PHILIP KARUHN
2015

Signal Transduction

Principles, Pathways, and Processes

ALSO FROM COLD SPRING HARBOR LABORATORY PRESS

ADVANCED TEXTBOOKS

Epigenetics, 2nd Edition
Mammalian Development: Networks, Switches, and Morphogenetic Processes
RNA Worlds: From Life's Origins to Diversity in Gene Regulation

SUBJECT COLLECTIONS FROM *COLD SPRING HARBOR PERSPECTIVES IN BIOLOGY*

Signaling by Receptor Tyrosine Kinases
Mitochondria
DNA Repair, Mutagenesis, and Other Responses to DNA Damage
Cell Survival and Cell Death
Immune Tolerance
Wnt Signaling
Protein Synthesis and Translational Control
The Synapse
Extracellular Matrix Biology
Protein Homeostasis
Calcium Signaling
Germ Cells
The Mammary Gland as an Experimental Model
The Biology of Lipids: Trafficking, Regulation, and Function
Auxin Signaling: From Synthesis to Systems Biology
Neuronal Guidance: The Biology of Brain Wiring
Cell–Cell Junctions
Generation and Interpretation of Morphogen Gradients
Immunoreceptor Signaling
NF-κB: A Network Hub Controlling Immunity, Inflammation, and Cancer
Symmetry Breaking in Biology
The p53 Family

SUBJECT COLLECTIONS FROM *COLD SPRING HARBOR PERSPECTIVES IN MEDICINE*

Bacterial Pathogenesis
Cystic Fibrosis: A Trilogy of Biochemistry, Physiology, and Therapy
Parkinson's Disease
Angiogenesis: Biology and Pathology
The Biology of Alzheimer Disease

OTHER RELATED TITLES

Genes & Signals
Means to an End: Apoptosis and Other Cell Death Mechanisms

Signal Transduction

Principles, Pathways, and Processes

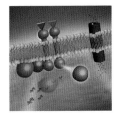

EDITED BY

Lewis C. Cantley
Harvard Medical School

Tony Hunter
Salk Institute for Biological Studies

Richard Sever
Cold Spring Harbor Laboratory

Jeremy Thorner
University of California at Berkeley

COLD SPRING HARBOR LABORATORY PRESS
Cold Spring Harbor, New York • www.cshlpress.org

Signal Transduction

Chapters online at cshperspectives.org and perspectivesinmedicine.org

Publisher	John Inglis
Director of Editorial Development	Jan Argentine
Project Manager	Inez Sialiano
Permissions Coordinator	Carol Brown
Production Editor	Diane Schubach
Production Manager/Cover Designer	Denise Weiss

Front cover artwork: Drawing by Nigel Hynes.

Library of Congress Cataloging-in-Publication Data

Signal transduction / edited by Lewis C. Cantley, Harvard Medical School, Tony Hunter, Salk Institute for Biological Studies, Richard Sever, Cold Spring Harbor Laboratory, Jeremy Thorner, University of California at Berkeley.
    p. cm.
  Summary: "This textbook provides a comprehensive view of signal transduction, covering both the fundamental mechanisms involved and their roles in key biological processes. It first lays out the basic principles of signal transduction, explaining how different receptors receive information and transmit it via signaling proteins, ions, and second messengers. It then surveys the major signaling pathways that operate in cells, before examining in detail how these function in processes such as cell growth and division, cell movement, metabolism, development, reproduction, the nervous system, and immune function"– Provided by publisher.
  Includes bibliographical references and index.
  ISBN 978-0-87969-901-7 (hardback)
1. Cellular signal transduction. 2. Developmental biology. 3. Pathology, Molecular. I. Cantley, Lewis, editor of compilation. II. Hunter, Tony, 1943- editor of compilation. III. Sever, Richard, editor of compilation. IV. Thorner, Jeremy W., editor of compilation.
  QP517.C45S534 2013
  571.7'4--dc23
                                        2013043753
10 9 8 7 6 5 4 3 2 1

For a complete catalog of all Cold Spring Harbor Laboratory Press publications, visit our website at www.cshlpress.org.

This book is dedicated to the memory of Tony Pawson (1952–2013). Tony was a giant in the field of signal transduction, who established principles of protein–protein interactions that have profoundly influenced our understanding of signal transduction. His enduring legacy will be the discovery that the Src homology 2 (SH2) domain of one protein can selectively interact with a tyrosine residue in a second protein, once it is phosphorylated in response to an upstream signal. This type of inducible protein–protein interaction can link intracellular signals generated in response to various upstream stimuli to downstream signaling events. This insight was the basis for his enormously influential idea that eukaryotic signaling systems involve modular and combinatorial interaction domains that propagate signals throughout the cell.

Contents

Contents

Preface

SIGNAL TRANSDUCTION PROCESSES CAN BE VIEWED as the higher command functions executed by cells on metabolic pathways (both catabolic and biosynthetic), macromolecular machinery and organellar compartments that allow an organism to maintain homeostasis and adjust cell number, cell behavior, and organismal physiology appropriately in response to internal cues and external stimuli. This book was conceived and organized as an instructional resource to introduce advanced students, investigators new to the field, and even researchers actively working in this general area to the underlying foundations and basic mechanisms of signal transduction in animal cells. Such a volume is needed because signaling impinges on every aspect of molecular and cellular biology—from biochemistry and structural biology to development and differentiation, endocrinology and systems biology, pharmacology and neuroscience, and immunology and cancer biology. Our objective is to explicate and illustrate the fundamental concepts, principles, and processes involved in signaling quite comprehensively, without necessarily being completely encyclopedic. We have taken a novel approach to conveying this large body of information and making it accessible, dividing the book up into distinct sections that describe principles, pathways, and processes.

The first four *principle* chapters set the stage, presenting molecular mechanisms and paradigms that are pertinent to all that follows. In Chapter 1, Carl-Henrik Heldin, Benson Lu, Ronald Evans, and Silvio Gutkind discuss signaling molecules and their receptors and downstream signaling events. In Chapter 2, Michael Lee and Michael Yaffe introduce the central role of proteins as transducers in signaling, describing the many ways by which signaling can control protein level, function, activity, and location. In Chapter 3, Alexandra Newton, Martin Bootman, and John Scott discuss the nature, generation, and action of intracellularly generated mediators ("second messengers"). In Chapter 4, Evren Azeloglu and Ravi Iyengar consider the circuit-like characteristics of signaling networks and systems, their emergent properties, and mathematical models we can use to describe them.

There follows a series of 14 *process* chapters that cover the roles of signaling in distinct biological processes and discuss how the general principles described in the four *principle* chapters apply in a specific context. Thus, the focus in these more specialized chapters is on the molecular basis of a particular aspect of signaling, its logic and its physiological consequences in biology, rather than a mere enumeration of pathway components and their interactions. Nonetheless, familiarity with signaling pathways used by cells is essential, and so separating the *principles* and *process* chapters are a series of *pathway* diagrams with short accompanying synopses written by other leaders in the field.

Different cell types possess a variety of mechanisms to sense and respond to diverse stimuli. Dedicated receptor cells, for example, respond to physical inputs from their surroundings, such as light, heat, and sound, as considered in the chapter by David Julius and Jeremy Nathans. The information is relayed via inorganic-ion-based electrical currents and release of and response to amino acids (glutamate and glycine), amino-acid-derived compounds, and other classes of substances that serve as neurotransmitters, as discussed in the chapters by Mary Kennedy and by Ivana Kuo and Barbara Ehrlich.

Cells respond to a plethora of other kinds of chemical signals, as disparate as inorganic substances (including gases) and a host of other organic molecules (from volatile substances to lipidic compounds to peptide hormones, growth factors, and morphogens), as presented in Chapter 1 and in the chapter by Norbert Perrimon, Chrysoula Pitsouli, and Ben-Zion Shilo. As discussed in Chapter 3, in many cases, the encounter with such extracellular ligands activates the production of second messengers, from phosphoinositides to cyclic nucleotides to less familiar, newly discovered metabolites. This allows amplification and spreading of the response by affecting the level, localization, and activity of numerous proteins and other cellular targets by mechanisms described in detail in Chapter 2. In addition to responses to native extracellular signals and normal internal cues, the specialized cells of our immune system must respond to attack by or internalization of potentially dangerous prokaryotic, viral and fungal pathogens, as reviewed in the chapters by Kim Newton and Vishva Dixit and by Doreen Cantrell. Microbes, in turn, have evolved an armamentarium of virulence factors and other effectors that they inject to specifically interdict signaling by lymphocytes and other cells, which also provide useful tools for experimentally interrogating signaling processes, as discussed in the chapter by Neal Alto and Kim Orth.

It is especially important that cells and tissues stay acutely attuned to their nutrient supply and adjust their metabolism accordingly. This aspect of signaling is described in the chapters by Patrick Ward and Craig Thompson and by Grahame Hardie. Cells also need to gauge their position in space and time and alter their morphology and adjust their movements in response to signals arising from cell–cell and cell–extracellular-matrix contacts, as presented in the chapters by Luke McCaffrey and Ian Macara and by Peter Devreotes and Rick Horowitz.

One reason for a cell to constantly gauge and integrate information about its nutrient supply, its developmental state, its neighboring cells, and demands of other tissues is to decide whether it should remain quiescent, grow and divide, or enter a developmental pathway leading to production of a highly specialized postmitotic cell type. The issue of how entry into the cell division cycle is controlled by signaling pathways is discussed in detail in the chapter by Robert Duronio and Yue Xiong. The internal, fail-safe signaling mechanisms (checkpoints) that ensure the proper spatial and temporal order of events in cell cycle progression, and act as delay timers to allow an adequate hiatus for any necessary repairs, are considered in the chapter by Nicholas Rhind and Paul Russell. When the normal signals that control the decision of cells to divide are subverted, and the negative controls on cell division are broken, malignant growth can occur. How defects in signaling lie at the heart of the molecular basis of cancers is discussed in the chapter by Richard Sever and Joan Brugge.

Concomitant with what may occur under optimal conditions, cells also have to cope with decisions about how to manage their resources and responses under more challenging and stressful conditions. Maybe the cell can overcome the problems, but, if it suffers irreversible harm to the integrity of its chromosomes, or to the functioning of a vital organelle, then alarm signals are in place to try to prevent any rogue or damaged cell from lingering. The signaling responses elicited by stressful conditions, and how those responses promote cell survival, are examined in the chapter by Gökhan Hotamisligil and Roger Davis. Conversely, how cells evoke and respond to the signals that lead to their own demise is described in the chapter by Douglas Green and Fabien Llambi.

Of course, most eukaryotes develop from multiplication of the single-celled zygote formed by the union of two germ cells, and how signaling is involved in gametogenesis and sexual reproduction is presented in the chapter by Sally Kornbluth and Rafael Fissore.

At the end of the book, we present an Outlook that provides some additional information and perspectives on recent developments (both methodological and conceptual) that further set the stage for future advances in the field of signal transduction. In it we discuss challenges and open questions that we hope will help point the way forward.

We would like to express our gratitude to all the authors who took time out of their busy schedules to contribute the fantastic chapters that make up this book. We also want to express our deep gratitude to the many investigators, too numerous to name individually here, who served as anonymous referees to evaluate the accuracy and effectiveness of the contents of this book. We would also like to thank Cell Signaling Technology, Inc., for financial support and for making available figures from which the pathway diagrams shown in the book were derived and adapted. Finally, we are indebted to Inez Sialiano, Diane Schubach, and Kathleen Bubbeo at Cold Spring Harbor Laboratory Press for all their hard work helping to get the book into print and online.

JEREMY THORNER
RICHARD SEVER
TONY HUNTER
LEWIS C. CANTLEY

Foreword

THIS TEXTBOOK ON *Signal Transduction*, edited by some of the foremost experts in this area, presents an encyclopedic view of a field that essentially did not exist 60 years ago. In those days, almost nothing was known about the mechanism by which enzymes and physiological processes were regulated, and terms such as "signaling" or "signal transduction" that are so commonly used today would not have been understood.

First, although endocrinology was already well established as a discipline, it remained purely at the phenomenological, mostly intact animal, level. The action of hormones stopped at the cell membrane and what happened next was totally unknown until Earl Sutherland and Ted Rall came along with their stunning discovery of cAMP, which served as a second, intracellular messenger for the action of epinephrine. Second, there was a fundamental difference in the way science was conducted. At that time and, in fact, since the days of Claude Bernard in the second half of the 19th century, one first observed a physiological phenomenon and then tried to identify the factors or enzymes involved. Whereas today, by and large, it is the other way around: new proteins are first identified mostly through genome sequencing projects and then, by overexpressing them or by knocking them in or out, one tries to define their function. Finally, essentially nothing was known about enzyme regulation. The prevailing idea was that they were regulated simply by the rate at which they were synthesized and degraded. But in the late 1940s/early 1950s, people began to realize that this could not be the case, that this would not work because protein synthesis and degradation are far too slow. Cells had to have ways of modulating the activity of their enzymes once they had been produced and liberated within the cells. They had to have the capability of adapting to their environment, of satisfying their metabolic needs, almost instantaneously in response to whatever internal or external demands are placed upon them. And this is where cell signaling and signal transduction came into play.

These fields did not originate from a single, explosive breakthrough or discovery. They grew step-by-step through successive small advances in the second half of the last century, originating perhaps with the finding that the control of glycogen phosphorylase, an enzyme shown by the Coris to catalyze the first step in the degradation of glycogen, occurs through a phosphorylation–dephosphorylation reaction. Since then, reversible protein phosphorylation has been found to be one of the most prevalent and versatile means by which cellular processes are regulated, being involved in the control of metabolism, gene expression, the immune response, cell development and differentiation, and what not. In fact, it would be difficult to find a physiological process that would not be, directly or indirectly, regulated by this kind of mechanism. It is implicated in innumerable hereditary diseases and pathological conditions, such as diabetes, Alzheimer's and Parkinson's diseases, and myelogenous leukemia, in viral diseases such as smallpox, and bacterial diseases such as cholera and plague.

Quantitatively, better than 99.9% of all these phosphorylation reactions occur on serine and threonine. But one of the most exciting developments in this field was the discovery, more than 30 years ago, that phosphorylation of proteins on tyrosyl residues was intimately implicated in cell transformation and oncogenesis, bringing into play a multitude of tyrosine kinases of cellular or viral origin, or linked to growth factor receptors.

Although reversible protein phosphorylation seemed to be for many years the main form of cellular regulation, a just as prevalent and far more complex regulatory mechanism has since been uncovered—namely, ubiquitylation. And it is very likely that other general regulatory systems might come to light, such as reversible protein acetylation, methylation, and oxidoreduction or the interaction of enzymes with their specific binding modules, anchors, and chaperones.

These advances could not have been possible without the development of sophisticated methodologies such as X-ray crystallography, nuclear magnetic resonance, mass spectrometry, and cryo-electron microscopy for protein structure determination and nanochemistry and the use of nanoparticles, monoclonal antibodies, and genetically encoded fluorescent marker proteins allowing one to monitor molecular processes without disrupting cell function.

Of course, the most spectacular advance occurred in genetic engineering with the cloning, manipulation, expression, and sequencing of genes, without which we would know essentially nothing about our genetic makeup or about a variety of hereditary and viral diseases. With the pervasive presence of the computer that allows one to display and analyze data and store and retrieve them at the touch of a button, today's investigators have at their disposal an array of technologies absolutely undreamed of just a few years ago.

Finally, what are some of the main problems that remain to be solved in signal transduction? Most of the major signaling pathways have probably been elucidated, and the structure, properties, regulation, and physiological function of the molecules involved have been well characterized. But these molecules are only the words the cells use to perform their daily chores. We know many of these words; we recognize probably bits and pieces of some of the sentences they spell out to elicit a particular response. But we are only just starting to understand the language the cell has to use to allow different receptors or pathways to speak with one another to coordinate all the reactions that take place. This communication often occurs through the formation of large macromolecular complexes comprising anchoring and scaffolding proteins and modules that link them to the cytoskeleton, providing those systems with the specificity and selectivity they require; however, how cells maintain and preserve the fidelity of signaling processes remains poorly understood.

The problem is further complicated by the fact that during the several billion years over which cells have evolved, they have had all the opportunities in the world to put in place the vast array of secondary or parallel pathways, shunts, compensatory mechanisms, feedback loops, and fail-safe systems they need to regulate their growth and development, to protect themselves against all sorts of adversity, and to program their own death when the time comes. And we do not know the myriads of signals that must exist to sort out all the reactions that take place.

Perhaps even more importantly, we do not understand the cross talk—the interactivity that must exist among cells and how they communicate with one another to synchronize their behavior in response to internal or external signals. This cross talk, this sharing of information, is crucial for the establishment of such sophisticated networks of communication as seen, for instance, during embryonic development and organogenesis, in the immune system, or in the infinitely more complex central nervous system, where a thousand billion cells speak with one another through more than a million billion synapses, leading ultimately to the generation of memory and thought and consciousness. Solving these problems will be one of the major challenges that will confront biologists in the years to come.

This textbook on signal transduction addresses most of these problems. It is directed toward future practitioners of biology and medicine: advanced graduate students, postdoctoral fellows, or researchers working in an academic, biotechnological, or pharmaceutical environment. It will be of enormous help to all those who would want to remain abreast of the field.

EDMOND FISCHER
University of Washington

GENERAL PRINCIPLES AND MECHANISMS

CHAPTER 1

Signals and Receptors

Carl-Henrik Heldin[1], Benson Lu[2], Ron Evans[2], and J. Silvio Gutkind[3]

[1]Ludwig Institute for Cancer Research, Science for Life Laboratory, Uppsala University, SE-75124 Uppsala, Sweden
[2]The Salk Institute for Biological Studies, Gene Expression Laboratory, La Jolla, California 92037
[3]National Institute of Dental and Craniofacial Research, National Institutes of Health, Bethesda, Maryland 20892-4340

Correspondence: c-h.heldin@licr.uu.se

SUMMARY

Communication between cells in a multicellular organism occurs by the production of ligands (proteins, peptides, fatty acids, steroids, gases, and other low-molecular-weight compounds) that are either secreted by cells or presented on their surface, and act on receptors on, or in, other target cells. Such signals control cell growth, migration, survival, and differentiation. Signaling receptors can be single-span plasma membrane receptors associated with tyrosine or serine/threonine kinase activities, proteins with seven transmembrane domains, or intracellular receptors. Ligand-activated receptors convey signals into the cell by activating signaling pathways that ultimately affect cytosolic machineries or nuclear transcriptional programs or by directly translocating to the nucleus to regulate transcription.

Outline

Cite this chapter as *Cold Spring Harb Perspect Biol* doi: 10.1101/cshperspect.a005900

1 INTRODUCTION

Cells within multicellular organisms need to communicate with each other to coordinate their growth, migration, survival, and differentiation. They do so by direct cell–cell contact and secretion or release of molecules that bind to and activate receptors on the surface of or inside target cells. Such factors can stimulate the producer cell itself (autocrine stimulation), cells in the immediate vicinity (paracrine stimulation), or cells in distant organs (endocrine stimulation). The signaling induced within target cells is important during embryonic development, as well as in the adult, where it controls cell proliferation, differentiation, the response to infection, and numerous organismal homeostatic mechanisms.

Many cell-surface receptors contain an extracellular ligand-binding region, a single transmembrane segment, and an intracellular effector region, which may or may not have an associated enzyme activity. Some receptors contain multiple subunits that together form the ligand-binding site. Others, including those encoded by the largest gene family in the human genome, consist of a polypeptide that spans the cell membrane seven times. Finally, there are receptors that are located intracellularly and are activated by ligands that cross the cell membrane, such as steroid hormones. Below, we describe the major families of ligands and receptors and the signal transduction mechanisms they activate.

2 CELL-SURFACE RECEPTORS

2.1 Receptors with Associated Protein Kinase Activity

Several types of cell-surface receptors contain or are associated with kinase activities that respond to the binding of a ligand. Perhaps best understood are receptors with intrinsic protein tyrosine kinase domains. This receptor tyrosine kinase (RTK) family has more than 50 human members (Lemmon and Schlessinger 2010). RTKs have important roles in the regulation of embryonic development, as well as in the regulation of tissue homeostasis in the adult. Each has an extracellular, ligand-binding region, which consists of different combinations of various domains, such as Ig-like, fibronectin type III, and cysteine-rich domains. This is linked to a single transmembrane segment and an intracellular region that includes a tyrosine kinase domain. Based on their structural features, RTKs can be divided into 20 subfamilies (Fig. 1), a well-studied example being the epidermal growth factor (EGF) receptors (EGFRs).

Members of the cytokine receptor family in contrast lack intrinsic kinase activity but associate with intracellular kinases. They have important roles in the regulation of the immune system and also promote cell differentiation. Cytokine receptors can be divided into two classes. The extracellular domains of class I cytokine receptors contain cytokine receptor homology domains (CHDs) consisting of two tandem fibronectin type III domains with a characteristic WSXWS motif in the second (Liongue and Ward 2007). Based on the number of CHDs and the presence of other types of domains, the class I cytokine receptors can be divided into five groups (Fig. 2), the growth hormone (GH) receptor being typical of the first group. Interferon receptors are typical of the 12-member class II cytokine receptor family, which also have extracellular regions based on tandem fibronectin domains but differ from those of class I receptors (Fig. 2) (Renauld 2003). Both classes of cytokine receptors have conserved box 1 and box 2 regions in their intracellular regions, which bind to JAK family tyrosine kinases that are activated upon ligand binding. The multisubunit antigen receptors on B cells and T cells (Zikherman and Weiss 2009) and the Fc receptors present on macrophages, mast cells, basophils, and other immune cells (Nimmerjahn and Ravetch 2008) are also associated with intracellular tyrosine kinases; activation of these receptors involves tyrosine phosphorylation by members of the Src family, followed by docking and activation of SH2-domain-containing Syk/Zap70 tyrosine kinases (see p. 125 [Samelson 2011]; Ch. 16 [Cantrell 2014]). Although receptors that have intrinsic tyrosine kinase activity and those associated with tyrosine kinases are structurally different and bind ligands of different kinds, the principles underlying their activation and the intracellular signals they initiate are similar (see below).

There is also a family of receptors that have intrinsic serine/threonine kinase domains, and these respond to members of the transforming growth factor β (TGFβ) family (see Moustakas and Heldin 2009; p. 113 [Wrana 2013]). The human genome has only 12 genes encoding receptors of this type (Fig. 3). These receptors have rather small cysteine-rich extracellular domains; their intracellular domains are most often also small and consist mainly of the kinase domains. TGFβ receptors mediate signaling events during embryonic development. Because they often inhibit cell growth, they also exert a controlling function on the immune system and other tissues.

2.2 Ligands

Each of these different receptor types responds to a subfamily of structurally similar ligands. The ligands are normally small monomeric, dimeric, or trimeric proteins, often derived by proteolytic processing from larger precursors, some of which are transmembrane proteins. There is not a strict specificity in ligand–receptor interactions with-

Cite this chapter as *Cold Spring Harb Perspect Biol* doi: 10.1101/cshperspect.a005900

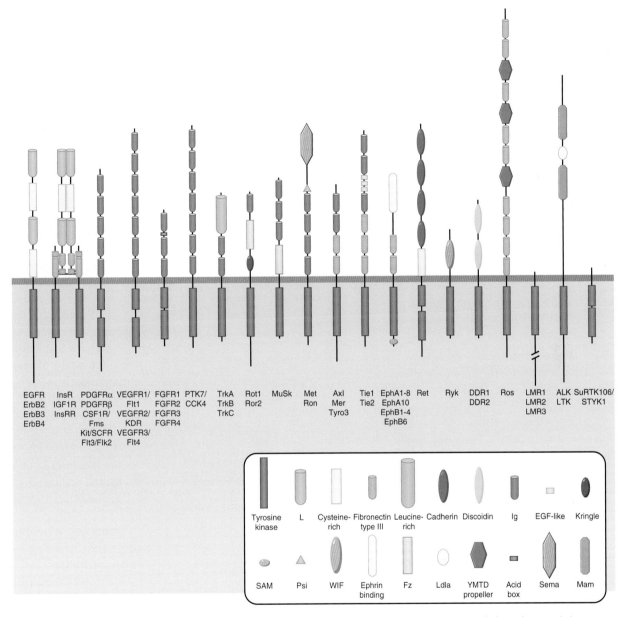

EGFR InsR PDGFRα VEGFR1/ FGFR1 PTK7/ TrkA Rot1 MuSk Met Axl Tie1 EphA1-8 Ret Ryk DDR1 Ros LMR1 ALK SuRTK106/
ErbB2 IGF1R PDGFRβ Flt1 FGFR2 CCK4 TrkB Ror2 Ron Mer Tie2 EphA10 DDR2 LMR2 LTK STYK1
ErbB3 InsRR CSF1R/ VEGFR2/ FGFR3 TrkC Tyro3 EphB1-4 LMR3
ErbB4 Fms KDR FGFR4 EphB6
Kit/SCFR VEGFR3/
Flt3/Flk2 Flt4

Tyrosine   L   Cysteine- Fibronectin Leucine- Cadherin Discoidin   Ig   EGF-like Kringle
kinase   rich   type III   rich

SAM   Psi   WIF   Ephrin   Fz   Ldla   YMTD   Acid   Sema   Mam
                    binding   propeller box

Figure 1. Receptor tyrosine kinase (RTK) families. The 20 subfamilies of human RTKs and their characteristic structural domains are shown. The individual members of each family are listed below. (From Lemmon and Schlessinger 2010; adapted, with permission.)

in the families; normally each ligand binds to more than one receptor, and each receptor binds more than one ligand. Although it is rare that ligands for completely different types of receptors bind to kinase-associated receptors, examples do exist.[4]

---

[4]The Ryk and Ror families of RTK, for example, bind members of the Wnt family. Ryk has a Wnt inhibitory factor-1 (WIF1) domain, and the two Ror receptors have cysteine-rich domains related to a domain in the Frizzled family of serpentine receptors to which ligands of the Wnt family bind (van Amerongen et al. 2008).

2.3 Activation by Dimerization

A common theme for activation of kinase-associated receptors is ligand-induced receptor dimerization or oligomerization (Heldin 1995). The juxtaposition of the intracellular kinase domains that occurs as a consequence allows autophosphorylation in *trans* within the complex. For RTKs, the autophosphorylation has two important consequences: it changes the conformation of the kinase domains, leading to an increase in their kinase activities; and it produces docking sites (phosphorylated sequences) for intracellular signaling molecules containing SH2 or PTB domains (see

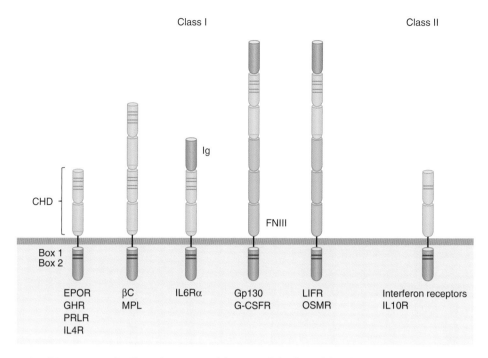

Figure 2. Cytokine receptor families. The structural features of the five subfamilies of class I and class II cytokine receptors are depicted. The characteristic cytokine homology domains (CHDs) with their four cysteine residues (blue lines) and a WW motif (green line), as well as the box 1 and box 2 regions (red bands), Ig-like domains, and FNIII domains are shown.

Ch. 2 [Lee and Yaffe 2014]). The autophosphorylation that controls the kinase activity occurs on residues in various regions of the receptors. Most RTKs are phosphorylated in the activation loops of the kinases, that is, a flexible loop of the carboxy-terminal domain, which can fold back and block the active site of the kinase when unphosphorylated. Exceptions are Ret and the four members of the EGFR family that can be activated without phosphorylation in their activation loops (Lemmon and Schlessinger 2010). Several RTKs are activated by phosphorylation in the juxtamem-

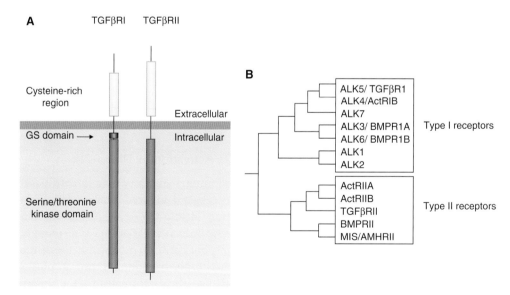

Figure 3. Serine/threonine kinase receptors. (A) The structural features of type I (TGFβRI) and type II (TGFβRII) serine/threonine kinase receptors. (B) The different members of the type I and type II receptor subfamilies and their evolutionary relations. Act, Activin; ALK, activin-receptor-like kinase.

Cite this chapter as *Cold Spring Harb Perspect Biol* doi: 10.1101/cshperspect.a005900

brane region, including MuSK, Flt3, Kit, and Eph family members. Moreover, Tie2, platelet-derived growth factor (PDGF) β receptor (PDGFRβ) and Ron have been shown to be activated by phosphorylation of their carboxy-terminal tails. In each case, autophosphorylation causes a change in conformation that opens up the active site of the kinase. Interestingly, a phosphorylation-independent mechanism for activation of EGFR kinase activity has been elucidated: the carboxy-terminal lobe of one kinase domain makes contact with the amino-terminal lobe of the other kinase domain in the dimer and thereby activates it (Zhang et al. 2006). The juxtamembrane region of the receptor, which in other RTKs has an inhibitory role, is needed to stabilize this dimeric interaction (Jura et al. 2009; Red Brewer et al. 2009).

Different ligands use different mechanisms to cause receptor dimerization. Thus, GH is a monomeric ligand that dimerizes its receptors by inducing formation of an asymmetric complex containing one ligand and two receptors (De Vos et al. 1992). EGF is also a monomeric factor, but each molecule binds to a single receptor molecule, causing a conformational change that promotes direct binding of two ligand-bound receptors (Fig. 4) (Burgess et al. 2003). Similarly, dimeric ligands can induce symmetric 2:2 ligand–receptor complexes. In the case of nerve growth factor (NGF), the ligand itself is solely responsible for the dimerization (Wehrman et al. 2007); in the case of Kit and the PDGFRs, the ligand-induced receptor dimerization is stabilized by direct interactions between the receptors (Yuzawa et al. 2007). In other cases, accessory molecules are needed to stabilize dimerization; that is, heparin or heparan sulfate stabilizes fibroblast growth factor (FGF)–induced dimers by interacting with both FGF and FGF-receptor subunits (Lemmon and Schlessinger 2010). Recent work on EGFR has shown that its activation requires an interaction between the amino-terminal parts of the transmembrane helices of the dimeric receptors, which promotes an antiparallel interaction between the cytoplasmic juxtamembrane segments and release of inhibition by the membrane (Arkhipov et al. 2013; Endres et al. 2013).

In addition to homodimerization of receptors, heterodimerization also often occurs. This is particularly common among cytokine receptors; a ligand-specific receptor subunit often interacts with one or more common subunits, such as gp130, β_c, or γ_c, to form a heterodimer, heterotrimer, or even more complicated multimer (Wang et al. 2009). Similarly, members of different RTK subfamilies often form heterodimers. For instance, one member of the EGFR subfamily, ErbB2, cannot bind to ligand itself, but acts in heterodimers with other members of the family (Yarden and Sliwkowski 2001). In the PDGFR family, different dimeric isoforms of the ligand induce formation of different dimeric complexes of PDGFα and PDGFβ recep-

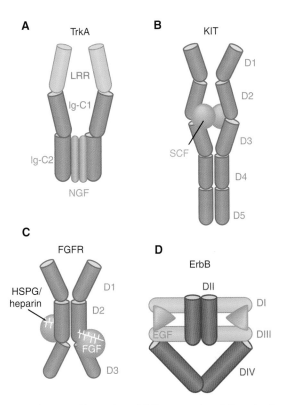

Figure 4. Schematic illustration of different modes of dimerization of RTKs. (A) Some dimeric ligands, such as nerve growth factor (NGF), bind to receptors in a symmetric manner, but the receptors do not contact each other. (B) Other dimeric ligands, such as stem cell factor (SCF), also bind to RTKs in a symmetric manner, but the receptor dimer is in addition stabilized by direct receptor–receptor interactions. (C) In the case of fibroblast growth factor (FGF), a ternary complex involving the ligand, the receptor, and heparin/heparin sulfate stabilizes the receptor dimer. (D) In the case of members of the epidermal growth factor (EGF) receptor family such as ErbB, ligand binding induces a conformational change in the extracellular domain of the receptor that promotes direct receptor–receptor interactions. (From Lemmon and Schlessinger 2010; adapted, with permission.)

tors (Heldin 1995). Because the downstream signaling pathways that are activated to a large extent depend on the specific docking of SH2- or PTB-domain-containing signaling molecules (see below and Ch. 2 [Lee and Yaffe 2014]), differences in the autophosphorylation patterns of homodimeric versus heterodimeric receptor complexes will give rise to different combinatorial signals.

Serine/threonine kinase receptors present another variation on this theme. These receptors are activated by ligand-induced assembly of two type I and two type II receptors into heterotetrameric receptor complexes, in which constitutively active type II receptors phosphorylate type I receptors in serine- and glycine-rich sequences just upstream of the kinase domains. In the TGFβ type I receptor, this causes a change in conformation that prevents the inhibitory interaction of the receptor with the immunophilin FK506-binding protein (FKBP12) and activates its kinase (Kang et al. 2009).

Note that certain receptors are present in dimeric complexes even in the absence of ligand; examples include the erythropoietin cytokine receptor, and the insulin and insulin-like growth factor-1 (IGF1) receptors, RTKs that actually occur as disulfide-bonded dimers. Here, ligand binding induces conformational changes that lead to activation of the receptor-associated kinase. Importantly, there are indications that reorientation of the intracellular regions of the receptors relative to each other in the dimer is important for their activation (Jiang and Hunter 1999).

2.4 Signaling Downstream from Tyrosine-Kinase-Associated Receptors via SH2- or PTB-Domain-Containing Molecules

An important aspect of signaling by tyrosine-kinase-associated receptors is the formation of multiprotein signaling complexes (Pawson 2004). As mentioned above, SH2- or PTB-domain-containing proteins bind to specific phosphorylated tyrosine residues in the receptors themselves or in scaffolding proteins associated with the receptors (see Ch. 2 [Lee and Yaffe 2014]). Examples of the latter include the four members of the insulin receptor substrate (IRS) family, which bind to insulin and IGF1 receptors; FRS2 family protein molecules, which bind to FGF and NGF receptors; Gab molecules, which bind to several RTKs; and SLP76 and LAT, which bind to antigen receptors.

Some of the proteins that bind to these sequences contain intrinsic enzymatic activities, for example, the tyrosine kinases of the Src family, the tyrosine phosphatases of the SHP family, GTPase-activating proteins (GAPs) that regulate Ras family GTPases, the ubiquitin ligase Cbl, and phospholipase C (PLC) γ. Others are proteins that form stable complexes with enzymes, for example, the regulatory p85 subunit of type 1 phosphoinositide 3-kinase (PI3K), which forms a complex with the catalytic p110 subunit, and Grb2, which forms a complex with SOS1, a guanine nucleotide exchange factor (GEF) that stimulates Ras. Cytokine receptors and some RTKs also activate STAT molecules, which translocate to the nucleus, where they act as transcription factors (see p. 117 [Harrison 2012]). Finally, SH2-/SH3-domain-containing adaptor molecules (e.g., Shc, Nck, and Crk) bind to certain tyrosine phosphorylated sites in the receptors or associated molecules and form bridges to other signaling molecules, including the enzymes mentioned above.

Many of the signaling proteins that bind to the receptors also contain domains that mediate interactions with other molecules. These include SH3 domains, which bind proline-rich motifs in proteins; PDZ domains, which

bind to the carboxy-terminal tails of proteins containing valine residues; and pleckstrin homology (PH) domains, which bind lipids present in membranes. Thus, large complexes of signaling proteins are transiently formed and mediate signaling to cytoplasmic machinery (e.g., enzymes controlling the cytoskeleton and cell migration), as well as to the nucleus, where transcription is affected.

Note that the kinase domains of certain RTKs (e.g., ErbB3, CCK4, EphB6, and Ryk) lack critical catalytic residues normally found in kinases and thus have no, or very low, kinase activity. However, these receptors still have important roles in signal transduction, acting as phosphorylatable scaffolds in heterodimers with other receptors.

There are also indications that the EGFRs can translocate to the nucleus to regulate transcription (Lin et al. 2001). However, the functional significance of this needs to be further clarified.

2.5 Downstream Signaling by Serine/Threonine Kinase Receptors

Serine/threonine kinase receptors activate members of the Smad transcription factor family (p. 113 [Wrana 2013]); similar to STATs, these molecules oligomerize after phosphorylation and activation and then translocate to the nucleus, where they induce transcription. In addition, serine/threonine kinase receptors can activate non-Smad pathways. The TGFβ type I receptor, for example, can autophosphorylate on tyrosine residues, which bind to the adaptor Shc, leading to activation of Ras and the ERK1/2 mitogen-activated protein kinase (MAPK) pathway (see p. 81 [Morrison 2012]). Moreover, the p38 MAPK pathway is activated via binding of the ubiquitin ligase TRAF6 to the TGFβ type I receptor (Kang et al. 2009). In addition to activating the type I receptor, the TGFβ type II receptor also contributes to signaling by phosphorylating the polarity complex protein PAR6 (Ozdamar et al. 2005; Ch. 9 [McCaffrey and Macara 2012]).

2.6 Interaction with Coreceptors

Several RTKs form complexes with nonkinase receptors, such as the hyaluronan receptor CD44 and integrins, which bind to extracellular matrix molecules. Such interactions can modulate their signaling. Similarly, the heparan sulfate proteoglycan agrin activates MuSK receptors by binding to low-density lipoprotein (LDL)-receptor-related protein 4 (LRP4); thus, LRP4 acts as a coreceptor for MuSK (Kim et al. 2008). Another example is glial-ceU-derived neurotrophic factor (GDNF) receptor α, a glycosylphosphatidylinositol (GPI)-anchored molecule needed for GDNF to bind to and activate Ret receptors (Schlee et al. 2006).

Cite this chapter as *Cold Spring Harb Perspect Biol* doi: 10.1101/cshperspect.a005900

2.7 Feedback and Amplification Mechanisms

Signaling via kinase-associated receptors is controlled by multiple feedback mechanisms to ensure an appropriate level. Thus, many tyrosine kinase and cytokine receptors bind protein tyrosine phosphatases (PTPs), including SHP1 and SHP2, that contain SH2 domains. Through dephosphorylation of specific tyrosine residues, PTPs modulate signaling quantitatively as well as qualitatively (Lemmon and Schlessinger 2010).

Another example is activation of the small G protein Ras, which occurs by docking of the GEF SOS1 in complex with the adaptor Grb2 (see Ch. 2 [Lee and Yaffe 2014]). PDGFRβ and possibly other receptors simultaneously bind the Ras-GAP molecules via another phosphorylated tyrosine residue, which counteracts Ras activation by promoting hydrolysis of bound GTP.

In addition, another downstream target of receptor signaling, protein kinase C (PKC), is involved in feedback control of certain RTKs, including EGFR, the insulin receptor, Met, and Kit. Activation of PLCγ leads to production of diacylglycerol (DAG) and increases in intracellular calcium levels, which activate the classical isoforms of PKC (see Ch. 3 [Newton et al. 2014]). Phosphorylation of the receptors by PKC subsequently inhibits the kinase activity of the receptors.

After the induction of immediate-early genes encoding various effector molecules by signaling to the nucleus, RTKs induce the expression of delayed-early genes, many of which encode proteins that suppress signaling. Examples include NAB2, which binds to and inhibits EGFR; FOSL1, which binds to and inhibits AP1 transcription factors; Id2, which inhibits the TCF transcription complex; DUSPs, which dephosphorylate and inactivate MAPKs; and ZFP36, which recognizes AU-motifs in the 3′ ends of mRNA molecules and causes their degradation (Amit et al. 2007). Similarly, TGFβ receptors and cytokine receptors induce Smad7 (Moustakas and Heldin 2009) and SOCS proteins (Yoshimura et al. 2003), respectively, which exert negative-feedback control by promoting ubiquitin-dependent degradation of receptors by targeting receptor-containing vesicles to lysosomes (see below).

2.8 Endocytosis of Kinase-Associated Receptors

After ligand binding, kinase-associated receptors are internalized into the cell via clathrin-dependent or clathrin-independent pathways (Zwang and Yarden 2009). Whereas internalization of RTKs is induced by ligand binding, internalization of serine/threonine receptors is constitutive and independent of ligand binding. Internalization has both positive and negative effects on signaling. Thus, when present in endosomes, the receptors are still active; in some cases, internalization is even necessary for the receptors to interact with signaling molecules on intracellular endosomes (Miaczynska et al. 2004). Examples include TGFβ receptors, which need to be internalized to interact with Smad molecules presented to the receptors by SARA and endofin proteins. These proteins reside on endosomes, their FYVE domains binding to phosphatidylinositol 3-phosphate (PI3P), a phospholipid that is enriched in endosomal membranes (Tsukazaki et al. 1998). Signaling is interrupted when the pH of the endosomes becomes so low that the ligand dissociates; at this stage, receptors are dephosphorylated and recycle back to the cell surface. Alternatively, the receptors are recognized by components of the endosomal sorting complex required for transport (ESCRT) machinery, which facilitates translocation to multivesicular bodies and degradation in lysosomes (Raiborg and Stenmark 2009). The latter route is promoted by ubiquitylation of the receptors by Cbl or other ubiquitin ligases.

2.9 Kinase-Associated Receptors and Disease

Because tyrosine-kinase-associated receptors often stimulate cell proliferation and survival, overactivity of these receptors is linked to the development of cancer and other diseases characterized by excess cell proliferation, such as inflammatory and fibrotic conditions. There are several examples of gain-of-function mutations in RTKs that occur in malignancies (Lengyel et al. 2007; Ch. 21 [Sever and Brugge 2014]). Point mutations in Kit and PDGFRα have been found in gastric intestinal stromal tumors, and the kinase domains of PDGFRs and FGFRs occur as constitutively active cytoplasmic fusion proteins in several rare leukemias. Moreover, the *ERBB2* gene is amplified in ~20% of breast cancers, and a mutated version of the *EGFR* gene is amplified in ~30% of glioblastoma cases.

Because overactivity of these receptors is common in malignancies, several antagonists have been developed, including inhibitory antibodies, ligand traps, and low-molecular-weight kinase inhibitors. Several of these are now used routinely in the clinic or are undergoing clinical trials.

The serine/threonine kinase receptors often relay growth inhibitory and apoptotic signals and therefore have tumor-suppressive effects. Thus, loss-of-function mutations of TGFβ type I and type II receptors have been observed in some cancers (e.g., colorectal carcinomas).

2.10 Receptors Activated by Proteolytic Cleavage

The highly conserved Notch family of receptors consists of four members, which have important roles during embryonic development and tissue renewal (Kopan and Ilagan 2009). The Notch receptors are single-pass transmembrane proteins with large extracellular domains containing multiple EGF-like repeats. They are cleaved extracellularly by furin-like proteases during transit to the plasma membrane to create a heterodimer held together by noncovalent interactions (p. 109 [Kopan 2012]).

There are five canonical Notch ligands (Delta-ligand-like 1, 3, and 4, and Jagged 1 and 2). These are also single-pass transmembrane proteins with extracellular EGF-like regions and other domains present on neighboring cells. The Notch receptor is normally triggered upon cell–cell contact. Notch activation involves a series of proteolytic events. Ligand binding induces cleavage of Notch by ADAM metalloproteases at a site about 12 amino acid residues outside the transmembrane domain. Removal of the extracellular domain allows an intramembrane cleavage by γ-secretase, which causes release of the intracellular domain (ICD) of Notch. Because the ICD has a nuclear localization sequence, it translocates to the nucleus, where it, together with the DNA-binding protein RBPjκ/CBF-1 and certain coactivators and corepressors (together referred to as CSL), regulates transcription of target genes via its transactivation domain. Because the Notch ligands are membrane-associated, signaling may also be induced in the ligand-bearing cells after binding to Notch.

Some RTK receptors (Ni et al. 2001) and the type I TGFβ receptor (Mu et al. 2011) are also subjected to sequential cleavage by metalloproteases and γ-secretase. The intracellular region of the receptors can then translocate to the nucleus to regulate transcription. Meanwhile, the extracellular portion of the receptor is liberated and can negatively regulate signaling by acting as a decoy for ligand (Ancot et al. 2009). In addition, some RTKs are classified as "dependence receptors." When they are not occupied by ligand, they can be cleaved by caspases (a group of proteases that control apoptosis) to generate fragments with apoptotic activity. For example, fragments of EGFR, ErbB2, Ret, Met, TrkC, ALK, and EphA4 have apoptotic effects, which contrast with the normal antiapoptotic effects of the full-length receptors stimulated by their ligands (Ancot et al. 2009).

2.11 G-Protein-Coupled Receptors

Approximately 2% of all genes in the human genome encode G-protein-coupled receptors (GPCRs), which represent the largest family of cell-surface molecules involved in signal transmission. They are so called because their signals are transduced by heterotrimeric G proteins; members of the GPCR family regulate a wide range of key physiological functions, including neurotransmission, blood pressure, cardiac activity, vascular integrity, hemostasis after tissue injury, glucose and lipid metabolism, sensory perception, regulation of endocrine and exocrine gland function, im-

mune responses, multiple developmental processes, and stem cell function and maintenance (Pierce et al. 2002; Dorsam and Gutkind 2007). Reflecting this remarkable multiplicity of activities, GPCR dysregulation contributes to some of the most prevalent human diseases. Indeed, GPCRs are the direct or indirect target of >50% of all available medicines (Flower 1999; Pierce et al. 2002).

2.12 A Shared Heptahelical Structure Transduces Signals Initiated by Highly Diverse Molecular Agonists

GPCRs are characterized by the presence of an extracellular amino terminus, an intracellular carboxy-terminal tail, and a shared structural core composed of seven transmembrane α helices that weave in and out of the membrane, thus forming three intracellular and three extracellular loops (Fig. 5) (Pierce et al. 2002). They are regulated by a diverse array of agonists, as small as photons, ions, nucleotides, amino acids, biogenic amines, bioactive lipids, and glucose metabolites, and as large as chemokines, glycoproteins, and proteases. They can be grouped into three classes and subfamilies of these according to the structural features involved in ligand–receptor recognition and subsequent stimulation (Fig. 5).

In class A GPCRs, the ligand-binding site is deep within the transmembrane domains in subfamily 1, which includes most receptors for small molecules, such as neurotransmitters, lipid mediators, and odorants. Subfamily 2 members are activated by protein ligands, such as chemokines and the tethered ligand resulting from thrombin-mediated cleavage of the protease-activated receptor 1 (PAR1) receptor. Subfamily 3 members have a very long extracellular domain, which binds to leuteinizing hormone (LH; also known as lutropin), thyroid-stimulating hormone (TSH), and follicle-stimulating hormone (FSH) (Ch. 17 [Kornbluth and Fissore 2014]). This subfamily also includes LGR5, LGR6, and LGR7, which are GPCR-like receptors involved in adult stem cell specification and function (Hsu et al. 2000; Leushacke and Barker 2011). The class B GPCRs are activated by high-molecular-weight hormones (e.g., glucagon, secretin, and vasoactive intestinal peptide [VIP]), whereas class C GPCRs include metabotropic glutamate receptors, calcium-sensing receptors, γ-amino butyric acid (GABA) B receptors, and receptors for taste compounds. Although many class A and class B GPCRs can form homo- and heterodimers (Terrillon and Bouvier 2004), dimer formation is obligatory for class C GPCRs (Kniazeff et al. 2011). The Frizzled family of receptors comprises the "Frizzled" and "Smoothened" subfamilies, which are structurally distinct and have complex mechanisms of agonist activation (p. 107 [Ingham 2012]; Lim and Nusse 2013).

Cite this chapter as *Cold Spring Harb Perspect Biol* doi: 10.1101/cshperspect.a005900

Class A

1

2

3

Class B

Class C

Frizzled

Smoothened

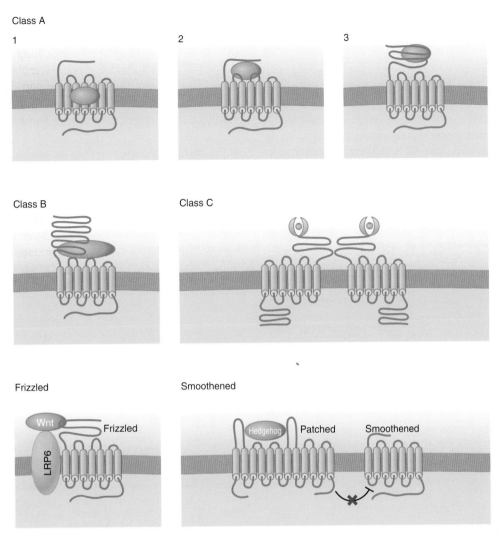

Figure 5. GPCRs use distinct structural features for ligand recognition. GPCRs have been classified based on their sequence similarities and ligand-binding properties. In class A GPCRs, the largest group, the ligand-binding site is deep within the transmembrane domains in subfamily 1. It involves interactions with the amino terminus, the extracellular loops, and the transmembrane domains in subfamily 2. The ligand-binding site is in the long extracellular domain in subfamily 3. Class B GPCRs are activated by high-molecular-weight hormones, which bind to the ligand-binding site within the long amino-terminal region, as well as some of the extracellular loops. Class C GPCRs are characterized by a very long amino terminus that shares some sequence similarity with periplasmic bacterial proteins; activation involves obligatory dimerization. The Frizzled family of receptors contains the Frizzled and "Smoothened" subfamilies, which are structurally distinct and have complex mechanisms of agonist activation. Wnt binds to and activates Frizzled through an interaction with a cysteine-rich amino-terminal region, whereas low-density lipoprotein-receptor-related protein 5 (LRP5) or LRP6 (single-transmembrane-span proteins) acts as a coreceptor (Lim and Nusse 2013). When Hedgehog binds to Patched, a negative regulatory effect of Patched on Smoothened activity is relieved, and Smoothened regulates both G-protein-dependent and -independent signals (p. 107 [Ingham 2012]).

2.13 GPCR Activation, Trafficking, and G-Protein-Coupling Specificity

The heterotrimeric G proteins that relay signals from GPCRs are associated with the underside of the plasma membrane and are composed of an α subunit and a $\beta\gamma$ dimer. Agonist-activated GPCRs act as GEFs that catalyze the exchange of GDP bound to the α subunit for GTP, causing release of G$\beta\gamma$ (Ch. 2 [Lee and Yaffe 2014]). A single ligand-bound GPCR can activate several G proteins, providing the first layer of signal amplification. The GTP-bound Gα subunits and G$\beta\gamma$ subunits can then promote the activation of a variety of downstream effectors, stimulating a network of signaling events that is highly dependent on the G-protein-coupling specificity of each receptor (Fig. 6). The human G-protein α subunits are encoded by 16 distinct genes and can be divided into four subfamilies: $G\alpha_s$ ($G\alpha_s$ and $G\alpha_{olf}$), $G\alpha_i$ ($G\alpha_t$, $G\alpha_{gust}$, $G\alpha_{i1-3}$, $G\alpha_o$, and $G\alpha_z$), $G\alpha_q$ ($G\alpha_q$, $G\alpha_{11}$, $G\alpha_{14}$, and $G\alpha_{15/16}$), and $G\alpha_{12}$ ($G\alpha_{12}$ and $G\alpha_{13}$) (Fig. 6) (Cabrera-Vera et al. 2003). A single GPCR can couple to either one or more than one family of Gα subunits. Five different β subunits and 12 γ subunits that form functional βγ dimers have been described.

Studies of the structural perturbations caused by ligand binding to class A GPCRs reveal that agonist binding causes a contraction of a ligand-binding pocket located within the transmembrane α-helical regions. In particular, upon ligand binding, the transmembrane (TM) α helix 6 moves outward from the center of the transmembrane bundle,

loses contact with TM3, and moves closer to TM5. The consequent conformational changes in the intracytoplasmic loops lead to the formation of a new pocket between TM3, TM5, and TM6, which binds to the carboxyl terminus of the Gα subunits, leading to G-protein activation by promoting the release of GDP and its exchange for GTP (Kobilka 2011). Ligands that bind and stabilize the receptors in conformations other than this fully activated state act as partial agonists, and they provoke a more limited G-protein activation and hence a restricted response. Ligands that stabilize the inactive conformation of the GPCR act as classical antagonists or inverse agonists. There is now a great interest in the development of novel drugs that act as allosteric modulators by binding at a site distinct from that to which the natural GPCR ligands bind. This new gener-

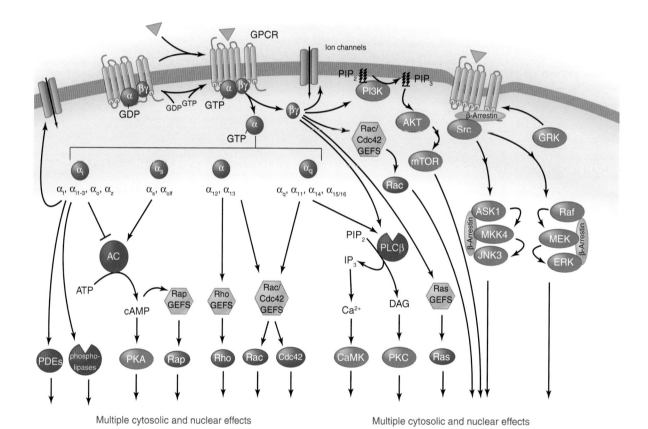

Figure 6. Regulation of classical second messenger systems and Ras and Rho GTPases by GPCRs. Agonist-activated GPCRs promote the dissociation of GDP bound to the α subunit pf heterotrimeric G proteins and its replacement by GTP. Gα and Gβγ subunits can then activate numerous downstream effectors. The 16 human G protein α subunits can be divided into the four subfamilies shown, and a single GPCR can couple to one or more families of Gα subunits. Downstream effectors regulated by their targets include a variety of second messenger systems, as well as members of the Ras and Rho families of small GTP-binding proteins, which, in turn, control the activity of multiple MAPKs, including ERK, JNK, p38, and ERK5. G-protein-dependent activation of these by GPCRs and β-arrestin-mediated G-protein-independent activation of ERK and JNK can have multiple effects in the cytosol. MAPKs also translocate to the nucleus, where they regulate gene expression. Activation of the PI3K–Akt and mTOR pathways plays a central role in the regulation of cell metabolism, migration, growth, and survival by GPCRs.

ation of pharmacological agents changes the receptor conformation, thereby modifying the affinity and/or efficacy of the endogenous ligands (May et al. 2007). Note that the nature of the agonist binding can impact receptor conformation, thus biasing the choice of G protein and hence the signaling output.

2.14 Regulation of Classical Second Messenger Systems by G Proteins and Their Linked GPCRs

Many of the immediate actions of GPCRs involve the rapid generation of second messengers (Fig. 6). $G\alpha_s$ stimulates adenylyl cyclases, which increases the cytosolic levels of cAMP, whereas $G\alpha_i$ inhibits adenylyl cyclases and hence decreases cAMP levels as a result of tonic phosphodiesterase activity. The different adenylyl cyclase isoforms appear to be distinctly regulated by $G\alpha_s$ and $G\alpha_i$, as well as by $G\beta\gamma$ subunits, intracellular calcium, and PKCs (Taussig and Gilman 1995). Thus, the effects of different GPCR agonists on the intracellular levels of cAMP are highly dependent on the adenylyl cyclases expressed in a given cell type. $G\alpha_t$ and $G\alpha_{gust}$ (also known as transducin and gustducin, respectively) activate phosphodiesterases in the visual system and gustative papillae, respectively, thus decreasing the cytoplasmic levels of cGMP by converting it to GMP (Cabrera-Vera et al. 2003). Members of the $G\alpha_q$ family bind to and activate phospholipase Cβ, which cleaves phosphatidylinositol 4,5-bisphosphate (PIP$_2$) into DAG and inositol 1,4,5-trisphosphate (IP$_3$). The latter causes an increase in cytosolic calcium levels by promoting the release of calcium from intracellular stores and also subsequent calcium influx into the cells, whereas DAG activates PKC (Hubbard and Hepler 2006; see Ch. 3 [Newton et al. 2014] and Ch. 11 [Julius and Nathans 2012] and p. 99 [Sassone-Corsi 2012]).

The released $G\beta\gamma$ dimers can independently activate many signaling molecules, including PLCβ, adenylyl cyclases, and ion channels (particularly the GIRK family of potassium channels). Although, in principle, the activation of any G protein by GPCRs should result in the release of $G\beta\gamma$ dimers and hence activation of their downstream targets, G_i and G_o are often the most highly expressed G proteins and represent the most frequent source of free $G\beta\gamma$ complexes.

The targets of diffusible second messengers activated by G proteins include a large number of ion channels, calcium-sensitive enzymes, and kinases such as PKA, PKC, cGMP-dependent kinase (PKG), and calcium-/calmodulin-dependent kinases (CaMKs), which are stimulated by cAMP, calcium/DAG, cGMP, and calcium, respectively. Heterotrimeric G proteins can also regulate other effectors, including signaling molecules that activate kinase cascades and small GTPases.

2.15 GPCRs Regulate a Network of Ras- and Rho-Related GTPases, MAPK Cascades, and PI3K-Regulated Signaling Circuits

GPCRs stimulate pathways that control cell migration, survival, and growth in part by activating MAPKs, a group of highly related proline-targeted serine/threonine kinases that link multiple cell-surface receptors to transcription factors. MAPKs include ERK1/2, JNK1-3, p38α-δ, and ERK5 MAPKs (Gutkind 1998; p. 81 [Morrison 2012]).

The small GTPase Ras, tyrosine kinases, PI3Ks, PKCs, and arrestins can act downstream from GPCRs to promote the activation of ERK1/2 in a cell-specific fashion (for review, see Gutkind 2000). The JNK cascade is activated downstream from the small G proteins Rac, Rho, and Cdc42 (Coso et al. 1995). Indeed, Rac and Cdc42 can mediate signaling of $G\beta\gamma$ dimers and $G\alpha_{12}$, $G\alpha_{13}$, $G\alpha_q$, and $G\alpha_i$ to JNK (Gutkind 2000; Yamauchi et al. 2000). Many GPCRs coupled to G_i activate Rac and JNK through the direct interaction of $G\beta\gamma$ subunits with the P-REX1/2 family of Rac-GEFs (Welch et al. 2002; Rosenfeldt et al. 2004). Similarly, $G\alpha_q$ activates Rho GTPases, and hence JNK, through direct interaction with p63-RhoGEF and Trio (Lutz et al. 2007). $G\alpha_{12}$ and $G\alpha_{13}$ bind to and act on three GEFs—p115-RhoGEF, PDZ-RhoGEF, and LARG—to promote the activation of Rho downstream from GPCRs (Hart et al. 1998; Fukuhara et al. 2001). Additional GEFs can also contribute to this network. How GPCRs activate p38 and ERK5 is much less clear, but, in general, these MAPKs are activated primarily by $G\alpha_q$, $G\alpha_{12/13}$, and $G\beta\gamma$ dimers (Gutkind 2000).

Although human cancer-associated viruses express constitutively active viral GPCRs, emerging data from deep sequencing studies have revealed that a large fraction of human malignancies harbour mutations in GPCRs and G-protein α subunits (O'Hayre et al. 2013). This has increased the interest in the molecular mechanisms by which G proteins and GPCRs control normal and cancer cell growth. Recent findings suggest that although GPCRs can stimulate multiple diffusible-second-messenger-generating systems, their ability to promote normal and aberrant cell proliferation often relies on the persistent activation of Rho GTPases and MAPK cascades based on the direct interaction of Gα subunits with RhoGEFs. The MAPKs regulate the activity of nuclear transcription factors and coactivators, such as Jun, Fos, and YAP (Yu et al. 2012; Vaqué et al. 2013).

Activation of the PI3K–Akt and mTOR pathways plays a central role in cell metabolism, migration, growth, and survival (Zoncu et al. 2011). PI3K generates 3′-phosphorylated inositol phosphates that participate in activation of the kinase Akt and mTOR, which relay downstream signals

(p. 87 [Hemmings and Restuccia 2012] and p. 91 [Laplante and Sabatini 2012]). PI3Kγ shows restricted tissue distribution and is activated specifically by GPCRs by the direct interaction of its catalytic (p110γ) and regulatory (p101) subunits with Gβγ subunits (Lopez-Ilasaca et al. 1997). PI3Kγ is involved in the chemokine-induced migration of leukocytes and plays significant roles in innate immunity (Costa et al. 2011). In cells lacking PI3Kγ expression, GPCRs can use PI3Kβ to stimulate phosphatidylinositol-(3,4,5)-tris-phosphate (PIP₃) synthesis (Ciraolo et al. 2008).

2.16 GPCR-Interacting Proteins in Receptor Compartmentalization, Trafficking, and G-Protein-Independent Signaling

GPCRs interact with a diverse array of proteins, which regulate compartmentalization to plasma membrane microdomains, endocytosis, trafficking between intracellular compartments and the plasma membrane, and G-protein-independent signaling (see below). These include receptor-activity-modifying proteins (RAMPs), GPCR-associated sorting proteins (GASPs), Homer, β-arrestins, arrestin-domain-containing proteins (ARRDCs), and DEP-domain proteins (Ballon et al. 2006; Magalhaes et al. 2012).

RAMPs are single-transmembrane-span proteins that associate with some class C GPCRs, such as the calcitonin receptor and calcitonin-like receptor. RAMPs facilitate the glycosylation of calcitonin family receptors in the endoplasmic reticulum, thereby facilitating their expression at the cell surface, and remain associated at the plasma membrane, where RAMPs contribute to ligand binding and receptor signaling (Bouschet et al. 2005). GASPs interact with the carboxy-terminal tail of many GPCRs and are primarily involved in their postendocytic sorting (Whistler et al. 2002). Homer 1a-c, Homer 2, and Homer 3 represent a class of proteins that harbor Enabled/VASP homology (EVH)–like domains and bind to metabotropic glutamate receptors (mGluRs) through a carboxy-terminal polyproline sequence (PPXXFP) (Bockaert and Pin 1999).

GPCRs can also associate with molecules containing protein–protein interaction domains, such as DEP, PDZ, WW, SH2, and SH3 domains (Ch. 2 [Lee and Yaffe 2014]), as well as polyproline-containing regions (Bockaert and Pin 1999; Brzostowski and Kimmel 2001). These interactions facilitate the localization of GPCR-initiated signaling to specific cellular structures or membrane microdomains, including the neuronal synapse, and also determine the signaling output by favoring the activation of a subset of GPCR targets by increasing their local accumulation in the vicinity of the GPCR.

β-Arrestins were initially described as adaptor proteins promoting the endocytosis of activated GPCRs (see below)

but are now believed to scaffold a wide variety of signaling complexes (Luttrell and Gesty-Palmer 2010; Rajagopal et al. 2010). In particular (Andreeva et al. 2007), they can interact with Src family kinases as well as multiple serine/threonine kinases, small GTPases and their GEFs, E3 ubiquitin ligases, phosphodiesterases, and transcription factors (Luttrell and Gesty-Palmer 2010; Rajagopal et al. 2010). β-Arrestins can act downstream from GPCRs within endocytic vesicles in a pathway leading to the activation of ERK1/2 and JNK, particularly in response to β-arrestin-biased agonists for some GPCRs, thus initiating intracellular signaling independently of the activation of heterotrimeric G proteins (Rajagopal et al. 2010). Interestingly, β-arrestins can form multimeric signaling complexes with ERK1/2 and JNK that are retained in the cytosol, thus restricting the nuclear translocation of these MAPKs, which instead act on cytosolic substrates (Fig. 6) (Rajagopal et al. 2010). Besides the best-studied β-arrestins, a family of α-arrestins that are conserved from budding yeast to humans has recently received increased attention because of their potential role in GPCR trafficking and degradation (Nabhan et al. 2010).

2.17 GPCR-Independent Activation of G Proteins

Heterotrimeric G proteins can be also activated in a GPCR-independent fashion by a family of proteins known as activators of G-protein-mediated signaling (AGS proteins) (Blumer et al. 2007). These proteins substitute for GPCRs by promoting nucleotide exchange on Gα subunits (e.g., AGS1), or can regulate the physical interaction and localization of G-protein subunits without affecting nucleotide exchange (e.g., AGS3, also known as LNG or PINS proteins) (Blumer et al. 2007). The latter play an important conserved role in cell polarity and polarized cell division and share the presence of a GoLoco motif by which they bind to Gα subunits to prevent nucleotide exchange (Willard et al. 2004).

2.18 GPCR Signal Termination

Considerable attention has focused on mechanisms of termination of GPCR signaling, because persistent activation occurs in many diseases (Pierce et al. 2002). This desensitization is highly regulated and occurs through several well-understood mechanisms, including GPCR-targeted kinases known as GPCR kinases (GRKs), and more general second-messenger-regulated kinases, such as PKC and PKA. PKC and PKA phosphorylation uncouples receptors from their respective G proteins, presumably by phosphorylating G-protein-interaction sites or by recruiting arrestins (Benovic et al. 1985), thereby forming a negative-feedback loop. Activation of PKA and PKC can also result in the heterologous desensitization of multiple GPCRs within a cell (Chuang et al. 1996). In contrast, GRKs phosphorylate only the activated or agonist-occupied form of the receptor, primarily in the carboxy-terminal tail, which then binds arrestin dimers that can prevent G-protein interaction and promote the removal of the receptor from the cell surface by endocytosis (Shenoy and Lefkowitz 2011). Internalized receptors can be recycled back to the plasma membrane or degraded in lysosomes, a process influenced by the ability of ligand-bound receptors to interact with ubiquitin ligases and a complex repertoire of sorting molecules (Hanyaloglu and von Zastrow 2008). Concomitantly, a family of regulators of GPCR signaling (RGS) molecules act as GAPs on GTP-bound G-protein α subunits, accelerating GTP hydrolysis and hence signal termination (Berman and Gilman 1998). Signaling and inactivation are intertwined, because molecules such as arrestins can also regulate GPCR signaling specificity and/or localization (Shenoy and Lefkowitz 2011), and many G-protein targets include RGS domains or act as GAPs, thus acting as direct effectors that concomitantly limit the duration of G-protein signaling.

2.19 GPCR Signal Integration

GPCRs are best known for their ability to control the activity of adenylyl cyclases, phosphodiesterases, phospholipases, ion channels, and ion transporters. The rapid regulation of these classical diffusible-second-messenger-generating systems and their direct molecular targets is now believed to represent a subset of the extensive repertoire of molecular mechanisms deployed by GPCRs in physiological and pathological contexts. Our recently gained understanding of GPCR signaling circuitries, including GEFs, Ras and Rho GTPases, MAPKs, PI3Ks, and their numerous downstream cytosolic and nuclear targets, provide a more global view of the general systems by which these receptors exert their numerous physiological and pathological roles. Indeed, the final biological outcome of GPCR activation most likely results from the integration of the network of GPCR-initiated biochemical responses in each cellular context. A new, systems-level understanding may provide a molecular framework for the development of novel approaches for therapeutic intervention in some of the most prevalent human diseases.

3 THE TNF RECEPTOR FAMILY

3.1 TNF Isoforms and TNF Receptors

Tumor necrosis factor (TNF) belongs to a 19-member family of structurally related factors that bind to the 29 members of the TNF receptor (TNFR) family. Most TNF and TNFR members are expressed in the immune system, where they regulate defense against infection (see Ch. 15 [Newton and Dixit 2012]); however, some members are expressed elsewhere, regulating hematopoiesis and morphogenesis. Perturbation of signaling by TNFRs is implicated in several diseases, including tumorigenesis, bone resorption, rheumatoid arthritis, and diabetes (Aggarwal 2003; Croft 2009).

Members of the TNFR family have an amino-terminal extracellular region consisting of one to four cysteine-rich domains, each of which contains three conserved intrachain disulfide bridges. These are linked to a single transmembrane segment and a cytoplasmic region that lacks enzymatic activity (Locksley et al. 2001). Several receptors contain binding motifs for members of the TRAF family of ubiquitin ligases in their intracellular regions, and eight of them contain death domains (DDs), which are involved in apoptotic signaling (Fig. 7). Some instead lack intracellular and/or transmembrane regions and act as decoy receptors.

The TNF family ligands are also transmembrane proteins but have extracellular carboxyl termini. Intriguingly, these can be shed following cleavage by proteases, and some inhibit the effects of the membrane-bound ligand (Suda et al. 1997). The ligands are characterized by a conserved TNF homology domain that mediates receptor binding. An exception is nerve growth factor (NGF), which in addition to binding to an RTK also binds to p75, a member of the TNFR family. Several ligands of this family bind to more than one receptor (Fig. 7).

3.2 Activation of TNFRs

Most TNF family ligands are noncovalent trimers that form symmetric 3:3 complexes with their receptors. Homomeric as well as heteromeric receptor complexes have been described (Schneider et al. 1997). Preformed receptor oligomers exist and TNFR1 and TNFR2 have a pre-ligand-binding assembly domain (PLAD) that is required for the assembly of TNFR complexes (Chan et al. 2000). Whereas juxtaposition of the receptors is important for activation, trimerization itself appears not to be necessary because there are examples of agonistic bivalent monoclonal antibodies directed against the extracellular domain, and because one of the ligands, NGF, is a dimer.

3.3 Signaling via TNFRs

The receptors that have DDs, including TNFR1, Fas (also known as CD95), death receptor (DR) 3, DR4, DR5, DR6, EDAR, and the p75 NGF receptor, induce apoptosis and necrosis of cells (Moquin and Chan 2010; Ch. 15 [Newton and Dixit 2012]; Ch. 19 [Green and Llambi 2014]). Ligand-induced receptor activation leads to the formation of complexes referred to as death-inducing signaling complexes

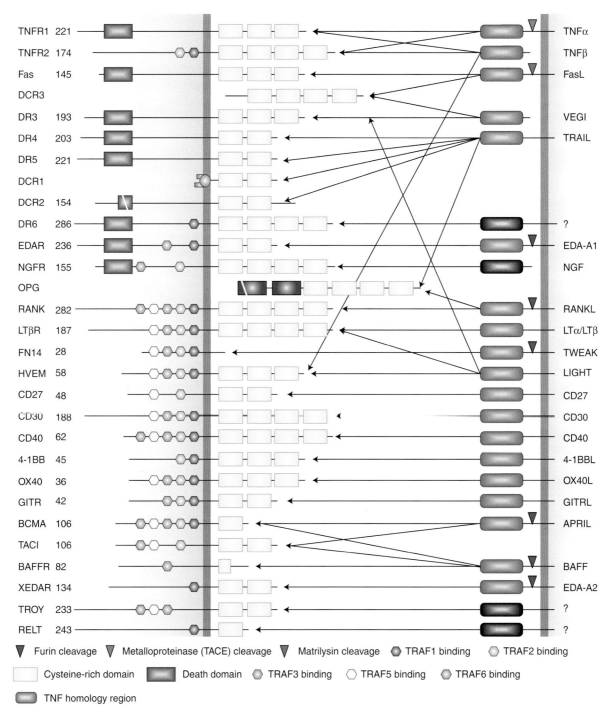

Figure 7. The tumor necrosis factor (TNF) receptor family. The structural features of the members of the TNF receptor family (*left*) and their ligands (*right*) are shown. Cysteine-rich domains, death domains, and interaction motifs for various members of the TRAF family are indicated. Cleavage sites for various proteases involved in processing of the ligands are also shown. (From Aggarwal 2003; adapted, with permission.)

(DISCs), in which the receptor DDs interact with DDs in adaptor molecules such as TRADD and FADD. The protease procaspase 8 is also recruited to the DISC, triggering a caspase cascade that results in apoptosis.

In addition to apoptosis, TNFRs control survival and differentiation. Many of their effects are mediated by the TRAF family of ubiquitin ligases, which stimulate NF-κB, PI3K, and JNK and p38 MAPK pathways (Faustman and

Davis 2010). Activation of NF-κB is of central importance and occurs downstream from all members of the TNFR family, except the decoy receptors. Moreover, certain members of the TNFR family stimulate cell proliferation, for example, by activation of the ERK1/2 MAPK.

3.4 TNFRs and Disease

Overactivity of members of the TNFR family is seen in immune-related diseases, and encouraging results have been obtained using inhibitory anti-TNF antibodies or ligand traps in the treatment of Crohn's disease (Suenaert et al. 2002) and rheumatoid arthritis (Feldmann and Maini 2001). Similarly, blocking signaling via the receptors OX40, 4-1BB, CD27, and DR3 is effective in various immune diseases (Croft 2009). Promising attempts have also been made to induce apoptosis of cancer cells by treatment with agonists of TNFR1, Fas, and TRAIL receptors (Fox et al. 2010).

4 CELL ADHESION RECEPTORS AND MECHANOTRANSDUCERS

4.1 Mechanosensing Signaling through Integrins

Cells deploy multiple signaling mechanisms to sense the biophysical properties of their surroundings and communicate with their neighboring cells. Integrins are perhaps the best known cell-surface receptors involved in these essential processes. They function primarily in cell adhesion to the extracellular matrix (ECM) or to other cells, the latter through a repertoire of cell-surface ligands involved in cell–cell interactions. Integrins accumulate at cell–ECM and cell–cell contact points and orchestrate the local assembly of multimolecular structures known as adhesion complexes, often referred to as focal adhesions (FAs) or adhesomes (Hynes 2002; Geiger and Yamada 2011).

Integrins provide a link between the extracellular environment and the intracellular cytoskeleton, and their dynamic engagement at cell–ECM and cell–cell adhesions results in the rapid activation of multiple intracellular signaling circuits, a process known as outside-in signaling (Hynes 2002; Miranti and Brugge 2002). In turn, the adhesive properties of integrins and hence the strength of the interactions with the ECM and other cells are regulated by a variety of signaling pathways, which impinge on the interaction of integrins with a key cytoskeletal molecule, talin (see below); this process is known as inside-out signaling (Hynes 2002). Integrins can be regulated in this way by signals from growth factors acting on RTKs and GPCRs, as well as by inflammatory cytokines and bioactive lipids (Hynes 2002; Miranti and Brugge 2002). The latter is of particular importance for immune cells and platelets, in which

integrins are maintained in an inactive state that enables cells to circulate in the bloodstream without interacting with endothelial cells in the blood vessel wall, but quickly deploy their adhesive properties in response to immune cell activation and the coagulation cascade (Moser et al. 2009). For example, cytokine-induced activation of the leukocyte integrin LFA1 (also known as $\alpha_L\beta_2$) leads to its rapid binding to intercellular adhesion molecule (ICAM)-1, an adhesion molecule expressed on the surface of activated endothelial cells, thereby stabilizing cell–cell interactions and facilitating the transmigration of circulating leukocytes through the endothelial cell layer into damaged tissues (Moser et al. 2009). The integrins also provide important survival signals, because many cells undergo anoikis, a form of programmed cell death, upon detaching from their surrounding ECM (Frisch and Screaton 2001). Overall, integrin-mediated cell adhesion controls cell migration, survival, growth, and differentiation, whereas at the organismal level, integrins play fundamental roles in tissue morphogenesis during development, and in the immune response, hemostasis, and tissue maintenance and repair (Hynes 2002; Miranti and Brugge 2002; Ch. 8 [Devreotes and Horwitz 2013]).

Each integrin is composed of a heterodimer of two transmembrane subunits (α and β). In humans, there are 18 α subunits (Fig. 8), which associate with eight different β subunits to form at least 24 heterodimeric complexes displaying distinct ligand-binding specificities and signaling capacities. Some of these dimers are widely expressed, whereas others show more restricted distributions and can therefore have more specific biological functions. Integrins typically have large extracellular regions, which interact with cell-surface ligands and with specific sequence motifs in ECM proteins, such as the tripeptide RGD motif originally described in fibronectin (Geiger et al. 2001; Hynes 2002). Some integrins are promiscuous and can bind to multiple ligands, for example, the ECM components vitronectin, fibrinogen, fibronectin, laminin, collagen, and tenascin. Other integrins show a more restricted binding pattern or bind to other cell adhesion receptors—for example, ICAMs, vascular cell adhesion molecule (VCAM), and E-cadherin (see Table 1). Most integrins possess a relatively short intracellular cytoplasmic domain (40–70 amino acids), with the exception of integrin β4, which has a long cytoplasmic domain.

The conformation of the integrin extracellular domains changes dramatically upon cell–ECM and cell–cell adhesions (Shattil et al. 2010). This results in the separation of the intracellular cytoplasmic tails, which lack enzymatic activity but instead act as platforms for the assembly of multimeric protein complexes that are involved in signal transduction (Shattil et al. 2010; Wehrle-Haller 2012) and link integrins to the cytoskeleton. The direct binding of talin to the cytoplas-

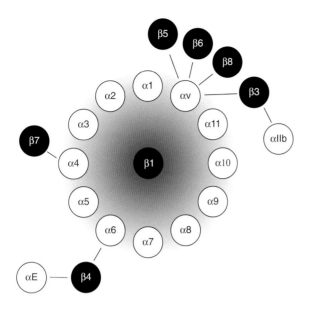

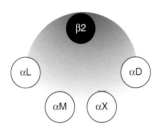

Figure 8. The integrin family of cell adhesion receptors. Integrins are composed of a heterodimer of two transmembrane α and β subunits. The 18 α subunits and eight β subunits can form at least 24 heterodimeric complexes displaying distinct binding specificity and signaling capacity. (Adapted from Hynes 2002.)

out signaling (Legate and Fassler 2009; Moser et al. 2009). Tensin is recruited to FAs and helps to establish additional connections between integrins and actin fibers. Ultimately, these multiple links between integrins and actin promote localized actin polymerization at sites of cell–ECM and cell–cell contact. This allows rapid assembly and disassembly of adhesion contacts and dynamic control of the cellular cytoskeleton during cell migration (Huttenlocher and Horwitz 2011; Ch. 8 [Devreotes and Horwitz 2013]).

FAs (Geiger and Yamada 2011) contain many cytoskeletal, adaptor, and signaling proteins. Among them, multiple nonreceptor tyrosine kinases, including focal adhesion kinase (FAK), the related proline-rich tyrosine kinase 2 (Pyk2), Src, and Src-family kinases act both as signal transducers and as scaffolds (Fig. 9) (Miranti and Brugge 2002; Geiger and Yamada 2011). Activation of these tyrosine kinases upon integrin ligation results in their autophosphorylation and cross-phosphorylation at several tyrosine residues, creating binding sites for multiple signaling proteins. These include the adaptor protein Grb2, which leads to recruitment of SOS (a GEF for Ras) and the consequent activation of Ras and the MAPK ERK1/2. In the case of FAK, the phosphotyrosines recruit Src, Grb2, and the p85 subunit of PI3K via their SH2 domains (Miranti and Brugge 2002; Legate et al. 2009; Shattil et al. 2010). The latter gen-

Table 1. Ligands for α β integrin heterodimers

Integrin	Representative ligands
α1β1	Collagen, laminin
α2β1	Collagen, laminin
α3β1	Laminin, fibronectin
α4β1	VCAM-1, fibronectin, thrombospondin
α4β7	VCAM-1, fibronectin, MAdCAM-1
α5β1	Fibronectin, neural adhesion molecule L1
α6β1	Laminin
α6β4	Laminin
α7β1	Laminin
α8β1	Osteopontin, fibronectin, vitronectin, tenascin
α9β1	Tenascin, osteopontin, VCAM-1
α10β1	Collagen
α11β1	Collagen
αEβ1	E-cadherin
αLβ2	ICAM-1, ICAM-2, ICAM-3
αMβ2	ICAM-1, ICAM-2, ICAM-4, fibrinogen
αXβ2	ICAM-1, fibrinogen
αDβ2	ICAM-3, VCAM-1
αIIbβ3	Fibrinogen, fibronectin, vitronectin, thrombospondin, von Willebrand factor
αVβ3	Vitronectin, fibrinogen, fibrinogen, osteopontin
αVβ5	Vitronectin
αVβ6	Vitronectin, fibronectin
αVβ8	Fibronectin, laminin, collagen, vitronectin
α6β4	Laminin
α4β7	VCAM-1, fibronectin

mic tails of most activated integrin β subunits is one of the earliest events after integrins bind to the ECM. This causes the local accumulation of PIP_2 upon recruitment of the lipid kinase phosphatidylinositol 4-phosphate 5-kinase via its interaction with talin and the subsequent recruitment of vinculin to the nascent adhesions (Legate and Fassler 2009; Moser et al. 2009; Shattil et al. 2010). This helps stabilize FAs, because integrin β subunits interact weakly with actin through talin in the absence of vinculin. Vinculin and talin also bind to α-actinin, which has a high affinity for actin and hence helps strengthen the interactions between β integrins and actin filaments. Other molecules linking the β subunits to actin include filamin, kindlins, migfilin, and integrin-linked kinase (ILK), a pseudokinase that bridges β integrins and parvin, another actin-binding protein (Hynes 2002; Legate et al. 2009; Geiger and Yamada 2011). ILK stabilizes the FA by retaining integrins in a clustered state and by reinforcing the connection with the cytoskeleton, whereas kindlins partner with talins in integrin inside-

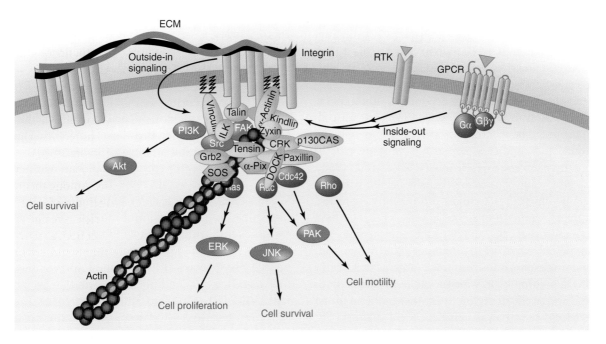

Figure 9. Integrin-based cell adhesion and signaling. Integrin engagement at cell-matrix adhesions or interaction with a repertoire of cell-surface ligands results in the rapid assembly of a multifunctional protein network (Geiger and Yamada 2011) containing many cytoskeletal, adaptor, and signaling proteins. This contributes to cell adhesion and activates multiple signaling events. The adhesive properties of integrins are, in turn, regulated by a variety of signaling pathways; this is known as inside-out signaling.

erates PIP$_3$ at the plasma membrane, which recruits ILK, Akt, and other proteins to the FA complex (Miranti and Brugge 2002; Shattil et al. 2010; Ch. 2 [Lee and Yaffe 2014]). Akt relays prosurvival and growth-promoting signals. A direct substrate of FAK, paxillin, acts as a multidomain adaptor protein that forms a scaffold organizing a variety of signaling molecules, including Src, ILK, Crk, and vinculin (Miranti and Brugge 2002; Geiger and Yamada 2011). Crk is an adaptor protein that binds to tyrosine-phosphorylated paxillin or another adaptor protein, p130Cas, and can then use its SH3 domain to recruit multiple additional proteins. These include the GEF DOCK180, which activates the small GTPase Rac1 (Miranti and Brugge 2002; Legate et al. 2009). The Rac/Cdc42-GEF α-Pix is also activated upon integrin ligation by binding to phosphorylated paxillin and by the local accumulation of PIP$_3$. Moreover, Src or FAK can phosphorylate and activate Vav, a Rac GEF that is recruited to the newly formed adhesive structures (Miranti and Brugge 2002; Legate et al. 2009). This results in the rapid remodeling of the actin cytoskeleton by Rho GTPases and their downstream effectors and the relay of the signals to the nucleus by JNK and p38 (Miyamoto et al. 1995; Miranti and Brugge 2002; p. 81 [Morrison 2012]).

The assembled protein network thus both contributes to adhesion to the ECM and other surrounding cells and orchestrates signaling events that enable the cells to respond appropriately to mechanical cues. In addition to outside-in and inside-out signaling, integrin activation can also induce the rapid clustering of multiple RTKs, such as EGFR, PDGFR, and FGFR (Miyamoto et al. 1996; Geiger and Yamada 2011). This enhances signaling in response to cell adhesion. Finally, note that rather than being rigid intracellular structures, most FAs are dynamic (Ch. 8 [Devreotes and Horwitz 2013]), and regulation of their multiple components and adhesive properties by mitogens and chemoattractants, for example, is essential for the rapid dissolution of preexisting adhesions and the establishment of new contact sites during cell migration.

4.2 Signaling by Cell Adhesion Molecules (CAMs)

The formation of adhesive structures between adjacent cells, including adherens junctions and tight junctions, contributes to the establishment of cell polarity, differentiation, and survival and consequently key morphogenetic processes involved in embryonic development, control of tissue homeostasis, and tissue repair in adults. A large family of cell adhesion molecules (CAMs) provides mechanical adhesion among cells, while initiating signaling via networks that control cellular behavior in response to the microenvironment. In general, cadherin molecules mediate cell–cell adhesion at adherens junctions, claudins contribute to the formation of tight junctions, and immuno-globulin-like CAMs (Ig-CAMs) accumulate throughout the intercellular boundary (Cavallaro and Dejana 2011). Other CAMs, known as nectins, can participate in both adherens and tight junctions (Takai et al. 2008).

The cadherins are calcium-dependent, homophilic, cell–cell adhesion molecules expressed in nearly all cells in solid tissues. These molecules participate in cell–cell recognition, and only cells expressing the same type of cadherins may adhere to each other (Nose et al. 1988). The "classical" cadherins were originally named based on the tissue in which they are most prominently expressed, for example, E-cadherin in epithelial cells, VE-cadherin in vascular endothelial cells, and N-cadherin in nervous system and mesenchymal cells (Gumbiner 2005). These cadherins are single-pass transmembrane proteins that form a core adhesion complex in which a cadherin dimer binds through its extracellular region to another cadherin dimer on an adjacent cell in a calcium-dependent manner. The cadherin intracellular region is anchored in the plasma membrane and linked to the cytoskeleton through a family of proteins known as catenins (Gumbiner 2005). β-Catenin interacts with the distal part of the cadherin cytoplasmic tail, and p120 catenin interacts with a more proximal region (Gumbiner 2005). α-Catenin binds primarily to cadherin-associated β-catenin and provides a physical link to the actin cytoskeleton, by binding actin filaments either directly or indirectly through other actin-binding proteins, such as vinculin, α-actinin, and formins (Kobielak and Fuchs 2004; Mege et al. 2006). p120 regulates cell movement through its ability to control both cell adhesion by governing the availability of cadherins at the plasma membrane and actin cytoskeleton organization by regulating Rho GTPases (Anastasiadis and Reynolds 2001; Grosheva et al. 2001; Yanagisawa and Anastasiadis 2006). p120 can also directly affect gene expression by repressing transcription by scaffolding a nuclear complex including the gene silencer Kaiso (Daniel and Reynolds 1999).

The formation of cadherin-dependent adhesions contributes to growth inhibition upon cell–cell contact. This has often been associated with inhibition of the canonical Wnt pathway by cadherins, which retain β-catenin at the plasma membrane and therefore limit the pool of free β-catenin available for nuclear signaling (Nelson and Nusse 2004). Recent evidence suggests that cadherin engagement can also activate the Hippo pathway, which results in the cytoplasmic retention of the transcriptional coactivator YAP that is necessary for cell growth (Kim et al. 2011). In epithelial cells, E-cadherin acts as a tumor and metastasis suppressor. Its expression and function are down-regulated or altered in many human cancers, and its reexpression decreases both the proliferative and invasive capacity of

tumor cells (Vleminckx et al. 1991; Thiery and Sleeman 2006). However, the engagement of cadherins in newly formed cell contacts promotes cell proliferation and survival through the activation of MAPKs, PI3K, and Rho GTPases (Pece et al. 1999; Pece and Gutkind 2000; Braga and Yap 2005). This process often involves the engagement of growth factor receptors such as EGFR, VEGFR2, FGFR, and PDGFR, promoting their ligand-independent clustering and activation and prolonging the activation by ligands by enhancing receptor recycling and limiting their degradation (Carmeliet et al. 1999; Pece and Gutkind 2000; Suyama et al. 2002; Cavallaro and Dejana 2011).

Cadherins can also control multiple cellular processes by associating with Src-family kinases, G proteins of the $G\alpha_{12/13}$ family, and phosphatases, such as density-enhanced phosphatase (RPTPη), the tyrosine phosphatase SHP2, and vascular endothelial protein tyrosine phosphatase (VE-PTP) (Cavallaro and Dejana 2011). Cadherins thus contribute to cell–cell recognition and adhesion while promoting cell survival and restricting cell motility/growth by regulating intracellular signaling at the plasma membrane. In some cases, the cadherin intracytoplasmic tail can translocate to the nucleus and regulate transcription after shedding of the ectodomain by cell-surface matrix and disintegrin family proteases and cleavage of the carboxy-terminal tail by intracellular proteases, such as γ-secretase (Cavallaro and Dejana 2011), which is reminiscent of Notch signaling.

Tight junctions involve numerous adhesion molecules, including occludin, junctional adhesion molecules (JAMs), and the claudin family of tetraspan transmembrane proteins, as well as intracellular adaptors, such as ZO1 and ZO2 (Tsukita et al. 2001). Claudins are the major adhesive proteins at tight junctions; whether they play a direct role in cell signaling is not clear yet.

Ig-CAMs are cell-surface glycoproteins that accumulate at the cell–cell boundary, and their homophilic interactions contribute to cell–cell recognition and adhesion in a calcium-independent fashion (Loers and Schachner 2007; Cavallaro and Dejana 2011). Although most Ig-CAMs have a transmembrane region and a cytoplasmic tail, some associate with the cell surface via a GPI anchor. In the former case, the cytoplasmic tail of Ig-CAMs can interact with cytoskeletal proteins such as actin, ankyrins, and spectrin and can also initiate signal transduction (Cavallaro and Dejana 2011). For example, the formation of NCAM-based adhesions results in the activation of a kinase cascade including CaMKIIα, the Src-family kinase Fyn, and FAK, and this promotes neurite outgrowth and neuronal survival (Bodrikov et al. 2008). NCAM also stimulates PKCβII by recruiting it to the membrane through the formation of a signalling complex involving a protein known as growth-associated protein 43 (GAP43) (Korshunova and Mosevitsky 2010). In common with cadherins, NCAM can interact with multiple growth factor receptors, including FGFR, regulating their signaling capacity.

Signaling by CAMs—either direct or indirect actions engaging and prolonging the activity of RTKs—is likely to play a key role during the formation of cell–cell contacts, particularly during embryonic development, morphogenesis, and tissue repair (Pece and Gutkind 2000; Dumstrei et al. 2002; Andl et al. 2003; Fedor-Chaiken et al. 2003). This may accelerate the growth rate without the need for elevated local levels of growth factors. The stabilization of cell–cell contacts may subsequently reduce signaling by ligand-activated RTKs by sequestering them in CAM-containing clusters, while favoring their ligand-independent activation of survival pathways, such as PI3K–Akt signaling (Pece and Gutkind 2000; Qian et al. 2004). Concomitantly, CAMs may restrict cell and organ overgrowth by limiting the availability of free β-catenin and YAP (Nelson and Nusse 2004; Kim et al. 2011). In these cases, CAMs may help prevent tumor formation while generating survival and differentiation signals. The microenvironment and RTK signaling networks can, in turn, modulate the localization, expression, and stability of CAMs and the proteins that they associate with, thus regulating their adhesive properties and signaling capacity. Conversely, CAMs can regulate RTK signaling, acting as rheostats governing the intensity and duration of their signals in response to environmental cues such as cell density, tissue architecture, and polarity.

5 NUCLEAR RECEPTORS

Nuclear receptors (NRs) comprise a large superfamily of intracellular transcription factors that can effect gene expression changes in response to a wide variety of lipophilic ligands (p. 129 [Sever and Glass 2013]). In this respect, they differ from most other receptors in that they do not reside in the plasma membrane, which many of their ligands can cross. Typical NR ligands include steroids, vitamins, dietary lipids, cholesterol metabolites, and xenobiotics. Ligands for 24 NRs have been identified; the remaining family members are considered orphan receptors (Mangelsdorf et al. 1995). There are 48 NRs in humans (49 in mice and 18 in *Drosophila*). NRs are believed to be only one of two transcription factor families that are metazoan specific (King-Jones et al. 2005; Degnan et al. 2009). Because many are endocrine hormone receptors, this suggests a potentially critical role in the evolution of animal physiology. Hormonal NRs are typically classified by the type of ligands to which they bind, whereas orphans have various different abbreviations, indicating, for example, similarity to known receptors (e.g., estrogen-related receptor, ERR).

Binding of their cognate ligands, in the most simplistic scenario, either changes the cellular localization of the NR or its interaction with repressive and activating cofactors in the nucleus. What distinguishes NRs from other genres of receptors is the ability to mediate transcription without intermediate signaling cascades. Instead, they directly bind to target genes. Nuclear–cytoplasmic cycling of some NRs allows them to have nongenomic effects and also to be targets of cytoplasmic signaling cascades (Wehling 1997).

5.1 Structure and Mechanisms of Receptor Activation

Nuclear receptors generally contain five functional domains. The A/B region at the amino terminus is divergent and in some NRs contains a ligand-independent transcriptional activation function domain (AF1). The highly conserved DNA-binding domain (DBD) is located in the central C domain. The carboxyl terminus is the E region, which contains the ligand-binding domain (LBD). A flexible hinge D region links the DBD and LBD (Fig. 10). The DBD contains tetracysteine (C4) zinc fingers that are unique to NRs and define membership in the superfamily. The DBD typically targets the NR to a specific DNA element termed a hormone-response element (HRE). The LBD is composed of 12 α helices and mediates ligand recognition, dimerization, interaction with coactivators/corepressors, and ligand-dependent transcriptional activation. The last helix, helix 12, within the LBD is the AF2 domain, which enables NRs to interact with short LxxLL motifs in coactivators or corepressors. These are termed the NR box or CoRNR box, respectively (Hu and Lazar 1999). Whereas AF1 domains vary greatly among all NRs and have a propensity to stay disordered, AF2 domains have similar structures (Warnmark et al. 2003).

5.2 Nuclear Receptor Classification

NRs can be classified based on their DNA-binding mechanism (Fig. 10) or their ligand (Table 2). On the basis of their mode of actions, nuclear receptors are classified into four types. Type I and type III NRs are normally sequestered in the cytoplasm by heat shock proteins. Ligand binding to type I NRs dissociates heat shock proteins and results in nuclear enrichment. Type I NRs bind to DNA as homodimers, and the response elements they recognize are inverted repeats. Type II receptors, unlike type I receptors, are normally enriched in the nucleus and are bound to DNA even in the absence of ligand. They generally bind to DNA as heterodimers with the retinoid X receptor (RXR). In the absence of ligand, type II receptors repress transcription via association with corepressors. In the presence of ligand,

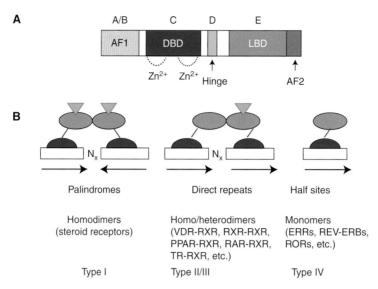

Figure 10. General structure and binding of nuclear receptors. (*A*) Domain organization of a typical nuclear receptor. (*B*) Three modes of signal transduction: as monomers, heterodimers, and homodimers. (From Sonoda et al. 2008: modified, with permission, © Elsevier.)

corepressors are dissociated, and coactivators including histone-modifying enzymes are recruited by the type II receptors for gene activation. Examples of type II receptors include the thyroid hormone receptor (TR) and retinoic acid receptor (RAR). Type IV receptors bind as monomers and recognize half-site response elements.

5.3 Three Modes of Binding to Promoter Regions Influence Transcription

Hormone-responsive NRs are sequestered in the cytoplasm or are bound to HREs in repressive complexes that include chromatin modifiers such as histone deacetylases (HDACs), nuclear corepressors (N-CoRs), and SMRT proteins (for "silencing mediator of retinoid and thyroid receptors") (Perissi et al. 2010). Ligand binding typically promotes nuclear translocation and/or recruitment of coactivators that displace corepressors to initiate transcription (Rosenfeld et al. 2006). Coactivators that mediate NR function include PGC1, SIRT1, p160, and p300 proteins (Sonoda et al. 2008). Mechanistically, NRs can mediate transcription (either repression or activation) in the following manner.

5.3.1 Homodimers

Type I and type III receptors bind to DNA as homodimers when bound to their cognate ligands. Type I receptors include all steroid receptors, and their response elements typically consist of two hexameric inverted (palindromic) repeat half-sites, for example, AGAACA (the glucocorticoid response element, GRE) or AGAACA (the estrogen response element, ERE). These sequences can mediate either activation or repression in response to hormone. The DBDs of steroid receptors are highly similar. Thus, the glucocorticoid receptor (GR), the mineralocorticoid receptor (MR), the androgen receptor (AR), and the progesterone receptor (PR) all bind to overlapping response elements. Type III receptors are similar to type I receptors except that they recognize direct repeats instead of inverted repeats.

5.3.2 Heterodimers

RXR forms heterodimers with various nonsteroidal members of the NR superfamily that do not bind to HREs efficiently by themselves. Depending on the type of NR, HREs have specific conformations. Unlike the type I receptors mentioned above, type II receptors form retinoid-acid-related-receptor (RXR)-containing heterodimers that recognize direct repeats of AGGTCA or similar sequences separated by specific spacing (Umesono et al. 1991). The RAR- and peroxisome proliferator-activated-receptor-γ (PPARγ)-containing dimers RAR–RXR and PPARγ-RXR bind to a direct repeat (DR-1), whereas the vitamin D_3 receptor (VDR)-, TR-, and RAR-containing dimers VDR–RXR, TR–RXR, and RAR–RXR bind to DR-3, DR-4, and DR-5, respectively (Evans 2005).

5.3.3 Monomers

Type IV NRs bind to just one AGGTCA site as monomers in a ligand-independent manner. These receptors use an extended DBD. There are currently three known types of

Table 2. The nuclear receptor superfamily

Steroid receptors	
GR	Glucococorticoid
MR	Mineralocorticoid
AR	Androgens
PR	Progesterone
TRα, TRβ	Thyroid hormone
ERα, ERβ	Estrogen
Vitamin receptors	
VDR	Vitamin D
RARα, RARβ, RARγ	Retinoic acid
Fatty acids and derivatives	
PPARα, PPARβ/δ, PPARγ	Fatty acids
LXRα, LXRβ	Oxysterol
FXR	Bile acids
RXRα, RXRβ, RXRγ	Rexinoids
Xenobiotic receptors	
CAR	Androstane metabolites
PXR	Pregnane derivatives
Adopted orphan receptors	
HNF4α, HNF4γ	Fatty acids
REV-ERBα, REV-ERBβ	Heme
RORα, RORβ, RORγ	Cholesterol, retinoic acid
SF1, LRH1	Phosphatidylinositols
ERRα, ERRβ, ERRγ	Estrogen
Orphan receptors	
SHP	
DAX1	
TLX	
PNR	
GCNF	
TR2, 4	
NR4A1, NR4A2, NR4A3	
COUP-TFα, COUP-TFβ, COUP-TFγ	

monomer-binding receptors, which are defined by unique sequences 5′ to the consensus site. REV-ERBs tend to bind to A(A/T)CTAGGTCA sequences, whereas ERRs bind to TNAAGGTCA sequences.

6 FUNCTIONS OF NUCLEAR RECEPTORS

6.1 Steroid and Thyroid Hormone Receptors

Steroid and thyroid receptors maintain homeostasis by acting as sensors for circulating hormones from the adrenals, ovaries, testes, and thyroid. GR, MR, AR, and PR are closely related and bind common DNA sequences. Following ligand treatment, these receptors translocate from the cytoplasm to the nucleus and bind to palindromes of the GR half-site AGAACA as homodimers. All other members of the superfamily, including ER and TR, bind to response elements that contain variations of the ER half-site AGGTCA.

6.2 Nonsteroidal Hormonal Receptors

Instead of acting as homodimers, all nonsteroidal hormone receptors, including TR, use RXR as a partner to form heterodimers. VDR and RAR are activated by calcitriol (vitamin D) and retinoic acid. RXR is conserved in invertebrates and may be an ancient member of the NR family. RXRs (α, β, and γ) are activated by the novel retinoid isomer 9-*cis* retinoic acid and other synthetic agents (rexinoids) (Mangelsdorf et al. 1992). RXR forms three different types of dimers. Other than the nonpermissive heterodimers mentioned above (e.g., RAR and VDR), which cannot be activated by the RXR ligand, RXR can form homodimers and permissive heterodimers (e.g., with PPAR and liver X receptor, LXR) that can be activated by RXR-ligand binding.

6.3 Receptors for Fatty Acids and Derivatives

The PPAR subfamily includes PPARα, PPARδ, and PPARγ. Each of the PPARs binds to common response elements as heterodimers with RXR, using fatty acid ligands, which function as metabolic sensors. PPARα is highly expressed in the liver and is the target of the fibrate class of drugs used to control hyperlipidemia (Issemann and Green 1990). PPARδ is universally expressed and promotes fatty-acid oxidation. PPARγ is required for adipogenesis, promotes fat storage, and has been widely studied for its role in the metabolic syndrome (Barak et al. 1999). LXRα and LXRβ and farnesoid X receptor (FXR) are sensors for cholesterol metabolites such as oxysterols and bile acids, respectively. Both LXRs and FXR form heterodimers with RXR. LXRs are important for maintaining cholesterol homeostasis, whereas FXR acts as the primary bile acid sensor by repressing expression of 7α-hydroxylase (CYP7A1), the rate-limiting enzyme in bile acid production, through up-regulation of two orphan NRs, SHP and LRH1, upon bile acid stimulation (Lu et al. 2001). SHP and LRH1 form inactivating heterodimers to repress transcription of *CYP7A1*, thus turning off bile acid production.

6.4 Xenobiotic Receptors

The xenobiotic receptors include the constitutive active/androstane receptor (CAR) and pregnane X receptor (PXR). Both CAR and PXR form heterodimers with RXR and bind promiscuously to different response elements (Lehmann et al. 1998; Sueyoshi et al. 1999). Their role is to induce expression of enzymes such as CYP3A that promote the metabolism of exogenous compounds (xenobiotics). Unlike typical NRs, PXR has a promiscuous LBD pocket that binds a wide variety of natural and synthetic compounds. Activation of xenobiotic receptors is not always beneficial to the host. For example, upon exposure to high doses of acetaminophen (a common component of pain killers and cocaine), CAR can mediate production of toxic by-products through up-regulation of three CYP enzymes that lead to acute hepatotoxicity (Zhang et al. 2002).

6.5 Adopted Orphan Receptors

Discovery of natural or synthetic ligands for several orphan NRs has led to their classification as "adopted" orphan receptors (e.g., REV-ERBα and REV-ERBβ). REV-ERBs are repressors that compete with retinoid-related orphan receptors (RORα, RORβ, and RORγ) for binding to (A/T)$_6$RGGTCA HREs (Forman et al. 1994). The REV-ERB ligand is neither a lipid nor a hormone but, rather, a single heme molecule (Reinking et al. 2005). Many heme-containing proteins also bind to diatomic gases, and REV-ERBs are also responsive to NO and CO (see below). RORs are proposed to bind cholesterol and retinoid acid, although whether this is important in antagonizing REV-ERBs is not clear (Kallen et al. 2002; Stehlin-Gaon et al. 2003). NR5A subfamily members SF1 (NR5A1) and LRH1 (NR5A2) were originally thought to be constitutively active. Recent analysis, however, reveals that these receptors can bind to phosphatidylinositols, including PIP$_2$ and PIP$_3$ (Ch. 3 [Newton et al. 2014]). They may therefore link transcription and phospholipid signaling (Krylova et al. 2005).

6.6 True Orphan Receptors

True orphans are constitutive activators or repressors that do not bind ligands. The NUR subfamily (NR4A1, NR4A2, and NR4A3) binds to DNA as monomers, homodimers, and heterodimers with RXR (or another NUR). NURs have no known ligands, and their activity is controlled by their expression levels. ERRα, ERRβ, and ERRγ interfere with ER and AR signaling. ERRs are activated by coactivators such as PGC1α, an important metabolic regulator (Ch. 14 [Hardie 2012]). Germ-cell nuclear receptor (GCNF) is an unusual member of the NR superfamily that binds to a DR-0 motif and may function as a transcriptional repressor in stem cells (Fuhrmann et al. 2001).

NRs thus represent an important metazoan strategy for regulating gene expression programs by integrating different intracellular and extracellular signals. Moreover, cross talk between NRs and other signal transduction pathways adds an additional layer of complexity to their already diverse functions.

7 GASES

The ability to sense and adapt to changes in levels of diatomic gases (O$_2$, NO, and CO) is crucial for an organism's survival. Like steroids, gases are able to cross the plasma membrane and bind to intracellular targets. Signal transduction via gases is predominantly mediated by heme-containing proteins. Binding of the gas to the sensor proteins via the heme moiety can affect their activity or stability. In the case of oxygen, the source of the gas is the environment. In contrast, nitric oxide and carbon monoxide can be generated in neighboring cells and function as bona fide paracrine signals (through covalent and noncovalent binding).

7.1 Signal Transduction via Oxygen

Levels of oxygen inside mammalian cells are mainly detected by hypoxia-inducible factor (HIF) (Wang et al. 1995). HIF is a heterodimer comprising a regulatory α subunit and a DNA-binding β subunit. At high oxygen levels, the HIFα subunit is hydroxylated by prolyl hydroxylase domain (PHD) proteins and targeted for proteosomal degradation by interacting with the von Hippel–Lindau (VHL) protein, which is the targeting subunit of a cullin RING ligase (CRL) 2 E3 ubiquitin ligase. The PHD proteins are dioxygenases that catalyze the incorporation of both atoms of molecular oxygen into their substrates. At low oxygen levels, HIFα degradation is inhibited and HIFα accumulates inside the cell. It associates with HIFβ, which binds to DNA elements termed hypoxia responsive elements (HREs) that have the core sequence RCGTG. HIF targets include erythropoietin, which stimulates red blood cell production; vascular endothelial factor (VEGF), which stimulates angiogenesis (growth of new blood vessels); and other enzymes in relevant pathways, including glycolytic enzymes.

7.2 Signal Transduction via Nitric Oxide

Nitric oxide (NO) was first discovered as an endothelial-derived relaxing factor (EDRF) and subsequently found to be a powerful mediator in other biological processes, including platelet aggregation, synaptic transmission, and inflammatory responses (Moncada et al. 1991). Vascular NO is a key regulator of endothelial function and stimulates blood vessel dilation, prevents platelet aggregation and adhesion, and inhibits proliferation of vascular smooth muscle cells. NO is produced by NO synthase (NOS) from the amino acid L-arginine and detected by the heme-containing enzyme soluble guanylyl cyclase (sGC). Synthesis of NO from arginine involves two mono-oxygenation steps and requires one O$_2$, one NADPH, and tetrahydrobiopterin (BH$_4$).

The electronic structure of NO makes it an excellent ligand for heme, allowing it to bind to sGC heme at a low and nontoxic concentration. sGC is a heterodimer composed of α1 and β1 subunits. H105 in the β1 amino terminus is the heme ligand (Wedel et al. 1994). In a two-step process, NO first binds to the sGC heme moiety, forming a 6-coordinate complex. In the second step, the heme-H105 bound breaks forming a 5-coordinate complex. The second step triggers a conformational change in the sGC catalytic domain, leading to its activation. When activated by NO, sGC converts GTP to cGMP, which then acts as a second messenger (Ch. 3 [Newton et al. 2014]). Besides sGC, NO can also interact with other heme-containing enzymes, including methionine synthase, cytochrome *c* oxidase, cytochrome P450, hemoglobin, and ferritin (Moncada and Palmer 1991; Brown and Cooper 1994; Danishpajooh et al. 2001). REV-ERBα and REV-ERBβ have also been shown to be able to bind NO through their heme-containing LBDs, which results in inhibition of the ability of REV-ERB to repress transcription (Pardee et al. 2009).

In *S*-nitrosylation, NO covalently links to the thiol side chain of cysteine residues in target proteins and modifies their activities. An example of NO-mediated *S*-nitrosylation is the modification of β-catenin in the adherens junction to increase endothelial permeability (Thibeault et al. 2010).

7.3 Signal Transduction via Carbon Monoxide

Endogenous carbon monoxide is generated by heme oxygenase (HO), an enzyme that degrades heme into iron, biliverdin, and CO in aging red blood cells and is induced by many cellular stressors and toxins, including UV, hydrogen peroxide, heavy metals, and arsinite. Like NO, CO has affinity for heme-containing proteins and activates sGC by binding to its heme motif, which stimulates similar biological processes, including intestinal smooth muscle relaxation and neural transmission (Rodgers 1999). In olfactory neurons, CO acts as a neurotransmitter via sGC to produce cGMPs (Verma et al. 1993). It can also bind to REV-ERBs, although at a lower affinity. In addition, CO can bind to other circadian-rhythm-regulating transcription factors, such as neuronal PAS domain protein 2 (NPAS2), which controls circadian gene expression. Upon exposure to CO, NPAS2 cannot form heterodimers with its partner Bmal (Dioum et al. 2002).

Gases are thus not only cell stress indicators, but also important endogenous signaling molecules. Their levels are also crucial and often disrupted in pathological processes, including cardiovascular disease, cancers, and CNS diseases. Future work on diatomic gas signaling thus has the potential lead to the development of entirely new classes of therapeutic agents.

8 CONCLUDING REMARKS

Signaling mechanisms controlling cell growth, migration, survival, and differentiation involve a plethora of different types of ligands, including proteins, peptides, and certain lipids, that bind to receptors at the cell surface, and steroids and gases that can pass across the plasma membrane and bind to intracellular receptors. Recent work has provided immense insights into how receptors are activated after ligand binding, how they initiate signaling inside the cell, and how signaling is terminated. A striking feature is that regulated dimerization or oligomerization is involved in the control of many plasma membrane receptors and intracellular signaling molecules. Moreover, signal transduction often occurs by the reversible formation of complexes and specific translocations of signaling molecules, rather than by random diffusion events. Another striking feature is that many different types of posttranslational modifications of proteins are involved in the regulation of the activity, stability, and subcellular localization of signaling molecules. Furthermore, different signaling inputs converge on a few well-conserved intracellular pathways, several of which link to transcriptional regulation of gene programs.

Whereas the initial phase of signal transduction often involves amplification mechanisms, such as inhibition of phosphatases in conjunction with activation of tyrosine kinases, subsequent phases contain different negative-feedback and termination mechanisms, acting at many different levels. There is also extensive cross talk between different signaling pathways. An important question that remains to be answered is exactly how specificity of signaling is achieved. Moreover, most of our knowledge about receptor signal transduction mechanisms comes from studies of cultured cells; much more work is needed before we understand signal transduction at the organismal level.

REFERENCES

*Reference is in this book.

Aggarwal BB. 2003. Signalling pathways of the TNF superfamily: A double-edged sword. *Nat Rev Immunol* **3**: 745–756.

Amit I, Citri A, Shay T, Lu Y, Katz M, Zhang F, Tarcic G, Siwak D, Lahad J, Jacob-Hirsch J, et al. 2007. A module of negative feedback regulators defines growth factor signaling. *Nat Genet* **39**: 503–512.

Anastasiadis PZ, Reynolds AB. 2001. Regulation of Rho GTPases by p120-catenin. *Curr Opin Cell Biol* **13**: 604.

Ancot F, Foveau B, Lefebvre J, Leroy C, Tulasne D. 2009. Proteolytic cleavages give receptor tyrosine kinases the gift of ubiquity. *Oncogene* **28**: 2185–2195.

Andl CD, Mizushima T, Nakagawa H, Oyama K, Harada H, Chruma K, Herlyn M, Rustgi AK. 2003. Epidermal growth factor receptor mediates increased cell proliferation, migration, and aggregation in esophageal keratinocytes in vitro and in vivo. *J Biol Chem* **278**: 1824–1830.

Andreeva AV, Kutuzov MA, Voyno-Yasenetskaya TA. 2007. Scaffolding proteins in G-protein signaling. *J Mol Signal* **2**: 13.

Arkhipov A, Shan Y, Das R, Endres NF, Eastwood MP, Wemmer DE, Kuriyan J, Shaw DE. 2013. Architecture and membrane interactions of the EGF receptor. *Cell* **152**: 557–569.

Ballon DR, Flanary PL, Gladue DP, Konopka JB, Dohlman HG, Thorner J. 2006. DEP-domain-mediated regulation of GPCR signaling responses. *Cell* **126:** 1079–1093.

Barak Y, Nelson MC, Ong ES, Jones YZ, Ruiz-Lozano P, Chien KR, Koder A, Evans RM. 1999. PPARγ is required for placental, cardiac, and adipose tissue development. *Mol Cell* **4:** 585–595.

Benovic JL, Pike LJ, Cerione RA, Staniszewski C, Yoshimasa T, Codina J, Caron MG, Lefkowitz RJ. 1985. Phosphorylation of the mammalian β-adrenergic receptor by cyclic AMP-dependent protein kinase. Regulation of the rate of receptor phosphorylation and dephosphorylation by agonist occupancy and effects on coupling of the receptor to the stimulatory guanine nucleotide regulatory protein. *J Biol Chem* **260:** 7094–7101.

Berman DM, Gilman AG. 1998. Mammalian RGS proteins: Barbarians at the gate. *J Biol Chem* **273:** 1269–1272.

Blumer JB, Smrcka AV, Lanier SM. 2007. Mechanistic pathways and biological roles for receptor-independent activators of G-protein signaling. *Pharmacol Ther* **113:** 488–506.

Bockaert J, Pin JP. 1999. Molecular tinkering of G protein-coupled receptors: An evolutionary success. *EMBO J* **18:** 1723–1729.

Bodrikov V, Sytnyk V, Leshchyns'ka I, den Hertog J, Schachner M. 2008. NCAM induces CaMKIIα-mediated RPTPα phosphorylation to enhance its catalytic activity and neurite outgrowth. *J Cell Biol* **182:** 1185–1200.

Bouschet T, Martin S, Henley JM. 2005. Receptor-activity-modifying proteins are required for forward trafficking of the calcium-sensing receptor to the plasma membrane. *J Cell Sci* **118:** 4709–4720.

Braga VMM, Yap AS. 2005. The challenges of abundance: Epithelial junctions and small GTPase signalling. *Curr Opin Cell Biol* **17:** 466.

Brown GC, Cooper CE. 1994. Nanomolar concentrations of nitric oxide reversibly inhibit synaptosomal respiration by competing with oxygen at cytochrome oxidase. *FEBS Lett* **356:** 295–298.

Brzostowski JA, Kimmel AR. 2001. Signaling at zero G: G-protein-independent functions for 7-TM receptors. *Trends Biochem Sci* **26:** 291–297.

Burgess AW, Cho HS, Eigenbrot C, Ferguson KM, Garrett TP, Leahy DJ, Lemmon MA, Sliwkowski MX, Ward CW, Yokoyama S. 2003. An open-and-shut case? Recent insights into the activation of EGF/ErbB receptors. *Mol Cell* **12:** 541–552.

Cabrera-Vera TM, Vanhauwe J, Thomas TO, Medkova M, Preininger A, Mazzoni MR, Hamm HE. 2003. Insights into G protein structure, function, and regulation. *Endocr Rev* **24:** 765–781.

* Cantrell D. 2014. Signaling in lymphocyte activation. *Cold Spring Harb Perspect Biol* doi: 10.1101/cshperspect.a018788.

Carmeliet P, Lampugnani MG, Moons L, Breviario F, Compernolle V, Bono F, Balconi G, Spagnuolo R, Oostuyse B, Dewerchin M, et al. 1999. Targeted deficiency or cytosolic truncation of the VE-cadherin gene in mice impairs VEGF-mediated endothelial survival and angiogenesis. *Cell* **98:** 147–157.

Cavallaro U, Dejana E. 2011. Adhesion molecule signalling: Not always a sticky business. *Nat Rev Mol Cell Biol* **12:** 189–197.

Chan FK, Chun HJ, Zheng L, Siegel RM, Bui KL, Lenardo MJ. 2000. A domain in TNF receptors that mediates ligand-independent receptor assembly and signaling. *Science* **288:** 2351–2354.

Chuang TT, Iacovelli L, Sallese M, De Blasi A. 1996. G protein-coupled receptors: Heterologous regulation of homologous desensitization and its implications. *Trends Pharmacol Sci* **17:** 416–421.

Ciraolo E, Iezzi M, Marone R, Marengo S, Curcio C, Costa C, Azzolino O, Gonella C, Rubinetto C, Wu H, et al. 2008. Phosphoinositide 3-kinase p110β activity: Key role in metabolism and mammary gland cancer but not development. *Sci Signal* **1:** ra3.

Coso OA, Chiariello M, Yu J-C, Teramoto H, Crespo P, Xu N, Miki T, Gutkind JS. 1995. The small GTP-binding proteins Rac1 and Cdc42 regulate the activity of the JNK/SAPK signaling pathway. *Cell* **81:** 1137–1146.

Costa C, Martin-Conte EL, Hirsch E. 2011. Phosphoinositide 3-kinase p110γ in immunity. *IUBMB Life* **63:** 707–713.

Croft M. 2009. The role of TNF superfamily members in T-cell function and diseases. *Nat Rev Immunol* **9:** 271–285.

Daniel JM, Reynolds AB. 1999. The catenin p120ctn interacts with Kaiso, a novel BTB/POZ domain zinc finger transcription factor. *Mol Cell Biol* **19:** 3614–3623.

Danishpajooh IO, Gudi T, Chen Y, Kharitonov VG, Sharma VS, Boss GR. 2001. Nitric oxide inhibits methionine synthase activity in vivo and disrupts carbon flow through the folate pathway. *J Biol Chem* **276:** 27296–27303.

Degnan BM, Vervoort M, Larroux C, Richards GS. 2009. Early evolution of metazoan transcription factors. *Curr Opin Genet Dev* **19:** 591–599.

De Vos AM, Ultsch M, Kossiakoff AA. 1992. Human growth hormone and extracellular domain of its receptor: Crystal structure of the complex. *Science* **255:** 306–312.

* Devreotes P, Horwitz AR. 2013. Signaling networks that regulate cell migration. *Cold Spring Harb Perspect Biol* doi: 10.1101/cshperspect.a005959.

Dioum EM, Rutter J, Tuckerman JR, Gonzalez G, Gilles-Gonzalez MA, McKnight SL. 2002. NPAS2: A gas-responsive transcription factor. *Science* **298:** 2385–2387.

Dorsam RT, Gutkind JS. 2007. G-Protein-coupled receptors and cancer. *Nat Rev Cancer* **7:** 79–94.

Dumstrei K, Wang F, Shy D, Tepass U, Hartenstein V. 2002. Interaction between EGFR signaling and DE-cadherin during nervous system morphogenesis. *Development* **129:** 3983–3994.

Endres NF, Das R, Smith AW, Arkhipov A, Kovacs E, Huang Y, Pelton JG, Shan Y, Shaw DE, Wemmer DE, et al. 2013. Conformational coupling across the plasma membrane in activation of the EGF receptor. *Cell* **152:** 543–556.

Evans RM. 2005. The nuclear receptor superfamily: A rosetta stone for physiology. *Mol Endocrinol* **19:** 1429–1438.

Faustman D, Davis M. 2010. TNF receptor 2 pathway: Drug target for autoimmune diseases. *Nat Rev Drug Discov* **9:** 482–493.

Fedor-Chaiken M, Hein PW, Stewart JC, Brackenbury R, Kinch MS. 2003. E-cadherin binding modulates EGF receptor activation. *Cell Commun Adhes* **10:** 105–118.

Feldmann M, Maini RN. 2001. Anti-TNFα therapy of rheumatoid arthritis: What have we learned? *Annu Rev Immunol* **19:** 163–196.

Flower DR. 1999. Modelling G-protein-coupled receptors for drug design. *Biochim Biophys Acta* **1422:** 207–234.

Forman BM, Chen J, Blumberg B, Kliewer SA, Henshaw R, Ong ES, Evans RM. 1994. Cross-talk among RORα1 and the Rev-erb family of orphan nuclear receptors. *Mol Endocrinol* **8:** 1253–1261.

Fox NL, Humphreys R, Luster TA, Klein J, Gallant G. 2010. Tumor necrosis factor-related apoptosis-inducing ligand (TRAIL) receptor-1 and receptor-2 agonists for cancer therapy. *Expert Opin Biol Ther* **10:** 1–18.

Frisch SM, Screaton RA. 2001. Anoikis mechanisms. *Curr Opin Cell Biol* **13:** 555–562.

Fuhrmann G, Chung AC, Jackson KJ, Hummelke G, Baniahmad A, Sutter J, Sylvester I, Scholer HR, Cooney AJ. 2001. Mouse germline restriction of Oct4 expression by germ cell nuclear factor. *Dev Cell* **1:** 377–387.

Fukuhara S, Chikumi H, Gutkind JS. 2001. RGS-containing RhoGEFs: The missing link between transforming G proteins and Rho? *Oncogene* **20:** 1661–1668.

Geiger B, Yamada KM. 2011. Molecular architecture and function of matrix adhesions. *Cold Spring Harb Perspect Biol* **3:** a005033.

Geiger B, Bershadsky A, Pankov R, Yamada KM. 2001. Transmembrane crosstalk between the extracellular matrix and the cytoskeleton. *Nat Rev Mol Cell Biol* **2:** 793–805.

* Green DR, Llambi F. 2014. Cell death signaling. *Cold Spring Harb Perspect Biol* doi: 10.1101/cshperspect.a006080.

Grosheva I, Shtutman M, Elbaum M, Bershadsky AD. 2001. p120 catenin affects cell motility via modulation of activity of Rho-family GTPases: A link between cell–cell contact formation and regulation of cell locomotion. *J Cell Sci* **114:** 695–707.

Gumbiner BM. 2005. Regulation of cadherin-mediated adhesion in morphogenesis. *Nat Rev Mol Cell Biol* **6:** 622–634.

Gutkind JS. 1998. The pathways connecting G protein-coupled receptors to the nucleus through divergent mitogen-activated protein kinase cascades. *J Biol Chem* **273:** 1839–1842.

Gutkind JS. 2000. Regulation of mitogen-activated protein kinase signaling networks by G protein-coupled receptors. *Sci STKE* **2000:** re1.

Hanyaloglu AC, von Zastrow M. 2008. Regulation of GPCRs by endocytic membrane trafficking and its potential implications. *Annu Rev Pharmacol Toxicol* **48:** 537–568.

* Hardie DG. 2012. Organismal carbohydrate and lipid homeostasis. *Cold Spring Harb Perspect Biol* **4:** a006031.

* Harrison DA. 2012. The JAK/STAT pathway. *Cold Spring Harb Perspect Biol* **4:** a011205.

Hart MJ, Jiang X, Kozasa T, Roscoe W, Singer WD, Gilman AG, Sternweis PC, Bollag G. 1998. Direct stimulation of the guanine nucleotide exchange activity of p115 RhoGEF by Gα13. *Science* **280:** 2112–2114.

Heldin C-H. 1995. Dimerization of cell surface receptors in signal transduction. *Cell* **80:** 213–223.

* Hemmings BA, Restuccia DF. 2012. PI3K-PKB/Akt pathway. *Cold Spring Harb Perspect Biol* **4:** a011189.

Hsu SY, Kudo M, Chen T, Nakabayashi K, Bhalla A, van der Spek PJ, van Duin M, Hsueh AJ. 2000. The three subfamilies of leucine-rich repeat-containing G protein-coupled receptors (LGR): Identification of LGR6 and LGR7 and the signaling mechanism for LGR7. *Mol Endocrinol* **14:** 1257–1271.

Hu X, Lazar MA. 1999. The CoRNR motif controls the recruitment of corepressors by nuclear hormone receptors. *Nature* **402:** 93–96.

Hubbard KB, Hepler JR. 2006. Cell signalling diversity of the Gqα family of heterotrimeric G proteins. *Cell Signal* **18:** 135–150.

Huttenlocher A, Horwitz AR. 2011. Integrins in cell migration. *Cold Spring Harb Perspect Biol* **3:** a005074.

Hynes RO. 2002. Integrins: Bidirectional, allosteric signaling machines. *Cell* **110:** 673–687.

* Ingham PW. 2012. Hedgehog signaling. *Cold Spring Harb Perspect Biol* **4:** a011221.

Issemann I, Green S. 1990. Activation of a member of the steroid hormone receptor superfamily by peroxisome proliferators. *Nature* **347:** 645–650.

Jiang G, Hunter T. 1999. Receptor signaling: When dimerization is not enough. *Curr Biol* **9:** R568–R571.

* Julius D, Nathans J. 2012. Signaling by sensory receptors. *Cold Spring Harb Perspect Biol* **4:** a005991.

Jura N, Endres NF, Engel K, Deindl S, Das R, Lamers MH, Wemmer DE, Zhang X, Kuriyan J. 2009. Mechanism for activation of the EGF receptor catalytic domain by the juxtamembrane segment. *Cell* **137:** 1293–1307.

Kallen JA, Schlaeppi JM, Bitsch F, Geisse S, Geiser M, Delhon I, Fournier B. 2002. X-ray structure of the hRORα LBD at 1.63 Å: Structural and functional data that cholesterol or a cholesterol derivative is the natural ligand of RORα. *Structure* **10:** 1697–1707.

Kang JS, Liu C, Derynck R. 2009. New regulatory mechanisms of TGFβ receptor function. *Trends Cell Biol* **19:** 385–394.

Kim N, Stiegler AL, Cameron TO, Hallock PT, Gomez AM, Huang JH, Hubbard SR, Dustin ML, Burden SJ. 2008. Lrp4 is a receptor for Agrin and forms a complex with MuSK. *Cell* **135:** 334–342.

Kim NG, Koh E, Chen X, Gumbiner BM. 2011. E-cadherin mediates contact inhibition of proliferation through Hippo signaling-pathway components. *Proc Natl Acad Sci* **108:** 11930–11935.

King-Jones K, Charles JP, Lam G, Thummel CS. 2005. The ecdysone-induced DHR4 orphan nuclear receptor coordinates growth and maturation in *Drosophila*. *Cell* **121:** 773–784.

Kniazeff J, Prezeau L, Rondard P, Pin JP, Goudet C. 2011. Dimers and beyond: The functional puzzles of class C GPCRs. *Pharmacol Ther* **130:** 9–25.

Kobielak A, Fuchs E. 2004. α-Catenin: At the junction of intercellular adhesion and actin dynamics. *Nat Rev Mol Cell Biol* **5:** 614–625.

Kobilka BK. 2011. Structural insights into adrenergic receptor function and pharmacology. *Trends Pharmacol Sci* **32:** 213–218.

* Kopan R. 2012. Notch signaling. *Cold Spring Harb Perspect Biol* **4:** a011213.

Kopan R, Ilagan MXG. 2009. The canonical Notch signaling pathway: Unfolding the activation mechanism. *Cell* **137:** 216–233.

* Kornbluth S, Fissore R. 2014. Vertebrate reproduction. *Cold Spring Harb Perspect Biol* doi: 10.1101/cshperspect.a006064.

Korshunova I, Mosevitsky M. 2010. Role of the growth-associated protein GAP-43 in NCAM-mediated neurite outgrowth. *Adv Exp Med Biol* **663:** 169–182.

Krylova IN, Sablin EP, Moore J, Xu RX, Waitt GM, MacKay JA, Juzumiene D, Bynum JM, Madauss K, Montana V, et al. 2005. Structural analyses reveal phosphatidyl inositols as ligands for the NR5 orphan receptors SF-1 and LRH-1. *Cell* **120:** 343–355.

* Laplante M, Sabatini DM. 2012. mTOR signaling. *Cold Spring Harb Perspect Biol* **4:** a011593.

* Lee MJ, Yaffe MB. 2014. Protein regulation in signal transduction. *Cold Spring Harb Perspect Biol* doi: 10.1101/cshperspect.a005918.

Legate KR, Fassler R. 2009. Mechanisms that regulate adaptor binding to β-integrin cytoplasmic tails. *J Cell Sci* **122:** 187–198.

Legate KR, Wickstrom SA, Fassler R. 2009. Genetic and cell biological analysis of integrin outside-in signaling. *Genes Dev* **23:** 397–418.

Lehmann JM, McKee DD, Watson MA, Willson TM, Moore JT, Kliewer SA. 1998. The human orphan nuclear receptor PXR is activated by compounds that regulate CYP3A4 gene expression and cause drug interactions. *J Clin Invest* **102:** 1016–1023.

Lemmon MA, Schlessinger J. 2010. Cell signaling by receptor tyrosine kinases. *Cell* **141:** 1117–1134.

Lengyel E, Sawada K, Salgia R. 2007. Tyrosine kinase mutations in human cancer. *Curr Mol Med* **7:** 77–84.

Leushacke M, Barker N. 2011. Lgr5 and Lgr6 as markers to study adult stem cell roles in self-renewal and cancer. *Oncogene* **31:** 3009–3022.

Lim X, Nusse R. 2013. Wnt signaling in skin development, homeostasis, and disease. *Cold Spring Harb Perspect Biol* **5:** a008029.

Lin S-Y, Makino K, Xia W, Matin A, Wen Y, Kwong KY, Bourguignon L, Hung M-C. 2001. Nuclear localization of EGF receptor and its potential new role as a transcription factor. *Nat Cell Biol* **3:** 802–808.

Liongue C, Ward AC. 2007. Evolution of Class I cytokine receptors. *BMC Evol Biol* **7:** 120.

Locksley RM, Killeen N, Lenardo MJ. 2001. The TNF and TNF receptor superfamilies: Integrating mammalian biology. *Cell* **104:** 487–501.

Loers G, Schachner M. 2007. Recognition molecules and neural repair. *J Neurochem* **101:** 865–882.

Lopez-Ilasaca M, Li W, Uren A, Yu J-C, Kazlauskas A, Gutkind JS, Heidaran MA. 1997. Requirement of phosphatidylinositol-3 kinase for activation of JNK/SAPKs by PDGF. *Biochem Biophys Res Commun* **232:** 273–277.

Lu TT, Repa JJ, Mangelsdorf DJ. 2001. Orphan nuclear receptors as eLiXiRs and FiXeRs of sterol metabolism. *J Biol Chem* **276:** 37735–37738.

Luttrell LM, Gesty-Palmer D. 2010. Beyond desensitization: Physiological relevance of arrestin-dependent signaling. *Pharmacol Rev* **62:** 305–330.

Lutz S, Shankaranarayanan A, Coco C, Ridilla M, Nance MR, Vettel C, Baltus D, Evelyn CR, Neubig RR, Wieland T, et al. 2007. Structure of Gαq-p63RhoGEF-RhoA complex reveals a pathway for the activation of RhoA by GPCRs. *Science* **318:** 1923–1927.

Magalhaes AC, Dunn H, Ferguson SS. 2012. Regulation of GPCR activity, trafficking and localization by GPCR-interacting proteins. *Br J Pharmacol* **165:** 1717–1736.

Mangelsdorf DJ, Borgmeyer U, Heyman RA, Zhou JY, Ong ES, Oro AE, Kakizuka A, Evans RM. 1992. Characterization of three RXR genes that mediate the action of 9-*cis* retinoic acid. *Genes Dev* **6:** 329–344.

Mangelsdorf DJ, Thummel C, Beato M, Herrlich P, Schutz G, Umesono K, Blumberg B, Kastner P, Mark M, Chambon P, et al. 1995. The nuclear receptor superfamily: The second decade. *Cell* **83**: 835–839.

May LT, Leach K, Sexton PM, Christopoulos A. 2007. Allosteric modulation of G protein-coupled receptors. *Annu Rev Pharmacol Toxicol* **47**: 1–51.

★ McCaffrey LM, Macara IG. 2012. Signaling pathways in cell polarity. *Cold Spring Harb Perspect Biol* **4**: a009654.

Mege RM, Gavard J, Lambert M. 2006. Regulation of cell–cell junctions by the cytoskeleton. *Curr Opin Cell Biol* **18**: 541–548.

Miaczynska M, Pelkmans L, Zerial M. 2004. Not just a sink: Endosomes in control of signal transduction. *Curr Opin Cell Biol* **16**: 400–406.

Miranti CK, Brugge JS. 2002. Sensing the environment: A historical perspective on integrin signal transduction. *Nat Cell Biol* **4**: E83–E90.

Miyamoto S, Teramoto H, Coso OA, Gutkind JS, Burbelo PD, Akiyama SK, Yamada KM. 1995. Integrin function: Molecular hierarchies of cytoskeletal and signaling molecules. *J Cell Biol* **131**: 791–805.

Miyamoto S, Teramoto H, Gutkind JS, Yamada KM. 1996. Integrins can collaborate with growth factors for phosphorylation of receptor tyrosine kinases and MAP kinase activation: Roles of integrin aggregation and occupancy of receptors. *J Cell Biol* **135**: 1633–1642.

Moncada S, Palmer RM. 1991. Inhibition of the induction of nitric oxide synthase by glucocorticoids: Yet another explanation for their anti-inflammatory effects? *Trends Pharmacol Sci* **12**: 130–131.

Moncada S, Palmer RM, Higgs EA. 1991. Nitric oxide: Physiology, pathophysiology, and pharmacology. *Pharmacol Rev* **43**: 109–142.

Moquin D, Chan FK-M. 2010. The molecular regulation of programmed necrotic cell injury. *Trends Biochem Sci* **35**: 434–441.

★ Morrison DK. 2012. MAP kinase pathways. *Cold Spring Harb Perspect Biol* **4**: a011254.

Moser M, Legate KR, Zent R, Fassler R. 2009. The tail of integrins, talin, and kindlins. *Science* **324**: 895–899.

Moustakas A, Heldin C-H. 2009. The regulation of TGFβ signal transduction. *Development* **136**: 3699–3714.

Mu Y, Sundar R, Thakur N, Ekman M, Gudey SK, Yakymovych M, Hermansson A, Dimitriou H, Bengoechea-Alonso MT, Ericsson J, et al. 2011. TRAF6 ubiquitinates TGFβ type I receptor to promote its cleavage and nuclear translocation in cancer. *Nat Commun* **2**: 330.

Nabhan JF, Pan H, Lu Q. 2010. Arrestin domain-containing protein 3 recruits the NEDD4 E3 ligase to mediate ubiquitination of the β2-adrenergic receptor. *EMBO Rep* **11**: 605–611.

Nelson WJ, Nusse R. 2004. Convergence of Wnt, β-catenin, and cadherin pathways. *Science* **303**: 1483–1487.

★ Newton K, Dixit VM. 2012. Signaling in innate immunity and inflammation. *Cold Spring Harb Perspect Biol* **4**: a006049.

★ Newton AC, Bootman MD, Scott JD. 2014. Second messengers. *Cold Spring Harb Perspect Biol* doi: 10.1101/cshperspect.a005926.

Ni C-Y, Murphy MP, Golde TE, Carpenter G. 2001. γ-Secretase cleavage and nuclear localization of ErbB-4 receptor tyrosine kinase. *Science* **294**: 2179–2181.

Nimmerjahn F, Ravetch JV. 2008. Fcγ receptors as regulators of immune responses. *Nat Rev Immunol* **8**: 34–47.

Nose A, Nagafuchi A, Takeichi M. 1988. Expressed recombinant cadherins mediate cell sorting in model systems. *Cell* **54**: 993–1001.

O'Hayre M, Vázquez-Prado J, Kufareva I, Stawiski EW, Handel TM, Seshagiri S, Gutkind JS. 2013. The emerging mutational landscape of G proteins and G-protein-coupled receptors in cancer. *Nat Rev Cancer* **13**: 412–424.

Ozdamar B, Bose R, Barrios-Rodiles M, Wang HR, Zhang Y, Wrana JL. 2005. Regulation of the polarity protein Par6 by TGFβ receptors controls epithelial cell plasticity. *Science* **307**: 1603–1609.

Pardee KI, Xu X, Reinking J, Schuetz A, Dong A, Liu S, Zhang R, Tiefenbach J, Lajoie G, Plotnikov AN, et al. 2009. The structural basis of gas-responsive transcription by the human nuclear hormone receptor REV-ERBβ. *PLoS Biol* **7**: e43.

Pawson T. 2004. Specificity in signal transduction: From phosphotyrosine–SH2 domain interactions to complex cellular systems. *Cell* **116**: 191–203.

Pece S, Gutkind JS. 2000. Signaling from E-cadherins to the MAPK pathway by the recruitment and activation of epidermal growth factor receptors upon cell–cell contact formation. *J Biol Chem* **275**: 41227–41233.

Pece S, Chiariello M, Murga C, Gutkind JS. 1999. Activation of the protein kinase Akt/PKB by the formation of E-cadherin-mediated cell–cell junctions. Evidence for the association of phosphatidylinositol 3-kinase with the E-cadherin adhesion complex. *J Biol Chem* **274**: 19347–19351.

Perissi V, Jepsen K, Glass CK, Rosenfeld MG. 2010. Deconstructing repression: Evolving models of co-repressor action. *Nat Rev Genet* **11**: 109–123.

Pierce KL, Premont RT, Lefkowitz RJ. 2002. Seven-transmembrane receptors. *Nat Rev Mol Cell Biol* **3**: 639–650.

Qian X, Karpova T, Sheppard AM, McNally J, Lowy DR. 2004. E-cadherin-mediated adhesion inhibits ligand-dependent activation of diverse receptor tyrosine kinases. *EMBO J* **23**: 1739–1748.

Raiborg C, Stenmark H. 2009. The ESCRT machinery in endosomal sorting of ubiquitylated membrane proteins. *Nature* **458**: 445–452.

Rajagopal S, Rajagopal K, Lefkowitz RJ. 2010. Teaching old receptors new tricks: Biasing seven-transmembrane receptors. *Nat Rev Drug Discov* **9**: 373–386.

Red Brewer M, Choi SH, Alvarado D, Moravcevic K, Pozzi A, Lemmon MA, Carpenter G. 2009. The juxtamembrane region of the EGF receptor functions as an activation domain. *Mol Cell* **34**: 641–651.

Reinking J, Lam MM, Pardee K, Sampson HM, Liu S, Yang P, Williams S, White W, Lajoie G, Edwards A, et al. 2005. The *Drosophila* nuclear receptor e75 contains heme and is gas responsive. *Cell* **122**: 195–207.

Renauld J-C. 2003. Class II cytokine receptors and their ligands: Key antiviral and inflammatory modulators. *Nat Rev Immunol* **3**: 667–676.

Rodgers KR. 1999. Heme-based sensors in biological systems. *Curr Opin Chem Biol* **3**: 158–167.

Rosenfeld MG, Lunyak VV, Glass CK. 2006. Sensors and signals: A coactivator/corepressor/epigenetic code for integrating signal-dependent programs of transcriptional response. *Genes Dev* **20**: 1405–1428.

Rosenfeldt H, Vazquez-Prado J, Gutkind JS. 2004. P-REX2, a novel PI-3-kinase sensitive Rac exchange factor. *FEBS Lett* **572**: 167–171.

★ Samelson LE. 2011. Immunoreceptor signaling. *Cold Spring Harb Perspect Biol* **3**: a011510.

★ Sassone-Corsi P. 2012. The cyclic AMP pathway. *Cold Spring Harb Perspect Biol* **4**: a011148.

Schlee S, Carmillo P, Whitty A. 2006. Quantitative analysis of the activation mechanism of the multicomponent growth-factor receptor Ret. *Nat Chem Biol* **2**: 636–644.

Schneider P, Thome M, Burns K, Bodmer JL, Hofmann K, Kataoka T, Holler N, Tschopp J. 1997. TRAIL receptors 1 (DR4) and 2 (DR5) signal FADD-dependent apoptosis and activate NF-κB. *Immunity* **7**: 831–836.

★ Sever R, Brugge JS. 2014. Signal transduction in cancer. *Cold Spring Harb Perspect Med* doi: 10.1101/cshperspect.a006098.

★ Sever R, Glass CK. 2013. Signaling by nuclear receptors. *Cold Spring Harb Perspect Biol* **5**: a016709.

Shattil SJ, Kim C, Ginsberg MH. 2010. The final steps of integrin activation: The end game. *Nat Rev Mol Cell Biol* **11**: 288–300.

Shenoy SK, Lefkowitz RJ. 2011. β-Arrestin-mediated receptor trafficking and signal transduction. *Trends Pharmacol Sci* **32**: 521–533.

Sonoda J, Pei L, Evans RM. 2008. Nuclear receptors: Decoding metabolic disease. *FEBS Lett* **582**: 2–9.

Stehlin-Gaon C, Willmann D, Zeyer D, Sanglier S, Van Dorsselaer A, Renaud JP, Moras D, Schule R. 2003. All-*trans* retinoic acid is a ligand for the orphan nuclear receptor RORβ. *Nat Struct Biol* **10**: 820–825.

Suda T, Hashimoto H, Tanaka M, Ochi T, Nagata S. 1997. Membrane Fas ligand kills human peripheral blood T lymphocytes, and soluble Fas ligand blocks the killing. *J Exp Med* **186:** 2045–2050.

Suenaert P, Bulteel V, Lemmens L, Noman M, Geypens B, Van Assche G, Geboes K, Ceuppens JL, Rutgeerts P. 2002. Anti-tumor necrosis factor treatment restores the gut barrier in Crohn's disease. *Am J Gastroenterol* **97:** 2000–2004.

Sueyoshi T, Kawamoto T, Zelko I, Honkakoski P, Negishi M. 1999. The repressed nuclear receptor CAR responds to phenobarbital in activating the human *CYP2B6* gene. *J Biol Chem* **274:** 6043–6046.

Suyama K, Shapiro I, Guttman M, Hazan RB. 2002. A signaling pathway leading to metastasis is controlled by N-cadherin and the FGF receptor. *Cancer Cell* **2:** 301–314.

Takai Y, Miyoshi J, Ikeda W, Ogita H. 2008. Nectins and nectin-like molecules: Roles in contact inhibition of cell movement and proliferation. *Nat Rev Mol Cell Biol* **9:** 603–615.

Taussig R, Gilman AG. 1995. Mammalian membrane-bound adenylyl cyclases. *J Biol Chem* **270:** 1 4.

Terrillon S, Bouvier M. 2004. Roles of G-protein-coupled receptor dimerization. *EMBO Rep* **5:** 30–34.

Thibeault S, Rautureau Y, Oubaha M, Faubert D, Wilkes BC, Delisle C, Gratton JP. 2010. S-Nitrosylation of β-catenin by eNOS-derived NO promotes VEGF-induced endothelial cell permeability. *Mol Cell* **39:** 468–476.

Thiery JP, Sleeman JP. 2006. Complex networks orchestrate epithelial–mesenchymal transitions. *Nat Rev Mol Cell Biol* **7:** 131.

Tsukazaki T, Chiang TA, Davison AF, Attisano L, Wrana JL. 1998. SARA, a FYVE domain protein that recruits Smad2 to the TGFβ receptor. *Cell* **95:** 779–791.

Tsukita S, Furuse M, Itoh M. 2001. Multifunctional strands in tight junctions. *Nat Rev Mol Cell Biol* **2:** 285–293.

Umesono K, Murakami KK, Thompson CC, Evans RM. 1991. Direct repeats as selective response elements for the thyroid hormone, retinoic acid, and vitamin D3 receptors. *Cell* **65:** 1255–1266.

van Amerongen R, Mikels A, Nusse R. 2008. Alternative wnt signaling is initiated by distinct receptors. *Sci Signal* **1:** re9.

Vaqué JP, Dorsam RT, Feng X, Iglesias-Bartolome R, Forsthoefel DJ, Chen Q, Debant A, Seeger MA, Ksander BR, Teramoto H, et al. 2013. A genome-wide RNAi screen reveals a Trio-regulated Rho GTPase circuitry transducing mitogenic signals initiated by G protein-coupled receptors. *Mol Cell* **49:** 94–108.

Verma A, Hirsch DJ, Glatt CE, Ronnett GV, Snyder SH. 1993. Carbon monoxide: A putative neural messenger. *Science* **259:** 381–384.

Vleminckx K, Vakaet L Jr, Mareel M, Fiers W, van Roy F. 1991. Genetic manipulation of E-cadherin expression by epithelial tumor cells reveals an invasion suppressor role. *Cell* **66:** 107–119.

Wang GL, Jiang BH, Rue EA, Semenza GL. 1995. Hypoxia-inducible factor 1 is a basic-helix–loop–helix-PAS heterodimer regulated by cellular O₂ tension. *Proc Natl Acad Sci* **92:** 5510–5514.

Wang X, Lupardus P, Laporte SL, Garcia KC. 2009. Structural biology of shared cytokine receptors. *Annu Rev Immunol* **27:** 29–60.

Warnmark A, Treuter E, Wright AP, Gustafsson JA. 2003. Activation functions 1 and 2 of nuclear receptors: Molecular strategies for transcriptional activation. *Mol Endocrinol* **17:** 1901–1909.

Wedel B, Humbert P, Harteneck C, Foerster J, Malkewitz J, Bohme E, Schultz G, Koesling D. 1994. Mutation of His-105 in the β₁ subunit yields a nitric oxide–insensitive form of soluble guanylyl cyclase. *Proc Natl Acad Sci* **91:** 2592–2596.

Wehling M. 1997. Specific, nongenomic actions of steroid hormones. *Annu Rev Physiol* **59:** 365–393.

Wehrle-Haller B. 2012. Assembly and disassembly of cell matrix adhesions. *Curr Opin Cell Biol* **24:** 569–581.

Wehrman T, He X, Raab B, Dukipatti A, Blau H, Garcia KC. 2007. Structural and mechanistic insights into nerve growth factor interactions with the TrkA and p75 receptors. *Neuron* **53:** 25–38.

Welch HC, Coadwell WJ, Ellson CD, Ferguson GJ, Andrews SR, Erdjument-Bromage H, Tempst P, Hawkins PT, Stephens LR. 2002. P-Rex1, a PtdIns(3,4,5)P₃- and Gβγ-regulated guanine-nucleotide exchange factor for Rac. *Cell* **108:** 809–821.

Whistler JL, Enquist J, Marley A, Fong J, Gladher F, Tsuruda P, Murray SR, Von Zastrow M. 2002. Modulation of postendocytic sorting of G protein-coupled receptors. *Science* **297:** 615–620.

Willard FS, Kimple RJ, Siderovski DP. 2004. Return of the GDI: The GoLoco motif in cell division. *Annu Rev Biochem* **73:** 925–951.

* Wrana JL. 2013. Signaling by the TGFβ superfamily. *Cold Spring Harb Perspect Biol* **5:** a011197.

Yamauchi J, Kawano T, Nagao M, Kaziro Y, Itoh H. 2000. Gᵢ-dependent activation of c-Jun N-terminal kinase in human embryonal kidney 293 cells. *J Biol Chem* **275:** 7633–7640.

Yanagisawa M, Anastasiadis PZ. 2006. p120 catenin is essential for mesenchymal cadherin-mediated regulation of cell motility and invasiveness. *J Cell Biol* **174:** 1087–1096.

Yarden Y, Sliwkowski MX. 2001. Untangling the ErbB signalling network. *Nat Rev Mol Cell Biol* **2:** 127–137.

Yoshimura A, Mori H, Ohishi M, Aki D, Hanada T. 2003. Negative regulation of cytokine signaling influences inflammation. *Curr Opin Immunol* **15:** 704–708.

Yu FX, Zhao B, Panupinthu N, Jewell JL, Lian I, Wang LH, Zhao J, Yuan H, Tumaneng K, Li H, et al. 2012. Regulation of the Hippo–YAP pathway by G-protein-coupled receptor signaling. *Cell* **150:** 780–791.

Yuzawa S, Opatowsky Y, Zhang Z, Mandiyan V, Lax I, Schlessinger J. 2007. Structural basis for activation of the receptor tyrosine kinase KIT by stem cell factor. *Cell* **130:** 323–334.

Zhang J, Huang W, Chua SS, Wei P, Moore D.D. 2002. Modulation of acetaminophen-induced hepatotoxicity by the xenobiotic receptor CAR. *Science* **298:** 422–424.

Zhang X, Gureasko J, Shen K, Cole PA, Kuriyan J. 2006. An allosteric mechanism for activation of the kinase domain of epidermal growth factor receptor. *Cell* **125:** 1137–1149.

Zikherman J, Weiss A. 2009. Antigen receptor signaling in the rheumatic diseases. *Arthritis Res Ther* **11:** 202.

Zoncu R, Efeyan A, Sabatini DM. 2011. mTOR: From growth signal integration to cancer, diabetes and ageing. *Nat Rev Mol Cell Biol* **12:** 21–35.

Zwang Y, Yarden Y. 2009. Systems biology of growth factor-induced receptor endocytosis. *Traffic* **10:** 349–363.

CHAPTER 2

Protein Regulation in Signal Transduction

Michael J. Lee and Michael B. Yaffe

David H. Koch Institute for Integrative Cancer Research at MIT, Department of Biology and Department of Biological Engineering, Massachusetts Institute of Technology, Cambridge, Massachusetts 02139

Correspondence: myaffe@mit.edu

SUMMARY

Cells must respond to a diverse, complex, and ever-changing mix of signals, using a fairly limited set of parts. Changes in protein level, protein localization, protein activity, and protein–protein interactions are critical aspects of signal transduction, allowing cells to respond highly specifically to a nearly limitless set of cues and also to vary the sensitivity, duration, and dynamics of the response. Signal-dependent changes in levels of gene expression and protein synthesis play an important role in regulation of protein levels, whereas posttranslational modifications of proteins regulate their degradation, localization, and functional interactions. Protein ubiquitylation, for example, can direct proteins to the proteasome for degradation or provide a signal that regulates their interactions and/or location within the cell. Similarly, protein phosphorylation by specific kinases is a key mechanism for augmenting protein activity and relaying signals to other proteins that possess domains that recognize the phosphorylated residues.

Outline

Cite this chapter as *Cold Spring Harb Perspect Biol* doi: 10.1101/cshperspect.a005918

1 INTRODUCTION

Signal transduction processes are, in many respects, protein-driven events. One common way to describe these circuits involves designating different proteins, or modular domains within an individual protein, as "readers," "writers," and "erasers." In this model, catalytic domains that add specific posttranslational modifications, such as kinases and acetyltransferases, are called writers because they leave behind a physical mark on the proteins they act on. Conversely, phosphatases, deacetylases, and other enzymes that remove these modifications are examples of erasers. In addition to writers and erasers, there are also domains that bind to specific posttranslationally modified or unmodified sequences of amino acids. These types of domains read the sequences produced by writers and erasers and are therefore called readers. These readers can be involved in protein–protein interactions or interactions between two parts of the same protein. Other readers can bind directly to specific phospho- and neutral lipids, or to specific ions, such as calcium, rather than to amino acid sequences. The targets of these readers, writers, and erasers are often short amino acid sequences (typically three to 15 amino acids in length) called motifs. Using these modular parts, cells dynamically encode information in response to environmental, chemical, or developmental stimuli, to transduce the signal (see Table 1).

For some cellular processes, the mere presence of a protein, monitored, for example, by the levels of its RNA transcript, is sufficient to provide insight into the current state of the cell. In contrast, this provides relatively limited information about the current state of a signaling network, primarily because an amazing amount of regulation occurs at the protein level, via control of the readers, writers, and erasers. Moreover, this occurs dynamically, reversibly, and sometimes very quickly. Below, we focus on the basic mechanisms of regulation that occur at the level of the protein, specifically focusing on how signal transduction can control protein number, protein localization, protein activity, and protein–protein interactions.

2 POSTTRANSLATIONAL MODIFICATIONS AND THE REGULATION OF PROTEIN ACTIVITY

Many signaling proteins have intrinsic enzymatic activities. For example, they may contain catalytic domains capable of phosphorylating or dephosphorylating proteins, lipids, or sugars (kinases) or hydrolyzing specific lipids or phospholipids (lipases and phospholipases, respectively). In addition to performing such covalent posttranslational

Table 1. Examples of modular domain readers of motifs and protein/lipid posttranslational modifications

Modular protein domain reader	Posttranslation modification recognized (if any)	Example motifs	Biological processes
SH2 domain	Phosphotyrosine	pYxxφ	Growth factor signaling
SH3 domain	None, usually recognizes proline-rich motifs	RxxPxxP, PxxPxR/K	Signaling scaffolds
14-3-3	Phosphoserine/phosphothreonine	RSx(pS/pT)xP, Rxx(pS/pT)xP	Chaperone/scaffold, subcellular localization
FHA domain	Phosphothreonine	pTxxφ, pTxxD	DNA damage response, gene expression, transport
BRCT domain	Phosphoserine/phosphothreonine	(pS/pT)xxφ	DNA damage response
Polo-box domain	Phosphoserine/phosphothreonine	S(pS/pT)P/X	Mitotic control by Polo-like kinases
WW domain	Most do not recognize any PTMS; a few recognize phosphoserine/phosphothreonine	PPxY, PPLP, PPR (pS/pT)P	Gene transcription, proline isomerization and cell-cycle control, development
Bromo domain	Acetyl-lysine	–	Gene expression, chromatin structure, and remodeling
Chromo domain	Methyl-lysine	–	Chromatin structure and remodeling
Tudor domain	Methyl-lysine	–	Chromatin structure and remodeling, DNA damage response
EVH1 domain	None, usually recognizes Pro-rich motifs	FPxφP, PPxxF	Actin cytoskeleton control, neurotransmission
PX domain	Phosphatidyl-3-phosphate, -3,4-bisphosphate, and -3,4,5-trisphosphate	(Not applicable)	Membrane targeting and trafficking, endosomal sorting
PH domain	Phosphatidyl-4,5-bisphosphate, -3,4-bisphosphate, and -3,4,5-trisphosphate	(Not applicable)	Membrane trafficking, cytoskeletal control
FYVE domain	Phosphatidyl-3-phosphate	(Not applicable)	Vacuolar protein sorting, endosomal sorting

Additional motifs beyond those shown are also recognized. (−) No distinct motif has emerged. (Not applicable) Lipid posttranslational modifications rather than protein sequence motifs.

pY, Phosphotyrosine; pS, phosphoserine; pT, phosphothreonine; F, hydrophobic amino acids.

Cite this chapter as *Cold Spring Harb Perspect Biol* doi: 10.1101/cshperspect.a005918

modifications, signaling proteins may also stimulate the exchange or hydrolysis of nucleotides in nucleotide-binding proteins such as small G proteins and heterotrimeric G proteins by acting as guanine-nucleotide exchange factors (GEFs) or GTPase-activating proteins (GAPs), respectively. Other types of catalytic domains participate in the formation of cyclic nucleotides such as cyclic AMP (cAMP) or cyclic GMP (cGMP) (Ch. 3 [Newton et al. 2014]), or synthesize specific types of gas molecules that are involved in signaling, such as nitric oxide, superoxide, and other reactive oxygen and nitrogen species.

The activity of these signaling molecules must be tightly controlled so that the products of the reactions that they catalyze—phospholipids, phosphoproteins, gases, activated or inactivated G proteins—are only produced at the appropriate time and place. Some of this control occurs through regulation of protein levels or protein localization, but much of it involves regulation of protein activity by posttranslational modification of the signaling molecule itself, intramolecular interactions between different domains within the protein, and/or intermolecular interactions with other proteins. In each case, this may involve a phenomenon known as allostery, in which the site where

the control is exerted (e.g., where the domains interact) is different from the site where catalysis occurs. Transmission of information via a conformational change allows the allosteric site to control the protein's catalytic activity, even though the two may be far removed in physical space (Fig. 1). Key examples in signaling include binding of calcium to calmodulin to activate calcium/calmodulin-dependent protein kinases (CaM kinases), and binding of cyclic nucleotides to activate protein kinase A (PKA) and protein kinase G (PKG) (see Ch. 3 [Newton et al. 2014]).

2.1 Posttranslational Modifications

Proteins can be exquisitely regulated by relatively small covalent changes to their basic chemical structure. These posttranslational modifications can profoundly alter a protein's activity, localization, stability, and/or binding partners, and therefore constitute the "front line" of many signaling systems within the cell. More than 350 different posttranslational modifications have been discovered, many of which are reversible.

One of the first types of protein posttranslational modifications to be identified was phosphorylation. Although

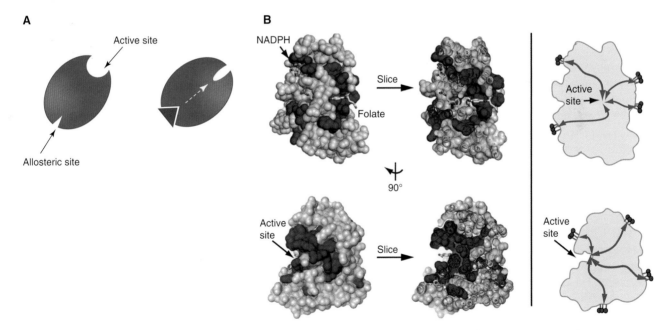

Figure 1. Allosteric modulation of protein activity. (*A*) Cartoon example of protein allostery. In this example, an allosteric-regulation site exists in a region of the protein that is spatially distinct from the active site. Modulation of the allosteric site (through posttranslational modification or the binding of a ligand, cofactor, or protein) causes changes in the active site. Importantly, these changes can be either activating or inhibiting. (*B*) Allosteric regulation of dihydrofolate reductase (DHFR). (*Left* panels) Two surface views of the DHFR enzyme, highlighting the active site (bound to folate), cofactor binding site (bound to NADPH), and residues involved in allosteric communication between the two sites (shown in blue). (*Middle* panels) A view of a slice through the protein core. (*Right* panels) A cartoon representation of the slice mappings shown in B. (Figure generously supplied by Rama Ranganathan and Cell Press.)

phosphorus was first noted as a trace element in purified egg white by Gerrit Mulder in 1835 (Mulder would subsequently invent the term protein to describe such materials), it was not until 1906 that an isolated cleavage fragment from a single protein, vitellin, was shown to contain a constant composition of ~10% phosphorus (Levene and Alsberg 1906). Subsequent studies in the 1930s revealed that the hydrolysis products of vitellin included serinephosphoric acid, indicating that the phosphorus was part of a covalent modification of the protein's constituent amino acids (Lipmann and Levene 1932). In the 1950s, George Burnett and Eugene Kennedy identified an enzymatic activity in mitochondrial lysates that was capable of transferring radioactive phosphate to exogenous substrates (Burnett and Kennedy 1954), but it was not until the pioneering work of Edwin Krebs and Edmond Fischer that protein phosphorylation was shown to reversibly regulate a key biological process (Krebs and Fischer 1956)—in this case, controlling whether glucose molecules are stored as polymers of glycogen or glycogen molecules are broken down to supply the body with glucose.

Using kinases as writers, phosphate groups are added by formation of esters with amino acids whose side chains contain alcohols. In mammals, the vast majority of protein phosphorylation occurs on serine residues (~85%), with a lesser amount on threonine (~15%) and only a tiny fraction on tyrosine residues (~0.4%). In addition to being relatively rare, tyrosine phosphorylation is restricted to higher eukaryotes, having evolved just before the origin of the metazoan lineage. Nevertheless, tyrosine phosphorylation seems to be particularly important in signaling and is frequently dysregulated in cancer. A handful of other residues can also be phosphorylated, including histidine, aspartic acid, arginine, and lysine. These modifications are typically labile and difficult to purify/study, but it is becoming increasingly clear that histidine phosphorylation, for example, is relatively common and an important player in many signaling systems.

Protein phosphorylation controls the functions of proteins through a variety of different molecular mechanisms. At physiological pH, the addition of a phosphate group adds about 1.5 electrostatic units of negative charge. In some cases, this negative charge can drive the formation of ionic bonds with positively charged residues, such as lysine and arginine, in other parts of the protein. In enzymes, these new bonds can shift the positions of α helices and loops in the protein to make the enzyme more or less active. For example, many protein kinases themselves are phosphorylated on residues in a region called the activation loop. Ionic interactions and hydrogen bonds involving these phosphorylated residues to nearby arginine and lysine residues lead to a dramatic rearrangement of the protein

that opens up the binding site for ATP and substrates, while simultaneously reorienting the catalytic residues into a conformation that allows the protein kinase to transfer a phosphate group from ATP to a serine, threonine, or tyrosine residue on the substrate (Fig. 2). Because protein kinases often both catalyze the phosphorylation of proteins and are themselves substrates for other, upstream protein kinases, they can be linked together into pathways where one protein kinase phosphorylates another, which then phosphorylates a third, which then phosphorylates some other type of protein. This creates a signal amplifier in which the activity of the first kinase is magnified by the catalytic activities of the other kinases downstream from it in the pathway and is a particularly prominent type of signaling circuit used by mitogen-activated protein kinases (MAPKs) (see p. 81 [Morrison 2012]). Phosphorylation is negatively regulated by a class of erasers called phosphatases, which are grouped into three main families based on sequence and structural similarity: phosphoprotein phosphatases, protein phosphatase metal-ion-dependent phosphatases, and protein tyrosine phosphatases. Each family is also further divided into subfamilies based on mechanisms of catalysis and substrate specificity.

Besides changing the enzymatic activity of a protein, phosphorylation can also disrupt the interactions between two or more proteins or cause two proteins to interact (see below), often changing the subcellular location of the phosphorylated protein.

A second common reversible posttranslational modification of proteins is acetylation. In this case, the positively charged ε amino groups on lysine residues are converted into neutral amides by the addition of acetate. This loss of positive charge prevents the acetylated lysines from making electrostatic interactions with phosphate groups and is therefore a prominent posttranslational modification found on DNA-binding histones. Because the DNA backbone is built from esters of phosphates and sugars, acetylation weakens binding of the histones in nucleosomes to DNA, allowing other DNA-binding proteins, such as transcription factors and RNA polymerase, to bind instead. This often results in changes in chromatin structure and transcriptional activity. Acetylation is also common in enzymes involved in metabolism (Wang et al. 2010) and probably functions, at least in part, by changing the activity of the enzyme, in a similar way to protein phosphorylation. Like phosphorylation, protein acetylation can also drive two proteins to bind to each other if one of the proteins has a domain that specifically recognizes acetyl-lysine (e.g., a bromo domain), or it can specifically enhance the dynamics of recruitment of other proteins to the acetylated protein relative to the unmodified form. For example, acetylation of the DNA damage kinase ATM by the acetyl-

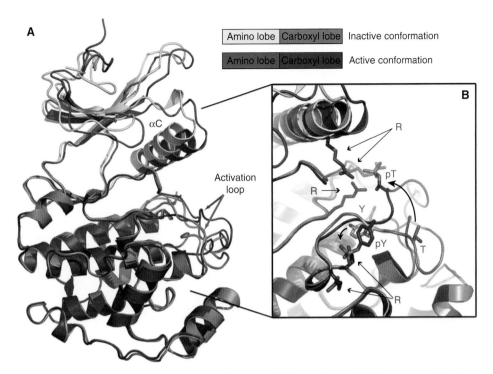

Figure 2. Mechanism of kinase activation. (*A*) Conformational changes in protein kinases upon phosphorylation enhance their catalytic ability. The structure of ERK2, an MAPK, is shown in its inactive nonphosphorylated state and its active phosphorylated state. The carboxy-terminal lobes of the kinase in both states (brown and red, respectively) have been superimposed. Note that following phosphorylation of the activation loop, there is a marked rotation and reorientation of the amino-terminal lobe (yellow and purple, respectively), bringing key catalytic residues, including those present in a critical α helix, αC, into position, converting the kinase into an active state that can now phosphorylate downstream substrates. (*B*) Close-up of phosphorylation-induced conformational changes in the activation loop. Two key residues in the activation loop of MAPKs, a threonine and a tyrosine residue, separated by a singe amino acid (i.e., a TXY motif) can interact with a network of surrounding arginine residues only when they are in their phosphorylated states. These interactions not only shift the positions of the threonine and tyrosine residues themselves (curved arrows), but drag the entire activation loop into a new conformation that communicates with the rest of the protein to move the entire amino-terminal lobe relative to the carboxy-terminal lobe, as shown in *A*.

transferase Tip60 (also known as Kat5) promotes recruitment of ATM to sites of DNA damage and ATM-dependent signaling. Similarly, microtubules containing acetylated tubulin are better able to recruit molecular motors that drive vesicular trafficking within the cell (Perdiz et al. 2011). As in the case of phosphorylation, acetylation is balanced by erasers, called deacetylases. Historically, histone proteins were thought to be the main targets of acetylation. Acetyl transferases and deacetylases are therefore commonly referred to as histone acetyltransferases (HATs) and histone deacetylases (HDACs), respectively. However, because neither of these classes of enzyme is specific for histone proteins, the terms lysine acetyltranferase (KAT) and lysine deacetylase (KDAC) are more appropriate.

Lysine and arginine residues can also be modified by methylation and/or demethylation (Fig. 3). Here, the amino acid side-chain nitrogen atoms have one or more of their hydrogen atoms replaced with methyl groups. A single lysine residue can contain one, two, or three methyl groups, whereas an arginine residue can contain one or two methyl groups distributed in different ways among the three side-chain guanidino nitrogens. Lysine methylation, like acetylation, can weaken interactions between histones and DNA (but can also lead to repression of transcription, depending on which histone lysine residue is methylated) and is therefore a major mechanism for the epigenetic control of gene expression. In addition, both lysine and arginine methylation can drive direct protein–protein interactions when the methylated lysine/arginine residues on one protein are recognized by modular domains of the Royal superfamily (e.g., Tudor, chromo, MBT, PWWP, and plant Agenet domains) on the other protein (Maurer-Stroh et al. 2003). For example, in response to DNA damage, kinases such as ATM and ATR phosphorylate a host of substrates to initiate cell-cycle arrest, DNA repair, and potentially cell death if the damage is too severe. One of these substrates

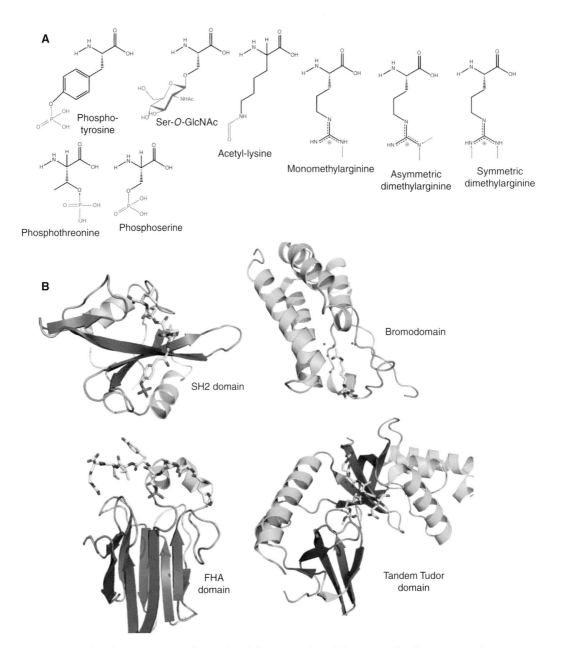

Figure 3. Examples of protein posttranslational modifications and modular protein-binding domains that recognize these modified amino acids. (*A*) Structures of common amino acid posttranslational modifications. The parent amino acid structure is shown in black; the modification is shown in red. (*B*) Cartoon representations of modular binding domains. α Helices are shown in cyan; β-strands are shown in purple; loops are shown in orange. SH2 domains recognize peptides containing phosphotyrosine, FHA domains recognize peptides containing phosphothreonine, Bromo domains recognize peptides containing acetyl-lysine, and tandem Tudor domains recognize dimethylarginine. In the examples shown, the SH2 domain is from Src kinase, the FHA domain is from Chk2, the bromo domain is from Brd4, and the tandem Tudor domains are from SND1.

is the methyltransferase MMSET (also known as NSD2 or WHSC1). Phosphorylated MMSET is recruited to sites of DNA damage, where it methylates histone H4 on lysine residue K20. Methylated H4K20 recruits the DNA repair protein 53BP1 through a Tudor domain in 53BP1, facilitating DNA repair (Pei et al. 2011).

Other types of posttranslational modifications include glycosylation, nitrosylation, and nitration. Glycosylation occurs when sugar residues are covalently attached to the amide nitrogens of asparagine (*N*-linked glycosylation) or to the hydroxyl groups of serine or threonine residues (*O*-linked glycosylation), usually as branched chains, in secreted proteins or the extracellular regions of transmembrane proteins. In most cases, these modifications help the protein to fold correctly or facilitate its transit and secretion, or insertion into the cell membrane; however, the addition of a single residue of a particular amino sugar—N-acetylglucosamine—to serine and threonine residues of cytoplasmic proteins may function in some cases by preventing those same residues from being phosphorylated (Dias et al. 2012).

Protein nitrosylation involves the covalent incorporation of nitric oxide into the thiol side chain of cysteine residues within proteins, whereas protein nitration involves the incorporation of nitric oxide and/or its reactive nitrogen species onto the ring –OH group of tyrosine residues to generate nitrotyrosine. Three isoforms of nitric oxide synthase (NOS), the enzymes that produce NO, are known, all of which appear to participate in protein nitrosylation and nitration. Although less well understood than protein phosphorylation, both S-nitrosylation and O-nitration can also regulate protein structure, catalytic activity, stability, localization, and protein–protein interactions. Protein nitration appears to occur primarily as a consequence of oxidative stress and is believed to affect tissue homeostasis (Radi 2013). In contrast, protein thiol nitrosylation is emerging as a prominent mechanism for regulating signal transduction pathways, particularly those within the cardiovascular system (Lima et al. 2010). Although the best-known role for NO in controlling vasodilation is through the generation of cGMP by activation of guanylyl cyclase (Ch. 3 [Newton et al. 2014]), many of the effects of NO are mediated by S-nitrosylation. For example, the chemokine SDF1 induces cell migration and angiogenesis by activating endothelial NOS, which S-nitrosylates and inactivates the MAPK phosphatase MKP7 to enhance downstream signaling. One of the most intriguing targets of protein nitrosylation is small G proteins of the Ras superfamily. Nitrosylation of a specific cysteine residue seems to facilitate their conversion from an inactive to an active form (see below) (Foster et al. 2009). Additional posttranslational modifications include ubiquitylation and lipidation (discussed in greater detail below).

2.2 Noncovalent Regulation of Protein Activity

Signals are not only transmitted in the form of protein posttranslational modifications. Ions, various lipids, and nucleotides can be produced or relocalized to function as second messengers that transmit a signal (Ch. 3 [Newton et al. 2014]). Another major class of signaling enzymes is guanine-nucleotide-binding proteins, also called G proteins, a large family of signaling proteins that control a wide array of cellular functions, including motility, hormone responses, sensory perception, and neurotransmission. G proteins function as molecular switches. Their activity is regulated by the intrinsic ability to bind and hydrolyze GTP to GDP.

In the basic GTPase cycle, G proteins exist in the "off" state bound to GDP (Fig. 4). G proteins have high affinity for GDP (and GTP); thus, the dissociation rates are very low. In addition, the nucleotide-free (i.e., empty) form of the protein is unstable, such that removal of GDP requires assistance from a guanine-nucleotide exchange factor (GEF). After dissociation of GDP, the empty G protein will favor binding to GTP because of the 10:1 ratio of GTP to GDP within the cell (Bos et al. 2007). The extra phosphate on GTP induces a conformational change in three switch regions near the nucleotide-binding pocket of the G protein, allowing substrate recognition. Signaling is promoted when G proteins are in the active state through binding to motifs in other proteins that specifically recognize the GTP-bound conformation of the G protein. Although the GTP-bound G protein is generally the active state, capable of transducing downstream signals, in some systems, GDP-bound G proteins transduce the active signal, particularly in plants (Temple and Jones 2007). G proteins have a slow intrinsic GTPase activity ($k_{cat} = 10^1 - 10^{-3}$ min^{-1}) that results in GTP hydrolysis and returns the protein to the "inactive" GDP-bound state; however, the rate of GTP hydrolysis can be dramatically enhanced (by a factor of $10^3 - 10^6$) by interaction with GTPase-activating proteins (GAPs). Another important class of regulators is guanine-nucleotide dissociation inhibitors (GDIs). These bind to and stabilize the inactive GDP-bound state, maintaining the G protein in the inactive conformation. Thus, coordinated actions of GEFs, GAPs, and GDIs are critical factors in determining the amplitude, dynamics, and duration of G-protein-transduced signals. The two major classes of G proteins are small G proteins and heterotrimeric G proteins.

2.2.1 Small G Proteins

Small G proteins are ~20–25 kDa in size and consist of a single monomeric subunit that has nucleotide binding and GTPase activities. More than 100 small G proteins exist in humans. The prototypic small G protein is Ras; thus, this class of proteins is sometimes referred to as the Ras superfamily. At least 10 distinct subfamilies exist within the Ras superfamily. Members within a single subfamily generally share similar sequence, structure, and functions (Wennerberg 2005). For example, members of the Rho subfamily (which include RhoA, Rac1, and Cdc42) are generally involved in cytoskeletal dynamics and cell morphology, whereas members of the Arf subfamily control vesicular transport. Targets for small G proteins are often themselves signaling proteins, which creates signaling cascades. For

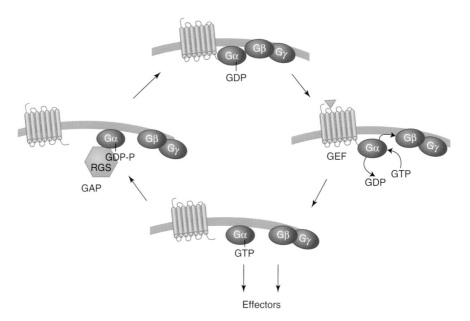

Figure 4. The GTPase cycle. G proteins can be small (Ras-like) or large (heterotrimeric). Depicted here is the nucleotide cycle for heterotrimeric G proteins, but the reactions are conceptually the same for small GTPases. G-protein-coupled receptors (GPCRs) bind extracellular ligands, and transmit signals to intracellular G proteins. The ligand-bound receptor functions as a guanine-nucleotide exchange factor (GEF), causing the Gα subunit to exchange GDP for GTP. GTP-bound Gα no longer interacts with the Gβγ dimer, and both entities are free to interact with downstream effector proteins. Gα controls the duration of the signal because it is a GTPase, whose activity can be stimulated by GTPase-activating proteins (GAPs) such as RGS proteins.

example, Ras activates the kinase Raf, and Rho activates the kinase ROCK.

The characteristics, mechanisms of action, and regulation of GEFs, GAPs, and GDIs differ between small G proteins and heterotrimeric G proteins. GEFs for small G proteins differ in structure and domain architecture for each of the Ras subfamilies, but their mechanisms of action generally involve interaction with the so-called switch regions of the GTPases and the coordination of a magnesium ion within the nucleotide-binding pocket (Sprang 1997). Small G proteins typically have only marginal GTPase activity (often 100-fold to 1000-fold slower than the Gα subunits of heterotrimeric G proteins). GAPs for this class of protein contain a catalytic "arginine finger" region that inserts into the binding pocket, greatly increasing the rate of hydrolysis (Kötting et al. 2008). GDIs generally stabilize the inactive state, but may also alter membrane association or stabilize the GTP-bound state. Members of the Ras subfamily—particularly H-Ras, K-Ras, and N-Ras—are important in many types of cancer because they control cell proliferation (see Ch. 21 [Sever and Brugge 2014]). Two constitutively active mutations are commonly seen in various cancers, including lung, colon, and pancreas. Mutation of the phosphate-binding loop glycine (G12) to valine or aspartic acid results in loss of sensitivity to GAPs, resulting in a prolonged activation period and a higher basal level

of signaling. Mutation of the catalytic glutamine residue (Q61), which normally coordinates a water molecule that attacks the β–γ bond in GTP to a leucine, results in an enzymatically inactive protein because the enzymatic transition state cannot be stabilized (Privé et al. 1992).

2.2.2 Heterotrimeric G Proteins

The second class of G proteins is heterotrimeric G proteins, which function downstream from G-protein-coupled receptors (GPCRs) (Hepler and Gilman 1992; Ch. 1 [Heldin et al. 2014]). The GTPase in this system is the Gα subunit. Roughly 20 different Gα subunits exist in humans, each of which contains two main domains: a GTPase domain that has sequence and structural similarity to Ras (also called the Ras-like domain) and an α-helical domain that can be posttranslationally modified and contributes to GTPase regulation (Dohlman and Jones 2012). The other components of the heterotrimer are the Gβ and Gγ subunits, which generally exist as an obligate dimer. In the basal state, Gα and Gβγ form a tripartite complex in which Gα is bound to GDP; however, the ligand-bound GPCR induces a conformational change in Gα, resulting in GDP-for-GTP exchange. In the GTP-bound state, Gα and Gβγ dissociate, and each is able to transduce a signal to downstream effector proteins. The GEFs for heterotrimeric G proteins are typically the GPCRs themselves (Weis and Kobilka 2008). Some non-GPCR GEFs have been identified, including Ric8a, Arr4, and various AGS family proteins, but their roles are still unclear. Gα subunits have GTP hydrolysis rates about 100 times higher than the Ras superfamily, but GAPs targeting them nevertheless exist. These proteins each contain a regulator of G-protein-signaling (RGS) domain. Some RGS proteins, like Sst2 in yeast or RGS2 and RGS4 in mammals, are relatively simple, essentially containing only an RGS box; others are more complex, containing numerous functional domains (Siderovski and Willard 2005). Unlike GAPs for small G proteins, RGS proteins function by interacting with the switch regions on Gα, and stabilizing the transition state between GDP- and GTP-bound forms (Tesmer et al. 1997).

2.2.3 The GPCR GEF Reaction

It has been estimated that 30%–50% of all drugs target GPCRs or some aspect of GPCR-mediated signal transduction (Wise et al. 2002). Despite this, our understanding of how GPCRs activate G proteins remains incomplete. Unlike GEFs for small GTPases, GPCRs do not make contact with the switch regions on Gα. Furthermore, other than the mobile switch regions, the Gα subunit was not thought to undergo large conformational changes upon activation. Recent high-resolution crystal structures of a GPCR–heterotrimeric G-protein-ternary complex, however, indicate that binding of ligand to the GPCR may induce a major displacement of the all-helical domain of Gα relative to the Ras-like GTPase domain during catalysis (Chung et al. 2011; Rasmussen et al. 2011). These structures suggest a mechanism by which binding of ligand to a GPCR induces nucleotide exchange.

3 REGULATION OF PROTEIN–PROTEIN INTERACTION

3.1 Protein Interaction Domains

What exactly is a protein domain, and what kind of amino acid sequences do reader domains involved in protein–protein interactions read? A domain is a segment of a protein, generally 50–400 amino acids in length, that folds independently into a stable three-dimensional structure and is capable of some type of independent function. Most signaling proteins are built of multiple domains. The sequences that connect the domains together are usually short and less structured parts of the protein. More than 1000 modular protein domains have been characterized using bioinformatics (Letunic et al. 2011).

A classic example of modular protein reader domains is SH2 domains, which bind to phosphotyrosine-containing sequences. Another example is SH3 domains, which bind to proline-containing sequences. Other domains, such as PTB domains, HYB domains, and some C2 domains, can also bind to phosphotyrosine-containing sequence motifs, whereas WW, EVH1, and GYF domains, like SH3 domains, bind to short proline-rich sequences.

A wide variety of domains recognize other posttranslational modifications. Domains such as 14–3–3, FHA domains, tandem BRCT domains, MH2 domains and Polo-box domains, for example, bind to short phosphoserine- and/or phosphothreonine-containing motifs (Yaffe and Smerdon 2004). Bromo domains recognize specific acetyl-lysine-containing sequences, and chromodomains, Tudor domains, MBT domains, and PWWP domains bind to sequences that contain methyl-lysine and methylarginine. Other domains, such as CUE, PAZ, UBA, and UEV domains, can bind to ubiquitin or specific types of ubiquitin chains. The ability of these domains to distinguish unmodified proteins from proteins containing these different types of posttranslational modifications ensures that protein–protein interactions occur only when one of the two proteins has been appropriately marked and modified by a writer. This use of readers allows protein–protein interactions and the assembly of multiprotein signaling machines, or even the activity of a single protein, to be precisely controlled by the actions of writers and erasers in response to a signal. For domains like SH3 domains that do not bind to modified sequences, their reader functions are typically regulated by conformational changes in other parts of the protein that modulate access of the domain to binding partners.

Reader domains are often found within proteins that also contain writer or eraser domains. Frequently, these interact with each other or with sequences that lie outside the domain, and this can have important functional consequences. This is perhaps best illustrated by the tyrosine kinase Src. Src contains a kinase writer domain, together with SH2 and SH3 reader domains. In the inactive state, the SH2 domain of Src is bound to a carboxy-terminal tyrosine residue that has been phosphorylated, while its SH3 domain is bound to a proline-containing linker region between the SH2 domain and the kinase domain. In this conformation, the SH2 and SH3 domains rest against the back surface of the kinase domain, holding it in a catalytically inactive form. Dephosphorylation of the carboxy-terminal tyrosine in Src by phosphatases such as SHP1 or SHP2, or displacement of the SH2 and SH3 domains through competitive binding to phosphotyrosine sequences (such as those found in platelet-derived growth factor receptor [PDGFR]) or focal adhesion kinase (FAK) and polyproline sequences (such as those found in the arrestin-bound β2 adrenergic receptor, NEF, or Sin) then releases the kinase domain to promote activation. Together with phosphorylation of a tyrosine residue in the activation loop, this results in full catalytic activity (Fig. 5).

A similar intramolecular interaction between reader and writer domains keeps the mitotic kinase Polo-like kinase 1 (Plk1) inactive until a certain level of cyclin-dependent kinase (CDK) activity has been reached (Ch. 6 [Rhind and Russell 2012]). Plk1 contains an amino-terminal kinase writer domain and a carboxy-terminal Polo-box reader domain that fold back against each other in interphase cells to keep the protein inactive (Lowery et al. 2005). When cells approach mitosis, CDKs phosphorylate substrates such as Cdc25C and Wee1 to generate phosphoserine/threonine motifs that can bind directly to the Polo-box domain, prying it away from the kinase domain. Again, together with phosphorylation of a threonine residue in the activation loop, this drives Plk1 into the fully active form required for cells to complete mitosis.

For many domains, portions that are not directly involved in motif recognition can also be instrumental in stabilizing protein–protein interactions. These types of domain–domain interactions are often important in assembly of multisubunit signaling complexes. For example, non-ligand-binding portions of the SH2 domain in phospholipase Cγ (PLCγ) are used to stabilize interactions with fibroblast growth factor 1 (FGFR1). Intriguingly, evolution has led to preferential cosegregation of particular protein domains within individual proteins (Jin and Pawson 2012). For example, SH2 domains and SH3 domains frequently co-occur, as do PX domains and SH3 domains. Presumably this is because specific combinations of domains have already mastered the molecular origami required for both productive interactions and allosteric control over additional domains with which they coassociate (see Table 1 for common domains, their partners, and motifs).

3.2 Motifs

For readers, writers, and erasers to function together to form signaling networks, they must operate on common short amino acid sequence motifs in their targets. Most such motifs contain ~15 amino acid residues and include particular amino acids that confer specificity to the writers and readers. For example, CDKs and MAPKs (writers) preferentially phosphorylate S-P and T-P motifs, whereas Polobox domains (readers) preferentially recognize motifs with the consensus S-pS/pT-P (where pS and pT are phosphorylated serine and threonine). Similarly, many tyrosine kinases (writers) phosphorylate tyrosine residues that are surrounded by acidic amino acid residues, whereas SH2 domains (readers) prefer bind to phosphotyrosine-con-

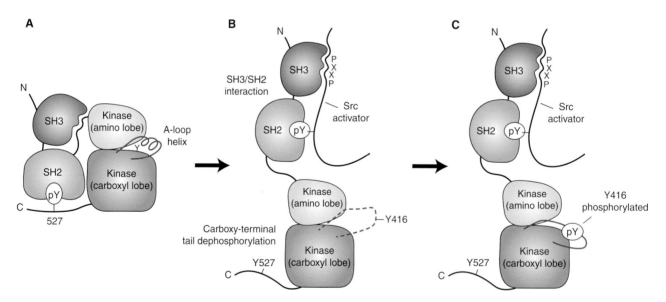

Figure 5. A multistep mechanism for maximal Src activation. (A) In the inactive state, Src is folded up as a consequence of multiple interactions between the reader domains and motifs in Src itself. The Src SH2 domain is bound to a phosphotyrosine residue (Y527) in the carboxyl terminus, while the SH3 domain binds to a polyproline-type helix in the linker that connects the SH2 domain to the kinase domain. (B) The kinase opens up into an active conformation when a ligand such as a growth factor receptor or an adaptor protein engages the SH2 and SH3 domains directly, usually accompanied by dephosphorylation of the Y527 site to prevent intramolecular reassociation into the closed form. (C) Autophosphorylation of Y416 in the activation loop of Src, or phosphoprylation of this site by another kinase, results in maximal activity. (From Xu et al. 1999; adapted, with permission, © Elsevier.)

taining sequences that contain specific patterns of hydrophobic or small amino acid residues carboxy terminal to the phosphotyrosine. Some domains can bind multiple sequence motifs or even bind ligands in multiple orientations, as is the case for SH3 domains and SUMO-SIM domains.

This type of limited overlap between the sequences that the writer domains generate and those that the reader domains recognize gives rise to the specificity we observe in signaling networks. That is, only a small fraction of the targets of any particular writer are then recognized by a particular reader. As mentioned above, not all motifs require posttranslational modification (e.g., proline-rich sequences are recognized by SH3 domains, WW domains, and EVH1 domains, among others), which leads to competition between a motif and multiple readers, depending on localization and timing. In contrast, some types of domains appear to recognize specific combinations of posttranslational modifications. Some bromo domains, for example, recognize specific phosphoacetyl-methyl motif combinations on histones (Filippakopoulos et al. 2012), whereas 14–3–3 proteins are able to recognize both phosphorylated and phosphoacetylated histone sequences (Macdonald et al. 2005). A similar effect can be achieved by adjacent protein–protein interaction domains that bind to different posttranslational modifications. These domains or domain combinations function as readers of a more complex code, essentially creating "AND gates," which can be used to dictate greater specificity and control, or functioning as integrators of signaling from multiple pathways.

Because most motifs are defined by the primary structure of a protein rather than a complicated 3D arrangement of noncontiguous elements, it is a relatively straightforward process to go motif hunting using protein sequences and bioinformatics search tools (Obenauer et al. 2003; Obenauer and Yaffe 2004). The small, "portable" nature of these motifs means that they have frequently moved around within the sequences of evolutionarily related proteins. Thus, a motif can sometimes be seen in one part of a protein sequence in a human protein and in another part of the sequence in a yeast ortholog.

4 REGULATION OF PROTEIN LOCATION

Signaling processes occur in the context of the multicompartment, 3D environment of the cell. Information transfer from the cell exterior, across lipid membranes, and through/within the cytosol must be tightly regulated in both space and time, and a key component to achieving specificity is the dynamic regulation of protein localization.

4.1 Compartmentalization

A common theme in signaling is modulating compartmentalization of signaling proteins within the cell. This can dictate access to substrates or environmental conditions that promote activation. Subcellular compartments, particularly organelles and cytoskeletal structures, can be very dynamic. In these cases, signaling can promote the formation of these compartments. The spatial segregation that organelles provide represents an important layer of regulation, allowing similar proteins to execute different functions in different environments. For example, mitotic kinases like Plk1, Aurora B, and Never in mitosis kinase 2 (Nek2) achieve substrate specificity through nonoverlapping subcellular localization, despite sharing partially overlapping substrate selection motifs (Alexander et al. 2011).

Such compartmentalization can be achieved in different ways. Some proteins have short sequences (also called signal peptides) that function as localization signals, allowing transport to a particular organelle. The best recognized of these are nuclear localization signals (NLSs), which typically feature one or more short sequences of positively charged lysine or arginine residues (Hung and Link 2011). Similar signals have been identified for many other organelles, including mitochondria, lysosomes, and peroxisomes. For proteins with localization sequences, regulated localization can be achieved through posttranslational modifications or conformational changes that mask or unmask a signal peptide. Type I nuclear receptors, for example, are typically retained in the cytoplasm in an inactive complex with HSP90 (see p. 129 [Sever and Glass 2013]; Ch. 1 [Heldin et al. 2014]). Ligand binding induces a conformational change, dissociation of HSP90, homodimerization, and active transport into the nucleus, promoting DNA binding and gene expression. Phosphorylation of the MAPK ERK on an SPS motif within the kinase-insert region results in recognition by the nuclear transport receptor importin β7 (Chuderland et al. 2008). Following nuclear import, ERK then phosphorylates nuclear substrates, including transcription factors Fos, Myc, and Elk1 (see p. 81 [Morrison 2012]). For proteins without localization sequences, protein–protein interactions or association with lipid membranes can localize them to specific subcellular regions.

4.2 Membrane Localization

Attachment to various cellular membranes can function as an anchor that restricts a protein or its activation to certain subcellular areas. Another benefit is that signaling is more efficient in two dimensions, because the local concentration near the membrane is greatly increased. Some estimates have suggested that membrane association increases the local protein concentration as much as 1000-fold (McLaughlin and Aderem 1995). Membrane-localized proteins can be either transmembrane proteins embedded in the membrane or peripheral membrane proteins that attach to membranes through protein–lipid interactions. Because they connect two physically separated environments, transmembrane proteins are well positioned to function as pumps, ion channels, or cell-surface receptors (Ch. 1 [Heldin et al. 2014]). Peripheral membrane proteins, which may interact with transmembrane receptors, typically function as signaling intermediaries. The most common method of membrane attachment for these proteins is through the covalent attachment of lipid groups (e.g., glycophosphatidylinositol [GPI] anchors and other lipid chains) (Casey 1995; Nadolski and Linder 2007) and reversible recruitment to the plasma membrane is a common method of signal regulation. Upon activation of certain receptor tyrosine kinases, for example, phosphoinositide 3-kinase (PI3K) is recruited to the receptor via interactions between the SH2 domains of the p85 subunit of PI3K and phosphorylated tyrosine residues on the activated receptor. Active PI3K then phosphorylates phosphatidylinositol 4,5-bisphosphate (PIP_2) to create phosphatidylinositol-3,4,5-trisphosphate (PIP_3). Downstream effectors like the kinase Akt then are recruited to the membrane through a PH-domain-dependent interaction with PIP_3. Once at the membrane, Akt can then be phosphorylated and activated by its activating kinase PDK1, which also has a PIP_3-binding PH domain. Similar mechanisms in which subcellular localization and activation are intrinsically coupled are also used for activation of Ras family GTPases. These involve recruitment of activation factors such as GEFs to the plasma membrane.

4.3 Lipid Modification

Membrane attachment can be induced by protein lipidation, owing to the hydrophobicity of the modifying lipid.

Many types of lipid modification exist, each unique in its regulation, chemical properties, and mechanism of linkage to proteins (Fig. 6). Myristoylation is the cotranslational addition of a saturated 14-carbon acyl chain to an amino-terminal glycine. The myristoyl moiety has a relatively low level of hydrophobicity. Thus, myristoylated proteins, such as Src-family kinases and some G proteins, typically achieve membrane attachment by using myristoylation in concert with other lipid modifications or nearby polybasic amino-acid stretches. Myristoylation can allow proteins to attach to membranes, but, in some cases, it alters protein conformation or protein–protein interaction (Boutin 1997). A myristoyl group added to the C subunit of PKA, for example, can insert into a lipid a bilayer, promoting membrane localization, or alternatively, fold into a hydrophobic pocket of the enzyme, allowing regulation of the protein–lipid conformational state by phosphorylation. Another common modification is prenylation, the post-translational addition of a 15-carbon farnesyl or 20-carbon geranylgeranyl chain attached to a carboxy-terminal cysteine. The enzymes responsible for prenylation have been identified, and inhibitors of these enzymes have received much attention as anticancer agents, owing to the importance of prenylation of GTPases in the Ras family (Resh 2012).

Another example is palmitoylation, the addition of a 16-carbon-chain fatty acid to cysteine residues through a thioester bond. Protein subcellular localization is often dictated by palmitoylation, through the localization of the palmitoyl acyltransferase (PAT). For example, proteins with only a myristoyl or prenyl group are thought to transiently interact with membranes, sampling many membrane surfaces within the cell. However, palmitoylation of a singly lipid-modified protein induces stable membrane attachment. Thus, through limited spatial expression of the PATs, palmitoylation can dictate protein subcellular

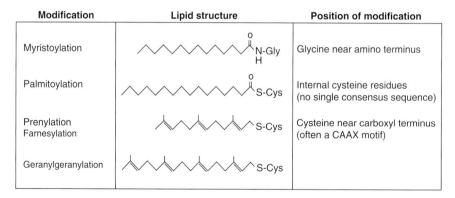

Figure 6. Common forms of protein lipidation. The table highlights four lipid moieties that are commonly used to modify proteins posttranslationally or (in the case of myristoylation) cotranslationally. These modifications differ in terms of their consensus sequences, position of the modification, hydrophobicity, and mechanism of regulation.

Cite this chapter as *Cold Spring Harb Perspect Biol* doi: 10.1101/cshperspect.a005918

localization. For example, the Ras family member H-Ras can localize to the plasma membrane or Golgi apparatus, having unique signaling roles at each location. H-Ras subcellular localization is dictated by palmitoylation on one of two carboxy-terminal cysteine residues (Roy et al. 2005). Furthermore, unlike myristoylation or prenylation, palmitoylation is a reversible process and can be dynamically regulated during signal transduction.

4.4 Membrane Microdomains and Spatially Restricted Signaling

Proteins and lipids in the plasma membrane are thought to be distributed heterogeneously, forming membrane microdomains (Maxfield 2002). This may help sequester low-abundance molecules, increasing their local concentrations or facilitating the formation of signaling machines. Examples of such microdomains include invaginations of the plasma membrane called caveolae and lipid rafts, cholesterol- and sphingolipid-rich portions of the membrane that potentially function as organizing centers for compartmentalized signaling. GPI-linked proteins are preferentially localized to such microdomains, which may play an important role in many signaling processes. Membrane microdomains can be used to promote spatially restricted signaling, such as those seen in the immunological synapse (the interface between an antigen-presenting cell and effector T cell) (see Ch. 16 [Cantrell 2014]).

Spatially regulated signal transduction is a well-established phenomenon, but an emerging paradigm is the positive and negative regulation of plasma-membrane-associated signals at intracellular locations. For example, H-Ras, well known to transduce signals from the plasma membrane, also exists at Golgi bodies, where it transmits a unique signal (Bivona et al. 2003). In addition, a growing body of evidence supports a positive role for the endocytic pathway in signal transduction (Zastrow and Sorkin 2007; Murphy et al. 2009). Numerous signals transmitted from endosomal locations have been identified, including those coupled to RTK- (Di Guglielmo et al. 1994) and GPCR-coupled signals (Lefkowitz and Shenoy 2005; Slessareva et al. 2006).

Signals can also be spatially restricted through binding to protein scaffolds. The co-occurrence of multiple motifs on a single molecule can lead to the formation of signaling scaffolds and signaling hubs. If a single protein contains multiple motifs and each is recognized by a different protein, then these will all converge. Such an arrangement is used, for example, to organize MAPK signaling in yeast and mammalian cells (Elion 2001; Engström et al. 2010; p. 81 [Morrison 2012]). By using scaffolds to bring two or more proteins together, cells overcome the diffusion problem and

no longer require the proteins to find each other in the crowded interior of the cell. As is the case for membrane attachment, the tethering of proteins to scaffolds greatly increases the effective concentration of signaling proteins. If the different proteins that are recruited to a multimotif protein do not communicate directly with each other, but instead act on other proteins, then the multimotif protein serves as a hub, where multiple signaling events converge in time and space. Such an arrangement is used, for example, by A-kinase-anchoring proteins (AKAPs) to coordinate signaling through PKA pathways (Ch. 3 [Newton et al. 2014]), and by the cytoplasmic tails of growth factor receptors to coordinate signaling by Ras, PI3K, PKC, and MAPKs (Ch. 1 [Heldin et al. 2014]).

5 REGULATION OF PROTEIN PRODUCTION

At its most basic level, signaling is controlled by the relative stoichiometry of positive and negative signaling intermediaries, which are frequently proteins. Any process that changes the balance of these proteins can, in turn, modulate the signal. One basic and widely used strategy is to change how much of the signaling intermediate—whether a ligand, second messenger, or protein—is available to transmit information. Cells use an incredibly diverse array of methods to regulate protein levels—so many, in fact, that an in-depth review of this topic would be a textbook unto itself. Here, we therefore focus on basic ways cells control signaling by modulating the rate of protein synthesis or protein degradation. The processes that can be targeted include transcription, RNA stability, RNA splicing, and translation. Of these, transcriptional regulation is probably the best studied and most commonly recognized target of signal transduction pathways.

5.1 Transcriptional Control

The output of many signaling pathways is transcription of genes that encode proteins necessary for the desired cellular response. The regulation can occur by modulating nuclear localization of transcription factors, coactivators, and other regulatory proteins by modulating the DNA-binding capabilities of these proteins or by posttranslationally modifying histones to modulate chromatin architecture. A good example is the transcriptional control of the tumor suppressor p53, which is a transcription factor, and its downstream targets in response to DNA damage signals (Brown et al. 2009; Ch. 5 [Duronio and Xiong 2013]). In unstressed cells, p53 expression remains low, owing to a low level of transcription and constitutive ubiquitin-dependent degradation of p53 promoted by an E3 ligase protein called MDM2 (Momand et al.

1992). In response to various cellular stressors, including ionizing radiation, UV, and other forms of genotoxic damage, the transcription of the p53 gene is dramatically increased, and the p53 protein is phosphorylated on one or more of its many phosphorylation sites. There are two important consequences of this phosphorylation event. The first is increased levels of p53 protein due to loss of interaction with MDM2, an E3 ubiquitin ligase that drives ubiquitin-mediated degradation (see below) of p53 in unstimulated cells. Second, as a result of MDM2 dissociation, p53 undergoes a conformational change that allows it to interact with DNA and drive transcription of new p53 target genes. Among the genes activated by p53 are those that control cell-cycle arrest, DNA repair, and apoptotic cell death, and even the *p53* gene itself (Bieging and Attardi 2012).

5.2 RNA Stability and miRNAs

Signal transduction can also involve modulation of mRNA processing or stability. Several splicing factors are phosphorylated in response to signaling, and these events may control alternative splicing of pre-mRNAs to give different mRNA isoforms (Lynch 2007). Cells can regulate mRNA stability by modulating mRNA maturation (via 5′ capping or 3′ polyadenylation) or interactions between RNA-binding proteins and mRNA transcripts (Wu and Brewer 2012). An emerging area of research is the control of signaling by cellular micro-RNAs (miRNAs) and other noncoding RNAs, such as long noncoding RNAs (lnc-RNAs). miRNAs are short (~21–23 nucleotide) untranslated RNAs (Ambros 2001) that typically induce degradation of target RNAs or block mRNA translation through sequence-specific base-pairing interactions. They are important regulators of normal developmental timing (Ambros 2011). miRNAs like Lin-4/mir-125 regulate the temporal transitions between pluripotent and differentiated states of numerous stem cell populations.

miRNAs are also common regulators of the dynamics, duration, and sensitivity of signaling processes. Many signaling pathways—including the Wnt, Notch, Hedgehog, and p53 pathways, for example—achieve specificity and sensitivity through active repression (i.e., basal repression or default repression). In this context, miRNAs target core pathway components or their transcriptional targets, maintaining cells in the inactive state, causing the system to require greater levels of stimulus for activation and also sharpening the response to activating stimuli. For example, miR-125 targets many components of p53 signaling, and loss of miR-125 causes spontaneous p53 activation.

5.3 Translational Control

Translation is also a target of signal transduction pathways. For example, the mammalian target of rapamycin (mTOR) and related pathways are now well known to control protein translation in response to nutrient, growth factor, and amino acid signals (Laplante and Sabatini 2012). A critical target involved in the regulation of translation by the mTORC1 complex is the eukaryotic translation initiation factor 4E (eIF4E)–binding protein 1 (4E-BP1) (Hara et al. 1997). 4E-BP1 inhibits cap-dependent translation by binding to eIF4E, the initiation factor that recognizes the 5′ cap on mRNA molecules. Phosphorylation of 4E-BP1 at T37 or T46 is thought to prime 4E-BP1 for subsequent phosphorylation at S65 and T70, leading to loss of interaction with eIF4E (Gingras et al. 1999), which results in a general increase in protein translation. The mTORC1 complex also regulates the activity of p70 S6 kinase 1 (S6K1), which has many targets whose phosphorylation activates translation (Pullen and Thomas 1997). Key among these is the S6 subunit of the 40S ribosome, which when phosphorylated by S6K1 promotes increased translation of mRNA transcripts containing an oligopyrimidine sequence in their 5′ untranslated region (5′-UTR). Additionally, translation of specific mRNAs can be controlled through feedback signaling at the level of translation initiation. For example, iron homeostasis is maintained in part by regulating the translation of proteins involved in iron import. In the presence of high concentrations of intracellular iron, the transferrin receptor mRNA is repressed by iron-induced binding of iron-response-element-binding protein (IREBP) to a 5′-UTR element in the transferrin receptor mRNA.

6 PROTEIN DEGRADATION

Every protein has a baseline level of expression resulting from the equilibrium between its synthesis and breakdown (see Box 1). A control mechanism widely used in signal transduction is regulation of the rate of protein degradation. Many methods exist for regulated degradation of proteins, including lysosomal and ER-mediated degradation (Ciechanover 2012), but the mechanism that seems to be most important in cell signaling is ubiquitin-dependent proteasomal degradation (Hershko and Ciechanover 1998). Ubiquitin is a small 76-amino-acid protein that can be conjugated through its carboxyl terminus to lysine residues on target proteins or to other ubiquitin molecules to form ubiquitin chains that serve as a marker for various cellular functions. Through diversity in chain length or the orientation of chain attachment, ubiquitin can regulate protein trafficking and protein–protein interactions, or

BOX 1. PROTEIN LEVELS DESCRIBED MATHEMATICALLY

Protein levels are best described using a deceptively simple equation:

$$d[\text{P}]/dt = k_1 - k_2[\text{P}],$$

where [P] is the protein concentration, k_1 is the synthesis rate, and k_2 is the degradation rate. Obviously, k_1 and k_2 reflect complex processes that are themselves regulated by signaling events at multiple levels. At steady state, when [P] is constant, $d[\text{P}]/dt = 0$ and $[\text{P}] = k_1/k_2$. Therefore, to increase [P], we need either to increase the synthesis rate or to decrease the degradation rate. The two key questions from a biologist's point of view are these: (1) what will the new level be? and (2) how fast will the changes happen—that is, how long until the new steady-state level is reached? Consider the case when the synthesis of a protein ceases entirely—that is, $k_1 = 0$. Because the protein is now subject only to degradation, the new steady-state level will be 0—that is, $[\text{P}] = 0/k_2$—but the rate of decline is given by k_2; that is, $t_{1/2} = \ln 2/k_2$. The somewhat surprising thing is that the same is true if we change the synthesis rate to some value other than zero. For example, if we double the synthesis rate, the new steady-state level will be twice the old steady-state level ($[\text{P}_{ss}] = k_1/k_2 \rightarrow 2k_1/k_2$), but the time it takes to reach the new steady-state level will be determined *not* by the change in synthesis rate but by the degradation rate, k_2. Thus, big changes in synthesis rate only cause rapid changes in steady-state protein levels if the degradation rate is fast, that is, if the protein is short lived. Similarly, minor changes in degradation rates are manifest with kinetics that depend *not* on the degradation rate but the synthesis rates—that is, rapid changes in degradation rate only manifest in changes in steady-state protein levels if the synthesis rate is fast.

function as a signaling scaffold (Muratani and Tansey 2003; Kirkin and Dikic 2007; Walczak et al. 2012).

Perhaps its major role, however, is to form polyubiquitin chains that target proteins for destruction by the proteasome. Recognition by the proteasome requires at least four ubiquitin monomers linked to each other through amide bonds involving the carboxyl terminus of one ubiquitin molecule and K48 on another ubiquitin molecule. Although chains of four monomers are required, longer chains are typical, presumably increasing the efficiency of proteasomal recognition. Ubiquitin chains are formed through a cascade of enzymatic processes performed by a ubiquitin-activating protein (E1), a ubiquitin-conjugating protein (E2), and a ubiquitin ligase (E3) (Fig. 7). The chains target proteins to the proteasome, a large protein complex containing multiple proteolytic enzymes, where ubiquitylated proteins are subsequently degraded and their parts recycled by the cell. A classic example of this process in action is the regulated expression of the cyclin proteins (Ch. 6 [Rhind and Russell 2012]), whose oscillating levels allow the periodic activation of cyclin-dependent kinases (CDKs) that regulates progression through the cell cycle (Malumbres and Barbacid 2005). These oscillations are achieved primarily by increased degradation of cyclins by ubiquitin-dependent proteasomal degradation, coupled with increased cyclin synthesis at different points in the cell cycle (Sudakin et al. 1995).

Other types of polyubiquitin chains, in which the individual ubiquitin molecules are connected through amide bonds involving residues in ubiquitin other than K48 can also be added to proteins (Fig. 7) (Ikeda and Dikic 2008). Seven such chain types can be formed, in which the lysine side-chain amino group in one of the seven lysine residues present in ubiquitin is linked to the carboxy-terminal diglycine motif of the next ubiquitin molecule through an isopeptide bond. Prominent examples of these other ubiquitin chains in signaling include the use of K63-linked chains to control protein–protein interaction rather than protein degradation. In the IKK/NF-κB pathway (see p. 121 [Lim and Staudt 2013]), for example, stimulation of the ubiquitin ligase activity of TRAF6, results in K63-linked polyubiquitylation of substrate proteins like NEMO and TRAF6 itself. K63-ubiquitylated TRAF6 binds to the NZF reader domain in TAB2/TAB3, which then recruits and activates TAK1, which, in turn, phosphorylates and activates IKK (Chen 2005).

In addition, an eighth chain type exists, linear ubiquitin chains, in which peptide bonds are formed between the amino-terminal methionine of a ubiquitin monomer and the carboxy-terminal carboxy group of another ubiquitin monomer. Linear ubiquitin chains are important for generating protein-binding surfaces and promote the activation of inflammation and stress-signaling pathways, such as those downstream from the tumor necrosis factor (TNF) receptor (Walczak et al. 2012). Furthermore, proteins can also be modified by the attachment of ubiquitin-like proteins, such as SUMO (for small ubiquitin-like molecule) and Nedd8 (for neural precursor cell expressed de-

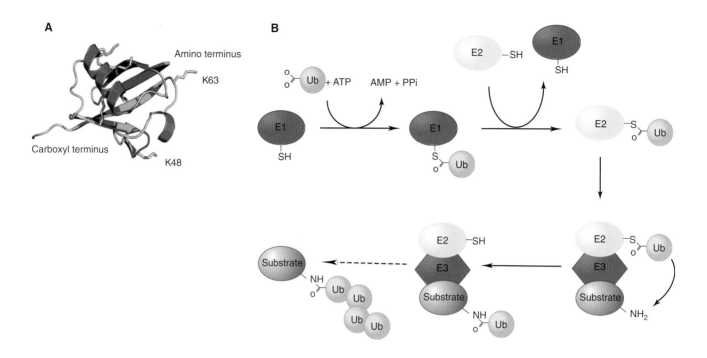

Figure 7. Protein ubiquitylation. (*A*) Structure of the ubiquitin monomer, highlighting the amino and carboxyl termini, as well as key lysine residues. (*B*) Schematic depiction of the posttranslational modification of substrate proteins with ubiquitin. Ubiquitin is added to proteins through a three-step enzymatic reaction featuring a ubiquitin-activating enzyme (E1), a ubiquin-conjugating enzyme (E2), and a ubiquitin ligase (E3). Subsequent rounds of ubiquitylation can result in the formation of ubiquitin chains. (*C*) Ubiquitin chain diversity. Ubiquitin contains seven lysine residues, each of which can be used as the anchorage point for subsequent ubiquitin monomers, in homotypic (same linkage throughout chain), heterotypic (mixed chains), or branched fashion (multiple ubiquitin monomers conjugated to a single ubiquitin). In addition, the amino-terminal methionine on a ubiquitin monomer linked to a substrate protein can be linked to the carboxy-terminal end of another ubiquin monomer (linear chains). Although all of these forms have been shown to exist in cells, physiological roles for many of these chains are still being elucidated.

Cite this chapter as *Cold Spring Harb Perspect Biol* doi: 10.1101/cshperspect.a005918

velopmentally down-regulated 8). Like K63 and linear ubiquitin chains, these molecules are thought to control protein–protein interactions rather than regulate protein degradation.

7 CONCLUDING REMARKS: WHAT DOES THE FUTURE HOLD

Modulation of proteins represents the frontline response of most signaling systems. Many proteins are exquisitely sensitive to even very small posttranslational chemical modifications, such as phosphorylation, methylation, and the other posttranslational modifications highlighted here. These modifications can create new protein–protein binding surfaces or alter the conformation or localization of the target protein (which can, in turn, offer new protein–protein interaction possibilities). Furthermore, these modifications can dramatically alter the abundance of a protein in both positive (as in the case of signaling driving transcription) and negative directions (as in the case of ubiquitin-mediated degradation).

An equally important question regarding posttranslational changes that modulate signal transduction is not "what" or "how," but rather "why." Why have these signaling systems used so much regulation at the protein level? Why are these complex regulatory mechanisms preferred? Many benefits have been proposed (and some validated) for the types of regulatory complexity that are seen in signaling biology, including signaling speed, robustness, reversibility, accuracy, and/or sensitivity that may be required for certain biological responses. The greatest benefit of these designs, however, is their diversity. Signaling processes were selected not to maximize efficiency but rather to maximize possibility. The modular organization of signaling components allows them to be imported into new biological contexts, where they may create new biological functions. This may be best illustrated by the roles of protein ubiquitylation in inactivating signals in some biological contexts and activating signals in others. Thus, using the tool kit of protein modules that function as readers, writers, and/or erasers, along with changes in protein abundance and location, signaling systems have evolved the potential to use a very limited list of parts to respond to a nearly limitless set of cues.

Although our knowledge of protein regulation in signal transduction is broad, it is important to stress that it is only deep at the detailed molecular level in a limited number of areas. Much remains to be learned, particularly with respect to how protein regulatory mechanisms work together at the systems level, and how protein regulatory behavior can be captured in the language of mathematics (Ch. 4 [Azeloglu and Iyengar 2014]). In addition, we can be sure that addi-tional types of posttranslational modifications, new modular binding domains, and new mechanisms of protein regulation will emerge.

REFERENCES

*Reference is in this book.

Alexander J, Lim D, Joughin BA, Hegemann B, Hutchins JRA, Ehrenberger T, Ivins F, Sessa F, Hudecz O, Nigg EA, et al. 2011. Spatial exclusivity combined with positive and negative selection of phosphorylation motifs is the basis for context-dependent mitotic signaling. *Sci Signal* **4:** ra42.

Ambros V. 2001. microRNAs: Tiny regulators with great potential. *Cell* **107:** 823–826.

Ambros V. 2011. MicroRNAs and developmental timing. *Curr Opin Genet Dev* **21:** 511–517.

* Azeloglu EU, Iyengar R. 2014. Signaling networks: Information flow, computation, and decision making. *Cold Spring Harb Perspect Med* doi: 10.1101/cshperspect.a005934.

Bieging KT, Attardi LD. 2012. Deconstructing p53 transcriptional networks in tumor suppression. *Trends Cell Biol* **22:** 97–106.

Bivona TG, Pérez De Castro I, Ahearn IM, Grana TM, Chiu VK, Lockyer PJ, Cullen PJ, Pellicer A, Cox AD, Philips MR. 2003. Phospholipase Cγ activates Ras on the Golgi apparatus by means of RasGRP1. *Nature* **424:** 694–698.

Bos JL, Rehmann H, Wittingshofer A. 2007. GEFs and GAPs: Critical elements in the control of small G proteins. *Cell* **129:** 865–877.

Boutin JA. 1997. Myristoylation. *Cell Signal* **9:** 15–35.

Brown CJ, Lain S, Verma CS, Fersht AR, Lane DP. 2009. Awakening guardian angels: Drugging the p53 pathway. *Nat Rev Cancer* **29:** 862–873.

Burnett G, Kennedy EP. 1954. The enzymatic phosphorylation of proteins. *J Biol Chem* **211:** 969–980.

* Cantrell D. 2014. Signaling in lymphocyte activation. *Cold Spring Harb Perspect Biol* doi: 10.1101/cshperspect.a018788.

Casey PJ. 1995. Protein lipidation in cell signaling. *Science* **268:** 221–225.

Chen ZJ. 2005. Ubiquitin signalling in the NF-κB pathway. *Nat Cell Biol* **7:** 758–765.

Chuderland D, Konson A, Seger R. 2008. Identification and characterization of a general nuclear translocation signal in signaling proteins. *Mol Cell* **31:** 850–861.

Chung KY, Rasmussen SGF, Liu T, Li S, Devree BT, Chae PS, Calinski D, Kobilka BK, Woods VL, Sunahara RK. 2011. Conformational changes in the G protein Gs induced by the β2 adrenergic receptor. *Nature* **477:** 611–615.

Ciechanover A. 2012. Intracellular protein degradation: From a vague idea thru the lysosome and the ubiquitin–proteasome system and onto human diseases and drug targeting. *Biochim Biophys Acta* **1824:** 3–13.

Dias WB, Cheung WD, Hart GW. 2012. O-GlcNAcylation of kinases. *Biochem Biophys Res Commun* **422:** 224–228.

Di Guglielmo GM, Baass PC, Ou WJ, Posner BI, Bergeron JJ. 1994. Compartmentalization of SHC, GRB2 and mSOS, and hyperphosphorylation of Raf-1 by EGF but not insulin in liver parenchyma. *EMBO J* **13:** 4269–4277.

Dohlman HG, Jones JC. 2012. Signal activation and inactivation by the Gα helical domain: A long-neglected partner in G protein signaling. *Sci Signal* **5:** re2–re2.

* Duronio RJ, Xiong Y. 2013. Signaling pathways that control cell proliferation. *Cold Spring Harb Perspect Biol* **5:** a008904.

Elion EA. 2001. The Ste5p scaffold. *J Cell Sci* **114:** 3967–3978.

Engström W, Ward A, Moorwood K. 2010. The role of scaffold proteins in JNK signalling. *Cell Prolif* **43:** 56–66.

Filippakopoulos P, Picaud S, Mangos M, Keates T, Lambert J-P, Barsyte-Lovejoy D, Felletar I, Volkmer R, Müller S, Pawson T, et al. 2012.

Histone recognition and large-scale structural analysis of the human bromodomain family. *Cell* **149:** 214–231.

Foster MW, Hess DT, Stamler JS. 2009. Protein *S*-nitrosylation in health and disease: A current perspective. *Trends Mol Med* **15:** 391–404.

Gingras AC, Gygi SP, Raught B, Polakiewicz RD, Abraham RT, Hoekstra MF, Aebersold R, Sonenberg N. 1999. Regulation of 4E-BP1 phosphorylation: A novel two-step mechanism. *Genes Dev* **13:** 1422–1437.

Hara K, Yonezawa K, Kozlowski MT, Sugimoto T, Andrabi K, Weng QP, Kasuga M, Nishimoto I, Avruch J. 1997. Regulation of eIF-4E BP1 phosphorylation by mTOR. *J Biol Chem* **272:** 26457–26463.

* Heldin C-H, Lu B, Evans R, Gutkind JS. 2014. Signals and receptors. *Cold Spring Harb Perspect Biol* doi: 10.1101/cshperspect.a005900.

Hepler JR, Gilman AG. 1992. G proteins. *Trends Biochem Sci* **17:** 383–387.

Hershko A, Ciechanover A. 1998. The ubiquitin system. *Annu Rev Biochem* **67:** 425–479.

Hung MC, Link W. 2011. Protein localization in disease and therapy. *J Cell Sci* **124:** 3381–3392.

Ikeda F, Dikic I. 2008. Atypical ubiquitin chains: New molecular signals. *EMBO Rep* **9:** 536–542.

Jin J, Pawson T. 2012. Modular evolution of phosphorylation-based signalling systems. *Philos Trans R Soc B Biol Sci* **367:** 2540–2555.

Kirklin V, Dikic I. 2007. Role of ubiquitin- and Ubl-binding proteins in cell signaling. *Curr Opin Cell Biol* **19:** 199–205.

Kötting C, Kallenbach A, Suveyzdis Y, Wittinghofer A, Gerwert K. 2008. The GAP arginine finger movement into the catalytic site of Ras increases the activation entropy. *Proc Natl Acad Sci* **105:** 6260–6265.

Krebs EG, Fischer EH. 1956. The phosphorylase b to a converting enzyme of rabbit skeletal muscle. *Biochim Biophys Acta* **20:** 150–157.

Laplante M, Sabatini DM. 2012. mTOR signaling in growth control and disease. *Cell* **149:** 274–293.

Lefkowitz RJ, Shenoy SK. 2005. Transduction of receptor signals by β-arrestins. *Science* **308:** 512–517.

Letunic I, Doerks T, Bork P. 2011. SMART 7: Recent updates to the protein domain annotation resource. *Nucleic Acids Res* **40:** D302–D305.

Levene PA, Alsberg CL. 1906. The cleavage products of vitellin. *J Biol Chem* **2:** 127–133.

* Lim KH, Staudt LM. 2013. Toll-like receptor signaling. *Cold Spring Harb Perspect Biol* **5:** a011247.

Lima B, Forrester MT, Hess DT, Stamler JS. 2010. *S*-Nitrosylation in cardiovascular signaling. *Circ Res* **106:** 633–646.

Lipmann FA, Levene PA. 1932. Serinephosphoric acid obtained on hydrolysis of vitellinic acid. *J Biol Chem* **98:** 109–114.

Lowery DM, Lim D, Yaffe MB. 2005. Structure and function of Polo-like kinases. *Oncogene* **24:** 248–259.

Lynch KW. 2007. Regulation of alternative splicing by signal transduction pathways. *Adv Exp Med Biol* **623:** 161–174.

Macdonald N, Welburn JPI, Noble MEM, Nguyen A, Yaffe MB, Clynes D, Moggs JG, Orphanides G, Thomson S, Edmunds JW, et al. 2005. Molecular basis for the recognition of phosphorylated and phospho-acetylated histone H3 by 14-3-3. *Mol Cell* **20:** 199–211.

Malumbres M, Barbacid M. 2005. Mammalian cyclin-dependent kinases. *Trends Biochem Sci* **30:** 630–641.

Maurer-Stroh S, Dickens NJ, Hughes-Davies L, Kouzarides T, Eisenhaber F, Ponting CP. 2003. The Tudor domain "Royal Family": Tudor, plant Agenet, Chromo, PWWP and MBT domains. *Trends Biochem Sci* **28:** 69–74.

Maxfield FR. 2002. Plasma membrane microdomains. *Curr Opin Cell Biol* **14:** 483–487.

McLaughlin S, Aderem A. 1995. The myristoyl-electrostatic switch: A modulator of reversible protein–membrane interactions. *Trends Biochem Sci* **20:** 272–276.

Momand J, Zambetti GP, Olson DC, George D, Levine AJ. 1992. The *mdm-2* oncogene product forms a complex with the p53 protein and inhibits p53-mediated transactivation. *Cell* **69:** 1237–1245.

* Morrison DK. 2012. MAPK kinase pathways. *Cold Spring Harb Perspect Biol* **4:** a011254.

Muratani M, Tansey WP. 2003. How the ubiquitin–proteasome system controls transcription. *Nat Rev Mol Cell Biol* **4:** 192–201.

Murphy JE, Padilla BE, Hasdemir B, Cottrell GS, Bunnett NW. 2009. Endosomes: A legitimate platform for the signaling train. *Proc Natl Acad Sci* **106:** 17615–17622.

Nadolski MJ, Linder ME. 2007. Protein lipidation. *FEBS J* **274:** 5202–5210.

* Newton AC, Bootman MD, Scott JD. 2014. Second messengers. *Cold Spring Harb Perspect Biol* doi: 10.1101/cshperspect.a005926.

Obenauer JC, Yaffe MB. 2004. Computational prediction of protein–protein interactions. *Methods Mol Biol* **261:** 445–468.

Obenauer JC, Cantley LC, Yaffe MB. 2003. Scansite 2.0: Proteome-wide prediction of cell signaling interactions using short sequence motifs. *Nucleic Acids Res* **31:** 3635–3641.

Pei H, Zhang L, Luo K, Qin Y, Chesi M, Fei F, Bergsagel PL, Wang L, You Z, Lou Z. 2011. MMSET regulates histone H4K20 methylation and 53BP1 accumulation at DNA damage sites. *Nature* **469:** 124–128.

Perdiz D, Mackeh R, Poüs C, Baillet A. 2011. The ins and outs of tubulin acetylation: More than just a post-translational modification? *Cell Signal* **23:** 763–771.

Privé GG, Milburn MV, Tong L, de Vos AM, Yamaizumi Z, Nishimura S, Kim SH. 1992. X-Ray crystal structures of transforming p21 *ras* mutants suggest a transition-state stabilization mechanism for GTP hydrolysis. *Proc Natl Acad Sci* **89:** 3649–3653.

Pullen N, Thomas G. 1997. The modular phosphorylation and activation of p70s6k. *FEBS Lett* **410:** 78–82.

Radi R. 2013. Protein tyrosine nitration: Biochemical mechanisms and structural basis of functional effects. *Acc Chem Res* **46:** 550–559.

Rasmussen SGF, Devree BT, Zou Y, Kruse AC, Chung KY, Kobilka TS, Thian FS, Chae PS, Pardon E, Calinski D, et al. 2011. Crystal structure of the β2 adrenergic receptor–Gs protein complex. *Nature* **477:** 549–555.

Resh MD. 2012. Targeting protein lipidation in disease. *Trends Mol Med* **18:** 206–214.

* Rhind N, Russell P. 2012. Signaling pathways that regulate cell division. *Cold Spring Harb Perspect Biol* **4:** a005942.

Roy S, Plowman S, Rotblat B, Prior IA, Muncke C, Grainger S, Parton RG, Henis YI, Kloog Y, Hancock JF. 2005. Individual palmitoyl residues serve distinct roles in H-ras trafficking, microlocalization, and signaling. *Mol Cell Biol* **25:** 6722–6733.

* Sever R, Brugge JS. 2014. Signal transduction in cancer. *Cold Spring Harb Perspect Med* doi: 10.1101/cshperspect.a006098.

* Sever R, Glass CK. 2013. Signaling by nuclear receptors. *Cold Spring Harb Perspect Biol* **5:** a016709.

Siderovski DP, Willard FS. 2005. The GAPs, GEFs, and GDIs of heterotrimeric G-protein α subunits. *Int J Biol Sci* **1:** 51.

Slessareva JE, Routt SM, Temple B, Bankaitis VA, Dohlman HG. 2006. Activation of the phosphatidylinositol 3-kinase Vps34 by a G protein α subunit at the endosome. *Cell* **126:** 191–203.

Sprang SR. 1997. G protein mechanisms: Insights from structural analysis. *Annu Rev Biochem* **66:** 639–678.

Sudakin V, Ganoth D, Dahan A, Heller H, Hershko J, Luca FC, Ruderman JV, Hershko A. 1995. The cyclosome, a large complex containing cyclin-selective ubiquitin ligase activity, targets cyclins for destruction at the end of mitosis. *Mol Biol Cell* **6:** 185–197.

Temple BRS, Jones AM. 2007. The plant heterotrimeric G-protein complex. *Annu Rev Plant Biol* **58:** 249–266.

Tesmer JJ, Berman DM, Gilman AG, Sprang SR. 1997. Structure of RGS4 bound to AlF$_4$-activated G$_i$α$_1$: Stabilization of the transition state for GTP hydrolysis. *Cell* **89:** 251–261.

Walczak H, Iwai K, Dikic I. 2012. Generation and physiological roles of linear ubiquitin chains. *BMC Biol* **10:** 23.

Wang Q, Zhang Y, Yang C, Xiong H, Lin Y, Yao J, Li H, Xie L, Zhao W, Yao Y, et al. 2010. Acetylation of metabolic enzymes coordinates carbon source utilization and metabolic flux. *Science* **327:** 1004–1007.

Weis WI, Kobilka BK. 2008. Structural insights into G-protein-coupled receptor activation. *Curr Opin Struct Biol* **18:** 734–740.

Wennerberg K. 2005. The Ras superfamily at a glance. *J Cell Sci* **118:** 843–846.

Wise A, Gearing K, Rees S. 2002. Target validation of G-protein coupled receptors. *Drug Discov Today* **7:** 235–246.

Wu X, Brewer G. 2012. The regulation of mRNA stability in mammalian cells: 2.0. *Gene* **500:** 10–21.

Xu W, Doshi A, Lei M, Eck MJ, Harrison SC. 1999. Crystal structures of c-Src reveal features of its autoinhibitory mechanism. *Mol Cell* **3:** 629–638.

Yaffe MB, Smerdon SJ. 2004. The use of in vitro peptide library screens in the analysis of phosphoserine/threonine-binding domain structure and function. *Annu Rev Biophys Biomol Struct* **33:** 225–244.

Zastrow von M, Sorkin A. 2007. Signaling on the endocytic pathway. *Curr Opin Cell Biol* **19:** 436–445.

CHAPTER 3

Second Messengers

Alexandra C. Newton[1], Martin D. Bootman[2], and John D. Scott[3]

[1]Department of Pharmacology, University of California at San Diego, La Jolla, California 92093

[2]Department of Life, Health and Chemical Sciences, The Open University, Walton Hall, Milton Keynes MK7 6AA, United Kingdom

[3]Department of Pharmacology, Howard Hughes Medical Institute, University of Washington School of Medicine, Seattle, Washington 98195

Correspondence: anewton@ucsd.edu

SUMMARY

Second messengers are small molecules and ions that relay signals received by cell-surface receptors to effector proteins. They include a wide variety of chemical species and have diverse properties that allow them to signal within membranes (e.g., hydrophobic molecules such as lipids and lipid derivatives), within the cytosol (e.g., polar molecules such as nucleotides and ions), or between the two (e.g., gases and free radicals). Second messengers are typically present at low concentrations in resting cells and can be rapidly produced or released when cells are stimulated. The levels of second messengers are exquisitely controlled temporally and spatially, and, during signaling, enzymatic reactions or opening of ion channels ensure that they are highly amplified. These messengers then diffuse rapidly from the source and bind to target proteins to alter their properties (activity, localization, stability, etc.) to propagate signaling.

Outline

1 INTRODUCTION

Signals received by receptors at the cell surface or, in some cases, within the cell are often relayed throughout the cell via generation of small, rapidly diffusing molecules referred to as second messengers. These second messengers broadcast the initial signal (the "first message") that occurs when a ligand binds to a specific cellular receptor (see Ch. 1 [Heldin et al. 2014]); ligand binding alters the protein conformation of the receptor such that it stimulates nearby effector proteins that catalyze the production or, in the case of ions, release or influx of the second messenger. The second messenger then diffuses rapidly to protein targets elsewhere within the cell, altering the activities as a response to the new information received by the receptor. Three classic second messenger pathways are illustrated in Figure 1: (1) activation of adenylyl cyclase by G-protein-coupled receptors (GPCRs) to generate the cyclic nucleotide second messenger $3'$-$5'$-cyclic adenosine monophosphate (cAMP); (2) stimulation of phosphoinositide 3-kinase (PI3K) by growth factor receptors to generate the lipid second messenger phosphatidylinositol 3,4,5-trisphosphate (PIP_3); and (3) activation of phospholipase C by GPCRs to generate the two second messengers membrane-bound messenger diacylglycerol (DAG) and soluble messenger inositol 1,4,5-trisphosphate (IP_3), which binds to receptors on subcellular organelles to release calcium into the cytosol. The activation of multiple effector pathways by a single plasma membrane receptor and the production of multiple second messengers by each effector can generate a high degree of amplification in signal transduction, and stimulate diverse, pleiotropic, responses depending on the cell type.

Second messengers fall into four major classes: cyclic nucleotides, such as cAMP and other soluble molecules that signal within the cytosol; lipid messengers that signal within cell membranes; ions that signal within and between cellular compartments; and gases and free radicals that can signal throughout the cell and even to neighboring cells. Second messengers from each of these classes bind to specific protein targets, altering their activity to relay downstream signals. In many cases, these targets are enzymes

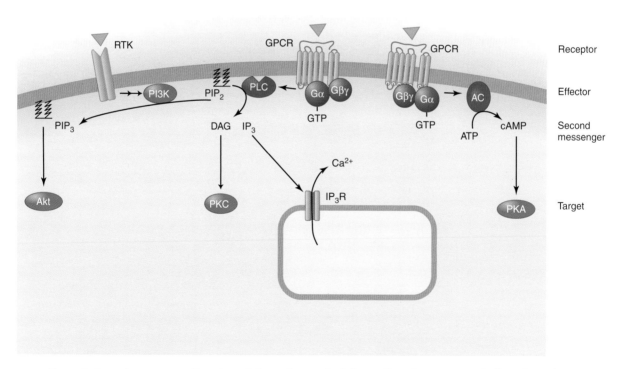

Figure 1. Second messengers disseminate information received from cell-surface receptors. Indicated are three examples of a receptor activating an effector to produce a second messenger that modulates the activity of a target. On the *right*, binding of agonists to a GPCR (the receptor) can activate adenylyl cyclase (the effector) to produce cAMP (the second messenger) to activate protein kinase A (PKA; the target). On the *left*, binding of growth factors to a receptor tyrosine kinase (RTK; the receptor) can activate PI3K (the effector) to generate PIP_3 (the second messenger), which activates Akt (the target). In the *center*, binding of ligands to a GPCR (receptor) activates phospholipase C (PLC; the effector) to clear the phospholipid PIP_2, to generate two second messengers, DAG and IP_3, which activate protein kinase C (PKC; the target) and release calcium from intracellular stores, respectively.

 Cite this chapter as *Cold Spring Harb Perspect Biol* doi: 10.1101/cshperspect.a005926

whose catalytic activity is modified by direct binding of the second messengers. The activation of multiple target enzymes by a single second messenger molecule further amplifies the signal.

Second messengers are not only produced in response to extracellular stimuli, but also in response to stimuli from within the cell. Moreover, their levels are exquisitely controlled by various homeostatic mechanisms to ensure precision in cell signaling. Indeed, dysregulation of the second messenger output in response to a particular agonist can result in cell/organ dysfunction and disease. For example, chronic exposure to cAMP in the heart results in an uncontrolled and asynchronous growth of cardiac muscle cells called pathological hypertrophy. This early stage heart disease presents as a thickening of the heart muscle (myocardium), a decrease in size of the chamber of the heart, and changes in contractility. Because such prolonged exposure to second messengers has deleterious effects, specific enzymes, channels, and buffering proteins exist to rapidly remove second messengers, either by metabolizing them or sequestering them away from target molecules.

2 CYCLIC NUCLEOTIDES

2.1 cAMP and a Major Target, PKA

The "fight or flight" mechanism, more accurately referred to as the adrenal response, prepares the body for situations of extreme stress. Release of the hormone epinephrine, also known as adrenaline, from the adrenal gland into the blood rapidly triggers vital cellular and physiological reactions that prepare the body for intense physical activity. One of the most important effects is the breakdown of glycogen into glucose to fuel muscles (Sutherland 1972).

Earl Sutherland first showed that epinephrine bound to cell-surface receptors stimulates membrane-associated adenylyl cyclases to synthesize the chemical messenger cAMP. Martin Rodbell and Alfred Gillman then discovered that G-protein subunits are the intermediates that shuttle between receptors and a family of eight adenylyl cyclase isoforms in the plasma membrane. Each G protein is a trimer consisting of $G\alpha$, $G\beta$, and $G\gamma$ subunits. The $G\alpha$ subunits in the G proteins G_s and G_i are distinct and provide the specificity for activation and inhibition of adenylyl cyclase, respectively. G_s and G_i can thus couple binding of ligands to GPCRs with either activation or inhibition of adenylyl cyclase, depending on the receptor type. This increases or reduces production of cAMP, which diffuses from the membrane into the cell. Edwin Krebs and Edmund Fischer later found that a principal task of cAMP is to stimulate protein phosphorylation (Fischer and Krebs 1955), ultimately showing cAMP-dependent protein ki-

nase (also known as protein kinase A, PKA) is responsible (Krebs 1993; Gilman 1995).

PKA is the major target for cAMP (see Fig. 2) (p. 99 [Sassone-Corsi 2012]). It is a heterotetramer consisting of two regulatory (R) subunits that maintain two catalytic (C) subunits in an inhibited state. When cAMP levels are low, the PKA holoenzyme is dormant; however, when cAMP levels are elevated, two molecules bind in a highly cooperative manner to each R subunit, causing a conformational change that releases the active C subunits (Taylor et al. 2012). In humans, four genes encode R subunits (RIα, RIβ, RIIα, and RIIβ) and two genes encode C subunits (Cα and Cβ), combinations of which are expressed in most, if not all, tissues. The C subunits of PKA phosphorylate serine or threonine residues on target substrates, typically within the sequence RRxS/TΦ, in which Φ is an aliphatic hydrophobic residue or an aromatic residue (Kemp et al. 1976). Around 300–500 distinct intracellular proteins can be phosphorylated by PKA in a typical cell. These include glycogen phosphorylase as part of the fight-or-flight mechanism. Active phosphorylase catalyzes the production of glucose 1-phosphate, a metabolic intermediate that is ultimately converted into other modified sugars that are used in various aspects of cellular catabolism and ATP production (see Ch. 14 [Hardie 2012]). On its release from the PKA holoenzyme, the C subunit of PKA can diffuse into the nucleus, where it phosphorylates transcription factors such as the cAMP-response-element-binding protein. cAMP can thereby ultimately influence transcriptional activation and reprogramming of the cell.

Additional layers of regulation ensure that PKA phosphorylates the correct proteins at the right time. For example, prostaglandin E1 and epinephrine both produce similar increases in cAMP and PKA activity in the heart, but only epinephrine increases glycogen phosphorylase activity. This is achieved in part through attachment of PKA to subcellular structures via interactions between R subunits and A-kinase-anchoring proteins (AKAPs) (Wong and Scott 2004). Anchoring of PKA to subcellular structures by AKAPs is a means to limit the range of action of PKA and avert the indiscriminate transmission of these responses throughout the cell.

A typical cell expresses 10 to 15 different AKAPs. These proteins all contain an amphipathic-helix motif that binds to a docking and dimerization domain formed by the R subunits of PKA and targeting domains that tether the AKAPs to intracellular membranes or organelles. AKAPs promote signaling efficacy by placing PKA near preferred substrates and insulating different anchored PKA complexes from one another (Scott and Pawson 2009). Most AKAPs also organize other signaling proteins, such as GPCRs, GTPases, protein phosphatases, phosphodiesterase (PDE), and other protein kinases.

A

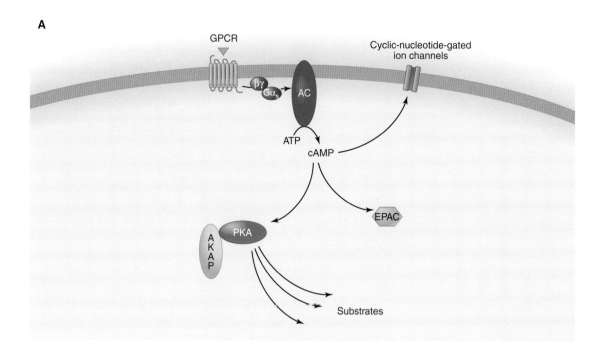

B

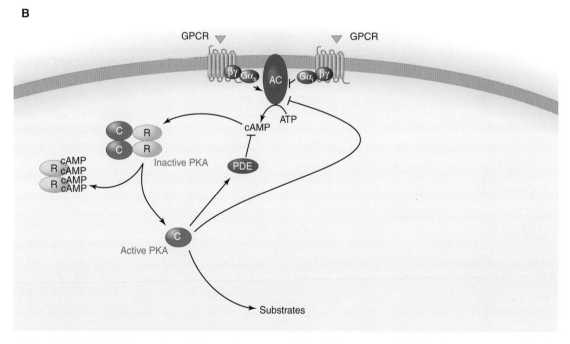

Figure 2. (*A*) cAMP is the archetypical second messenger. Its levels increase rapidly following receptor-mediated activation of adenylyl cyclase (AC), which catalyzes the conversion of adenosine monophosphate (AMP) to cAMP. This small second messenger activates PKA at specific cellular locations as a result of anchoring of PKA to A-kinase-anchoring proteins (AKAPs). In addition, cAMP activates EPACs (exchange proteins directly activated by cAMP). (*B*) AC activity is controlled by the opposing actions of the G_s and G_i proteins. The cAMP produced by AC activates PKA by binding to its R subunit. This releases the C subunit. The signal can be terminated by the action of phosphodiesterase (PDE) enzymes.

2.2 3′-5′-Cyclic Guanosine Monophosphate and cGMP-Dependent Protein Kinase

3′-5′-cyclic guanosine monophosphate (cGMP) is another cyclic nucleotide that serves as a second messenger. cGMP is often synthesized by receptor guanylyl cyclases, which have their catalytic domains close to the inner face of the plasma membrane and are stimulated by small peptide factors that bind to their extracellular domains. It can also be produced by soluble cytoplasmic guanylyl cyclases, which are stimulated by nitric oxide. This latter signaling pathway is involved in a number of important physiological processes, including smooth muscle relaxation and neurotransmission (see Ch. 1 [Heldin et al. 2014]). cGMP targets cGMP-dependent protein kinase (PKG), a dimeric enzyme that has an R and C domain within the same polypeptide chain. When cGMP levels are low, PKG is dormant; however, when cGMP levels are elevated, two molecules bind to each R domain in the dimer, exposing the active catalytic domains. There are two PKG isozymes: the type I enzyme is soluble and predominantly cytoplasmic, whereas the type II enzyme is particulate and is attached to a variety of biological membranes. PKG phosphorylates peptide sites with the general consensus—RRXS/TX. Hence, PKA and PKG have overlapping substrate specificities. However, a major difference is that PKG expression is restricted to the brain, lungs, and vascular tissue, where this enzyme controls important physiological processes, including the regulation of smooth muscle relaxation, platelet function in the blood, cell division, and aspects of nucleic acid synthesis. PKG causes relaxation of vascular smooth-muscle cells, for example, by phosphorylating several structural proteins that are vital for this process.

2.3 Other cAMP and cGMP Targets

Other cAMP and cGMP targets also play key roles in cellular physiology. For example, cyclic nucleotide-gated channels (nonselective channels that allow many ions to flow into or out of a cell) have important functions in retinal photoreceptors and olfactory receptor neurons (see Ch. 11 [Julius and Nathans 2012]). Most notably, cGMP is a key second messenger in vision: photoisomerization of the chromophore in the light receptor, rhodopsin, produces a conformational change in the receptor that allows it to activate the G protein transducin, which, in turn, activates a cGMP PDE. The resulting reduction in the concentration of cGMP leads to the closure of sodium and calcium channels and, thus, hyperpolarization of photoreceptor cells, leading to changes in neurotransmitter release. Mutations in many components of this pathway, including the PDE, cause blindness. Regulation of ion channels by cAMP is particularly important in the sinoatrial node of the heart, in which cAMP-responsive hyperpolarization-activated cyclic nucleotide-gated (HCNs) channels help to generate pacemaker currents that control cardiac contractility (see Ch. 13 [Kuo and Ehrlich 2014]). Other classes of HCN channel have analogous functions in the brain and nervous system. cAMP also controls the cAMP-responsive guanine nucleotide exchange factor EPAC1, a protein that promotes activation of the Rap1 GTPase to regulate cell adhesion by stimulating integrin molecules in the plasma membrane (Bos 2003).

2.4 Cyclic Nucleotide PDEs

Termination of cAMP and cGMP signals is mediated by PDEs. These enzymes represent a vast gene family of 11 distinct subtypes and more than 100 isoforms that can break the phosphoester bond of either cyclic nucleotide to liberate AMP or GMP (Beavo and Brunton 2002). PDE activity can be regulated in a variety of ways. For example, calcium-dependent processes control the activity of the PDE1 and PDE2 isoforms, PKA phosphorylation attenuates the activity of PDE3 and PDE4 isoforms, and PKA or PKG phosphorylation participates in the control of PDE5. Less is known about the mechanisms of regulation for PDE6-11. More recently, there has been considerable interest in the development and clinical application of small-molecule PDE inhibitors. Selective PDE inhibitors that produce elevated levels of cAMP/cGMP have been used clinically to alleviate chronic obstructive pulmonary disease, asthma, and combat certain immune disorders, but their most celebrated therapeutic application has been in the treatment of male erectile dysfunction. Sildenafil citrate (Viagra) and its relatives act by inhibiting cGMP-specific PDE5 (Beavo and Brunton 2002) in the arterial wall smooth muscle of the penis, which elevates cGMP and increases blood flow.

3 LIPID AND LIPID-DERIVED SECOND MESSENGERS

Two seemingly unrelated discoveries half a century ago provided the first insights into how cells use lipids to signal. Hokin and Hokin discovered that cholinergic stimulation of pancreatic slices promotes incorporation of ^{32}P from radiolabeled ATP into phospholipids, which led to an explosion of studies showing that numerous extracellular signals promote hydrolysis of phosphoinositides in cells and eventually the identification of DAG as a major second messenger. During this period, Yasutomi Nishizuka purified the enzyme PKC and showed it is the target for this lipid second messenger (Nishizuka 1992).

A

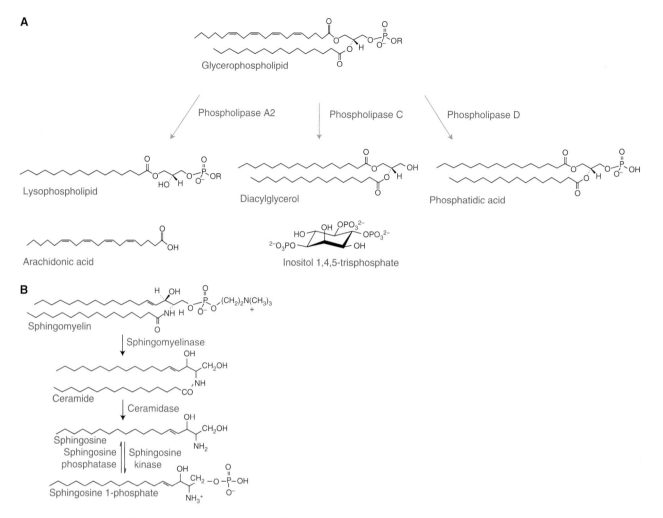

B

Figure 3. Lipid-derived second messengers. (*A*) Hydrolysis of glycerophospholipids results in the production of lysophospholipids, free fatty acids (arachidonic acid is shown), diacylglycerol, phosphatidic acid, and, in the case of hydrolysis of PIP$_2$, the water soluble inositol 3,4,5-trisphosphate. (*B*) Hydrolysis of sphingolipids yields ceramide, sphingosine, and sphingosine 1-phosphate.

We now know that many extracellular signals trigger hydrolysis of two classes of lipids to generate second messengers: glycerolipids, whose hydrophobic backbone is DAG (e.g., phosphatidylinositol 4,5-bisphosphate, PIP$_2$), and sphingolipids, whose hydrophobic backbone is ceramide (e.g., sphingomyelin). Both produce a variety of products that are involved in cell signaling (Fig. 3). Binding of ligands to receptors that couple to lipid-modifying enzymes (see below) results in activation of enzymes that hydrolyze specific acyl chains or polar head groups from each lipid class to generate lipid second messengers. Specifically, hydrolysis of acyl chains generates lysophospholipids (e.g., lysophosphatidylcholine and lysophosphatidic acid through hydrolysis catalyzed by phospholipase A2) and lysophingolipids (e.g., sphingosine through hydrolysis catalyzed by ceramidases). It also produces free fatty acids,

such as arachidonic acid, which are important signaling entities in their own right.

Hydrolysis of polar head groups from glycerophospholipids by phospholipase D and PLC results in generation of phosphatidic acid and DAG, respectively. The polar head group released following PIP$_2$ hydrolysis, IP$_3$, is a key second messenger itself. IP$_3$ can also be further phosphorylated to form polyphosphoinositols that serve as additional second messengers. Hydrolysis of the polar head group from sphingomyelin, catalyzed by sphingomyelinase, produces ceramide. Further hydrolysis of the acyl chain by ceramidase produces sphingosine, and the subsequent phosphorylation of this by sphingosine kinase produces sphingosine 1-phosphate.

Lipid second messengers that retain two acyl chains (e.g., phosphatidic acid, DAG, and ceramide) remain mem-

Cite this chapter as *Cold Spring Harb Perspect Biol* doi: 10.1101/cshperspect.a005926

brane associated, but the decreased lipophilicity of second messengers that retain only one acyl chain (lysolipids such as lysophosphatidic acid and sphingosine) allows them to dissociate from membranes and serve as soluble second messengers. Finally, note that in addition to its hydrolysis, PIP_2 can be phosphorylated to produce another key second messenger, PIP_3 (see below).

3.1 IP$_3$ and DAG

Signals from both GPCRs (e.g., histamine receptors) and RTKs (e.g., epidermal growth factor [EGF] receptors) can result in activation of PLC, which cleaves phospholipids to generate DAG and the lipid headgroup. In the case of GPCR signaling, lipid hydrolysis is mediated by binding of G proteins (notably Gq) to phospholipases (PLCβ); in the case of RTKs, lipid hydrolysis is mediated by recruitment of phospholipases (PLCγ) to tyrosine-phosphorylated proteins at the plasma membrane (tyrosine phosphorylation of PLCγ by RTKs also stimulates its activity directly). If the lipid cleaved is PIP_2 (rather than phosphatidylcholine [PC]), then water-soluble IP_3 is also produced. IP_3 binds to IP_3 receptors on the endoplasmic reticulum (ER), and other organelles, causing release of calcium into the cytosol (Fig. 1). It can be further modified to yield additionally phosphorylated phosphoinositols, including diphosphoryl inositol phosphates (Tsui and York 2010). Inositol 1,3,4,5-tetrakisphosphate (IP_4) activates chloride channels. Inositol 1,2,3,4,5,6-hexaphosphate (IP_6) is a compound found in plants (hence, its name phytic acid), but it is also present in yeast and animal cells, along with the enzymes that synthesize and metabolize it, and is gaining interest as a potential anticancer agent because of its apoptosis-promoting properties. The DAG sensor is a small globular domain, called the C1 domain, originally identified in PKC. Approximately 30 other proteins contain a C1 domain (Colon-Gonzalez and Kazanietz 2006), including protein kinase D, Ras, gastrin-releasing polypeptides, DAG kinase, and *n*-chimaerin.

PKC represents a family of nine genes grouped into three families (Newton 2009). The so-called conventional PKC isozymes require the coordinated presence of both calcium (sensed by the C2 domain) and DAG (sensed by the adjacent C1 domain) for activation and thus transduce signals that trigger PIP_2 hydrolysis, but not those that trigger hydrolysis of other phospholipids, such as PC (because these do not mobilize calcium via IP_3). Novel isozymes of PKC do not have a calcium sensor, but because they bind DAG with an affinity two orders of magnitude higher than conventional PKC isozymes, they are efficiently activated by DAG alone. Thus, they can transduce signals triggered by PC hydrolysis. A third class of PKC isozymes, the atyp-

ical PKCs, do not respond to DAG or calcium. Signaling by conventional and novel PKC isozymes is terminated by phosphorylation of DAG by DAG kinase, which removes the second messenger. Note that PKC isozymes are constitutively phosphorylated so, unlike many other kinases, these phosphorylations do not acutely regulate activity.

DAG- or calcium-dependent translocation of conventional and novel PKC isozymes to the membrane is a hallmark of their activation (see Fig. 4). In unstimulated cells, PKC localizes to the cytosol, where it may be concentrated on specific protein scaffolds. For example, the conventional PKCα binds to the PDZ-domain scaffold DLG1. PIP_2 hydrolysis provides calcium, which binds to the C2 domain and thus recruits cytosolic PKC to the plasma membrane; there, PKC binds to the membrane-embedded second messenger DAG via its C1 domain. Binding of both the C2 and C1 domains to membranes provides the energy to release an autoinhibitory pseudosubstrate segment from the substrate-binding cavity, allowing PKC to phosphorylate substrates and relay signals. Interestingly, a function common to many of the PKC isozymes is that their phosphorylation of certain substrates terminates signaling pathways. For example, phosphorylation of receptors such as the EGF and insulin receptors by PKC promotes their internalization and degradation, thus acting as a negative-feedback loop to attenuate signaling via these pathways.

PKC has a unique signature of activation depending on its cellular location (Gallegos and Newton 2008). In general, its activation kinetics mirror those for the increase in intracellular calcium concentration, and deactivation kinetics mirror the decay of DAG. Thus, membranes such as the Golgi in which DAG production is sustained serve as a platform for sustained PKC activity, whereas membranes such as the plasma membrane where DAG is more rapidly removed by phosphorylation serve as platforms for transient PKC activity. Although PKC can phosphorylate cytosolic targets, these events likely occur at the membrane. Reduction in cytoplasmic calcium levels and phosphorylation of DAG by DAG kinase effectively terminates PKC signaling.

3.2 PIP$_3$ and Akt Signaling

Binding of growth factors to receptor tyrosine kinases results in the activation of PI3K isoforms (Fig. 4B), which catalyze the phosphorylation of PIP_2 at the 3′ position to generate the very minor, but highly efficacious, lipid second messenger PIP_3 in the plasma membrane (Cantley 2002). PI3K also phosphorylates phosphatidylinositol and phosphatidylinositol 4-phosphate (PIP) at the 3′ position to generate corresponding 3′-phosphoinositides that also serve as second messengers. PIP_2 is itself a minor compo-

A

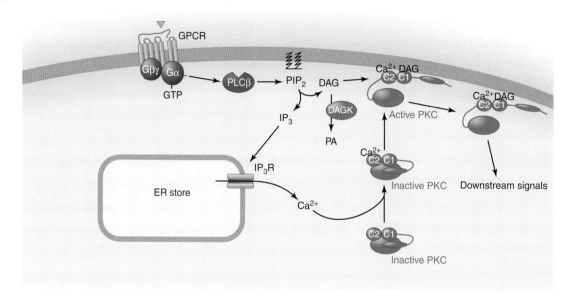

B

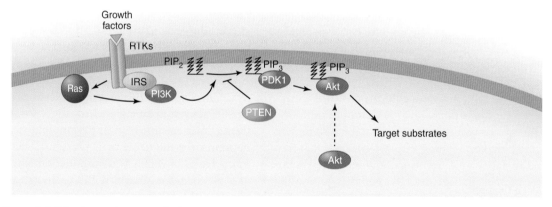

Figure 4. (*A*) PKC transduces signals that promote phospholipid hydrolysis. Signals that cause phospholipase-C-mediated hydrolysis of PIP$_2$ activate conventional PKC isozymes by a two-step mechanism. First, calcium (whose levels increase following IP$_3$ production) binds to the C2 domain of PKC. This increases the affinity of the module for the plasma membrane, causing PKC to translocate to the membrane. Here, it binds its allosteric activator, DAG. This binding produces a conformational change that expels the autoinhibitory pseudosubstrate segment from the active site, allowing substrate phosphorylation and downstream signaling. (*B*) Activation of PI3K following engagement of growth factor receptors such as insulin receptor generates the phospholipid PIP$_3$, which recruits the kinase Akt to the membrane where it is phosphorylated by the upstream kinase PDK1. Subsequent phosphorylation events lead to activation of Akt.

nent of the plasma membrane, accounting for ~0.05% of the total phospholipid, yet the phosphorylation of 10% of this pool effectively transduces growth factor signals. The reason for this highly effective signaling is the use of protein modules that bind with high affinity and specificity to the various phosphorylated species of the inositol head group (e.g., inositol 3,4,5-trisphosphate versus inositol 3,4-bisphosphate) that is exposed to the cytosol. These

include pleckstrin homology (PH), FYVE, and PX domains (Lemmon 2007; p. 87 [Hemmings and Restuccia 2012]).

The best defined target for PIP$_3$ is the prosurvival kinase Akt (Manning and Cantley 2007). Like PKC, Akt is present in the cytosol in unstimulated cells. PIP$_3$ produced following PI3K activation recruits Akt to the plasma membrane via its PH domain. This triggers the phosphorylation of Akt at two sites, leading to its activation and

Cite this chapter as *Cold Spring Harb Perspect Biol* doi: 10.1101/cshperspect.a005926

downstream signaling. The first phosphorylation is catalyzed by the phosphoinositide kinase PDK1 (also recruited to the membrane by PIP_3) at a segment near the entrance to the active site, which, in turn, leads to rapid phosphorylation at a carboxy-terminal site. Autophosphorylation and the mTORC2 kinase complex have been proposed to regulate this second site (see p. 91 [Laplante and Sabatini 2012]). Once phosphorylated, Akt is locked in an active and signaling-competent conformation and can be released from the plasma membrane to signal throughout the cell, including the nucleus. Signaling is terminated by dephosphorylation of PIP_3 and dephosphorylation of Akt. The former is catalyzed by the tumor suppressor PTEN, a lipid phosphatase, which removes the lipid second messenger by dephosphorylating the $3'$ phosphate of PIP_3. Dephosphorylation of the two Akt sites is catalyzed by protein phosphatases such as the tumor suppressor PHLPP.

3.3 Sphingolipid Hydrolysis and Protein Targets that Control Apoptosis

Sphingolipids are a class of bioactive lipids that play important roles in diverse cellular responses (Hannun and Obeid 2008). The major signaling lipids derived from sphingomyelin are ceramide, sphingosine, and sphingosine 1-phosphate, which are produced by the action of sphingomyelinase, ceramidase, and sphingosine kinase, respectively (Fig. 3).

Ceramide propagates information in stress-response pathways; it is produced following activation of receptors by stimuli such as cytokines or death ligands and also by nonreceptor signals in response to radiation, cytotoxic insult, or infection by pathogens. Sphingosine also has apoptotic roles, although its direct targets in cells are unclear. In contrast to ceramide and sphingosine, sphingosine 1-phosphate promotes prosurvival signaling. When secreted, it binds to a class of GPCRs that promote diverse cellular effects, including cell-survival, differentiation, inflammation, and angiogenesis (Pitson 2011). Intracellular sphingosine 1-phosphate acts as a second messenger, activating enzymes such as the TRAF2 E3 ligase and some histone deacetylases.

3.4 Lysolipids, Prostaglandins, and Other Eicosanoids

Oxidation of 20-carbon fatty acids produces eicosanoids, thus signaling molecules that bind to GPCRs to mediate inflammatory and immune responses (Wymann and Schneiter 2008). Phospholipase-A2-mediated hydrolysis of phospholipids containing arachidonic acid (a 20-carbon unsaturated fatty acid) in the *sn*-2 position of the glycerol backbone (see Fig. 3) yields a fatty-acid product that can be oxidized by cyclooxygenases to yield prostaglandins, or by lipooxygenases to yield leukotrienes. These signaling molecules bind to specific receptors that couple to diverse pathways. Many of these are involved in inflammatory responses. Notably, leukotrienes bind to GPCRs that cause inflammatory responses common in asthma, and, as such, leukotriene pathways serve as targets for antiasthma drugs. Aspirin inhibits cyclooxygenases and thus the production of prostaglandins, which are inflammatory (Venteclef et al. 2011). As mentioned above, arachidonic acid itself also serves as a second messenger.

4 IONS AS INTRACELLULAR MESSENGERS

Ions such as sodium, potassium, calcium, magnesium, protons, chloride, iron, and copper play essential roles in cell function. Many ions act as cofactors for structural proteins and enzymes. Moreover, cellular processes are sensitive to changes in their ambient ionic environment; changes in pH, for example, can alter enzymatic activity and the behavior of cellular ion transporters (Vaughan-Jones et al. 2009). In addition, ions control cellular activity by spreading electrical signals, for example, action potentials in heart and neurons.

Ions such as calcium and magnesium can also play direct roles as dynamic intracellular messengers that regulate specific protein targets during signal transduction (Fig. 5A). These ions are imported from the extracellular milieu or mobilized from intracellular stores and control the activity of protein targets by binding to specific motifs on the protein itself or upstream regulators (e.g., calmodulin) (Berridge et al. 2000). Note that most ions should not be considered intracellular messengers, however. For example, the sodium ions that enter cardiac myocytes during an action potential serve only to depolarize the cell membrane. This causes the opening of voltage-operated calcium channels (VOCCs), allowing influx of the real messenger—calcium ions—which control contraction by binding to proteins such as calmodulin and troponin C (Ch. 13 [Kuo and Ehrlich 2014]). Similarly, to reverse an action potential, potassium ions are rapidly extruded and the membrane potential is restored, but it is the reversion of the calcium concentration to the prestimulated state through the action of the sodium/calcium exchanger (NCX) that represents signal termination (sodium and potassium levels are returned to the original state by the sodium/potassium ATPase [Na^+/K^+-ATPase], so that another action potential can be evoked).

A key reason for using ions as messengers is speed of response. Cells use energy to maintain gradients of ions across their lipid membranes. By activating channels or transporters, cells can use the potential energy established

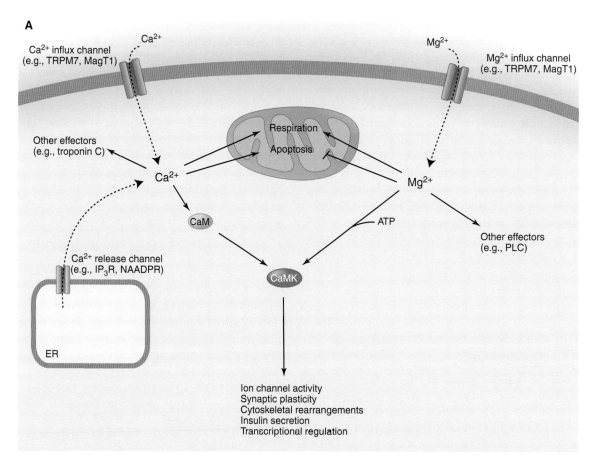

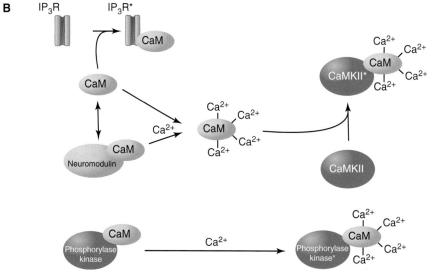

Figure 5. (A) Calcium and magnesium signals. Both ions can enter cells via channels in the plasma membrane. In addition, calcium is stored in organelles such as the ER. Calcium exerts its effects by binding to numerous cellular protein targets, including calmodulin, whereas magnesium may function as a calcium mimetic or have magnesium-specific effects. (B) The various ways in which calmodulin can function to alter cellular targets. It is generally thought that cellular calmodulin is largely bound to proteins even when the calcium concentration is low, and that there is a relatively small pool of cytosolic calcium-free calmodulin (apocalmodulin). However, even apocalmodulin can regulate specific cellular processes (e.g., IP_3 receptors, IP_3Rs). When the calcium concentration is elevated, calcium ions bind to calmodulin. This causes calmodulin to be displaced from some targets and associate with others. In some cases (e.g., phosphorylase kinase), calmodulin is a constitutively bound subunit that binds calcium and activates the enzyme when the calcium concentration is elevated. Other calcium-binding proteins, such as neuronal calcium sensors, may also display complex interactions with their various targets. *CaM-dependent inhibition of IP3R channels.

by the electrochemical gradient of an ion to rapidly generate a cellular signal (Clapham 2007). Unlike other intracellular messengers, ionic signals can be generated with no enzymatic steps. The speed of the response depends on the rate at which the intracellular concentration of the ion changes and the proximity of the ions to their cellular targets. For example, action potentials cause the fast release of neurotransmitters at nerve terminals because the cytosolic concentration of calcium ions just beneath the plasma membrane increases from $\sim$100 nM to $>$10 μM within milliseconds (Berridge 2006). In situations in which ions have to diffuse further before encountering their target(s), the response will be slower. As we discuss below, the ability to generate different spatiotemporal patterns, such as waves and oscillations, is another important advantage of ionic intracellular messengers.

4.1 Calcium

Calcium is an extremely versatile intracellular messenger that controls a wide range of cellular functions by regulating the activity of a vast number of target proteins. The cellular effects of calcium are mediated either by direct binding to a target protein, or stimulation of calcium sensors that detect changes in calcium concentration and then activate different downstream responses (Berridge 2004). The multitude of sensors that mediate effects of calcium can be characterized by the nature of their calcium-binding site(s). The most common calcium-binding motifs are EF-hands and C2 domains. Synaptotagmin and troponin C are examples of proteins with C2 domains and EF-hands, respectively. Calcium sensors also act by recruiting a range of intermediary effectors, such as calcium-sensitive enzymes—for example, calcium/calmodulin-dependent protein kinases (CaMKs), calcineurin (also known as protein phosphatase 2B), myosin light chain kinase, and phosphorylase kinase (see Ch. 13 [Kuo and Ehrlich 2014]).

4.2 Control of Calcium Levels

Intracellular calcium levels are controlled by an assortment of channels, pumps, transporters, buffers, and effector moieties. At rest, cytosolic calcium is maintained at $\sim$100 nM. In most cases, its levels are elevated by the opening of channels located either on various organellar stores or in the plasma membrane. Release of calcium from internal stores represents a major source of signal calcium for many cells. The principal calcium stores are the ER, sarcoplasmic reticulum (SR), Golgi, and acidic organelles of the endolysosomal system (Bootman et al. 2002). Calcium release channels are present on the membranes of these or-

ganelles and gate the flux of calcium from the ER/SR lumen into the cytosol. Ubiquitous calcium-release channels include the IP$_3$Rs that respond to IP$_3$ produced by hydrolysis of PIP$_2$ (Parys and De Smedt 2012), ryanodine receptors (Belevych et al. 2013), and two-pore channels that respond to nicotinic acid adenine dinucleotide phosphate (Galione 2011).

Channels that permit the influx of calcium across the plasma membrane are typically characterized by their activation mechanism. Receptor-operated calcium channels and second messenger-operated calcium channels are opened by the binding of an external or internal ligand. Examples of these are N-methyl-D-aspartate (NMDA) receptors that respond to the neurotransmitter glutamate (see Ch. 12 [Kennedy 2013]), and Orai channels regulated by the intracellular messenger arachidonic acid. VOCCs that are activated by a change in membrane potential are widely expressed in excitable tissues and can be divided into three families [Ca$_V$1 (L-type), Ca$_V$2 (N, P/Q, and R type), and Ca$_V$3 (T type)], which have specific characteristics and functions (Catterall 2011). The transient receptor potential (TRP) family includes a number of calcium-permeable channels with distinct activation mechanisms (Gees et al. 2012). Depletion of ER or SR calcium activates "store-operated" calcium entry. This pathway for calcium influx has been known for more than two decades, and the store-operated current (I$_{crac}$, for calcium-release-activated current) has been well characterized. However, the molecular players, STIM (stromal interaction molecule) and Orai, have only been identified relatively recently through investigation of patients with severe combined immunodeficiency caused by a lack of calcium influx into their T lymphocytes (Hogan et al. 2010).

4.3 Calcium Buffers

Cells express a number of calcium-binding proteins that buffer calcium changes within various cellular compartments and modulate calcium signals. The rapid binding of calcium by the cytosolic buffers parvalbumin and calbindin D-28k shapes both the spatial and temporal properties of intracellular calcium signals (Schwaller 2010). ER/SR calcium-binding proteins (e.g., calsequestrin, calreticulin, GRP78, and GRP94) facilitate the accumulation of large amounts of calcium, which is necessary for rapid cell signaling (Prins and Michalak 2011). Mitochondria also play a key buffering role in that they express a calcium uniporter capable of taking up substantial amounts of calcium whenever the cytosolic calcium concentration increases during signaling (Rizzuto and Pozzan 2006). The uptake of calcium into the mitochondrial matrix stimulates the citric acid cycle to produce more ATP. This increased oxidation also enhances mitochondrial reactive oxygen species (ROS) formation, which contributes to redox signaling pathways. Loading of calcium into mitochondria is believed to be essential to avoid induction of autophagy. However, exaggerated mitochondrial calcium loading can precipitate apoptosis (Baumgartner et al. 2009).

4.4 Signal Termination

Calcium pumps and exchangers are responsible for pumping calcium out of the cell or back into the ER/SR to terminate a calcium signal. These pumps and exchangers operate at different times during the recovery process. NCX proteins have low affinities for calcium, but have high capacities that enable them to rapidly remove large quantities of calcium and are especially evident in excitable cells (Nicoll et al. 2013). The plasma membrane calcium-ATPase and SR/ER calcium-ATPase enzymes have lower transport rates but higher affinities, which means that they operate at relatively low cytosolic calcium concentrations (Vandecaetsbeek et al. 2011). Calcium/hydrogen exchangers are important for the loading of calcium into endo/lysosomal compartments.

4.5 Spatiotemporal Organization of the Calcium Signaling Machinery

The substantial diversity of cellular calcium signals in different cell types reflects their use of distinct combinations and arrangements of calcium-transport mechanisms. In neurons, for example, microscopic calcium signals trigger the release of neurotransmitter-containing vesicles at presynaptic terminals (Berridge 2014). The calcium signal is a plume of ions around the mouth of an open channel that dissipates within milliseconds and can only activate processes within a few tens of nanometers. The receiver for the calcium signal is synaptotagmin, a calcium-binding protein that promotes fusion of the neurotransmitter-containing vesicle with the neuronal plasma membrane. The rapidity of this response depends on the calcium channels and exocytotic machinery (SNARE complexes) being linked together within a highly localized microdomain. A key component of the SNARE complex is SNAP-25, a membrane-bound protein that interacts with both the VOCCs and synaptotagmin. In contrast, the regular beating of the heart relies on the sequential elevation of calcium levels within all the myocytes of the atrial and ventricular chambers. In this case, the calcium signal is global, occurring across the whole volume of each myocyte to evenly activate troponin C, thereby allowing actin and myosin to engage and cause contraction (Ch. 13 [Kuo and Ehrlich 2014]).

An important driver for global calcium signaling is regenerative calcium elevation—that is, the autocatalytic increase in cellular calcium concentration. Regenerative calcium signals arise because several components of the calcium signaling toolkit, such as PLC, RYRs, and IP$_3$Rs, are activated by calcium (Roderick et al. 2003). The opening of one calcium channel can therefore promote a positive feed-forward reaction in which more calcium ions enter the cytosol. This regenerative calcium-induced calcium release (CICR) process can override cellular calcium buffers, enabling rapid coordinated surges in calcium concentration. The triggering of CICR is typically the critical event in causing the transition from local to global calcium signaling. Without CICR, the global calcium signals in cardiac myocytes that trigger contraction (described above) would not occur. CICR leads to a rapid increase in calcium concentration. Within an individual cell, each spike of calcium caused by CICR is evident as calcium waves, such as the calcium waves evoked by fertilization of an oocyte (see Wakai et al. 2011). Microscopic analysis and modeling has shown that such waves propagate in a salutatory manner, involving successive rounds of CICR and diffusion, as the wave jumps from a cluster of calcium channels to the next (Thul et al. 2012). Such waves can engulf the entirety of the cellular cytoplasm, and pervade into the nucleus to activate or inhibit transcription (Bootman et al. 2009).

If a cellular stimulus is not sufficient to trigger a global response, then only local calcium elevations will occur. The calcium ions emanating from channels during local calcium signaling only switch on processes within their immediate vicinity. The sphere of action of such microdomains depends on how long the channels are open, their ability to conduct calcium, and the capacity of surrounding calcium buffers and homeostatic mechanisms that remove calcium (Parekh 2008). Calcium buffers and pumps are highly effective at restricting the diffusion of calcium so that the calcium concentration declines exponentially with distance from a calcium source. If local calcium events do not turn into global calcium signals, the proximity of particular effectors and their affinity for calcium determines the response. For example, relatively low-affinity effector molecules such as synaptotagmin (described above) and BK potassium channels (big calcium-activated potassium channels, also called maxi-K channels) depend on submicrometer proximity to a channel to sense an activating calcium concentration. Sensor proteins with relatively higher affinity, such as calmodulin and SK potassium channels (small calcium-activated potassium channels), can be activated at greater distances.

4.6 Magnesium

Magnesium can also be considered an intracellular messenger because its concentration can change dynamically in response to cellular stimulation (Li et al. 2011). Given that magnesium binds to nucleotides, oligonucleotides, and hundreds of enzymes, it is reasonable to conclude that it is an intracellular messenger in its own right. Of its potential binding sites, ATP is particularly important. Cellular processes typically use ATP complexed with magnesium as an energy source. Moreover, magnesium has been shown to cause prolonged inhibition of potassium channels in neurons following muscarinic acetylcholine receptor activation (an effect that is not mimicked by calcium), thereby regulating neuronal excitability (Chuang et al. 1997).

In addition, magnesium deserves consideration because it influences the effects of calcium. Indeed, magnesium and calcium are typically thought to have antagonistic actions. Magnesium frequently inhibits the transport and cellular activities of calcium and can prevent pathological consequences of increases in calcium levels (Romani 2013). Like calcium, there is an electrochemical gradient of magnesium across the plasma membrane that can serve as a reservoir for signal generation. The extracellular concentrations of magnesium and calcium are similar (1.1–1.5 mM), and magnesium can also act as a "calcimimetic" (e.g., by binding to the calcium-sensing receptor), a GPCR that has pleiotropic actions. Within cells, the magnesium concentration (0.3–1.5 mM) is several orders of magnitude higher than that of calcium. It has been suggested that mitochondria might serve as a store of magnesium and that magnesium can potentially regulate cellular respiration (Wolf and Trapani 2012). Moreover, magnesium in the mitochondrial matrix inhibits permeability transition pore (PTP) activation, an increase in the leakiness of the inner mitochondrial membrane that allows solutes <1500 Da to pass, and can precipitate mitochondrial swelling, apoptosis, and cell death. Calcium promotes PTP, so magnesium acts as a counteracting antagonist.

Like calcium, magnesium has a plethora of transport pathways. Of these, TRPM6 and TRPM7 (members of the TRP family) are relatively well understood. TRPM6 is restricted to kidney tubules and the intestinal epithelium, and plays an important role in magnesium (re)adsorption (defective TRPM6 function leads to hypomagnesemia), whereas TRPM7 is ubiquitously expressed in mammals. Both TRPM6 and TRPM7 are "chanzymes": ion channels that incorporate a kinase domain. Interestingly, TRPM7 is regulated by PIP$_2$, the source of the calcium-mobilizing messenger IP$_3$. Moreover, it binds to PLC, the enzyme that hydrolyses PIP$_2$. Consequently, hormonal stimulation of cells will lead to both a calcium signal (via IP$_3$ production) and a change in magnesium flux (via loss of PIP$_2$-mediated regulation of TRPM7) (Langeslag et al. 2007). MagT1 is also believed to be a key cellular magnesium channel. Like TRPM6 and TRPM7, it is located on the plasma membrane of mammalian cells. However, whereas TRPM6 and TRPM7 allow both calcium and magnesium fluxes, it is believed that MagT1 is a specific pathway for magnesium. Individuals bearing mutations in the *MAGT1* gene show high levels of Epstein–Barr virus infection and a predisposition to lymphoma (Chaigne-Delalande et al. 2013).

5 CONCLUDING REMARKS

Second messengers disseminate information received by cellular receptors rapidly, faithfully, and efficaciously. They are small, nonprotein organic molecules or ions that bind to specific target proteins, altering their activities in a variety of ways that allow them to respond appropriately to the information received by receptors.

A key advantage of second messengers over proteins is that, unlike proteins, second messenger levels are controlled with rapid kinetics. Thus, whereas it may take tens of minutes for the levels of a protein to increase significantly, most second messenger levels increase within microseconds (e.g., ions) to seconds (e.g., DAG), They are often produced from precursors that are abundant in cells or released from stores that contain high concentrations of the second messenger; so, their generation is not rate limiting. Thus, when the appropriate signal is received, second messengers are rapidly generated, diffuse rapidly, and alter target protein function highly efficiently.

Second messengers vary significantly in size and chemical character: from ions to hydrophilic molecules such as cyclic nucleotides to hydrophobic molecules such as diacylglycerol. Moreover, the continuing discovery of new second messengers is expanding the repertoire of molecules known to convey information within the cell. Indeed, only very recently, cyclic guanosine monophosphate-adenosine monophosphate was shown to be a second messenger that is synthesized by the enzyme cGAS in response to HIV infection and binds to and activates a protein called STING, leading to induction of interferon (Wu et al. 2013). The ability to respond rapidly to information thus depends on an expanding library of small molecules.

REFERENCES

*Reference is in this book.

Baumgartner HK, Gerasimenko JV, Thorne C, Ferdek P, Pozzan T, Tepikin AV, Petersen OH, Sutton R, Watson AJ, Gerasimenko OV. 2009. Calcium elevation in mitochondria is the main Ca^{2+} requirement for mitochondrial permeability transition pore (mPTP) opening. *J Biol Chem* **284**: 20796–20803.

Beavo JA, Brunton LL. 2002. Cyclic nucleotide research—Still expanding after half a century. *Nat Rev Mol Cell Biol* **3**: 710–718.

Belevych AE, Radwanski PB, Carnes CA, Gyorke S. 2013. "Ryanopathy": Causes and manifestations of RyR2 dysfunction in heart failure. *Cardiovasc Res* **98**: 240–247.

Berridge MJ. 2004. Calcium signal transduction and cellular control mechanisms. *Biochim Biophys Acta* **1742**: 3–7.

Berridge MJ. 2006. Calcium microdomains: Organization and function. *Cell Calcium* **40**: 405–412.

Berridge MJ. 2014. Calcium regulation of neural rhythms, memory and Alzheimer's disease. *J Physiol* **592**: 281–293.

Berridge MJ, Lipp P, Bootman MD. 2000. The versatility and universality of calcium signalling. *Nat Rev Mol Cell Biol* **1**: 11–21.

Bootman MD, Berridge MJ, Roderick HL. 2002. Calcium signalling: More messengers, more channels, more complexity. *Curr Biol* **12**: R563–R565.

Bootman MD, Fearnley C, Smyrnias I, MacDonald F, Roderick HL. 2009. An update on nuclear calcium signalling. *J Cell Sci* **122**: 2337–2350.

Bos JL. 2003. Epac: A new cAMP target and new avenues in cAMP research. *Nat Rev Mol Cell Biol* **4**: 733–738.

Cantley LC. 2002. The phosphoinositide 3-kinase pathway. *Science* **296**: 1655–1657.

Catterall WA. 2011. Voltage-gated calcium channels. *Cold Spring Harb Perspect Biol* **3**: a003947.

Chaigne-Delalande B, Li FY, O'Connor GM, Lukacs MJ, Jiang P, Zheng L, Shatzer A, Biancalana M, Pittaluga S, Matthews HF, et al. 2013. Mg²⁺ regulates cytotoxic functions of NK and CD8 T cells in chronic EBV infection through NKG2D. *Science* **341**: 186–191.

Chuang H, Jan YN, Jan LY. 1997. Regulation of IRK3 inward rectifier K⁺ channel by m1 acetylcholine receptor and intracellular magnesium. *Cell* **89**: 1121–1132.

Clapham DE. 2007. Calcium signaling. *Cell* **131**: 1047–1058.

Colon-Gonzalez F, Kazanietz MG. 2006. C1 domains exposed: From diacylglycerol binding to protein–protein interactions. *Biochim Biophys Acta* **1761**: 827–837.

Fischer EH, Krebs EG. 1955. Conversion of phosphorylase *b* to phosphorylase *a* in muscle extracts. *J Biol Chem* **216**: 121–132.

Galione A. 2011. NAADP receptors. *Cold Spring Harb Perspect Biol* **3**: a004036.

Gallegos LL, Newton AC. 2008. Spatiotemporal dynamics of lipid signaling: Protein kinase C as a paradigm. *IUBMB Life* **60**: 782–789.

Gees M, Owsianik G, Nilius B, Voets T. 2012. TRP channels. *Compr Physiol* **2**: 563–608.

Gilman AG. 1995. Nobel Lecture. G proteins and regulation of adenylyl cyclase. *Biosci Rep* **15**: 65–97.

Hannun YA, Obeid LM. 2008. Principles of bioactive lipid signalling: Lessons from sphingolipids. *Nat Rev Mol Cell Biol* **9**: 139–150.

* Hardie DG. 2012. Organismal carbohydrate and lipid homeostasis. *Cold Spring Harb Perspect Biol* **4**: a006031.

* Heldin C-H, Lu B, Evans R, Gutkind JS. 2014. Signals and receptors. *Cold Spring Harb Perspect Biol* doi: 10.1101/cshperspect.a005900.

* Hemmings BA, Restuccia DF. 2012. PI3K-PKB/Akt pathway. *Cold Spring Harb Perspect Biol* **4**: a011189.

Hogan PG, Lewis RS, Rao A. 2010. Molecular basis of calcium signaling in lymphocytes: STIM and ORAI. *Annu Rev Immunol* **28**: 491–533.

* Julius D, Nathans J. 2012. Signaling by sensory receptors. *Cold Spring Harb Perspect Biol* **4**: a005991.

Kemp BE, Benjamini E, Krebs EG. 1976. Synthetic hexapeptide substrates and inhibitors of 3′:5′-cyclic AMP-dependent protein kinase. *Proc Natl Acad Sci* **73**: 1038–1042.

* Kennedy MB. 2013. Synaptic signaling in learning and memory. *Cold Spring Harb Perspect Biol* doi: 10.1101/cshperspect.a016824.

Krebs EG. 1993. Nobel Lecture. Protein phosphorylation and cellular regulation I. *Biosci Rep* **13**: 127–142.

* Kuo IY, Ehrlich BE. 2014. Signaling in muscle contraction. *Cold Spring Harb Perspect Biol* doi: 10.1101/cshperspect.a006023.

Langeslag M, Clark K, Moolenaar WH, van Leeuwen FN, Jalink K. 2007. Activation of TRPM7 channels by phospholipase C-coupled receptor agonists. *J Biol Chem* **282**: 232–239.

* Laplante M, Sabatini DM. 2012. mTOR signaling. *Cold Spring Harb Perspect Biol* **4**: a011593.

Lemmon MA. 2007. Pleckstrin homology (PH) domains and phosphoinositides. *Biochem Soc Symp* **2007**: 81–93.

Li FY, Chaigne-Delalande B, Kanellopoulou C, Davis JC, Matthews HF, Douek DC, Cohen JI, Uzel G, Su HC, Lenardo MJ. 2011. Second messenger role for Mg²⁺ revealed by human T-cell immunodeficiency. *Nature* **475**: 471–476.

Manning BD, Cantley LC. 2007. AKT/PKB signaling: Navigating downstream. *Cell* **129**: 1261–1274.

Newton AC. 2009. Lipid activation of protein kinases. *J Lipid Res* **50**: S266–S271.

Nicoll DA, Ottolia M, Goldhaber JI, Philipson KD. 2013. 20 years from NCX purification and cloning: Milestones. *Adv Exp Med Biol* **961**: 17–23.

Nishizuka Y. 1992. Intracellular signaling by hydrolysis of phospholipids and activation of protein kinase C. *Science* **258**: 607–614.

Parekh AB. 2008. Ca²⁺ microdomains near plasma membrane Ca²⁺ channels: Impact on cell function. *J Physiol* **586**: 3043–3054.

Parys JB, De Smedt H. 2012. Inositol 1,4,5-trisphosphate and its receptors. *Adv Exp Med Biol* **740**: 255–279.

Pitson SM. 2011. Regulation of sphingosine kinase and sphingolipid signaling. *Trends Biochem Sci* **36**: 97–107.

Prins D, Michalak M. 2011. Organellar calcium buffers. *Cold Spring Harb Perspect Biol* **3**: a004069.

Rizzuto R, Pozzan T. 2006. Microdomains of intracellular Ca²⁺: Molecular determinants and functional consequences. *Physiol Rev* **86**: 369–408.

Roderick HL, Berridge MJ, Bootman MD. 2003. Calcium-induced calcium release. *Curr Biol* **13**: R425.

Romani AM. 2013. Magnesium homeostasis in mammalian cells. *Metal Ions life Sci* **12**: 69–118.

* Sassone-Corsi P. 2012. The cyclic AMP pathway. *Cold Spring Harb Perspect Biol* **4**: a011148.

Schwaller B. 2010. Cytosolic Ca²⁺ buffers. *Cold Spring Harb Perspect Biol* **2**: a004051.

Scott JD, Pawson T. 2009. Cell signaling in space and time: Where proteins come together and when they're apart. *Science* **326**: 1220–1224.

Sutherland EW. 1972. Studies on the mechanism of hormone action. *Science* **171**: 401–408.

Taylor SS, Ilouz R, Zhang P, Kornev AP. 2012. Assembly of allosteric macromolecular switches: Lessons from PKA. *Nat Rev Mol Cell Biol* **13**: 646–658.

Thul R, Coombes S, Roderick HL, Bootman MD. 2012. Subcellular calcium dynamics in a whole-cell model of an atrial myocyte. *Proc Natl Acad Sci* **109**: 2150–2155.

Tsui MM, York JD. 2010. Roles of inositol phosphates and inositol pyrophosphates in development, cell signaling and nuclear processes. *Adv Enzyme Regul* **50**: 324–337.

Vandecaetsbeek I, Vangheluwe P, Raeymaekers L, Wuytack F, Vanoevelen J. 2011. The Ca²⁺ pumps of the endoplasmic reticulum and Golgi apparatus. *Cold Spring Harb Perspec Biol* **3**: a004184.

Vaughan-Jones RD, Spitzer KW, Swietach P. 2009. Intracellular pH regulation in heart. *J Mol Cell Cardiol* **46**: 318–331.

Venteclef N, Jakobsson T, Steffensen KR, Treuter E. 2011. Metabolic nuclear receptor signaling and the inflammatory acute phase response. *Trends Endocrinol Metab* **22**: 333–343.

Wakai T, Vanderheyden V, Fissor RA. 2011. Ca²⁺ signaling during mammalian fertilization: Requirements, players, and adaptations. *Cold Spring Harb Perspect Biol* **3**: a006767.

Wolf FI, Trapani V. 2012. Magnesium and its transporters in cancer: A novel paradigm in tumour development. *Clin Sci (Lond)* **123**: 417–427.

Wong W, Scott JD. 2004. AKAP Signalling complexes: Focal points in space and time. *Nat Rev Mol Cell Biol* **5**: 959–971.

Wu J, Sun L, Chen X, Du F, Shi H, Chen C, Chen ZJ. 2013. Cyclic GMP-AMP is an endogenous second messenger in innate immune signaling by cytosolic DNA. *Science* **339**: 826–830.

Wymann MP, Schneiter R. 2008. Lipid signalling in disease. *Nat Rev Mol Cell Biol* **9**: 162–176.

CHAPTER 4

Signaling Networks: Information Flow, Computation, and Decision Making

Evren U. Azeloglu and Ravi Iyengar

Department of Pharmacology and Systems Therapeutics and Systems Biology Center New York, Mount Sinai School of Medicine, New York, New York 10029

SUMMARY

Signaling pathways come together to form networks that connect receptors to many different cellular machines. Such networks not only receive and transmit signals but also process information. The complexity of these networks requires the use of computational models to understand how information is processed and how input–output relationships are determined. Two major computational approaches used to study signaling networks are graph theory and dynamical modeling. Both approaches are useful; network analysis (application of graph theory) helps us understand how the signaling network is organized and what its information-processing capabilities are, whereas dynamical modeling helps us determine how the system changes in time and space upon receiving stimuli. Computational models have helped us identify a number of emergent properties that signaling networks possess. Such properties include ultrasensitivity, bistability, robustness, and noise-filtering capabilities. These properties endow cell-signaling networks with the ability to ignore small or transient signals and/or amplify signals to drive cellular machines that spawn numerous physiological functions associated with different cell states.

Outline

1 INTRODUCTION

Coordinated regulation of cellular processes allows cells to maintain homeostatic balance and make decisions as to whether to divide, differentiate, or die. In each case, the cell responds to chemical, mechanical, or electrical signals, including hormones, neurotransmitters, mechanical stretch and shear, and ion currents. The signaling pathways activate relay information to effectors that alter subcellular processes, but they also process this information. Information processing involves computation in which the network of connected signaling molecules recognizes the amplitude and duration of the incoming signal and produces an output signal of appropriate strength and duration. This depends on the organization (topology) of the signaling network and can change the relationship between inputs and outputs. For example, the cell can limit responses to maintain homeostatic balance or trigger a set of changes that takes it to another state (e.g., differentiation or division). The inherent computational abilities of signaling networks thus provide the cell with decision-making capabilities.

In all signaling pathways, information flows through coupled biochemical reactions and molecular interactions to the cellular machines that control its output: biochemical (e.g., metabolic enzymes), mechanical (e.g., actin cytoskeleton contractility), or electrical (e.g., ion channel activity in excitable cells). Most of these pathways are part of wider networks (Weng et al. 1999). Studying the flow of information (signal) through these networks (see Box 1) is like studying traffic patterns, which depend on the population density of the towns through which the highways pass, the state of local roads, the time of the day, the weather, and other factors. Similarly, to understand the flow of signals that regulate cellular machines, we must know how the pathways involved are connected to form networks, the characteristics of the connections, and how information is processed as it flows through these interconnected pathways. Below, we discuss the organization of these networks and explain how they can be modeled quantitatively.

1.1 Versatility of Signaling Components Enables Pathways to Form Networks

Most signaling pathways in mammalian cells interact with one another (Jordan et al. 2000). This so-called "cross talk" starts at the level of receptors. Growth factor receptors with intrinsic tyrosine kinase activity, for example, interact with multiple effector pathways, including the Ras-ERK-kinase pathway (p. 81 [Morrison 2012]), the phosphoinositide 3-kinase (PI3K) pathway (p. 87 [Hemmings and Restuccia 2012]), and the phospholipase-C γ pathway (Fig. 1A) (p. 95 [Bootman 2012]). Similarly, in the case of receptors coupled to heterotrimeric G proteins, the ability of both the Gα and Gβγ subunits to transmit signals allows the receptors to couple to multiple downstream effectors (Fig. 1B).

Small G proteins, such as Ras, Rho, Rap, and Cdc42, are also major loci of interconnectivity (Bar-Sagi and Hall 2000). Their guanine nucleotide exchange factors (GEFs) and GTPase-activating proteins (GAPs) can be regulated by numerous mechanisms, including binding of ligands such as cAMP, calcium, and diacylglycerol (DAG), or posttranslational modifications such as phosphorylation and acylation. Multiple receptors feed into these GTPase regulators, and the small G proteins themselves can modulate the activity of multiple signaling pathways by targeting several different effectors (Fig. 1C).

Protein kinases and protein phosphatases are additional loci at which multiple signaling pathways interact. Typically, a single protein kinase or phosphatase can be activated by multiple receptors and has multiple substrates. Together, all these components make the signaling pathways within a cell extensively interconnected.

Most mammalian proteins are present as multiple isoforms. These can arise from alternative splicing of a single mRNA or use of alternative initiation codons, or they can be products of different genes altogether. Typically, these isoforms have different characteristics that alter both upstream and downstream interactions (connectivity) and/or their intracellular localization. This further increases connectivity. An early example is adenylyl cyclases (AC1-AC8), which allow signals from multiple types of receptors and ion channels to feed into the cAMP pathway (Pieroni et al. 1993; p. 99 [Sassone-Corsi 2012]). These enzymes can be activated or inhibited by signals relayed via different types of receptors (Fig. 2), thus enabling cAMP levels to provide an integrated measure of the information coming into the cell from various sources. cAMP in turn works through multiple effectors (e.g., protein kinase A, the EPAC Rap-GEF, and the cyclic-nucleotide-gated channels) to regulate numerous cellular targets to evoke physiological responses.

Why is one class of signaling component often responsible for regulating many cellular machines? We do not yet know the answer. The cellular machinery probably evolved to produce distinct activity patterns in response to specific signals, whose amplitude, duration, and subcellular location evoke signature responses from only a subset of cellular machines. There is suggestive evidence to support this idea but no definitive proof, and this is an area of ongoing research in systems biology.

BOX 1. INFORMATION AND SIGNALS IN CELL-SIGNALING PATHWAYS

In the context of intracellular-signaling pathways and networks, the terms information and signal are used interchangeably. These terms are most often used to distinguish the activity state of a signaling component. Biochemical information, or signal, flows from component A to component B as the activity state of A controls the activity state of component B. The activity state of a given molecule is defined as the fraction of active (e.g., A* or B*) to total molecules (e.g., A or B). A typical node A in this example could be a receptor, such as the adrenergic receptor, that has an interaction (i.e., edge) with a target node B, such as the G protein Gs. Communication through this edge results in node A activating node B (i.e., β adrenergic receptor-activating Gs). It is important to note that information content at any given time depends on both the local concentration of interacting components and the reaction kinetics that govern change in activity state. Green arrows in the figure denote information flow.

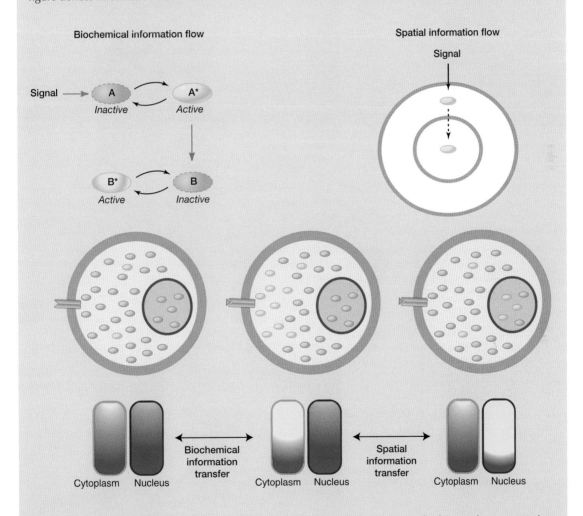

Biochemical information transfer is established through coupled reactions, which are either noncovalent reversible binding reactions or enzymatic reactions that involve covalent modification. However, such information flow sometimes results in the movement of components from one subcellular compartment to another. The most common examples of such movements are the translocation of protein kinases or transcription factors from the cytoplasm to the nucleus in response to hormone or growth factor signals, where they interact with other proteins or DNA to affect gene expression. This is considered spatial information transfer.

E.U. Azeloglu and R. Iyengar

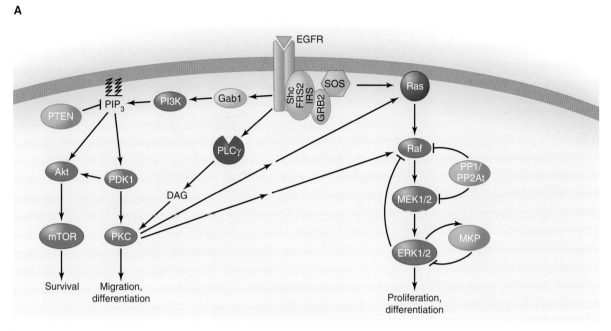

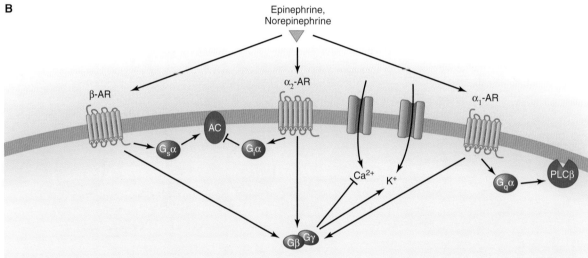

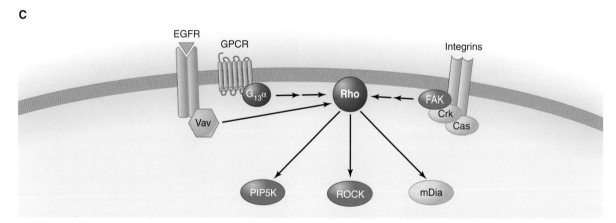

Figure 1. Interaction of multiple components with receptors leads to signal flow within multiple signaling pathways. Good examples are the interplay of PI3K, PLCγ, and Ras ERK signaling following activation of epidermal growth factor receptor (EGFR) (*A*), heterotrimeric G proteins linked to adrenergic receptors (ARs) (*B*), and the small G protein Rho (*C*).

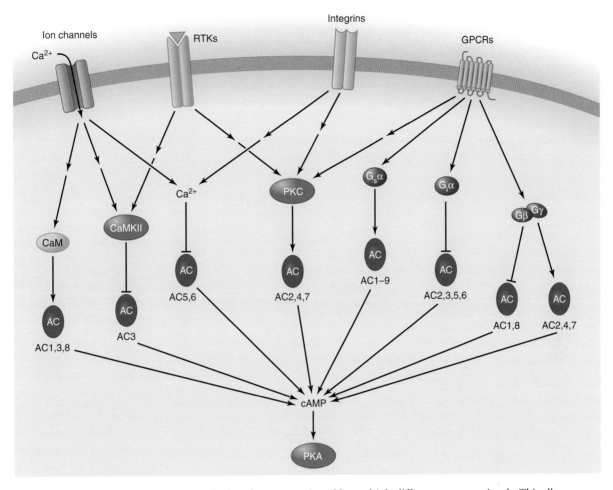

Figure 2. Different adenylyl cyclase (AC) isoforms are activated by multiple different upstream signals. This allows the second messenger cAMP to provide an integrated measure of information from several different sources.

1.2 Graph-Theory-Based Models Allow Us to Infer the Information-Processing Ability of Signaling Networks

The complexity of signaling networks requires that we have an orderly approach if we are to understand what these networks do, and how they do it. For this, in addition to experiments, we need a theoretical framework. Networks are often studied using a branch of mathematics called graph theory that analyzes systems composed of objects (called nodes or vertices) and binary relationships (called edges). First developed by Euler in 1736 (Biggs et al. 1986), graph theory focuses on the combinatorial study of multiple interactions between numerous components, and it has been very successfully used in many disciplines, including social sciences. A classic experiment by Stanley Milgram in 1967 using chain letters showed that any two persons in the United States are connected by a median of five interconnected individuals and led to the concept of "six degrees of separation" (Pool et al. 1989). Statistical analyses have since defined metrics that provide estimates of overall connectivity and organization within networks. In general, these focus on computing the relationships between edges and nodes within the network (see Box 2).

Signaling networks, like many other networks, are scale-free. This means the ratio of edges to nodes follows a power law rather than a Gaussian distribution (Barabasi and Albert 1999). In a typical network, there are a small number of nodes with many edges, called *hubs*. In contrast, most nodes have only few edges. Distribution of highly and sparsely connected nodes is not evenly spaced. Interconnectivity between nodes is also not evenly distributed with the overall networks. A metric called the clustering coefficient can be used to compute the extent of interconnectivity. This is used to identify highly or sparsely connected regions within a network, which allows us to identify clusters or subnetworks within the larger network (see Box 2).

BOX 2. COMPUTATIONAL MODELS OF SIGNALING NETWORKS

The complexity that arises from having many isoforms of signaling components, and their differential connectivity, makes it hard to intuitively understand the organization of signaling networks and to make accurate predictions of their outputs a priori. However, computational analysis based on experimental observations could yield accurate predictions while improving our understanding of the signaling landscape. Two types of computational approaches are useful: (1) network models, and (2) dynamical models.

1. Network Models

Network analysis, application of a field of mathematics called graph theory, typically involves the use of statistical methods to identify characteristics of a network. So-called network models incorporate an additional level of complexity, the relationship between the different entities. Although statistical and network models share many features in terms of methods used for analysis, the outcomes are different. Statistical models provide knowledge about relationships within a system by analyzing the experimental background for significant correlative changes. Network models provide knowledge about the organization of the system, in addition to identifying the underlying relationships. The term used to describe network organization is topology. The organization of networks can be categorized based on how the components are connected to one another. There are different types of networks. In part B in the figure below, three different networks are shown in decreasing order of clustering coefficients. Small-world networks have densely connected topologies, in which the typical distance between two randomly chosen nodes is proportional to the logarithm of the number of nodes (Barabasi and Albert 1999).

The connectivity between components may be direct (i.e., component A chemically interacts with component B) or indirect (wherein changes in component A modulate changes in component C without direct chemical interactions). Direct binary relationships allow tracking of pathways from receptors to effectors, and prediction of distal relationships. The use of models of network topology to predict the information-processing capability of the system relies on identification of small organizing units, called network motifs, such as feedback and feedforward loops. The smallest fundamental motifs have two to four interacting components; some of the motifs that are prominently featured in biological networks are outlined in part B in the figure below.

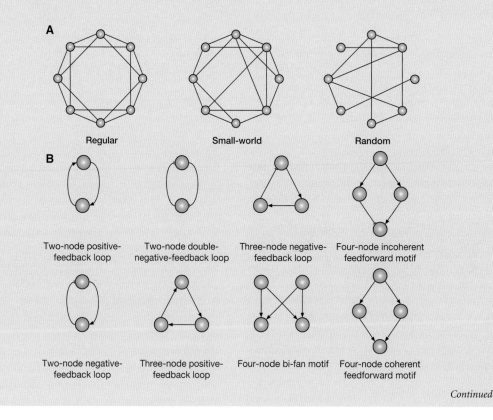

A

Regular   Small-world   Random

B

Two-node positive-feedback loop   Two-node double-negative-feedback loop   Three-node negative-feedback loop   Four-node incoherent feedforward motif

Two-node negative-feedback loop   Three-node positive-feedback loop   Four-node bi-fan motif   Four-node coherent feedforward motif

Continued

Cite this chapter as *Cold Spring Harb Perspect Biol* doi: 10.1101/cshperspect.a005934

A limitation of graph-theory-based analysis is that it provides a static view of the system. It tells us how the system is organized, but does not tell us how such organization enables the system to change with respect to time on receiving a signal. In signaling networks, every connection between components in a signaling reaction represents one or more chemical reactions, whose dynamics are regulated by both the concentration of the reactants and the rates of the forward and reverse reactions. Accordingly, connectivity between components does not automatically imply that signals are propagated through them. To understand how spatiotemporal reaction dynamics affect signal propagation, we need a second type of analysis called dynamical modeling.

2. Dynamical Models

Dynamical models describe how a system changes with respect to time, based on reaction rates between components. They can be divided into two classes: (1) deterministic models, in which the time-dependent change in the state of the system can be predicted from the initial concentrations of the reactants and the reaction rates; and (2) stochastic models, in which the trajectory of the system is not fully predictable but depends on the probability of a set of reactions occurring. Typically, stochastic models involve systems in which the concentrations of reactants or the volumes occupied are very small, which decrease or increase the certainty of interactions, respectively.

Deterministic models usually use ordinary-differential-equation (ODE)-based representations of biochemical or biophysical interactions. ODE systems are based on the assumption that system dynamics depend only on time. These offer a detailed and quantitative picture of interactions. Using the known parameters of reaction stoichiometry, a system of ODEs can be formulated to represent the flow of information within a signaling network. These equations can then be simultaneously solved as a function of time to reveal the behavior of individual components as the reactions proceed (see figure above). A typical ODE model can be used to describe signal flow upon receptor ligand activation.

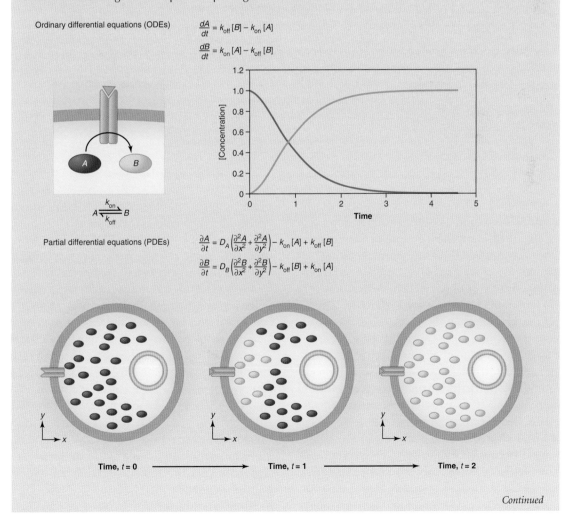

Ordinary differential equations (ODEs)

$$\frac{dA}{dt} = k_{off}[B] - k_{on}[A]$$

$$\frac{dB}{dt} = k_{on}[A] - k_{off}[B]$$

$$A \underset{k_{off}}{\overset{k_{on}}{\rightleftharpoons}} B$$

Partial differential equations (PDEs)

$$\frac{\partial A}{\partial t} = D_A\left(\frac{\partial^2 A}{\partial x^2} + \frac{\partial^2 A}{\partial y^2}\right) - k_{on}[A] + k_{off}[B]$$

$$\frac{\partial B}{\partial t} = D_B\left(\frac{\partial^2 B}{\partial x^2} + \frac{\partial^2 B}{\partial y^2}\right) - k_{off}[B] + k_{on}[A]$$

Time, $t = 0$ Time, $t = 1$ Time, $t = 2$

Continued

Although ODE-based models are common and very valuable as a first approximation, they suffer from an inherent simplifying assumption that all components are evenly distributed within the reaction space (i.e., time is the only independent variable). This is usually not true within cells. Simple ODE models can be improved by introducing spatial information transfer via multiple subcellular compartments such that the rates at which components move between compartments are specified. Typically, for a signal flowing from the cell surface to the nucleus, the cell can be thought of as a three-compartment system comprising the plasma membrane, cytoplasm, and the nucleus.

A more sophisticated, and complete, way of improving deterministic models is to use partial differential equations, in which each component has two associated attributes: specific reaction rates (k) and a diffusion coefficient (D), which define system behavior as a function of time (t) and space (x,y), respectively (see figure above). For example, a model of signal flow from receptor tyrosine kinases to extracellular-regulated kinase (ERK) to transcription factors involves information flow from the cell surface membrane through the cytoplasm to the nucleus, and it is best described by a partial-differential-equation (PDE)-based model. Although such models provide a more realistic picture of the cell, they have some drawbacks. For example, diffusion coefficients for most cellular components have not been measured and they need to be approximated from molecular weights. In addition, numerical simulations of PDE-based models remain computationally expensive; so such models are not widely used. Nevertheless, to fully understand how signaling reactions occur in localized regions within the cell (especially for cells with nonstandard geometries) one must use PDE-based models

Sometimes reactions in cells involve components that are present in very small numbers or do not react in an easily predictable manner. The most common example is the binding of a transcription factor to the promoter of a gene. To model such an event accurately, we need to consider the probability of the interaction occurring as a function of time, volume, and number of molecules. Stochastic models incorporate these parameters, and use Gillespie's algorithm as the standard approach for computations (Gillespie 2007). These models combine individual probabilities of interactions at the molecular level under given circumstances (e.g., probability of binding of a transcription factor to a promoter given its concentration and volume constraints). In theory, they can provide highly accurate representations of biochemical events; however, because of the need for a large number of operations per unit time, they tend to be computationally expensive even with small spatiotemporal constraints. In addition, because of their random nature, in most cases, they require Monte Carlo simulations that use random sampling to determine how the system changes with respect to time or space. Such models are also often difficult to develop owing to a paucity of experimental data. Hence, stochastic models are not as commonly used as deterministic models to describe large systems but could become more popular in the future as more information becomes available.

The local topology within a signaling network can have numerous effects on its input–output relationships. Signals may be amplified, dampened, prolonged or shortened; such behavior is controlled by ubiquitously repeating patterns of connectivity called regulatory motifs. Several classes of regulatory motifs are observed in biological networks (Alon 2007). These include feedforward and feedback loops and bi-fan motifs (see Box 2).

The feedforward motif has two arms from a starting node to a target node. A good example is the signal from the MAP kinase ERK to the transcription factor Fos (Fig. 3A). ERK stimulates transcription of the Fos gene and consequently production of Fos protein, which it can also phosphorylate. Because the phosphorylated form of Fos is protected from degradation, the feedforward loop prolongs the signal from the ERK pathway. Extending signal output is one feature of a feedforward motif. Another is redundancy, which can make a system robust against perturbations (see below). Feedforward loops in which all the relationships are of the same sign (e.g., all stimulatory) are called coherent feedforward loops, and they can be used for additive computation within the signaling network. When two arms of the starting node have opposing effects, the motif is called an incoherent feedforward loop. Incoherent feedforward loops can induce accelerated transient output following a step input (Mangan and Alon 2003), which could be useful for rapid cellular response.

Feedback loops are common regulatory features of biochemical pathways and networks. They can be both positive and negative. In negative-feedback loops, a downstream component inhibits an upstream stimulator. This type of regulation is very common in metabolic pathways, where a downstream metabolite will allosterically inhibit an upstream enzyme to reduce flux. It is also found in signaling. Phosphorylation and deactivation of the β-adrenergic receptor by PKA to limit the duration of responses to the hormone epinephrine is a classic example (Fig. 3B). Positive-feedback loops are also found in signaling pathways.

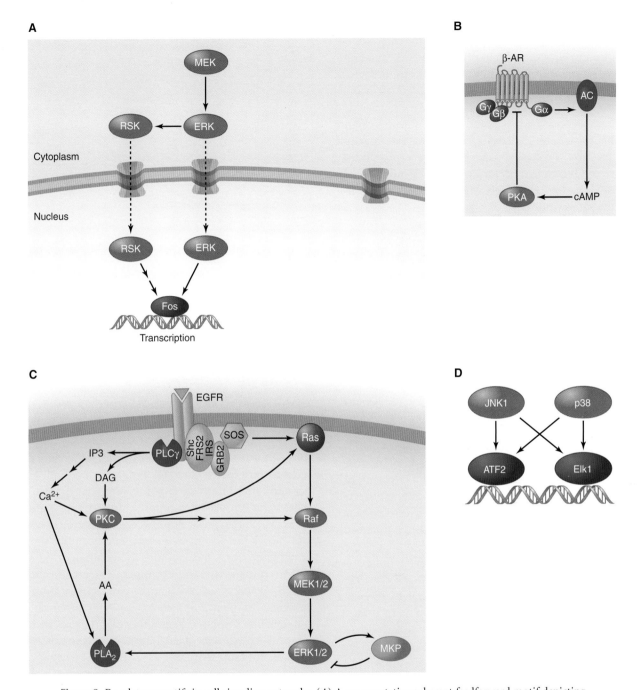

Figure 3. Regulatory motifs in cell signaling networks. (*A*) A representative coherent feedforward motif depicting signal flow from ERK to the transcription factor Fos, where both transcriptional activation and posttranslational stabilization lead to prolonged signal duration. (*B*) Negative feedback through β-adrenergic receptor (β-AR)-activated PKA limits the duration of response. (*C*) Positive-feedback loops within ERK-PLA2-PKC pathways can lead to bistability. (*D*) Kinases JNK and p38 phosphorylate and activate both of the transcription factors ATF2 and Elk1, forming a bi-fan motif that provides tight regulation, including temporal control and coincidence detection.

An example is the interaction between ERK and phospholipase A2 (PLA2), which results in the activation of PLA2 (probably via activation of MNK1/2, which phosphorylates PLA2 at S727), leading to the increased production of arachidonic acid. Diacylglycerol and arachidonic acid together can activate protein kinase C (PKC). This positive-feedback loop results in prolonged activation of PKC as well as ERK (Fig. 3C). Such positive-feedback loops can function as switches to move the signaling network from an inactive to persistently active state.

Bi-fan motifs form another class of abundant motifs found in signaling networks, especially at the interface between protein kinases and transcription factors. Most protein kinases phosphorylate multiple transcription factors, and many transcription factors are phosphorylated by multiple protein kinases. The JNK and p38 MAP kinases and the transcription factors ATF2 and Elk1 are shown in Figure 3D. This crisscrossing allows a cell to integrate multiple signals to give a cohesive pattern of gene expression. In the case of Elk1 and ATF2, for example, the motif enables upstream kinases to effectively synchronize the activity levels of the transcription factors.

2 EMERGENT PROPERTIES OF SIGNALING NETWORKS

The organization of a signaling pathway and its regulatory motifs can endow a signaling network with emergent properties that the individual components by themselves do not possess. These are difficult to predict from examination of any one of the components.

2.1 Ultrasensitivity

Sometimes a stimulus, such as receptor activation, can produce a switchlike response in a downstream signaling component. In such a system, at a certain point called the threshold, a small change in the ligand/receptor can cause a large change in the activity of a downstream effector. Such responses are called ultrasensitive. This was first observed with coupled enzymes switching between activity states (Goldbeter and Koshland 1981). Subsequently, similar ultrasensitivity has been observed in control of the cell cycle by the ERK pathway, where small changes in stimulus cause large changes in ERK activity (Ferrell and Machleder 1998). Ultrasensitivity can be produced by several mechanisms such as cooperativity, multistep regulation as seen in the ERK pathway, and by changing the levels of activators to inhibitors of a signaling component such as a GTPase (Lipshtat et al. 2010).

2.2 Bistability

The ability of positive-feedback loops like the one involving PKC, ERK, and PLA2 (Fig. 3C) to function as switches is an emergent property called bistability. ODE-based models (see Box 2) of the ERK-PLA2-PKC pathway (Bhalla and Iyengar 1999) allow one to plot the activity of ERK as a function of protein kinase C activity and vice versa (Fig. 4). The two curves intersect three times. In this system, if the initial stimulus is above the threshold level (the middle intersection point) needed to increase activity of either protein kinase C or ERK, the system will move from

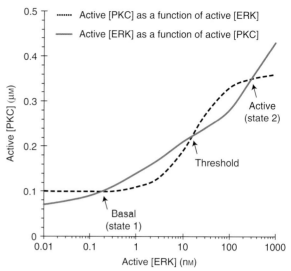

Figure 4. Activity states of components in a positive-feedback loop can be plotted as functions of each other to study the characteristics of a bistable system that switches from an inactive *basal* state (state 1) to an *active* state (state 2) once a threshold is reached. For example, when the concentration of active MAPK or PKC is above the threshold levels upon growth factor receptor activation, the system moves to state 2. The plot is obtained by computational modeling of the positive-feedback loop in Figure 3C, and recording the steady-L-state levels of active PKC for a given fixed concentration of active MAPK (black dashed line) upon stimulation, and vice versa (red solid line).

the lower intersection point (basal state) to the upper intersection point (active state) for a prolonged period. The upper and lower intersection points represent the two stable states. The middle intersection point represents the level above which the initial stimulus engages the feedback loop and moves the system to the activated equilibrium state. Otherwise, when the stimulus is withdrawn the system will revert to its basal equilibrium state. The computational modeling analysis shows that the connectivity alone is insufficient for the switching behavior. The concentrations of the signaling components and the reaction rates are critical in ensuring that a feedback loop with this topology functions as a bistable switch.

Bistability is involved in triggering numerous biological processes. The positive ERK-PLA2-PKC-feedback loop described above functions as a switch in cerebellar neurons, where it converts brief electrical signals into long-lasting changes in cellular responsiveness called long-term depression (Tanaka and Augustine 2008). In another example, during cell replication, a positive-feedback loop confers bistability on cyclin B expression, enabling the cell to enter mitosis (Sha et al. 2003). Similarly, programmed cell death is controlled by two cooperative positive-feedback loops, which confer bistability on mitochondrial permeability initiated by caspase-3 (Bagci et al. 2006; Ch. 19 [Green and Llambi 2014]).

 Cite this chapter as *Cold Spring Harb Perspect Biol* doi: 10.1101/cshperspect.a005934

2.3 Redundancy and Robustness

Because most signaling pathways are critical for survival, unsurprisingly, cells often use multiple pathways to connect a single input with an output. For example, activated epidermal growth factor receptors (EGFRs) simultaneously activate PKC (via tyrosine phosphorylation of PLCγ and generation of diacylglycerol [Ch. 3 (Newton et al. 2014)]) and the small G protein Ras (via formation of Grb2-SOS adapter-GEF complexes), both of which feed into the ERK pathway (p. 81 [Morrison 2012]). As shown in Figure 1A, this redundancy provides a safety net for activation of key processes that are necessary for cell survival or homeostasis. It further allows signaling networks to withstand perturbations that can alter input–output relationships, and it can ensure that cells respond only to stimuli of appropriate duration and magnitude, increasing robustness.

Network topology plays an important role in the robustness of signaling systems. Coherent feedforward motifs (see Box 2), for example, can be used in transcriptional regulatory networks as "persistence monitors" that respond to only sustained stimuli (Mangan and Alon 2003). Biochemical properties of signaling components such as a high catalytic rate or the ability to be altered by posttranslational modification are also major contributors to robustness of signaling networks.

2.4 Oscillatory Behavior

Oscillatory phenomena occur in cellular events from cell division to circadian rhythms. They generally involve negative-feedback loops; coupled positive- and negative-feedback loops can also lead to sustained oscillations (Tyson and Novak 2010). The cell cycle, for example, requires the cyclic expression and degradation of cyclins (Ch. 5 [Duronio and Xiong 2013]). Control of the cell cycle in *Xenopus* oocytes was one of the first oscillatory biological systems to be modeled with numerical simulations (Novak and Tyson 1993). Two feedback loops, one on top of the other, operate in the *Xenopus* oocyte cell cycle. In this case, a negative-feedback loop is necessary and sufficient for the oscillations of the driver, cyclin-CDK complex, but a positive-feedback loop dampens and synchronizes the cycles and is necessary for systems-level physiological operations (Pomerening et al. 2005).

3 INFORMATION FLOW AND PROCESSING

3.1 Information in Signaling Network Is Contextual

Although most cells share a multitude of common signaling components, identical signals can lead to diverse biological responses from different cell types. For instance, a pathway could have different outputs in two different cell types because they have different numbers or types of receptors.

A good example is a small peptide hormone known by two names: vasopressin (VP) and antidiuretic hormone (ADH). The two names arise because the hormone has very different physiological effects in two different cell types. In smooth muscle cells, it binds to V1 receptors that couple through the G protein Gq and calcium to regulate contractility, leading to contraction of blood vessels. In kidney cells, the same hormone binds to V2 receptors that couple through Gs and cAMP to regulate water uptake, leading to an antidiuretic effect. Thus, information in the hormone molecule is very differently interpreted by the two different cell types in different tissues in the body.

Such contextual dependence of information can occur within the cell as well. For example, in neurons, cAMP can bind to and activate PKA, leading to effects on metabolic enzymes, gene expression, and channel activity. cAMP can also directly bind to and gate a class of cation channels called HCNs. Because the affinity of cAMP for the HCN channels is substantially less than its affinity for PKA and because these channels are largely present in the distal dendrites, the information in cAMP can be selectively transmitted to metabolic enzymes and gene regulation without substantially affecting the electrical behavior of the neurons. In effect, cAMP at low concentrations represents information for PKA but no information for HCN channels, because the lower affinity precludes the channels from binding cAMP. In considering information flow within the signaling network, we therefore always need to know the characteristics of both the transmitter of information and the receiver.

3.2 Information Transmitted Can Be Proportional to the Stimulus, Processed, or Dissipated

Often the subcellular localization of the components of a signaling network and their relative affinities ensure that information flows proportionally (e.g., linearly) from receptor to effector. Typically, when the strength of the extracellular signal is such that only a small proportion of receptors are activated, the resultant downstream responses are proportional to the level of receptor activation. In some cases, however, network topology, spatial localization of the signaling components, and biochemical characteristics of individual components can lead to information processing. Enzymatic reactions, for example, often amplify signals, depending on the catalytic rate of the enzyme and its ability to activate multiple targets (Fig. 1B). Conversely, signal flow that involves adaptor proteins or other binding reactions such as scaffold proteins is less likely to amplify information because these interactions are stoichiometric. Adaptors and scaffolds play a critical role in signal processing, however, adding spatial constraints to information flow and defining bidirectional specificity. Network topology such as positive-feedback loops and coherent feedforward loops can increase the amplitude as well as the duration of output. Amplification by any of these mechanisms can increase sensitivity of the response or reduce responses to small or transient stimuli.

For certain signaling pathways that are ultrasensitive, dissipation of biological information is just as important as amplification. Prolonged signaling may lead to cell death or disease states. Drug-induced calcium excitotoxicity, which kills neurons, and mutated signaling proteins that lead to neoplasia, provide good examples of the deleterious effects of aberrant signals. Hence, signal dissipation is critical for proper functioning of signaling pathways. A range of negative regulators act on receptors, GTPases, effectors, and protein kinase signals to control and dissipate them before they can reach downstream effectors and produce physiological responses. Receptors are negatively regulated by protein kinases, phosphatases, and binding proteins called arrestins. Almost all receptors undergo desensitization when the cell is exposed to extracellular signals for prolonged periods (Ch. 1 [Heldin et al. 2014]). Desensitization often involves these proteins. Similarly, GTPases are deactivated by GAPs, and the effects of protein kinases are tightly controlled by protein phosphatases that antagonize them. The activities of effectors such as ACs are controlled by degradation of their products (in the case of AC, the second messenger cAMP by phosphodiesterases). There is a large family of phosphodiesterases that is subject to differential regulation by protein kinases, scaffolds, and by calcium/calmodulin. Both GAPs and phosphatases can be regulated by protein kinases and scaffolds as well. Thus, every positive signal generator in the cell has a negative counterpart that allows signal dissipation as needed.

3.3 Signaling Pathways and Networks Can Filter Noise

The coupling of positive and negative components allows signaling networks to filter noise in their inputs. Thus, low-level or transient signals may activate a few upstream signaling reactions but generate no cellular response (e.g., gene expression). A number of regulatory motifs improve signal-to-noise properties of networks. As described above, positive-feedback loops that function as switches can filter out subthreshold stimuli as noise. Feedforward motifs and interlinked feedback loops can also be configured so that they respond to sustained stimuli but ignore transient stimuli as noise (Brandman et al. 2005). Bi-fans that require

activation of both upstream components to produce downstream effects function as AND gates (Fig. 3D). Such a network motif can function as a double coincidence detector that requires multiple upstream signals to activate downstream components. Such a configuration can help prevent activation of downstream events such as transcription when the transcription factor is phosphorylated by only one protein kinase (i.e., only one signal is received) (Lipshtat et al. 2008).

4 CONCLUDING REMARKS

Signaling networks in cells produce outputs that are manifested as decisions to perform physiological functions in response to biochemical, electrical, or mechanical stimuli. These outputs are most often controlled by protein kinases in the cytoplasm and near the plasma membrane. These protein kinases phosphorylate and regulate metabolic enzymes, channels, transporters, and components of the cytoskeletal machinery. In the nucleus, the targets are typically transcription factors and proteins that control chromosomal organization and dynamics. The outputs, whether they are production of glucose from glycogen by liver cells in response to epinephrine, firing of neurons in response to neurotransmitters, or hormone-driven changes in gene expression of ovarian cells, reflect both the characteristics (amplitude and duration) of the external signals and information processing within signaling networks. The ability to balance these allows the cell to respond to varying stimuli and return to homeostatic balance.

When signaling components are inappropriately activated or inactivated (e.g., by mutations), however, the information-processing capability of the networks is also altered. This may lead to sustained changes in gene expression that push the cell into a different state, for example, an enhanced rate of proliferation, which is often detrimental and can cause diseases such as cancer (Ch. 21 [Sever and Brugge 2014]). Similarly, bacterial toxins can cause disease by altering components within signaling networks (Ch. 20 [Alto and Orth 2012]). Both examples show how critical it is for us to understand the organization and information-processing capability of signaling networks.

REFERENCES

*Reference is in this book.

Alon U. 2007. Network motifs: Theory and experimental approaches. *Nat Rev Genet* **8:** 450–461.
* Alto NM, Orth K. 2012. Subversion of cell signaling by pathogens. *Cold Spring Harb Perspect Biol* **4:** a006114.
Bagci EZ, Vodovotz Y, Billiar TR, Ermentrout GB, Bahar I. 2006. Bistability in apoptosis: Roles of bax, bcl-2, and mitochondrial permeability transition pores. *Biophys J* **90:** 1546–1559.

Barabasi AL, Albert R. 1999. Emergence of scaling in random networks. *Science* **286:** 509–512.

Bar-Sagi D, Hall A. 2000. Ras and Rho GTPases: A family reunion. *Cell* **103:** 227–238.

Bhalla US, Iyengar R. 1999. Emergent properties of networks of biological signaling pathways. *Science* **283:** 381–387.

Biggs N, Lloyd EK, Wilson RJ. 1986. *Graph theory 1736–1936.* Clarendon, Oxford.

* Bootman M. 2012. Calcium signaling. *Cold Spring Harb Perspect Biol* **4:** a011171.

Brandman O, Ferrell JE Jr, Li R, Meyer T. 2005. Interlinked fast and slow positive feedback loops drive reliable cell decisions. *Science* **310:** 496–498.

* Duronio RJ, Xiong X. 2013. Signaling pathways that control cell proliferation. *Cold Spring Harb Perspect Biol* **5:** a008904.

Ferrell JE Jr, Machleder EM. 1998. The biochemical basis of an all-or-none cell fate switch in *Xenopus* oocytes. *Science* **280:** 895–898.

Gillespie DT. 2007. Stochastic simulation of chemical kinetics. *Annu Rev Phys Chem* **58:** 35–55.

Goldbeter A, Koshland DE Jr. 1981. An amplified sensitivity arising from covalent modification in biological systems. *Proc Natl Acad Sci* **78:** 6840–6844.

* Green DR, Llambi F. 2014. Cell death signaling. *Cold Spring Harb Perspect Biol* doi: 10.1101/cshperspect.a006080.

* Heldin C-H, Lu B, Evans R, Gutkind JS. 2014. Signals and receptors. *Cold Spring Harb Perspect Biol* doi: 10.1101/cshperspect.a005900.

* Hemmings BA, Restuccia DF. 2012. The PI3K-PKB/AKT pathway. *Cold Spring Harb Perspect Biol* **4:** 011189.

Jordan JD, Landau EM, Iyengar R. 2000. Signaling networks: The origins of cellular multitasking. *Cell* **103:** 193–200.

Lipshtat A, Purushothaman SP, Iyengar R, Ma'ayan A. 2008. Functions of bifans in context of multiple regulatory motifs in signaling networks. *Biophys J* **94:** 2566–2579.

Lipshtat A, Jayaraman G, He JC, Iyengar R. 2010. Design of versatile biochemical switches that respond to amplitude, duration, and spatial cues. *Proc Natl Acad Sci* **107:** 1247–1252.

Mangan S, Alon U. 2003. Structure and function of the feed-forward loop network motif. *Proc Natl Acad Sci* **100:** 11980–11985.

* Morrison D. 2012. MAPK kinase pathways. *Cold Spring Harb Perspect Biol* **4:** a011254.

* Newton AC, Bootman MD, Scott JD. 2014. Second messengers. *Cold Spring Harb Perspect Biol* doi: 10.1101/cshperspect.a005926.

Novak B, Tyson JJ. 1993. Numerical analysis of a comprehensive model of M-phase control in *Xenopus* oocyte extracts and intact embryos. *J Cell Sci* **106:** 1153–1168.

Pieroni JP, Jacobowitz O, Chen J, Iyengar R. 1993. Signal recognition and integration by Gs-stimulated adenylyl cyclases. *Curr Opin Neurobiol* **3:** 345–351.

Pomerening JR, Kim SY, Ferrell JE Jr. 2005. Systems-level dissection of the cell-cycle oscillator: Bypassing positive feedback produces damped oscillations. *Cell* **122:** 565–578.

Pool IdS, Milgram S, Newcomb T. 1989. *The small world.* Ablex, Norwood, NJ.

* Sassone-Corsi P. 2012. The cyclic AMP pathway. *Cold Spring Harb Perspect Biol* **4:** a011148.

* Sever R, Brugge JS. 2014. Signal transduction in cancer. *Cold Spring Harb Perspect Med* doi: 10.1101/cshperspect.a006098.

Sha W, Moore J, Chen K, Lassaletta AD, Yi CS, Tyson JJ, Sible JC. 2003. Hysteresis drives cell-cycle transitions in *Xenopus laevis* egg extracts. *Proc Natl Acad Sci* **100:** 975–980.

Tanaka K, Augustine GJ. 2008. A positive feedback signal transduction loop determines timing of cerebellar long-term depression. *Neuron* **59:** 608–620.

Tyson JJ, Novak B. 2010. Functional motifs in biochemical reaction networks. *Annu Rev Phys Chem* **61:** 219–240.

Weng G, Bhalla US, Iyengar R. 1999. Complexity in biological signaling systems. *Science* **284:** 92–96.

SIGNALING PATHWAYS

MAP Kinase Pathways

Deborah K. Morrison

Laboratory of Cell and Developmental Signaling, National Cancer Institute, Frederick, Maryland 21702

Correspondence: morrisod@mail.nih.gov

Mitogen-activated protein kinase (MAPK) modules containing three sequentially activated protein kinases are key components of a series of vital signal transduction pathways that regulate processes such as cell proliferation, cell differentiation, and cell death in eukaryotes from yeast to humans (Fig. 1) (Qi and Elion 2005; Raman et al. 2007; Keshet and Seger 2010). Each cascade is initiated by specific extracellular cues and leads to activation of a particular MAPK following the successive activation of a MAPK kinase kinase (MAPKKK) and a MAPK kinase (MAPKK) (Fig. 1). The MAPKKK is typically activated by interactions with a small GTPase and/or phosphorylation by protein kinases downstream from cell surface receptors (Cuevas et al. 2007). The MAPKKK directly phosphorylates and activates the MAPKK, which, in turn, activates the MAPK by

dual phosphorylation of a conserved tripeptide TxY motif in the activation segment. Once activated, the MAPK phosphorylates diverse substrates in the cytosol and nucleus to bring about changes in protein function and gene expression that execute the appropriate biological response. MAPKs generally contain docking sites for MAPKKs and substrates, which allow high-affinity protein–protein interactions to ensure both that they are activated by a particular upstream MAPKK (Bardwell and Thorner 1996) and that they recognize specific downstream targets (Tanoue and Nishida 2003).

The MAP kinases can be grouped into three main families. In mammals, these are ERKs (extracellular-signal-regulated kinases), JNKs (Jun amino-terminal kinases), and p38/SAPKs (stress-activated protein kinases). ERK family

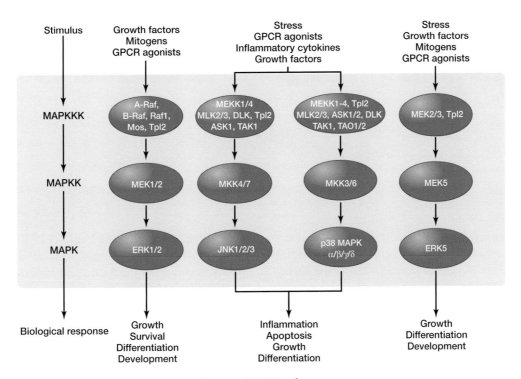

Figure 1. MAPK pathways.

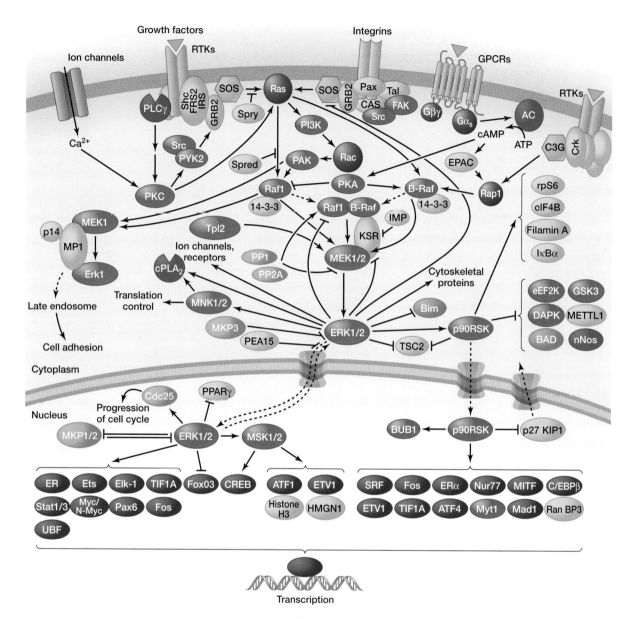

Figure 2. The ERK MAPK pathway.

members possess a TEY motif in the activation segment and can be subdivided into two groups: the classic ERKs that consist mainly of a kinase domain (ERK1 and ERK2) and the larger ERKs (such as ERK5) that contain a much more extended sequence carboxy-terminal to their kinase domain (Zhang and Dong 2007). The classic ERK1/2 module (Fig. 2) responds primarily to growth factors and mitogens to induce cell growth and differentiation (McKay and Morrison 2007; Shaul and Seger 2007). Important upstream regulators of this module include cell surface receptors, such as receptor tyrosine kinases (RTKs), G-protein-coupled receptors (GPCRs), and integrins, as well as

the small GTPases Ras and Rap. MAPKKs for the classic ERK1/2 module are MEK1 and MEK2, and the MAPKKKs include members of the Raf family, Mos, and Tpl2.

JNK family members contain a TPY motif in the activation segment and include JNK1, JNK2, and JNK3. The JNK module (Fig. 3) is activated by environmental stresses (ionizing radiation, heat, oxidative stress, and DNA damage) and inflammatory cytokines, as well as growth factors, and signaling to the JNK module often involves the Rho family GTPases Cdc42 and Rac (Johnson and Nakamura 2007). The JNK module plays an important role in apoptosis, inflammation, cytokine production, and

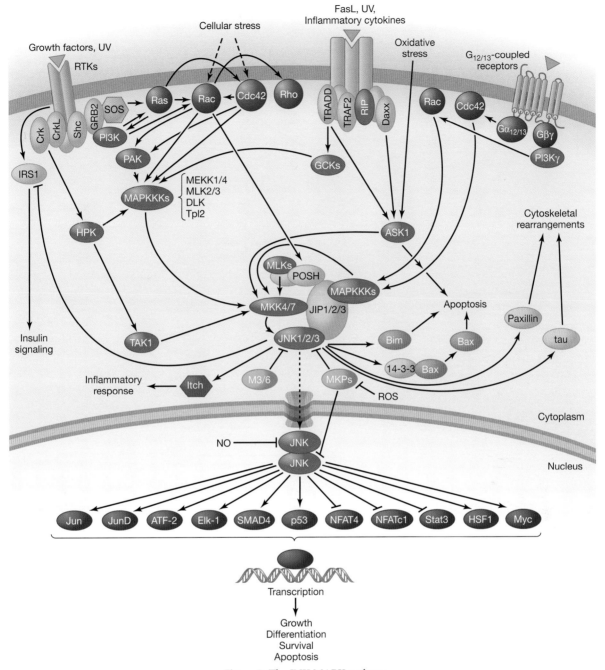

Figure 3. The JNK MAPK pathway.

metabolism (Dhanasekaran and Reddy 2008; Huang et al. 2009; Rincon and Davis 2009). MAPKKs for the JNK module are MKK4 and MKK7, and the MAPKKKs include MEKK1 and MEKK4, MLK2 and MLK3, ASK1, TAK1, and Tpl2.

p38 family members possess a TGY motif in the activation segment and include p38α, p38β, p38γ, and p38δ. Like JNK modules, p38 modules (Fig. 4) are strongly activated by environmental stresses and inflammatory cyto-

kines. p38 activation contributes to inflammation, apoptosis, cell differentiation, and cell cycle regulation (Cuenda and Rousseau 2007; Cuadrado and Nebreda 2010). The primary MAPKKs for p38 modules are MKK3 and MKK6, and the MAPKKKs include MLK2 and MLK3, MEKKs, ASKs, TAK1, and TAO1 and TAO2. Important substrates in p38 signaling include the downstream kinases MK2/3, PRAK, and MSK1 and MSK2, as well as various transcription factors.

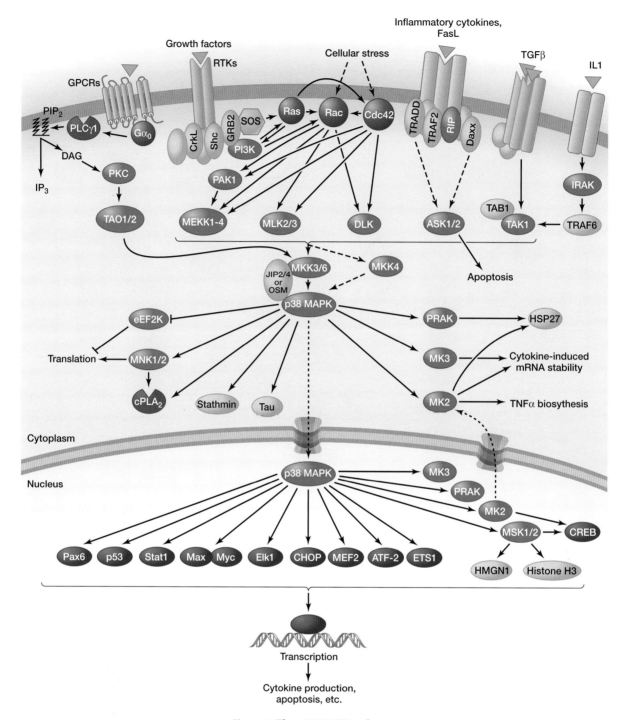

Figure 4. The p38 MAPK pathway.

For all of the MAPK modules, specific scaffold proteins (Good et al. 2011) have been identified that dock at least two of the core kinases of the module. These scaffolds contribute to MAPK signaling by increasing the local concentration of the components, providing spatial temporal regulation of cascade activation, and/or localizing the module to specific cellular sites or substrates. Scaffold proteins involved in MAPK cascade signaling include KSR and MP1 for the ERK module; JIP1, JIP2, JIP3, JIP4, and POSH for the JNK module; and JIP2, JIP4, and OSM for the p38 module (Dhanasekaran et al. 2007).

Figure 1 adapted, with permission, from Cell Signaling Technology (http://www.cellsignal.com).

Cite this chapter as *Cold Spring Harb Perspect Biol* doi: 10.1101/cshperspect.a011254

REFERENCES

Bardwell L, Thorner J. 1996. A conserved motif at the amino termini of MEKs might mediate high-affinity interaction with the cognate MAPKs. *Trends Biochem Sci* **21**: 373–374.

Cuadrado A, Nebreda AR. 2010. Mechanisms and functions of p38 MAPK signalling. *Biochem J* **429**: 403–417.

Cuenda A, Rousseau S. 2007. p38 MAP-kinases pathway regulation, function and role in human diseases. *Biochim Biophys Acta* **1773**: 1358–1375.

Cuevas BD, Abell AN, Johnson GL. 2007. Role of mitogen-activated protein kinase kinase kinases in signal integration. *Oncogene* **26**: 3159–3171.

Dhanasekaran DN, Reddy EP. 2008. JNK signaling in apoptosis. *Oncogene* **27**: 6245–6251.

Dhanasekaran DN, Kashef K, Lee CM, Xu H, Reddy EP. 2007. Scaffold proteins of MAP-kinase modules. *Oncogene* **26**: 3185–3202.

Good MC, Zalatan JG, Lim WA. 2011. Scaffold proteins: Hubs for controlling the flow of cellular information. *Science* **332**: 680–686.

Huang G, Shi LZ, Chi H. 2009. Regulation of JNK and p38 MAPK in the immune system: Signal integration, propagation and termination. *Cytokine* **48**: 161–169.

Johnson GL, Nakamura K. 2007. The c-Jun kinase/stress-activated pathway: Regulation, function and role in human disease. *Biochim Biophys Acta* **1773**: 1341–1348.

Keshet Y, Seger R. 2010. The MAP kinase signaling cascades: A system of hundreds of components regulates a diverse array of physiological functions. *Methods Mol Biol* **661**: 3–38.

McKay MM, Morrison DK. 2007. Integrating signals from RTKs to ERK/MAPK. *Oncogene* **26**: 3113–3121.

Qi M, Elion EA. 2005. MAP kinase pathways. *J Cell Sci* **118**: 3569–3572.

Raman M, Chen W, Cobb MH. 2007. Differential regulation and properties of MAPKs. *Oncogene* **26**: 3100–3112.

Rincon M, Davis RJ. 2009. Regulation of the immune response by stress-activated protein kinases. *Immunol Rev* **228**: 212–224.

Shaul YD, Seger R. 2007. The MEK/ERK cascade: From signaling specificity to diverse functions. *Biochim Biophys Acta* **1773**: 1213–1226.

Tanoue T, Nishida E. 2003. Molecular recognitions in MAP kinase cascades. *Cell Signal* **15**: 455–462.

Zhang Y, Dong C. 2007. Regulatory mechanisms of mitogen-activated kinase signaling. *Cell Mol Life Sci* **64**: 2771–2789.

The PI3K-PKB/Akt Pathway

Brian A. Hemmings and David F. Restuccia

Friedrich Miescher Institute for Biomedical Research, Basel 4058, Switzerland

Correspondence: david.restuccia@fmi.ch

Identification of the phosphoinositide-3-kinase–protein kinase B/Akt (PI3K-PKB/Akt) pathway and activating receptor tyrosine kinases (RTKs) began in earnest in the early 1980s through vigorous attempts to characterize insulin receptor signaling (for review, see Alessi 2001; Brazil and Hemmings 2001). These humble beginnings led to the identification of the components and mechanism of insulin receptor signaling via insulin receptor substrate (IRS) proteins to PI3K and consequent PKB/Akt-mediated activation by 3-phosphoinositide-dependent protein kinase 1 (PDK1). With the discovery of the potent contribution of PI3K and PKB/Akt activation to tumorigenesis, intense research into the regulation of this pathway led to the discovery of the negative regulators, the protein phosphatase 2 (PP2A), phosphatase and tensin homolog (PTEN), and the PH-domain leucine-rich-repeat-containing protein phosphatases (PHLPP1/2). More recently, the elusive PKB/

Akt hydrophobic motif kinases—i.e., the mammalian target of rapamycin (mTOR), when associated with the mTOR complex 2 (mTORC2), and the DNA-dependent protein kinase (DNA-PK)—were identified, as was the ability of Ras to affect the PI3K-PKB/Akt pathway via PI3K, completing our current model of the PI3K-PKB/Akt pathway.

The PI3K-PKB/Akt pathway is highly conserved, and its activation is tightly controlled via a multistep process (as shown in Fig. 1) Activated receptors directly stimulate class 1A PI3Ks bound via their regulatory subunit or adapter molecules such as the insulin receptor substrate (IRS) proteins. This triggers activation of PI3K and conversion by its catalytic domain of phosphatidylinositol (3,4)-bis-phosphate (PIP_2) lipids to phosphatidylinositol (3,4,5)-tris-phosphate (PIP_3). PKB/Akt binds to PIP_3 at the plasma membrane, allowing PDK1 to access and phosphorylate

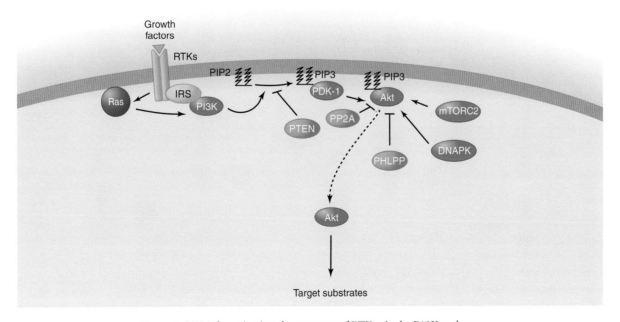

Figure 1. PKB/Akt activation downstream of RTKs via the PI3K pathway.

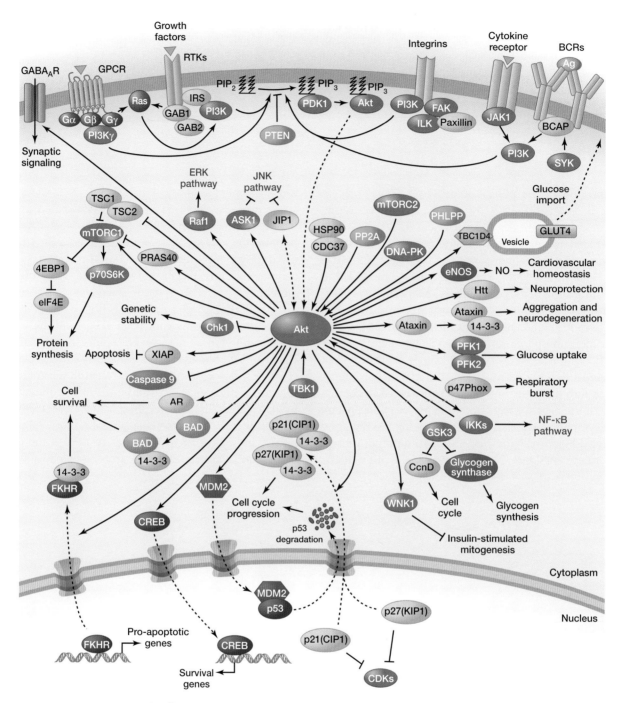

Figure 2. Signalling events activating PKB/Akt and cellular functions regulated by PKB/Akt.

T308 in the "activation loop," leading to partial PKB/Akt activation (Alessi et al. 1997). This PKB/Akt modification is sufficient to activate mTORC1 by directly phosphorylating and inactivating proline-rich Akt substrate of 40 kDa (PRAS40) and tuberous sclerosis protein 2 (TSC2) (Vander Haar et al. 2007). mTORC1 substrates include the eukaryotic translation initiation factor 4E binding protein 1 (4EBP1),

and ribosomal protein S6 kinase, 70 kDa, polypeptide 1 (S6K1), which, in turn, phosphorylates the ribosomal protein S6 (S6/RPS6), promoting protein synthesis and cellular proliferation.

Phosphorylation of Akt at S473 in the carboxy-terminal hydrophobic motif, either by mTOR (Sarbassov et al. 2005) or by DNA-PK (Feng et al. 2004), stimulates full Akt activity.

Full activation of Akt leads to additional substrate-specific phosphorylation events in both the cytoplasm and nucleus, including inhibitory phosphorylation of the pro-apoptotic FOXO proteins (Guertin et al. 2006). Fully active PKB/Akt mediates numerous cellular functions including angiogenesis, metabolism, growth, proliferation, survival, protein synthesis, transcription, and apoptosis (as shown in Fig. 2). Dephosphorylation of T308 by PP2A (Andjelković et al. 1996), and S473 by PHLPP1/2 (Brognard et al. 2007), and the conversion of PIP_3 to PIP_2 by PTEN (Stambolic et al. 1998) antagonize Akt signaling.

Figures adapted, with kind permission, from Cell Signaling Technology (http://www.cellsignal.com).

REFERENCES

Alessi DR. 2001. Discovery of PDK1, one of the missing links in insulin signal transduction. Colworth Medal Lecture. *Biochem Soc Trans* **29:** 1–14.

Alessi DR, James SR, Downes CP, Holmes AB, Gaffney PR, Reese CB, Cohen P. 1997. Characterization of a 3-phosphoinositide-dependent protein kinase which phosphorylates and activates protein kinase Bα. *Curr Biol* **7:** 261–269.

Altomare DA, Testa JR. 2005. Perturbations of the AKT signaling pathway in human cancer. *Oncogene* **24:** 7455–7464.

Andjelković M, Jakubowicz T, Cron P, Ming XF, Han JW, Hemmings BA. 1996. Activation and phosphorylation of a pleckstrin homology domain containing protein kinase (RAC-PK/PKB) promoted by serum and protein phosphatase inhibitors. *Proc Natl Acad Sci* **93:** 5699–5704.

Bozulic L, Hemmings BA. 2009. PIKKing on PKB: Regulation of PKB activity by phosphorylation. *Curr Opin Cell Biol* **21:** 256–261.

Brazil DP, Hemmings BA. 2001. Ten years of protein kinase B signalling: A hard Akt to follow. *Trends Biochem Sci* **26:** 657–664.

Brognard J, Sierecki E, Gao T, Newton AC. 2007. PHLPP and a second isoform, PHLPP2, differentially attenuate the amplitude of Akt signaling by regulating distinct Akt isoforms. *Mol Cell* **25:** 917–931.

Feng J, Park J, Cron P, Hess D, Hemmings BA. 2004. Identification of a PKB/Akt hydrophobic motif Ser-473 kinase as DNA-dependent protein kinase. *J Biol Chem* **279:** 41189–41196.

Guertin DA, Stevens DM, Thoreen CC, Burds AA, Kalaany NY, Moffat J, Brown M, Fitzgerald KJ, Sabatini DM. 2006. Ablation in mice of the mTORC components raptor, rictor, or mLST8 reveals that mTORC2 is required for signaling to Akt-FOXO and PKCα, but not S6K1. *Dev Cell* **11:** 859–871.

Manning BD, Cantley LC. 2007. AKT/PKB signaling: Navigating downstream. *Cell* **129:** 1261–1274.

Sarbassov DD, Guertin DA, Ali SM, Sabatini DM. 2005. Phosphorylation and regulation of Akt/PKB by the rictor–mTOR complex. *Science* **307:** 1098–1101.

Stambolic V, Suzuki A, de la Pompa JL, Brothers GM, Mirtsos C, Sasaki T, Ruland J, Penninger JM, Siderovski DP, Mak TW. 1998. Negative regulation of PKB/Akt-dependent cell survival by the tumor suppressor PTEN. *Cell* **95:** 29–39.

Vander Haar E, Lee SI, Bandhakavi S, Griffin TJ, Kim DH. 2007. Insulin signalling to mTOR mediated by the Akt/PKB substrate PRAS40. *Nat Cell Biol* **9:** 316–323.

mTOR Signaling

Mathieu Laplante and David M. Sabatini

Centre de Recherche de l'Institut Universitaire de Cardiologie et de Pneumologie de Québec (CRIUCPQ), Université Laval, Québec G1V 4G5, Canada

Correspondence: mathieu.laplante@criucpq.ulaval.ca

The scarcity of nutrients/energy interspersed with sporadic periods of abundance means that cells must transition between anabolic and catabolic states. An important protein that has evolved to respond to this need is the target of rapamycin (TOR). TOR is a well-conserved serine/ threonine kinase that plays an important role in the signaling network that controls growth and metabolism in response to environmental cues. As its name indicates, TOR is the target of a molecule named rapamycin, an anti-fungal macrolide produced by the bacterial species *Streptomyces hygroscopicus*, which was isolated from a soil sample from the Eastern Islands in the 1970s (Vezina et al. 1975). In addition to its anti-fungal properties, rapamycin strongly inhibits cell growth and proliferation, making this molecule a valuable tool to study cell growth control. In the early 1990s, yeast genetic screens led to the identification of

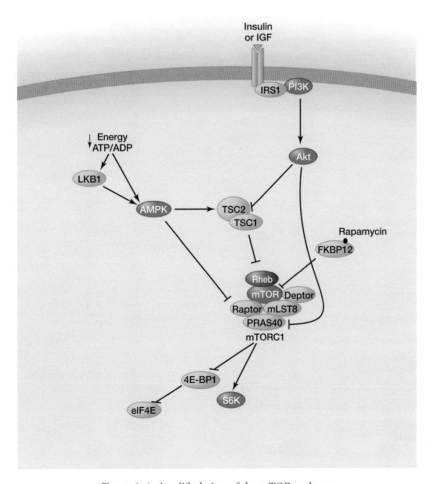

Figure 1. A simplified view of the mTOR pathway.

Cite this chapter as *Cold Spring Harb Perspect Biol* doi: 10.1101/cshperspect.a011593

TOR as one mediator of the toxic effect of rapamycin in yeast (Heitman et al. 1991; Kunz et al. 1993). Shortly after, the mammalian TOR (mTOR), now officially known as the mechanistic TOR, was identified as the physical target of rapamycin (Fig. 1) (Brown et al. 1994; Sabatini et al. 1994; Sabers et al. 1995).

mTOR interacts with many proteins to form at least two distinct multiprotein complexes: mTOR complex 1 (mTORC1) and mTOR complex 2 (mTORC2) (Zoncu et al. 2011). In addition to their different protein compositions, the mTOR complexes have important differences in their sensitivities to rapamycin, in the upstream signals

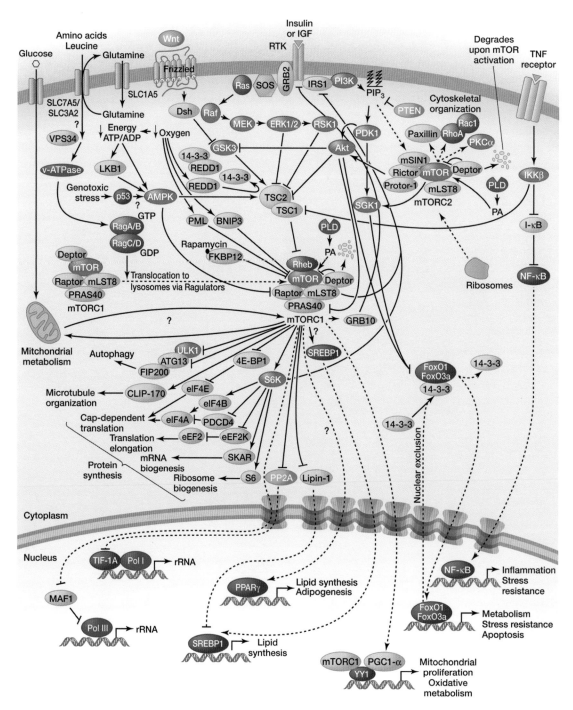

Figure 2. The mTOR pathway.

Cite this chapter as *Cold Spring Harb Perspect Biol* doi: 10.1101/cshperspect.a011593

they integrate, in the substrates they regulate, and in the biological processes they control (Laplante and Sabatini 2009b). mTORC1 is sensitive to rapamycin and integrates inputs from at least five major signals—growth factors, genotoxic stress, energy status, oxygen, and amino acids—to regulate many processes involved in the promotion of cell growth and proliferation (Fig. 2). With the exception of amino acids, all of these inputs regulate mTORC1 activity by modulating the activity of TSC1–TSC2. TSC1–TSC2 negatively regulates mTORC1 activity by converting Rheb into its inactive GDP-bound state. Growth factors block TSC1–TSC2 activity and activate mTORC1, whereas genotoxic stress, energy deficit, and oxygen deprivation promote TSC1–TSC2 action and inhibit mTORC1. Conversely, amino acids activate mTORC1 through the action of the Rag GTPases. In the presence of amino acids, the Rag GTPases interact with mTORC1, which promotes its translocation from the cytoplasm to lysosomal membranes, where Rheb is thought to reside.

When activated, mTORC1 promotes protein synthesis mainly by phosphorylating the kinase S6K and the translation regulator 4E-BP1 (Ma and Blenis 2009). This complex also induces lipid biogenesis by activating SREBP1 and PPARγ transcription factors (Laplante and Sabatini 2009a). In addition to promoting anabolism, mTORC1 also inhibits catabolism by blocking autophagy through the phosphorylation of the ULK1–Atg13–FIP200 complex. Importantly, sustained mTORC1 activation blocks growth factor signaling by activating many negative-feedback loops that inhibit PI3-kinase (PI3K) signaling.

Compared with mTORC1, our knowledge of mTORC2 biology is still very limited. mTORC2 is activated by growth factors through a mechanism that is not well understood.

It is insensitive to acute treatment with rapamycin and regulates cell survival, metabolism, and cytoskeletal organization by phosphorylating many AGC kinases, including Akt, SGK1, and PKCα (Laplante and Sabatini 2009b; Zoncu et al. 2011).

Figures adapted from Laplante and Sabatini (2009b).

REFERENCES

Brown EJ, Albers MW, Shin TB, Ichikawa K, Keith CT, Lane WS, Schreiber SL. 1994. A mammalian protein targeted by G1-arresting rapamycin-receptor complex. *Nature* **369:** 756–758.

Heitman J, Movva NR, Hall MN. 1991. Targets for cell cycle arrest by the immunosuppressant rapamycin in yeast. *Science* **253:** 905–909.

Kunz J, Henriquez R, Schneider U, Deuter-Reinhard M, Movva NR, Hall MN. 1993. Target of rapamycin in yeast, TOR2, is an essential phosphatidylinositol kinase homolog required for G₁ progression. *Cell* **73:** 585–596.

Laplante M, Sabatini DM. 2009a. An emerging role of mTOR in lipid biosynthesis. *Curr Biol* **19:** R1046–R1052.

Laplante M, Sabatini DM. 2009b. mTOR signaling at a glance. *J Cell Sci* **122:** 3589–3594.

Ma XM, Blenis J. 2009. Molecular mechanisms of mTOR-mediated translational control. *Nat Rev Mol Cell Biol* **10:** 307–318.

Sabatini DM, Erdjument-Bromage H, Lui M, Tempst P, Snyder SH. 1994. RAFT1: A mammalian protein that binds to FKBP12 in a rapamycin-dependent fashion and is homologous to yeast TORs. *Cell* **78:** 35–43.

Sabers CJ, Martin MM, Brunn GJ, Williams JM, Dumont FJ, Wiederrecht G, Abraham RT. 1995. Isolation of a protein target of the FKBP12–rapamycin complex in mammalian cells. *J Biol Chem* **270:** 815–822.

Vezina C, Kudelski A, Sehgal SN. 1975. Rapamycin (AY-22,989), a new antifungal antibiotic. I. Taxonomy of the producing streptomycete and isolation of the active principle. *J Antibiot (Tokyo)* **28:** 721–726.

Zoncu R, Efeyan A, Sabatini DM. 2011. mTOR: From growth signal integration to cancer, diabetes and ageing. *Nat Rev Mol Cell Biol* **12:** 21–35.

Calcium Signaling

Martin D. Bootman

The Babraham Institute Babraham Research Campus, Cambridge CB22 3AT, United Kingdom

Correspondence: martin.bootman@bbsrc.ac.uk

Calcium ions regulate processes as diverse as cell motility, gene transcription, muscle contraction, and exocytosis (Berridge et al. 2000). The first realization that they are critical for cellular function is often attributed to Sydney Ringer, who discovered in 1883 that saline solution made up using London tap water (which contained calcium) supported the contraction of isolated frog hearts, whereas saline made up using distilled water (which lacked calcium) could not. Subsequent work revealed that numerous cell biological processes are controlled by calcium (Carafoli

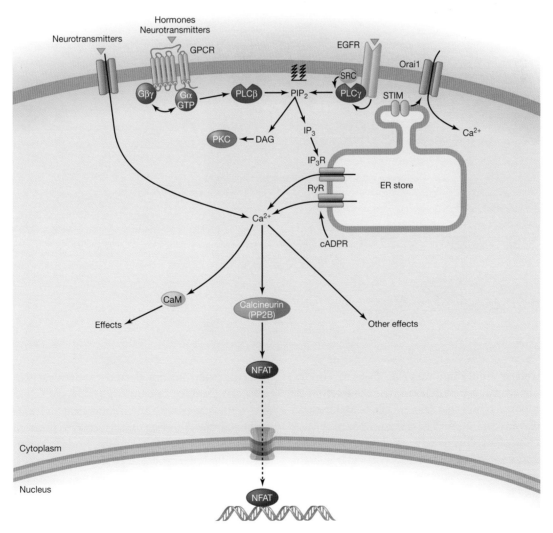

Figure 1. Calcium signaling (simplified view).

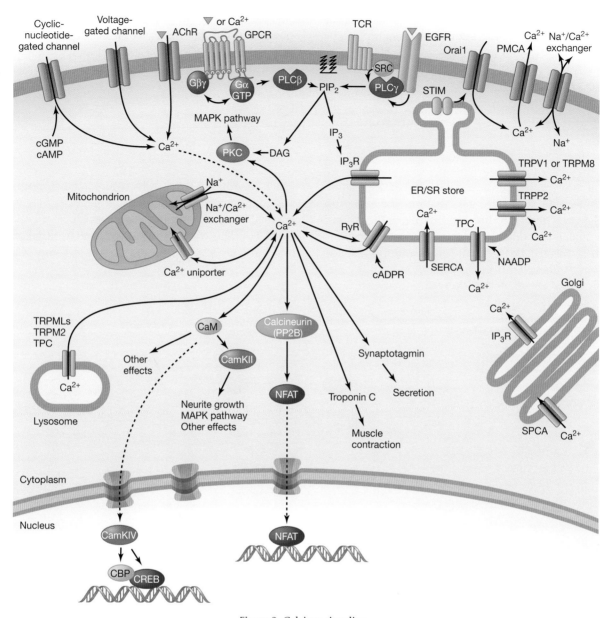

Figure 2. Calcium signaling.

2003). Particularly important was the discovery in the 1950s that calcium triggers skeletal muscle contraction by binding to troponin C and that calcium can be sequestered in the sarcoplasmic reticulum. These studies led to the notion that calcium signals inside cells oscillate: the cytoplasmic calcium concentration increases, a particular effector is activated, and then the calcium signal is reversed to reset the system (see Figs. 1 and 2). Cells use a "toolkit" of channels, pumps, and cytosolic buffers to control calcium levels (Berridge et al. 2000). Numerous proteins are modulated directly or indirectly by calcium. These include kinases and phosphatases, transcription factors such as NF-AT, and the ubiquitous calcium-binding protein calmodulin (CaM).

Electrical, hormonal, and mechanical stimulation of cells can produce calcium signals by causing entry of the ion across the plasma membrane or its release from intracellular stores. Binding of hormones to G-protein-coupled receptors (GPCRs), for example, leads to generation of the second messenger inositol 1,4,5-trisphosphate (IP_3), which releases calcium from intracellular stores such as the endoplasmic reticulum. By contrast, electrical or neurotransmitter stimulation of neurons causes calcium to enter cells from outside via channels in the plasma membrane. This can increase the average cytosolic calcium concentration from around 100 nM to around 1 μM. Close to an active channel the calcium concentration can reach tens of micromolar. Such local hot-spots of calcium are used by cells to activate specific process that are generally not sensitive to the bulk cytosolic calcium concentration (Bootman et al. 2001).

Cellular calcium signaling proteomes are tissue-specific, producing unique calcium signals that suit a tissue's physiology (Berridge et al. 2003). For example, cardiac myocytes require a rapid (hundreds of milliseconds) whole-cell calcium transient to trigger contraction every second (Bers 2002), whereas cells that are not electrically excitable typically display calcium oscillations that last for tens of seconds, and can have a periodicity of several minutes, to control gene expression and metabolism (Dupont et al. 2011). The rapid calcium signals within myocytes are caused by calcium entering through voltage-activated calcium channels in the plasma membrane, which then triggers calcium release via ryanodine receptors on the sarcoplasmic reticulum. The slower calcium signals in nonexcitable cells typically rely on IP_3, which binds to channels ($InSP_3$-Rs) on the endoplasmic reticulum, or potentially nicotinic acid adenine dinucleotide phosphate-gated calcium channels (two pore channels) on acidic organelles, leading to release of calcium into the cytoplasm (Galione 2011). Calcium signals can also pass through gap junctions to coordinate activities of neighboring cells (Sanderson et al. 1994).

The actions of calcium can be mediated by direct binding of calcium to effectors, such as the phosphatase calcineurin (Berridge 2006). Alternatively, it can act via the ubiquitous calcium-binding protein CaM. The interaction of calcium with CaM leads to a rearrangement of the protein that allows it to bind and allosterically regulate target molecules such as the calcium/calmodulin-dependent kinases CaMKII and CaMKIV. CaM is mobile within cells and can associate with its targets after binding calcium. However, in some cases, it is prebound to its target, which provides rapid control. Ultimately, calcium signals are reversed by the action of pumps such as the sarco/endoplasmic reticulum ATPases (SERCA) that return it from the cytosol to intracellular stores or the external milieu.

REFERENCES

Berridge MJ. 2006. Calcium microdomains: Organization and function. *Cell Calcium* **40**: 405–412.

Berridge MJ, Lipp P, Bootman MD. 2000. The versatility and universality of calcium signalling. *Nat Rev Mol Cell Biol* **1**: 11–21.

Berridge MJ, Bootman MD, Roderick HL. 2003. Calcium signalling: Dynamics, homeostasis and remodelling. *Nat Rev Mol Cell Biol* **4**: 517–529.

Bers DM. 2002. Cardiac excitation-contraction coupling. *Nature* **415**: 198–205.

Bootman MD, Lipp P, Berridge MJ. 2001. The organisation and functions of local Ca^{2+} signals. *J Cell Sci* **114**: 2213–2222.

Carafoli E. 2003. The calcium-signalling saga: Tap water and protein crystals. *Nat Rev Mol Cell Biol* **4**: 326–332.

Dupont G, Combettes L, Bird GS, Putney JW. 2011. Calcium oscillations. *Cold Spring Harb Perspect Biol* **3**: a004226.

Galione A. 2011. NAADP receptors, *Cold Spring Harb Perspect Biol* **3**: a004036.

Sanderson MJ, Charles AC, Boitano S, Dirksen ER. 1994. Mechanisms and function of intercellular calcium signaling. *Mol Cell Endocrinol* **98**: 173–187.

The Cyclic AMP Pathway

Paolo Sassone-Corsi

Center for Epigenetics and Metabolism, School of Medicine, University of California, Irvine, California 92697

Correspondence: psc@uci.edu

Cyclic adenosine 3′,5′-monophosphate (cAMP) was the first second messenger to be identified and plays fundamental roles in cellular responses to many hormones and neurotransmitters (Sutherland and Rall 1958). The intra- cellular levels of cAMP are regulated by the balance between the activities of two enzymes (see Fig. 1): adenylyl cyclase (AC) and cyclic nucleotide phosphodiesterase (PDE). Different isoforms of these enzymes are encoded by a large

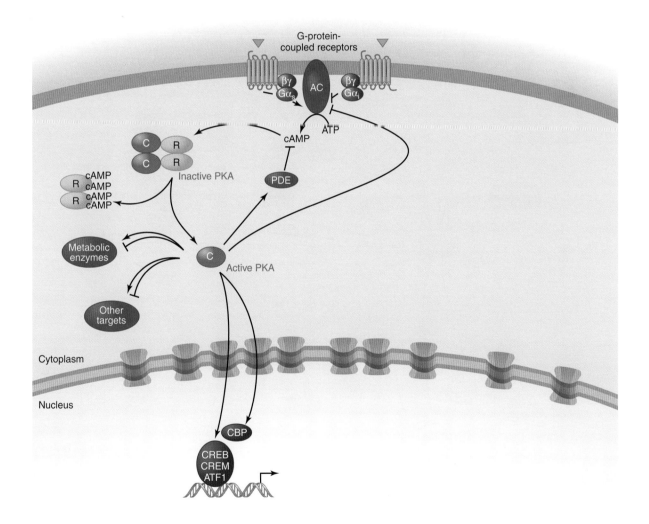

Figure 1. PKA regulation.

P. Sassone-Corsi

number of genes, which differ in their expression patterns and mechanisms of regulation, generating cell-type and stimulus-specific responses (McKnight 1991).

Most ACs (soluble bicarbonate-regulated ACs are the exception) are activated downstream from G-protein-coupled receptors (GPCRs) such as the β adrenoceptor by interactions with the α subunit of the G$_s$ protein (α_s). α_s is released from heterotrimeric αβγ G-protein complexes following binding of agonist ligands to GPCRs (e.g., epinephrine in the case of β adrenoceptors) and binds to and activates AC. The βγ subunits can also stimulate some AC isoforms. cAMP generated as a consequence of AC activation can activate several effectors, the most well studied of

which is cAMP-dependent protein kinase (PKA) (Pierce et al. 2002).

Alternatively, AC activity can be inhibited by ligands that stimulate GPCRs coupled to G$_i$ and/or cAMP can be degraded by PDEs. Indeed both ACs and PDEs are regulated positively and negatively by numerous other signaling pathways (see Fig. 2), such as calcium signaling (through calmodulin [CaM], CamKII, CamKIV, and calcineurin [also know as PP2B]), subunits of other G proteins (e.g., α_i, α_o, and α_q proteins, and the βγ subunits in some cases), inositol lipids (by PKC), and receptor tyrosine kinases (through the ERK MAP kinase and PKB) (Yoshimasa et al. 1987; Bruce et al. 2003; Goraya and Cooper 2005).

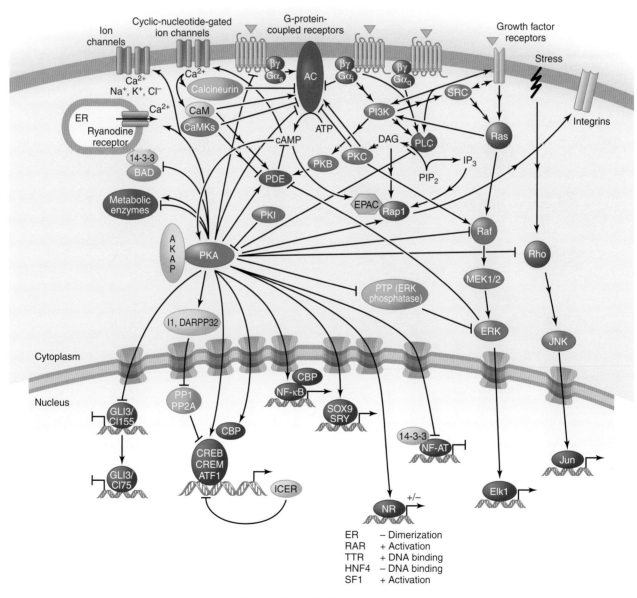

Figure 2. The cAMP/PKA pathway.

Cite this chapter as *Cold Spring Harb Perspect Biol* doi: 10.1101/cshperspect.a011148

Cross talk with other pathways provides further modulation of the signal strength and cell-type specificity, and feedforward signaling by PKA itself stimulates PDE4.

There are three main effectors of cAMP: PKA, the guanine-nucleotide-exchange factor (GEF) EPAC and cyclic-nucleotide-gated ion channels. Protein kinase (PKA), the best-understood target, is a symmetrical complex of two regulatory (R) subunits and two catalytic (C) subunits (there are several isoforms of both subunits). It is activated by the binding of cAMP to two sites on each of the R subunits, which causes their dissociation from the C subunits (Taylor et al. 1992). The catalytic activity of the C subunit is decreased by a protein kinase inhibitor (PKI), which can also act as a chaperone and promote nuclear export of the C subunit, thereby decreasing nuclear functions of PKA. PKA-anchoring proteins (AKAPs) provide specificity in cAMP signal transduction by placing PKA close to specific effectors and substrates. They can also target it to particular subcellular locations and anchor it to ACs (for immediate local activation of PKA) or PDEs (to create local negative-feedback loops for signal termination) (Wong and Scott 2004).

A large number of cytosolic and nuclear proteins have been identified as substrates for PKA (Tasken et al. 1997). PKA phosphorylates numerous metabolic enzymes, including glycogen synthase and phosphorylase kinase, which inhibits glycogen synthesis and promotes glycogen breakdown, respectively, and acetyl CoA carboxylase, which inhibits lipid synthesis. PKA also regulates other signaling pathways. For example, it phosphorylates and thereby inactivates phospholipase C (PLC) β2. In contrast, it activates MAP kinases; in this case, PKA promotes phosphorylation and dissociation of an inhibitory tyrosine phosphatase (PTP). PKA also decreases the activities of Raf and Rho and modulates ion channel permeability. In addition, it regulates the expression and activity of various ACs and PDEs.

Regulation of transcription by PKA is mainly achieved by direct phosphorylation of the transcription factors cAMP-response element-binding protein (CREB), cAMP-responsive modulator (CREM), and ATF1. Phosphorylation is a crucial event because it allows these proteins to interact with the transcriptional coactivators CREB-binding protein (CBP) and p300 when bound to cAMP-response elements (CREs) in target genes (Mayr and Montminy 2001). The *CREM* gene also encodes the powerful repressor ICER, which negatively feeds back on cAMP-induced transcription (Sassone-Corsi 1995). Note, however, that the picture is more complex, because CREB, CREM, and ATF1 can all be phosphorylated by many different kinases, and PKA can also influence the activity of other transcription factors, including some nuclear receptors.

In addition to the negative regulation by signals that inhibit AC or stimulate PDE activity, the action of PKA is counterbalanced by specific protein phosphatases, including PP1 and PP2A. PKA in turn can negatively regulate phosphatase activity by phosphorylating and activating specific PP1 inhibitors, such as I1 and DARPP32. PKA-promoted phosphorylation can also increase the activity of PP2A as part of a negative-feedback mechanism.

Another important effector for cAMP is EPAC, a GEF that promotes activation of certain small GTPases (e.g., Rap1). A major function of Rap1 is to increase cell adhesion via integrin receptors (how this occurs is unclear) (Bos 2003).

Finally, cAMP can bind to and modulate the function of a family of cyclic-nucleotide-gated ion channels. These are relatively nonselective cation channels that conduct calcium. Calcium stimulates CaM and CaM-dependent kinases and, in turn, modulates cAMP production by regulating the activity of ACs and PDEs (Zaccolo and Pozzan 2003). The channels are also permeable to sodium and potassium, which can alter the membrane potential in electrically active cells.

Figure 2 adapted from Fimia and Sassone-Corsi (2001).

REFERENCES

Bos JL. 2003. Epac: A new cAMP target and new avenues in cAMP research. *Nat Rev Mol Cell Biol* **4:** 733–738.

Bruce JI, Straub SV, Yule DI. 2003. Crosstalk between cAMP and Ca^{2+} signaling in non-excitable cells. *Cell Calcium* **34:** 431–444.

Fimia GM, Sassone-Corsi P. 2001. Cyclic AMP signaling. *J Cell Sci* **114:** 1971–1972.

Goraya TA, Cooper DMF. 2005. Ca^{2+}-calmodulin-dependent phosphodiesterase (PDE1): Current perspectives. *Cell Signal* **17:** 789–797.

Mayr B, Montminy M. 2001. Transcriptional regulation by the phosphorylation-dependent factor CREB. *Nat Rev Mol Cell Biol* **2:** 599–609.

McKnight GS. 1991. Cyclic AMP second messenger systems. *Curr Opin Cell Biol* **3:** 213–217.

Pierce KL, Premont RT, Lefkowitz RJ. 2002. Seven-transmembrane receptors. *Nat Rev Mol Cell Biol* **3:** 639–650.

Sassone-Corsi P. 1995. Transcription factors responsive to cAMP. *Annu Rev Cell Dev Biol* **11:** 355–377.

Sutherland EW, Rall TW. 1958. Fractionation and characterization of a cyclic adenine ribonucleotide formed by tissue particles. *J Biol Chem* **232:** 1077–1091.

Tasken K, Skalhegg BS, Tasken KA, Solberg R, Knutsen HK, Levy FO, Sandberg M, Orstavik S, Larsen T, Johansen AK, et al. 1997. Structure, function and regulation of human cAMP-dependent protein kinases. *Adv Second Messenger Phosphoprotein Res* **31:** 191–203.

Taylor SS, Knighton DR, Zheng J, Ten Eyck LF, Sowadski JM. 1992. Structural framework for the protein kinase family. *Annu Rev Cell Biol* **8:** 429–462.

Wong W, Scott JD. 2004. AKAP signaling complexes: Focal points in space and time. *Nat Rev Mol Cell Biol* **5:** 959–970.

Yoshimasa T, Sibley DR, Bouvier M, Lefkowitz RJ, Caron MG. 1987. Cross-talk between cellular signaling pathways suggested by phorbol-ester induced adenylate cyclase phosphorylation. *Nature* **327:** 67–70.

Zaccolo M, Pozzan T. 2003. cAMP and Ca^{2+} interplay: A matter of oscillation patterns. *Trends Neurosci* **26:** 53–55.

Wnt Signaling

Roel Nusse

Howard Hughes Medical Institute, Stanford University Medical Center, Stanford, California 94305-5428

Correspondence: rnusse@stanford.edu

Members of the Wnt family are secreted ligands that regulate numerous developmental pathways (Cadigan and Peifer 2009; Van Amerongen and Nusse 2009; and see Ch. 10 [Perrimon et al. 2013]). Wnt binds to members of the Frizzled family, activating a canonical signaling pathway that targets members of the LEF/TCF transcription factor family (Fig. 1). These control gene expression programs that regulate cell fate and morphogenesis (Van Amerongen and Nusse 2009). Wnt also activates so-called non-canonical pathways (Fig. 2), which regulate planar cell

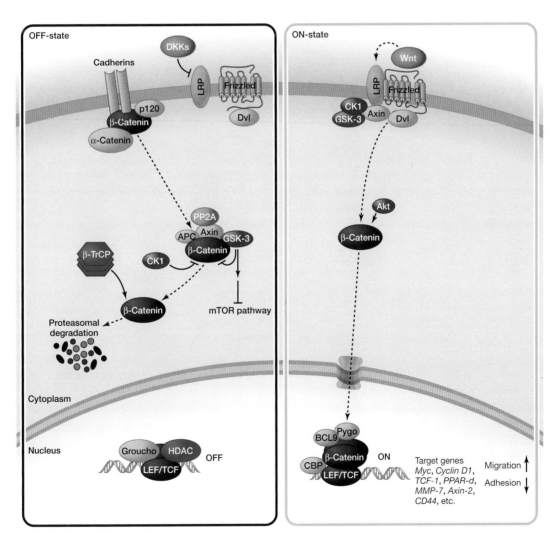

Figure 1. Canonical Wnt signaling.

Cite this chapter as *Cold Spring Harb Perspect Biol* doi: 10.1101/cshperspect.a011163

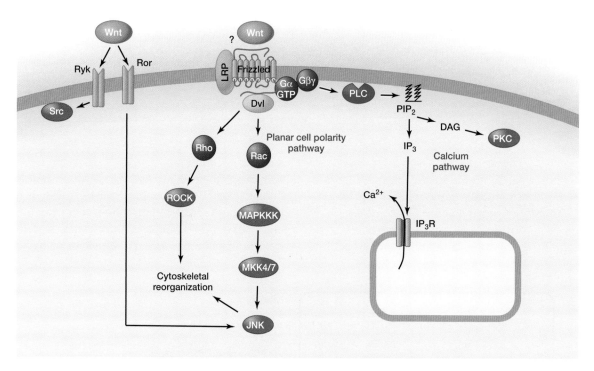

Figure 2. Non-canonical Wnt signaling pathways.

polarity by stimulating cytoskeletal reorganization and can also lead to calcium mobilization (Veeman et al. 2003).

The canonical Wnt signaling pathway is regulated at many levels, including by extensive negative control steps. In cells not exposed to Wnt signals, the major signaling components including the receptors and the β-catenin protein are kept in an off state (Fig. 1, left). Active Wnt signaling rearranges these complexes (Fig. 1, right). In the inactive state, β-catenin levels are kept low through interactions with the protein kinases GSK-3b and CK1, the adenomatous polyposis coli tumor suppressor protein (APC), and the scaffolding protein axin. β-Catenin is degraded, after phosphorylation by GSK-3 and CK1 through the ubiquitin pathway, which involves interactions with β-TrCP. β-Catenin is also regulated by adhesion complexes containing cadherins and α-catenin. At the level of receptors, the negative regulator DKK can bind to the LRP receptor and inhibit Wnt signaling.

During signaling (right), Wnt proteins interact with Frizzled receptors; the transmembrane protein LRP is also required for Wnt signaling. When Wnt proteins bind, the receptors presumably rearrange, leading to the activation of β-catenin. The cytoplasmic tail of LRP binds to axin in a Wnt- and phosphorylation-dependent manner. Phosphorylation of the tail of LRP is regulated by CK1, and Dishevelled (Dvl) and Frizzled also have roles in this process. In a current model, Wnt signaling initially leads to

formation of a complex involving Dvl, axin, and GSK3. The DIX domain in axin is similar to the NH2 terminus in Dvl and promotes interactions between Dvl and axin. As a consequence, GSK does not phosphorylate β-catenin anymore, releasing it from the axin complex and allowing it to accumulate. The stabilized β-catenin then enters the nucleus to interact with TCF/LEF transcription factors. Note that GSK3 also participates in other pathways, such as the mTor and Akt pathway (see p. 87 [Hemmings and Restuccia 2012] and p. 91 [Laplante and Sabatini 2012]).

In the nucleus, in the absence of the Wnt signal, TCF/LEF acts as a repressor of Wnt target genes, in a complex with Groucho. β-Catenin can convert TCF/LEF into a transcriptional activator of the same genes that are repressed by TCF/LEF alone. Three other key players in this complex are BCL9, Pygopos, and CBP. There are many target genes for the canonical Wnt pathway. Most of these genes are cell type specific, with the possible exception of axin 2, which acts as a negative-feedback regulator (Grigoryan 2008).

In non-canonical Wnt signaling, Wnt stimulates the planar cell polarity pathway by activating the small GTPases Rho and Rac. These induce cytoskeletal rearrangements that lead to the development of lateral asymmetry in epithelial sheets and other structures. Wnt can also provoke release of calcium from intracellular stores, probably via heterotrimeric G-proteins. A less-well-understood

Cite this chapter as *Cold Spring Harb Perspect Biol* doi: 10.1101/cshperspect.a011163

mechanism involves activation of the Ror and Ryk tyrosine kinase receptors, which control the activities of the JNK and Src kinases, respectively (van Amerongen et al. 2008).

Figure 1 adapted, with permission, from Cell Signaling Technology (http://www.cellsignal.com).

REFERENCES

*Reference is in this book.

Cadigan KM, Peifer M. 2009. Wnt signaling from development to disease: Insights from model systems. *Cold Spring Harb Perspect Biol* **1:** a002881.

Grigoryan T, Wend P, Klaus A, Birchmeier W. 2008. Deciphering the function of canonical Wnt signals in development and disease: Condi-tional loss- and gain-of-function mutations of β-catenin in mice. *Genes Dev* **22:** 2308–2341.

⋆ Hemmings BA, Restuccia DF. 2012. The P13K-PKB/Akt pathway. *Cold Spring Harb Perspect Biol* **4:** a011189.

⋆ Laplante M, Sabatini DM. 2012. mTOR signaling. *Cold Spring Harb Perspect Biol* **4:** a011593.

⋆ Perrimon N, Pitsouli C, Shilo B-Z. 2012. Signaling mechanisms controlling cell fate and embryonic patterning. *Cold Spring Harb Perspect Biol* **4:** a005975.

van Amerongen R, Nusse R. 2009. Towards an integrated view of Wnt signaling in development. *Development* **136:** 3205–3214.

van Amerongen R, Mikels A, Nusse R. 2008. Alternative Wnt signaling is initiated by distinct receptors. *Sci Signal* **1:** re9.

Veeman MT, Axelrod JD, Moon RT. 2003. A second canon. Functions and mechanisms of β-catenin-independent Wnt signaling. *Dev Cell* **5:** 367–377.

Hedgehog Signaling

Philip W. Ingham

Institute of Molecular and Cell Biology, Singapore 138673

Correspondence: pingham@imcb.a-star.edu.sg

The Hedgehog (Hh) proteins belong to one of a small number of families of secreted signals that play a central role in the development of most metazoans. *hh* itself was originally identified as a mutation that causes a "segment polarity" phenotype in *Drosophila*—as were most of the core components of the Hh signal transduction pathway (Ingham et al. 2011). These components have been highly conserved between flies and vertebrates (Figs. 1 and 2); however, mammals express three related proteins, Sonic hedgehog (Shh), Indian hedgehog, and Desert hedgehog (Dhh), and genetic analysis in mice has uncovered a vertebrate-specific role of the primary cilium in Hh signaling (Goetz et al. 2009).

An unusual feature of Hh proteins is their covalent coupling to cholesterol, which occurs during autocleavage of the proprotein to yield the signaling moiety, HhN (Beachy et al. 1997). HhN is also palmitoylated at its amino terminus by an acyl transferase encoded by the *Drosophila skinny hedgehog* gene (Skn). Secretion of lipidated HhN requires the function of a large multipass transmembrane protein, Dispatched, that is structurally related to the Hh receptor Patched (Ptch1) (Ingham et al. 2011). Binding of HhN to Ptc1 is promoted by two other transmembrane proteins, CDO and BOC (known as IHOG and BOI in *Drosophila*), which act redundantly to bind HhN via one of several fibronectin III (FnIII) motifs in their extracellular domains (Beachy et al. 2010). In its unbound state, Ptc1 localizes to the primary cilium (Rohatgi and Scott 2007), where it acts via a poorly characterized mechanism, thought to involve the transport of one or more lipids, to suppress the activity of the G-protein-coupled receptor (GPCR)-like protein Smoothened (Smo) (Ayers and Thérond 2010).

The principal response of cells to HhN is the activation of target genes by the Gli zinc finger proteins. Gli2 and Gli3 (but not Gli1) are bifunctional transcription factors: their full-length forms function as transcriptional activators, but they can be converted into lower-molecular-weight transcriptional repressors. This is promoted by the phosphorylation of a series of motifs within the carboxy-terminal domain of the protein. These motifs are sequentially phosphorylated by PKA, GSK3, and CKI to generate recognition signals for the F-box protein bTrCP, a component of the SCF complex. This in turn catalyzes the ubiquitylation of the carboxyl terminus, targeting it for degradation by the proteasome to yield the truncated amino-terminal repressor forms, Gli2/3R (Ingham et al. 2011). These bind to Hh target genes to repress their transcription. Gli2/3 proteins appear to shuttle up and down the primary cilium in association with the Cos/Kif7 and SuFu proteins; in the absence of Smo activity, this association seems to promote their processing at the base of the primary cilium. Inactivation of Ptc by HhN results in Smo being transported to the tip of the primary cilium (a process that requires Kif3a and β-arrestin activity), where its activity promotes the dissociation of the Gli2/3-Cos-SuFu complex (Tukachinsky et al. 2010), releasing the full-length highly labile forms

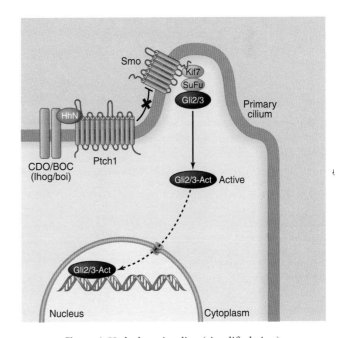

Figure 1. Hedgehog signaling (simplified view).

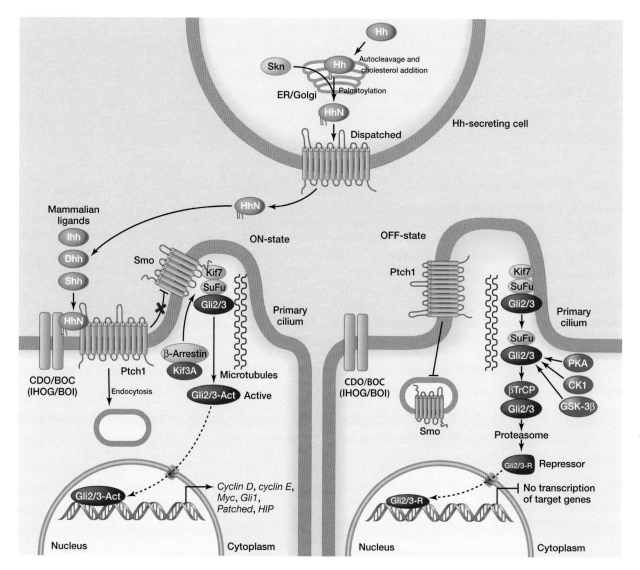

Figure 2. Hedgehog signaling.

of the Gli proteins. These translocate to the nucleus and activate transcription of target genes, such as *Ptch1*, which attenuates the signal, *Gli1*, which amplifies the signal, and the genes encoding the cell cycle regulators, Myc, cyclin D, and E.

Figures 1 and 2 adapted, with permission, from Cell Signaling Technology (http://www.cellsignal.com).

REFERENCES

Ayers KL, Thérond PP. 2010. Evaluating smoothened as a G-protein-coupled receptor for Hedgehog signalling. *Trends Cell Biol* **20:** 287–298.

Beachy PA, Cooper MK, Young KE, von Kessler DP, Park WJ, Hall TM, Leahy DJ, Porter JA. 1997. Multiple roles of cholesterol in hedgehog protein biogenesis and signaling. *Cold Spring Harb Symp Quant Biol* **62:** 191–204.

Beachy PA, Hymowitz SG, Lazarus RA, Leahy DJ, Siebold C. 2010. Interactions between Hedgehog proteins and their binding partners come into view. *Genes Dev* **24:** 2001–2012.

Goetz SC, Ocbina PJ, Anderson KV. 2009. The primary cilium as a Hedgehog signal transduction machine. *Methods Cell Biol* **94:** 199–222.

Ingham PW, Nakano Y, Seger C. 2011. Mechanisms and functions of Hedgehog signalling across the metazoa. *Nat Rev Genet* **12:** 393–406.

Rohatgi R, Scott MP. 2007. Patching the gaps in Hedgehog signalling. *Nat Cell Biol* **9:** 1005–1009.

Tukachinsky H, Lopez LV, Salic A. 2010. A mechanism for vertebrate Hedgehog signaling: Recruitment to cilia and dissociation of SuFu-Gli protein complexes. *J Cell Biol* **191:** 415–428.

Cite this chapter as *Cold Spring Harb Perspect Biol* doi: 10.1101/cshperspect.a011221

Notch Signaling

Raphael Kopan

Department of Developmental Biology, and Department of Medicine (Division of Dermatology), Washington University, St. Louis, Missouri 63110

Correspondence: kopan@wustl.edu

The Notch pathway regulates cell proliferation, cell fate, differentiation, and cell death in all metazoans. Notch itself is a cell-surface receptor that transduces short-range signals by interacting with transmembrane ligands such as Delta (termed Delta-like in humans) and Serrate (termed Jagged in humans) on neighboring cells (Fig. 1). Some soluble ligands have also been identified in *Caenorhabditis elegans*, but these bind to Notch together with transmembrane adaptors (Komatsu et al. 2008). Ligand binding leads to cleavage and release of the Notch intracellular domain

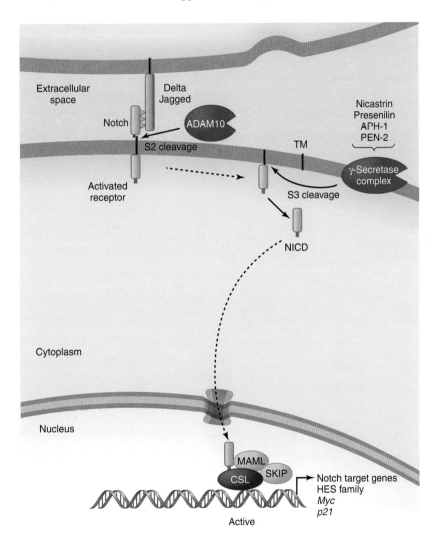

Figure 1. Notch signaling (simplified view).

(NICD), which then travels to the nucleus to regulate transcriptional complexes containing the DNA-binding protein CBF1/RBPjk/Su(H)/Lag1 (CSL).

Following their synthesis, Notch receptors are cleaved by protein convertases during exocytosis at site 1 (S1), which regulates their trafficking and signaling activity (Logeat et al. 1998; Gordon et al. 2009). During passage through the Golgi, they can be glycosylated by glycosyl-transferases such as Fringe, which determines the subsequent response to different subfamilies of ligands. These and other posttranslational modifications of the receptors and ligands tune the amplitude and timing of Notch activity to generate context-specific signals. Several proteins, including E3 ubiquitin ligases (e.g., Deltex and Nedd4), Numb, and α-adaptin, regulate the steady-state levels of the Notch receptor at the cell surface. In signal-sending

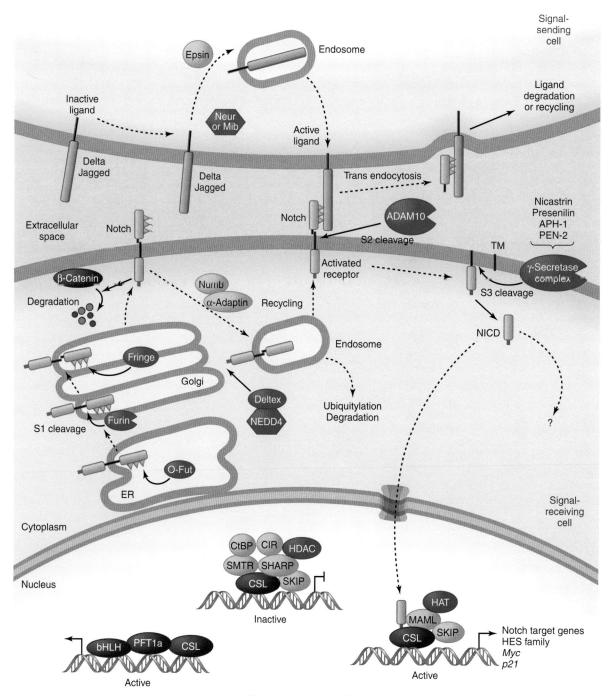

Figure 2. Notch signaling.

Cite this chapter as *Cold Spring Harb Perspect Biol* doi: 10.1101/cshperspect.a011213

cells, E3 ubiquitin ligases (Neur and MIB) similarly ubiquitylate the intracellular domain of the ligand to promote epsin-mediated endocytosis, which is associated with ligand activation (Fig. 2).

Following ligand binding, signaling is initiated when endocytosis of ligand–receptor complexes induces unfolding of a juxtamembrane negative control region (NRR) unique to Notch proteins. Unfolding of the NRR allows access by the protease ADAM10 (also known as KUZ), which removes the Notch extracellular domain by cleaving at site 2 (S2); γ-secretase then cleaves Notch within its transmembrane domain at site 3 (S3) to release various forms of the NICD. Those that have valine or methionine residues at the amino terminus escape the N-end-rule degradation pathway (Tagami et al. 2008) and are stable enough to impact transcription (see below). Interestingly, productive interactions between Notch and its ligands occur when these are present on neighboring cells (i.e., in trans); when receptor and ligand are present on the same cell (i.e., interactions are in cis), activation is inhibited. cis interactions thus determine whether a cell will signal (the ligand is more abundant than Notch) or receive (Notch is more abundant than the ligand) (Sprinzak et al. 2010). Alternatively, in some cases ligand and receptors can be segregated into different subdomains to allow simultaneous transmission and reception of signals (Luty et al. 2007).

Only one nuclear protein is known to mediate the bulk of Notch signals: CSL (Kopan and Ilagan 2009). CSL is a DNA-binding adaptor that interacts with many proteins to build either repressor complexes, which include histone deacetylases (HDACs) that preserve a closed chromatin conformation, or activating complexes, which contain NICD, along with other proteins including histone acetyltransferases (HATs) that open up chromatin. In canonical, CSL-mediated Notch signaling, NICD translocates to the nucleus, binds to CSL, and helps recruit the adaptor protein Mastermind-like (MAML) (Kopan and Ilagan 2009). MAML recruits the HAT p300 and components of the transcription machinery. Thus, every cleaved Notch molecule generates one signaling unit, and tuning the effectiveness of receptor–ligand interaction directly determines the amount of NICD in the nucleus. During the transcriptional activation process, NICD is phosphorylated on a degron within its PEST domain by kinases such as cyclin-dependent kinase 8 (CDK8) and targeted for proteasome-mediated degradation by E3 ubiquitin ligases such as Sel10 (also known as Fbw7). This limits the half-life of a canonical Notch signal and resets the cell for the next pulse of signaling.

In addition to the canonical signals, mounting evidence indicates that CSL-independent activities of Notch also regulate vertebrate (Rangarajan et al. 2001; Demehri et al. 2008) and invertebrate (Ramain et al. 2001) development,

but the biochemical details of this aspect of the pathway are yet to be uncovered. In the absence of ligand, Notch may also be involved in other cellular processes, such as regulating the stability of β-catenin (Sanders et al. 2009), a component of the Wnt signaling pathway (p. 103 [Nusse 2012]).

Under most physiological conditions, unbound Notch receptors simply recycle or are targeted for lysosomal degradation. With one exception (Mukherjee et al. 2011), only pathological or experimental conditions are known to lead to receptor activation without ligand. These include mutations in the NRR domain (Weng et al. 2004), overexpression of Notch with ADAM proteases, and exposure to calcium chelators (Bozkulak and Weinmaster 2009; van Tetering et al. 2009), all of which expose Notch to shedding by ADAM17 (also known as TACE). Notch can also become activated when ESCRT components are mutated (Moberg et al. 2005; Thompson et al. 2005; Vaccari and Bilder 2005), which delays entry into the lysosome and permits ligand-independent activation. The frequent activation of Notch by mutations in T-cell acute lymphoblastic leukemia and its frequent inactivation in head and neck squamous cell carcinomas (Agrawal et al. 2011; Stransky et al. 2011) illustrate the importance of the pathway for control of cell fate and proliferation and the severe consequences of its dysregulation.

Figures adapted with kind permission of Cell Signaling Technology (http://www.cellsignal.com).

REFERENCES

*Reference is in this book.

Agrawal N, Frederick MJ, Pickering CR, Bettegowda C, Chang K, Li RJ, Fakhry C, Xie TX, Zhang J, Wang J, et al. 2011. Exome sequencing of head and neck squamous cell carcinoma reveals inactivating mutations in NOTCH1. Science 333: 1154–1157.

Bozkulak EC, Weinmaster G. 2009. Selective use of ADAM10 and ADAM17 in activation of Notch1 signaling. Mol Cell Biol 29: 5679–5695.

Demehri S, Liu Z, Lee J, Lin MH, Crosby SD, Roberts CJ, Grigsby PW, Miner JH, Farr AG, Kopan R. 2008. Notch-deficient skin induces a lethal systemic B-lymphoproliferative disorder by secreting TSLP, a sentinel for epidermal integrity. PLoS Biol 6: e123.

Gordon WR, Vardar-Ulu D, L'Heureux S, Ashworth T, Malecki MJ, Sanchez-Irizarry C, McArthur DG, Histen G, Mitchell JL, Aster JC, et al. 2009. Effects of S1 cleavage on the structure, surface export, and signaling activity of human Notch1 and Notch2. PLoS ONE 4: e6613.

Komatsu H, Chao MY, Larkins-Ford J, Corkins ME, Somers GA, Tucey T, Dionne HM, White JQ, Wani K, Boxem M, et al. 2008. OSM-11 facilitates LIN-12 Notch signaling during C. elegans vulval development. PLoS Biol 6: e196.

Kopan R, Ilagan MX. 2009. The canonical Notch signaling pathway: Unfolding the activation mechanism. Cell 137: 216–233.

Logeat F, Bessia C, Brou C, Lebail O, Jarriault S, Seidah NG, Israël A. 1998. The Notch1 receptor is cleaved constitutively by a furin-like convertase. Proc Natl Acad Sci 95: 8108–8112.

Luty WH, Rodeberg D, Parness J, Vyas YM. 2007. Antiparallel segregation of Notch components in the immunological synapse directs reciprocal signaling in allogeneic Th:DC conjugates. *J Immunol* **179:** 819–829.

Moberg KH, Schelble S, Burdick SK, Hariharan IK. 2005. Mutations in *erupted*, the *Drosophila* ortholog of mammalian tumor susceptibility gene 101, elicit non-cell-autonomous overgrowth. *Dev Cell* **9:** 699–710.

Mukherjee T, Kim WS, Mandal L, Banerjee U. 2011. Interaction between Notch and Hif-α in development and survival of *Drosophila* blood cells. *Science* **332:** 1210–1213.

* Nusse R. 2012. Wnt signaling. *Cold Spring Harb Perspect Biol* **4:** a011163.

Ramain P, Khechumian K, Seugnet L, Arbogast N, Ackermann C, Heitzler P. 2001. Novel Notch alleles reveal a Deltex-dependent pathway repressing neural fate. *Curr Biol* **11:** 1729–1738.

Rangarajan A, Talora C, Okuyama R, Nicolas M, Mammucari C, Oh H, Aster JC, Krishna S, Metzger D, Chambon P, et al. 2001. Notch signaling is a direct determinant of keratinocyte growth arrest and entry into differentiation. *EMBO J* **20:** 3427–3436.

Sanders PG, Munoz-Descalzo S, Balayo T, Wirtz-Peitz F, Hayward P, Arias AM. 2009. Ligand-independent traffic of Notch buffers activated armadillo in *Drosophila*. *PLoS Biol* **7:** e1000169.

Sprinzak D, Lakhanpal A, LeBon L, Santat LA, Fontes ME, Anderson GA, Garcia-Ojalvo J, Elowitz MB. 2010. *Cis*-interactions between Notch and Delta generate mutually exclusive signaling states. *Nature* **465:** 86–90.

Stransky N, Egloff AM, Tward AD, Kostic AD, Cibulskis K, Sivachenko A, Kryukov GV, Lawrence MS, Sougnez C, McKenna A, et al. 2011. The mutational landscape of head and neck squamous cell carcinoma. *Science* **333:** 1157–1160.

Tagami S, Okochi M, Yanagida K, Ikuta A, Fukumori A, Matsumoto N, Ishizuka-Katsura Y, Nakayama T, Itoh N, Jiang J, et al. 2008. Regulation of Notch signaling by dynamic changes in the precision in S3 cleavage of Notch-1. *Mol Cell Biol* **28:** 165–176.

Thompson BJ, Mathieu J, Sung HH, Loeser E, Rorth P, Cohen SM. 2005. Tumor suppressor properties of the ESCRT-II complex component Vps25 in *Drosophila*. *Dev Cell* **9:** 711–720.

Vaccari T, Bilder D. 2005. The *Drosophila* tumor suppressor *vps25* prevents nonautonomous overproliferation by regulating Notch trafficking. *Dev Cell* **9:** 687–698.

van Tetering G, van Diest P, Verlaan I, van der Wall E, Kopan R, Vooijs M. 2009. Metalloprotease ADAM10 is required for Notch1 site 2 cleavage. *J Biol Chem* **284:** 31018–31027.

Weng AP, Ferrando AA, Lee W, Morris JPIV, Silverman LB, Sanchez-Irizarry C, Blacklow SC, Look AT, Aster JC. 2004. Activating mutations of *NOTCH1* in human T cell acute lymphoblastic leukemia. *Science* **306:** 269–271.

Signaling by the TGFβ Superfamily

Jeffrey L. Wrana

Samuel Lunenfeld Research Institute, Mount Sinai Hospital, Toronto, Ontario M5G 1X5, Canada

Correspondence: wrana@lunenfeld.ca

The transforming growth factor β (TGFβ) superfamily was discovered in a hunt for autocrine factors secreted from cancer cells that promote transformation (Roberts et al. 1981). However, it soon became clear that TGFβ and the related bone morphogenetic proteins (BMPs) regulate diverse developmental and homeostatic processes and are mutated in numerous human diseases. Furthermore, TGFβ-superfamily members such as activins, Nodal, and growth differentiation factors (GDFs) were shown to control cell fate as a function of concentration, thus defining them as a key class of secreted morphogens (Green and Smith 1990).

TGFβ-superfamily members are highly conserved across animals and comprise the largest family of secreted morphogens. Ligands are produced by cleavage of a prodomain that releases active disulfide-linked homo- or heterodimers. TGFβ (plus activin, Nodal, and GDFs) and BMPs both signal through transmembrane serine/threonine kinase receptors. These are typically presented as using distinct downstream pathways (see Figs. 1 and 2), although some cross talk in certain cell types occurs. In each pathway there are two kinds of receptors: type I and type II (Massague 1998). Ligand binding induces formation of heterotetramers containing two type II and two type I receptors, which allows the constitutively active type II receptor to phosphorylate a glycine-serine (GS)-rich region in the type I receptor. This initiates signaling through the Smad pathway (Fig. 1), with the phosphorylated GS region providing a docking site for receptor-regulated Smad proteins (R-Smads). In TGFβ signaling, this is promoted by Smad anchor for receptor activation (SARA) in the endosomal compartment (Attisano and Wrana 2002).

Despite the large TGFβ superfamily (>30 members in humans), the receptor repertoire is limited: only five type II and seven type I receptors are encoded in mammalian genomes, with the type I receptors funneling signaling into one of two distinct R-Smad pathways: the TGFβ-Smad pathway (R-Smad2/3) or the BMP-Smad pathway (R-Smad1/5/8) (Massague 2012). Docking of R-Smads allows

phosphorylation of the last two serines in their carboxyl termini by the type I kinase, which induces dissociation from the receptor, binding to the common Smad, Smad4, and nuclear accumulation of the Smad complex. In the nucleus, most Smads regulate transcriptional responses by direct binding of their MH1 domain to DNA (Smad2 only binds to DNA indirectly) in cooperation with various DNA-binding partners that include sequence-specific transcription factors. These Smad DNA-binding partners typically bind to the R-Smad, thus maintaining specificity of the transcriptional response. Smads also interact with transcriptional coactivators or corepressors that modulate the transcriptional output. Interestingly, the remaining two Smads, Smad6 and Smad7, are transcriptional targets of R-Smads and act in a negative-feedback loop to inhibit signaling by interacting with the receptors and recruiting Smurf and related ubiquitin ligases of the Nedd4 family to induce receptor degradation. SnoN and the related Ski, as well as Arkadia, a ring-finger ubiquitin ligase (also known as RNF111), are important examples of negative and positive regulators of Smad2/3-dependent transcription, respectively (Stroschein et al. 1999; Niederlander et al. 2001). SnoN and Ski negatively regulate the pathway as corepressors of Smads, whereas Arkadia promotes signaling by degrading negative regulators of the pathway, such as Smad7 and SnoN/Ski.

The Smad pathways are the major mediators of transcriptional responses induced by the TGFβ family, which control cell-fate determination, cell-cycle arrest, apoptosis, and actin rearrangements. However, receptor signaling is not restricted to R-Smad activation (Fig. 2). The type II receptor phosphorylates Par6 polarity proteins bound to the type I receptor to dissolve tight junctions in epithelial cells (Ozdamar et al. 2005) and specify axons (Yi et al. 2010). Furthermore, protein phosphatase 2A (PP2A) is regulated by the receptors (Griswold-Prenner et al. 1998), signaling via SHC-Grb2 has been reported (Lee et al. 2007), and interactions between the kinase PAK and the TGFβ receptor link it to cytoskeletal and focal adhesion dynamics

Copyright © 2014 Cold Spring Harbor Laboratory Press; all rights reserved
Cite this chapter as *Cold Spring Harb Perspect Biol* doi: 10.1101/cshperspect.a011197

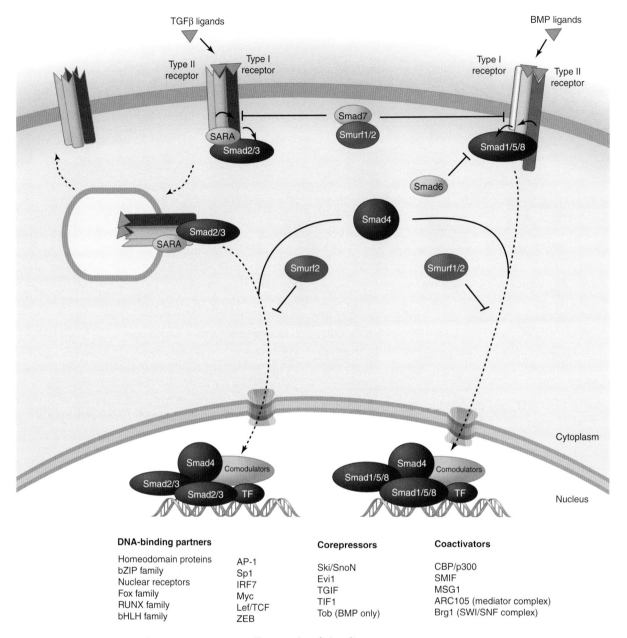

Figure 1. Smad signaling.

<table>
<thead>
<tr><th>DNA-binding partners</th><th></th><th>Corepressors</th><th>Coactivators</th></tr>
</thead>
<tbody>
<tr><td>Homeodomain proteins</td><td>AP-1</td><td>Ski/SnoN</td><td>CBP/p300</td></tr>
<tr><td>bZIP family</td><td>Sp1</td><td>Evi1</td><td>SMIF</td></tr>
<tr><td>Nuclear receptors</td><td>IRF7</td><td>TGIF</td><td>MSG1</td></tr>
<tr><td>Fox family</td><td>Myc</td><td>TIF1</td><td>ARC105 (mediator complex)</td></tr>
<tr><td>RUNX family</td><td>Lef/TCF</td><td>Tob (BMP only)</td><td>Brg1 (SWI/SNF complex)</td></tr>
<tr><td>bHLH family</td><td>ZEB</td><td></td><td></td></tr>
</tbody>
</table>

(Wilkes et al. 2009). BMPRII is of particular interest because it has a unique carboxy-terminal tail that serves as a docking site for binding of the kinases LIMK and JNK, thus linking BMP signaling to the actin cytoskeleton and the microtubule network, respectively (Miyazono et al. 2010; Podkowa et al. 2010). These pathways are critical in neuronal dendritogenesis and may be important in familial pulmonary hypertension, in which mutations in BMPRII are a major cause of disease (Morrell 2011). Numerous kinase cascades, including the ERK, JNK, and p38 MAPK pathways, are also regulated by TGFβ signaling (Mu et al. 2012), with signaling to TRAF6 being one important mechanism

of MAPK regulation. These non-Smad pathways combine with the gene-expression programs controlled by Smads to yield an integrated response to TGFβ signals.

Finally, TGFβ signaling is embedded in a higher-order network of interactions with other signaling pathways. For example, Smads interact with the Wnt pathway (via Lef/TCF transcription factors and β-catenin), the Hedgehog pathway (via Gli transcriptional regulators), and the Hippo pathway (via TAZ and YAP). This provides an integrated signaling network that allows contextual interpretation of morphogen signals in diverse biological settings (Attisano and Wrana 2013).

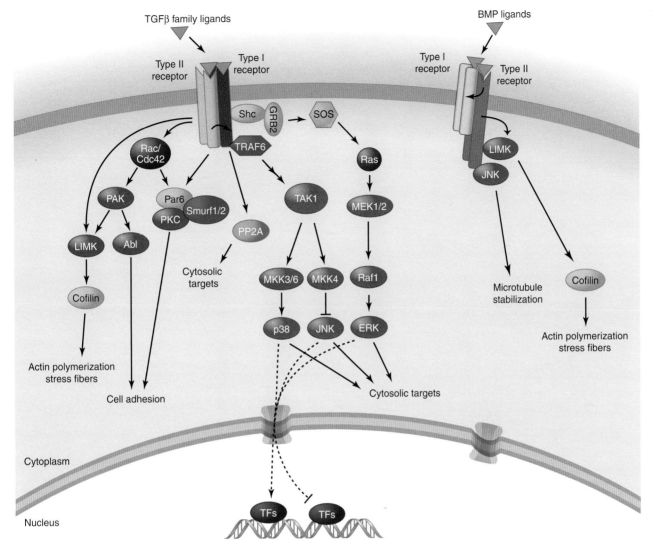

Figure 2. Non-Smad signals.

REFERENCES

Attisano L, Wrana JL. 2002. Signal transduction by the TGF-β superfamily. *Science* **296:** 1646–1647.

Attisano L, Wrana JL. 2013. Signal integration in TGF-β, WNT and Hippo pathways. *F1000Prime Rep* **5:** 17.

Green JB, Smith JC. 1990. Graded changes in dose of a *Xenopus* activin A homologue elicit stepwise transitions in embryonic cell fate. *Nature* **347:** 391–394.

Griswold-Prenner I, Kamibayashi C, Maruoka EM, Mumby MC, Derynck R. 1998. Physical and functional interactions between type I transforming growth factor β receptors and Bα, a WD-40 repeat subunit of phosphatase 2A. *Mol Cell Biol* **18:** 6595–6604.

Lee MK, Pardoux C, Hall MC, Lee PS, Warburton D, Qing J, Smith SM, Derynck R. 2007. TGF-β activates Erk MAP kinase signalling through direct phosphorylation of ShcA. *EMBO J* **26:** 3957–3967.

Massague J. 1998. TGF-β signal transduction. *Annu Rev Biochem* **67:** 753–791.

Massague J. 2012. TGFβ signalling in context. *Nat Rev Mol Cell Biol* **13:** 616–630.

Miyazono K, Kamiya Y, Morikawa M. 2010. Bone morphogenetic protein receptors and signal transduction. *J Biochem* **147:** 35–51.

Morrell NW. 2011. Role of bone morphogenetic protein receptors in the development of pulmonary arterial hypertension. *Adv Exp Med Biol* **661:** 251–264.

Mu Y, Gudey SK, Landström M. 2012. Non-Smad signaling pathways. *Cell Tissue Res* **347:** 11–20.

Niederlander C, Walsh JJ, Episkopou V, Jones CM. 2001. Arkadia enhances nodal-related signalling to induce mesendoderm. *Nature* **410:** 830–834.

Ozdamar B, Bose R, Barrios-Rodiles M, Wang HR, Zhang Y, Wrana JL. 2005. Regulation of the polarity protein Par6 by TGF-β receptors controls epithelial cell plasticity. *Science* **307:** 1603–1609.

Podkowa M, Zhao X, Chow CW, Coffey ET, Davis RJ, Attisano L. 2010. Microtubule stabilization by bone morphogenetic protein receptor-mediated scaffolding of c-Jun N-terminal kinase promotes dendrite formation. *Mol Cell Biol* **30:** 2241–2250.

Roberts AB, Anzano MA, Lamb LC, Smith JM, Sporn MB. 1981. New class of transforming growth factors potentiated by epidermal growth factor: Isolation from non-neoplastic tissues. *Proc Natl Acad Sci* **78:** 5339–5343.

Stroschein SL, Wang W, Zhou S, Zhou Q, Luo K. 1999. Negative feedback regulation of TGF-β signaling by the SnoN oncoprotein. *Science* **286:** 771–774.

Wilkes MC, Repellin CE, Hong M, Bracamonte M, Penheiter SG, Borg JP, Leof EB. 2009. Erbin and the NF2 tumor suppressor Merlin cooper-atively regulate cell-type-specific activation of PAK2 by TGF-β. *Dev Cell* **16:** 433–444.

Yi JJ, Barnes AP, Hand R, Polleux F, Ehlers MD. 2010. TGF-β signaling specifies axons during brain development. *Cell* **142:** 144–157.

The JAK/STAT Pathway

Douglas A. Harrison

Department of Biology, University of Kentucky, Lexington, Kentucky 40506

Correspondence: DougH@uky.edu

Cellular responses to dozens of cytokines and growth factors are mediated by the evolutionarily conserved Janus kinase/signal transducers and activators of transcription (JAK/STAT) signaling pathway (Fig. 1). These responses include proliferation, differentiation, migration, apoptosis, and cell survival, depending on the signal, tissue, and cellular context. JAK/STAT signaling is essential for numerous developmental and homeostatic processes, including hematopoiesis, immune cell development, stem cell maintenance, organismal growth, and mammary gland development (Ghoreschi et al. 2009).

Janus kinases (JAKs) were identified through sequence comparisons as a unique class of tyrosine kinases that contain both a catalytic domain and a second kinase-like

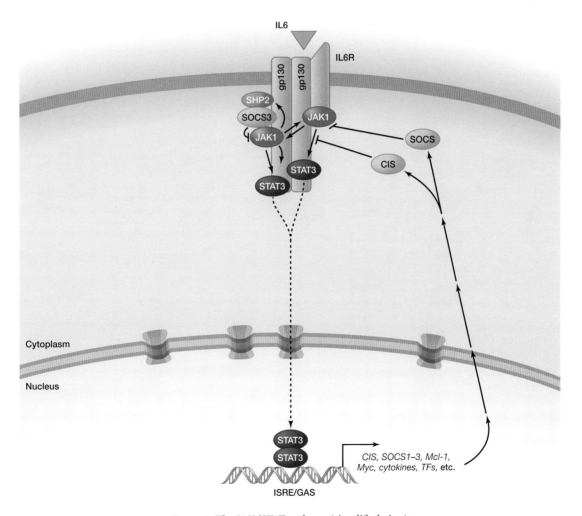

Figure 1. The JAK/STAT pathway (simplified view).

Cite this chapter as *Cold Spring Harb Perspect Biol* doi: 10.1101/cshperspect.a011205

domain that serves an autoregulatory function, hence the homage to the two-faced Roman god. They were functionally linked to STATs and interferon signaling in powerful somatic cell genetic screens (Darnell et al. 1994; Schindler and Plumlee 2008). The JAK/STAT cascade is among the simplest of the conserved metazoan signaling pathways. The binding of extracellular ligand leads to pathway activation via changes to the receptors that permit the intracellular JAKs associated with them to phosphorylate one another. Trans-phosphorylated JAKs then phosphorylate downstream substrates, including both the receptor and the STATs. Activated STATs enter the nucleus and bind as dimers or as more complex oligomers to specific enhancer sequences in target genes, thus regulating their transcription (Fig. 2).

In mammals, there are four members of the JAK family and seven STATs. Different JAKs and STATs are recruited based on their tissue specificity and the receptors engaged in the signaling event (Schindler and Plumlee 2008). In

invertebrates, the *Drosophila* JAK/STAT pathway has been extensively studied and comprises only one JAK and one STAT (Arbouzova and Zeidler 2006). Although the canonical JAK/STAT pathway is simple and direct, pathway components regulate or are regulated by members of other signaling pathways, including those involving the ERK MAP kinase, PI 3-kinase (PI3K), and others. Furthermore, non-canonical JAK and STAT activities influence the global transcriptional state through modification of chromatin structure (Li 2008; Dawson et al. 2009).

Human JAK mutations cause numerous diseases, including severe combined immune deficiency, hyperIgE syndrome, certain leukemias, polycythemia vera, and other myeloproliferative disorders (Jatiani et al. 2010). Because of the causative role in these diseases and their central significance in immune response, JAKs have become attractive targets for development of therapeutics for a variety of hematopoietic and immune system disorders (Pesu et al. 2008; Haan et al.

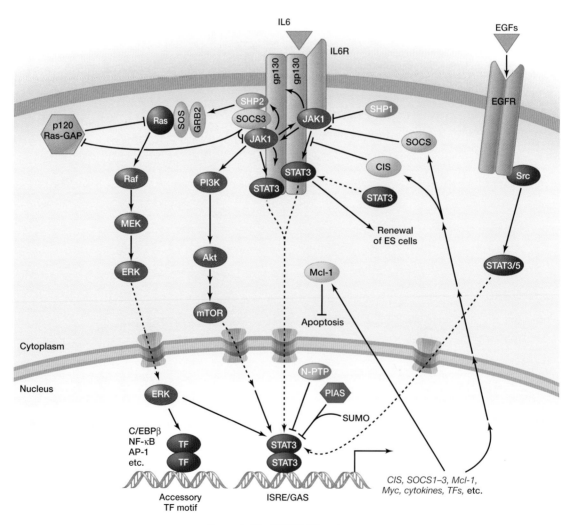

Figure 2. The JAK/STAT pathway.

2010). Owing to the pleiotropy of the JAK/STAT pathway, agents that selectively perturb specific family members are being sought.

Figures adapted by kind permission of Cell Signaling Technology (http://cellsignal.com).

REFERENCES

Arbouzova NI, Zeidler MP. 2006. JAK/STAT signalling in *Drosophila*: Insights into conserved regulatory and cellular functions. *Development* **133:** 2605–2616.

Darnell JE Jr, Kerr IM, Stark GR. 1994. Jak-STAT pathways and transcriptional activation in response to IFNs and other extracellular signaling proteins. *Science* **264:** 1415–1421.

Dawson MA, Bannister AJ, Gottgens B, Foster SD, Bartke T, Green AR, Kouzarides T. 2009. JAK2 phosphorylates histone H3Y41 and excludes HP1a from chromatin. *Nature* **461:** 819–822.

Ghoreschi K, Laurence A, O'Shea JJ. 2009. Janus kinases in immune cell signaling. *Immunol Rev* **228:** 273–287.

Haan C, Behrmann I, Haan S. 2010. Perspectives for the use of structural information and chemical genetics to develop inhibitors of Janus kinases. *J Cell Mol Med* **14:** 504–527.

Jatiani SS, Baker SJ, Silverman LR, Reddy EP. 2010. JAK/STAT pathways in cytokine signaling and myeloproliferative disorders: Approaches for targeted therapies. *Genes Cancer* **1:** 979–993.

Li WX. 2008. Canonical and non-canonical JAK-STAT signaling. *Trends Cell Biol* **18:** 545–551.

Pesu M, Laurence A, Kishore N, Zwillich SH, Chan G, O'Shea JJ. 2008. Therapeutic targeting of Janus kinases. *Immunol Rev* **223:** 132–142.

Schindler C, Plumlee C. 2008. Inteferons pen the JAK-STAT pathway. *Semin Cell Dev Biol* **19:** 311–318.

Toll-Like Receptor Signaling

Kian-Huat Lim and Louis M. Staudt

Metabolism Branch, Center for Cancer Research, National Cancer Institute, National Institutes of Health, Bethesda, Maryland 20892

Correspondence: lstaudt@mail.nih.gov

Toll-like receptors (TLRs) are protective immune sentries that sense pathogen-associated molecular patterns (PAMPs) such as unmethylated double-stranded DNA (CpG), single-stranded RNA (ssRNA), lipoproteins, lipopolysaccharide (LPS), and flagellin. In innate immune myeloid cells, TLRs induce the secretion of inflammatory cytokines (Ch. 15 [Newton and Dixit 2012]), thereby engaging lymphocytes to mount an adaptive, antigen-specific

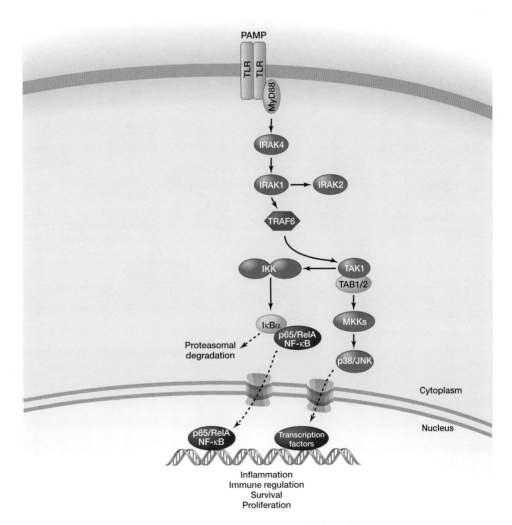

Figure 1. TLR signaling (simplified view).

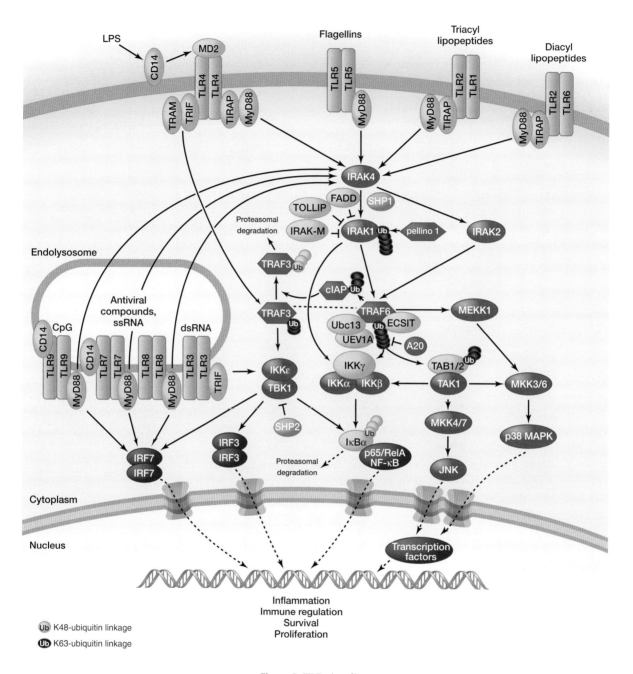

Figure 2. TLR signaling.

immune response (see Fig. 1) that ultimately eradicates the invading microbes (Kawai and Akira 2010).

Identification of TLR innate immune function began with the discovery that *Drosophila* mutants in the *Toll* gene are highly susceptible to fungal infection (Lemaitre et al. 1996). This was soon followed by identification of a human *Toll* homolog, now known as TLR4 (Medzhitov et al. 1997). To date, 10 TLR family members have been identified in humans, and at least 13 are present in mice. All TLRs

consist of an amino-terminal domain, characterized by multiple leucine-rich repeats, and a carboxy-terminal TIR domain that interacts with TIR-containing adaptors. Nucleic acid–sensing TLRs (TLR3, TLR7, TLR8, and TLR9) are localized within endosomal compartments, whereas the other TLRs reside at the plasma membrane (Blasius and Beutler 2010; McGettrick and O'Neill 2010). Trafficking of most TLRs from the endoplasmic reticulum (ER) to either the plasma membrane or endolysosomes is orchestrated by

Cite this chapter as *Cold Spring Harb Perspect Biol* doi: 10.1101/cshperspect.a011247

ER-resident proteins such as UNC93B (for TLR3, TLR7, TLR8, and TLR9) and PRAT4A (for TLR1, TLR2, TLR4, TLR7, and TLR9) (Blasius and Beutler 2010). Once in the endolysosomes, TLR3, TLR7, and TLR9 are subject to stepwise proteolytic cleavage, which is required for ligand binding and signaling (Barton and Kagan 2009). For some TLRs, ligand binding is facilitated by coreceptors, including CD14 and MD2.

Following ligand engagement, the cytoplasmic TIR domains of the TLRs recruit the signaling adaptors MyD88, TIRAP, TRAM, and/or TRIF (see Fig. 2). Depending on the nature of the adaptor that is used, various kinases (IRAK4, IRAK1, IRAK2, TBK1, and IKKε) and ubiquitin ligases (TRAF6 and pellino 1) are recruited and activated, culminating in the engagement of the NF-κB, type I interferon, p38 MAP kinase (MAPK), and JNK MAPK pathways (Kawai and Akira 2010; p. 81 [Morrison 2012]). TRAF6 is modified by K63-linked autoubiquitylation, which enables the recruitment of IκB kinase (IKK) through a ubiquitin-binding domain of the IKKγ (also known as NEMO) subunit. In addition, a ubiquitin-binding domain of TAB2 recognizes ubiquitylated TRAF6, causing activation of the associated TAK1 kinase, which then phosphorylates the IKKβ subunit. Pellino 1 can modify IRAK1 with K63-linked ubiquitin, allowing IRAK1 to recruit IKK directly. TLR4 signaling via the TRIF adaptor protein leads to K63-linked polyubiquitylation of TRAF3, thereby promoting the type I interferon response via interferon regulatory factor (IRFs) (Hacker et al. 2011). Alternatively, TLR4 signaling via MyD88 leads to the activation of TRAF6, which modifies cIAP1 or cIAP2 with K63-linked polyubiquitin (Hacker et al. 2011). The cIAPs are thereby activated to modify TRAF3 with K48-linked polyubiquitin, causing its proteasomal degradation. This allows a TRAF6–TAK1 complex to activate the p38 MAPK pathway and promote inflammatory cytokine production (Hacker et al. 2011). TLR signaling is turned off by various negative regulators: IRAK-M and MyD88 short (MyD88s), which antagonize IRAK1 activation; FADD, which antagonizes MyD88 or IRAKs; SHP1 and SHP2, which dephosphorylate IRAK1 and TBK1, respectively; and A20, which deubiquitylates TRAF6 and IKK (Flannery and Bowie 2010; Kawai and Akira 2010).

Deregulation of the TLR signaling cascade causes several human diseases. Patients with inherited deficiencies of MyD88, IRAK4, UNC93B1, or TLR3 are susceptible to recurrent bacterial or viral infections (Casanova et al. 2011). Chronic TLR7 and/or TLR9 activation in autoreactive B cells, in contrast, underlies systemic autoimmune diseases (Green and Marshak-Rothstein 2011). Furthermore, oncogenic activating mutations of MyD88 occur frequently in the activated B-cell-like subtype of diffuse large B-cell lymphoma and in other B-cell malignancies (Ngo et al. 2011). Inhibitors of various TLRs or their associated kinases are currently being developed for autoimmune or inflammatory diseases and also hold promise for the treatment of B-cell malignancies with oncogenic MyD88 mutations. Many TLR7 and TLR9 agonists are currently in clinical trials as adjuvants to boost host antitumor responses in cancer patients (Hennessy et al. 2010).

Figures adapted with kind permission of Cell Signaling Technology (http://www.cellsignal.com).

REFERENCES

*Reference is in this book.

Barton GM, Kagan JC. 2009. A cell biological view of Toll-like receptor function: Regulation through compartmentalization. *Nat Rev Immunol* **9:** 535–542.

Blasius AL, Beutler B. 2010. Intracellular Toll-like receptors. *Immunity* **32:** 305–315.

Casanova JL, Abel L, Quintana-Murci L. 2011. Human TLRs and IL-1Rs in host defense: Natural insights from evolutionary, epidemiological, and clinical genetics. *Annu Rev Immunol* **29:** 447–491.

Flannery S, Bowie AG. 2010. The interleukin-1 receptor-associated kinases: Critical regulators of innate immune signalling. *Biochem Pharmacol* **80:** 1981–1991.

Green NM, Marshak-Rothstein A. 2011. Toll-like receptor driven B cell activation in the induction of systemic autoimmunity. *Semin Immunol* **23:** 106–112.

Hacker H, Tseng PH, Karin M. 2011. Expanding TRAF function: TRAF3 as a tri-faced immune regulator. *Nat Rev Immunol* **11:** 457–468.

Hennessy EJ, Parker AE, O'Neill LA. 2010. Targeting Toll-like receptors: Emerging therapeutics? *Nat Rev Drug Discov* **9:** 293–307.

Kawai T, Akira S. 2010. The role of pattern-recognition receptors in innate immunity: Update on Toll-like receptors. *Nat Immunol* **11:** 373–384.

Lemaitre B, Nicolas E, Michaut L, Reichhart JM, Hoffmann JA. 1996. The dorsoventral regulatory gene cassette spatzle/Toll/cactus controls the potent antifungal response in *Drosophila* adults. *Cell* **86:** 973–983.

McGettrick AF, O'Neill LA. 2010. Localisation and trafficking of Toll-like receptors: An important mode of regulation. *Curr Opin Immunol* **22:** 20–27.

Medzhitov R, Preston-Hurlburt P, Janeway CA Jr. 1997. A human homologue of the *Drosophila* Toll protein signals activation of adaptive immunity. *Nature* **388:** 394–397.

* Morrison DK. 2012. MAP kinase pathways. *Cold Spring Harb Perspect Biol* **4:** a011254.

* Newton K, Dixit VM. 2012. Signaling in innate immunity and inflammation. *Cold Spring Harb Perspect Biol* **4:** a006049.

Ngo VN, Young RM, Schmitz R, Jhavar S, Xiao W, Lim KH, Kohlhammer H, Xu W, Yang Y, Zhao H, et al. 2011. Oncogenically active MYD88 mutations in human lymphoma. *Nature* **470:** 115–119.

Immunoreceptor Signaling

Lawrence E. Samelson

Center for Cancer Research, National Cancer Institute, National Institutes of Health, Bethesda, Maryland 20892-4256

Correspondence: lsamelson@comcast.net

T cells and B cells are stimulated when antigens bind to T-cell receptors (TCRs) and B-cell receptors (BCRs) in their respective plasma membranes (Fig. 1). The antigens presented to T cells are in the form of short peptides that have been processed in infected cells and are displayed on the surface bound to major histocompatibility complex (MHC) class I or class II molecules. BCRs, in contrast, recognize free antigens in their native form present in the extracellular milieu or on cell surfaces.

Despite superficial differences, the responses in the two cell types are remarkably similar. For both, signaling begins with engagement of a complex receptor composed of antigen receptor subunits with immunoglobulin or immunoglobulin-like domains. Both receptors also contain non-polymorphic subunits, the TCRζ chain and CD3 subunits (γ, δ, and ε) for the TCR, and the α and β chains of the BCR (Reth 1995; Wucherpfennig et al. 2010). Coreceptors are important in both T cells (CD4 for helper T cells; CD8 for cytotoxic T cells) and B cells (CD19). For both receptors, members of two protein tyrosine kinase (PTK) families, Src (Lck for T cells and Lyn for B cells) and Syk (Zap-70 for T cells and Syk for B cells) (Bradshaw 2010; Liu et al. 2010; Wang et al. 2010), are critical signaling molecules closely associated with the receptors and cor-

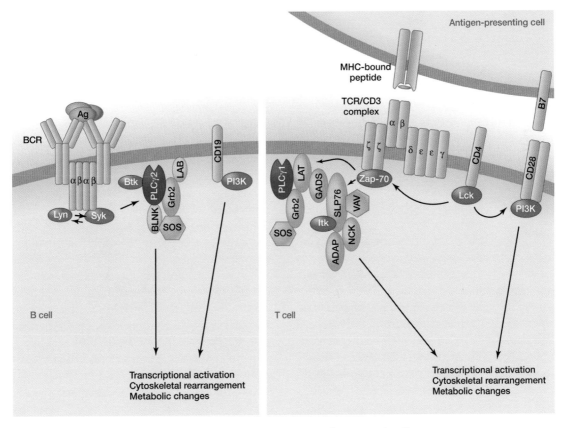

Figure 1. Early events in T cell and B cell receptor signaling.

Cite this chapter as *Cold Spring Harb Perspect Biol* doi: 10.1101/cshperspect.a011510

eceptors. Activation of tyrosine kinases is the first biochemical change that follows receptor engagement.

Following antigen engagement, activated tyrosine kinases phosphorylate several adapter proteins and signaling enzymes, which then interact. For T cells, this is exemplified by the phosphorylation of the adapters LAT and SLP-76 (Fig. 2), the formation of multiprotein complexes nucleated at these adapters, and the inclusion of such enzymes as phospholipase Cγ1 (PLCγ1) and VAV in the protein assemblies (Balagopalan et al. 2010; Jordan and

Koretzky 2010). Activation of an additional protein tyrosine kinase, Itk in T cells and Btk in B cells, occurs in these complexes (Andreotti et al. 2010). Similarly in B-cells, phosphorylation of BLNK and CD19 leads to recruitment of adapters and signaling enzymes. In both cell types these events occur at the plasma membrane, where critical lipid substrates of enzymes such as PLCγ1 and PI3 kinase (PI3K) are located.

Downstream from these initial events, a number of other protein complexes are formed, and various signaling

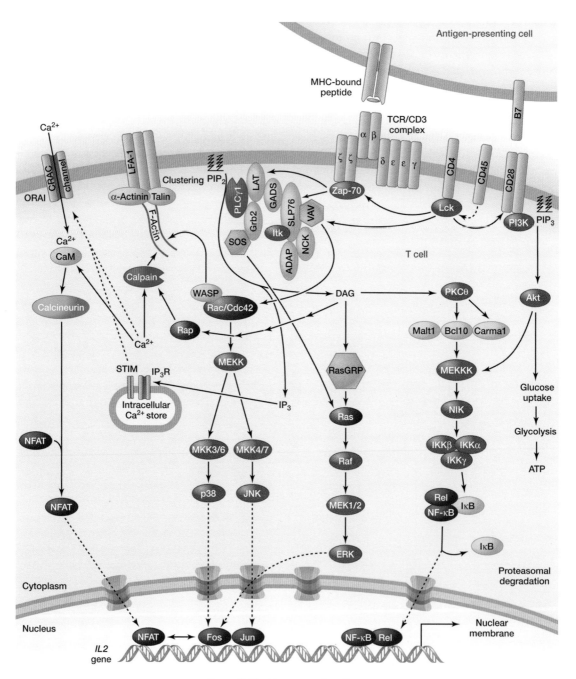

Figure 2. T cell receptor signaling.

Cite this chapter as *Cold Spring Harb Perspect Biol* doi: 10.1101/cshperspect.a011510

pathways are activated. Many of these involve activation of protein serine kinases (Finlay and Cantrell 2010). In T cells and B cells, activation of protein kinase C (PKC) leads to formation and activation of the CARMA1–Bcl10–MALT1 complex (Thome et al. 2010). Stimulation of PI3K leads to activation of the kinase Akt (Huang and Sauer 2010). Downstream from the TCR and BCR, one also observes activation of small G proteins such as Ras, which, in turn, leads to activation of the Raf/MEK and ERK kinases. Finally, in both cell types, these various events lead to cytoskeletal changes and gene expression induced by transcriptional activators such as NF-AT and NF-κB. T cells and B cells thus activated can proliferate, differentiate, and synthesize the cytokines and effector molecules that allow them to fulfill their roles in the adaptive immune response.

Figures adapted with kind permission of Cell Signaling Technology (http://cellsignal.com).

REFERENCES

Andreotti AH, Schwartzberg PL, Joseph RE, Berg LJ. 2010. T-cell signaling regulated by the Tec family kinase, Itk. *Cold Spring Harb Perspect Biol* **2**: a002287.

Balagopalan L, Coussens NP, Sherman E, Samelson LE, Sommers CL. 2010. The LAT story: A tale of cooperativity, coordination, and choreography. *Cold Spring Harb Perspect Biol* **2**: a005512.

Bradshaw JM. 2010. The Src, Syk, and Tec family kinases: Distinct types of molecular switches. *Cell Signal* **22**: 1175–1184.

Finlay D, Cantrell D. 2010. The coordination of T-cell function by serine/threonine kinases. *Cold Spring Harb Perspect Biol* **3**: a002261.

Huang YH, Sauer K. 2010. Lipid signaling in T-cell development and function. *Cold Spring Harb Perspect Biol* **2**: a002428.

Jordan MS, Koretzky GA. 2010. Coordination of receptor signaling in multiple hematopoietic cell lineages by the adaptor protein SLP-76. *Cold Spring Harb Perspect Biol* **2**: a002501.

Liu W, Sohn HW, Tolar P, Pierce SK. 2010. It's all about change: The antigen-driven initiation of B-cell receptor signaling. *Cold Spring Harb Perspect Biol* **2**: a002295.

Reth M. 1995. The B-cell antigen receptor complex and co-receptors. *Immunol Today* **16**: 310–313.

Thome M, Charton JE, Pelzer C, Hailfinger S. 2010. Antigen receptor signaling to NF-κB via CARMA1, BCL10, and MALT1. *Cold Spring Harb Perspect Biol* **2**: a003044.

Wang H, Kadlecek TA, Au-Yeung BB, Goodfellow HE, Hsu LY, Freedman TS, Weiss A. 2010. ZAP-70: An essential kinase in T-cell signaling. *Cold Spring Harb Perspect Biol* **2**: a002279.

Wucherpfennig KW, Gagnon E, Call MJ, Huseby ES, Call ME. 2010. Structural biology of the T-cell receptor: Insights into receptor assembly, ligand recognition, and initiation of signaling. *Cold Spring Harb Perspect Biol* **2**: a005140.

Signaling by Nuclear Receptors

Richard Sever[1] and Christopher K. Glass[2]

[1]Cold Spring Harbor Laboratory, Cold Spring Harbor, New York 11724

[2]University of California San Diego, La Jolla, California 92093

Correspondence: ckg@ucsd.edu

Nuclear receptors are a family of ligand-regulated transcription factors that are activated by steroid hormones, such as estrogen and progesterone, and various other lipid-soluble signals, including retinoic acid, oxysterols, and thyroid hormone (Mangelsdorf et al. 1995). Unlike most intercellular messengers, the ligands can cross the plasma membrane and directly interact with nuclear receptors inside the cell (Fig. 1), rather than having to act via cell surface

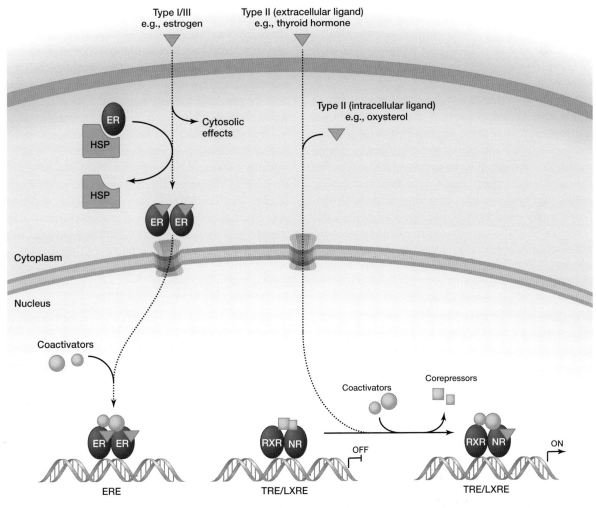

Figure 1. Nuclear receptor signaling.

receptors. Once activated, nuclear receptors directly regulate transcription of genes that control a wide variety of biological processes, including cell proliferation, development, metabolism, and reproduction. Although nuclear receptors primarily function as transcription factors, some have also been found to regulate cellular functions within the cytoplasm. For example, estrogens act through the estrogen receptor in the cytoplasm of endothelial cells to rapidly activate signaling pathways that control vascular tone and endothelial cell migration (Wu et al. 2011).

Studies of the salivary glands of insect larva in the 1960s first indicated that steroid hormones regulate transcription. Subsequent work showed that estrogen can selectively activate the genes encoding egg-white and yolk proteins, leading to the cloning of the estrogen, glucocorticoid, and thyroid hormone receptors in the 1980s (Hollenberg et al. 1985; Green et al. 1986; Miesfeld et al. 1986; Sap et al. 1986; Weinberger et al. 1986). We now know that 48 nuclear receptors are encoded in the human genome (Mangelsdorf et al. 1995). In many cases, ligands for these have been identified, but several "orphan receptors" remain (Burris et al. 2012). Whether all of these have bona fide ligands is unclear, because some nuclear receptors can act in the absence of a ligand (Table 1).

Nuclear receptors share a common structure, comprising a highly variable amino-terminal domain that includes several distinct transactivation regions (the A/B domain; also referred to as AF1 for activation function 1), a central conserved DNA-binding domain that includes two Zn fingers (the C domain), a short region responsible for nuclear localization (the D domain), and a large fairly well-conserved carboxy-terminal ligand-binding domain (the E domain, or LBD) that also contributes to interactions of the subset of nuclear receptors that form heterodimers (Mangelsdorf et al. 1995). Some also possess a highly variable

carboxy-terminal tail (the F domain) that in most cases has unknown functions.

The receptors can exist as monomers, homodimers, or heterodimers and recognize DNA sequences termed hormone response elements (HREs) derived from pairs of sequences with the consensus RGGTCA (R is a purine). They can be grouped into four subtypes based on their mode of action. Type I receptors, such as the androgen receptor, the estrogen receptor, and the progesterone receptor, are anchored in the cytoplasm by chaperone proteins (e.g., HSP90) (Echeverria and Picard 2010). Ligand binding frees the receptor from the chaperone, allowing homodimerization, exposure of the nuclear localization sequence, and entry into the nucleus (Fig. 1). Once in the nucleus, the ligand–receptor complex associates with transcriptional coactivators that facilitate binding to and activation of target genes (Glass and Rosenfeld 2000; Bulynko and O'Malley 2011). Recent genome-wide location analysis indicates that most nuclear-receptor binding sites in the genome are located in enhancer elements that are far away from the transcriptional start site, as first documented for the estrogen receptor (Carroll et al. 2006). Studies of the glucocorticoid receptor suggest that the ligand-bound receptor rapidly exchanges with its binding sites and that increases and decreases in receptor activity follow changes in the concentration of endogenous glucocorticoids.

Type II receptors, such as the thyroid hormone receptor and the retinoic acid receptor, in contrast, reside in the nucleus bound to their specific DNA response elements even in the absence of ligand. They generally form heterodimers with the retinoid X receptor (RXR) and in the absence of ligand exert active repressive functions through interactions with NCoR and SMRT corepressor complexes (Chen and Evans 1995; Horlein et al. 1995) that are associated with histone deacetylases (HDACs) (Watson et al. 2012). Binding of ligand to the LBD leads to dissociation of corepressors and their replacement with coactivator complexes. Coactivator complexes typically contain proteins with enzymatic functions, including histone acetyltransferases, that help open up chromatin and facilitate activation of target genes (Glass and Rosenfeld 2000). Note that several type II receptors bind to ligands produced in the same cell (e.g., LXR responses to oxysterols), which allows cell-autonomous feedback regulation.

Type III receptors function similarly to type I receptors except that the organization of the HRE differs (it is a direct repeat rather than inverted) and type IV receptors instead bind as monomers to half-site HREs (Mangelsdorf et al. 1995).

Ligands allosterically control the interactions of nuclear receptors with coactivators and corepressors by influencing the conformation of a short helix, referred to as AF2

Table 1. Common nuclear receptors and their ligands

Receptor	Abbreviation	Ligand
Androgen receptor	AR	Testosterone
Estrogen receptor	ER	Estrogen
Estrogen-related receptor	ERR	?
Glucocorticoid receptor	GR	Cortisol
Mineralocorticoid receptor	MR	Aldosterone
Progesterone receptor	PR	Progesterone
Retinoic acid receptor	RAR	Retinoic acid
Retinoid orphan receptor	ROR	?
Retinoic acid-related receptor	RXR	Rexinoids
Liver X receptor	LXR	Oxysterols
Peroxisome proliferator-activated receptor γ	PPARγ	Fatty acid metabolites
Thyroid hormone receptor	TR	Thyroid hormone
Vitamin D$_3$ receptor	VDR	Vitamin D$_3$

(activation function 2), at the carboxy-terminal end of the LBD (Glass and Rosenfeld 2000). In the absence of ligand, the AF2 helix is in an open conformation that enables binding of corepressors to type II receptors. Upon agonist binding, the AF2 helix adopts a conformation in which it forms one side of a charge clamp that grips the ends of a short helix of consensus sequence LxxLL present in coactivator proteins that interact directly with the LBD (Nolte et al. 1998). Selective modulation of nuclear receptor activities can be achieved by synthetic ligands that differentially alter the AF2 conformation (Glass and Rosenfeld 2000). For example, the estrogen receptor modulator tamoxifen prevents AF2 from adopting a charge-clamp conformation, thereby blocking AF2-dependent transcriptional activity.

The functions of nuclear receptors can also be modulated by posttranslational modifications that include phosphorylation, ubiquitylation, and SUMOylation (Berrabah et al. 2011; Treuter and Venteclef 2011; Lee and Lee 2012). Phosphorylation can activate some nuclear receptors independently of ligand binding and function as the major mechanism regulating activities of orphan receptors (Berrabah et al. 2011). Receptor ubiquitylation can occur in response to ligand binding and may contribute to termination of hormonal signaling (Lee and Lee 2012). SUMOylation typically reduces the activation function of nuclear receptors and/or promotes repressor activity (Treuter and Venteclef 2011).

A characteristic feature of nuclear receptors with respect to their integrative roles in development and homeostasis is their ability to regulate different genes in different cell types. For example, estrogen receptors regulate different sets of genes in the brain, breast, and uterus that contribute to the distinct functions of those organs. Recent studies indicate that tissue-specific responses are a consequence of binding of nuclear receptors to enhancer elements that are selected in a cell-specific manner. Cell-specific enhancer selection is conferred by the key lineage-determining factors for each cell type, which interact in a collaborative manner to generate open regions of chromatin that provide access points for signal-dependent transcription factors (Fig. 2) (Heinz et al. 2010). In the case of LXRs, for example, macrophage-specific binding sites are established by interactions between macrophage-lineage-determining factors that include PU.1 and AP-1, whereas in liver (Heinz et al. 2010) LXR-binding sites occur in association with the hepatocyte-lineage-determining factors HNF4 and C/EBPα (Boergesen et al. 2012). In each case, a complex multistep process involving numerous coactivator proteins is involved in building a functional enhancer, and the tissue-specific responses can be further tailored by expression of distinct coactivator/corepressor complexes (Fig. 2) (Bulynko and O'Malley 2011).

Given the wide variety of processes controlled by nuclear receptors, their dysregulation can contribute to nu-

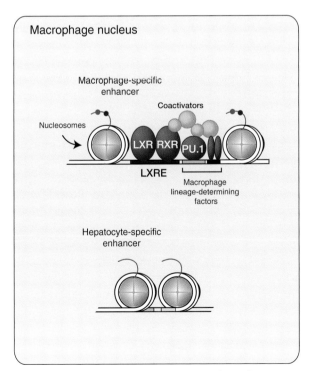 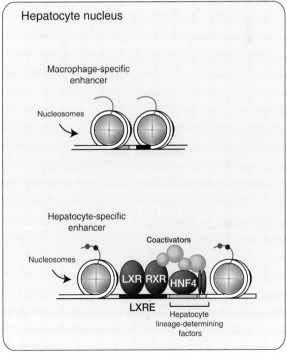

Figure 2. Tissue-specific nuclear receptor signaling in hepatocytes versus macrophages.

merous diseases, including cancer, diabetes, and infertility. However, because they bind to small molecules, they represent promising therapeutic targets for which selective agonists and antagonists can be engineered (Burris et al. 2012). Tamoxifen, for example, is an estrogen receptor antagonist currently used to treat breast cancer, and thiazolidinediones that target peroxisome proliferator-activated receptor γ (PPARγ) are used to treat type 2 diabetes. Because nuclear receptors regulate many genes in many tissues, synthetic ligands usually show beneficial therapeutic effects and unwanted side effects that limit clinical use. Major goals in the nuclear receptor field therefore include attaining a better understanding of the mechanisms underlying their actions in specific cell types and ways in which to selectively modulate their activities (Burris et al. 2012).

REFERENCES

Berrabah W, Aumercier P, Lefebvre P, Staels B. 2011. Control of nuclear receptor activities in metabolism by post-translational modifications. *FEBS Lett* **585:** 1640–1650.

Boergesen M, Pedersen TA, Gross B, van Heeringen SJ, Hagenbeek D, Bindesboll C, Caron S, Lalloyer F, Steffensen KR, Nebb HI, et al. 2012. Genome-wide profiling of liver X receptor, retinoid X receptor, and peroxisome proliferator-activated receptor α in mouse liver reveals extensive sharing of binding sites. *Mol Cell Biol* **32:** 852–867.

Bulynko YA, O'Malley BW. 2011. Nuclear receptor coactivators: Structural and functional biochemistry. *Biochemistry* **50:** 313–328.

Burris TP, Busby SA, Griffin PR. 2012. Targeting orphan nuclear receptors for treatment of metabolic diseases and autoimmunity. *Chem Biol* **19:** 51–59.

Carroll JS, Meyer CA, Song J, Li W, Geistlinger TR, Eeckhoute J, Brodsky AS, Keeton EK, Fertuck KC, Hall GF, et al. 2006. Genome-wide analysis of estrogen receptor binding sites. *Nat Genet* **38:** 1289–1297.

Chen JD, Evans RM. 1995. A transcriptional co-repressor that interacts with nuclear hormone receptors. *Nature* **377:** 454–457.

Echeverria PC, Picard D. 2010. Molecular chaperones, essential partners of steroid hormone receptors for activity and mobility. *Biochim Biophys Acta* **1803:** 641–649.

Glass CK, Rosenfeld MG. 2000. The coregulator exchange in transcriptional functions of nuclear receptors. *Genes Dev* **14:** 121–141.

Green S, Walter P, Kumar V, Krust A, Bornert JM, Argos P, Chambon P. 1986. Human oestrogen receptor cDNA: Sequence, expression and homology to v-erb-A. *Nature* **320:** 134–139.

Heinz S, Benner C, Spann N, Bertolino E, Lin YC, Laslo P, Cheng JX, Murre C, Singh H, Glass CK. 2010. Simple combinations of lineage-determining transcription factors prime *cis*-regulatory elements required for macrophage and B cell identities. *Mol Cell* **38:** 576–589.

Hollenberg SM, Weinberger C, Ong ES, Cerelli G, Oro A, Lebo R, Thompson EB, Rosenfeld MG, Evans RM. 1985. Primary structure and expression of a functional human glucocorticoid receptor cDNA. *Nature* **318:** 635–641.

Horlein AJ, Naar AM, Heinzel T, Torchia J, Gloss B, Kurokawa R, Ryan A, Kamei Y, Soderstrom M, Glass CK, et al. 1995. Ligand-independent repression by the thyroid hormone receptor mediated by a nuclear receptor co-repressor. *Nature* **377:** 397–404.

Lee JH, Lee MJ. 2012. Emerging roles of the ubiquitin–proteasome system in the steroid receptor signaling. *Arch Pharm Res* **35:** 397–407.

Mangelsdorf DJ, Thummel C, Beato M, Herrlich P, Schutz G, Umesono K, Blumberg B, Kastner P, Mark M, Chambon P, et al. 1995. The nuclear receptor superfamily: The second decade. *Cell* **83:** 835–839.

Miesfeld R, Rusconi S, Godowski PJ, Maler BA, Okret S, Wikstrom AC, Gustafsson JA, Yamamoto KR. 1986. Genetic complementation of a glucocorticoid receptor deficiency by expression of cloned receptor cDNA. *Cell* **46:** 389–399.

Nolte RT, Wisely GB, Westin S, Cobb JE, Lambert MH, Kurokawa R, Rosenfeld MG, Willson TM, Glass CK, Milburn MV. 1998. Ligand binding and co-activator assembly of the peroxisome proliferator-activated receptor-γ. *Nature* **395:** 137–143.

Sap J, Munoz A, Damm K, Goldberg Y, Ghysdael J, Leutz A, Beug H, Vennstrom B. 1986. The c-erb-A protein is a high-affinity receptor for thyroid hormone. *Nature* **324:** 635–640.

Treuter E, Venteclef N. 2011. Transcriptional control of metabolic and inflammatory pathways by nuclear receptor SUMOylation. *Biochim Biophys Acta* **1812:** 909–918.

Watson PJ, Fairall L, Schwabe JW. 2012. Nuclear hormone receptor co-repressors: Structure and function. *Mol Cell Endocrinol* **348:** 440–449.

Weinberger C, Thompson CC, Ong ES, Lebo R, Gruol DJ, Evans RM. 1986. The *c-erb-A* gene encodes a thyroid hormone receptor. *Nature* **324:** 641–646.

Wu Q, Chambliss K, Umetani M, Mineo C, Shaul PW. 2011. Non-nuclear estrogen receptor signaling in the endothelium. *J Biol Chem* **286:** 14737–14743.

Cite this chapter as *Cold Spring Harb Perspect Biol* doi: 10.1101/cshperspect.a016709

The Hippo Pathway

Kieran F. Harvey[1] and Iswar K. Hariharan[2]

[1]Peter MacCallum Cancer Centre, East Melbourne 3002, Australia

[2]University of California, Berkeley, Berkeley, California 94720

Correspondence: ikh@berkeley.edu

The Hippo pathway (Fig. 1), also known as the Salvador-Warts-Hippo pathway, regulates tissue growth in a wide variety of organisms (Harvey and Tapon 2007; Grusche et al. 2010; Oh and Irvine 2010; Pan 2010; Halder and Johnson 2011; Zhao et al. 2011). Many components of the pathway were identified as a result of mutations in the fruit fly *Drosophila melanogaster* that resulted in tissue overgrowth (Table 1). The pathway is conserved in

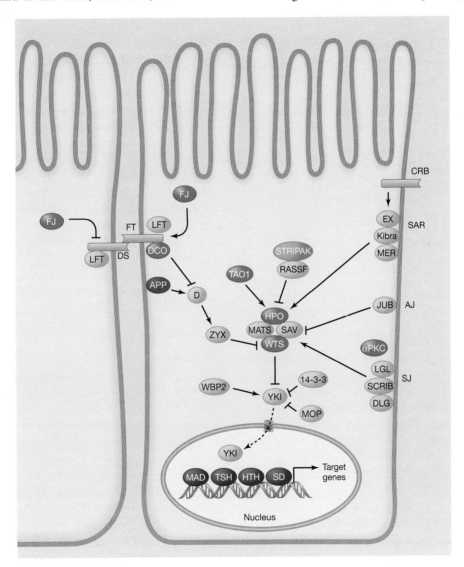

Figure 1. The *Drosophila* Hippo pathway.

Table 1. Components of the *Drosophila melanogaster* Salvador-Warts-Hippo pathway and their human homologues

Drosophila	Human
Upstream	
Fat (FT)	FAT1—FAT4
Dachsous (DS)	DCHS1 and DCHS2
Discs overgrown (DCO)	CK1ε and CK1δ
Lowfat (LFT)	LIX1 and LIX1L
Four-jointed (FJ)	FJX1
Dachs (D)	?
Approximated (APP)	ZDHHC9, ZDHHC14 and ZDHHC18
Zyxin (ZYX)	Zyxin, LPP and TRIP6
Merlin (MER)	Merlin (also known as NF2)
Expanded (EX)	Willin/FRMD6 and FRMD1
Kibra	Kibra
Crumbs (CRB)	CRB1—CRB3
Lethal giant larvae (LGL)	LGL1 and LGl2
Discs large (DLG)	DLG1—DLG4
Scribble (SCRIB)	SCRIB
aPKC	aPKCι and aPKCζ
STRIPAK (PP2A)	PP2A (STRIPAK)
RASSF	RASSF1-RASSF6
Myopic (MOP)	HD-PTP
JUB	Ajuba, LIMD1, WTIP
TAO1	TAO1—TAO3
Core	
Hippo (HPO)	MST1 and MST2
Salvador (SAV)	SAV1
Mats (MTS)	MOBKL1A and MOBKL1AB
Warts (WTS)	LATS1 and LATS2
Downstream	
Yorkie (YKI)	YAP and TAZ
WBP2	WBP2
Scalloped (SD)	TEAD1—TEAD4
MAD	SMADs
TSH	TSHZ1—TSHZ3
HTH	MEIS1—MEIS3

vertebrates, including mammals (Fig. 2), and changing the activity of the pathway can result in dramatic changes in the size of certain organs, most notably the liver (Pan 2010; Halder and Johnson 2011). In addition to its role in regulating tissue growth, the pathway has been implicated in the control of other biological processes, such as cell-fate determination, mitosis, and pluripotency. Deregulation of Hippo pathway activity has been reported in many human cancers. The human homolog of *D. melanogaster* Merlin (MER), also known as Neurofibromatosis Type 2 (NF2) is a bona fide tumor suppressor, while altered activity of several Hippo pathway components has been implicated in human tumorigenesis (Harvey and Tapon 2007).

At the core of the pathway is a module composed of two kinases—Hippo (HPO) (Harvey et al. 2003; Jia et al. 2003; Pantalacci et al. 2003; Udan et al. 2003; Wu et al. 2003) and

Warts (WTS; also known as LATS) (Justice et al. 1995; Xu et al. 1995)—and two other proteins—Salvador (SAV) (Kango-Singh et al. 2002; Tapon et al. 2002) and Mob as Tumor Suppressor (MATS) (Lai et al. 2005). HPO functions upstream of WTS and can directly phosphorylate it. Mutations that inactivate any of these four proteins result in tissue overgrowth. The first indication that some of these proteins might function in a pathway was the observation that *sav* and *wts* mutants display similar phenotypic abnormalities and that the two proteins can interact with each other (Tapon et al. 2002). More recently it has been shown that activity of this module can be regulated by RASSF, a scaffold protein that promotes tissue growth by recruiting the serine-threonine phosphatase complex STRIPAK to inhibit HPO autophosphorylation, and hence HPO activity (Ribeiro et al. 2010).

The main output of the module involves the transcriptional coactivator Yorkie (YKI) (Huang et al. 2005). Phosphorylation of YKI by WTS induces binding of 14-3-3 proteins to YKI that limit YKI activity by preventing nuclear accumulation. Phosphatases that counter the activity of WTS have not been discovered but the Myopic (MOP) tyrosine phosphatase regulates YKI activity, repressing it (Gilbert et al. 2011). YKI promotes tissue growth by increasing expression of positive regulators of cell growth and inhibitors of apoptosis. YKI, itself does not bind DNA but functions together with several transcription factors, including Scalloped (SD; the homolog of TEAD transcription factors in vertebrates), Homothorax (HTH), Teashirt (TSH), and Mothers against DPP (MAD). Transcriptional regulatory proteins such as WBP2 also control Hippo-pathway-dependent tissue growth (Zhang et al. 2011). WBP2 and other as-yet-unidentified proteins have been predicted to interact with YKI via its WW domains, which are important for YKI's transcription activation function (Oh and Irvine 2010).

The HPO and WTS kinases appear to receive multiple inputs. The first upstream regulators to be discovered were the Band 4.1 proteins Expanded (EX) and MER (Hamaratoglu et al. 2006). These function together with the WW-domain-containing protein Kibra to activate the core kinase cassette by an unknown mechanism. EX is also thought to repress YKI by physical interaction and sequestration. The Fat/Dachsous branch of the pathway consists of the atypical cadherins Fat (FT) and Dachsous (DS) as well as the downstream effector proteins Discs overgrown (DCO, a serine-threonine kinase also known as casein kinase 1ε), Dachs (D, an atypical myosin), Approximated (APP, a palmitoyltransferase), Lowfat (LFT), and Zyxin (ZYX) (Grusche et al. 2010; Rauskolb et al. 2011). The Fat/Dachsous branch impinges on pathway activity by modulating the abundance of WTS and also modulates

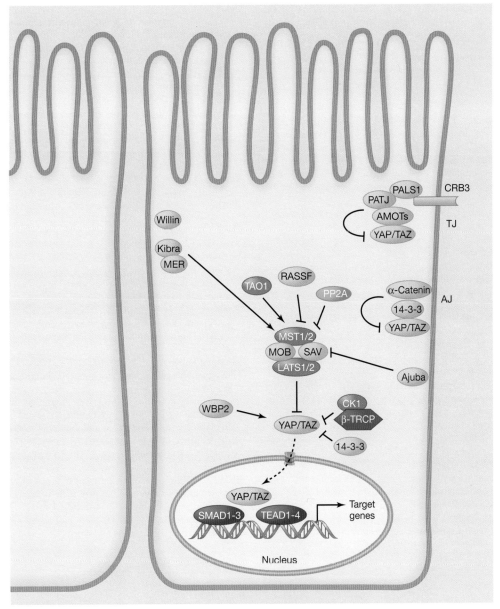

Figure 2. The mammalian Hippo pathway.

the Kibra-EX-MER branch by regulating EX levels. The sterile 20-like kinase, TAO1, phosphorylates and activates HPO (MST1/2 in mammals) although it is unclear whether TAO1 activity is regulated (Boggiano et al. 2011; Poon et al. 2011).

Increasing evidence underlines the importance of cell junctions for regulation of Hippo pathway activity. In *D. melanogaster* epithelial cells, many Hippo pathway proteins reside, at least partially, at the sub-apical region (SAR), adherens junction (AJ) or septate junction (SJ). Examples of such junctional proteins include the AJ protein Jub and the apical-basal polarity proteins Discs large (DLG), Lethal giant larvae (LGL), Scribble (SCRIB), Crumbs (CRB), and

atypical protein kinase C (aPKC). In mammalian epithelial cells, several other junctional proteins regulate Hippo pathway activity (see Table 2), including angiomotin and

Table 2. Components of the human Salvador-Warts-Hippo pathway and their *Drosophila melanogaster* homologs

Human	*Drosophila*
Upstream	
α-Catenin	α-Catenin
PATJ	Discs lost
PALS1	Stardust
AMOTs	?
β-TRCP	Slimb

α-catenin (Schlegelmilch et al. 2011; Zhao et al. 2011). The Hippo pathway may therefore help couple tissue growth to mechanical stresses or cell–cell contact, which might be important for organ size regulation.

REFERENCES

Boggiano JC, Vanderzalm PJ, Fehon RG. 2011. Tao-1 phosphorylates Hippo/MST kinases to regulate the Hippo-Salvador-Warts tumor suppressor pathway. *Dev Cell* **21:** 888–895.

Gilbert MM, Tipping M, Veraksa A, Moberg KH. 2011. A screen for conditional growth suppressor genes identifies the *Drosophila* homolog of HD-PTP as a regulator of the oncoprotein Yorkie. *Dev Cell* **20:** 700–712.

Grusche FA, Richardson HE, Harvey KF. 2010. Upstream regulation of the hippo size control pathway. *Curr Biol* **20:** R574–582.

Halder G, Johnson RL. 2011. Hippo signaling: Growth control and beyond. *Development* **138:** 9–22.

Hamaratoglu F, Willecke M, Kango-Singh M, Nolo R, Hyun E, Tao C, Jafar-Nejad H, Halder G. 2006. The tumour-suppressor genes NF2/Merlin and Expanded act through Hippo signalling to regulate cell proliferation and apoptosis. *Nat Cell Biol* **8:** 27–36.

Harvey K, Tapon N. 2007. The Salvador-Warts-Hippo pathway—an emerging tumour-suppressor network. *Nat Rev Cancer* **7:** 182–191.

Harvey KF, Pfleger CM, Hariharan IK. 2003. The *Drosophila* Mst ortholog, hippo, restricts growth and cell proliferation and promotes apoptosis. *Cell* **114:** 457–467.

Huang J, Wu S, Barrera J, Matthews K, Pan D. 2005. The Hippo signaling pathway coordinately regulates cell proliferation and apoptosis by inactivating Yorkie, the *Drosophila* homolog of YAP. *Cell* **122:** 421–434.

Jia J, Zhang W, Wang B, Trinko R, Jiang J. 2003. The *Drosophila* Ste20 family kinase dMST functions as a tumor suppressor by restricting cell proliferation and promoting apoptosis. *Genes Dev* **17:** 2514–2519.

Justice RW, Zilian O, Woods DF, Noll M, Bryant PJ. 1995. The *Drosophila* tumor suppressor gene warts encodes a homolog of human myotonic dystrophy kinase and is required for the control of cell shape and proliferation. *Genes Dev* **9:** 534–546.

Kango-Singh M, Nolo R, Tao C, Verstreken P, Hiesinger PR, Bellen HJ, Halder G. 2002. Shar-pei mediates cell proliferation arrest during imaginal disc growth in *Drosophila*. *Development* **129:** 5719–5730.

Lai ZC, Wei X, Shimizu T, Ramos E, Rohrbaugh M, Nikolaidis N, Ho LL, Li Y. 2005. Control of cell proliferation and apoptosis by mob as tumor suppressor, mats. *Cell* **120:** 675–685.

Oh H, Irvine KD. 2010. Yorkie: The final destination of Hippo signaling. *Trends Cell Biol* **20:** 410–417.

Pan D. 2010. The hippo signaling pathway in development and cancer. *Dev Cell* **19:** 491–505.

Pantalacci S, Tapon N, Leopold P. 2003. The Salvador partner Hippo promotes apoptosis and cell-cycle exit in *Drosophila*. *Nat Cell Biol* **5:** 921–927.

Poon CL, Lin JI, Zhang X, Harvey KF. 2011. The sterile 20-like kinase Tao-1 controls tissue growth by regulating the Salvador-Warts-Hippo pathway. *Dev Cell* **21:** 896–906.

Rauskolb C, Pan G, Reddy BV, Oh H, Irvine KD. 2011. Zyxin links fat signaling to the hippo pathway. *PLoS Biol* **9:** e1000624.

Ribeiro PS, Josue F, Wepf A, Wehr MC, Rinner O, Kelly G, Tapon N, Gstaiger M. 2010. Combined functional genomic and proteomic approaches identify a PP2A complex as a negative regulator of Hippo signaling. *Mol Cell* **39:** 521–534.

Schlegelmilch K, Mohseni M, Kirak O, Pruszak J, Rodriguez JR, Zhou D, Kreger BT, Vasioukhin V, Avruch J, Brummelkamp TR, et al. 2011. Yap1 acts downstream of alpha-catenin to control epidermal proliferation. *Cell* **144:** 782–795.

Tapon N, Harvey KF, Bell DW, Wahrer DC, Schiripo TA, Haber DA, Hariharan IK. 2002. *salvador* promotes both cell cycle exit and apoptosis in *Drosophila* and is mutated in human cancer cell lines. *Cell* **110:** 467–478.

Udan RS, Kango-Singh M, Nolo R, Tao C, Halder G. 2003. Hippo promotes proliferation arrest and apoptosis in the Salvador/Warts pathway. *Nat Cell Biol* **5:** 914–920.

Wu S, Huang J, Dong J, Pan D. 2003. hippo encodes a Ste-20 family protein kinase that restricts cell proliferation and promotes apoptosis in conjunction with salvador and warts. *Cell* **114:** 445–456.

Xu T, Wang W, Zhang S, Stewart RA, Yu W. 1995. Identifying tumor suppressors in genetic mosaics: The *Drosophila* lats gene encodes a putative protein kinase. *Development* **121:** 1053–1063.

Zhang X, Milton CC, Poon CL, Hong W, Harvey KF. 2011. Wbp2 cooperates with Yorkie to drive tissue growth downstream of the Salvador-Warts-Hippo pathway. *Cell Death Differ* **18:** 1346–1355.

Zhao B, Tumaneng K, Guan KL. 2011. The Hippo pathway in organ size control, tissue regeneration and stem cell self-renewal. *Nat Cell Biol* **13:** 877–883.

SIGNALING PROCESSES

CHAPTER 5

Signaling Pathways that Control Cell Proliferation

Robert J. Duronio[1,2,3] and Yue Xiong[2,3,4]

[1]Department of Biology and Genetics, University of North Carolina, Chapel Hill, North Carolina 27599

[2]Program in Molecular Biology and Biotechnology, University of North Carolina, Chapel Hill, North Carolina 27599

[3]Lineberger Comprehensive Cancer Center, University of North Carolina, Chapel Hill, North Carolina 27599

[4]Department of Biochemistry and Biophysics, University of North Carolina, Chapel Hill, North Carolina 27599

Correspondence: duronio@med.unc.edu; yxiong@email.unc.edu

SUMMARY

Cells decide to proliferate or remain quiescent using signaling pathways that link information about the cellular environment to the G_1 phase of the cell cycle. Progression through G_1 phase is controlled by pRB proteins, which function to repress the activity of E2F transcription factors in cells exiting mitosis and in quiescent cells. Phosphorylation of pRB proteins by the G_1 cyclin-dependent kinases (CDKs) releases E2F factors, promoting the transition to S phase. CDK activity is primarily regulated by the binding of CDK catalytic subunits to cyclin partners and CDK inhibitors. Consequently, both mitogenic and antiproliferative signals exert their effects on cell proliferation through the transcriptional regulation and ubiquitin-dependent degradation of cyclins and CDK inhibitors.

Outline

R.J. Duronio and Y. Xiong

1 INTRODUCTION

Control of cell proliferation generally occurs during the first gap phase (G_1) of the eukaryotic cell division cycle (see Box 1). Multiple signals, ranging from growth factors to DNA damage to developmental cues, influence the decision to enter S phase, when DNA is replicated (Fig. 1). Hence, G_1 phase cell cycle control is intrinsically linked with a diverse set of pathways controlling differentiation, stem and progenitor cell quiescence, senescence, and responses to a variety of stresses. The decision to enter S phase from G_1 represents a point of no return that, in the absence of stress such as DNA damage, commits cells to complete the cell cycle and divide, and is therefore tightly controlled. This decision is made at what is called the "restriction point" in mammalian cells and "START" in yeast, after which cells become largely refractory to extracellular signals and will complete S phase and proceed through a second gap phase (G_2 phase) and then mitosis. In multicellular organisms, most differentiated cells exit the active cell cycle during G_1 phase and enter G_0 phase, in which they remain metabolically active for days or even years, performing specialized functions. Postmitotic nerve and skeletal muscle cells provide good examples. Some G_0 cells, such

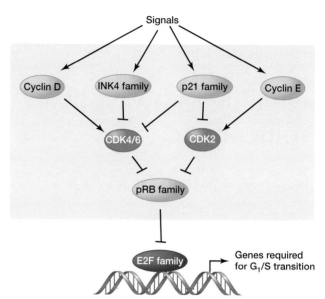

Figure 1. G_1 cell cycle control by the pRB pathway. Many cellular signaling events are intrinsically linked to G_1 phase of the cell cycle, which is controlled by the RB pathway. Signaling to the RB pathway, and thus G_1 control by different cellular processes, is achieved mainly through the regulation of cyclins and CDK inhibitors (CKIs). In mammalian cells, mitogenic signals first induce the synthesis of D-type cyclins, leading to activation of cyclin-D-dependent CDK4 and CDK6, and then induce E-type cyclins to activate CDK2. Cyclin-D-CDK4/6 and cyclin-E-CDK2 cooperatively phosphorylate RB-family proteins, derepressing E2F to allow transcription of E2F-target genes, thereby promoting the G_1/S transition. The INK4 proteins specifically inhibit CDK4 and CDK6, whereas the p21 (CIP/KIP) family of CKIs inhibits multiple CDKs. Although the schematic illustration is based on mammalian cells, the regulation of both G_1 cyclins and CDK inhibitors is evolutionarily conserved.

BOX 1. THE EUKARYOTIC CELL CYCLE

The classical cell cycle comprises four phases—G_1, S, G_2, and M—and is controlled by cyclin-dependent kinases (CDKs) and their cyclin partners. The commitment to divide occurs in G_1 phase, which is controlled by cyclin-D-CDK4/6 and cyclin-E-CDK2 at the so-called G_1/S transition. DNA is then replicated in S phase. This is followed by a second gap phase, G_2, at the end of which cyclin-B-CDK1 controls entry into M phase (mitosis), when the cell divides. Cells can exit the cell cycle in G_1 phase and enter G_0 phase (quiescence). In some cases, they can reenter the cell cycle and begin dividing again (see main text).

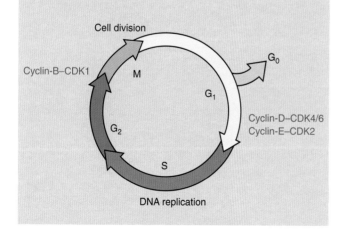

as quiescent T cells, can be stimulated by mitogenic signals to reenter the cell cycle.

The restriction point is primarily controlled in mammalian cells by the RB pathway, named after the first tumor suppressor identified, the retinoblastoma protein (pRB) (Weinberg 1995). pRB is a member of a highly conserved family of proteins, encoded by a single gene in the single-celled green alga *Chlamydomonas* (*MAT3*), *Caenorhabditis elegans* (*LIN-35*), and *Arabidopsis* (*RBR1*); two genes in *Drosophila* (*RBF1 and RBF2*); and three genes in mammalian cells (*RB1*; *p107*, also known as *RBL1*; and *p130*, also known as *RBL2*) (Weinberg 1995; van den Heuvel and Dyson 2008). Budding yeast cells contain a protein (Whi5) that, although it does not share sequence similarity with pRB, functions at START in a similar manner (Costanzo et al. 2004; de Bruin et al. 2004). pRB proteins are present as hypophosphorylated, active forms in cells exiting mitosis and in quiescent cells, where they use a conserved pocket to bind to LxCxE motifs in numerous chromatin-

Cite this chapter as *Cold Spring Harb Perspect Biol* doi: 10.1101/cshperspect.a008904

associated proteins and transcription factors, particularly members of the E2F family. pRB proteins negatively regulate the expression of E2F-target genes, many of which are required for entry into and progression through S phase, by recruiting various repressive chromatin regulatory complexes and histone-modifying enzymes or by blocking the transactivation function of E2F proteins. Phosphorylation of the pRB family proteins by CDKs during G_1 phase causes pRB to dissociate from E2Fs, allowing the transcription of target genes that stimulate progression into S phase (Fig. 1) (Dyson 1998).

The principal kinases that phosphorylate pRB family proteins during G_1 phase in mammalian cells are three cyclin-dependent kinases (CDKs)[5]: cyclin-D-dependent CDK4 and CDK6 (Ewen et al. 1993; Kato et al. 1993) and cyclin-E-dependent CDK2 (Akiyama et al. 1992; Hinds et al. 1992). As many as eight distinct mammalian G_1 CDK–cyclin complexes can be formed from combinatorial association of three D-type cyclins (cyclins D1, D2, and D3) with CDK4 and CDK6 and two E-type cyclins (cyclins E1 and E2) with CDK2, and these phosphorylate as many as 16 sites in pRB proteins (Akiyama et al. 1992; Kitagawa et al. 1996). Regulation of pRB-E2F by G_1 CDKs has been evolutionarily conserved in plants, worms, flies, and mammals (Inze 2005; van den Heuvel and Dyson 2008). The complexity of the pRB pathway reflects the need to meet the demand to integrate diverse signals from different signaling pathways into a central G_1 control mechanism. Disruption of this mechanism results in a wide range of developmental defects and human diseases, particularly cancer. Indeed, disruption of G_1 control probably represents a common event in the development of most types of human cancer (Sherr 1996).

The critical role of pRB and G_1 CDKs in controlling the G_1/S transition is further illustrated by the studies of three DNA tumor viruses: adenovirus, human papilloma virus (HPV), and simian virus 40 (SV40). Although evolutionarily distant from each other, these viruses encode unrelated proteins (E1A in adenovirus, E7 in HPV, and large T in SV40) that bind to and inactivate pRB via an LxCxE motif to promote cell proliferation and viral replication. Primate herpesvirus saimiri and human Kaposi's sarcoma virus encode cyclin D homologs (v-cyclins) that preferentially bind to and activate CDK6, creating complexes that are resistant to CDK inhibitors (CKIs; see below).

The steady-state levels of CDK2, CDK4, and CDK6 proteins remain relatively constant during the normal cell cycle and in quiescent, aging, and even terminally differentiated cells. Signaling pathways that affect G_1 phase

progression thus do not affect CDK levels and instead act mainly through regulation of CDK activity by controlling the abundance of their cyclin partners and a group of CKIs. Although both cyclins and CKIs can be regulated at the level of mRNA stability, translational control, and subcellular localization, the two major control mechanisms are transcriptional regulation and ubiquitin-dependent proteolysis. We discuss these mechanisms below, focusing on the regulation of expression and ubiquitylation of G_1 cyclins and CKIs by different signal transduction pathways.

2 TRANSCRIPTIONAL REGULATION OF G_1 CYCLINS BY MITOGENIC SIGNALS

2.1 D-Type Cyclins

D-type cyclins were simultaneously isolated initially from mammalian cells in a genetic screen for genes capable of complementing G_1 cyclin deficiency in yeast, as the product of a gene whose expression is induced by colony-stimulating factor (CSF1), and as the product of the potential oncogene *BCL1* that is clonally rearranged and overexpressed in a subset of parathyroid tumors (Matsushime et al. 1991; Motokura et al. 1991; Xiong et al. 1991). These findings provided early evidence linking the activation of a G_1 cyclin with mitogenic growth factors and implicating abnormal expression of G_1 cyclins in tumorigenesis. However, subsequent genetic analyses revealed only a relatively minor role of cyclin-D-dependent CDK activity in cell proliferation and development (Meyer et al. 2000; Kozar et al. 2004; Malumbres et al. 2004), although mouse embryonic fibroblasts (MEFs) from mice lacking CDK4 and CDK6 do have a reduced rate of exiting from quiescence in response to mitogenic stimulation. Hence, the D-type cyclins, although not an obligate component of the cell cycle machinery, couple extracellular mitogenic signals to the G_1/S transition (Sherr and Roberts 2004).

The canonical Ras–Raf–MEK–ERK mitogen-activated protein kinase (MAPK) pathway is the best characterized pathway for the activation of cyclin D transcription (p. 81 [Morrison 2012]). It stimulates the expression of AP1 transcription factors such as the proto-oncogene products Jun and Fos, which bind directly to an AP1 site in the cyclin D1 promoter (Albanese et al. 1995). D-type cyclins can also be induced by other signaling pathways, including mitogen-stimulated Rac and NF-κB signaling, cytokine signaling, signaling by receptors for extracellular matrix (ECM) proteins (e.g., integrins), and the Wnt and Notch pathways (p. 109 [Kopan 2012] and p. 103 [Nusse 2012]). Multiple transcription factors directly regulate cyclin D genes, including Jun, Fos, STAT3, β-catenin, and NF-κB (Fig. 2A). Cyclin D genes are expressed at very low levels in most differentiated

---

[5]CDKs are a family of kinases that regulate the cell cycle and that require binding to noncatalytic partner proteins termed cyclins for activity.

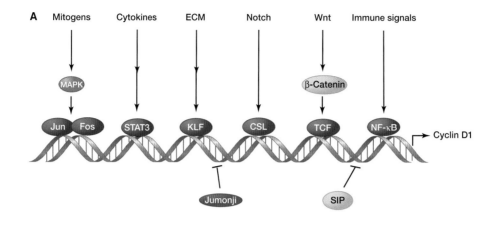

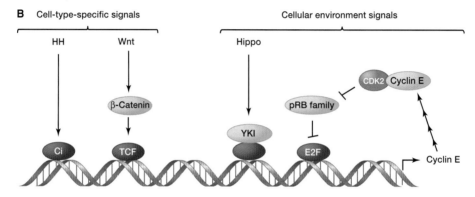

Figure 2. Transcriptional regulation of G_1 cyclins. (*A*) The expression of cyclin genes is tightly regulated at the level of transcription by different signals, including many mitogens. The figure uses human cyclin D1 as an example. (*B*) Cyclin E expression is also highly regulated and responds to two types of developmental signals, those that are cell-type specific and those that all cells use to control proliferation in response to their environment. MAPK, Mitogen-activated protein kinases; ECM, extracellular matrix; STAT, signal transducers and activators of transcription; KLF, Krüppel-like factor; CSL, CBF-1/suppressor of hairless/LAG-1; TCF, ternary complex factor; NF-κB, nuclear factor-κB; SIP1, SMAD interacting protein 1; HH, Hedgehog; Ci, *Drosophila* cubitus interruptus; YKI, Yorkie.

tissues, in part because of transcriptional repression by proteins such as Jumonji and SIP (Klein and Assoian 2008). Repression of G_1 cyclin expression is an important part of cell cycle exit and terminal differentiation, and inappropriate reactivation of D- or E-type cyclins can drive differentiated cells back into S phase (Buttitta et al. 2007; Korzelius et al. 2011).

In contrast to cyclin D repression, inappropriate cyclin-D-dependent CDK4/6 activity represents the most frequent alteration of human cyclins in cancer and bears clear pathological significance. Human cyclin D1 is amplified in an estimated 13% of neoplasms of different types, including breast cancer, esophageal cancer, and lymphoma (Bates and Peters 1995). Mice transgenically expressing cyclin D1 develop mammary gland tumors and conversely are protected against mammary tumors if cyclin D1 is deleted (Wang et al. 1994; Yu et al. 2001). Likewise, CDK4 and CDK6 are also frequently amplified in diverse human cancers. Mouse cells lacking either combination of the three

cyclin D proteins or CDK4/6 are more resistant to oncogenic transformation (Sherr and Roberts 2004; Malumbres and Barbacid 2009). These observations indicate that whereas a low level of G_1 CDK activity is sufficient to support cell proliferation in response to normal physiological levels of mitogens, significantly higher levels of G_1 CDK activity are required to sustain hyperproliferative stimulation, such as those elicited by activated oncogenes.

2.2 Cyclin E Expression

Cyclin E is encoded by a single gene in *C. elegans* (*CYE-1*) and *Drosophila* (*CycE*) and by two genes in mammalian cells (*E1* and *E2*). The worm and fly cyclin E genes are essential for cell cycle progression and development (Knoblich et al. 1994; Fay and Han 2000). In contrast, mice lacking both cyclin E1 and E2 or CDK2 are viable and display relatively minor defects late in development, owing to compensation by other CDKs (Berthet et al. 2003; Geng et al. 2003; Ortega et al. 2003; Parisi et al. 2003). In well-fed proliferating cells, cyclin E expression is cyclical, peaking at the G_1/S transition and being low or absent at other times in the cell cycle (Lew et al. 1991; Dulic et al. 1992; Koff et al. 1992). Conversely, MEFs lacking both cyclins E1 and E2 proliferate more slowly than normal cells and have a significantly reduced response to mitogenic stimulation, and cyclin E gene expression is repressed in serum-deprived cells, all of which suggest that cyclin E responds to growth factors (Herrera et al. 1996; Geng et al. 2003). This regulation is important, because forced overexpression of cyclin E can shorten G_1 phase and drive cells into S phase, in part by causing phosphorylation of pRB family proteins (Hinds et al. 1992; Ohtsubo and Roberts 1993; Resnitzky et al. 1994). In vivo, transgenic expression of cyclin E under the control of the β-lactoglobulin promoter in mice results in mammary tumorigenesis (Smith et al. 2006), and overexpression of cyclin E is frequently observed in various human cancers and correlates with increased tumor aggression (Hwang and Clurman 2005). Hence, tight control of the levels of cyclin E is critically important for normal cell physiology and for preventing a neoplastic cell cycle. This notion is supported by biochemical and genetic analyses of the regulation of cyclin E by phosphorylation and by its regulatory protein FBW7 (see below).

Cyclin E transcription is directly controlled by E2F (Duronio and O'Farrell 1995; Ohtani et al. 1995; Geng et al. 1996). Thus, one important way that signaling regulates cyclin E is through the pRB/E2F pathway, which also integrates the output from the growth factor signals that control D-type-cyclin-dependent CDK activity. Indeed, if the mouse cyclin E gene is engineered to respond to the signals that control cyclin D1 gene expression, then cyclin D1 is no longer needed (Geng et al. 1999). Because cyclin-E–CDK2 can phosphorylate and inactivate pRB, resulting in E2F activity, a positive-feedback amplification is an important part of G_1/S control (Fig. 1B). This helps produce the switch-like behavior needed for unidirectional decisions like the G_1/S transition (Xiong and Ferrell 2003; Ferrell et al. 2009).

Control of cyclin E transcription via E2F is a cornerstone of G_1/S cell cycle control, but the cyclin E gene also responds directly to signaling pathways. This often occurs when developmental programs coordinate cell cycle progression with cell differentiation. In the *Drosophila* eye, for example, Hedgehog signaling induces cyclin E at the G_1/S transition of the last cell cycle before differentiation of specialized cell types such as photoreceptors (p. 107 [Ingham 2012]). The *Drosophila CycE* gene contains multiple enhancer elements that respond to and integrate various signals (Fig. 2B), including those from the pRB/E2F, Hedgehog and Wnt signaling pathways, in different cell types at different stages of

development (Jones et al. 2000; Deb et al. 2008; p. 107 [Ingham 2012]).

In *Drosophila*, *CycE* is also a target of the growth-inhibitory Hippo pathway (p. 133 [Harvey and Hariharan 2012]), whose main target is the inactivation of the transcriptional coactivator Yorkie (YKI) (Huang et al. 2005). Tissue overgrowth upon disruption of the Hippo pathway is accompanied by increased expression of cyclin E, probably through direct regulation of *CycE* transcription by transcription factors associated with YKI. In vertebrates, the Hippo-pathway-mediated regulation of cell proliferation appears to be largely mediated by cyclin D1 (Cao et al. 2008).

Transcription of the cyclin E gene thus responds to two types of developmental signals: those that are cell type specific and essential for cell cycle progression (e.g., Hedgehog and Wnt signals), and those that are not cell type specific or strictly essential for cell cycle progression but instead modulate the rate of growth and cell proliferation in response to the cellular environment (e.g., E2F-mediated responses and Hippo) (Fig. 2B).

2.3 Posttranscriptional Regulation of CDKs

Posttranscriptional mechanisms also regulate CDK activity in response to various signals. The mitotic CDK, CDK1 (also known as CDC2), is inhibited during interphase by phosphorylation at two adjacent residues within its catalytic pocket, T14 and Y15, and is activated by CDC25-mediated dephosphorylation to bring about a sudden burst of CDK1 activity that triggers mitosis (Ch. 6 [Rhind and Russel 2012]). Both CDK2 and CDK4 are also phosphorylated at analogous residues to mediate the responses to different signals: phosphorylation of T14 and Y15 of CDK2 is important for regulating the timing of DNA replication and centrosome duplication (Zhao et al. 2012), and phosphorylation of Y17 of CDK4 is required for G_1 arrest upon UV irradiation, which could cause DNA damage that should be repaired before entry into S phase (Terada et al. 1995).

3 TRANSCRIPTIONAL REGULATION OF CDK INHIBITORS

CKIs play an important role in arresting the cell cycle in G_1 phase in response to a variety of stimuli, ranging from growth factor deprivation to DNA damage, cellular stress, differentiation, and senescence. Failure to arrest the cell cycle resulting from loss of function of a CKI can cause developmental defects or hyperplasia and tumorigenesis. The first CKI characterized was mammalian p21 (also known as CDKN1A, CIP1, or WAF1), which binds to and inhibits the activity of multiple CDK–cyclin complexes (Xiong et al. 1992, 1993a; Harper et al. 1993). The p21 family (also known as the CIP/KIP family) includes three related proteins: p21, p27 (also known as CDKN1B or KIP1), and p57 (also known as CDKN1C or KIP2). A distinct CKI, p16 (also known as INK4A), was isolated around the same time and is a specific inhibitor of CDK4 (Serrano et al. 1993). p16 is the founding member of a separate family of INK4 CKIs that includes three additional proteins: p15 (also known as INK4B), p18 (also known as INK4C), and p19 (also known as INK4D) (Sherr and Roberts 1995).

These two families of CKIs inhibit CDK via different mechanisms. The INK4 proteins bind selectively to the catalytic subunits of two CDKs, CDK4 and CDK6, preventing cyclin binding; and the p21 CKIs bind to the cyclin–CDK complex by contacting both subunits via different motifs to block kinase activity and substrate binding. CKIs of both families are localized predominantly in the nucleus in most tissues, but p21 family CKIs have also been frequently observed in the cytoplasm, where they have been linked to CDK-independent functions and tumor development. In particular, reduced nuclear p27 and accumulation of cytoplasmic p27 have been observed in multiple types of human cancers and are associated with poor prognosis of breast cancer (Wander et al. 2011).

The two separate families of multiple CDK inhibitors evolved to meet the increasing need to integrate numerous different antiproliferative signals that can arrest cells in G_1 phase. Mice lacking CKI genes have various phenotypes, ranging from a compromised DNA damage response (p21 mutants) to widespread hyperplastic cell proliferation and organomegaly (p18- and p27-null mice), spontaneous tumor development (p16-null mice), and perinatal lethality and widespread developmental defects (in p57-null mice) (Ortega et al. 2002). Furthermore, genetic studies of p21-type CKIs in worms and flies have revealed various functions from control of cell cycle progression to cell cycle exit in specific cell types at various times in development (de Nooij et al. 1996; Lane et al. 1996; Hong et al. 1998; Firth and Baker 2005).

One major difference between the two CKI families is their stability. The p21 family inhibitors are intrinsically unstable ($t_{1/2} < 30$ min) as a result of ubiquitin-dependent, and in most cases phosphorylation-promoted, proteasomal degradation, and cause a rapid and transient cell cycle arrest, for example, following DNA damage. In contrast, the INK4 proteins are stable ($t_{1/2} > 4$–6 h) and are subject to minimal posttranslational regulation. INK4 proteins therefore maintain a long-term or permanent cell cycle arrest in stem, progenitor, senescent, and postmitotic cells. Accordingly, whereas p21 family CKIs are regulated both

transcriptionally and posttranscriptionally, the INK4 members are regulated primarily at the level of transcription.

3.1 p21 Transcription Regulation by p53-Dependent and -Independent Mechanisms

Cells use signaling pathways to respond to a variety of exogenous and intrinsic stresses that have the potential to damage the genome. The tumor suppressor p53 functions as a transcription factor to activate the expression of many genes involved in stress responses, and defects in p53-mediated stress responses are associated with most types of human cancer. p53-mediated transcriptional activation of p21 following DNA damage was the first identified example of G_1-phase regulation of a CKI gene (El-Deiry et al. 1993; Xiong et al. 1993b). Given that none of the other six CKI genes is a direct target of p53, the p53–p21–CDK regulatory module constitutes a major mechanism for DNA-damage-induced cell cycle arrest. Indeed, knocking out the p21 gene compromises the DNA damage response despite having little effect on overall mouse development (Brugarolas et al. 1995; Deng et al. 1995).

Transcriptional regulation of the p21 gene has also been linked to p53-independent cell cycle exit during development. In the Drosophila embryonic epidermis, activation of the dacapo (dap) gene, which encodes a p21-type CKI, triggers cell cycle exit (de Nooij et al. 1996; Lane et al. 1996). In Caenorhabditis elegans, the insulin-like growth factor signaling pathway similarly induces p21 expression in response to starvation, which results in cell cycle arrest in stem cells (Baugh and Sternberg 2006), and Ras/MAPK signaling activates p21 to control cell cycle exit in vulval precursor cells (Clayton et al. 2008). This diversity of responses probably relies on the existence of multiple, modular enhancers for the p21 gene that respond to different signaling pathways (Liu et al. 2002; Meyer et al. 2002).

3.2 INK4 Repression in Stem and Progenitor Cells

INK4 genes have distinct expression patterns during development in adult tissues and in response to different conditions (Roussel 1999). p16 is a target of Polycomb group (PcG) transcriptional repressors: deletion of the Polycomb gene Bmi1 retards cell proliferation, and this is associated with up-regulation of p16 and can be partially rescued by deletion of p16 (van Lohuizen et al. 1991; Jacobs et al. 1999). Furthermore, both PcG repression complexes (PRC1 and PRC2) collaborate with pRB proteins to bind to the p16 locus and trimethylate histone H3 lysine 27 (H3K27) to repress the expression of p16 (Bracken et al. 2007; Kotake et al. 2007). These findings explain how the up-regulation of p16 in aging stem cells results from de-

Cite this chapter as Cold Spring Harb Perspect Biol doi: 10.1101/cshperspect.a008904

creased expression of Polycomb genes and reveal a negative-feedback loop between p16 and pRB.[6] In many different types of human tumors, *p16* expression is silenced by promoter DNA methylation (Merlo et al. 1995).

Unlike *p16* mRNA, which is undetectable in young tissues and is induced during aging, *p18* mRNA is present early in embryogenesis and maintains a high level throughout life in many adult tissues (Zindy et al. 1997). Deletion of *p18* in mice results in spontaneous development of various tumors (Franklin et al. 1998; Pei et al. 2009) and increases self-renewing division of hematopoietic stem cells and expansion of mammary luminal progenitor cells (Yuan et al. 2004; Pei et al. 2009). *p18* thus seems to suppress tumorigenesis by maintaining a quiescent state in stem and progenitor cells of different organs. GATA3, a transcription factor specifying mammary luminal cell fate, binds to the *p18* locus and represses *p18* transcription (Pei et al. 2009). It provides an example of a lineage-specifying factor that regulates cell differentiation in part by repressing the expression of an *INK4* gene to allow quiescent progenitor cells to exit G_0/G_1 arrest, reenter the cell cycle, and proliferate.

4 CONTROL OF G_1 CYCLINS BY THE UBIQUITIN–PROTEASOME SYSTEM

Like their mitotic counterparts, G_1 cyclins undergo rapid turnover and are degraded by the ubiquitin–proteasome pathway. This process is tightly regulated through the phosphorylation of cyclins and, in some cases, by proteins that target cyclins to E3 ubiquitin ligases, which provide mechanisms for extracellular factors to signal to the G_1-phase cell cycle control machinery.

The level of cyclin E, and associated CDK2 activity, oscillates during the cell cycle (Dulic et al. 1992; Koff et al. 1992). Cyclin E begins to accumulate during the middle of G_1 phase (as a result of E2F-mediated transcriptional activation), peaks at the G_1/S transition, and then is destroyed during S phase following ubiquitylation. FBW7 (also known as Cdc4 or Ago) is an F-box protein that is the substrate-recognition component of the E3 ubiquitin ligase SCF (also known as CRL1) and recognizes two phosphodegrons in cyclin E: a carboxy-terminal degron centered on T380 and an amino-terminal degron centered

---

[6]Linked to p16, both structurally in the genome and through regulation by Polycomb group proteins, is the product of the *ARF* tumor suppressor gene, which is transcribed from an alternative promoter and translated in an alternative reading frame from *p16*. ARF does not share any amino acid sequence similarity with INK4 proteins and instead acts as a p53 activator by binding to and inhibiting the activity of MDM2, the principle E3 ubiquitin ligase for and negative regulator of p53. As a result, any signal, such as oncogenic stimulation, that induces the expression of *ARF* will stabilize p53 and activate p21, leading to G_1 cell cycle arrest.

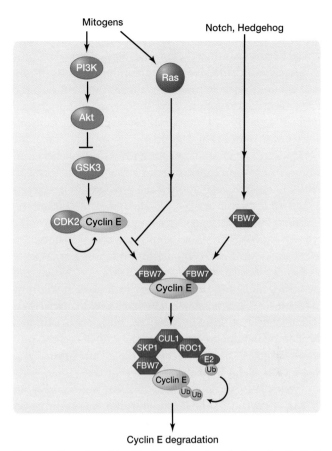

Figure 3. Targeting ubiquitin-dependent degradation of cyclin E. F-box protein FBW7 specifically recognizes two separate phosphodegrons in cyclin E and targets cyclin E for ubiquitin-dependent proteasome degradation by the SCF-FBW7 E3 ligase complex. The phosphorylation of both amino- and carboxy-terminal degrons in cyclin E is catalyzed by GSK3 and CDK2 and creates two separate binding sites for FBW7. Both mitogenic and antiproliferative signals exert their effect on the cell cycle through cyclin E ubiquitylation by inhibiting the activity of GSK3 or stimulating the expression of FBW7, respectively.

on T62 (Fig. 3) (Welcker and Clurman 2008). Both cyclin E degrons are phosphorylated by GSK3 and CDK2 itself, creating two independent FBW7-binding sites. Cyclin-E–CDK2 is thought to phosphorylate cyclin E first at T384, creating a "priming phosphate" that is needed for GSK3 to phosphorylate T380 upstream, thus generating the doubly phosphorylated phosphodegron that is specifically recognized by the FBW7 targeting subunit of SCF-FBW7.

Because GSK3 plays critical roles in diverse signals, including those activated by insulin, mitogenic growth factors, Wnts, Hedgehog, and cytokines, GSK3 activity can link the regulation of cyclin E and thus G_1 progression to different signaling pathways. For example, GSK3 is regulated by the phosphoinositide 3 kinase (PI3K)–AKT pathway, which allows a major mitogen signaling pathway (p. 87 [Hemmings and Restuccia 2012]) to couple cell growth to G_1 regulation. Transgenic expression of mutant cyclin E (T380A) in mammary glands causes more widespread hyperplasia than that of wild-type cyclin E and promotes p53 loss of heterozygosity and tumorigenesis (Smith et al. 2006). Knock-in mutations that ablate both T62 and T380 result in disruption of cyclin E periodicity, increased cyclin E activity, and abnormal proliferation in multiple cell types (Minella et al. 2008).

Studies of *Fbw7*-mutant mice and loss-of-function mutations of *FBW7* in human cancer support a role for SCF-FBW7 in negative regulation of cell proliferation by targeting cyclin E, as well as Myc, Notch, and Jun (Welcker and Clurman 2008). Mitogen signaling can also influence the activity of FBW7 itself. In mammalian cells, activated Ras increases cyclin E levels by inhibiting binding of cyclin E to FBW7 (Welcker and Clurman 2008), and Notch and Hedgehog signaling suppresses cyclin E accumulation by inducing FBW7 expression in *Drosophila* eye imaginal discs (Nicholson et al. 2011). Therefore, both oncogenic and developmental signals can control the level of cyclin E protein by regulating components of the E3 ubiquitin ligase that targets cyclin E for destruction (Fig. 3).

Cyclin D is phosphorylated at T286, a site analogous to T380 in cyclin E, and T286 phosphorylation promotes cyclin D destruction (Diehl et al. 1998). Multiple F-box proteins, such as Fbxo41, Fbxw8, SKP2, and Fbxo31, have been implicated in targeting cyclin D for destruction, but the E3 ligase responsible remains to be definitively identified (Kanie et al. 2012). Promoting the destruction of both D- and E-type G_1 cyclins by GSK3-mediated phosphorylation, however, could allow cells to effectively couple the PI3K–AKT pathway to G_1 cell cycle control. T286-phosphorylated cyclin D1 can also be recognized and stabilized in the nucleus by Pin1, a prolyl isomerase that regulates the function of proteins by causing conformational change of their S/T-phosphorylated forms (Liou et al. 2002).

Progression through G_1 phase is also controlled by other E3 ligases. In particular, the anaphase-promoting complex (APC), which promotes the ubiquitin-dependent proteasomal degradation of multiple mitotic regulatory proteins, remains active in G_1 phase to suppress accumulation of mitotic cyclins until cyclin-E–CDK2 is activated at the G_1/S transition.

5 CONTROL OF G_1 CDK INHIBITORS BY THE UBIQUITIN–PROTEASOME SYSTEM

Some CKIs are also regulated by the ubiquitin–proteasome pathway. Again, this regulation involves phosphorylation of these CKIs, which provides a mechanism linking extracellular signaling to the G_1 cell cycle control machinery.

5.1 Phosphorylation-Dependent Ubiquitylation and Degradation of a Yeast CKI

In *Saccharomyces cerevisiae*, a single CDK, Cdc28, forms multiple B-type cyclin–CDK complexes to drive both S phase and mitosis. Cdc28 is inhibited by Sic1, a CKI that is unrelated in sequence to either the p21 or INK4 family of CKIs. Sic1 is targeted for ubiquitylation (Fig. 4) following phosphorylation by the G_1 cyclin–CDK complex Cln–Cdc28 (Schwob et al. 1994). Inactivation of Sic1 rescues the inviability of yeast cells lacking the G_1 cyclins Cln1, Cln2, and Cln3 (Schneider et al. 1996), and mutation of CDK phosphorylation sites in Sic1 causes stabilization of Sic1 and blocks DNA replication. These observations indicate that the primary function of these three G_1 cyclins, once mitogenically activated, is to promote Sic1 ubiquitylation to bring about the G_1/S transition. Phosphorylated, but not unmodified, Sic1 binds to the F-box protein Cdc4, which, through a linker protein, Skp1, brings Sic1 to the Cul1 (also known as Cdc53)–Roc1 (also known as Rbx or Hrt1) E3 ligase complex for ubiquitylation by the E2 enzyme Cdc34 (Feldman et al. 1997; Skowyra et al. 1997). Nine sites in Sic1 are phosphorylated, and each contributes to Cdc4 binding, with any six being required (Nash et al. 2001). This multisite phosphorylation requirement makes Sic1 ubiquitylation ultrasensitive to the level of G_1 CDK activity, enabling cells to measure the strength of mitogens and set the level of CDK activity that determines the timing of DNA replication. It transforms a gradual accumulation process, such as protein synthesis during G_1 phase, into an irreversible switch for the onset of DNA replication. Sic1 is also phosphorylated by its target, the B-type cyclin–CDK complex Clb5–CDK1, which may ensure irreversibility of the G_1/S transition once DNA replication has been initiated.

In response to mating pheromones, budding yeast cells arrest their cycle in G_1 phase and fuse cytoplasms and nuclei to generate a diploid cell. This G_1 cell cycle arrest is regulated by the Fus3 MAPK pathway, which leads to phosphorylation and activation of Far1, a second budding yeast CDK inhibitor that is unrelated to Sic1 and other CKIs in sequence. Far1 selectively inhibits G_1 cyclin–Cdc28, leading to the inhibition of Cln–Cdc28-induced Sic1 degradation and G_1 arrest.

The distantly related fission yeast, *Schizosaccharomyces pombe*, contains a single CKI, Rum1, that is unrelated to Sic1, p21, or INK4 CKIs in sequence. Rum1 inhibits the cyclin B–CDK complex Cdc13–Cdc2 and is an essential G_1 regulator whose deletion causes premature S-phase initiation immediately after mitosis (Correa-Bordes and Nurse 1995). Rum1 is degraded following ubiquitylation by the SCF-Pop1 ligase, which uses Pop1, an ortholog of

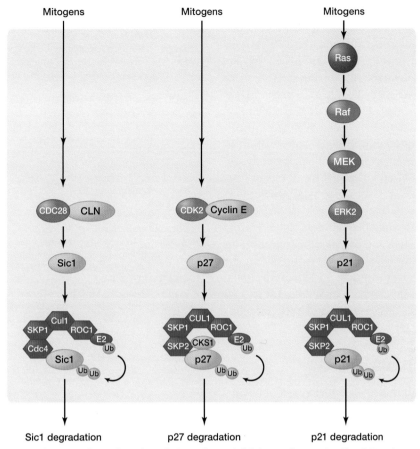

Figure 4. Targeting ubiquitin-dependent degradation of CDK inhibitors. The p21 family of CKIs is regulated by the ubiquitin–proteasome pathway. In many cases, this involves phosphorylation of these CKIs. Phosphorylated CKIs are recognized by F-box proteins such as Cdc4 in budding yeast or SKP2 in human cells, which, through the SKP1 linker protein, recruits the CKI substrate to the SCF E3 ligase for ubiquitylation.

budding yeast Cdc4, to target Rum1 (Kominami and Toda 1997). Hence, the mechanism for targeting G_1 CDK inhibitors for ubiquitylation has been conserved between two yeast species that are as evolutionarily divergent from each other as either is from animals.

5.2 Regulation of Mammalian CIP/KIP by E3 Ligases

The mammalian CKI p27 is also regulated by ubiquitin-dependent proteolysis (Pagano et al. 1995). p27 and its close relative p57 are phosphorylated by cyclin-E−CDK2 at analogous sites (T187 in p27 and T310 in p57), which promotes their binding to the F-box protein SKP2 and subsequent ubiquitylation by the SCF-SKP2 E3 ligase. The recognition of T187-phosphorylated p27 by SKP2 requires CKS1, a small evolutionarily conserved protein whose function is essential for yeast cell viability and normal mouse development (Fig. 3). A second p27 E3 ligase, KIP1-ubiquitylation-promoting complex (KPC), preferentially recognizes free p27 and is competed off by the

binding of cyclin-E−CDK2 (Kamura et al. 2004). Mitogen-stimulated cyclin E expression and thus the formation of the cyclin-E−CDK2 complex may switch cells from KPC-mediated degradation of p27 during early G_0/G_1 transitions to SCF-mediated degradation at the G_1/S transition. Likewise, p57, which plays important roles in development, is also ubiquitylated by the SCF-SKP2 E3 ligase and a second E3 ligase, SCF-FBL12, containing FBL12. FBL12 is induced by TGFβ1 and binds only to p57, providing a mechanism for TGFβ1-induced degradation of p57, but not p27 or p21 (Kim et al. 2008a).

p21 expression oscillates twice during each cell cycle: it is high in G_1 phase, decreases during S phase, reaccumulates during G_2 phase, and then decreases at early mitosis. The protein has a very short half-life (<30 min) and is rapidly turned over by ubiquitin-dependent proteolysis. Several E3 ligases can target p21 ubiquitylation at different phases of the cell cycle in both phosphorylation-dependent and phosphorylation-independent manners. During G_1 phase, sustained activation of the ERK2 MAPK by mitogenic stimuli such as epidermal growth factor (EGF) results in T57 and S130 phosphorylation on p21, leading to its ubiquitin-dependent degradation (Fig. 3) (Hwang et al. 2009). During S phase, WD40 protein CDT2 and the F-box protein SKP2 target p21 for ubiquitylation by the CRL4-CDT2 and SCF-SKP2 E3 ligases to prevent DNA rereplication (Bornstein et al. 2003; Abbas et al. 2008; Kim et al. 2008b; Nishitani et al. 2008). The SCF-SKP2-mediated p21 ubiquitylation requires S130 phosphorylation by cyclin-E–CDK2 (Bornstein et al. 2003). During early mitosis, Cdc20 binds to p21 and targets it for ubiquitylation by APC. CRL4 also targets p21 for ubiquitylation after low-dose UV irradiation, to delay the cell cycle, allowing time for optimal DNA repair (Bendjennat et al. 2003; Havens and Walter 2011; Starostina and Kipreos 2012). Hence, the mechanism for targeting G_1 CKIs for ubiquitylation has been conserved from yeast to animals and links the regulation of CKI stability to signals from different pathways via the phosphorylation of CKI proteins and their targeting molecules.

6 CONCLUDING REMARKS

Precise cell cycle regulation is an essential aspect of normal development and adult homeostasis. To achieve this, cells in G_1 phase integrate inputs from major cellular signaling pathways to decide whether or not to enter S phase, which is an irreversible cell cycle step. This integration of signals is transformed into an appropriate level of CDK activity in large part via changes in the level of cyclins and CKIs achieved through the regulation of both transcription and protein stability. One challenge for the future is to understand how multiple signaling pathways cooperate to precisely regulate cyclin and CKI activity in various cell types, particularly stem cells, in intact tissues. Another is to use this information to develop novel therapeutics for the treatment of cancer, which arises in part because of disruptions to signaling pathways that affect cell cycle regulation.

ACKNOWLEDGMENTS

We thank Alan Diehl, Andrew Koff, Kun-Liang Guan, Michele Pagano, DJ Pan, and Xin-Hai Pei for discussions, and Tadashi Nakagawa and Ruiting Lin for helping with figure preparation.

REFERENCES

*Reference is in this book.

Abbas T, Sivaprasad U, Terai K, Amador V, Pagano M, Dutta A. 2008. PCNA-dependent regulation of p21 ubiquitylation and degradation via the CRL4Cdt2 ubiquitin ligase complex. *Genes Dev* **22:** 2496–2506.

Akiyama T, Ohuchi T, Sumida S, Matsumoto K, Toyoshima K. 1992. Phosphorylation of the retinoblastoma protein by cdk2. *Proc Natl Acad Sci* **89:** 7900–7904.

Albanese C, Johnson J, Watanabe G, Eklund N, Vu D, Arnold A, Pestell RG. 1995. Transforming p21^{ras} mutants and c-Ets-2 activate the cyclin D1 promoter through distinguishable regions. *J Biol Chem* **270:** 23589–23597.

Bates S, Peters G. 1995. Cyclin D1 as a cellular proto-oncogene. *Semin Cancer Biol* **6:** 73–82.

Baugh LR, Sternberg PW. 2006. DAF-16/FOXO regulates transcription of cki-1/Cip/Kip and repression of lin-4 during *C. elegans* L1 arrest. *Curr Biol* **16:** 780–785.

Bendjennat M, Boulaire J, Jascur T, Brickner H, Barbier V, Sarasin A, Fotedar A, Fotedar R. 2003. UV irradiation triggers ubiquitin-dependent degradation of p21^{WAF1} to promote DNA repair. *Cell* **114:** 599–610.

Berthet C, Aleem E, Coppola V, Tessarollo L, Kaldis P. 2003. Cdk2 knock-out mice are viable. *Curr Biol* **13:** 1775–1785.

Bornstein G, Bloom J, Sitry-Shevah D, Nakayama K, Pagano M, Hershko A. 2003. Role of the SCFSkp2 ubiquitin ligase in the degradation of p21^{Cip1} in S phase. *J Biol Chem* **278:** 25752–25757.

Bracken AP, Kleine-Kohlbrecher D, Dietrich N, Pasini D, Gargiulo G, Beekman C, Theilgaard-Monch K, Minucci S, Porse BT, Marine JC, et al. 2007. The Polycomb group proteins bind throughout the INK4A-ARF locus and are disassociated in senescent cells. *Genes Dev* **21:** 525–530.

Brugarolas J, Chandrasekaran C, Gordon JI, Beach D, Jacks T, Hannon GJ. 1995. Radiation-induced cell cycle arrest compromised by p21 deficiency. *Nature* **377:** 552–557.

Buttitta LA, Katzaroff AJ, Perez CL, de la Cruz A, Edgar BA. 2007. A double-assurance mechanism controls cell cycle exit upon terminal differentiation in *Drosophila*. *Dev Cell* **12:** 631–643.

Cao X, Pfaff SL, Gage FH. 2008. YAP regulates neural progenitor cell number via the TEA domain transcription factor. *Genes Dev* **22:** 3320–3334.

Clayton JE, van den Heuvel SJ, Saito RM. 2008. Transcriptional control of cell-cycle quiescence during *C. elegans* development. *Dev Biol* **313:** 603–613.

Correa-Bordes J, Nurse P. 1995. p25^{rum1} orders S phase and mitosis by acting as an inhibitor of the p34^{cdc2} mitotic kinase. *Cell* **83:** 1001–1009.

Costanzo M, Nishikawa JL, Tang X, Millman JS, Schub O, Breitkreuz K, Dewar D, Rupes I, Andrews B, Tyers M. 2004. CDK activity antagonizes Whi5, an inhibitor of G_1/S transcription in yeast. *Cell* **117:** 899–913.

Deb DK, Tanaka-Matakatsu M, Jones L, Richardson HE, Du W. 2008. Wingless signaling directly regulates cyclin E expression in proliferating embryonic PNS precursor cells. *Mech Dev* **125:** 857–864.

de Bruin RA, McDonald WH, Kalashnikova TI, Yates J 3rd, Wittenberg C. 2004. Cln3 activates G_1-specific transcription via phosphorylation of the SBF bound repressor Whi5. *Cell* **117:** 887–898.

Deng C, Zhang P, Harper JW, Elledge SJ, Leder P. 1995. Mice lacking p21$^{CIP1/WAF1}$ undergo normal development, but are defective in G_1 checkpoint control. *Cell* **82:** 675–684.

de Nooij JC, Letendre MA, Hariharan IK. 1996. A cyclin-dependent kinase inhibitor, Dacapo, is necessary for timely exit from the cell cycle during *Drosophila* embryogenesis. *Cell* **87:** 1237–1247.

Diehl JA, Cheng M, Roussel M, Sherr CJ. 1998. Glycogen synthase kinase-3β regulates cyclin D1 proteolysis and subcellular localization. *Genes Dev* **12:** 3499–3511.

Dulic V, Lees E, Reed SI. 1992. Association of human cyclin E with a periodic G_1–S phase protein kinase. *Science* **257:** 1958–1961.

Duronio RJ, O'Farrell PH. 1995. Developmental control of the G_1 to S transition in *Drosophila*: Cyclin E is a limiting downstream target of E2F. *Genes Dev* **9:** 1456–1468.

Dyson N. 1998. The regulation of E2F by pRB-family proteins. *Genes Dev* **12:** 2245–2262.

El-Deiry WS, Tokino T, Velculescu VE, Levy DB, Parsons R, Lin DM, Mercer WE, Kinzler KWV, Vogelstein B. 1993. WAF1, a potential mediator of p53 tumor suppression. *Cell* **75:** 817–825.

Ewen ME, Sluss HK, Sherr CJ, Matsushime H, Kato J, Livingston DM. 1993. Functional interactions of the retinoblastoma protein with mammalian D-type cyclins. *Cell* **73:** 487–497.

Fay DS, Han M. 2000. Mutations in *cye-1*, a *Caenorhabditis elegans* cyclin E homolog, reveal coordination between cell-cycle control and vulval development. *Development* **127:** 4049–4060.

Feldman RMR, Correll CC, Kaplan KB, Deshaies RJ. 1997. A complex of Cdc4p, Skp1p, and Cdc53p/Cullin catalyzes ubiquitination of the phosphorylated CDK inhibitor Sic1p. *Cell* **91:** 221–230.

Ferrell JE Jr, Pomerening JR, Kim SY, Trunnell NB, Xiong W, Huang CY, Machleder EM. 2009. Simple, realistic models of complex biological processes: Positive feedback and bistability in a cell fate switch and a cell cycle oscillator. *FEBS Lett* **583:** 3999–4005.

Firth LC, Baker NE. 2005. Extracellular signals responsible for spatially regulated proliferation in the differentiating *Drosophila* eye. *Dev Cell* **8:** 541–551.

Franklin DS, Godfrey VL, Lee H, Kovalev GI, Schoonhoven R, Chen-Kiang S, Su L, Xiong Y. 1998. CDK inhibitors p18^{INK4c} and p27^{KIP1} mediate two separate pathways to collaboratively suppress pituitary tumorigenesis. *Genes Dev* **12:** 2899–2911.

Geng Y, Eaton EN, Picon M, Roberts JM, Lundberg AS, Gifford A, Sardet C, Weinberg RA. 1996. Regulation of cyclin E transcription by E2Fs and retinoblastoma protein. *Oncogene* **12:** 1173–1180.

Geng Y, Whoriskey W, Park MY, Bronson RT, Medema RH, Li T, Weinberg RA, Sicinski P. 1999. Rescue of cyclin D1 deficiency by knockin cyclin E. *Cell* **97:** 767–777.

Geng Y, Yu Q, Sicinska E, Das M, Schneider JE, Bhattacharya S, Rideout WM, Bronson RT, Gardner H, Sicinski P. 2003. Cyclin E ablation in the mouse. *Cell* **114:** 431–443.

Harper JW, Adami GR, Wei N, Keyomarsi K, Elledge SJ. 1993. The p21 Cdk-interacting protein Cip1 is a potent inhibitor of G1 cyclin-dependent kinases. *Cell* **75:** 805–816.

* Harvey KF, Hariharan IK. 2012. The Hippo pathway. *Cold Spring Harb Perspect Biol* **4:** a011288.

Havens CG, Walter JC. 2011. Mechanism of CRL4(Cdt2), a PCNA-dependent E3 ubiquitin ligase. *Genes Dev* **25:** 1568–1582.

* Hemmings BA, Restuccia DF. 2012. The PI3K-PKB/Akt pathway. *Cold Spring Harb Perspect Biol* **4:** a011189.

Herrera RE, Sah VP, Williams BO, Makela TP, Weinberg RA, Jacks T. 1996. Altered cell cycle kinetics, gene expression, and G$_1$ restriction point regulation in *Rb*-deficient fibroblasts. *Mol Cell Biol* **16:** 2402–2407.

Hinds PW, Mittnacht S, Dulic V, Arnold A, Reed SI, Weinberg RA. 1992. Regulation of retinoblastoma protein functions by ectopic expression of human cyclins. *Cell* **70:** 993–1006.

Hong Y, Roy R, Ambros V. 1998. Developmental regulation of a cyclin-dependent kinase inhibitor controls postembryonic cell cycle progression in *Caenorhabditis elegans*. *Development* **125:** 3585–3597.

Huang J, Wu S, Barrera J, Matthews K, Pan D. 2005. The Hippo signaling pathway coordinately regulates cell proliferation and apoptosis by inactivating Yorkie, the *Drosophila* homolog of YAP. *Cell* **122:** 421–434.

Hwang HC, Clurman BE. 2005. Cyclin E in normal and neoplastic cell cycles. *Oncogene* **24:** 2776–2786.

Hwang CY, Lee C, Kwon KS. 2009. Extracellular signal-regulated kinase 2-dependent phosphorylation induces cytoplasmic localization and degradation of p21^{Cip1}. *Mol Cell Biol* **29:** 3379–3389.

* Ingham PW. 2012. Hedgehog signaling. *Cold Spring Harb Perspect Biol* **4:** a011221.

Inze D. 2005. Green light for the cell cycle. *EMBO J* **24:** 657–662.

Jacobs JJ, Kieboom K, Marino S, DePinho RA, van Lohuizen M. 1999. The oncogene and Polycomb-group gene *bmi-1* regulates cell proliferation and senescence through the *ink4a* locus. *Nature* **397:** 164–168.

Jones L, Richardson H, Saint R. 2000. Tissue-specific regulation of cyclin E transcription during *Drosophila melanogaster* embryogenesis. *Development* **127:** 4619–4630.

Kamura T, Hara T, Matsumoto M, Ishida N, Okumura F, Hatakeyama S, Yoshida M, Nakayama K, Nakayama KI. 2004. Cytoplasmic ubiquitin ligase KPC regulates proteolysis of p27^{Kip1} at G$_1$ phase. *Nat Cell Biol* **6:** 1229–1235.

Kanie T, Onoyama I, Matsumoto A, Yamada M, Nakatsumi H, Tateishi Y, Yamamura S, Tsunematsu R, Matsumoto M, Nakayama KI. 2012. Genetic reevaluation of the role of F-box proteins in cyclin D1 degradation. *Mol Cell Biol* **32:** 590–605.

Kato J-Y, Matsushime H, Hiebert SW, Ewen M, Sherr CJ. 1993. Direct binding of cyclin D to the retinoblastoma gene product (pRb) and pRb phosphorylation by the cyclin D-dependent kinase CDK4. *Genes Dev* **7:** 331–342.

Kim M, Nakamoto T, Nishimori S, Tanaka K, Chiba T. 2008a. A new ubiquitin ligase involved in p57^{KIP2} proteolysis regulates osteoblast cell differentiation. *EMBO Rep* **9:** 878–884.

Kim Y, Starostina NG, Kipreos ET. 2008b. The CRL4Cdt2 ubiquitin ligase targets the degradation of p21^{Cip1} to control replication licensing. *Genes Dev* **22:** 2507–2519.

Kitagawa M, Higashi H, Jung HK, Suzuki-Takahashi I, Ikeda M, Tamai K, Kato J, Segawa K, Yoshida E, Nishimura S, et al. 1996. The consensus motif for phosphorylation by cyclin D1–Cdk4 is different from that for phosphorylation by cyclin A/E–Cdk2. *EMBO J* **15:** 7060–7069.

Klein EA, Assoian RK. 2008. Transcriptional regulation of the cyclin D1 gene at a glance. *J Cell Sci* **121:** 3853–3857.

Knoblich JA, Sauer K, Jones L, Richardson H, Saint R, Lehner CF. 1994. Cyclin E controls S phase progression and its down-regulation during *Drosophila* embryogenesis is required for the arrest of cell proliferation. *Cell* **77:** 107–120.

Koff A, Giordano A, Desai D, Yamashita K, Harper JW, Elledge S, Nishimoto T, Morgan DO, Franza BR, Roberts JM. 1992. Formation and activation of a cyclin E–cdk2 complex during the G$_1$ phase of the human cell cycle. *Science* **257:** 1689–1694.

Kominami K-I, Toda T. 1997. Fission yeast WD-repeat protein Pop1 regulates genome ploidy through ubiquitin–proteasome-mediated degradation of the CDK inhibitor Rum1 and the S-phase initiator Cdc18. *Genes Dev* **11:** 1548–1560.

* Kopan R. 2012. Notch signaling. *Cold Spring Harb Perspect Biol* **4:** a011213.

Korzelius J, The I, Ruijtenberg S, Prinsen MB, Portegijs V, Middelkoop TC, Groot Koerkamp MJ, Holstege FC, Boxem M, van den Heuvel S. 2011. *Caenorhabditis elegans* cyclin D/CDK4 and cyclin E/CDK2 induce distinct cell cycle re-entry programs in differentiated muscle cells. *PLoS Genet* **7:** e1002362.

Kotake Y, Cao R, Viatour P, Sage J, Zhang Y, Xiong Y. 2007. pRB family proteins are required for H3K27 trimethylation and Polycomb repression complexes binding to and silencing p16^{INK4a} tumor suppressor gene. *Genes Dev* **21:** 49–54.

Kozar K, Ciemerych MA, Rebel VI, Shigematsu H, Zagozdzon A, Sicinska E, Geng Y, Yu Q, Bhattacharya S, Bronson RT, et al. 2004. Mouse development and cell proliferation in the absence of D-cyclins. *Cell* **118:** 477–491.

Lane ME, Sauer K, Wallace K, Jan YN, Lehner CF, Vaessin H. 1996. Dacapo, a cyclin-dependent kinase inhibitor, stops cell proliferation during *Drosophila* development. *Cell* **87:** 1225–1235.

Lew D, Dulic V, Reed SI. 1991. Isolation of three novel human cyclins by rescue of G$_1$ cyclin (*Cln*) function in yeast. *Cell* **66:** 1197–1206.

Liou YC, Ryo A, Huang HK, Lu PJ, Bronson R, Fujimori F, Uchida T, Hunter T, Lu KP. 2002. Loss of Pin1 function in the mouse causes phenotypes resembling cyclin D1-null phenotypes. *Proc Natl Acad Sci* **99:** 1335–1340.

Liu TH, Li L, Vaessin H. 2002. Transcription of the *Drosophila* CKI gene *dacapo* is regulated by a modular array of *cis*-regulatory sequences. *Mech Dev* **112:** 25–36.

Malumbres M, Barbacid M. 2009. Cell cycle, CDKs and cancer: A changing paradigm. *Nat Rev Cancer* **9:** 153–166.

Malumbres M, Sotillo R, Santamaria D, Galan J, Cerezo A, Ortega S, Dubus P, Barbacid M. 2004. Mammalian cells cycle without the D-type cyclin-dependent kinases Cdk4 and Cdk6. *Cell* **118:** 493–504.

Matsushime H, Roussel MF, Ashmum RA, Sherr CJ. 1991. Colony-stimulating factor 1 regulates a novel gene (*CYL1*) during the G_1 phase of the cell cycle. *Cell* **65:** 701–713.

Merlo A, Herman JG, Mao L, Lee DJ, Gabrielson E, Burger PC, Baylin SB, Sidransky D. 1995. 5′ CpG island methylation is associated with transcriptional silencing of the tumour suppressor p16/CDKN2/MTS1 in human cancers. *Nat Med* **1:** 686–692.

Meyer CA, Jacobs HW, Datar SA, Du W, Edgar BA, Lehner CF. 2000. *Drosophila* Cdk4 is required for normal growth and is dispensable for cell cycle progression. *EMBO J* **19:** 4533–4542.

Meyer CA, Kramer I, Dittrich R, Marzodko S, Emmerich J, Lehner CF. 2002. *Drosophila* p27^{Dacapo} expression during embryogenesis is controlled by a complex regulatory region independent of cell cycle progression. *Development* **129:** 319–328.

Minella AC, Loeb KR, Knecht A, Welcker M, Varnum-Finney BJ, Bernstein ID, Roberts JM, Clurman BE. 2008. Cyclin E phosphorylation regulates cell proliferation in hematopoietic and epithelial lineages in vivo. *Genes Dev* **22:** 1677–1689.

* Morrison DK. 2012. MAP kinase pathways. *Cold Spring Harb Perspect Biol* **4:** 011254.

Motokura T, Bloom T, Kim HG, Juppner H, Ruderman JV, Kronenberg HM, Arnold A. 1991. A novel cyclin encoded by a bcl1-linked candidate oncogene. *Nature* **350:** 512–515.

Nash P, Tang X, Orlicky S, Chen Q, Gertler FB, Mendenhall MD, Sicheri F, Pawson T, Tyers M. 2001. Multisite phosphorylation of a CDK inhibitor sets a threshold for the onset of DNA replication. *Nature* **414:** 514–521.

Nicholson SC, Nicolay BN, Frolov MV, Moberg KH. 2011. Notch-dependent expression of the archipelago ubiquitin ligase subunit in the *Drosophila* eye. *Development* **138:** 251–260.

Nishitani H, Shiomi Y, Iida H, Michishita M, Takami T, Tsurimoto T. 2008. CDK inhibitor p21 is degraded by a proliferating cell nuclear antigen-coupled Cul4–DDB1Cdt2 pathway during S phase and after UV irradiation. *J Biol Chem* **283:** 29045–29052.

* Nusse R. 2012. Wnt signaling. *Cold Spring Harb Perspect Biol* **4:** a011163.

Ohtani K, Degregori J, Nevins JR. 1995. Regulation of the cyclin E gene by transcription factor E2F1. *Proc Natl Acad Sci* **92:** 12146–12150.

Ohtsubo M, Roberts JM. 1993. Cyclin-dependent regulation of G_1 in mammalian fibroblasts. *Science* **259:** 1908–1912.

Ortega S, Malumbres M, Barbacid M. 2002. Cyclin D-dependent kinases, INK4 inhibitors and cancer. *Biochim Biophys Acta* **1602:** 73–87.

Ortega S, Prieto I, Odajima J, Martin A, Dubus P, Sotillo R, Barbero JL, Malumbres M, Barbacid M. 2003. Cyclin-dependent kinase 2 is essential for meiosis but not for mitotic cell division in mice. *Nat Genet* **35:** 25–31.

Pagano M, Tam SW, Theodoras AM, Beer-Romero P, Del Sal G, Chau V, Yew PR, Draetta GF, Rolfe M. 1995. Role of the ubiquitin–proteasome pathway in regulating abundance of the cyclin-dependent kinase inhibitor p27. *Science* **269:** 682–685.

Parisi T, Beck AR, Rougier N, McNeil T, Lucian L, Werb Z, Amati B. 2003. Cyclins E1 and E2 are required for endoreplication in placental trophoblast giant cells. *EMBO J* **22:** 4794–4803.

Pei XH, Bai F, Smith MD, Usary J, Fan C, Pai S-Y, Ho IC, Perou CM, Xiong Y. 2009. CDK inhibitor p18^{INK4c} is a downstream target of GATA3 and restrains mammary luminal progenitor cell proliferation and tumorigenesis. *Cancer Cell* **15:** 389–401.

Resnitzky D, Gossen M, Bujard H, Reed S. 1994. Acceleration of the G_1/S phase transition by expression of cyclins D1 and E with an inducible system. *Mol Cell Biol* **14:** 1669–1679.

* Rhind N, Russel P. 2012. Signaling pathways that regulate cell division. *Cold Spring Harb Perspect Biol* **4:** a005942.

Roussel MF. 1999. The INK4 family of cell cycle inhibitors in cancer. *Oncogene* **18:** 5311–5317.

Schneider BL, Yang QH, Futcher AB. 1996. Linkage of replication to start by the Cdk inhibitor Sic1. *Science* **272:** 560–562.

Schwob E, Bohm T, Mendenhall MD, Nasmyth K. 1994. The B-type cyclin kinase inhibitor p40^{SIC1} controls the G_1 to S transition in *S. cerevisiae*. *Cell* **79:** 233–244.

Serrano M, Hannon GJ, Beach D. 1993. A new regulatory motif in cell cycle control causing specific inhibition of cyclin D/CDK4. *Nature* **366:** 704–707.

Sherr CJ. 1996. Cancer cell cycle. *Science* **274:** 1672–1677.

Sherr CJ, Roberts JM. 1995. Inhibitors of mammalian G_1 cyclin-dependent kinases. *Genes Dev* **9:** 1149–1163.

Sherr CJ, Roberts JM. 2004. Living with or without cyclins and cyclin-dependent kinases. *Genes Dev* **18:** 2699–2711.

Skowyra D, Craig K, Tyers M, Elledge SJ, Harper JW. 1997. F-box proteins are receptors that recruit phosphorylated substrates to the SCF ubiquitin-ligase complex. *Cell* **91:** 209–219.

Smith AP, Henze M, Lee JA, Osborn KG, Keck JM, Tedesco D, Bortner DM, Rosenberg MP, Reed SI. 2006. Deregulated cyclin E promotes p53 loss of heterozygosity and tumorigenesis in the mouse mammary gland. *Oncogene* **25:** 7245–7259.

Starostina NG, Kipreos ET. 2012. Multiple degradation pathways regulate versatile CIP/KIP CDK inhibitors. *Trends Cell Biol* **22:** 33–41.

Terada Y, Tatsuka M, Jinno S, Okayama H. 1995. Requirement for tyrosine phosphorylation of Cdk4 in G_1 arrest induced by ultraviolet irradiation. *Nature* **376:** 358–362.

van den Heuvel S, Dyson NJ. 2008. Conserved functions of the pRB and E2F families. *Nat Rev Mol Cell Biol* **9:** 713–724.

van Lohuizen M, Verbeek S, Scheijen B, Wientjens E, van der Gulden H, Berns A. 1991. Identification of cooperating oncogenes in E$^{\mu}$-*myc* transgenic mice by provirus tagging. *Cell* **65:** 737–752.

Wander SA, Zhao D, Slingerland JM. 2011. p27: A barometer of signaling deregulation and potential predictor of response to targeted therapies. *Clin Cancer Res* **17:** 12–18.

Wang TC, Cardiff RD, Zukerberg L, Lees E, Arnold A, Schmidt EV. 1994. Mammary hyperplasia and carcinoma in MMTV–cyclin D1 transgenic mice. *Nature* **369:** 669–671.

Weinberg RA. 1995. The retinoblastoma protein and cell cycle control. *Cell* **81:** 323–330.

Welcker M, Clurman BE. 2008. FBW7 ubiquitin ligase: A tumour suppressor at the crossroads of cell division, growth and differentiation. *Nat Rev Cancer* **8:** 83–93.

Xiong W, Ferrell JE Jr. 2003. A positive-feedback-based bistable "memory module" that governs a cell fate decision. *Nature* **426:** 460–465.

Xiong Y, Connolly T, Futcher B, Beach D. 1991. Human D-type cyclin. *Cell* **65:** 691–699.

Xiong Y, Zhang H, Beach D. 1992. D-type cyclins associate with multiple protein kinases and the DNA replication and repair factor PCNA. *Cell* **71:** 505–514.

Xiong Y, Hannon G, Zhang H, Casso D, Kobayashi R, Beach D. 1993a. p21 is a universal inhibitor of the cyclin kinases. *Nature* **366:** 701–704.

Xiong Y, Zhang H, Beach D. 1993b. Subunit rearrangement of cyclin-dependent kinases is associated with cellular transformation. *Genes Dev* **7:** 1572–1583.

Yu Q, Geng Y, Sicinski P. 2001. Specific protection against breast cancers by cyclin D1 ablation. *Nature* **411:** 1017–1021.

Yuan Y, Shen H, Franklin DS, Scadden DT, Cheng T. 2004. In vivo self-renewing divisions of haematopoietic stem cells are increased in the absence of the early G_1-phase inhibitor, p18^{INK4C}. *Nat Cell Biol* **6:** 436–442.

Zhao H, Chen X, Gurian-West M, Roberts JM. 2012. Loss of cyclin-dependent kinase 2 (CDK2) inhibitory phosphorylation in a CDK2AF knock-in mouse causes misregulation of DNA replication and centrosome duplication. *Mol Cell Biol* **32:** 1421–1432.

Zindy F, Quelle DE, Roussel MF, Sherr CJ. 1997. Expression of the p16^{INK4a} tumor suppressor versus other INK4 family members during mouse development and aging. *Oncogene* **15:** 203–211.

CHAPTER 6

Signaling Pathways that Regulate Cell Division

Nicholas Rhind[1] and Paul Russell[2]

[1]Department of Biochemistry and Molecular Pharmacology, University of Massachusetts Medical School, Worcester, Massachusetts 01605

[2]Department of Molecular Biology, Department of Cell Biology, The Scripps Research Institute, La Jolla, California 92037

Correspondence: nick.rhind@umassmed.edu

SUMMARY

Cell division requires careful orchestration of three major events: entry into mitosis, chromosomal segregation, and cytokinesis. Signaling within and between the molecules that control these events allows for their coordination via checkpoints, a specific class of signaling pathways that ensure the dependency of cell-cycle events on the successful completion of preceding events. Multiple positive- and negative-feedback loops ensure that a cell is fully committed to division and that the events occur in the proper order. Unlike other signaling pathways, which integrate external inputs to decide whether to execute a given process, signaling at cell division is largely dedicated to completing a decision made in G_1 phase—to initiate and complete a round of mitotic cell division. Instead of deciding if the events of cell division will take place, these signaling pathways entrain these events to the activation of the cell-cycle kinase cyclin-dependent kinase 1 (CDK1) and provide the opportunity for checkpoint proteins to arrest cell division if things go wrong.

Outline

1 INTRODUCTION

The cell cycle (see Fig. 1) consists of DNA synthesis (S) and mitosis (M) phases separated by gap phases in the order G_1–S–G_2–M (Murray and Hunt 1993; Nurse 2000; Morgan 2006). Cell division involves two connected processes triggered at the end of G_2 phase: mitosis itself (segregation of the chromosomes, which duplicate in S phase) and cytokinesis (division of the cell, per se). Mitosis can be subdivided into six distinct phases (see Box 1): (1) prophase, in which the spindle begins to assemble in the cytoplasm and chromosomes begin to condense in the nucleus; (2) prometaphase, in which the nuclear envelope breaks down and chromosomes attach to the spindle; (3) metaphase, in which chromosomes align at the spindle midzone; (4) anaphase A, in which chromosomes move to the centrosomes, which form the spindle poles; (5) anaphase B, in which the spindle elongates; and (6) telophase, in which the nuclear envelope reforms around the new daughter nuclei. Mitosis in yeasts differs in that the nuclear envelope does not break down; instead the spindle-pole body, the yeast equivalent of the centrosome, spans the nuclear envelope, allowing the spindle to access both the nucleus and the cytoplasm. Signals during telophase trigger cytokinesis, which separates the daughter nuclei into two daughter cells.

The signaling pathways that operate in G_2 phase to control the onset of mitosis, and those that operate during mitosis to control chromosome segregation and the initiation of cytokinesis, are somewhat different from other cellular signaling pathways. Instead of being involved in decisions about potential cell fates or responses to varying environmental conditions, they are involved in executing a decision that is made in G_1 phase—to enter and complete another cell cycle. These signaling pathways have two key roles: to coordinate potentially independent events, such as mitosis and cytokinesis; and to provide quality-control checkpoints that arrest the process when things go wrong. The cell division signaling pathways are the archetypical checkpoints, defined as signaling pathways that ensure a dependency for the execution of later cell-cycle events on the successful completion of preceding events (Hartwell and Weinert 1989).

Two major transitions are required for cell division: the G_2/M transition and the metaphase/anaphase transition. These are regulated by the protein kinase cyclin-dependent kinase 1 (CDK1) and the anaphase-promoting complex (APC, an E3 ubiquitin ligase), respectively (Murray and Hunt 1993; Nurse 2000; Morgan 2006). During G_2 phase CDK1 is maintained at a low level of activity by mechanisms described below. Activation of CDK1 is necessary and sufficient to trigger entry into mitosis. Thus, CDK1 activation is the focal point of many signaling pathways that control the commitment to cell division. These pathways have both positive effects, such as activating CDK1 when cells reach a critical size, and negative effects, such as delaying CDK1 activation in the presence of DNA damage. In general, the negative regulatory pathways, known as checkpoints, are much better understood than the positive regulatory pathways. Activation of CDK1 and entry into mitosis lead to activation of the APC. The APC in turn causes separation of sister chromatids, which allows them to segregate to opposite spindle poles at anaphase and the complete inactivation of CDK1, which allows exit from mitosis and the resetting of the cell cycle to G_1 phase. Being necessary and sufficient for the initiation of anaphase, the

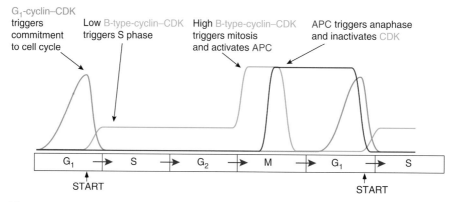

Figure 1. The major events of the cell cycle. The major events of the cell cycle are regulated by successive waves of kinase and ubiquitin ligase activity. G_1-cyclin–CDK activity is required to initiate the cell cycle and activate B-type-cyclin–CDK activity. Low levels of B-type-cyclin–CDK activity are sufficient to trigger S phase, but tyrosine phosphorylation by Wee1 prevents full activation, preventing premature mitosis. Full CDK activation triggers mitosis and activates APC, which triggers anaphase and feeds back to inactivate CDK activity. Inactivation of CDK allows exit from mitosis and the reestablishment of interphase chromosome and nuclear structure in G_1 phase. See Box 1 for description of the stages of mitosis.

Cite this chapter as *Cold Spring Harb Perspect Biol* doi: 10.1101/cshperspect.a005942

BOX 1. THE MAIN STAGES OF MITOSIS

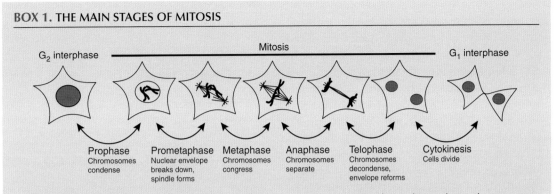

Prophase Triggered by activation of CDK1, CDK1–cyclin-B is imported into the nucleus, chromosomes condense from their diffuse interphase state to compact rods, and the centrosomes begin to separate, forming a spindle between them.

Prometaphase The nuclear envelope breaks down, the mature spindle is formed with centrosomes on either side of the cell, and so-called kinetochore microtubules from the spindle poles interact with the kinetochore protein complexes that form at chromosome centromeres, moving chromosomes to the spindle midzone. In yeasts, which do not break down their nuclear envelopes for mitosis, the spindle forms inside the nucleus, and the spindle poles span the nuclear envelope to connect nuclear and cytoplasmic microtubules.

Metaphase When each of the kinetochores on a pair of sister chromosomes is attached to microtubules from opposite spindle poles, the opposing force pulls the pair to the metaphase plate at the middle of the spindle in a "bi-oriented" configuration. Until all the chromosomes are bi-oriented, unattached kinetochores produce a checkpoint signal that prevents the metaphase/anaphase transition.

Anaphase Once the mitotic checkpoint has been satisfied, the APC ubiquitin ligase is activated. Ubiquitin-dependent proteolysis of the inhibitor securin leads to activation of separase—a protease that cleaves the cohesin proteins that hold sister chromatids together. This cleavage causes a loss of chromosome cohesion, allowing the chromosomes to separate. They do so in two movements: anaphase A, in which chromosomes are pulled toward the spindle poles by contraction of the kinetochore microtubules; and anaphase B, in which the spindles are pushed away from each other by the elongation of inter-spindle-pole microtubules.

Telophase After the chromosomes have been segregated to opposite sides of the cell, CDK1 activity is inhibited by APC-mediated destruction of cyclin B and activation of the CDK1-antagonizing phosphatase Cdc14. The reduction in CDK1 activity allows for reformation of the nuclear envelope, decondensation of chromosomes, and entry into G_1.

APC is the major target of the checkpoints that arrest cells in metaphase (Fig. 1).

Below we examine these G_2 and mitotic signaling pathways and the checkpoints that regulate them, explaining how they ensure that events take place in the correct order even in the face of cell-cycle disruptions.

2 CDK1 AND ENTRY INTO MITOSIS

The onset of mitosis is brought about by the activation of CDK1, a 34-kDa proline-directed serine/threonine protein kinase (Nurse 1990). CDK1, also known as Cdc2 (cell division cycle 2), phosphorylates hundreds of different substrates at an S/TPx(x)R/K consensus, including downstream effector kinases, to initiate the events required for progression through mitosis. Unsurprisingly, it is tightly regulated at many levels. CDK1 is maintained at a constant concentration throughout the cell cycle; however, its periodic activation and inactivation is achieved by interactions

with specific regulatory subunits and by dephosphorylation of critical residues.

As their name indicates, CDKs such as CDK1 require binding of a regulatory cyclin subunit (Murray and Hunt 1993; Morgan 1995; Bloom and Cross 2007), which activates the kinase by causing conformational changes near the active site. These structural changes allow the kinase to bind ATP and substrates in an orientation that promotes the transfer of the terminal γ phosphate of ATP to the target serine or threonine residue in the substrate. Different CDK–cyclin heterodimers regulate the different cell-cycle transitions (Morgan 2006). The G_1 cyclins, in association with CDK4, CDK6, and CDK2, regulate entry into the cell cycle (Duronio and Xiong 2012), whereas S-phase events and the G_2/M transition are primarily regulated by CDK1 (CDK2 also functions in S phase) bound to a members of the B-type family of cyclins (the CDK1–cyclin-B complex is also known as M-phase promoting factor [MPF]).

The periodic accumulation and disappearance of cyclins has a central role in regulating CDK1 activity during the cell cycle. B-type cyclin genes are generally expressed during S and G_2 phases, leading to the accumulation of CDK1–cyclin complexes during these phases of the cell cycle. At the metaphase/anaphase transition, APC ubiquitylates B-type cyclins, which targets them for rapid destruction by the proteasome. The destruction of cyclins causes a precipitous drop in CDK activity, resetting the cell cycle as cells enter G_1 phase.

B-type cyclins comprise a large and quickly evolving protein family (Table 1). Most species express multiple B-type cyclins, which are believed to confer different substrate specificities on CDK1–cyclin complexes. However, these differences do not appear to be crucial. For example, cyclin A (a mammalian B-type cyclin) is expressed during S phase and is responsible for triggering DNA replication in conjunction with CDK2, but in its absence cyclin B can provide this function (Kalaszczynska et al. 2009). More dramatically, in the fission yeast *Schizosaccharomyces pombe*, the cell cycle can be regulated by a single cyclin–CDK1 complex, demonstrating that differential B-type-cyclin–CDK1 specificities are not required for S-phase and M-phase events (Coudreuse and Nurse 2010). Likewise, a single B-type cyclin can drive S phase and M phase in budding yeast (Haase and Reed 1999).

These results have led to a quantitative model of cell-cycle regulation, in which relatively high CDK1 activity is required to bring about the onset of M phase, whereas a much lower level of CDK1 activity is sufficient to catalyze the events of S phase (Stern and Nurse 1996; Coudreuse and Nurse 2010). Nonetheless, different cyclins are expressed at different times and confer differential substrate specificities on CDK1, allowing layers of fine-tuning to the basic quantitative model (Uhlmann et al. 2011).

Cyclin binding is necessary but not sufficient for robust CDK1 activity. Full activation of CDK1 also requires phosphorylation of a threonine residue near the active site, which causes further realignment of active-site residues into an active conformation. This phosphorylation is catalyzed by a CDK-activating kinase (CAK), which requires the CDK to be cyclin bound (Morgan 1995). Curiously, the identity of CAK varies, depending on the species, and in some organisms there are multiple CAKs. CAK activity does not change during the cell cycle, nor is the dephosphorylation of the CAK-phosphorylated CDK–cyclin complexes regulated in a cell-cycle-dependent manner. Thus, phosphorylation of CDK1 by CAK is dependent on cyclin binding and is essential for cell division, but it does not control when CDK1 is activated.

Full activation of CDK1, and thus entry into M phase, is restrained by Wee1 and related protein kinases (Fig. 2),

Table 1. Key proteins in cell division control

Protein	Budding yeast	Fission yeast	Human
Cyclin-dependent kinase 1	Cdc28	Cdc2	**CDK1**[a]
S-phase-expressed B-type cyclin	Clb5,6	Cig2	**Cyclin A**
M-phase-expressed B-type cyclin	Clb1,2	Cdc13	**Cyclin B**
CDK-inhibitory kinase	Swe1	**Wee1**, Mik1	**WEE1**, MYT1
CDK-activating phosphatase	Mih1	**Cdc25**	CDC25B, CDC25C
Checkpoint kinase	Tel1	Tel1	**ATM**
Checkpoint kinase	Mec1	Rad3	**ATR**
Checkpoint effector kinase	**Chk1**	**Chk1**	**CHK1**
Checkpoint effector kinase	Rad53, Dun1	Cds1	**CHK2**
MRN nuclease	**Mre11**	Rad32	**MRE11**
MRN scaffold	**Rad50**	**Rad50**	**RAD50**
MRN regulator	Xrs2	**Nbs1**	**NBS1**
ATR targeting subunit	Ldc1	Rad26	**ATRIP**
9-1-1 checkpoint clamp	Ddc1	**Rad9**	**RAD9**
9-1-1 checkpoint clamp	Rad17	**Rad1**	**RAD1**
9-1-1 checkpoint clamp	Rad24	**Hus1**	**HUS1**
Checkpoint mediator	–[b]	–	**MDC1**
Checkpoint mediator	Rad9	Crb2	
Checkpoint mediator	Dpb11	Rad4/Cut5	**TopBP1**
Checkpoint mediator	Mrc1	Mrc1	Claspin
Fork protection complex	Tof1	Swi1	Timeless
Fork protection complex	Csm3	Swi3	Tipin
Mitotic kinase	Cdc5	Plo1	Plk1-4
Mitotic kinase	Ipl1	Ark1	**Aurora** A/B
APC regulator	**Cdc20**	Slp1	**CDC20**
APC regulator	**Cdh1**	Srw1	**CCH1**
Mitotic checkpoint regulator	**Mad1**	**Mad1**	**MAD1**
Mitotic checkpoint regulator	**Mad2**	**Mad2**	**MAD2**
Mitotic checkpoint regulator	**Mad3**	**Mad3**	BUBR1
Mitotic checkpoint kinase	**Bub1**	**Bub1**	**BUB1**
Mitotic checkpoint regulator	**Bub3**	**Bub3**	BUB3
MEN/SIN scaffold	Nud1	Cdc11	?[c]
MEN/SIN GTPase	Tem1	Spg1	?
MEN/SIN GAP	Bub2	Cdc16	?
MEN/SIN GAP cofactor	Byr4	Byr4	?
MEN/SIN GEF	Lte1	Etd1	?
MEN/SIN kinase	Cdc15	Cdc7	MST1/2
MEN/SIN kinase	–	Sid1	MST1/2
MEN/SIN kinase regulator	–	Cdc14	?
MEN/SIN kinase	Dbf2	Sid2	LATS1/2
MEN/SIN kinase regulator	Mob1	Mob1	MOB1A,B
Phosphatase	**Cdc14**	Clp1	**CDC14**

[a]Names used in the text are in bold.
[b]–, No ortholog is believed to exist.
[c]?, An ortholog has not been identified.

Cite this chapter as *Cold Spring Harb Perspect Biol* doi: 10.1101/cshperspect.a005942

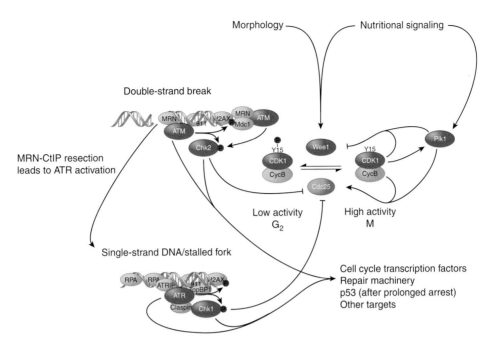

Figure 2. Signaling at the G_2/M transition. The rate-limiting step for the transition from G_2 to mitosis is the dephosphorylation of CDK1 on Y15, and in some organisms T14. This phosphorylation is catalyzed by the Wee1 family of dual-specificity kinases and the phosphate is removed by Cdc25 phosphatases. Most of the many signaling pathways that affect the G_2/M transition regulate Wee1 or Cdc25. The DNA damage and replication checkpoints inactivate Cdc25; the morphogenesis and nutritional checkpoints activate Wee1. CDK1 regulates its own activation as part of a feedback loop by directly phosphorylating Wee1 and Cdc25 or doing so indirectly through Plk1.

which inhibit the activity of CDK1–cyclin complexes that accumulate during S and G_2 phases (Russell and Nurse 1987; Lundgren et al. 1991; Mueller et al. 1995). Wee1 family kinases resemble serine/threonine protein kinases, but actually phosphorylate CDK1 on a tyrosine residue (Y15), as well as the adjacent threonine residue (T14) once it is bound to a cyclin (Gould and Nurse 1989; Featherstone and Russell 1991; Parker and Piwnica-Worms 1992; Mueller et al. 1995). These phosphorylations probably inhibit CDK1 kinase activity by interfering with substrate binding (Welburn et al. 2007). Once CDK1 is bound to a cyclin and phosphorylated by CAK and Wee1, it is primed and ready to be activated through the dephosphorylation of T14 and Y15. Cdc25, a dual-specificity protein phosphatase, catalyzes this dephosphorylation reaction to bring about the G_2/M transition (Gautier et al. 1991; Kumagai and Dunphy 1991; Strausfeld et al. 1991; Beausoleil et al. 2006).

Studies of fission yeast established that Wee1 and Cdc25 together determine when cells initiate mitosis (Nurse 1975; Russell and Nurse 1986, 1987). Mutants lacking Wee1 divide prematurely at about half the size of wild-type cells. Strains that have extra copies of *wee1* divide at progressively larger sizes that directly correlate with increasing *wee1* gene dosage. In contrast, elimination of Cdc25 activity generates cells that grow progressively larger and cannot initiate

mitosis. Strains that overexpress Cdc25 look similar to *wee1* mutants.

The opposing activities of Wee1 and Cdc25 underlie the rapid "switchlike" activation of CDK1 that occurs at the G_2/M transition (Fig. 2). This behavior is controlled by feedback regulation from CDK1 (Pomerening et al. 2003). CDK1 can directly phosphorylate Wee1 and Cdc25. It can also activate Polo-like kinase 1 (Plk1), which in turn leads to the degradation of Wee1 and stimulates Cdc25 phosphatase activity (Kumagai and Dunphy 1996). These feedback loops create a bi-stable regulatory mechanism in which passage through a tipping point ensures the rapid and stable transition from a low-activity CDK1 state to a high-CDK1-activity state.

In addition to the feedback loops that activate CDK1, a feedback loop in animal cells inactivates the protein phosphatase 2A (PP2A), which antagonizes CDK phosphorylation (Wurzenberger and Gerlich 2011). CDK1 activates the Greatwall kinase, which in turn activates Arpp19 and α-endosulfine, two small inhibitors of the B55 isoform of PP2A (Fig. 3) (Castilho et al. 2009; Gharbi-Ayachi et al. 2010; Mochida et al. 2010). Inhibition of PP2A-B55 increases the effective activity of CDK1 by reducing the rate at which CDK1 substrates are dephosphorylated. These interacting feedback loops are believed to be critical for

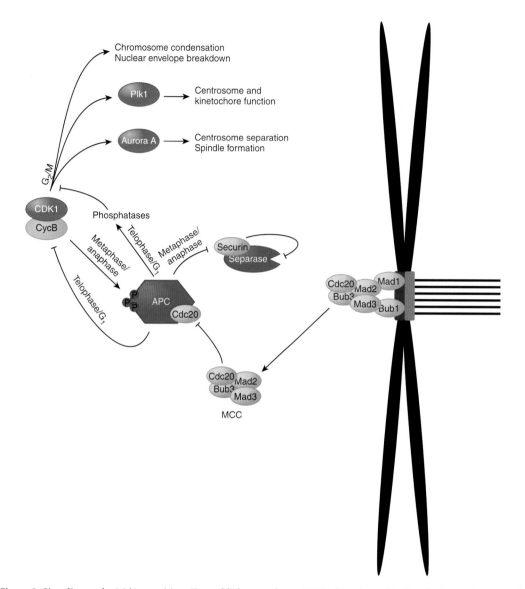

Figure 3. Signaling at the M/A transition. To establish metaphase, CDK1 directly, and indirectly through a network of kinases including Plk1 and Aurora A, phosphorylates substrates that trigger nuclear envelope breakdown, chromosome condensation, centrosome separation, and spindle assembly. In addition, in animal cells CDK1 activates the Greatwall kinase, which indirectly inactivates the CDK-antagonizing PP2A-B55 phosphatase via phosphorylation of the small phosphatase inhibitors Arpp19 and endosulfine. The rate-limiting step for the transition from metaphase to anaphase is the activation of the APC ubiquitin ligase. The APC is activated by binding of the Cdc20 regulatory subunit (and later Cdh1) and by CDK1 phosphorylation. Active APC targets securin for destruction, activating separase to release chromatid cohesion and trigger anaphase. In the presence of unattached kinetochores, anaphase is delayed by the mitotic checkpoint complex (MCC), a soluble inhibitor of the APC produced at unattached kinetochores. The APC also feeds back to inactivate CDK1 by targeting cyclin B for degradation and activates phosphatases (CDC14 in yeasts and PP1 and PP2A-B55 in animals) that oppose CDK activity, resetting the cell cycle to a CDK-free G_1 state.

ensuring the commitment to mitosis that occurs at the G_2/M transition (Domingo-Sananes et al. 2011).

The regulation of protein subcellular localization, in addition to the regulation of protein activity, is important in the control of the G_2/M transition. In particular, CDK1–cyclin-B complexes, which have critical nuclear

targets, are maintained in the cytoplasm during G_2 phase, where they can be inhibited by Wee1, but sequestered away from activation by nuclear Cdc25. One of the early events of the G_2/M transition is the nuclear localization of CDK1–cyclin-B (Porter and Donoghue 2003). However, it is unclear whether nuclear localization of CDK1–cyclin-B

Cite this chapter as *Cold Spring Harb Perspect Biol* doi: 10.1101/cshperspect.a005942

causes its activation or is a consequence of activation that enforces the CDK1–Cdc25 positive-feedback loop. Subcellular localization of regulatory proteins also plays a crucial role in the control of cytokinesis, as described below.

3 REGULATION OF THE G_2/M TRANSITION

The G_2/M transition that commits cells to division is, in general, a default consequence of initiating the cell cycle at the G_1/S transition; examples of cells voluntarily arresting in G_2 phase are rare. However, what ultimately triggers the activation of CDK1 is unclear. Plk1 may do so (Pomerening et al. 2003; Barr et al. 2004), but such a model just begs the question of what activates Plk1. Moreover, the involvement of Plk1 in the feedback loops that regulate CDK1 makes it difficult to disentangle cause and effect.

One important input is cell size (Kellogg 2003; Tzur et al. 2009). In steady-state growth conditions, cell division is coordinated with cellular growth in such a way that a newly born cell doubles its mass before undergoing division. This coordination can be achieved by linking the activation of CDK1 at the G_2/M transition to attainment of a particular cell size (Jorgensen and Tyers 2004). A number of potential mechanisms have been explored over the years (Turner et al. 2012), including size-dependent accumulation of activators (Tyson 1983), spatial gradients (Martin and Berthelot-Grosjean 2009; Moseley et al. 2009), and translational capacity (Polymenis and Schmidt 1997). Nonetheless, the mechanism by which cell size is measured is still unknown.

Another regulator, and one that is intimately entwined with cell size, is cell morphology. In budding yeast, defects in cell morphology or cytoskeletal architecture delay cell division (Howell and Lew 2012). This cell-cycle delay is enforced by the accumulation of Swe1, the budding yeast Wee1 ortholog. Swe1 is normally degraded at the end of S phase, which allows the dephosphorylation of CDK1 by Mih1, the budding yeast Cdc25 homolog. Swe1 degradation requires its localization to the bud neck; thus, disruption of bud-neck structure leads to Swe1 accumulation and cell-cycle arrest. In addition, disruption of the actin cytoskeleton also leads to Swe1 stabilization, further tying cell-cycle progression to cell morphology.

An unrelated signaling pathway required for proper morphology in fission yeast also affects the timing of division (Martin and Berthelot-Grosjean 2009; Moseley et al. 2009). Wee1 localizes at cortical nodes near the middle of cells. Pom1, a serine/threonine protein kinase, forms a concentration gradient, spreading out from the ends of cells, and indirectly activates Wee1 via inhibitory phosphorylation of Cdr2, a kinase that colocalizes with, and negatively regulates, Wee1 (Hachet et al. 2011). In newly born cells, which are short, the concentration of Pom1 near

the center of the cells is reasonably high, which results in higher Wee1 kinase activity. As cells grow, the concentration of Pom1 decreases near the middle of the cells, decreasing Wee1 activity and tipping the Wee1–Cdc25 balance in favor of Cdc25 and activation of CDK1. This model explains how Pom1 can influence cell size at division; however, cells lacking Pom1 divide at a well-defined length, albeit about 15% shorter than that of wild type, which implies that Pom1 is not required to maintain accurate size control. Because Pom1-deficient cells frequently display asymmetrical division, instead of controlling cell size per se, Pom1 appears to function in a morphogenesis checkpoint that arrests cell-cycle progression if the incipient division site is too close to one of the cell tips.

The size at which cells initiate mitosis is also linked to nutrient conditions (Fantes and Nurse 1977). Although nutritional signaling via the PI3K and TOR pathways has profound effects on G_1 cell-cycle progression (Ch. 5 [Duronio and Xiong 2012] and Ch. 7 [Ward and Thompson 2012]), it can also affect the G_2/M transition. For instance, in fission yeast, poor nitrogen availability activates the Spc1/Sty1 mitogen-associated protein kinase (MAPK), which in turn activates Plk1, thereby inhibiting Wee1 and activating Cdc25 to delay CDK1 activation (Shiozaki and Russell 1995; Petersen and Nurse 2007). Note that this nutritional signaling may be considered part of the G_2 checkpoints discussed below.

4 CHECKPOINT REGULATION OF THE G_2/M TRANSITION

Although arrest in G_2 phase in normal, unperturbed cells is uncommon, it occurs in response to activation of a variety of quality-control checkpoints. Because Y15 dephosphorylation of CDK1 is the rate-limiting step for entry into mitosis, these checkpoints target the regulators of this step: Wee1 and Cdc25. The best-understood checkpoints are those activated by DNA damage and problems with DNA replication.

5 THE DNA DAMAGE AND REPLICATION CHECKPOINTS

The purpose of the cell division cycle is to distribute complete and accurate copies of the genome to daughter cells. Genome instability arises if cells initiate mitosis when chromosomes are only partially replicated or are damaged by a double-strand DNA break (DSB). The consequences of genome instability can be cell death or neoplastic transformation. DNA damage and replication checkpoints ensure that the onset of nuclear division is delayed when chromosomes are broken or incompletely replicated. The cellular responses to DNA damage and replication-fork stalling are controlled by ATM (ataxia telangiectasia mutated), ATR (ATM and Rad3-related), and DNA-PK (DNA-dependent protein kinase) (Shiloh 2003; Cimprich and Cortez 2008). These protein kinases belong to the PIKK (phosphoinositide-3-like-kinase kinase) family, whose members include mTOR (p. 91 [Laplante and Sabatini 2012]). PIKKs strongly prefer to phosphorylate serine and threonine residues followed by glutamine (S/TQ). Of the checkpoint kinases, ATM and ATR are conserved from yeasts to humans, whereas DNA-PK is found only in metazoans (but not all).

ATM, ATR, and DNA-PK are nuclear proteins that rapidly localize to sites of DNA damage through interactions with targeting factors. Acting with accessory proteins, these kinases set up signaling platforms on the chromatin that flanks DNA lesions. DNA-PK associates with the Ku70–Ku80 heterodimeric protein complex that binds to DNA ends and promotes repair by nonhomologous end joining. DNA-PK does not have a clearly established role in regulating cell-cycle progression. ATM and ATR, in contrast, control multiple responses to DNA damage, including regulation of DNA synthesis and repair proteins, regulation of transcription of the genes involved in DNA synthesis and repair, and regulation of cell-cycle progression. ATM and ATR regulate some of these activities directly, whereas others are controlled by their effector kinases: Chk1 and Chk2. ATM functions specifically in the signaling downstream from DSBs, whereas ATR is activated in response to formation of single-stranded DNA (ssDNA), which can occur through resection of DSBs, by endonucleolytic removal of other types of DNA lesions, or by the stalling or collapse of replication forks.

6 THE DNA DAMAGE CHECKPOINT TRIGGERED BY DSBs

DSBs are especially dangerous DNA lesions because they disrupt the physical continuity of chromosomes. Initiating mitosis with broken chromosomes results in gross genome instability, which is a feature of cancer in humans and is usually lethal in single-celled microorganisms (Hartwell and Weinert 1989; Harper and Elledge 2007; Jackson and Bartek 2009). DSBs therefore generate a potent checkpoint response. Indeed, in yeasts a single unrepaired DSB is sufficient to induce a robust checkpoint response that prevents the onset of mitosis even as cells continue to grow beyond the size at which they normally divide (Harrison and Haber 2006).

ATM initiates the checkpoint response to DSBs (Shiloh 2003). It associates with DSBs by interacting with the MRN protein complex, which consists of Mre11 (a nuclease), Rad50 (a dimeric scaffolding protein with ATPase activity), and Nbs1 (an adaptor protein) subunits (Fig. 2). Mre11 and Rad50, which are conserved in eubacteria and archaea, directly engage DNA ends. Nbs1, which is only found in eukaryotes, directly binds ATM (Stracker and Petrini 2011).

Once localized at DNA ends, ATM phosphorylates proteins involved in cell-cycle checkpoints, DNA repair, and chromatin structure (Fig. 2). Phosphoproteomic studies suggest that it has hundreds of substrates, but only a few have been studied in detail (Matsuoka et al. 2007). One of these is Chk2, which is activated by ATM. Another ATM target is an SQ motif in the exposed carboxy-terminal tail of the histone variant H2AX. Phospho-H2AX (also known as γH2AX) serves as a marker for DNA damage in large multi-kilobase domains of chromatin flanking DSBs (Bonner et al. 2008). In mammals, γH2AX establishes a recruitment platform for mediator of DNA-damage checkpoint protein 1 (Mdc1), which is critical for amplifying and maintaining the checkpoint signal (Stewart et al. 2003). The carboxy-terminal region of Mdc1 contains a pair of breast cancer susceptibility protein 1 (BRCA1) carboxy-terminal (BRCT) domains that form a phosphopeptide-binding pocket for the phosphorylated γH2AX tail (Stucki et al. 2005). Once Mdc1 binds to γH2AX, it recruits more MRN complex through an interaction involving phosphorylated motifs in Mdc1 and the amino-terminal forkhead-associated (FHA) domain of Nbs1 (Melander et al. 2008; Spycher et al. 2008). These constitutive phosphorylations of Mdc1 appear to be catalyzed by the kinase CK2. The MRN complex that binds to Mdc1 recruits additional ATM, thus establishing a checkpoint signal amplification loop that activates Chk2.

In addition to recruiting ATM to DSBs, the MRN complex functions with the DNA-end-processing factor CtIP (known as Sae2 in budding yeast and Ctp1 in fission yeast) to initiate the 5′-3′ resection of DSBs (Mimitou and Symington 2011). This resection generates ssDNA tails that are extended through further resection by Exo1 exonuclease and the Dna2–Sgs1 (also known as BLM) DNA endonuclease–helicase. These ssDNA tails are rapidly bound by heterotrimeric replication protein A (RPA). RPA is essential for homology-directed repair of DNA damage, but it is also a recruitment factor for ATR (Cimprich and Cortez 2008). ATR associates with RPA through its targeting subunit ATRIP (Fig. 2). ATM promotes CtIP recruitment to DSBs in mammals, thereby enhancing resection; hence, ATM is required for recruitment and activation of ATR at DSBs (You et al. 2009). Recruitment of the CtIP orthologs in budding yeast (Sae2) or fission yeast (Ctp1) does not require the ATM ortholog Tel1, which is a key reason why the ATR orthologs have a more dominant checkpoint role in these organisms.

DNA damage is recognized independently by a hetero-trimeric ring-shaped protein complex known as the 9-1-1 (Rad9–Hus1–Rad1) checkpoint clamp. Rad17 and four subunits of replication factor C (RFC) form a protein complex that loads the 9-1-1 checkpoint clamp onto resected DNA ends. The 9-1-1 complex associates with TopBP1, which also interacts with ATR and enhances its kinase activity (Navadgi-Patil and Burgers 2009). Both ATR and 9-1-1 are required for checkpoint signaling. Furthermore, in budding yeast, ectopically targeting both complexes to the same locus is sufficient for checkpoint activation, which suggests that recruitment of the two complexes to sites of damage is necessary and sufficient for checkpoint signaling (Bonilla et al. 2008).

ATM and ATR share multiple substrates, including histone H2AX. In fission yeast the checkpoint mediator protein Crb2 has carboxy-terminal BRCT domains that bind to γH2AX. In addition, Crb2 also has Tudor domains that bind to histone H4 dimethylated on K20. Crb2 can also localize to DSBs by binding to the TopBP1 ortholog, which interacts with the 9-1-1 checkpoint clamp. Either Crb2-recruitment pathway is sufficient to mount a partial checkpoint response, but loss of both pathways abrogates the checkpoint because Crb2 localization at DSBs is required for activation of Chk1 by the ATR ortholog (Du et al. 2006).

7 THE DNA REPLICATION CHECKPOINT TRIGGERED BY STALLED REPLICATION FORKS

During DNA replication, ssDNA is exposed at the replication fork when DNA lesions, DNA-bound protein complexes, or limiting supplies of deoxynucleotide triphosphates (dNTPs) cause replicative DNA polymerases to slow down or stall. This ssDNA is bound by RPA, which elicits a checkpoint response by ATR/ATRIP and their orthologs (Bartek et al. 2004; Branzei and Foiani 2010). This checkpoint also requires the 9-1-1 checkpoint clamp and TopBP1, similarly to the DNA damage checkpoint activated by DSBs, but the checkpoint mediators are different. In budding and fission yeast Mrc1 (mediator of the replication checkpoint 1) travels with the replication fork and is required for activation of Chk2. In addition to other functions, the activated Chk2 arrests cell-cycle progression and maintains the S-phase pattern of gene expression (de Bruin et al. 2008; Dutta et al. 2008), which is critical for completing DNA replication. A similar checkpoint pathway operates in mammals, which use an Mrc1-related protein known as claspin to mediate ATR-dependent activation of Chk1 in response to stalled replication forks (Kumagai and Dunphy 2000). Another protein complex, the fork protection complex,

comprising Tof1–Csm3 in budding yeast, Swi1–Swi3 in fission yeast, and Timeless–Tipin in mammals, also associates with stalled forks. The fork protection complex stabilizes stalled forks in a manner that promotes the replication checkpoint response (McFarlane et al. 2010).

8 Chk1 AND Chk2 TARGET Cdc25

Chk1 and Chk2, the two (unrelated) key effectors of the DNA damage and replication checkpoints, share Cdc25 as a primary target (Fig. 2). This mechanism was first discovered in fission yeast, where Chk1 is essential for reducing the rate of CDK1 Y15 dephosphorylation in response to DNA damage (Rhind and Russell 2000). Chk2 similarly regulates Y15 dephosphorylation in response to replication fork arrest (in mammals this is done by Chk1). Both kinases directly phosphorylate Cdc25, the key effects being inhibition of its protein phosphatase activity and exclusion from the nucleus (Karlsson-Rosenthal and Millar 2006). The latter involves binding of 14-3-3 proteins to Cdc25.

9 THE MAPK-DEPENDENT STRESS CHECKPOINT

In addition to DNA damage, a variety of other cellular stresses, such as osmotic shock, oxidative stress, and microtubule depolymerization, can delay cells in G_2 phase. Many of these stresses activate the p38 MAPK pathway (p. 81 [Morrison 2012]; Ch. 18 [Hotamisligil and Davis 2014]). Although the details are not well understood, p38 MAPK is believed to inactivate Cdc25 (Shiozaki and Russell 1995; Karlsson-Rosenthal and Millar 2006).

10 RESOLUTION OF G_2 CHECKPOINT ARRESTS

G_2 checkpoint arrests are generally reversible. Once the problem is resolved, the checkpoint is inactivated and the cell can proceed with mitosis. This inactivation appears to be a passive process in which resolution of the checkpoint-initiating event (e.g., the repair of the DNA damage) removes the signal that activates the checkpoint kinases, allowing constitutive phosphatases to remove the phosphates from their substrates and reset the system. In situations in which the problem cannot be resolved, cells can adapt to the checkpoint signal and enter mitosis in the presence of ongoing checkpoint signaling. Such adaptation generally appears to be a failure of checkpoint function, as opposed to a regulated attenuation of checkpoint signaling (Harrison and Haber 2006). As cells continue to grow during a checkpoint arrest, it is likely that the mitosis-promoting activities that link the onset of mitosis to attainment of a sufficient cell size eventually overcome mitosis-inhibiting activity of the checkpoint. Alternatively, in metazoans, transcriptional circuits involving p53 and p21 can be activated to drive cells into senescence or apoptosis in response to prolonged G_2 arrest (Kastan and Bartek 2004). A failure to activate these circuits can lead to premature resumption of cell division in the presence of DNA damage (Bunz et al. 1998).

11 REGULATION OF MITOSIS

If the G_2 checkpoints are not triggered, cells fully activate CDK1 and proceed through the G_2/M transition into mitosis. This decision to commit to cell division is implemented by signaling pathways that regulate the various processes of cell division, in particular mitosis and cytokinesis. They also regulate organelles and other cytoplasmic components in ways that are less well understood.

The transition from G_2 phase to mitosis involves reorganization of the nucleus, the condensation of the chromosomes, and the formation of the mitotic spindle (see Box 1 and Fig. 3). These events are triggered by CDK1 and culminate with the mitotic chromosomes aligned on the metaphase plate (Morgan 2006). How it coordinates these processes is a long-standing question in the field.

12 DISSECTING THE NETWORK

The challenge of dissecting this regulatory network is twofold. First, CDK1 has dozens, if not hundreds, of mitotic substrates, many of which are not essential for mitosis. This apparent redundancy is presumably because the transition to metaphase is driven by many phosphorylation events, each one of which slightly biases a single protein toward its mitotic state, but no one of which is necessary or sufficient to drive mitosis, per se. This complexity makes validating the importance of any single phosphorylation difficult.

Second, CDK1 activates a network of other kinases that are involved in various steps in the G_2/M transition, and some of these feed back to ensure full activation of CDK1 (Fig. 3). The major mitotic kinases are the Polo and Aurora families (Ma and Poon 2011). In addition to functioning in feedback loops that stimulate CDK1, Plk1 is essential for executing mitotic progression, especially centrosome separation and formation of a bipolar spindle (Archambault and Glover 2009). Plk1 phosphorylates numerous substrates at the centrosome and kinetochore complexes that link them to spindle microtubules, many of which are also CDK1 substrates. Furthermore, CDK1 primes Plk1 substrates. The Polo-box domain of Plk1 binds to phospho-S/TP sites (the preferred CDK1 phosphorylation site), and therefore CDK1 phosphorylation of a protein can increase its affinity for Plk1, priming it for Plk1 phosphorylation.

The Aurora A and Aurora B kinases are also essential for mitosis. Aurora A acts earlier in CDK1 activation,

centromere separation, and spindle formation. Aurora B acts later to detect and correct improperly attached kinetochores (see below). In addition to Plk1 and Aurora kinases, the NIMA and Greatwall families of kinases are involved in executing the G_2/M transition and in the feedback loops that ensure and maintain the full activation of CDK1.

13 ANAPHASE ENTRY

Once chromosomes are properly aligned on the metaphase plate, anaphase is triggered via the activation of the APC (Fig. 3) (Morgan 2006). The APC is a large multisubunit E3 ubiquitin ligase that regulates the stability of a range of mitotic proteins by targeting them for ubiquitin-dependent proteolysis. It is regulated by the binding of one of two substrate-selectivity subunits: Cdc20 at the metaphase/anaphase transition and Cdh1 during telophase and into G_1 phase (Pesin and Orr-Weaver 2008). APC–Cdc20 promotes the degradation of several key substrates that trigger the irreversible transition from metaphase to anaphase and the subsequent exit from mitosis. One key substrate is securin, a stoichiometric inhibitor of separase, the protease that cleaves the cohesin complexes that hold sister chromatids together during metaphase. Another key substrate is cyclin B, degradation of which inactivates CDK1. In addition, APC activation leads to the indirect activation of the Cdc14 phosphatase, which dephosphorylates many CDK1 targets, further enforcing the inactivation of CDK1 (Clifford et al. 2008). PP1 and PP2A also play a role (Wurzenberger and Gerlich 2011).

In yeasts, there appears to be an intrinsic delay between the activation of CDK1 and the activation of APC, which usually gives the cells enough time to set up the metaphase plate. APC is activated by CDK1 phosphorylation (Kraft et al. 2003). The lag may give chromosomes adequate time to align. If not, a checkpoint signal (described below) prevents APC activation until the problem is resolved. In most metazoans, metaphase is rarely achieved in time, presumably because the metazoan spindle is larger and more complicated, and the checkpoint is activated in most cell cycles to delay anaphase until the chromosomes are properly aligned. In effect, the regulation of anaphase has gone from being a quality-control checkpoint in yeast to being a central signaling pathway in metazoans that triggers anaphase upon the successful completion of metaphase.

14 CHECKPOINT REGULATION OF MITOSIS

The signaling pathway that delays initiation of anaphase until the successful alignment of metaphase chromosomes is called the spindle assembly checkpoint or mitotic checkpoint (Fig. 3). The checkpoint acts to prevent action of

APC–Cdc20, thus preventing the degradation of securin and cyclin B (Musacchio and Salmon 2007). While securin remains stable, sister-chromatid cohesion is maintained, preventing separation of chromosomes; while cyclin B remains stable, CDK1 stays active, maintaining the mitotic phosphorylation state of the nucleus and preventing mitotic exit.

The spindle assembly checkpoint monitors chromosome alignment during the stages leading up to metaphase. In so-called bi-oriented attachment, the two sister chromatids are attached to spindle microtubules emanating from opposite poles; the checkpoint is triggered by sister-chromatid pairs lacking such attachments. The identity of the trigger that senses this lack of bi-orientation and activates the checkpoint has been controversial. Early experiments showed that a single unattached chromosome can activate the checkpoint, and that pulling on such a chromosome with a glass needle is sufficient to stop the checkpoint signal (Li and Nicklas 1995; Rieder et al. 1995). The interpretation of this result was that tension across the kinetochore is sufficient to stop activation of the checkpoint and thus that it monitors kinetochore tension. However, subsequent data were difficult to reconcile with the tension model and instead supported an occupancy model in which the checkpoint monitors proper binding of microtubules to the kinetochore, independently of the tension they generate (Khodjakov and Pines 2010).

These two models can be reconciled by the existence of an Aurora-B-dependent mechanism that recognizes chromosomes that are not bi-oriented (including those with kinetochores lacking tension) and destabilizes their kinetochore–microtubule attachments, allowing proper attachments to reform (Tanaka et al. 2002). In such a model, during reattachment cycles, lack of occupancy at the kinetochore activates the checkpoint. The checkpoint thus monitors kinetochore occupancy, but stable occupancy requires kinetochore tension (Khodjakov and Pines 2010).

A single unattached kinetochore can activate the spindle assembly checkpoint and inhibit all of the APC–Cdc20 in the cell, which suggests that unattached kinetochores produce a diffusible inhibitor of APC–Cdc20. That diffusible inhibitor is believed to be the mitotic checkpoint complex (MCC), which contains three checkpoint proteins—Mad2, Mad3/BubR1, and Bub3—as well as Cdc20 itself (Musacchio and Salmon 2007). The components of MCC interact dynamically with unattached kinetochores in a manner dependent on the more stable interaction of other checkpoint proteins, including Mad1 and Bub1. A crucial step in MCC formation is believed to be the conversion of the soluble open form of Mad2 to a closed form, which can bind stably to Cdc20. This conformational conversion of Mad2 is catalyzed by binding to Mad1 at unattached kinetochores and facilitates the assembly of MCC, which is then released to bind to and inhibit the APC. In metazoans MCC formation is inhibited in the absence of unattached kinetochores by p31-commet, which is believed to prevent open-Mad2 from binding to closed-Mad2 (Musacchio and Salmon 2007). The MCC-dependent checkpoint signal prevents anaphase until all kinetochores are properly attached. Once all kinetochores are occupied, the production of MCC ceases, the MCC-dependent checkpoint inhibition of the APC is relieved, and anaphase ensues.

Another signaling pathway that can block the metaphase/anaphase transition is the DNA damage checkpoint. In most organisms studied, the damage checkpoint arrests cells in G_2 phase by preventing the activation of CDK1, as described above. However, in budding yeast the DNA damage checkpoint directly targets the metaphase/anaphase transition by preventing separase activity via Chk1 phosphorylation and stabilization of securin (Sanchez et al. 1999). Metazoan cells also appear to regulate metaphase progression in response to DNA damage (Rieder 2011). However, because these cells also display robust G_2 damage arrest, the significance of the response is less clear.

Following the completion of anaphase, the CDK1 phosphorylation events that established the mitotic state are reversed and the cell returns to an interphase state in a process known as mitotic exit (Clifford et al. 2008; Rieder 2011; Wurzenberger and Gerlich 2011). CDK1 activity is reduced by APC-mediated proteolysis of cyclin B. In addition, the rate of CDK1 substrate dephosphorylation is increased by the activation of phosphatases that antagonize CDK1 phosphorylation. In yeast, Cdc14 is activated to reverse Cdk1-catalyzed phosphorylations, as described below. In animal cells, PP1 and PP2A appear to be the major phosphatases antagonizing CDK phosphorylation (Wurzenberger and Gerlich 2011). One mechanism for activation of PP2A-B55 at mitotic exit is the inactivation of Greatwall, but the discovery of other regulatory loops involving phosphatases seems likely. The process of mitotic exit is intimately connected to the final stage of cell division, cytokinesis.

15 REGULATION OF CYTOKINESIS

The proper coordination of cytokinesis with mitosis is essential to ensure faithful chromosome segregation and avoid aneuploidy or polyploidy. This coordination requires the septation initiation network (SIN) in fission yeast and the mitotic exit network (MEN) in budding yeast (McCollum and Gould 2001; Goyal et al. 2011; Meitinger et al. 2012). Cytokinesis is entrained to mitosis; thus, the signaling pathways that regulate cytokinesis are not involved so much in deciding when cytokinesis should occur as in coordinating cytokinesis with mitosis and providing opportunities for checkpoint regulation.

The MEN and SIN pathways are GTPase-triggered kinase cascades that culminate in the activation of the Sid2 kinase in fission yeast and the Dbf2 kinase in budding yeast (Goyal et al. 2011; Meitinger et al. 2012). Components of the MEN/SIN pathways are organized on the spindle-pole body, the yeast equivalent of the centrosome, making it a nexus for signaling pathways that control cell division. Sid2 is necessary and sufficient for initiating cytokinesis in fission yeast, although its exact targets are not known. Therefore, it is essential to restrain SIN signaling until after the successful completion of anaphase. Initiation of SIN signaling by the Spg1 GTPase is inhibited by the GTPase-activating protein Cdc16. Full activation of SIN signaling is antagonized by CDK1 activity (Guertin et al. 2000), which prevents cytokinesis until after activation of the APC and ensures that activation of the spindle assembly checkpoint will also delay cell division. In budding yeast, Dbf2 regulates cytokinesis by promoting localization of the chitin synthase Chs2 and the cytokinesis regulator Hof1 to the bud neck; this activity is antagonized by CDK1 activity (Meitinger et al. 2012).

In return, the MEN/SIN pathways antagonize the activity of CDK1 (Goyal et al. 2011; Meitinger et al. 2012). This reciprocal regulation allows cytokinesis errors, which prolong SIN signaling, to restrain CDK1 activity in the subsequent cell cycle, thus arresting cells in G_2 phase and preventing the next mitosis until the previous cytokinesis is successfully completed. An important component of the MEN/SIN pathways is the Cdc14 (Clp1 in fission yeast) phosphatase (Clifford et al. 2008). Cdc14 directly dephosphorylates CDK1 targets, facilitating mitotic exit and resetting the cell to an interphase state at the beginning of G_1 phase.

Many proteins in the MEN/SIN pathways are conserved in metazoans. In particular, the LATS kinases, relatives of Sid2/Dbf2 that also function in the Hippo pathway (p. 133 [Harvey and Hariharan 2012]), appear to have roles in the regulation of cytokinesis (Yang et al. 2004). Although the details have yet to be established, similar signaling pathways probably coordinate mitosis and cytokinesis in metazoans.

In addition to coordinating the timing of cytokinesis, signaling during cell division is required to determine the location of cytokinesis. Although the details differ between organisms, the location and orientation of the cytokinesis furrow is generally determined by the mitotic spindle, except in fission yeast, in which the cleavage plane is determined directly by the location of the nucleus (Almonacid and Paoletti 2010). How the signal is transmitted from the spindle or the nucleus to the cortex to establish the site of cytokinesis has yet to be established. Budding yeast is unusual in this context because the cleavage plane (the bud neck) is established before mitosis. Therefore, instead of using the spindle to orient cell division, budding yeast uses cell division to orient the spindle. Specifically, the MEN GTPase Tem1 is localized to the spindle-pole body, whereas Tem1's inhibitors, Bub2, Bfa1, and Kin4, are localized to the mother cell, and its activator Lte1, a guanine nucleotide exchange factor (GEF) relative, is localized to the daughter cell (Bardin et al. 2000). Thus, the MEN is only activated once the spindle is oriented such that one end of the spindle is through the bud neck and in the daughter cell. However, this strategy of triggering cytokinesis as a spatial consequence of spindle elongation may be general, as a similar mechanism functions in fission yeast (Garcia-Cortes and McCollum 2009).

16 CONCLUDING REMARKS

The major cell-cycle transitions that constitute cell division—the G_2/M transition, the metaphase/anaphase transition, and cytokinesis—provide important decision points that are regulated by a number of signaling pathways. These pathways ensure that the critical events of cell division occur in the proper order and provide the quality controls that prevent cells from dividing with damaged DNA or misaligned chromosomes. As such, they are instrumental in maintaining genomic integrity and, in metazoans, preventing cancer.

These negative regulatory signaling pathways were the original inspiration for the checkpoint paradigm of active negative regulation of cell-cycle events (Hartwell and Weinert 1989). The initial model posited that checkpoints delayed cell-cycle transitions to allow time for checkpoint-independent processes to fix whatever problem had triggered the checkpoint. Since then, these same pathways have been shown to regulate many other aspects of cell metabolism, such as DNA repair, and the term "checkpoint" is now used much more broadly than originally intended. Nonetheless, these signaling pathways serve as prime examples of how cells reorganize their metabolism and cell cycle to damage and other perturbations.

Although the well-studied signaling pathways that regulate cell division inhibit transitions in response to signals of damage or other problems, there is evidence for at least one positive signaling pathway, the one that regulates cell size. One of the enduring mysteries of cell biology is how cells measure size and how that information is used to regulate cell-cycle transitions such as cell division. Notwithstanding the current lack of mechanistic insight, the way size is measured and the pathways that transduce that signal are poised to be areas of future progress in the field.

The signal transduction pathways that regulate cell division continue to be the focus of significant experimental effort. That effort looks set to continue as work continues on the discovery of new regulators of cell division, the increasingly detailed mechanistic understanding of the major checkpoint pathways, and the translation of our understanding of these pathways into diagnostic and therapeutic advances in fields such as fertility, cancer, and aging.

ACKNOWLEDGMENTS

We thank Dan McCollum for valuable insight. N.R. is supported by NIH R01-GM069957 and an ACS Research Scholar Grant. P.R. is supported by NIH R01-GM59447, CA77325, and CA117638.

REFERENCES

*Reference is in this book.

Almonacid M, Paoletti A. 2010. Mechanisms controlling division-plane positioning. *Semin Cell Dev Biol* 21: 874–880.

Archambault V, Glover DM. 2009. Polo-like kinases: Conservation and divergence in their functions and regulation. *Nat Rev Mol Cell Biol* 10: 265–275.

Bardin AJ, Visintin R, Amon A. 2000. A mechanism for coupling exit from mitosis to partitioning of the nucleus. *Cell* 102: 21–31.

Barr FA, Sillje HH, Nigg EA. 2004. Polo-like kinases and the orchestration of cell division. *Nat Rev Mol Cell Biol* 5: 429–440.

Bartek J, Lukas C, Lukas J. 2004. Checking on DNA damage in S phase. *Nat Rev Mol Cell Biol* 5: 792–804.

Beausoleil SA, Villen J, Gerber SA, Rush J, Gygi SP. 2006. A probability-based approach for high-throughput protein phosphorylation analysis and site localization. *Nat Biotechnol* 24: 1285–1292.

Bloom J, Cross FR. 2007. Multiple levels of cyclin specificity in cell-cycle control. *Nat Rev Mol Cell Biol* 8: 149–160.

Bonilla CY, Melo JA, Toczyski DP. 2008. Colocalization of sensors is sufficient to activate the DNA damage checkpoint in the absence of damage. *Mol Cell* 30: 267–276.

Bonner WM, Redon CE, Dickey JS, Nakamura AJ, Sedelnikova OA, Solier S, Pommier Y. 2008. γH2AX and cancer. *Nat Rev Cancer* 8: 957–967.

Branzei D, Foiani M. 2010. Maintaining genome stability at the replication fork. *Nat Rev Mol Cell Biol* 11: 208–219.

Bunz F, Dutriaux A, Lengauer C, Waldman T, Zhou S, Brown JP, Sedivy JM, Kinzler KW, Vogelstein B. 1998. Requirement for p53 and p21 to sustain G₂ arrest after DNA damage. *Science* 282: 1497–1501.

Castilho PV, Williams BC, Mochida S, Zhao Y, Goldberg ML. 2009. The M phase kinase Greatwall (Gwl) promotes inactivation of PP2A/B55δ, a phosphatase directed against CDK phosphosites. *Mol Biol Cell* 20: 4777–4789.

Cimprich KA, Cortez D. 2008. ATR: An essential regulator of genome integrity. *Nat Rev Mol Cell Biol* 9: 616–627.

Clifford DM, Chen CT, Roberts RH, Feoktistova A, Wolfe BA, Chen JS, McCollum D, Gould KL. 2008. The role of Cdc14 phosphatases in the control of cell division. *Biochem Soc Trans* 36: 436–438.

Coudreuse D, Nurse P. 2010. Driving the cell cycle with a minimal CDK control network. *Nature* 468: 1074–1079.

de Bruin RA, Kalashnikova TI, Aslanian A, Wohlschlegel J, Chahwan C, Yates JR, Russell P, Wittenberg C. 2008. DNA replication checkpoint promotes G1-S transcription by inactivating the MBF repressor Nrm1. *Proc Natl Acad Sci* 105: 11230–11235.

Domingo-Sananes MR, Kapuy O, Hunt T, Novak B. 2011. Switches and latches: A biochemical tug-of-war between the kinases and phosphatases that control mitosis. *Philos Trans R Soc Lond B Biol Sci* 366: 3584–3594.

Du LL, Nakamura TM, Russell P. 2006. Histone modification-dependent and -independent pathways for recruitment of checkpoint protein Crb2 to double-strand breaks. *Genes Dev* 20: 1583–1596.

*Duronio RJ, Xiong Y. 2012. Signaling pathways that control cell proliferation. *Cold Spring Harb Perspect Biol* 4: a008904.

Dutta C, Patel PK, Rosebrock A, Oliva A, Leatherwood J, Rhind N. 2008. The DNA replication checkpoint directly regulates MBF-dependent G1/S transcription. *Mol Cell Biol* 28: 5977–5985.

Fantes P, Nurse P. 1977. Control of cell size at division in fission yeast by a growth-modulated size control over nuclear division. *Exp Cell Res* 107: 377–386.

Featherstone C, Russell P. 1991. Fission yeast p107^wee1 mitotic inhibitor is a tyrosine/serine kinase. *Nature* 349: 808–811.

Garcia-Cortes JC, McCollum D. 2009. Proper timing of cytokinesis is regulated by *Schizosaccharomyces pombe* Etd1. *J Cell Biol* 186: 739–753.

Gautier J, Solomon MJ, Booher RN, Bazan JF, Kirschner MW. 1991. cdc25 is a specific tyrosine phosphatase that directly activates p34^cdc2. *Cell* 67: 197–211.

Gharbi-Ayachi A, Labbe JC, Burgess A, Vigneron S, Strub JM, Brioudes E, Van-Dorsselaer A, Castro A, Lorca T. 2010. The substrate of Greatwall kinase, Arpp19, controls mitosis by inhibiting protein phosphatase 2A. *Science* 330: 1673–1677.

Gould KL, Nurse P. 1989. Tyrosine phosphorylation of the fission yeast cdc2+ protein kinase regulates entry into mitosis. *Nature* 342: 39–45.

Goyal A, Takaine M, Simanis V, Nakano K. 2011. Dividing the spoils of growth and the cell cycle: The fission yeast as a model for the study of cytokinesis. *Cytoskeleton (Hoboken)* 68: 69–88.

Guertin DA, Chang L, Irshad F, Gould KL, McCollum D. 2000. The role of the Sid1p kinase and Cdc14p in regulating the onset of cytokinesis in fission yeast. *EMBO J* 19: 1803–1815.

Haase SB, Reed SI. 1999. Evidence that a free-running oscillator drives G1 events in the budding yeast cell cycle. *Nature* 401: 394–397.

Hachet O, Berthelot-Grosjean M, Kokkoris K, Vincenzetti V, Moosbrugger J, Martin SG. 2011. A phosphorylation cycle shapes gradients of the DYRK family kinase Pom1 at the plasma membrane. *Cell* 145: 1116–1128.

Harper JW, Elledge SJ. 2007. The DNA damage response: Ten years after. *Mol Cell* 28: 739–745.

Harrison JC, Haber JE. 2006. Surviving the breakup: The DNA damage checkpoint. *Annu Rev Genet* 40: 209–235.

Hartwell LH, Weinert TA. 1989. Checkpoints: Controls that ensure the order of cell cycle events. *Science* 246: 629–634.

*Harvey KF, Hariharan IK. 2012. Hippo signaling. *Cold Spring Harb Perspect Biol* 4: a011288.

*Hotamisligil GS, Davis RJ. 2014. Cell signaling and stress responses. *Cold Spring Harb Perspect Biol* doi: 10.1101/cshperspect.a006072.

Howell AS, Lew DJ. 2012. Morphogenesis and the cell cycle. *Genetics* 190: 51–77.

Jackson SP, Bartek J. 2009. The DNA-damage response in human biology and disease. *Nature* 461: 1071–1078.

Jorgensen P, Tyers M. 2004. How cells coordinate growth and division. *Curr Biol* 14: R1014–R1027.

Kalaszczynska I, Geng Y, Iino T, Mizuno S, Choi Y, Kondratiuk I, Silver DP, Wolgemuth DJ, Akashi K, Sicinski P. 2009. Cyclin A is redundant in fibroblasts but essential in hematopoietic and embryonic stem cells. *Cell* 138: 352–365.

Karlsson-Rosenthal C, Millar JB. 2006. Cdc25: Mechanisms of checkpoint inhibition and recovery. *Trends Cell Biol* 16: 285–292.

Kastan MB, Bartek J. 2004. Cell-cycle checkpoints and cancer. *Nature* 432: 316–323.

Kellogg DR. 2003. Wee1-dependent mechanisms required for coordination of cell growth and cell division. *J Cell Sci* 116: 4883–4890.

Khodjakov A, Pines J. 2010. Centromere tension: A divisive issue. *Nat Cell Biol* **12**: 919–923.

Kraft C, Herzog F, Gieffers C, Mechtler K, Hagting A, Pines J, Peters JM. 2003. Mitotic regulation of the human anaphase-promoting complex by phosphorylation. *EMBO J* **22**: 6598–6609.

Kumagai A, Dunphy WG. 1991. The cdc25 protein controls tyrosine dephosphorylation of the cdc2 protein in a cell-free system. *Cell* **64**: 903–914.

Kumagai A, Dunphy WG. 1996. Purification and molecular cloning of Plx1, a Cdc25-regulatory kinase from *Xenopus* egg extracts. *Science* **273**: 1377–1380.

Kumagai A, Dunphy WG. 2000. Claspin, a novel protein required for the activation of Chk1 during a DNA replication checkpoint response in *Xenopus* egg extracts. *Mol Cell* **6**: 839–849.

* Laplante M, Sabatini DM. 2012. mTOR signaling. *Cold Spring Harb Perspect Biol* **4**: a011593.

Li X, Nicklas RB. 1995. Mitotic forces control a cell-cycle checkpoint. *Nature* **373**: 630–632.

Lundgren K, Walworth N, Booher R, Dembski M, Kirschner M, Beach D. 1991. Mik1 and Wee1 cooperate in the inhibitory tyrosine phosphorylation of Cdc2. *Cell* **64**: 1111–1122.

Ma HT, Poon RY. 2011. How protein kinases co-ordinate mitosis in animal cells. *Biochem J* **435**: 17–31.

Martin SG, Berthelot-Grosjean M. 2009. Polar gradients of the DYRK-family kinase Pom1 couple cell length with the cell cycle. *Nature* **459**: 852–856.

Matsuoka S, Ballif BA, Smogorzewska A, McDonald ER III, Hurov KE, Luo J, Bakalarski CE, Zhao Z, Solimini N, Lerenthal Y, et al. 2007. ATM and ATR substrate analysis reveals extensive protein networks responsive to DNA damage. *Science* **316**: 1160–1166.

McCollum D, Gould KL. 2001. Timing is everything: Regulation of mitotic exit and cytokinesis by the MEN and SIN. *Trends Cell Biol* **11**: 89–95.

McFarlane RJ, Mian S, Dalgaard JZ. 2010. The many facets of the Tim-Tipin protein families' roles in chromosome biology. *Cell Cycle* **9**: 700–705.

Meitinger F, Palani S, Pereira G. 2012. The power of MEN in cytokinesis. *Cell Cycle* **11**: 219–228.

Melander F, Bekker-Jensen S, Falck J, Bartek J, Mailand N, Lukas J. 2008. Phosphorylation of SDT repeats in the MDC1 N terminus triggers retention of NBS1 at the DNA damage-modified chromatin. *J Cell Biol* **181**: 213–226.

Mimitou EP, Symington LS. 2011. DNA end resection—unraveling the tail. *DNA Repair (Amst)* **10**: 344–348.

Morgan DO. 1995. Principles of CDK regulation. *Nature* **374**: 131–134.

Morgan DO. 2006. *The cell cycle: Principles of control.* Oxford University Press, Sunderland, MA.

* Morrison DK. 2012. MAP kinase pathways. *Cold Spring Harb Perspect Biol* **4**: a011254.

Moseley JB, Mayeux A, Paoletti A, Nurse P. 2009. A spatial gradient coordinates cell size and mitotic entry in fission yeast. *Nature* **459**: 857–860.

Mueller PR, Coleman TR, Kumagai A, Dunphy WG. 1995. Myt1: A membrane-associated inhibitory kinase that phosphorylates Cdc2 on both threonine-14 and tyrosine-15. *Science* **270**: 86–90.

Murray AW, Hunt T. 1993. *The cell cycle: An introduction.* W.H. Freeman, New York.

Musacchio A, Salmon ED. 2007. The spindle-assembly checkpoint in space and time. *Nat Rev Mol Cell Biol* **8**: 379–393.

Navadgi-Patil VM, Burgers PM. 2009. A tale of two tails: Activation of DNA damage checkpoint kinase Mec1/ATR by the 9-1-1 clamp and by Dpb11/TopBP1. *DNA Repair (Amst)* **8**: 996–1003.

Nurse P. 1975. Genetic control of cell size at cell division in yeast. *Nature* **256**: 547–551.

Nurse P. 1990. Universal control mechanism regulating onset of M-phase. *Nature* **344**: 503–508.

Nurse P. 2000. A long twentieth century of the cell cycle and beyond. *Cell* **100**: 71–78.

Parker LL, Piwnica-Worms H. 1992. Inactivation of the p34cdc2–cyclin B complex by the human WEE1 tyrosine kinase. *Science* **257**: 1955–1957.

Pesin JA, Orr-Weaver TL. 2008. Regulation of APC/C activators in mitosis and meiosis. *Annu Rev Cell Dev Biol* **24**: 475–499.

Petersen J, Nurse P. 2007. TOR signalling regulates mitotic commitment through the stress MAP kinase pathway and the Polo and Cdc2 kinases. *Nat Cell Biol* **9**: 1263–1272.

Polymenis M, Schmidt EV. 1997. Coupling of cell division to cell growth by translational control of the G1 cyclin CLN3 in yeast. *Genes Dev* **11**: 2522–2531.

Pomerening JR, Sontag ED, Ferrell JE Jr. 2003. Building a cell cycle oscillator: Hysteresis and bistability in the activation of Cdc2. *Nat Cell Biol* **5**: 346–351.

Porter LA, Donoghue DJ. 2003. Cyclin B1 and CDK1: Nuclear localization and upstream regulators. *Prog Cell Cycle Res* **5**: 335–347.

Rhind N, Russell P. 2000. Chk1 and Cds1: Linchpins of the DNA damage and replication checkpoint pathways. *J Cell Sci* **113**: 3889–3896.

Rieder CL. 2011. Mitosis in vertebrates: The G2/M and M/A transitions and their associated checkpoints. *Chromosome Res* **19**: 291–306.

Rieder CL, Cole RW, Khodjakov A, Sluder G. 1995. The checkpoint delaying anaphase in response to chromosome monoorientation is mediated by an inhibitory signal produced by unattached kinetochores. *J Cell Biol* **130**: 941–948.

Russell P, Nurse P. 1986. *cdc25*[+] functions as an inducer in the mitotic control of fission yeast. *Cell* **45**: 145–153.

Russell P, Nurse P. 1987. Negative regulation of mitosis by *wee1*[+], a gene encoding a protein kinase homolog. *Cell* **49**: 559–567.

Sanchez Y, Bachant J, Wang H, Hu F, Liu D, Tetzlaff M, Elledge SJ. 1999. Control of the DNA damage checkpoint by Chk1 and Rad53 protein kinases through distinct mechanisms. *Science* **286**: 1166–1171.

Shiloh Y. 2003. ATM and related protein kinases: Safeguarding genome integrity. *Nat Rev Cancer* **3**: 155–168.

Shiozaki K, Russell P. 1995. Cell-cycle control linked to extracellular environment by MAP kinase pathway in fission yeast. *Nature* **378**: 739–743.

Spycher C, Miller ES, Townsend K, Pavic L, Morrice NA, Janscak P, Stewart GS, Stucki M. 2008. Constitutive phosphorylation of MDC1 physically links the MRE11–RAD50–NBS1 complex to damaged chromatin. *J Cell Biol* **181**: 227–240.

Stern B, Nurse P. 1996. A quantitative model for the cdc2 control of S phase and mitosis in fission yeast. *Trends Genet* **12**: 345–350.

Stewart GS, Wang B, Bignell CR, Taylor AM, Elledge SJ. 2003. MDC1 is a mediator of the mammalian DNA damage checkpoint. *Nature* **421**: 961–966.

Stracker TH, Petrini JH. 2011. The MRE11 complex: Starting from the ends. *Nat Rev Mol Cell Biol* **12**: 90–103.

Strausfeld U, Labbe JC, Fesquet D, Cavadore JC, Picard A, Sadhu K, Russell P, Doree M. 1991. Dephosphorylation and activation of a p34[cdc2]/cyclin B complex *in vitro* by human CDC25 protein. *Nature* **351**: 242–245.

Stucki M, Clapperton JA, Mohammad D, Yaffe MB, Smerdon SJ, Jackson SP. 2005. MDC1 directly binds phosphorylated histone H2AX to regulate cellular responses to DNA double-strand breaks. *Cell* **123**: 1213–1226.

Tanaka TU, Rachidi N, Janke C, Pereira G, Galova M, Schiebel E, Stark MJ, Nasmyth K. 2002. Evidence that the Ipl1-Sli15 (Aurora kinase-INCENP) complex promotes chromosome bi-orientation by altering kinetochore-spindle pole connections. *Cell* **108**: 317–329.

Tyson JJ. 1983. Unstable activator models for size control of the cell cycle. *J Theor Biol* **104**: 617–631.

Tzur A, Kafri R, LeBleu VS, Lahav G, Kirschner MW. 2009. Cell growth and size homeostasis in proliferating animal cells. *Science* **325**: 167–171.

Uhlmann F, Bouchoux C, Lopez-Aviles S. 2011. A quantitative model for cyclin-dependent kinase control of the cell cycle: Revisited. *Philos Trans R Soc Lond B Biol Sci* **366:** 3572–3583.

* Ward PS, Thompson CB. 2012. Signaling in control of cell growth and metabolism. *Cold Spring Harb Perspect Biol* **4:** a006783.

Welburn JP, Tucker JA, Johnson T, Lindert L, Morgan M, Willis A, Noble ME, Endicott JA. 2007. How tyrosine 15 phosphorylation inhibits the activity of cyclin-dependent kinase 2-cyclin A. *J Biol Chem* **282:** 3173–3181.

Wurzenberger C, Gerlich DW. 2011. Phosphatases: Providing safe passage through mitotic exit. *Nat Rev Mol Cell Biol* **12:** 469–482.

Yang X, Yu K, Hao Y, Li DM, Stewart R, Insogna KL, Xu T. 2004. LATS1 tumour suppressor affects cytokinesis by inhibiting LIMK1. *Nat Cell Biol* **6:** 609–617.

You Z, Shi LZ, Zhu Q, Wu P, Zhang YW, Basilio A, Tonnu N, Verma IM, Berns MW, Hunter T. 2009. CtIP links DNA double-strand break sensing to resection. *Mol Cell* **36:** 954–969.

CHAPTER 7

Signaling in Control of Cell Growth and Metabolism

Patrick S. Ward[1,2] and Craig B. Thompson[1]

[1]Cancer Biology and Genetics Program, Memorial Sloan-Kettering Cancer Center, New York, New York 10065

[2]Department of Cancer Biology, Perelman School of Medicine at the University of Pennsylvania, Philadelphia, Pennsylvania 19104

Correspondence: thompsonc@mskcc.org

SUMMARY

Mammalian cells require growth-factor-receptor-initiated signaling to proliferate. Signal transduction not only initiates entry into the cell cycle, but also reprograms cellular metabolism. This instructional metabolic reprogramming is critical if the cell is to fulfill the anabolic and energetic requirements that accompany cell growth and division. Growth factor signaling mediated by the PI3K/Akt pathway plays a major role in regulating the cellular uptake of glucose, as well as the incorporation of this glucose carbon into lipids for membrane synthesis. Tyrosine-kinase-based regulation of key glycolytic enzymes such as pyruvate kinase also plays a critical role directing glucose carbon into anabolic pathways. In addition, the Myc transcription factor and mTOR kinase regulate the uptake and utilization of amino acids for protein and nucleic acid synthesis, as well as for the supply of intermediates to the mitochondrial Krebs cycle. However, the relationship between cellular signaling and metabolism is not unidirectional. Cells, by sensing levels of intracellular metabolites and the status of key metabolic pathways, can exert feedback control on signal transduction networks through multiple types of metabolite-derived protein modifications. These mechanisms allow cells to coordinate growth and division with their metabolic activity.

Outline

1 INTRODUCTION

Unicellular organisms have evolved to grow and divide rapidly when nutrients are abundant, and they take up nutrients in a cell-autonomous manner. The macromolecular precursors and free energy derived from metabolism of these nutrients are used to synthesize the new biomass required for cell growth and division. When the nutrient supply dwindles, anabolic metabolism in these organisms decreases. The cells then shift to catabolic pathways that maximize the efficiency of energy production to survive periods of nutrient limitation (Vander Heiden et al. 2009).

In multicellular organisms, cells are generally surrounded by sufficient nutrients to engage in continuous cell growth and proliferation. However, organismal integrity requires that proliferation not be a cell-autonomous process dictated by available nutrients. Mammalian cells require receptor-mediated signal transduction initiated by extracellular growth factors to leave the quiescent state and enter the cell cycle. The onset of cell growth and division introduces a metabolic requirement for sufficient carbon, nitrogen, and free energy to support synthesis of the new proteins, lipids, and nucleic acids needed by a proliferating cell. Recent studies have shown that this additional uptake of nutrients is regulated by signal transduction pathways (Fig. 1). This growth-factor-directed uptake of nutrients is critical to supporting a rate of macromolecular synthesis sufficient for growth (DeBerardinis et al. 2008; Vander Heiden et al. 2009).

Mammalian cells instructed to proliferate via signal transduction are generally successful at avoiding metabolic collapse. Assuming that extracellular nutrients are abundant, these signaling-instructed cells will increase both uptake of nutrients and nutrient flux through anabolic pathways. However, if the regulation of cell growth by signaling pathways goes unchecked, problems can rapidly develop. The availability of a key extracellular nutrient could be limited in a particular context, or an important enzyme in a critical anabolic pathway may, for some reason, be deficient. Thus, to ensure that cell growth is properly coordinated with both the availability of key nutrients and with the cellular capacity to use them effectively, cells need a way to slow their own growth if their metabolic state cannot support biomass production. Such a brake on anabolic metabolism must be able to function even in the presence of growth-factor-initiated signaling. Metabolically sensitive posttranslational modifications, including acetylation and glycosylation, of signaling proteins provide an important mechanism by which cellular metabolism can exert feedback control on the output of signal transduction cascades. The relationship between cell signaling and metabolism is thus bidirectional.

2 PI3K/AKT SIGNALING CONTROLS GLUCOSE METABOLISM AND THE INCORPORATION OF CARBON INTO MACROMOLECULES

A highly conserved signal transduction pathway initiated by extracellular growth factors is the phosphoinositide 3-kinase (PI3K)/Akt pathway, whose components are conserved throughout metazoan species (p. 87 [Hemmings and Restuccia 2012]). In mammals, the pathway plays a particularly critical role downstream from insulin signaling to facilitate glucose uptake in insulin-dependent tissues such as fat and muscle. In these tissues, the PI3K/Akt pathway promotes the trafficking of the glucose transporter GLUT4 to the cell surface (Kohn et al. 1996; Ch. 14 [Hardie 2012]). However, this pathway plays multiple other roles in glucose metabolism, and its activity is not limited to those tissues classically described as insulin dependent.

In the normal, non-cancerous, setting, PI3K is activated in cells when cell membrane receptor tyrosine kinases (RTKs), as well as G-protein-coupled receptors (GPCRs) and cytokine receptors, are stimulated by extracellular growth factors. Following activation, PI3K phosphorylates membrane phosphatidylinositol lipids, which, in turn, leads to the recruitment and activation of additional kinases, most notably Akt (Fig. 2). One of the major effects of Akt on glucose metabolism is at the level of glucose uptake through increased expression of GLUT1 on the cell surface (Rathmell et al. 2003). GLUT1 is the major glucose transporter in most cell types. The translation of GLUT1 mRNA is also increased by Akt signaling through mTORC1 and 4EBP (Taha et al. 1999). In addition, Akt signaling stimulates the activity of several glycolytic enzymes. Hexokinase, which performs the first enzymatic reaction of glycolysis— glucose $\rightarrow$ glucose 6-phosphate (G6P)—is more active when associated with mitochondria, and this association is promoted by Akt (Gottlob et al. 2001). Akt also directly phosphorylates phosphofructokinase 2, and the resulting increase in fructose 2,6-bisphosphate levels enhances the activity of the glycolytic enzyme phosphofructokinase 1 (PFK1) (Deprez et al. 1997).

Akt signaling may also increase glycolytic flux indirectly by transcriptionally up-regulating the endoplasmic reticulum enzyme ENTPD5 (Fang et al. 2010). ENTPD5 promotes proper protein glycosylation and folding and does so through a cycle of reactions coupled to ATP hydrolysis. Thus, Akt-mediated up-regulation of ENTPD5 can increase the cellular ADP:ATP ratio. Increasing the ADP:ATP ratio can relieve allosteric inhibition of PFK1 by ATP. It can also lead to increased activity of downstream glycolytic enzymes that require ADP as a cofactor. Enhanced ATP consumption caused by PI3K/Akt signaling may therefore enhance the proliferating cell's ability to rapidly metabolize glucose carbon.

A

Absence of growth factor

RTKs

Quiescent cell metabolism in absence of extracellular growth factor

Little nutrient uptake

Mitchondrion

Krebs cycle

ATP production

Carbon dioxide

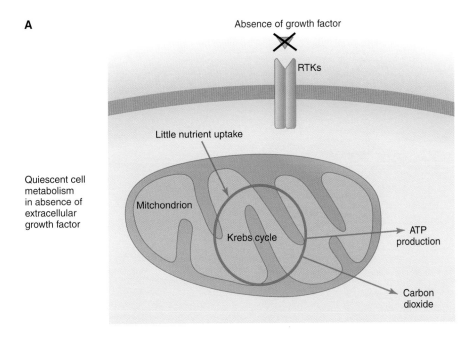

B

Glutamine

Growth factor

Glucose

RTKs

Proliferating cell metabolism in presence of extracellular growth factor

Biomass

ADP

ATP

ATP consumption to maintain glycolytic flux

Mitchondrion

Krebs cycle

Lactate

ATP

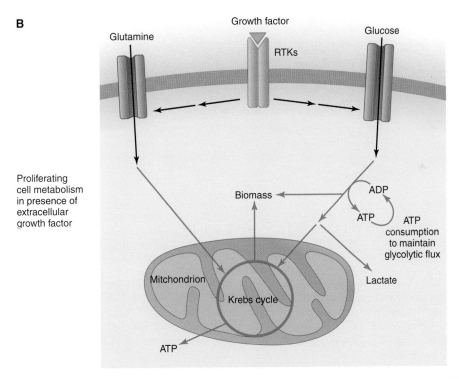

Figure 1. Growth-factor-initiated signaling reprograms metabolism in proliferating cells. (A) In multicellular organisms, cells that are not instructed to proliferate by extracellular growth factors are generally quiescent. In these cells, glucose carbon is predominantly metabolized to carbon dioxide in the mitochondrial Krebs cycle when oxygen is available. This mitochondrial oxidation maximizes free-energy generation in the form of ATP. (B) When cells are instructed to proliferate by growth factor signaling, they increase their nutrient uptake, particularly that of glucose and glutamine. Much of this increased nutrient uptake is used to fulfill the lipid, protein, and nucleotide synthesis (biomass) required for cell growth, and the excess carbon is secreted as lactate. Proliferating cells also may adopt strategies to increase their ATP consumption to maintain glycolytic flux. Metabolic pathways are indicated by green arrows.

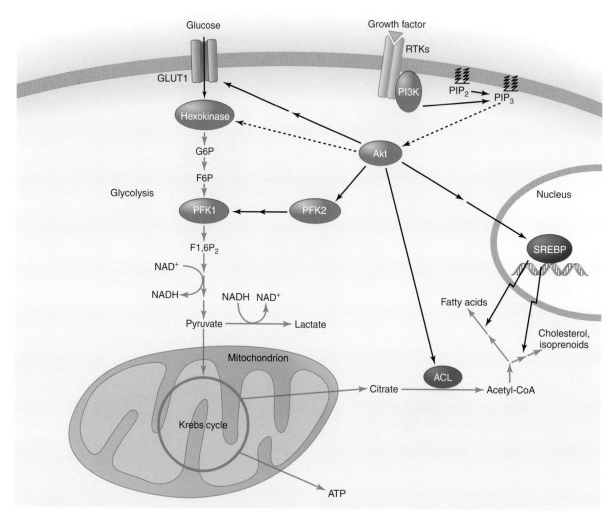

Figure 2. Glucose carbon flux is regulated by PI3K/Akt signaling. PI3K generates phosphatidylinositol 3,4,5-tris-phosphate (PIP$_3$) from phosphatidylinositol 4,5-bisphosphate (PIP$_2$). PIP$_3$ recruits Akt to the cell membrane, permitting its activation by upstream kinases. Akt can then activate a variety of proteins that increase glycolytic metabolism and direct glucose carbon flux toward biosynthesis. Akt enhances glucose transporter 1 (GLUT1) translation and localization to the cell surface to increase glucose uptake and increases the activity of the glycolytic enzymes hexokinase and phosphofructokinase 1 (PFK1). Akt also promotes lipid synthesis from glucose carbon by multiple mechanisms. It can activate ATP-citrate lyase (ACL) to increase the availability of cytosolic acetyl-CoA. Akt activation can also indirectly enhance the proteolytic release of the sterol regulatory element-binding protein transcription factors (SREBPs), thereby promoting the expression of genes involved in the pathways for fatty acid and cholesterol/isoprenoid synthesis. Abbreviations: G6P, glucose 6-phosphate; F6P, fructose 6-phosphate; F1,6P$_2$, fructose 1,6-bisphosphate.

Through these multiple effects on glycolytic gene expression and enzyme activity, activated Akt is able to increase the overall glycolytic rate in the cell. Even under aerobic conditions, cells with activated Akt display a greatly increased rate of glycolysis and secrete a high proportion of glucose carbon from the cell in the form of lactate (Elstrom et al. 2004). Originally described by Otto Warburg as a phenomenon specific to tumor cells, "aerobic glycolysis" is now known to be a general characteristic of rapidly

proliferating cells, both cancerous and non-cancerous (Bauer et al. 2004).

The PI3K/Akt pathway also regulates the fate of glucose carbon beyond glycolysis for biosynthetic purposes. When glycolysis metabolizes glucose to pyruvate, the pyruvate can enter mitochondria and be converted to acetyl-CoA. This acetyl-CoA can then condense with oxaloacetate to form citrate. The best-understood fate of mitochondrial citrate is oxidation within the mitochondrial matrix through the Krebs citric acid cycle for ATP production. Indeed, this fate predominates in tissues such as skeletal muscle and heart, where the demand for free energy in the form of ATP is high. However, proliferating cells with high levels of glucose uptake do not oxidize all of their mitochondrial citrate, but instead transport some to the cytosol. This cytosolic citrate can be converted to cytosolic acetyl-CoA by ATP-citrate lyase (ACL), another enzyme stimulated by Akt (Berwick et al. 2002; Bauer et al. 2005).

Cytosolic acetyl CoA can initiate and/or elongate fatty acid chains as part of de novo lipid synthesis. By activating ACL, Akt signaling thus increases the availability of acetyl-CoA precursors for lipid synthesis. It also induces lipogenic genes involved in isoprenoid, cholesterol, and fatty acid biosynthesis. These actions depend on mTOR-mediated activation of protein synthesis. Enhanced synthesis of ER-exported proteins leads to depletion of ER-Golgi lipids. This, in turn, promotes ER-to-Golgi transport and proteolytic release of the sterol regulatory element-binding protein (SREBP) transcription factors (Porstmann et al. 2005; Bobrovnikova-Marjon et al. 2008; Ch. 14 [Hardie 2012]). The additional lipids synthesized as a consequence of PI3K/Akt activation play a critical role as components of the plasma and organelle membranes that must be synthesized if cells are to grow and proliferate. Some of them obviously also play important signaling roles within the cell.

3 HIF1 SIGNALING PROVIDES ADDITIONAL REGULATION OF GLUCOSE METABOLISM IN RESPONSE TO BOTH OXYGEN AVAILABILITY AND NUTRIENT STATUS

The HIF1 (hypoxia inducible factor 1) signaling pathway also plays a central role in the regulation of cellular glucose metabolism. HIF1 is a transcription factor that was initially identified through its role in the adaptive cellular response to hypoxia (low oxygen tension). When oxygen is limited, the Krebs cycle in the mitochondria will cause mitochondrial redox stress if ATP continues to be produced solely by oxidative phosphorylation. Under these conditions, HIF1 promotes the expression of genes whose products are involved in anaerobic glycolysis (Fig. 3). This leads to increased generation of ATP in the cytosol from the conversion of glucose to pyruvate (Semenza et al. 1994).

To maintain a high rate of glycolysis, cells must also have a high rate of pyruvate → lactate conversion, which regenerates NAD^+ from NADH. This is important for maintaining a sufficiently high cytoplasmic NAD^+:NADH ratio to facilitate flux through the glycolytic enzyme glyceraldehyde-3-phosphate dehydrogenase. HIF1 promotes lactate production and NAD^+ regeneration by positively regulating the expression of the enzyme lactate dehydrogenase A. HIF1

also reduces the rate of pyruvate entry into the mitochondria by promoting the expression of pyruvate dehydrogenase kinase, an enzyme that phosphorylates and inactivates the mitochondrial pyruvate dehydrogenase complex (PDH) responsible for the formation of mitochondrial acetyl CoA from pyruvate (Kim et al. 2006; Papandreou et al. 2006). HIF1-mediated regulation of PDH flux serves to limit the flow of glucose-derived carbon into the mitochondrial, oxidative, Krebs cycle when excessive levels of pyruvate are generated. If pyruvate is oxidized in the Krebs cycle and increases electron transport beyond the cell's ability to assimilate the resulting electron flux using molecular oxygen and/or the electrochemical potential of hydrogen in ATP generation, a considerable increase in the levels of potentially damaging reactive oxygen species (ROS) will result. HIF1-mediated regulation of PDH also limits the flow of glucose carbon into biosynthetic pathways that emanate from mitochondrial Krebs cycle intermediates (Lum et al. 2007).

HIF1 is a heterodimer composed of a labile α-subunit and a stable β-subunit (Majmundar et al. 2010). At normal oxygen levels, the α-subunit of HIF1 (HIF1α) is posttranslationally modified by hydroxylation of specific proline residues by a prolyl hydroxylase (PHD). This proline hydroxylation reaction requires both α-ketoglutarate and O_2 as reactants, and generates succinate and CO_2 as products. Proline hydroxylation of HIF1α promotes its association with a ubiquitin E3 ligase complex that is composed of the von Hippel–Lindau (VHL) tumor suppressor as the substrate specificity subunit together with a cullin-Rbx1 ligase (CRL). Ubiquitylation of HIF1α by the CRL–VHL complex then targets it for degradation by the proteasome. At low oxygen levels, the prolyl hydroxylation of HIF1α does not occur, because the PHD reaction requires molecular oxygen (O_2) as a substrate. This permits stabilization of HIF1α during hypoxia and enhanced expression of HIF1 target genes.

In proliferating cells, several mechanisms increase HIF1 signaling even at normal oxygen tensions. Activated Akt, through mTORC1, enhances translation of HIF1α mRNA (Zhong et al. 2000; Hudson et al. 2002; p. 91 [Laplante and Sabatini 2012]). Increased cellular nutrient uptake can also lead to inhibition of the proline hydroxylation reactions that would otherwise target HIF1α for degradation. Excess nutrient metabolism can lead to increased cellular levels of reactive oxygen species (ROS) if reactions in the Krebs cycle occur at a rate exceeding the capacity of the electron transport chain to capture the electrons generated (Wellen and Thompson 2010). ROS are potent inhibitors of the HIF-targeting PHDs (Shatrov et al. 2003). Levels of succinate and fumarate may also increase with signaling-induced elevations in nutrient metabolism; both of these substrates

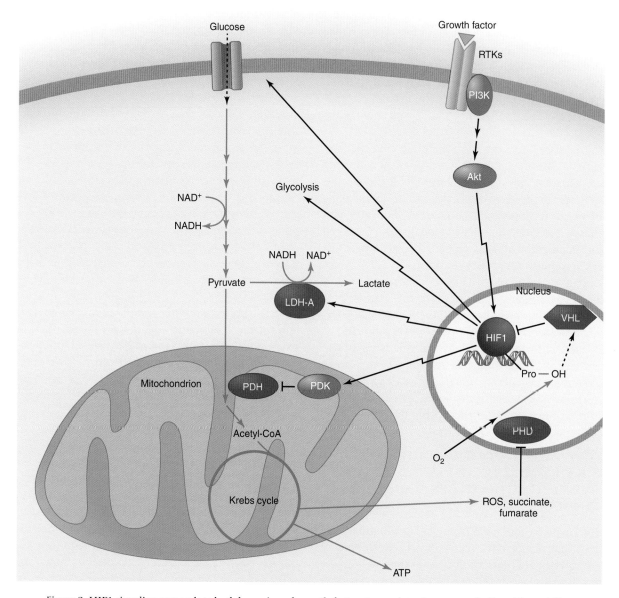

Figure 3. HIF1 signaling responds to both hypoxia and growth factors to regulate glucose metabolism. The stability of the transcription factor HIF1 is regulated through hydroxylation of proline residues on the HIF1α-subunit by prolyl hydroxylase enzymes (PHDs). Molecular oxygen (O_2) is a substrate for PHD activity, which targets HIF1α for recognition by the Von Hippel–Lindau (VHL) ubiquitin E3 ligase and subsequent degradation by the proteasome. Inhibition of PHD activity and HIF1 degradation occurs in the presence of elevated levels of reactive oxygen species (ROS) or the Krebs cycle intermediates succinate and fumarate. HIF1 can also be positively regulated transcriptionally and translationally downstream from growth factor signaling. When not degraded, HIF1 can promote the expression of genes encoding glucose transporters and most enzymes of glycolysis. It also inhibits glucose carbon flow into the mitochondria by promoting the expression of pyruvate dehydrogenase kinase (PDK), which acts to inhibit pyruvate dehydrogenase (PDH). Furthermore, HIF1 up-regulates lactate dehydrogenase A (LDH-A) expression to increase the conversion of pyruvate to lactate. This conversion regenerates the ratio of $NAD^+/NADH$ necessary for continued flux through glycolysis.

can also inhibit HIF-targeting PHDs (Isaacs et al. 2005; Selak et al. 2005). PHD inhibition by succinate is an example of product inhibition. The result of PHD inhibition by ROS, succinate, and/or fumarate is a feedback mechanism that decreases the flow of glucose carbon into the mitochondria and thus inhibits further generation of ROS through mitochondrial metabolism.

4 TYROSINE KINASE SIGNALING AND PYRUVATE KINASE: REGULATION OF A METABOLIC SWITCH IN PROLIFERATING CELLS

A unique aspect of glycolysis in proliferating cells is the isoform specificity of pyruvate kinase, the enzyme that converts phosphoenolpyruvate (PEP) to pyruvate with the concomitant generation of ATP. Several alternatively spliced isoforms of pyruvate kinase exist, but most cells in the adult predominantly express the M1 isoform (PKM1). However, in early development, embryonic tissues predominantly express an alternatively spliced M2 isoform (PKM2). Furthermore, in all cancer cells examined to date, PKM2 is the predominant isoform found (Christofk et al. 2008a). The splicing factors PTB, hnRNPA1, and hnRNPA2 that regulate the alternative splicing of PKM mRNA to favor PKM2 over PKM1 expression are themselves up-regulated by the Myc transcription factor, a proto-oncogene product (Clower et al. 2010; David et al. 2010). Myc-regulated transcription of these splicing factors thereby ensures a high PKM2/ PKM1 ratio in cells with activated growth factor signaling. Myc also regulates the expression of several glycolytic enzymes (Shim et al. 1997; Osthus et al. 2000).

Unlike the constitutively active PKM1 and the other pyruvate kinase isoforms, PKM2 is uniquely sensitive to regulation by tyrosine kinase signaling pathways downstream from growth factor receptors (Fig. 4). PKM2 can bind to the phosphorylated tyrosine residues in tyrosine-phosphorylated proteins and is also a target for tyrosine phosphorylation on itself (Christofk et al. 2008b; Hitosugi et al. 2009). These events lead to inhibition of PKM2 enzymatic activity, at least partly by promoting the release of PKM2's allosteric activator fructose 1,6-bisphosphate. Binding of the latter to PKM2 causes it to form an active tetramer, but tyrosine-phosphorylated peptides displace fructose 1,6-bisphosphate and cause the PKM2 tetramer to dissociate, resulting in enzyme inactivation. The sensitivity of PKM2 to inhibition by tyrosine kinase signaling allows it to act as a gatekeeper for the metabolic fate of glucose carbon. When growth factors are present, PKM2 is inhibited, which can promote the channeling of upstream glycolytic intermediates into anabolic pathways such as nucleotide biosynthesis. When growth factors are absent, PKM2 is active and can generate pyruvate for subsequent

catabolism in the Krebs cycle (Vander Heiden et al. 2009). All of this is consistent with the growth advantage that PKM2-expressing cells show in vivo compared with cells exclusively expressing PKM1 (Christofk et al. 2008a).

The full range of effects that PKM2 may have on cell growth and metabolism, however, remains incompletely characterized. Recently, an alternative glycolytic pathway has been proposed to be present in PKM2-expressing cells (Vander Heiden et al. 2010). When inactive PKM2 cannot effectively convert PEP to pyruvate, PEP can donate its high-energy phosphate to H11 of the upstream glycolytic enzyme phosphoglycerate mutase. This pathway may allow the generation of pyruvate from PEP in a step that is independent of ATP generation. Although further work is needed to characterize and purify an enzyme that can catalyze this proposed activity, the pathway may serve as an additional mechanism, besides Akt-mediated up-regulation of ENTPD5, by which proliferating cells can decrease their ATP:ADP ratio. The potential importance of PKM2 as a binding partner for a variety of signaling kinases and transcription factors has also received attention. For example, PKM2 translocates to the nucleus and potentiates the transcriptional activity of the Oct4 transcription factor involved in maintaining pluripotency (Lee et al. 2008). Many additional PKM2-interacting proteins have been reported, most recently HIF1 and β-catenin, but the significance of each of these interactions for facilitating cell growth and proliferation requires further study.

The regulation of PKM2 activity by metabolites is also continuing to be investigated. For example, ROS have recently been proposed to cause oxidation, dissociation, and inactivation of the PKM2 tetramer (Anastasiou et al. 2011). Conversely, the non-essential amino acid serine allosterically activates PKM2 (Eigenbrodt et al. 1983). Ongoing work is addressing whether inactivation of PKM2 and the resultant accumulation of glycolytic intermediates can lead to enhanced flux through the serine synthetic pathway and thereby promote a feedback loop through serine synthesis that can modulate the flux of glycolytic intermediates to support cellular biosynthetic requirements.

5 AMINO ACID METABOLISM IS ALSO REGULATED BY CELLULAR SIGNALING CASCADES

Proliferating cells require a nitrogen source for protein and nucleotide biosynthesis. Mammalian cells acquire nitrogen through the uptake and metabolism of amino acids, using mechanisms that are also highly regulated by growth factor signaling. Myc, which is downstream from several signaling pathways, including those involving Ras and Hedgehog (see p. 81 [Morrison 2012] and p. 107 [Ingham 2012]),

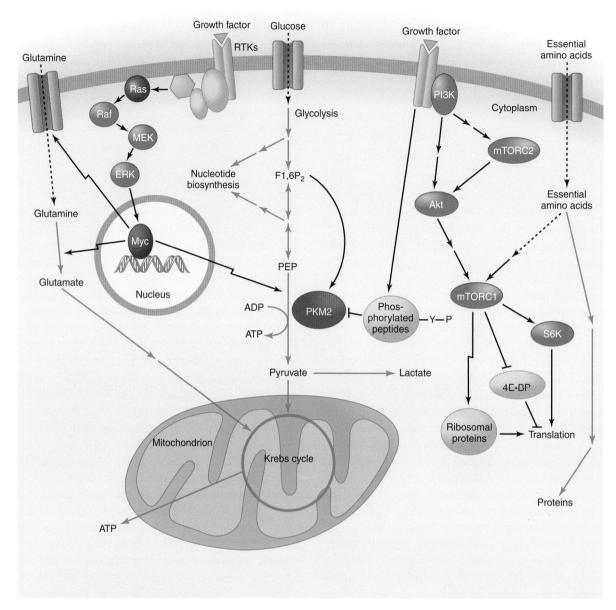

Figure 4. Pyruvate kinase as a glycolytic switch, and mTOR and Myc regulate amino acid uptake and metabolism. Pyruvate kinase catalyzes the late step of glycolysis that converts phosphoenolpyruvate (PEP) to pyruvate, with the concomitant generation of ATP. The M2 isoform of pyruvate kinase (PKM2) is specific to proliferating cells. Its activity is allosterically activated by the upstream glycolytic intermediate fructose 1,6-bisphosphate (F1,6P$_2$). Tyrosine kinase signaling downstream from growth factor receptors causes the release of F1,6P$_2$ from PKM2, which results in decreased pyruvate kinase enzyme activity. This enzymatic inhibition can promote the redistribution of upstream glycolytic intermediates into anabolic pathways like nucleotide biosynthesis. The transcription factor Myc positively regulates the expression of PKM2 versus the constitutively active PKM1 isoform by up-regulating several RNA splicing factors. Glutamine uptake and metabolism are also regulated by the Myc transcription factor, which can be activated downstream from growth factor signaling pathways, such as those involving Ras. Myc positively regulates the expression of glutamine transporters as well as the enzyme glutaminase. Glutamate, the product of glutaminase activity, can be further metabolized to α-ketoglutarate to supply intermediates for the mitochondrial Krebs cycle. Another major regulator of amino acid metabolism is mTOR complex 1 (mTORC1), which can be activated downstream from PI3K/Akt signaling as well as through the sensing of essential amino acids. mTORC1 positively regulates protein synthesis in response to these inputs by activating S6 kinase (S6K), inhibiting eukaryotic initiation factor 4E binding protein (4E-BP), and promoting the expression of ribosomal proteins. A distinct mTOR complex, mTORC2, lies upstream of Akt and is positively regulated by PI3K signaling.

plays an especially important role in regulating metabolism of the amino acid glutamine (Yuneva et al. 2007; Wise et al. 2008; Gao et al. 2009). It is perhaps surprising that glutamine metabolism is highly regulated by growth factor signaling, because it is traditionally considered a non-essential amino acid that can be synthesized from other amino acids through enzymes such as glutamine synthetase. However, observations made by Eagle and others revealed more than 50 years ago that glutamine is uniquely critical for the proliferation of mammalian cells in culture (Wise and Thompson 2010). Myc stimulates glutamine uptake by promoting the expression of glutamine transporter proteins. It can also promote the intracellular metabolism of glutamine by increasing the expression of the enzyme glutaminase, which converts glutamine to glutamate and ammonia (Fig. 4).

Glutamine's carbon backbone is important for replenishing the supply of mitochondrial Krebs cycle intermediates (anaplerosis). This is accomplished by the conversion of glutamine to glutamate and then to the Krebs cycle intermediate α-ketoglutarate (DeBerardinis et al. 2007). Importantly, treating cells that are glutamine-addicted because they express oncogenic Myc with a cell-permeable form of α-ketoglutarate can maintain cell viability in the absence of glutamine. However, for the cells to proliferate, the complete glutamine molecule that includes nitrogenous groups is required (Wise et al. 2008). Myc-induced glutamine catabolism can provide both the amide and amino groups needed for non-essential amino acid synthesis. It can also generate precursors for the biosynthesis of nucleotides and nicotinamide. Myc provides additional stimulation of nucleotide biosynthesis by promoting the expression of several key enzymes including TS, IMPDH2, and PRPS2 (Tong et al. 2009).

Growth factor signaling also regulates amino acid metabolism through the mTOR kinase. mTORC1, one of two protein complexes containing mTOR (see p. 91 [Laplante and Sabatini 2012]), is stimulated by PI3K/Akt, which phosphorylates and consequently disrupts the tuberous sclerosis complex (TSC). When disrupted, the TSC complex can no longer hydrolyze the GTP-binding protein Rheb to its GDP-bound form, thus leaving the GTP-bound Rheb free to interact with and activate mTORC1 (Sengupta et al. 2010). mTORC1 inhibition can also be relieved by Akt phosphorylation of PRAS40 at T246 (Sancak et al. 2007).

mTORC1 promotes protein synthesis by phosphorylating and inhibiting the activity of 4E-BPs, which normally sequester eIF4E and thus block cap-dependent mRNA translation (Hara et al. 1997). mTORC1 can also phosphorylate and activate S6Ks to enhance translational elongation by relieving the suppression of eEF2 (Wang et al. 2001). In addition, mTORC1 regulates the translation of the 5′TOP mRNAs, an abundant class of mRNAs that includes many encoding components of the translational apparatus (Tang et al. 2001). mTORC1 also increases ribosome synthesis by activating RNA polymerase I to transcribe rRNAs (Mayer et al. 2004).

Through its responsiveness to PI3K/Akt signaling, mTORC1 activity enhances protein synthesis when a cell is directed to grow. However, mTORC1 is also highly sensitive to inputs that reflect the cell's metabolic state. For example, upon depletion of cellular ATP levels, AMP-activated protein kinase (AMPK) can inhibit mTORC1 through multiple mechanisms, including activation of the TSC complex and inhibition of the mTOR binding partner Raptor (Corradetti et al. 2004; Gwinn et al. 2008). This provides a means for the cell to down-regulate the energetically costly process of protein synthesis when ATP is limiting. In contrast, mTORC1 can be activated by the availability of essential amino acids including leucine. Essential amino acids induce the Rag GTPase complex to assume an active conformation on lysosomal membranes. Active Rag can then recruit mTORC1 to the lysosomal surface, where it may be more amenable to activation by Rheb (Sancak et al. 2008; Sancak et al. 2010). Interestingly, recent work shows that cellular leucine uptake is coupled with glutamine efflux (Nicklin et al. 2009), linking Myc-regulated glutamine uptake with regulation of mTORC1 by essential amino acid availability.

The much less well understood mTORC2 complex also has important roles in growth factor signaling and metabolic regulation. Like mTORC1, it is activated by growth factors such as insulin. But, in contrast to mTORC1, mTORC2 lies upstream of Akt. Recent data indicate that growth-factor-stimulated activation of mTORC2 involves a direct association with ribosomes, which may ensure that mTORC2 is active only in cells that are growing and undergoing protein synthesis (Oh et al. 2010; Zinzalla et al. 2011). Once active, mTORC2 can phosphorylate Akt at S473, which is considered important both for enhancing the strength of Akt activation downstream from PI3K and for widening the range of effective Akt substrates.

6 METABOLICALLY SENSITIVE PROTEIN MODIFICATIONS LINK GROWTH FACTOR SIGNALING TO CELLULAR RESPONSES

Although many metabolic pathways that facilitate cell growth and proliferation are up-regulated in response to growth-factor-initiated signaling, the cell is not merely a passive recipient of instructions from growth factors. Rather, intracellular metabolites can exert feedback control on signaling initiated by growth factors through posttranslational modifications of critical signaling proteins (Metallo and Vander Heiden 2010).

One area where this regulation occurs is at the level of the growth factor receptor. Recent studies of cells whose ability to die by apoptosis upon nutrient withdrawal has been eliminated have shown that cells normally directed to take up nutrients upon stimulation by the growth factor interleukin 3 (IL3) can no longer take up glutamine when glucose is withdrawn from the medium (Wellen et al. 2010). This deficiency is due to the cell's inability to continue displaying the IL3 receptor at the cell surface, necessary for IL3-dependent regulation of glutamine uptake and metabolism. Metabolism of glucose to UDP-*N*-acetylglucosamine (UDP-GlcNac) via the hexosamine biosynthetic pathway, which branches off glycolysis, is necessary for the proper *N*-linked glycosylation of the IL3 receptor α-subunit. This glycosylation is critical for the proper folding of the IL3 receptor and its localization to the cell surface. The production of UDP-GlcNac through the hexosamine pathway also requires glutamine as a nitrogen donor. Thus, the dependency of IL3 receptor signaling on receptor glycosylation ensures that this pathway remains active only when there are adequate sources of both glucose and glutamine, as well as intact enzymatic pathways for their metabolism (Fig. 5). Appearance of the TGFβ, epidermal growth factor (EGF), insulin-like growth factor, and Her2 receptors at the cell surface also responds to glucose availability and/or *N*-glycosylation (Wu and Derynck 2009; Fang et al. 2010).

Histone acetylation is another metabolically sensitive protein modification that influences signaling outputs. Acetylation of histone lysine residues promotes open chromatin and increased gene expression (Li et al. 2007). In mammalian cells, the acetyl donor for histone acetylation, acetyl-CoA, is predominantly generated from glucose-derived citrate through the enzyme ACL (Wellen et al. 2009). Through ACL activity, increased levels of glucose availability and metabolism are thus able to facilitate open chromatin formation via increased histone acetylation. Further evidence for the concept that physiological variations in acetyl-CoA levels can alter the degree to which histones are acetylated has been found in yeast, where the transcription of cell growth genes is linked to increased histone acetylation that occurs following a surge in acetyl-CoA production upon entry into a growth phase (Cai and Tu 2011). In yeast, the acetylation of all proteins is not regulated by variations in acetyl-CoA levels; this regulation appears limited to substrates of the histone acetyltransferase Gcn5. Gcn5 has an in vitro K_d of 8.5 μM and K_m of 2.5 μM, within the range of yeast intracellular acetyl-CoA concentrations that have been estimated to vary from 3 to 30 μM over the metabolic cycle (Cai et al. 2011). As in yeast, in mammalian cells, not all protein acetylation varies in response to the acetyl-CoA concentration. Although the addition of acetyl groups to histone proteins has been shown

to vary with acetyl-CoA availability, the acetylation of tubulin remains relatively constant (Wellen et al. 2009). The removal of acetylation marks from histones can also be metabolically responsive. When glucose availability and metabolism are diminished, cellular NAD^+ levels and the NAD^+:NADH ratio are increased. This activates the sirtuin enzymes, class III histone deacetylases that remove the acetylation mark from histone lysine residues (Haigis and Sinclair 2010).

Note that cells use multiple other mechanisms to modify signaling pathways and outputs based on the metabolic state of the cell. As previously discussed, mTORC1 activity is regulated in part by the availability of essential amino acids, and cells also sense their metabolic state through AMPK (Ch. 14 [Hardie 2012]). AMPK generally functions to activate/up-regulate ATP-producing pathways and to down-regulate ATP-consuming activities when cellular free-energy levels are low. Other targets of AMPK activity, in addition to mTORC1, include key enzymes for fatty acid synthesis and cholesterol synthesis: acetyl-CoA carboxylase and HMG-CoA reductase.

7 PERTURBATIONS OF CELLULAR METABOLISM IN DISEASE

In disease, particularly in cancer, the control of cellular metabolism by growth-factor-receptor-initiated signaling pathways often becomes dysregulated. As discussed previously, the preferential expression of PKM2 over PKM1 has been found in all cancer cells examined to date, and this selectivity is promoted by the Myc oncogene product. Loss-of-function mutations in the PTEN tumor suppressor that antagonizes PI3K are also common. These mutations impair negative regulation of PI3K/Akt signaling, thereby enhancing glucose uptake and the flux of glucose into lipid synthesis. Overexpression and/or constitutive activation of growth factor receptors, such as the EGF receptor and Her2, are also found frequently and can promote increased nutrient uptake and anabolic metabolism for cancer cells.

Not all disease-associated mutations affecting cellular metabolism are so clearly linked to enhanced anabolic pathways. For example, recent investigations have discovered that specific mutations in the active site of $NADP^+$-linked cytosolic isocitrate dehydrogenase 1 (IDH1), or in its mitochondrial relative IDH2, facilitate a neomorphic enzyme activity. This neomorphic activity converts the Krebs cycle intermediate α-ketoglutarate to a rare metabolite not found at high levels in mammalian cells under normal conditions, 2-hydroxyglutarate (2HG) (Dang et al. 2009; Ward et al. 2010, 2011). IDH1 and IDH2 mutations occur in cancers including glioma, acute myeloid leukemia, and chondrosarcoma, and in a large percentage of patients

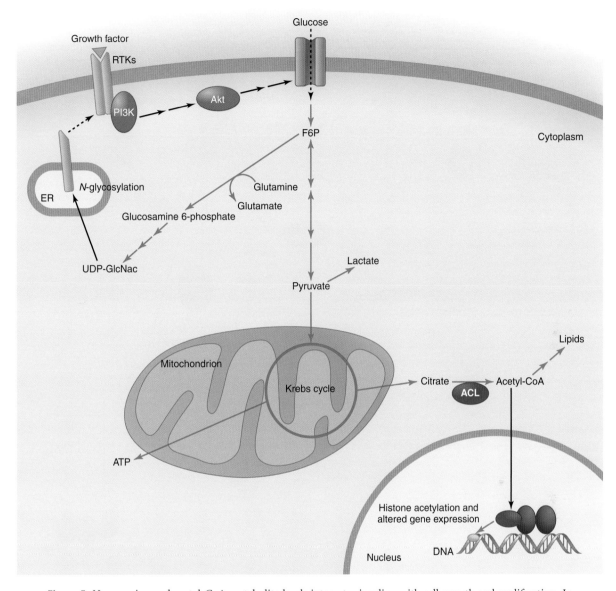

Figure 5. Hexosamine and acetyl-CoA metabolite levels integrate signaling with cell growth and proliferation. In addition to being metabolized through glycolysis, fructose 6-phosphate can act as a nitrogen acceptor from glutamine and be converted to glucosamine 6-phosphate. Further metabolism through the hexosamine biosynthetic pathway results in UDP-N-acetylglucosamine (UDP-GlcNac). UDP-GlcNac can be used for N-linked-glycosylation reactions, including that involving growth factor receptor subunits as they are being processed in the ER. This glycosylation promotes the expression and localization of receptor subunits at the cell membrane, where they can respond to growth factor. Another metabolically sensitive protein modification occurs when acetyl-CoA is used as a substrate for the acetylation of histones in the nucleus, a modification that alters gene expression. The supply of acetyl-CoA for acetylation reactions depends on the activity of ATP-citrate lyase (ACL).

with the inborn error of metabolism 2HG aciduria (Mardis et al. 2009; Yan et al. 2009; Kranendijk et al. 2010; Ward et al. 2010; Amary et al. 2011). Although the function of the "oncometabolite" 2HG in the context of various cell types and tissues remains under active investigation, current evidence suggests that its major role is to competitively

inhibit α-ketoglutarate-dependent enzymes that modify chromatin, particularly the TET family of DNA 5-methyl-cytosine hydroxylases and Jumonji-C domain histone demethylases (Figueroa et al. 2010; Chowdhury et al. 2011). Unlike other cancer-associated mutations, direct evidence for IDH mutations promoting cell-autonomous growth

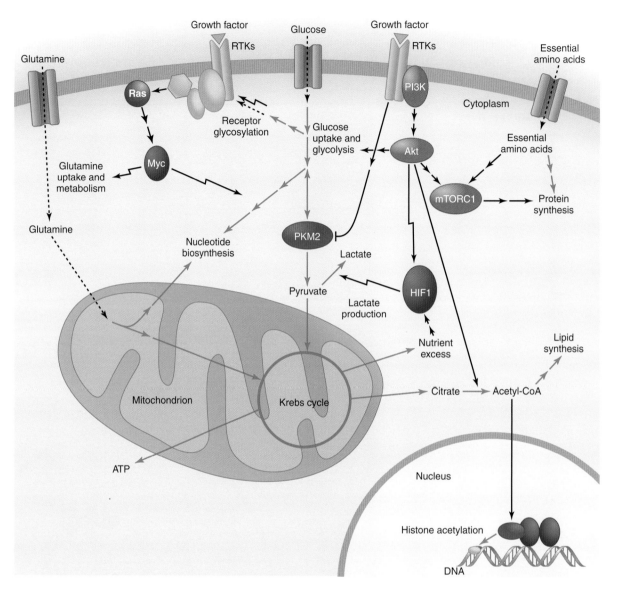

Figure 6. Growth factor signaling reprograms cellular metabolism to promote biosynthesis but is sensitive to feedback control by metabolically sensitive protein modifications. Signaling downstream from PI3K/Akt enhances glucose uptake, glycolysis, and the flux of glucose carbon into cytosolic acetyl-CoA and lipids. HIF1 signaling further promotes glycolysis while also enhancing the flux of pyruvate into lactate under conditions of O_2 limitation or nutrient excess. The Myc transcription factor, activated downstream from Ras, enhances glutamine uptake and metabolism as well as nucleotide biosynthesis. mTORC1 signaling responds to both upstream PI3K/Akt signaling and the levels of essential amino acids to promote protein synthesis. PKM2 is a pyruvate kinase isoform specific to proliferating cells that is uniquely sensitive to inhibition by tyrosine kinase signaling downstream from growth factors. Metabolically sensitive protein modifications, such as receptor glycosylation and nuclear histone acetylation, provide a way for the cell to exert feedback control on the output of growth factor signaling.

and activation of anabolic pathways is lacking, which is consistent with IDH mutations predominantly arising in lower-grade, more indolent lesions. This has led to the alternative proposal that the effects of mutant IDH and 2HG in the tumor and its microenvironment ultimately lead to a block in cellular differentiation (Ward et al. 2010).

8 CONCLUDING REMARKS

In multicellular organisms, cell growth and proliferation are normally not cell autonomous. Receptor-mediated signal transduction, initiated by extracellular growth factors, promotes entry into the cell cycle and reprograms cellular metabolism to fulfill the biosynthetic needs of cell growth and division (Fig. 6). However, despite having become highly dependent on instruction from extracellular growth factors, mammalian cells have retained the ability to sense their internal metabolic reserves and adjust their growth and biosynthetic activities accordingly. Much of this feedback control occurs at the level of posttranslational modifications of signal transduction proteins by key cellular metabolites. Moreover, intracellular metabolites can also regulate chromatin accessibility to control gene expression.

The evolution of the ability to regulate chromatin accessibility by specific metabolites may have preceded the ability of growth factor signaling to reprogram metabolism in multicellular organisms. Nonetheless, many aspects of how variations in cellular metabolism influence chromatin accessibility remain to be fully characterized. In addition to intracellular glucose metabolism directly altering the acetylation state of histones, the methylation state of histone lysine and DNA cytosine may also be metabolically responsive. These methylation states were once thought to be irreversible, but recent work has described demethylating enzymes for both histones and DNA that are sustained by α-ketoglutarate production downstream from glutamine metabolism. As discussed above, these α-ketoglutarate-dependent enzymes can be inhibited by the 2HG produced by IDH1/IDH2 mutations.

Another area of continuing investigation is how minimizing ATP production and enhancing ATP consumption can facilitate the rapid nutrient metabolism of proliferating cells (Israelsen and Vander Heiden 2010). Minimizing cellular ATP accumulation runs counter to the metabolic strategy of quiescent cells, which completely oxidize the majority of glucose carbon in the mitochondria to maximize ATP production. But proliferating cells with activated growth factor signaling pathways usually take up nutrients far in excess of the level required to maintain ATP levels and avoid AMPK activation. Moreover, glycolysis in proliferating cells is limited by the rate of ATP consumption, not ATP production (Scholnick et al. 1973). Increasing cellular ATP consumption and the ADP:ATP ratio may be critical for relieving the inhibition of glycolytic enzymes that can occur when ATP levels are high, inhibition that would otherwise prevent high glycolytic flux and enhanced macromolecular biosynthesis from glycolytic intermediates.

REFERENCES

*Reference is in this book.

Amary MF, Bacsi K, Maggiani F, Damato S, Halai D, Berisha F, Pollock R, O'Donnell P, Grigoriadis A, Diss T, et al. 2011. IDH1 and IDH2 mutations are frequent events in central chondrosarcoma and central and periosteal chondromas but not in other mesenchymal tumours. *J Pathol* 224: 334–343.

Anastasiou D, Poulogiannis G, Asara JM, Boxer MB, Jiang JK, Shen M, Bellinger G, Sasaki AT, Locasale JW, Auld DS, et al. 2011. Inhibition of pyruvate kinase M2 by reactive oxygen species contributes to antioxidant responses. *Science* 334: 1278–1283.

Bauer DE, Harris MH, Plas DR, Lum JJ, Hammerman PS, Rathmell JC, Riley JL, Thompson CB. 2004. Cytokine stimulation of aerobic glycolysis in hematopoietic cells exceeds proliferative demand. *FASEB J* 18: 1303–1305.

Bauer DE, Hatzivassiliou G, Zhao F, Andreadis C, Thompson CB. 2005. ATP citrate lyase is an important component of cell growth and transformation. *Oncogene* 24: 6314–6322.

Berwick DC, Hers I, Heesom KJ, Moule SK, Tavare JM. 2002. The identification of ATP-citrate lyase as a protein kinase B (Akt) substrate in primary adipocytes. *J Biol Chem* 277: 33895–33900.

Bobrovnikova-Marjon E, Hatzivassiliou G, Grigoriadou C, Romero M, Cavener DR, Thompson CB, Diehl JA. 2008. PERK-dependent regulation of lipogenesis during mouse mammary gland development and adipocyte differentiation. *Proc Natl Acad Sci* 105: 16314–16319.

Cai L, Tu BP. 2011. On acetyl-CoA as a gauge of cellular metabolic state. *Cold Spring Harb Symp Quant Biol* doi: 10.1101/sqb.2011.76.010769.

Cai L, Sutter BM, Li B, Tu BP. 2011. Acetyl-CoA induces cell growth and proliferation by promoting the acetylation of histones at growth genes. *Mol Cell* 42: 426–437.

Chowdhury R, Yeoh KK, Tian YM, Hillringhaus L, Bagg EA, Rose NR, Leung IK, Li XS, Woon EC, Yang M, et al. 2011. The oncometabolite 2-hydroxyglutarate inhibits histone lysine demethylases. *EMBO Rep* 12: 463–469.

Christofk HR, Vander Heiden MG, Harris MH, Ramanathan A, Gerszten RE, Wei R, Fleming MD, Schreiber SL, Cantley LC. 2008a. The M2 splice isoform of pyruvate kinase is important for cancer metabolism and tumour growth. *Nature* 452: 230–233.

Christofk HR, Vander Heiden MG, Wu N, Asara JM, Cantley LC. 2008b. Pyruvate kinase M2 is a phosphotyrosine-binding protein. *Nature* 452: 181–186.

Clower CV, Chatterjee D, Wang Z, Cantley LC, Vander Heiden MG, Krainer AR. 2010. The alternative splicing repressors hnRNP A1/A2 and PTB influence pyruvate kinase isoform expression and cell metabolism. *Proc Natl Acad Sci* 107: 1894–1899.

Corradetti MN, Inoki K, Bardeesy N, DePinho RA, Guan KL. 2004. Regulation of the TSC pathway by LKB1: Evidence of a molecular link between tuberous sclerosis complex and Peutz-Jeghers syndrome. *Genes Dev* 18: 1533–1538.

Dang L, White DW, Gross S, Bennett BD, Bittinger MA, Driggers EM, Fantin VR, Jang HG, Jin S, Keenan MC, et al. 2009. Cancer-associated IDH1 mutations produce 2-hydroxyglutarate. *Nature* 462: 739–744.

David CJ, Chen M, Assanah M, Canoll P, Manley JL. 2010. HnRNP proteins controlled by c-Myc deregulate pyruvate kinase mRNA splicing in cancer. *Nature* 463: 364–368.

DeBerardinis RJ, Mancuso A, Daikhin E, Nissim I, Yudkoff M, Wehrli S, Thompson CB. 2007. Beyond aerobic glycolysis: Transformed cells can engage in glutamine metabolism that exceeds the requirement for protein and nucleotide synthesis. *Proc Natl Acad Sci* 104: 19345–19350.

DeBerardinis RJ, Lum JJ, Hatzivassiliou G, Thompson CB. 2008. The biology of cancer: Metabolic reprogramming fuels cell growth and proliferation. *Cell Metab* 7: 11–20.

Deprez J, Vertommen D, Alessi DR, Hue L, Rider MH. 1997. Phosphorylation and activation of heart 6-phosphofructo-2-kinase by protein kinase B and other protein kinases of the insulin signaling cascades. *J Biol Chem* 272: 17269–17275.

Eigenbrodt E, Leib S, Kramer W, Friis RR, Schoner W. 1983. Structural and kinetic differences between the M2 type pyruvate kinases from lung and various tumors. *Biomed Biochim Acta* 42: S278–S282.

Elstrom RL, Bauer DE, Buzzai M, Karnauskas R, Harris MH, Plas DR, Zhuang H, Cinalli RM, Alavi A, Rudin CM, et al. 2004. Akt stimulates aerobic glycolysis in cancer cells. *Cancer Res* 64: 3892–3899.

Fang M, Shen Z, Huang S, Zhao L, Chen S, Mak TW, Wang X. 2010. The ER UDPase ENTPD5 promotes protein *N*-glycosylation, the Warburg effect, and proliferation in the PTEN pathway. *Cell* **143:** 711–724.

Figueroa ME, Abdel-Wahab O, Lu C, Ward PS, Patel J, Shih A, Li Y, Bhagwat N, Vasanthakumar A, Fernandez HF, et al. 2010. Leukemic IDH1 and IDH2 mutations result in a hypermethylation phenotype, disrupt TET2 function, and impair hematopoietic differentiation. *Cancer Cell* **18:** 553–567.

Gao P, Tchernyshyov I, Chang TC, Lee YS, Kita K, Ochi T, Zeller KI, De Marzo AM, Van Eyk JE, Mendell JT, et al. 2009. c-Myc suppression of miR-23a/b enhances mitochondrial glutaminase expression and glutamine metabolism. *Nature* **458:** 762–765.

Gottlob K, Majewski N, Kennedy S, Kandel E, Robey RB, Hay N. 2001. Inhibition of early apoptotic events by Akt/PKB is dependent on the first committed step of glycolysis and mitochondrial hexokinase. *Genes Dev* **15:** 1406–1418.

Gwinn DM, Shackelford DB, Egan DF, Mihaylova MM, Mery A, Vasquez DS, Turk BE, Shaw RJ. 2008. AMPK phosphorylation of raptor mediates a metabolic checkpoint. *Mol Cell* **30:** 214–226.

Haigis MC, Sinclair DA. 2010. Mammalian sirtuins: Biological insights and disease relevance. *Annu Rev Pathol* **5:** 253–295.

Hara K, Yonezawa K, Kozlowski MT, Sugimoto T, Andrabi K, Weng QP, Kasuga M, Nishimoto I, Avruch J. 1997. Regulation of eIF-4E BP1 phosphorylation by mTOR. *J Biol Chem* **272:** 26457–26463.

* Hardie DG. 2012. Organismal carbohydrate and lipid homeostasis. *Cold Spring Harb Perspect Biol* **4:** a006031.

* Hemmings BA, Restuccia DF. 2012. The PI3K-PKB/Akt pathway. *Cold Spring Harb Perspect Biol* **4:** a011189.

Hitosugi T, Kang S, Vander Heiden MG, Chung TW, Elf S, Lythgoe K, Dong S, Lonial S, Wang X, Chen GZ, et al. 2009. Tyrosine phosphorylation inhibits PKM2 to promote the Warburg effect and tumor growth. *Sci Signal* **2:** ra73.

Hudson CC, Liu M, Chiang GG, Otterness DM, Loomis DC, Kaper F, Giaccia AJ, Abraham RT. 2002. Regulation of hypoxia-inducible factor 1α expression and function by the mammalian target of rapamycin. *Mol Cell Biol* **22:** 7004–7014.

* Ingham P. 2012. Hedgehog signaling. *Cold Spring Harb Perspect Biol* **4:** a011221.

Isaacs JS, Jung YJ, Mole DR, Lee S, Torres-Cabala C, Chung YL, Merino M, Trepel J, Zbar B, Toro J, et al. 2005. HIF overexpression correlates with biallelic loss of fumarate hydratase in renal cancer: Novel role of fumarate in regulation of HIF stability. *Cancer Cell* **8:** 143–153.

Israelsen WJ, Vander Heiden MG. 2010. ATP consumption promotes cancer metabolism. *Cell* **143:** 669–671.

Kim JW, Tchernyshyov I, Semenza GL, Dang CV. 2006. HIF-1-mediated expression of pyruvate dehydrogenase kinase: A metabolic switch required for cellular adaptation to hypoxia. *Cell Metab* **3:** 177–185.

Kohn AD, Summers SA, Birnbaum MJ, Roth RA. 1996. Expression of a constitutively active Akt Ser/Thr kinase in 3T3-L1 adipocytes stimulates glucose uptake and glucose transporter 4 translocation. *J Biol Chem* **271:** 31372–31378.

Kranendijk M, Struys EA, van Schaftingen E, Gibson KM, Kanhai WA, van der Knaap MS, Amiel J, Buist NR, Das AM, de Klerk JB, et al. 2010. IDH2 mutations in patients with D-2-hydroxyglutaric aciduria. *Science* **330:** 336.

* Laplante M, Sabatini DM. 2012. mTOR signaling. *Cold Spring Harb Perspect Biol* **4:** a011593.

Lee J, Kim HK, Han YM, Kim J. 2008. Pyruvate kinase isozyme type M2 (PKM2) interacts and cooperates with Oct-4 in regulating transcription. *Int J Biochem Cell Biol* **40:** 1043–1054.

Li B, Carey M, Workman JL. 2007. The role of chromatin during transcription. *Cell* **128:** 707–719.

Lum JJ, Bui T, Gruber M, Gordan JD, DeBerardinis RJ, Covello KL, Simon MC, Thompson CB. 2007. The transcription factor HIF-1α plays a critical role in the growth factor-dependent regulation of both aerobic and anaerobic glycolysis. *Genes Dev* **21:** 1037–1049.

Majmundar AJ, Wong WJ, Simon MC. 2010. Hypoxia-inducible factors and the response to hypoxic stress. *Mol Cell* **40:** 294–309.

Mardis ER, Ding L, Dooling DJ, Larson DE, McLellan MD, Chen K, Koboldt DC, Fulton RS, Delehaunty KD, McGrath SD, et al. 2009. Recurring mutations found by sequencing an acute myeloid leukemia genome. *N Engl J Med* **361:** 1058–1066.

Mayer C, Zhao J, Yuan X, Grummt I. 2004. mTOR-dependent activation of the transcription factor TIF-IA links rRNA synthesis to nutrient availability. *Genes Dev* **18:** 423–434.

Metallo CM, Vander Heiden MG. 2010. Metabolism strikes back: Metabolic flux regulates cell signaling. *Genes Dev* **24:** 2717–2722.

* Morrison DK. 2012. MAP kinase pathways. *Cold Spring Harb Perspect Biol* **4:** a011254.

Nicklin P, Bergman P, Zhang B, Triantafellow E, Wang H, Nyfeler B, Yang H, Hild M, Kung C, Wilson C, et al. 2009. Bidirectional transport of amino acids regulates mTOR and autophagy. *Cell* **136:** 521–534.

Oh WJ, Wu CC, Kim SJ, Facchinetti V, Julien LA, Finlan M, Roux PP, Su B, Jacinto E. 2010. mTORC2 can associate with ribosomes to promote cotranslational phosphorylation and stability of nascent Akt polypeptide. *EMBO J* **29:** 3939–3951.

Osthus RC, Shim H, Kim S, Li Q, Reddy R, Mukherjee M, Xu Y, Wonsey D, Lee LA, Dang CV. 2000. Deregulation of glucose transporter 1 and glycolytic gene expression by c-Myc. *J Biol Chem* **275:** 21797–21800.

Papandreou I, Cairns RA, Fontana L, Lim AL, Denko NC. 2006. HIF-1 mediates adaptation to hypoxia by actively downregulating mitochondrial oxygen consumption. *Cell Metab* **3:** 187–197.

Porstmann T, Griffiths B, Chung YL, Delpuech O, Griffiths JR, Downward J, Schulze A. 2005. PKB/Akt induces transcription of enzymes involved in cholesterol and fatty acid biosynthesis via activation of SREBP. *Oncogene* **24:** 6465–6481.

Rathmell JC, Fox CJ, Plas DR, Hammerman PS, Cinalli RM, Thompson CB. 2003. Akt-directed glucose metabolism can prevent Bax conformation change and promote growth factor-independent survival. *Mol Cell Biol* **23:** 7315–7328.

Sancak Y, Thoreen CC, Peterson TR, Lindquist RA, Kang SA, Spooner E, Carr SA, Sabatini DM. 2007. PRAS40 is an insulin-regulated inhibitor of the mTORC1 protein kinase. *Mol Cell* **25:** 903–915.

Sancak Y, Peterson TR, Shaul YD, Lindquist RA, Thoreen CC, Bar-Peled L, Sabatini DM. 2008. The Rag GTPases bind raptor and mediate amino acid signaling to mTORC1. *Science* **320:** 1496–1501.

Sancak Y, Bar-Peled L, Zoncu R, Markhard AL, Nada S, Sabatini DM. 2010. Ragulator–Rag complex targets mTORC1 to the lysosomal surface and is necessary for its activation by amino acids. *Cell* **141:** 290–303.

Scholnick P, Lang D, Racker E. 1973. Regulatory mechanisms in carbohydrate metabolism. IX. Stimulation of aerobic glycolysis by energy-linked ion transport and inhibition by dextran sulfate. *J Biol Chem* **248:** 5175–5182.

Selak MA, Armour SM, MacKenzie ED, Boulahbel H, Watson DG, Mansfield KD, Pan Y, Simon MC, Thompson CB, Gottlieb E. 2005. Succinate links TCA cycle dysfunction to oncogenesis by inhibiting HIF-α prolyl hydroxylase. *Cancer Cell* **7:** 77–85.

Semenza GL, Roth PH, Fang HM, Wang GL. 1994. Transcriptional regulation of genes encoding glycolytic enzymes by hypoxia-inducible factor 1. *J Biol Chem* **269:** 23757–23763.

Sengupta S, Peterson TR, Sabatini DM. 2010. Regulation of the mTOR complex 1 pathway by nutrients, growth factors, and stress. *Mol Cell* **40:** 310–322.

Shatrov VA, Sumbayev VV, Zhou J, Brune B. 2003. Oxidized low-density lipoprotein (oxLDL) triggers hypoxia-inducible factor-1α (HIF-1α) accumulation via redox-dependent mechanisms. *Blood* **101:** 4847–4849.

Shim H, Dolde C, Lewis BC, Wu CS, Dang G, Jungmann RA, Dalla-Favera R, Dang CV. 1997. c-Myc transactivation of LDH-A: Implications for tumor metabolism and growth. *Proc Natl Acad Sci* **94:** 6658–6663.

Taha C, Liu Z, Jin J, Al-Hasani H, Sonenberg N, Klip A. 1999. Opposite translational control of GLUT1 and GLUT4 glucose transporter mRNAs in response to insulin. Role of mammalian target of rapamycin, protein kinase b, and phosphatidylinositol 3-kinase in GLUT1 mRNA translation. *J Biol Chem* **274:** 33085–33091.

Tang H, Hornstein E, Stolovich M, Levy G, Livingstone M, Templeton D, Avruch J, Meyuhas O. 2001. Amino acid–induced translation of TOP mRNAs is fully dependent on phosphatidylinositol 3-kinase-mediated signaling, is partially inhibited by rapamycin, and is independent of S6K1 and rpS6 phosphorylation. *Mol Cell Biol* **21:** 8671–8683.

Tong X, Zhao F, Thompson CB. 2009. The molecular determinants of de novo nucleotide biosynthesis in cancer cells. *Curr Opin Genet Dev* **19:** 32–37.

Vander Heiden MG, Cantley LC, Thompson CB. 2009. Understanding the Warburg effect: The metabolic requirements of cell proliferation. *Science* **324:** 1029–1033.

Vander Heiden MG, Locasale JW, Swanson KD, Sharfi H, Heffron GJ, Amador-Noguez D, Christofk HR, Wagner G, Rabinowitz JD, Asara JM, et al. 2010. Evidence for an alternative glycolytic pathway in rapidly proliferating cells. *Science* **329:** 1492–1499.

Wang X, Li W, Williams M, Terada N, Alessi DR, Proud CG. 2001. Regulation of elongation factor 2 kinase by p90 (RSK1) and p70 S6 kinase. *EMBO J* **20:** 4370–4379.

Ward PS, Patel J, Wise DR, Abdel-Wahab O, Bennett BD, Coller HA, Cross JR, Fantin VR, Hedvat CV, Perl AE, et al. 2010. The common feature of leukemia-associated IDH1 and IDH2 mutations is a neomorphic enzyme activity converting α-ketoglutarate to 2-hydroxyglutarate. *Cancer Cell* **17:** 225–234.

Ward PS, Cross JR, Lu C, Weigert O, Abel-Wahab O, Levine RL, Weinstock DM, Sharp KA, Thompson CB. 2011. Identification of additional IDH mutations associated with oncometabolite R(−)-2-hydroxyglutarate production. *Oncogene* doi: 10.1038/onc.2011.416.

Wellen KE, Thompson CB. 2010. Cellular metabolic stress: Considering how cells respond to nutrient excess. *Mol Cell* **40:** 323–332.

Wellen KE, Hatzivassiliou G, Sachdeva UM, Bui TV, Cross JR, Thompson CB. 2009. ATP-citrate lyase links cellular metabolism to histone acetylation. *Science* **324:** 1076–1080.

Wellen KE, Lu C, Mancuso A, Lemons JM, Ryczko M, Dennis JW, Rabinowitz JD, Coller HA, Thompson CB. 2010. The hexosamine biosynthetic pathway couples growth factor-induced glutamine uptake to glucose metabolism. *Genes Dev* **24:** 2784–2799.

Wise DR, Thompson CB. 2010. Glutamine addiction: A new therapeutic target in cancer. *Trends Biochem Sci* **35:** 427–433.

Wise DR, DeBerardinis RJ, Mancuso A, Sayed N, Zhang XY, Pfeiffer HK, Nissim I, Daikhin E, Yudkoff M, McMahon SB, et al. 2008. Myc regulates a transcriptional program that stimulates mitochondrial glutaminolysis and leads to glutamine addiction. *Proc Natl Acad Sci* **105:** 18782–18787.

Wu L, Derynck R. 2009. Essential role of TGF-β signaling in glucose-induced cell hypertrophy. *Dev Cell* **17:** 35–48.

Yan H, Parsons DW, Jin G, McLendon R, Rasheed BA, Yuan W, Kos I, Batinic-Haberle I, Jones S, Riggins GJ, et al. 2009. IDH1 and IDH2 mutations in gliomas. *N Engl J Med* **360:** 765–773.

Yuneva M, Zamboni N, Oefner P, Sachidanandam R, Lazebnik Y. 2007. Deficiency in glutamine but not glucose induces MYC-dependent apoptosis in human cells. *J Cell Biol* **178:** 93–105.

Zhong H, Chiles K, Feldser D, Laughner E, Hanrahan C, Georgescu MM, Simons JW, Semenza GL. 2000. Modulation of hypoxia-inducible factor 1α expression by the epidermal growth factor/phosphatidylinositol 3-kinase/PTEN/AKT/FRAP pathway in human prostate cancer cells: Implications for tumor angiogenesis and therapeutics. *Cancer Res* **60:** 1541–1545.

Zinzalla V, Stracka D, Oppliger W, Hall MN. 2011. Activation of mTORC2 by association with the ribosome. *Cell* **144:** 757–768.

CHAPTER **8**

Signaling Networks that Regulate Cell Migration

Peter Devreotes[1] and Alan Rick Horwitz[2]

[1]Department of Cell Biology, Johns Hopkins University School of Medicine, Baltimore, Maryland 21205
[2]Department of Cell Biology, University of Virginia School of Medicine, Charlottesville, Virginia 22908

Correspondence: pnd@jhmi.edu a005959

SUMMARY

Stimuli that promote cell migration, such as chemokines, cytokines, and growth factors in metazoans and cyclic AMP in *Dictyostelium*, activate signaling pathways that control organization of the actin cytoskeleton and adhesion complexes. The Rho-family GTPases are a key convergence point of these pathways. Their effectors include actin regulators such as formins, members of the WASP/WAVE family and the Arp2/3 complex, and the myosin II motor protein. Pathways that link to the Rho GTPases include Ras GTPases, TorC2, and PI3K. Many of the molecules involved form gradients within cells, which define the front and rear of migrating cells, and are also established in related cellular behaviors such as neuronal growth cone extension and cytokinesis. The signaling molecules that regulate migration can be integrated to provide a model of network function. The network displays biochemical excitability seen as spontaneous waves of activation that propagate along the cell cortex. These events coordinate cell movement and can be biased by external cues to bring about directed migration.

Outline

1 INTRODUCTION

Cell migration plays a pivotal role in a wide variety of phenomena throughout phylogeny (Trinkaus 1969; Ridley et al. 2003; Lee et al. 2005; Cai et al. 2010). In the amoeba *Dictyostelium,* it functions in nutrient seeking, cell–cell aggregation, and the morphogenesis of multicellular structures. In metazoans, cells migrate both throughout embryogenesis and in the adult. In early developmental events, such as gastrulation or dorsal closure, large cell sheets migrate and fold; at later stages, precursor cells that reside in the neural crest, somites, brain ventricles, and other stem cell regions leave epithelial sheets and migrate to their target destinations. In the adult, cell migrations are critical for immune cell trafficking, wound healing, and stem cell homing, among other processes. A closely related phenomenon is the directed growth of specialized cellular extensions (e.g., yeast mating "shmoos" [Slessareva and Dohlman 2006], pollen tubes [Takeuchi and Higashiyama 2011], and the dendrites, axons, and spines of neurons [Tada and Sheng 2006; Geraldo and Gordon-Weeks 2009]).

Most migrating cells and cellular extensions have an internal compass that enables them to sense and move along gradients of soluble attractants and repellents, a process referred to as chemotaxis (Devreotes and Janetopoulos 2003). Many chemoattractants act through G-protein-coupled receptors (GPCRs). Examples include cAMP acting on cAMP receptors in *Dictyostelium* and chemokines such as SDF1 acting on chemokine receptors in metazoans. Growth factors acting on receptor tyrosine kinases and cytokines such as transforming growth factor β (TGFβ) also function as chemoattractants. Like migrating cells, the growth of axons can be guided by a series of extracellular protein attractants and repellents (e.g., nephrins). Cells can

also be guided by gradients of immobilized signaling molecules (haplotaxis), substrate rigidity (durotaxis), electric fields (galvanotaxis), and shear force (mechanotaxis). Importantly, many diseases involve defective or unregulated cell migration or protrusive growth (Ridley et al. 2003). For example, tumor invasion and metastasis occur as a consequence of the movement of both individual cells and large collectives (Friedl and Gilmour 2009; Friedl and Alexander 2011), arthritis and asthma result from excessive migration of inflammatory cells (Montoya et al. 2002; Vicente-Manzanares et al. 2002), and several cognitive disorders are accompanied by abnormal neuronal extensions (Newey et al. 2005; van Galen and Ramakers 2005).

Cell migration requires coordination of cytoskeletal dynamics and reorganization, cell adhesion, and signal transduction, and takes a variety of forms (see Box 1) (Lauffenburger and Horwitz 1996; Mitchison and Cramer 1996; Ridley et al. 2003). Here, we first examine the machinery that drives migration—the actin cytoskeleton, cell adhesions, and their regulators. We then discuss signaling networks that control the migration machinery, starting with those closest to the cytoskeleton then adding upstream components. Finally, we address how chemotactic cues regulate motility. There are, of course, other kinds of motility, such as sperm and cilial motility, but they use microtubule-based mechanisms and are not addressed here.

2 THE MIGRATION MACHINERY

2.1 Actin Polymerization and Myosin-Mediated Contraction

Polymerization of globular (G) actin monomers to form filamentous (F) actin is critical for cell migration (Pollard

---

BOX 1. THE SPECTRUM OF CELL MIGRATION BEHAVIORS

Cells can move in a variety of different ways, depending on the differentiated cell type, the surrounding environment, and the organism. The "mesenchymal" migration of fibroblasts, which have large actin filament bundles and prominent adhesions, is slow, for example. Similarly, keratocytes have an actin-rich lamellipodium, but these move more rapidly than fibroblasts. The amoeboid movements of neutrophils and *Dictyostelium* are instead characterized by the presence of rapid, efficient pseudopodial extensions and low adhesion. Cells such as primordial germ cells and some leukocytes and tumor cells can move by "blebbing," a contraction-mediated squeezing from the rear that produces a protrusion in regions lacking highly organized actomyosin filaments (Charras and Paluch 2008; Friedl and Wolf 2010; Schmidt and Friedl 2010). These migration modes are related, residing along a continuum, and can interconvert depending on cell state, the extracellular environment, and the relative activation of different pathways; but they are distinct from the "swimming" driven by beating of flagella or cilia that is observed in some cells. Migration can result in the movement of single cells, small collectives, or large sheets. It can also occur over a variety of substrata that include other cells and extracellular matrix components. Tumor cells can adapt to their environment by using diverse migration modes that include mesenchymal, amoeboid, and blebbing modes. They can also use specialized adhesion structures like invadopodia, which localize proteolytic activity that degrades the local matrix (Linder et al. 2011).

Cite this chapter as *Cold Spring Harb Perspect Biol* doi: 10.1101/cshperspect.a005959

and Borisy 2003; Ridley 2011). It produces oriented filaments that grow at the so-called barbed end and push the
front (the leading edge) of the cell forward, driving cell
migration. In cells that migrate by blebbing, actin stabilizes
the blebs following their protrusion (Charras and Paluch
2008; Fackler and Grosse 2008). Actin filaments arise and

grow through a complex but well-understood process (Fig.
1). Actin nucleation and polymerization are regulated by
formins (e.g., mDia1 and mDia2) and the Arp2/3 complex
(Insall and Machesky 2009; Chesarone et al. 2010; Ridley
2011). The formins nucleate and regulate the growth of
linear actin filaments (Goode and Eck 2007; Paul and Pol-

Figure 1. Regulation of actin dynamics by formins and Arp2/3 in cellular protrusions. The Rho GTPases Rac, RhoA,
and Cdc42 regulate actin dynamics at the leading edge via their effects on the activities of formins (mDia), Arp2/3
complex, and LIM kinase (LIMK). Arp2/3 nucleates actin branches that are seen in broad protrusions. Its activity is
regulated by Cdc42 and Rac1, which act on WASP/WAVE-containing protein complexes. Rac and Cdc42 also act on
PAK, which phosphorylates LIM kinase, which in turn regulates cofilin, a severing protein. Finally, RhoA acts on
mDia1 and Cdc42 acts on mDia2 to promote actin polymerization using a processive capping mechanism. RhoA
also activates profilin, which binds to actin monomers and increases the rate of polymerization. These GTPases are
activated in a clear temporal sequence near the leading edge (Machacek et al. 2009). AID, autoinhibitory domain;
FH, formin homology domains; RBD, Rho-GTPase-binding domain.

lard 2009). These processive capping proteins sequentially add actin monomers while remaining weakly bound to the rapidly growing (barbed) end of the filaments, a process termed processive elongation. The Arp2/3 complex nucleates branches from existing actin filaments at a 70° angle and thereby produces the dendritic actin network that is prominent near the leading edge of broad protrusions and appears to stabilize them (Insall and Machesky 2009).

These mediators of actin branching and polymerization are highly regulated. In fibroblasts and many other cells, mDia1 and mDia2 are regulated by the Rho-family small GTPases RhoA and Cdc42, respectively, which relieve an autoinhibitory state. Two other proteins, ABI1 and Gα12/13, appear to direct mDia1 to actin filaments and adhesions. The Arp2/3 complex contains six subunits, including two actin-related proteins, Arp2 and Arp3, which nucleate new actin filaments by binding to the side of existing filaments. WASP family members (WAVE [also known as Scar] and the WASPs) are targets of the Rho-family GTPases Rac1 (WAVE) and Cdc42 (WASP) and in turn interact with the Arp2/3 complex and regulate its activity (Pollitt and Insall 2009; Padrick and Rosen 2010).

Actin filaments are capped at the barbed end by capping proteins, which inhibits depolymerization and thereby stabilizes them. Anticapping proteins of the Mena/Vasp family, in turn, antagonize capping proteins and thereby regulate capping (Krause et al. 2003; Bear and Gertler 2009). Cofilin is another major regulator of actin filament stability that severs actin filaments; it also binds to G-actin, increasing the off-rate of actin monomers at the pointed (nonpolymerizing) end (Condeelis 2001; Bamburg and Bernstein 2008). LIM kinase, which is activated by Cdc42 and Rac1, stimulates cofilin activity by phosphorylation. LIM kinase is itself regulated by Cdc42 and Rac1, which act via the kinase PAK (Yamaguchi and Condeelis 2007; Bamburg and Bernstein 2008). Finally, profilin binds to actin monomers and increases the polymerization rate.

Contraction forces generated by myosin II motor proteins (Bugyi and Carlier 2010) are coordinated with actin polymerization at the leading edge and have several roles in migration (Small and Resch 2005; Vicente-Manzanares et al. 2009). First, in fibroblasts and epithelial cells, myosin II promotes retrograde movement of actin filaments away from the zone of active actin polymerization in the lamellipodium. This retrograde flow essentially subtracts from actin polymerization at the leading edge and can reduce the net protrusion rate (Ponti et al. 2004). The forces from both retrograde flow and actin polymerization can be "shunted" to the substratum via integrin-based adhesions linked to actin filaments (Mitchison and Kirschner 1988; Jay 2000). This shunting inhibits retrograde flow and enhances the protrusion rate, because the full force of actin polymer-

ization acts at adhesions and the membrane at the leading edge. However, the transmission of force from actin through adhesions to the substratum is not always complete and can lead to varying rates of retrograde flow (Brown et al. 2006; Hu et al. 2007; Wang 2007; Chen et al. 2012). Second, the pressure from myosin-mediated contractions in the rear and sides can produce blebs in regions depleted of actomyosin filaments (Charras and Paluch 2008). In cells that move by blebbing, myosin-based contraction alone drives migration; however, the blebs are stabilized by the formation of a dendritic actin meshwork.

Myosin II activity is regulated by phosphorylation of myosin's regulatory light chain (RLC), and its assembly into filaments is regulated by phosphorylation in the tail region of the heavy chain (Vicente-Manzanares et al. 2009). A number of kinases can phosphorylate the RLC. Among the best studied are myosin light-chain kinase (MLCK) and Rho-associated protein kinase (ROCK); myosin phosphatase hydrolyzes the phosphate. In contrast to RLC regulation, the kinases regulating filament assembly are not well understood (Vicente-Manzanares et al. 2009).

2.2 Adhesion

Adhesion to the substrate is common to most migrating cells. Integrin-mediated adhesions are the best studied (Hynes 2002). The integrins are a large family of heterodimeric transmembrane receptors that link to actin via a specialized set of molecules that include talin, vinculin, and α-actinin. The adhesions in which these components reside are large assemblies containing >150 different molecules that mediate intracellular signaling in addition to adhesion to proteins in the extracellular matrix, such as fibronectin and laminin (Zaidel-Bar et al. 2007; Parsons et al. 2010; Zaidel-Bar and Geiger 2010). The affinity of integrins is regulated by the binding of talin and kindlin, cytoplasmic proteins that bind directly to the cytoplasmic domain of the integrin β subunit, and also by phosphatidylinositol 4,5-bisphosphate (PIP_2) and other adhesion-associated molecules (Moser et al. 2009; Shattil et al. 2010).

Adhesions serve both as traction points and as signaling centers during cell migration (Parsons et al. 2010). As traction points, they transmit forces to the substrate so that actin polymerization causes protrusion at the cell front. These traction points are released at the cell rear as it retracts and the cell moves forward. Although this release is efficient in some cells and substrates, it is not in others and can be rate limiting for migration (Lauffenburger and Horwitz 1996). Thus, there is an optimum strength of attachment that allows sufficient adhesion for traction at the cell front and yet allows for efficient release at the rear (Palecek et al. 1997). As signaling centers, adhesions in protrusions

Cite this chapter as *Cold Spring Harb Perspect Biol* doi: 10.1101/cshperspect.a005959

regulate actin polymerization and myosin II activity through Rho-family GTPases (Parsons et al. 2010). Although adhesions can vary considerably in size, location, and presumably function, they have not yet been classified clearly and meaningfully based on differences in composition and function. Adhesions in vivo tend to be small and dynamic in migrating cells; however, highly elongated adhesions have also been observed (Harunaga and Yamada 2011; Kubow and Horwitz 2011).

The cytoskeleton in turn regulates adhesions via an incompletely understood feedback loop involving actin polymerization and myosin-II-mediated contraction (Fig. 2). Nascent adhesions form in the region of dendritic actin, and their formation is coupled to actin polymerization (Alexandrova et al. 2008; Choi et al. 2008). At the interface of dendritic actin in the lamellipodium and the actin bundles in the adjacent lamellum, adhesions elongate along actin filament bundles (Small et al. 2002; Choi et al. 2008; Geiger and Yamada 2011; Oakes et al. 2012).

The fraction of adhesions that grow, as well as the extent of maturation, is determined at least in part by myosin II activity (Vicente-Manzanares et al. 2009; Oakes et al. 2012). Proteases such as calpain, a calcium-activated protease, mediate adhesion disassembly by acting on adhesion proteins such as talin, which link actin and integrins (Franco et al. 2004; Chan et al. 2010; Cortesio et al. 2011). The repeated direct contact between microtubule tips and adhesions and endocytosis of integrins driven by the GTPase dynamin also contribute to disassembly (Kaverina et al. 1999; Broussard et al. 2008; Ezratty et al. 2009; Gerisch et al. 2011).

The formation, dynamics, and function of adhesions are highly regulated. In migrating cells, they form in protrusions near the leading edge. In rapidly migrating, amoeboid-like cells, adhesions in protrusions are small and tend to form and turnover rapidly, making them difficult to visualize. Few, if any, undergo significant maturation into large, elongated structures. Cells undergoing slower, mes-

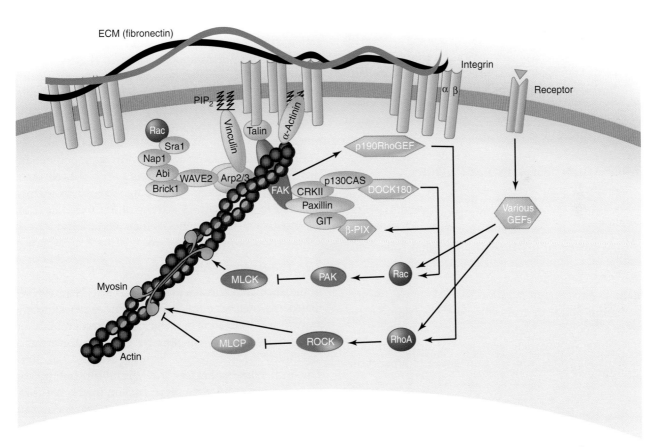

Figure 2. Adhesions serve as contact points and signaling centers. Integrin-based adhesions are large, complex assemblies that link the substratum to actin and generate signals that regulate Rho GTPases and cell migration. The structural linkage to actin is thought to be mediated by talin, vinculin, and perhaps α-actinin. The signaling is mediated by adhesion-associated complexes. The paxillin/FAK module and its link to some Rac GEFs and Rho GEFs is shown as an example. The Arp2/3 complex and myosin II, whose activity is regulated by Rho and Rac, are also shown.

enchymal migration tend to have larger adhesions with a significant number maturing to large, elongate adhesions (Parsons et al. 2010). These large adhesions do not appear to generate signals that drive actin polymerization.

2.3 Polarization

The presence of a distinct front and rear is a key feature of cell migration. Some cells can polarize spontaneously and migrate in a directionally persistent manner. The machinery that establishes polarity is incompletely understood but microtubules, vesicle cycling, and actomyosin filaments appear to be the drivers. In epithelial cells and astrocytes, polarity is established through a signaling pathway involving Cdc42, Par3/6, and atypical protein kinase C (aPKC) that targets microtubules (Etienne-Manneville and Hall 2002; Etienne-Manneville et al. 2005). This pathway orients the microtubule-organizing center (MTOC) and Golgi apparatus (Ch. 9 [McCaffrey and Macara 2012]). In fibroblasts, activated myosin II creates a region of actomyosin filament bundles that terminate in adhesions that do not contain guanine nucleotide exchange factors (GEFs) and therefore do not support Rac or Cdc42 signaling and actin polymerization (Vicente-Manzanares et al. 2008, 2011). This region becomes the rear and sides, and zones of active Rac generate protrusions that elongate the cell to form the front; the actomyosin system appears to set up the initial polarity, which is then refined by the microtubule system (Vicente-Manzanares et al. 2008).

3 MIGRATION SIGNALING NETWORKS

A complex signaling network regulates the cytoskeleton and adhesion in the context of migration. Below, we focus on Rho GTPases, integrins, and phosphoinositides, which have been extensively investigated, although it is clear that Ras proteins, calcium, cyclic nucleotides, numerous kinases, and other components are also involved. These networks contain positive- and negative-feedback loops, redundancies, and points of cross talk often involving synergy between adhesions, chemotactic receptors, and growth factor receptors. We speculate below on the different roles of these signaling events, which are integrated to bring about migration.

3.1 Rho-Family GTPases Regulate Cytoskeletal Activity

Rho-family small GTPases are a major convergence point of migration-associated signaling (Heasman and Ridley 2008). Protrusion, adhesion, and polarization are all regulated by Rho-family GTPases, and many receptor-initiated signaling pathways link to the Rho GTPases (Ridley et al.

2003; Zaidel-Bar et al. 2007; Parsons et al. 2010). These include chemokine receptors and growth factor receptors, such as the epidermal growth factor (EGF) receptor. The response depends on the spatial and temporal segregation of the activities of the different GTPases and involves multiple, parallel, redundant, and synergistic pathways that form a complex network.

The pathways involved regulate Rho GTPase activity by acting on the many GEFs and GTPase-activating proteins (GAPs) that control their activity (Etienne-Manneville and Hall 2002; DerMardirossian and Bokoch 2005). The GTPases are also regulated by Rho GDP dissociation inhibitors (GDIs), which remove them from the membrane (Garcia-Mata et al. 2011). The Rho family has several members, whose functions, in the context of migration, are represented by Rac, RhoA, and Cdc42. These act on a number of effectors that control the cytoskeletal machinery (see above) (Ridley 2006, 2011; Heasman and Ridley 2008). For example, Rac and Cdc42 regulate actin polymerization by acting on the WASP family and consequently Arp2/3: Cdc42 regulates mDia2, and Rac regulates cofilin through LIMK (Ridley 2011). RhoA also regulates actin polymerization by acting on mDia1. In addition, it regulates adhesion and actin organization via myosin II activity: RhoA activates ROCK, which phosphorylates the myosin RLC and inhibits myosin phosphatase, both of which stimulate myosin II. Cdc42 also regulates microtubule dynamics, which in turn affects the turnover of some adhesions (Kraynov et al. 2000; Nalbant et al. 2004; Machacek et al. 2009). The front-back polarity required for migration requires that actin polymerization localizes to specific cellular regions (i.e., the leading edge of migrating cells). This is reflected in the polarized activity of Rac and Cdc42 and the intricate relative kinetics of their activation.

3.2 The Paxillin/FAK Signaling Module

In mesenchymal cells at least, Rho-family GTPases are regulated by signaling complexes that reside in adhesions; they are activated by ligation of integrins to matrix proteins such as fibronectin, whose signaling synergizes with growth factor and chemokine receptor pathways (Ridley et al. 2003). The focal adhesion kinase (FAK)/paxillin signaling module is the best-studied example. Paxillin is among the earliest molecules to enter adhesions as they form and remains present until they disassemble (Webb et al. 2004). It functions as a signaling adapter that binds to numerous molecules involved in Rho-family GTPase signaling (Brown and Turner 2004; Deakin and Turner 2008).

The amino terminus of paxillin has a series of LD regions. The first two of these include two SH2-domain-binding sites (around Y31 and Y118). They are phosphorylated by FAK and Src and bind to a number of molecules, including Crk/p130Cas, FAK/Src, and Ras GAP. The phosphatase PTP PEST binds near the carboxyl terminus of paxillin and dephosphorylates these sites. p130Cas recruits a Cas-Crk-Dock180-Elmo complex, in which Dock180 functions as a GEF that activates Rac. Two other LD regions of paxillin bind GIT1 and GIT2 (Turner et al. 2001; Hoefen and Berk 2006). These adapters bind to Pix, a GEF for Cdc42 and Rac (Deakin and Turner 2008). Pix also binds to the kinase PAK, a Rac effector, creating a Rac and Cdc42 activator-effector signaling module (Bokoch 2003). The p85 subunit of phosphoinositide 3-kinase (PI3K) also binds to this region of paxillin and is involved in signaling to Vav, another Rac GEF (see below) (Tybulewicz et al. 2003). This list of interactions is not exhaustive but serves to illustrate the central role of paxillin as a regulator of protrusion through its action on Rac and Cdc42. Vinculin, a tension-sensitive structural molecule implicated in the integrin-actin linkage, also binds to the LD region of paxillin (Deakin and Turner 2008).

Nearly all of these binding interactions are regulated by phosphorylation of paxillin on Y31 and Y118, which creates the two SH2-binding sites as well as inducing a major conformational change. Conformational regulation is a general theme in adhesion signaling (Parsons et al. 2010; Zaidel-Bar and Geiger 2010). Src, FAK, paxillin, and p130 Cas are all conformationally activated, at least in part, by phosphorylation events, which often serve to release an autoinhibitory state (Cohen et al. 2006; Sawada et al. 2006; Parsons et al. 2010). Tension can also activate or regulate the activities of adhesion molecules. For example, p130Cas, talin, and vinculin are all tension sensitive (Sawada et al. 2006; del Rio et al. 2009; Grashoff et al. 2010). These sensitivities are thought to regulate adhesion-generated signals (Bershadsky et al. 2003; Schwartz 2010).

FAK binds to paxillin following tyrosine phosphorylation of paxillin Y31 and Y118 (Choi et al. 2011). The phosphorylation occurs after both molecules are in nascent adhesions and appears to result from a conformational activation of paxillin, because FAK binds to the carboxyl end of paxillin. FAK also recruits molecules that regulate Rho-family GTPases, functions as a tyrosine kinase, and possesses tyrosine phosphorylation sites, most of which are phosphorylated by Src (Parsons 2003; Mitra et al. 2005; Frame et al. 2010). Activated FAK binds to Src (via its SH2 domain), to p130Cas (via an SH3 domain), and to a p120RasGAP-p190RhoGAP complex, which negatively regulates RhoA activity. FAK binds to two Rho GEFs, p190RhoGEF and PDZRhoGEF, which activate RhoA (Tomar and Schlaepfer 2009). p190RhoGEF and p190RhoGAP do not appear to bind to FAK at the same time, and in spreading cells, binding of p190RhoGAP precedes that of

p190RhoGEF. This provides a potential mechanism for the transient, local and sequential activation of RhoA seen at the leading edge of migrating cells. Thus, RhoA activity appears to be controlled by antagonistic regulators that interact with FAK. In addition, the activity of p190RhoGAP, for example, depends on phosphorylation, which suggests that the GAPs (and probably the GEFs) are regulated by phosphorylation (Tomar and Schlaepfer 2009).

3.3 Other Rho-Regulating Modules

The Pax/FAK model reveals the importance and function of signaling complexes that localize the activity of Rho-family GTPases; but other complexes do this as well. For example, the ILK-pinch-parvin complex is another adhesion-associated system that signals to Rac. The pseudokinase scaffold protein ILK binds to parvin, which binds to Pix (Sepulveda et al. 2005; Legate et al. 2006). The SH2/SH3 adapter NCK is also implicated in Rac signaling in adhesions (Ruusala and Aspenstrom 2008). Finally, note that Rho-family GTPases are regulated by their association with the plasma membrane. They are targeted to lipid rafts and regulated by phosphorylation-dependent interactions with Rho GDI and endocytic events, which thereby regulate their activity (Grande-Garcia et al. 2005).

3.4 The PI3K Signaling Module

The PI3K signaling module (p. 87 [Hemmings and Restuccia 2012]) acts dynamically at the leading edge of the cell to regulate the accumulation of phosphatidylinositol 3,4,5-trisphosphate (PIP$_3$) and in turn cytoskeletal activities. The links between PIP$_3$ and the cytoskeleton are a subject of intense investigation. PIP$_3$ targets include substrates of Akt, including PAK, as well as a series of PH-domain-containing proteins (Kamimura et al. 2008; Tang et al. 2011). In randomly migrating cells, patches of PIP$_3$ are generated spontaneously and appear at the tips of protrusions. In cells migrating in gradients of chemoattractants such as chemokines or growth factors, receptors and G proteins are distributed uniformly around the cell perimeter. However, PIP$_3$ accumulation as well as other signaling events are dynamically localized at the tips of pseudopodia at the front (Parent and Devreotes 1999). PIP$_3$ accumulates at the leading edge of the pseudopodia because, in *Dictyostelium* at least, PI3K is recruited to, and PTEN (a 3′-specific phosphatase that hydrolyzes PIP$_3$) is lost from, these regions (Funamoto et al. 2002; Iijima and Devreotes 2002; Arai et al. 2010). In neutrophils, the 5′-phosphatase Ship1 also degrades PIP$_3$ (Nishio et al. 2007; Mondal et al. 2012).

Asymmetric PIP$_3$ accumulation has been conserved throughout evolution in migrating cells and also operates where cells undergo related morphological changes (Fig. 3)—for example, in the extending membranes of neuronal growth cones (Wang et al. 2002; Lacalle et al. 2004; Chadborn et al. 2006; Evans and Falke 2007; Yoo et al. 2010), during cytokinesis (a process which resembles two cells migrating away from each other) (Janetopoulos et al. 2005; Janetopoulos and Devreotes 2006), in phagocytosis (Clarke et al. 2006), and in mammary and prostate epithelia (in which the basal lateral and apical portions of stationary cells are akin to the front and rear of a migrating cell, PIP$_3$ being localized to the basal–lateral region and PTEN being localized to the apical region [Shewan et al. 2011]).

It is now clear that the asymmetrical accumulation of PIP$_3$ is sufficient to promote actin polymerization and produce cellular projections; but it is one of a number of parallel pathways. Alterations in PIP$_3$ levels lead to defects in cell migration, cytokinesis, phagocytosis, and epithelial architecture. In *Dictyostelium* cells lacking PTEN or neutrophils lacking Ship1, for example, excessive amounts of PIP$_3$ are generated at the front of the cell and this diffuses

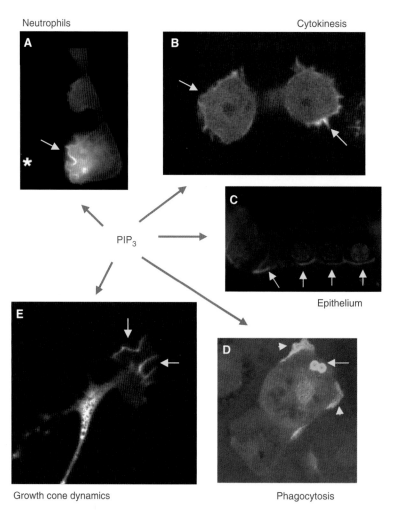

Figure 3. Asymmetric accumulation of PIP$_3$ is a feature of a spectrum of cell morphological changes. Panels show snapshots of the dynamic distribution of PIP$_3$ in cells undergoing various morphological changes. (*A*) Human neutrophils expressing a biosensor for PIP$_3$ (PHakt-GFP). The cells have been exposed to a gradient formed by a micropipette filled with the chemoattractant C5a (position indicted by *). The arrow shows recruitment of PHakt-GFP to the membrane, indicating an elevated level of PIP$_3$. (*B*) A dividing *Dictyostelium* cell expressing PHCrac-GFP as a biosensor for PIP$_3$. Arrows point to the accumulation of PIP$_3$ at the poles of the dividing cell. (*C*) Prostate epithelial cells expressing PHakt-GFP. (Image courtesy of Tamara Lotan.) Arrows point to the accumulation of PIP$_3$ on the basal–lateral membranes. (*D*) *Dictyostelium* cell expressing PHCrac-GFP phagocytizing latex beads. The arrow indicates accumulation of PIP$_3$ around two beads; arrowheads point to PIP$_3$-labeled pseudopods in the same cell. (Image courtesy of Margaret Clarke.) (*E*) The growth cone of rat dorsal root ganglion expressing PHakt-GFP. Arrows indicate the accumulation of PIP$_3$ at the leading edge. (Image courtesy of Britta Eickholt.)

Cite this chapter as *Cold Spring Harb Perspect Biol* doi: 10.1101/cshperspect.a005959

along most of the cell perimeter (Funamoto et al. 2002; Iijima and Devreotes 2002; Nishio et al. 2007). This additional PIP$_3$ elicits ectopic pseudopodia at lateral regions outside the leading edge. If the PTEN-deficient cells are treated with inhibitors of PI3K, the morphological defects are suppressed, and the cells again display a single anterior pseudopod (Chen et al. 2003). Using a synthetic PI3K activation system in neutrophils, Inoue and Meyers showed that elevation of PIP$_3$ alone is sufficient to initiate pseudopodial extensions (Inoue and Meyer 2008). However, because asymmetrical generation of PIP$_3$ is only one of several parallel pathways, it is not essential for a directional response. Inhibition of PI3K blocks migration in zebrafish neutrophils and fibroblasts and random migration in *Dictyostelium* but, under certain conditions, does not block chemoattractant-driven migration in amoebae or human neutrophils (Chen et al. 2003; Ferguson et al. 2007; Hoeller and Kay 2007). Furthermore, primordial germ cells in zebrafish have persistent, uniformly distributed membrane PIP$_3$ levels even as they migrate directionally (Dumstrei et al. 2004). These observations confirm that there are parallel pathways that allow cells to receive directional cues from chemoattractant receptors in the absence of PIP$_3$.

3.5 Genetic Analysis of a Signaling Network

The multiple parallel pathways that drive migration form a complex network of coordinated events that are triggered by chemoattractants but can also occur spontaneously as cells migrate. A successful explanation of cell migration has to integrate the information from initially independent studies of pathways believed to directly regulate the cytoskeleton, such as those involving Rho GTPases, with others thought to transduce signals from receptors, such as PI3K signaling. Interestingly, an emerging theme is that downstream and upstream events are probably linked through multiple feedback loops.

Genetic analyses in *Dictyostelium* have implicated about 95 nonlethal genes in chemotaxis and about 40 can be organized into an internally consistent "wiring diagram" (Swaney et al. 2010). Some of the major features of the network include the presence of parallel pathways defined by cyclic GMP, myosin heavy-chain kinase (MHCK), the kinase Tor complex 2 (TorC2), PIP$_3$, and phospholipase A2 (PLA2) (Veltman et al. 2008). Four isoforms of the small G protein Ras are activated by chemoattractant and seem to act early in these pathways (Kae et al. 2004; Sasaki and Firtel 2009). A portion of the network involving PIP3 and TorC2 is examined in more detail below. Interestingly TorC2 is defined by subunits Pianissimo and Rip3. These highly conserved genes were first identified as causing chemotaxis defects in *Dictyostelium* and later renamed as Rictor and

Sin1, respectively (see p. 91 [Laplante and Sabatini 2012]). Remarkably, the basic elements of this network appear to be conserved in human neutrophils, although further analysis of each pathway is needed. Similarities include the rapid activation of K-, H-, and N-Ras by chemoattractants, localization of PIP$_3$ at the leading edge of the cell, and the critical role for mTorC2 (Bokoch 2003; Van Keymeulen et al. 2006; Liu and Parent 2011). One apparent difference is that a role for cyclic GMP has not been described in neutrophils.

There is a spatiotemporal pattern to many elements of the network. Biosensors for activation/formation/recruitment of Ras, PI3K, PIP$_3$, HSPC300 (a subunit of the WAVE complex), and LimE (an actin-binding protein) serve as dynamic markers for the front of the cell whereas others, such as PTEN and myosin, define the back. The front markers reside in the cytosol, but as protrusions form they are recruited to the tips of pseudopodia (Van Haastert and Devreotes 2004). In contrast, the back markers reside uniformly in the cortex and dissociate from regions where protrusions form (Funamoto et al. 2002; Iijima and Devreotes 2002; Robinson and Spudich 2004). For example, a biosensor for the collective activation of Ras proteins (the Ras-binding domain of the kinase Raf fused to GFP) moves to protrusions at the leading edge of a migrating cell, whereas PTEN-GFP falls off. When cells are stimulated with a uniform chemoattractant, all of the front components are recruited from the cytosol to the cell periphery, whereas the back components fall off and move to the cytosol (Swaney et al. 2010). These changes are transient and the components reestablish their original locations within a few minutes.

Examining a small portion of the network involving PIP$_3$ and TorC2 in detail reveals how genetic analyses have helped delineate signaling mechanisms involved in cell migration (Fig. 4). As outlined above, cells with elevated PIP$_3$ levels owing to loss of PTEN display a "migration" phenotype in which ectopic protrusions form over most of the cell perimeter. This can be reversed by disrupting the Akt ortholog PKBA (Tang et al. 2011). Interestingly, the cells lacking both PTEN and PKBA have elevated PIP$_3$ levels around the perimeter but still respond effectively to chemotactic cAMP gradients. Parallel pathways must therefore mediate the directional response despite the uniform PIP$_3$ distribution. One of these involves a second PKB isoform, PKBR1, which can be activated independently of PIP$_3$; unlike Akt and PKBA, PKBR1 lacks a PH domain and is instead tethered to the membrane by myristoylation (Kamimura et al. 2008). The activation of PKBR1 is mediated by phosphorylation of a hydrophobic motif by TorC2. Phosphorylation of serines/threonines within conserved hydrophobic motifs in the carboxy-terminal region of

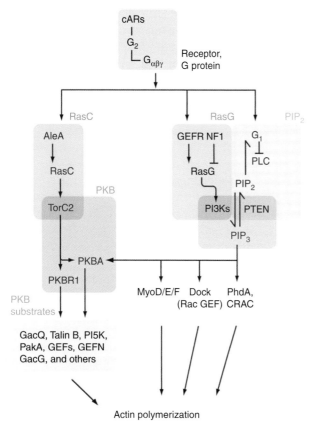

Figure 4. A portion of the *Dictyostelium* migration signaling network involving PIP₃ and TorC2. Colored blocks delineate modules. The overlapping of blocks indicates that some components belong to several modules. CARE, cystic AMP receptors.

many ACG-family kinases (which included PKA, PKC, and PKG) can be required for their action. The phosphorylation of PKBR1 is part of a pathway that leads from chemoattractant-mediated activation of Ras through TorC2 to regulation of the cytoskeleton (Chen et al. 1997; Lee et al. 2005; Cai et al. 2010). In cells lacking RasC or Aimless, a Ras GEF for RasC, TorC2 activation is greatly reduced, whereas in cells expressing a constitutively active RasC (Q62L), TorC2 activation is elevated and ectopic sites of actin polymerization appear around the cell perimeter. However, in cells lacking Pianissimo, a key subunit of TorC2, there is no effect of expressing RasC Q62L and no phosphorylation of PKBR1. These studies focus attention on PKB substrates. In *Dictyostelium*, there are at least nine substrates that are rapidly, transiently phosphorylated in response to chemoattractant. These include signaling and cytoskeletal proteins, such as Ras GEFs and Rac GAPs, talin, PakA, and phosphatidylinositol 5-kinase (PI5K) (Kamimura et al. 2008).

The PIP₃-independent role of TorC2 is another example of an element of the network that is conserved in neutrophils. Neutrophils lacking PI3K or exposed to PI3K

inhibitors migrate along chemoattractant gradients when plated on certain extracellular matrices (Ferguson et al. 2007). Knocking down the Rictor subunit of mTorC2 causes a severe defect in neutrophil chemotaxis (Liu et al. 2010; Wang, pers. comm.). Although Akt is a substrate of mTorC2 in neutrophils, other targets are probably more important. Because PKC and other ACG kinases are phosphorylated on their hydrophobic motifs, it is possible that the mTorC2 targets are these kinases in neutrophils and other cells.

4 BIASED EXCITABLE BIOCHEMICAL NETWORKS IN CHEMOTAXIS

4.1 Excitability of Signaling Networks Linked to Migration

The signaling network controlling cell migration displays behavior, including oscillations and cortical wave propagation, which suggest the system is excitable. Excitability typically arises when a system contains opposing positive- and negative-feedback loops. These systems are in a resting state until a threshold is crossed and an all-or-nothing response ensues. Total internal reflection microscopy (TIRF) reveals that subunits of the WAVE complex and actin-binding proteins participate in wavelike phenomena that propagate along the basal surface of the cell. When these waves reach the edge of the cell, they appear to push the perimeter outward. Essentially similar phenomena are observed in human neutrophils and *Dictyostelium* amoebae (Gerisch et al. 2004, 2011; Weiner et al. 2007; Bretschneider et al. 2009; Gerisch 2010). Actin and integrin waves have also been observed in fibroblasts (Giannone et al. 2004; Döbereiner et al. 2006; Case and Waterman 2011). The waves form in the absence of stimulation and in mutants lacking G proteins, indicating that this is an intrinsic behavior of motile cells. Furthermore, activation of Ras and accumulation of PIP₃ also occur in flashes and bursting waves, which are propagated across the cortex (Arai et al. 2010; Xiong et al. 2010).

These waves can be modeled mathematically, like action potentials in neurons (Levine et al. 2006; Insall and Machesky 2009; Xiong et al. 2010; Hecht et al. 2011), by linking components of a hypothetical network in positive- and negative-feedback loops. Some feedback loops have been described in the real network. For example, a positive-feedback loop appears to link cytoskeletal events and PIP₃ because inhibition of either reduces the spontaneous activation of the other (Weiner et al. 2002; Inoue and Meyer 2008). Second, a negative-feedback loop involving the phosphorylation of upstream Ras GEF by downstream PKB has been described in *Dictyostelium* (Charest et al. 2010).

Cell migration may thus involve a mechanism in which guidance cues differentially alter the excitability of the network on one side of the cell. That is, if a cue decreases the threshold for excitability on the side of the cell closer to the source and increases the threshold on the distal side, the cell will be attracted to it. Such theoretical models are referred to as biased excitable networks (BENs). Because they are typically excitable within very narrow parameter ranges, BENs can provide extreme sensitivity to external signals (Iglesias and Devreotes 2012). This mechanism may explain how professional chemotactic cells such as leukocytes and *Dictyostelium* are able to move up gradients of chemoattractant that differ by <2% over their body length.

4.2 Adaptation to Chemotactic Signaling and Local Excitation–Global Inhibition Models

The influence of an external signal on a BEN can depend on the absolute or relative amount of the signal. The chemotactic response of 3T3 fibroblasts to platelet-derived growth factor (PDGF), for example, depends on the absolute concentration and diminishes as the concentration increases and the fractional difference in receptor occupancy across the cell decreases (Schneider and Haugh 2006). In contrast, responses to chemoattractants that signal via GPCRs, such as FMLP in human leukocytes or cAMP in *Dictyostelium*, depend primarily on the relative steepness rather than the absolute concentration of the gradient (Devreotes and Zigmond 1988). The "relative" systems can maintain sensitivity over a wide range of concentrations.

Studies of cells treated with inhibitors of the cytoskeleton, which allow the direction-sensing system to be examined in isolation, have shown these receptors produce rapid but transient signaling events, such as PIP_3 accumulation and Ras activation. Further responses can only be elicited if the stimulus is increased or removed and then reapplied (i.e., the cells adapt when receptor occupancy is held constant). In contrast, when cells are exposed to a chemotactic gradient, the biosensors form a crescent toward the high side. The crescent is maintained persistently at steady state but can immediately adjust if the chemoattractant gradient is shifted to a new direction. How do they adapt to uniform stimuli yet respond persistently to a gradient? The differential response to uniform versus gradient stimuli can be explained by local excitation–global inhibition (LEGI) models (Fig. 5) (Parent and Devreotes 1999; Janetopoulos et al. 2004; Levine et al. 2006). In a LEGI model, an increase in receptor occupancy triggers a rapid excitatory process, such as the dissociation of G-protein subunits, and a slower inhibitory process that balances excitation. Whenever excitation exceeds inhibition, a response regulator is generated; when inhibition catches up with excitation, the

response regulator returns to its basal level. Because excitation is local, whereas the inhibitor is more global, in a gradient there is a persistent deflection of the response regulator above and below its basal level at the front and back of the cell, respectively.

The LEGI model is a useful conceptual device that allows one to predict the response to any combination of applied temporal and spatial stimuli, but further studies are needed to define the underlying biochemical events and to link the model to cell migration. First, the excitatory process likely corresponds to G-protein activation. When cells are exposed to chemoattractant, the G-protein subunits dissociate within a few seconds and all of the biochemical responses in the network are triggered. During the next several minutes, the responses gradually subside even though the G protein does not reassociate. The mechanism that offsets the activity of the G-protein and causes the responses to subside remains to be determined. Second, LEGI schemes can account for all of the behavior of immobilized cells but fail to explain migration or polarity. However, the output of LEGI could enhance excitability at the front and suppress it at the rear (Xiong et al. 2010). This would ensure that the system responds to the steepness of a gradient but is independent of its midpoint concentration. LEGI-BEN schemes are capable of extraordinary sensitivity, and computer simulations show that this model can produce realistic temporal and spatial chemotactic responses.

5 CONCLUDING REMARKS

Many of the principles described in this chapter appear to be general and apply to cellular behaviors analogous to migration. For example, dendritic spines, small extensions along dendrites of neurons in the central nervous system, contain a highly organized postsynaptic density that receives excitatory signals. Like protrusions in migrating cells, dendritic spines are highly dynamic, they undergo complex morphologic changes, and they contain a highly organized adhesion associated with the postsynaptic density. Actin polymerization and actomyosin activity play a major role in spine and postsynaptic density (PSD) organization (Oertner and Matus 2005; Hodges et al. 2011). Rho-family GTPases have emerged as major regulators of spine organization and dynamics and are implicated in human cognitive diseases. Indeed, mutations in regulators of Rho-family GTPases are implicated in spine-related diseases including autism, schizophrenia, and nonsyndromic mental retardation. With respect to the latter, α-Pix (a Rac GEF), PAK3 (a Rac/Cd42 effector), and oligphrenin 1 (a Rho GAP) are all associated with nonsyndromic mental retardation in humans—a disease characterized by spine defects (van Galen and Ramakers 2005). Pix and PAK are localized

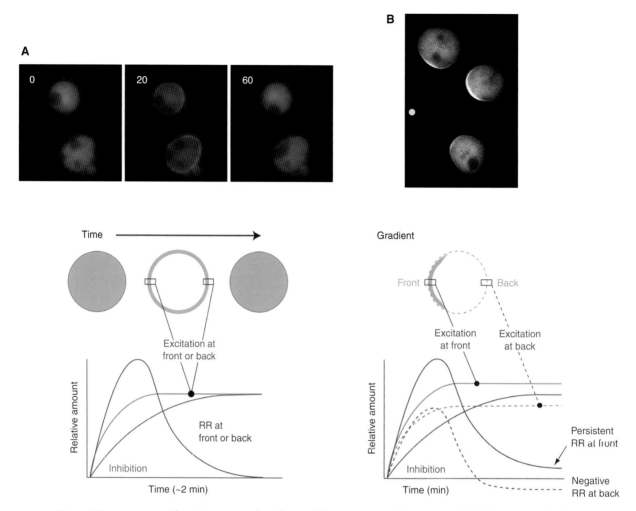

Figure 5. Responses to uniform increases and gradients of chemoattractants in a LEGI model. (*A*) Micrographs show translocation of a biosensor for PIP$_3$ (PHCrac-GFP) to the membrane. PIP$_3$ levels rise transiently during persistent stimulation with a uniform chemoattractant. The schematic depicts the response of a "front" marker such as PIP$_3$ to uniform stimulation. A LEGI model assumes that the level of a response regulator is controlled by the difference between rapid excitatory and slower inhibitory processes. The response regulator (RR, blue line) rises when excitation (green line) is higher than inhibition (red line) and then falls as inhibition catches up. (*B*) The micrograph shows that the steady-state accumulation of PIP$_3$ forms a crescent facing the high side of the gradient produced by a micropipette releasing chemoattractant. The schematic depicts the behavior of a "front" marker such as PIP$_3$ in response to a gradient of chemoattractant. In the LEGI model, the response regulator (blue line) rises when excitation (green line) is higher than inhibition (red line) and then falls to a new steady state. Because inhibition is more global than the excitation the differences generate a response regulator that has a higher concentration than basal at the front and a lower concentration than basal at the back.

to dendritic spines via GIT1 and thought to spatially restrict their formation (Zhang et al. 2005). Asymmetric localization of PIP$_3$ is observed in many migrating cells. Oncogenic mutations leading to overproduction of PIP$_3$ are typically assumed to increase growth rates; but it is likely that many of the cancer-causing effects can also be attributed to alterations in the cytoskeleton (Kim et al. 2011). Distortion of cell migration signaling networks plays a critical role in migration-related diseases such as invasive and metastatic cancer (Ch. 21 [Sever and Brugge 2014]).

The plethora of signaling pathways that converge on Rho GTPases means there is a very large potential set of loci for the misregulation of migration. It also suggests that drugs directed against any particular pathway may not be effective for long given the selection that occurs in the tumor environment. However, the convergence on Rho-family GTPases and the limited migration machinery on which it acts hold promise for therapeutic strategies targeting these GTPases and their downstream effectors or diagnostic routes to identifying cells with invasive potential.

Another emerging area of research is the study of migration in 3D. Until recently, most migration studies focused on migration on planar substrates using integrin-mediated adhesion. Analogous mechanisms are probably used by cells migrating in 3D, or using other receptors (e.g., in the central nervous system). However, different pathways might play more prominent roles in each case. For example, in 3D, cell protrusions, adhesion, and cell morphology all appear to differ from that generally seen on rigid planar 2D substrates (Even-Ram and Yamada 2005; Provenzano et al. 2009; Friedl and Wolf 2010; Sanz-Moreno and Marshall 2010). In 3D, the cells are more elongated and possess narrower protrusions and smaller adhesions (Harunaga and Yamada 2011). Moreover, there is evidence that different signaling pathways are indeed involved. For example, depleting paxillin produces a mesenchymal phenotype in 3D environments, whereas depleting the paxillin relative Hic5 produces an amoeboid morphology (Deakin and Turner 2008, 2011). The particular signaling pathway used seems to depend on the cellular microenvironment, which can differ between normal and tumor cells.

Research into cell migration has clearly made enormous progress. The basic machines that drive migration have been described, and many of the pathways that regulate them have been identified. However, we have only scratched the surface and much remains to be understood. The interactions and regulation of the complex signaling networks that orchestrate migration and the mechanism by which extracellular forces affect these networks are not understood. Furthermore, new modes of migration are being uncovered, including blebbing-mediated migration and the newly described lobopodia migration (Petrie et al. 2012). In addition, migration in complex in vivo environments differs from that seen on rigid planar substrates and its study presents unexpected challenges. Finally, integrative, quantitative models of migration that conjoin the plethora of regulatory networks are only now beginning to be developed.

REFERENCES

Reference is in this book.

Alexandrova AY, Arnold K, Schaub S, Vasiliev JM, Meister JJ, Bershadsky AD, Verkhovsky AB. 2008. Comparative dynamics of retrograde actin flow and focal adhesions: Formation of nascent adhesions triggers transition from fast to slow flow. *PLoS ONE* **3**: e3234.
Arai Y, Shibata T, Matsuoka S, Sato MJ, Yanagida T, Ueda M. 2010. Self-organization of the phosphatidylinositol lipids signaling system for random cell migration. *Proc Natl Acad Sci* **107**: 12399–12404.
Bamburg JR, Bernstein BW. 2008. ADF/cofilin. *Curr Biol* **18**: R273–R275.
Bear JE, Gertler FB. 2009. Ena/VASP: Towards resolving a pointed controversy at the barbed end. *J Cell Sci* **122**: 1947–1953.

Bershadsky AD, Balaban NQ, Geiger B. 2003. Adhesion-dependent cell mechanosensitivity. *Annu Rev Cell Dev Biol* **19**: 677–695.
Bokoch GM. 1995. Chemoattractant signaling and leukocyte activation. *Blood* **86**: 1649–1660.
Bokoch GM. 2003. Biology of the p21-activated kinases. *Annu Rev Biochem* **72**: 743–781.
Bretschneider T, Anderson K, Ecke M, Muller-Taubenberger A, Schroth-Diez B, Ishikawa-Ankerhold HC, Gerisch G. 2009. The three-dimensional dynamics of actin waves, a model of cytoskeletal self-organization. *Biophys J* **96**: 2888–2900.
Broussard JA, Webb DJ, Kaverina I. 2008. Asymmetric focal adhesion disassembly in motile cells. *Curr Opin Cell Biol* **20**: 85–90.
Brown MC, Turner CE. 2004. Paxillin: Adapting to change. *Physiol Rev* **84**: 1315–1339.
Brown CM, Hebert B, Kolin DL, Zareno J, Whitmore L, Horwitz AR, Wiseman PW. 2006. Probing the integrin-actin linkage using high-resolution protein velocity mapping. *J Cell Sci* **119**: 5204–5214.
Bugyi B, Carlier MF. 2010. Control of actin filament treadmilling in cell motility. *Annu Rev Biophys* **39**: 449–470.
Cai H, Das S, Kamimura Y, Comer FI, Parent DA, Devreotes PN. 2010. Ras-mediated activation and inactivation of the TorC2-PKB pathway are critical for chemotaxis. *J Cell Biol* **190**: 233–245.
Case LB, Waterman CM. 2011. Adhesive F-actin waves: A novel integrin-mediated adhesion complex coupled to ventral actin polymerization. *PLoS ONE* **6**: e26631.
Chadborn NH, Ahmed AI, Holt MR, Prinjha R, Dunn GA, Jones GE, Eickholt BJ. 2006. PTEN couples Sema3A signalling to growth cone collapse. *J Cell Sci* **119**: 951–957.
Chan KT, Bennin DA, Huttenlocher A. 2010. Regulation of adhesion dynamics by calpain-mediated proteolysis of focal adhesion kinase (FAK). *J Biol Chem* **285**: 11418–11426.
Charest PG, Shen Z, Lakoduk A, Sasaki AT, Briggs SP, Firtel RA. 2010. A Ras signaling complex controls the RasC-TORC2 pathway and directed cell migration. *Dev Cell* **18**: 737–749.
Charras G, Paluch E. 2008. Blebs lead the way: How to migrate without lamellipodia. *Nat Rev Mol Cell Biol* **9**: 730–736.
Chen M-Y, Long Y, Devreotes PN. 1997. A novel cytosolic regulator, Pianissimo, is required for chemoattractant receptor and G protein-mediated activation of the twelve transmembrane domain adenylyl cyclase in *Dictyostelium. Genes Dev* **11**: 3218–3231.
Chen L, Janetopoulos C, Huang YE, Iijima M, Borleis J, Devreotes PN. 2003. Two phases of actin polymerization display different dependences on PI(3,4,5)P$_3$ accumulation and have unique roles during chemotaxis. *Mol Biol Cell* **14**: 5028–5037.
Chen L, Iijima M, Tang M, Landree MA, Huang YE, Xiong Y, Iglesias PA, Devreotes PN. 2007. PLA$_2$ and PI3K/PTEN pathways act in parallel to mediate chemotaxis. *Dev Cell* **12**: 603–614.
Chen L, Vicente-Manzanares M, Potvin-Trottier L, Wiseman PW, Horwitz AR. 2012. The integrin-ligand interaction regulates adhesion and migration through a molecular clutch. *PLoS ONE* **7**: e40202.
Chesarone MA, DuPage AG, Goode BL. 2010. Unleashing formins to remodel the actin and microtubule cytoskeletons. *Nat Rev Mol Cell Biol* **11**: 62–74.
Choi CK, Vicente-Manzanares M, Zareno J, Whitmore LA, Mogilner A, Horwitz AR. 2008. Actin and α-actinin orchestrate the assembly and maturation of nascent adhesions in a myosin II motor-independent manner. *Nat Cell Biol* **10**: 1039–1050.
Choi CK, Zareno J, Digman MA, Gratton E, Horwitz AR. 2011. Cross-correlated fluctuation analysis reveals phosphorylation-regulated paxillin-FAK complexes in nascent adhesions. *Biophys J* **100**: 583–592.
Clarke M, Muller-Taubenberger A, Anderson KI, Engel U, Gerisch G. 2006. Mechanically induced actin-mediated rocketing of phagosomes. *Mol Biol Cell* **17**: 4866–4875.
Cohen DM, Kutscher B, Chen H, Murphy DB, Craig SW. 2006. A conformational switch in vinculin drives formation and dynamics of a talin-vinculin complex at focal adhesions. *J Biol Chem* **281**: 16006–16015.

Comer FI, Parent CA. 2007. Phosphoinositides specify polarity during epithelial organ development. *Cell* 128: 239–240.

Condeelis J. 2001. How is actin polymerization nucleated in vivo? *Trends Cell Biol* 11: 288–293.

Cortesio CL, Boateng LR, Piazza TM, Bennin DA, Huttenlocher A. 2011. Calpain-mediated proteolysis of paxillin negatively regulates focal adhesion dynamics and cell migration. *J Biol Chem* 286: 9998–10006.

Deakin NO, Turner CE. 2008. Paxillin comes of age. *J Cell Sci* 121: 2435–2444.

Deakin NO, Turner CE. 2011. Distinct roles for paxillin and Hic-5 in regulating breast cancer cell morphology, invasion, and metastasis. *Mol Biol Cell* 22: 327–341.

del Rio A, Perez-Jimenez R, Liu R, Roca-Cusachs P, Fernandez JM, Sheetz MP. 2009. Stretching single talin rod molecules activates vinculin binding. *Science* 323: 638–641.

DerMardirossian C, Bokoch GM. 2005. GDIs: Central regulatory molecules in Rho GTPase activation. *Trends Cell Biol* 15: 356–363.

Devreotes P, Janetopoulos C. 2003. Eukaryotic chemotaxis: Distinctions between directional sensing and polarization. *J Biol Chem* 278: 20445–20448.

Devreotes PN, Zigmond SH. 1988. Chemotaxis in eucaryotic cells: A focus on leukocytes and *Dictyostelium*. *Ann Rev Cell Biol* 4: 649–686.

Döbereiner HG, Dubin-Thaler BJ, Hofman JM, Xenias HS, Sims TN, Giannone G, Dustin ML, Wiggins CH, Sheetz MP. 2006. Lateral membrane waves constitute a universal dynamic pattern of motile cells. *Phys Rev Lett* 97: 038102.

Dumstrei K, Mennecke R, Raz E. 2004. Signaling pathways controlling primordial germ cell migration in zebrafish. *J Cell Sci* 117: 4787–4795.

Etienne-Manneville S. 2004. Cdc42—The centre of polarity. *J Cell Sci* 117: 1291–1300.

Etienne-Manneville S, Hall A. 2002. Rho GTPases in cell biology. *Nature* 420: 629–635.

Etienne-Manneville S, Hall A. 2003. Cell polarity: Par6, aPKC and cytoskeletal crosstalk. *Curr Opin Cell Biol* 15: 67–72.

Etienne-Manneville S, Manneville JB, Nicholls S, Ferenczi MA, Hall A. 2005. Cdc42 and Par6-PKCζ regulate the spatially localized association of Dlg1 and APC to control cell polarization. *J Cell Biol* 170: 895–901.

Evans JH, Falke JJ. 2007. Ca²⁺ influx is an essential component of the positive-feedback loop that maintains leading-edge structure and activity in macrophages. *Proc Natl Acad Sci* 104: 16176–16181.

Even-Ram S, Yamada KM. 2005. Cell migration in 3D matrix. *Curr Opin Cell Biol* 17: 524–532.

Ezratty EJ, Partridge MA, Gundersen GG. 2005. Microtubule-induced focal adhesion disassembly is mediated by dynamin and focal adhesion kinase. *Nat Cell Biol* 7: 581–590.

Ezratty EJ, Bertaux C, Marcantonio EE, Gundersen GG. 2009. Clathrin mediates integrin endocytosis for focal adhesion disassembly in migrating cells. *J Cell Biol* 187: 733–747.

Fackler OT, Grosse R. 2008. Cell motility through plasma membrane blebbing. *J Cell Biol* 181: 879–884.

Ferguson GJ, Milne L, Kulkarni S, Sasaki T, Walker S, Andrews S, Crabbe T, Finan P, Jones G, Jackson S, et al. 2007. PI(3)Kγ has an important context-dependent role in neutrophil chemokinesis. *Nat Cell Biol* 9: 86–91.

Frame MC, Patel H, Serrels B, Lietha D, Eck MJ. 2010. The FERM domain: Organizing the structure and function of FAK. *Nat Rev Mol Cell Biol* 11: 802–814.

Franco SJ, Rodgers MA, Perrin BJ, Han J, Bennin DA, Critchley DR, Huttenlocher A. 2004. Calpain-mediated proteolysis of talin regulates adhesion dynamics. *Nat Cell Biol* 6: 977–983.

Friedl P, Alexander S. 2011. Cancer invasion and the microenvironment: Plasticity and reciprocity. *Cell* 147: 992–1009.

Friedl P, Gilmour D. 2009. Collective cell migration in morphogenesis, regeneration and cancer. *Nat Rev Mol Cell Biol* 10: 445–457.

Friedl P, Wolf K. 2010. Plasticity of cell migration: A multiscale tuning model. *J Cell Biol* 188: 11–19.

Funamoto S, Meili R, Lee S, Parry L, Firtel RA. 2002. Spatial and temporal regulation of 3-phosphoinositides by PI 3-kinase and PTEN mediates chemotaxis. *Cell* 109: 611–623.

Garcia-Mata R, Boulter E, Burridge K. 2011. The "invisible hand": Regulation of Rho GTPases by Rho GDIs. *Nat Rev Mol Cell Biol* 12: 493–504.

Geiger B, Yamada KM. 2011. Molecular architecture and function of matrix adhesions. *Cold Spring Harb Perspect Biol* 3: a005033.

Geraldo S, Gordon-Weeks PR. 2009. Cytoskeletal dynamics in growth-cone steering. *J Cell Sci* 122: 3595–3604.

Gerisch G. 2010. Self-organizing actin waves that simulate phagocytic cup structures. *PMC Biophys* 3: 7.

Gerisch G, Bretschneider T, Müller-Taubenberger A, Simmeth E, Ecke M, Diez S, Anderson K. 2004. Mobile actin clusters and traveling waves in cells recovering from actin depolymerization. *Biophys J* 87: 3493–3503.

Gerisch G, Ecke M, Wischnewski D, Schroth-Diez B. 2011. Different modes of state transitions determine pattern in the Phosphatidylinositide-Actin system. *BMC Cell Biol* 12: 42.

Giannone G, Dubin-Thaler BJ, Döbereiner HG, Kieffer N, Bresnick AR, Sheetz MP. 2004. Periodic lamellipodial contractions correlate with rearward actin waves. *Cell* 116: 431–443.

Goode BL, Eck MJ. 2007. Mechanism and function of formins in the control of actin assembly. *Annu Rev Biochem* 76: 593–627.

Grande-Garcia A, Echarri A, Del Pozo MA. 2005. Integrin regulation of membrane domain trafficking and Rac targeting. *Biochem Soc Trans* 33: 609–613.

Grashoff C, Hoffman BD, Brenner MD, Zhou R, Parsons M, Yang MT, McLean MA, Sligar SG, Chen CS, Ha T, et al. 2010. Measuring mechanical tension across vinculin reveals regulation of focal adhesion dynamics. *Nature* 466: 263–266.

Harunaga JS, Yamada KM. 2011. Cell-matrix adhesions in 3D. *Matrix Biol* 30: 363–368.

Heasman SJ, Ridley AJ. 2008. Mammalian Rho GTPases: New insights into their functions from in vivo studies. *Nat Rev Mol Cell Biol* 9: 690–701.

Hecht I, Skoge ML, Charest PG, Ben-Jacob E, Firtel RA, Loomis WF, Levine H, Rappel WJ. 2011. Activated membrane patches guide chemotactic cell motility. *PLoS Comput Biol* 7: e1002044.

* Hemmings BA, Restuccia DF. 2012. PI3K-PKB/Akt pathway. *Cold Spring Harb Perspect Biol* 4: a011189.

Hodges JL, Newell-Litwa K, Asmussen H, Vicente-Manzanares M, Horwitz AR. 2011. Myosin IIb activity and phosphorylation status determines dendritic spine and post-synaptic density morphology. *PLoS ONE* 6: e24149.

Hoefen RJ, Berk BC. 2006. The multifunctional GIT family of proteins. *J Cell Sci* 119: 1469–1475.

Hoeller O, Kay RR. 2007. Chemotaxis in the absence of PIP3 gradients. *Curr Biol* 17: 813–817.

Hu K, Ji L, Applegate KT, Danuser G, Waterman-Storer CM. 2007. Differential transmission of actin motion within focal adhesions. *Science* 315: 111–115.

Huang YE, Iijima M, Parent CA, Funamoto S, Firtel RA, Devreotes P. 2003. Receptor-mediated regulation of PI3Ks confines PI(3,4,5)P3 to the leading edge of chemotaxing cells. *Mol Biol Cell* 14: 1913–1922.

Huttenlocher A, Horwitz AR. 2011. Integrins in cell migration. *Cold Spring Harb Perspect Biol* 3: a005074.

Huttenlocher A, Palecek SP, Lu Q, Zhang W, Mellgren RL, Lauffenburger DA, Ginsberg MH, Horwitz AF. 1997. Regulation of cell migration by the calcium-dependent protease calpain. *J Biol Chem* 272: 32719–32722.

Hynes RO. 2002. Integrins: Bidirectional, allosteric signaling machines. *Cell* 110: 673–687.

Cite this chapter as *Cold Spring Harb Perspect Biol* doi: 10.1101/cshperspect.a005959

Iglesias PA, Devreotes PN. 2012. Biased excitable networks: How cells direct motion in response to gradients. *Curr Opin Cell Biol* **757**: 451–468.

Iglesias PA, Levchenko A. 2002. Modeling the cell's guidance system. *Sci STKE* **2002**: re12.

Iijima M, Devreotes PN. 2002. Tumor suppressor PTEN mediates sensing of chemoattractant gradients. *Cell* **109**: 599–610.

Inoue T, Meyer T. 2008. Synthetic activation of endogenous PI3K and Rac identifies an AND-gate switch for cell polarization and migration. *PLoS ONE* **3**: e3068.

Insall RH, Machesky LM. 2009. Actin dynamics at the leading edge: From simple machinery to complex networks. *Dev Cell* **17**: 310–322.

Janetopoulos C, Devreotes PN. 2006. Phosphoinositide signaling plays a key role in cytokinesis. *J Cell Biol* **174**: 485–490.

Janetopoulos C, Ma L, Iglesias PA, Devreotes PN. 2004. Chemoattractant-induced temporal and spatial PI(3,4,5)P$_3$ accumulation is controlled by a local excitation, global inhibition mechanism. *Proc Natl Acad Sci* **101**: 8951–8956.

Janetopoulos C, Borleis J, Vazquez F, Iijima M, Devreotes P. 2005. Temporal and spatial regulation of phosphoinositide signaling mediates cytokinesis. *Dev Cell* **8**: 467–477.

Jay DG. 2000. The clutch hypothesis revisited: Ascribing the roles of actin-associated proteins in filopodial protrusion in the nerve growth cone. *J Neurobiol* **44**: 114–125.

Kae H, Lim CJ, Spiegelman GB, Weeks G. 2004. Chemoattractants-induced Ras activation during Dictyostelium aggregation. *EMBO Rep* **5**: 602–606.

Kamimura Y, Xiong Y, Iglesias PA, Hoeller O, Bolourani P, Devreotes PN. 2008. PIP3-independent activation of TorC2 and PKB at the cell's leading edge mediates chemotaxis. *Curr Biol* **18**: 1034–1043.

Kaverina I, Krylyshkina O, Small JV. 1999. Microtubule targeting of substrate contacts promotes their relaxation and dissociation. *J Cell Biol* **146**: 1033–1044.

Kim EK, Yun SJ, HA JM, Kim YW, Jin IH, Yun J, Shin HK, Song SH, Kim JH, Lee JS, et al. 2011. Selective activation of Akt1 by mammalian target of rapamycin complex 2 regulates cancer cell migration, invasion, and metastasis. *Oncogene* **30**: 2954–2963.

Krause M, Dent EW, Bear JE, Loureiro JJ, Gertler FB. 2003. Ena/VASP proteins: Regulators of the actin cytoskeleton and cell migration. *Annu Rev Cell Dev Biol* **19**: 541–564.

Kraynov VS, Chamberlain C, Bokoch GM, Schwartz MA, Slabaugh S, Hahn KM. 2000. Localized Rac activation dynamics visualized in living cells. *Science* **290**: 333–337.

Kubow KE, Horwitz AR. 2011. Reducing background fluorescence reveals adhesions in 3D matrices. *Nat Cell Biol* **13**: 3–5.

Lacalle RA, Gómez-Moutón C, Barber DF, Jiménez-Baranda S, Mira E, Martínez-A C, Carrera AC, Mañes S. 2004. PTEN regulates motility but not directionality during leukocyte chemotaxis. *J Cell Sci* **117**: 6207–6215.

* Laplante M, Sabatini DM. 2012. mTOR signaling. *Cold Spring Harb Perspect Biol* **4**: a011593.

Lauffenburger DA, Horwitz AF. 1996. Cell migration: A physically integrated molecular process. *Cell* **84**: 359–369.

Lee S, Comer FI, Sasaki A, McLeod IX, Duong Y, Okumura K, Yates JR 3rd, Parent CA, Firtel RA. 2005. TOR complex 2 integrates cell movement during chemotaxis and signal relay in Dictyostelium. *Mol Biol Cell* **16**: 4572–4583.

Legate KR, Montañez E, Kudlacek O, Fässler R. 2006. ILK, PINCH and parvin: The tIPP of integrin signalling. *Nat Rev Mol Cell Biol* **7**: 20–31.

Levine H, Kessler DA, Rappel WJ. 2006. Directional sensing in eukaryotic chemotaxis: A balanced inactivation model. *Proc Natl Acad Sci* **103**: 9761–9766.

Linder S, Wiesner C, Himmel M. 2011. Degrading devices: Invadosomes in proteolytic cell invasion. *Annu Rev Cell Dev Biol* **27**: 185–211.

Liu L, Parent CA. 2011. TOR kinase complexes and cell migration. *J Cell Biol* **194**: 815–824.

Liu L, Das S, Losert W, Parent CA. 2010. mTORC2 regulates neutrophil chemotaxis in a cAMP- and RhoA-dependent fashion. *Dev Cell* **19**: 845–857.

Machacek M, Hodgson L, Welch C, Elliott H, Pertz O, Nalbant P, Abell A, Johnson GL, Hahn KM, Danuser G. 2009. Coordination of Rho GTPase activities during cell protrusion. *Nature* **461**: 99–103.

* McCaffrey LM, Macara IG. 2012. Signaling pathways in cell polarity. *Cold Spring Harb Perspect Biol* **4**: a009654.

Meili R, Ellsworth C, Lee S, Reddy TB, Ma H, Firtel RA. 1999. Chemoattractant-mediated transient activation and membrane localization of Akt/PKB is required for efficient chemotaxis to cAMP in *Dictyostelium*. *EMBO J* **18**: 2092–2105.

Mitchison TJ, Cramer LP. 1996. Actin-based cell motility and cell locomotion. *Cell* **84**: 371–379.

Mitchison T, Kirschner M. 1988. Cytoskeletal dynamics and nerve growth. *Neuron* **1**: 761–772.

Mitra SK, Hanson DA, Schlaepfer DD. 2005. Focal adhesion kinase: In command and control of cell motility. *Nat Rev Mol Cell Biol* **6**: 56–68.

Mondal S, Subramanian KK, Sakai J, Bajrami B, Luo HR. 2012. Phosphoinositide lipid phosphatase SHIP1 and PTEN coordinate to regulate cell migration and adhesion. *Mol Biol Cell* **23**: 1219–1230.

Montoya MC, Sancho D, Vicente-Manzanares M, Sánchez-Madrid F. 2002. Cell adhesion and polarity during immune interactions. *Immunol Rev* **186**: 68–82.

Moser M, Legate KR, Zent R, Fässler R. 2009. The tail of integrins, talin, and kindlins. *Science* **324**: 895–899.

Nalbant P, Hodgson L, Kraynov V, Toutchkine A, Hahn KM. 2004. Activation of endogenous Cdc42 visualized in living cells. *Science* **305**: 1615–1619.

Newey SE, Velamoor V, Govek EE, Van Aelst L. 2005. Rho GTPases, dendritic structure, and mental retardation. *J Neurobiol* **64**: 58–74.

Nishio M, Watanabe K, Sasaki J, Taya C, Takasuga S, Iizuka R, Balla T, Yamazaki M, Watanabe H, Itoh R, et al. 2007. Control of cell polarity and motility by the PtdIns(3,4,5)P3 phosphatase SHIP1. *Nat Cell Biol* **9**: 36–44.

Oakes PW, Beckham Y, Stricker J, Gardel ML. 2012. Tension is required but not sufficient for focal adhesion maturation without a stress fiber template. *Cell Biol* **196**: 363–374.

Oertner TG, Matus A. 2005. Calcium regulation of actin dynamics in dendritic spines. *Cell Calcium* **37**: 477–482.

Padrick SB, Rosen MK. 2010. Physical mechanisms of signal integration by WASP family proteins. *Annu Rev Biochem* **79**: 707–735.

Palecek SP, Loftus JC, Ginsberg MH, Lauffenburger DA, Horwitz AF. 1997. Integrin-ligand binding properties govern cell migration speed through cell-substratum adhesiveness. *Nature* **385**: 537–540.

Parent CA, Devreotes PN. 1999. A cell's sense of direction. *Science* **284**: 765–770.

Parent CA, Blacklock BJ, Froehlich WM, Murphy DB, Devreotes PN. 1998. G protein signaling events are activated at the leading edge of chemotactic cells. *Cell* **95**: 81–91.

Parsons JT. 2003. Focal adhesion kinase: The first ten years. *J Cell Sci* **116**: 1409–1416.

Parsons JT, Horwitz AR, Schwartz MA. 2010. Cell adhesion: Integrating cytoskeletal dynamics and cellular tension. *Nat Rev Mol Cell Biol* **11**: 633–643.

Paul AS, Pollard TD. 2009. Review of the mechanism of processive actin filament elongation by formins. *Cell Motil Cytoskeleton* **66**: 606–617.

Petrie RJ, Gavara N, Chadwick RS, Yamada KM. 2012. Nonpolarized signaling reveals two distinct modes of 3D cell migration. *J Cell Biol* **197**: 439–455.

Pollard TD, Borisy GG. 2003. Cellular motility driven by assembly and disassembly of actin filaments. *Cell* **112**: 453–465.

Pollitt AY, Insall RH. 2009. WASP and SCAR/WAVE proteins: The drivers of actin assembly. *J Cell Sci* **122**: 2575–2578.

Ponti A, Machacek M, Gupton SL, Waterman-Storer CM, Danuser G. 2004. Two distinct actin networks drive the protrusion of migrating cells. *Science* **305:** 1782–1786.

Provenzano PP, Eliceiri KW, Keely PJ. 2009. Shining new light on 3D cell motility and the metastatic process. *Trends Cell Biol* **19:** 638–648.

Ridley AJ. 2006. Rho GTPases and actin dynamics in membrane protrusions and vesicle trafficking. *Trends Cell Biol* **16:** 522–529.

Ridley AJ. 2011. Life at the leading edge. *Cell* **145:** 1012–1022.

Ridley AJ, Schwartz MA, Burridge K, Firtel RA, Ginsberg MH, Borisy G, Parsons JT, Horwitz AR. 2003. Cell migration: Integrating signals from front to back. *Science* **302:** 1704–1709.

Robinson DN, Spudich JA. 2004. Mechanics and regulation of cytokinesis. *Curr Opin Cell Biol* **16:** 182–188.

Ruusala A, Aspenstrom P. 2008. The atypical Rho GTPase Wrch1 collaborates with the nonreceptor tyrosine kinases Pyk2 and Src in regulating cytoskeletal dynamics. *Mol Cell Biol* **28:** 1802–1814.

Sasaki AT, Firtel RA. 2009. Spatiotemporal regulation of Ras-GTPases during chemotaxis. *Methods Mol Biol* **571:** 333–348.

Sawada Y, Tamada M, Dubin-Thaler BJ, Cherniavskaya O, Sakai R, Tanaka S, Sheetz MP. 2006. Force sensing by mechanical extension of the Src family kinase substrate p130Cas. *Cell* **127:** 1015–1026.

Schmidt S, Friedl P. 2010. Interstitial cell migration: Integrin-dependent and alternative adhesion mechanisms. *Cell Tissue Res* **339:** 83–92.

Schneider IC, Haugh JM. 2006. Mechanisms of gradient sensing and chemotaxis: Conserved pathways, diverse regulation. *Cell Cycle* **5:** 1130–1134.

Schwartz MA. 2010. Integrins and extracellular matrix in mechanotransduction. *Cold Spring Harb Perspect Biol* **2:** a005066.

Sepulveda JL, Gkretsi V, Wu C. 2005. Assembly and signaling of adhesion complexes. *Curr Top Dev Biol* **68:** 183–225.

* Sever R, Brugge JS. 2014. Signal transduction in cancer. *Cold Spring Harb Perspect Med* doi: 10.1101/cshperspect.a006098.

Shattil SJ, Kim C, Ginsberg MH. 2010. The final steps of integrin activation: The end game. *Nat Rev Mol Cell Biol* **11:** 288–300.

Shewan A, Eastburn DJ, Mostov K. 2011. Phosphoinositides in cell architecture. *Cold Spring Harb Perspect Biol* doi: 10.1101/cshperspect.a004796.

Slessareva JE, Dohlman HG. 2006. G protein signaling in yeast: New components, new connections, new compartments. *Science* **314:** 1412–1413.

Small JV, Resch GP. 2005. The comings and goings of actin: Coupling protrusion and retraction in cell motility. *Curr Opin Cell Biol* **17:** 517–523.

Small JV, Stradal T, Vignal E, Rottner K. 2002. The lamellipodium: Where motility begins. *Trends Cell Biol* **12:** 112–120.

Swaney KF, Huang CH, Devreotes PN. 2010. Eukaryotic chemotaxis: A network of signaling pathways controls motility, directional sensing, and polarity. *Annu Rev Biophys* **278:** 20445–20448.

Takeuchi H, Higashiyama T. 2011. Attraction of tip-growing pollen tubes by the female gametophyte. *Curr Opin Plant Biol* **14:** 614–621.

Tada T, Sheng M. 2006. Molecular mechanisms of dendritic spine morphogenesis. *Curr Opin Neurobiol* **16:** 95–101.

Tang M, Iijima M, Kamimura Y, Chen L, Long Y, Devreotes PN. 2011a. Disruption of PKB signaling restores polarity to cells lacking tumor suppressor PTEN. *Mol Biol Cell* **22:** 437–447.

Tang M, Iijima M, Devreotes P. 2011b. Generation of cells that ignore the effects of PIP3 on cytoskeleton. *Cell Cycle* **10:** 2817–2818.

Tomar A, Schlaepfer DD. 2009. Focal adhesion kinase: Switching between GAPs and GEFs in the regulation of cell motility. *Curr Opin Cell Biol* **21:** 676–683.

Trinkaus JP. 1969. *Cells into organs: The forces that shape the embryo*, p. 215. Prentice-Hall, Englewood Cliffs, NJ.

Turner CE, West KA, Brown MC. 2001. Paxillin-ARF GAP signaling and the cytoskeleton. *Curr Opin Cell Biol* **13:** 593–599.

Tybulewicz VL, Ardouin L, Prisco A, Reynolds LF. 2003. Vav1: A key signal transducer downstream of the TCR. *Immunol Rev* **192:** 42–52.

van Galen EJ, Ramakers GJ. 2005. Rho proteins, mental retardation and the neurobiological basis of intelligence. *Prog Brain Res* **147:** 295–317.

Van Haastert PJ, Devreotes PN. 2004. Chemotaxis: Signalling the way forward. *Nat Rev Mol Cell Biol* **5:** 626–634.

Van Keymeulen A, Wong K, Knight ZA, Govaerts C, Hahn KM, Shokat KM, Bourne HR. 2006. To stabilize neutrophil polarity, PIP3 and Cdc42 augment RhoA activity at the back as well as signals at the front. *J Cell Biol* **174:** 437–445.

Veltman DM, Keizer-Gunnik I, Van Haastert PJ. 2008. Four key signaling pathways mediating chemotaxis in *Dictyostelium discoideum*. *J Cell Biol* **180:** 747–753.

Vicente-Manzanares M, Sancho D, Yáñez-Mó M, Sánchez-Madrid F. 2002. The leukocyte cytoskeleton in cell migration and immune interactions. *Int Rev Cytol* **216:** 233–289.

Vicente-Manzanares M, Koach MA, Whitmore L, Lamers ML, Horwitz AF. 2008. Segregation and activation of myosin IIB creates a rear in migrating cells. *J Cell Biol* **183:** 543–554.

Vicente-Manzanares M, Ma X, Adelstein RS, Horwitz AR. 2009. Nonmuscle myosin II takes centre stage in cell adhesion and migration. *Nat Rev Mol Cell Biol* **10:** 778–790.

Vicente-Manzanares M, Newell-Litwa K, Bachir AI, Whitmore LA, Horwitz AR. 2011. Myosin IIA/IIB restrict adhesive and protrusive signaling to generate front-back polarity in migrating cells. *J Cell Biol* **193:** 381–396.

Wang Y-L. 2007. Flux at focal adhesions: Slippage clutch, mechanical gauge, or signal depot. *Sci STKE* **2007:** e10.

Wang F, Herzmark P, Weiner OD, Srinivasan S, Servant G, Bourne HR. 2002. Lipid products of PI(3)Ks maintain persistent cell polarity and directed motility in neutrophils. *Nature Cell Biology* **4:** 513–518.

Webb DJ, Donais K, Whitmore LA, Thomas SM, Turner CE, Parsons JT, Horwitz AF. 2004. FAK-Src signalling through paxillin, ERK and MLCK regulates adhesion disassembly. *Nat Cell Biol* **6:** 154–161.

Weiner OD, Neilsen PO, Prestwich GD, Kirschner MW, Cantley LC, Bourne HR. 2002. A PtdInsP(3)- and Rho GTPase-mediated positive feedback loop regulates neutrophil polarity. *Nat Cell Biol* **4:** 509–513.

Weiner OD, Marganski WA, Wu LF, Altschuler SJ, Kirschner MW. 2007. An actin-based wave generator organizes cell motility. *PLoS Biol* **5:** e221.

Welch HC, Coadwell WJ, Ellson CD, Ferguson GJ, Andrews SR, Erdjument-Bromage H, Tempst P, Hawkins PT, Stephens LR. 2002. P-Rex1, a PtdIns(3,4,5)P3- and Gβγ-regulated guanine-nucleotide exchange factor for Rac. *Cell* **108:** 809–821.

Xiong Y, Huang C-H, Iglesias PA, Devreotes PN. 2010. Cells navigate with a local-excitation, global-inhibition-biased excitable network. *Proc Natl Acad Sci* **107:** 17079–17086.

Yamaguchi H, Condeelis J. 2007. Regulation of the actin cytoskeleton in cancer cell migration and invasion. *Biochim Biophys Acta* **1773:** 642–652.

Yoo SK, Deng Q, Cavnar PJ, Wu YI, Hahn KM, Huttenlocher A. 2010. Differential regulation of protrusion and polarity by PI3K during neutrophil motility in live zebrafish. *Dev Cell* **18:** 226–236.

Zaidel-Bar R, Geiger B. 2010. The switchable integrin adhesome. *J Cell Sci* **123:** 1385–1388.

Zaidel-Bar R, Itzkovitz S, Ma'ayan A, Iyengar R, Geiger B. 2007a. Functional atlas of the integrin adhesome. *Nat Cell Biol* **9:** 858–867.

Zaidel-Bar R, Milo R, Kam Z, Geiger B. 2007b. A paxillin tyrosine phosphorylation switch regulates the assembly and form of cell-matrix adhesions. *J Cell Sci* **120:** 137–148.

Zhang H, Webb D, Asmussen J, Niu H, Horwitz AF. 2005. A GIT1/PIX/Rac/PAK signaling module regulates spine morphogenesis and synapse formation through MLC. *J Neurosci* **25:** 3379–3388.

CHAPTER 9

Signaling Pathways in Cell Polarity

Luke Martin McCaffrey[1] and Ian G. Macara[2]

[1]Department of Oncology, Rosalind and Morris Goodman Cancer Research Centre, McGill University, Montreal, Quebec, Canada

[2]Department of Microbiology, Center for Cell Signaling, University of Virginia School of Medicine, Charlottesville, Virginia 22908

Correspondence: lgm9c@virginia.edu

SUMMARY

A key function of signal transduction during cell polarization is the creation of spatially segregated regions of the cell cortex that possess different lipid and protein compositions and have distinct functions. Polarity can be initiated spontaneously or in response to signaling inputs from adjacent cells or soluble factors and is stabilized by positive-feedback loops. A conserved group of proteins, the Par proteins, plays a central role in polarity establishment and maintenance in many contexts. These proteins generate and maintain their distinct locations in cells by actively excluding one another from specific regions of the plasma membrane. The Par signaling pathway intersects with multiple other pathways that control cell growth, death, and organization.

Outline

1 INTRODUCTION

The asymmetric distribution of proteins, lipids, and RNAs is necessary for cell fate determination, differentiation, and a multitude of specialized cell functions that underlie morphogenesis (St Johnston 2005; Gonczy 2008; Knoblich 2008; Macara and Mili 2008; Martin-Belmonte and Mostov 2008). The establishment of cell polarity can be dissected into three primary processes: (1) breaking symmetry, either through extrinsic cues or stochastically; (2) establishing spatial organization through signal transduction; and (3) amplifying and maintaining the polarized state through feedback loops (Fig. 1). Even single-celled organisms such as budding yeast are polarized and engage sophisticated signaling mechanisms to initiate and organize asymmetric cell divisions. Higher organisms use polarity to build diverse cell types, such as neurons and epithelial cells in animals or stomatal cells in plants. Polarity spatially segregates important cellular functions from one another—for instance, in neurons, it separates synaptic inputs (along dendrites) from signaling outputs (along the axons). Epithelial cell polarity separates the apical membrane, which is specialized for interactions with the external environment, from the baso-lateral membrane, which contacts extracellular matrix or other cell types. In some epithelia a barrier called the tight junction separates the two membrane regions and prevents the intercellular diffusion of material across the epithelial sheet. Once established, cell polarity is often stable for the lifetime of the cell, as in neurons, but it can also be dynamic, for example, during development, when neural crest cells lose their epithelial character and become mesenchymal (this is termed the epithelial mesenchymal transition, EMT).

A conserved set of proteins called the Par polarity proteins is used in many contexts throughout the animal kingdom—for example, to polarize epithelia, to specify axons versus dendrites in a neuron, and to drive the asymmetric division of a nematode zygote (Fig. 2). These proteins are components of signal transduction pathways and include kinases, GTPases, adaptor proteins, and scaffolds. Additional pathways have evolved that play more specific roles—for instance, in epithelial apical/basal polarization or in planar polarity of epithelial sheets. However, our knowledge of the inputs to and outputs from these pathways and their intersection with other signaling networks remains incomplete. Moreover, despite the high level of conservation at the sequence level, the regulation and cross talk between the polarity proteins and other signaling components vary from one context to another and from one species to another, which complicates the task of dissecting polarity protein function. Nonetheless, rapid progress is being made in our understanding of polarity signaling, which we outline here, with an emphasis on Cdc42 and the Par proteins.

2 THE POLARIZATION MACHINERY

2.1 Symmetry Breaking and Positive-Feedback Loops

Symmetry breaking has been studied most intensively in the budding yeast *Saccharomyces cerevisiae*, which during the cell cycle switches from isotopic growth as a spherical cell to the polarized growth of bud formation before cell division, or to the "schmoo" formation necessary for mating (Slaughter et al. 2009). The key signaling pathway involves the small GTP-binding protein Cdc42 and its various regulators. A positive-feedback loop generates a high local enrichment of GTP-bound Cdc42 at the cell cortex that nucleates actin cables, which recruit more Cdc42 to the site, which, in turn, leads to further actin nucleation (Fig. 3). The initial local enrichment of Cdc42 relies on a separate feedback loop involving a complex of a guanine nucleotide exchange factor (GEF) for Cdc42, called Cdc24, and an adaptor protein (Bem1) (Fig. 3). Cdc42-GTP recruits the adaptor, which, in turn, recruits the GEF, which generates more Cdc42-GTP locally (Butty et al. 2002). Note that wild-type haploid yeast cells are never entirely unpolarized, and the new bud always forms adjacent to the bud scar left over from the previous cell cycle.

A pair of membrane-associated proteins near the scar function as the landmark for the new bud and recruit the GEF for a different GTPase, Rsr1, which then recruits Bem1

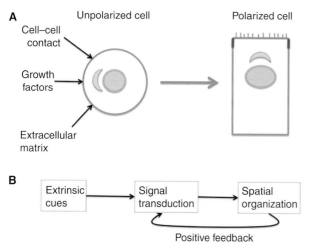

Figure 1. (*A*) Extrinsic signals normally are responsible for driving cell polarization, although it can also occur spontaneously under certain conditions. (*B*) Cell polarization is established by signal transduction pathways that spatially segregate different regions of the cell, especially the cell cortex, and this organization is reinforced and maintained by positive-feedback loops.

 Cite this chapter as *Cold Spring Harb Perspect Biol* doi: 10.1101/cshperspect.a009654

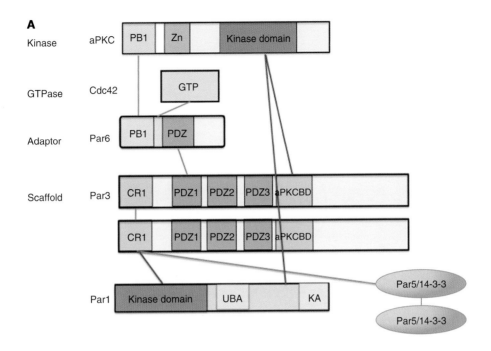

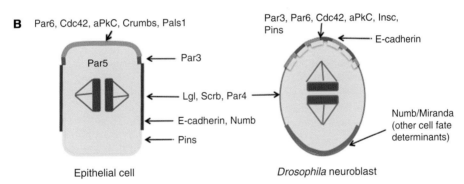

Figure 2. (*A*) Schematic showing domain structures of Par polarity proteins and their interactions. Phox and Bem1 domain (PB1), forms homodimers and heterodimers; zinc finger domain (Zn); PSD95, Dlg1, ZO-1 domain (PDZ), binds other PDZ domains and carboxy terminal peptide motifs; conserved region domain (CR1), forms homooligomers; atypical protein kinase C binding domain (aPKCBD); ubiquitin-binding-associated domain (UBA); kinase-associated domain (KA). (*B*) The different distributions of these polarity proteins in an epithelial cell and a neuroblast stem cell, together with the localization of other interacting proteins. Note that whereas in neuroblasts all of the polarity proteins form a complex (the "Par complex") at the apical cortex, this is not the case in epithelial cells, in which Par3 is not associated with Par6 and aPKC but is associated instead with the tight junction complex. The orientation of the mitotic spindle is controlled by the Par proteins and is different in neuroblasts (vertical) versus epithelial cells (horizontal). This difference reflects the distinct functions of polarity in the two cell types: segregation of cell fate determinants into only one daughter cell in the neuroblast versus formation of a polarized sheet of cells by the epithelium.

and the Cdc42-specific GEF Cdc24. This GTPase cascade is thought to amplify and stabilize the initial local cue. Interestingly, a GTPase-activating protein (GAP) that inactivates Cdc42 is localized to the old bud site and prevents its reuse in the next cycle (Tong et al. 2007). Other GAPs, which are not localized, inactivate any Cdc42-GTP that diffuses away from the bud site, thereby helping to maintain a focused spot of active Cdc42 at the correct cortical location. Therefore, local activation and global inactivation of a GTPase, plus cortical landmarks and organization of the actin cytoskeleton, are all used by budding yeast to ensure correct polarization for bud formation.

Do other organisms or cell types use the same mechanisms to drive polarization? Although the details differ, the general concept of reinforcing initial polarity cues with positive-feedback loops is widespread. Perhaps the best

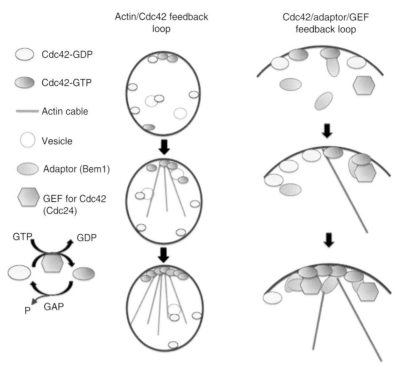

Figure 3. Positive-feedback loops that drive polarization of budding yeast during cell division. Cdc42 can cycle between GDP- and GTP-bound states, catalyzed by guanine nucleotide exchange factors (GEFs) that load GTP onto the protein, and GTPase-activating proteins (GAPs) that stimulate hydrolysis of the bound GTP. In the actin/Cdc42 loop, local enrichment of Cdc42-GTP at the cell cortex triggers nucleation of actin cables, along which vesicles recruit more Cdc42, which nucleates more actin cables. In the Cdc42/adaptor/GEF loop, local enrichment of Cdc42-GTP recruits an adaptor protein (Bem1), which, in turn, recruits a GEF for Cdc42 (Cdc24), which produces more Cdc42-GTP, which can, in turn, recruit more Bem1 and Cdc24.

understood examples are the migration of cells up gradients of a chemical attractant (chemotaxis) in *Dictyostelium* and in mammalian neutrophils (see Ch. 8 [Devreotes and Horwitz 2012]). These cells use positive feedback to reinforce polarization in the direction of the gradient, but instead of Cdc42, the loop involves PI-3 kinase, the phosphatase PTEN, and a protein kinase, Akt (Charest and Firtel 2006). As in yeast, local activation coupled with global inactivation seems to play an important role in stabilizing polarity (Xiong et al. 2010). Interestingly, *Dictyostelium* spontaneously and transiently polarizes in random directions even in the absence of any external gradient. This suggests that the detection of the chemotactic signal functions primarily to reinforce and stabilize a preexisting polarity rather than to break symmetry.

2.2 Par Proteins

Perhaps the clearest example of symmetry breaking and cell polarization is the fertilization of the *Caenorhabditis elegans* oocyte. Here the entry of the sperm into the egg breaks symmetry, driving a wave of acto-myosin contractions across the cell cortex and the establishment of an

anterior/posterior polarity, exemplified by the distribution of the Par proteins. The *par* (for "partition-defective") genes were identified in an elegant screen by Jim Priess and Ken Kemphues for maternal-effect genes that are embryonically lethal in *C. elegans* (Kemphues et al. 1988). Seven genes were identified in the screen, and they are all essential for the first asymmetric cell division of the zygote. Par1 and Par4 (also known as LKB1) are serine/threonine kinases (Guo and Kemphues 1995; Watts et al. 2000); Par2 is a RING-finger domain protein that may function as an E3 ubiquitin ligase (Levitan et al. 1994); Par3 and Par6 are PDZ-domain-containing proteins that have scaffolding or adaptor functions (Etemad-Moghadam et al. 1995; Hung and Kemphues 1999); Par5 is a 14-3-3 protein that binds to phosphorylated serine and threonine residues (Morton et al. 2002); and PKC-3 is an atypical protein kinase C (aPKC) (Fig. 2). With the exception of Par2, all of the Par proteins and aPKC are conserved throughout the Metazoa.

Strikingly, most of these polarity proteins show a polarized distribution within the zygote (Tabuse et al. 1998). Par1 and Par2 are restricted to the posterior of the zygote cortex, whereas Par3, Par6, and aPKC are restricted to the anterior cortex (although they are also present in the

 Cite this chapter as *Cold Spring Harb Perspect Biol* doi: 10.1101/cshperspect.a009654

cytoplasm) (Schneider and Bowerman 2003; Munro 2006). The segregation of the two cortical groups of Par proteins depends on their mutual antagonism, and loss of one Par protein results in escape of the others from their respective domains. In addition, the worm homolog of Lethal Giant Larvae (Lgl), a polarity protein originally discovered in *Drosophila*, localizes to the posterior cortex of the worm zygote and helps exclude the Par3–Par6–aPKC complex (Beatty et al. 2010; Hoege et al. 2010). Only Par4 and Par5 are non-polarized, and they are distributed diffusely throughout the cytoplasm, but they are required for the asymmetric distribution of the cortical Par proteins. The mechanisms that underlie this mutual antagonism are described below.

Par3, Par6, and aPKC can form a physical complex (Fig. 2), sometimes called the Par complex (Joberty et al. 2000; Lin et al. 2000; Wodarz et al. 2000), that has been identified in all animal cells that have been examined (Goldstein and Macara 2007). Par6 acts as a regulatory subunit for aPKC. The two proteins are attached to one another through their amino-terminal PB1 domains (Hirano et al. 2005), and this association inhibits the basal activity of aPKC. Par6 can also recruit substrates for phosphorylation (Yamanaka et al. 2001). Interaction of Par6 with Cdc42-GTP induces a conformational switch that relieves the inhibition, enabling the kinase to phosphorylate its substrates. One of these substrates is Par3. Atypical PKC binds through its kinase domain directly to a small region in the carboxy-terminal half of Par3 and phosphorylates S827 within this region (Nagai-Tamai et al. 2002), which causes the disassociation of the kinase from Par3. However, an additional interaction, between the PDZ domains of Par6 and Par3, can indirectly tether aPKC to Par3 even after S827 phosphorylation. The function of this rather complicated set of interactions is necessary for delivery of aPKC and Par6 to the apical surface of epithelial cells.

Importantly, Par3, Par6, and aPKC do not form a constitutive complex. Their interactions are regulated by multiple protein kinases, by small GTPases, and by competition for other binding partners, including other polarity proteins. These regulators determine the subcellular distribution of the Par proteins. Two striking examples are *Drosophila* neuroblasts and epithelial cells (Fig. 2). In the neuroblasts, Par3, Par6, and aPKC all localize together at the apical crescent, in a complex with two other proteins, Inscuteable and Partner of Inscuteable (Pins), which control spindle orientation during mitosis. This clustering of polarity proteins is independent of the phosphorylation of Par3 by aPKC. In contrast, only aPKC and Par6 are apical in epithelial cells, whereas Par3 segregates to the lateral/apical boundary (or to tight junctions in mammalian epithelial cells) (Fig. 2) (Izumi et al. 1998; Joberty et al. 2000). Moreover, the

localization of aPKC and Par6 to the apical cortex depends on the ability of aPKC to phosphorylate Par3. As described above, the phosphorylation partially disengages aPKC from Par3; but the two proteins remain attached through Par6. An epithelium-specific polarity protein at the apical membrane, called Crumbs, outcompetes Par6, displacing Par3 (Morais-de-Sa et al. 2010). In this way, Par3 is completely disengaged, and the Par6–aPKC complex is retained at the apical cortex.

2.3 Intercellular Junctions

Intercellular junctions are a universal feature of multicellular organisms. They provide the glue that binds cells together into tissues and organs, but also provide for the transmission of signals between adjacent cells. Many types of adhesive proteins have evolved, but the most widespread are the cadherins—transmembrane proteins that form calcium-dependent homophilic interactions between adjacent cells. The intracellular domains of cadherins bind to catenins, which perform multiple functions, including stabilizing adhesive clusters at cell–cell interfaces, interacting with the actin cytoskeleton, and serving as signaling platforms that are coupled to many of the known signal transduction networks within the cell, including the Par polarity proteins. Vertebrate epithelia and endothelia also possess tight junctions, which form both a barrier between cells and a fence between the apical and lateral domains of the plasma membrane within each cell. The fence prevents the free diffusion of membrane proteins and lipids between these two domains, thereby helping to maintain apical/baso-lateral polarity. Tight junctions are composed principally of transmembrane proteins called claudins, but—as is true for adhesive junctions—there are numerous additional proteins that associate with the claudins to form the junctional structures.

Despite the important role of intercellular junctions in polarity, cell polarization is not dependent on their existence. Baas et al. (2004) have shown that single intestinal cells can be induced to polarize simply by the activation of Par4, in the complete absence of attachment to any neighboring cell. Moreover, *Drosophila* epithelial cells do not possess tight junctions, yet are able to segregate apical from lateral proteins as efficiently as do their mammalian counterparts.

3 FACTORS THAT CONTROL POLARITY PROTEIN LOCALIZATION

The Par proteins provide critical spatial information during polarization, to identify different regions of the cell cortex. The localization of polarity proteins is therefore central to their biological functions. Protein localization often involves distinct steps that can include transport, delivery, anchoring at the destination, and active exclusion from other areas of the cell. Transport can simply involve passive diffusion, or directed movement along the cytoskeleton. Alternatively, the mRNA encoding the protein might be transported to the destination, where it is locally translated. Anchors can include phospholipids, cytoskeletal elements, or more specific protein complexes.

3.1 Membrane Attachment via Phospholipids

All of the Par proteins except for Par4 and Par5, plus other polarity proteins including Lgl, Scribble, Dlg, Pals1, Patj, and Crumbs, are found predominantly at the cell cortex. Crumbs is a transmembrane protein that tethers Patj, Pals1, and Lin7 to the apical cortex. Par3 contains a conserved basic amino acid motif in the carboxy-terminal half of the protein that can bind directly to phosphoinositides (Krahn et al. 2010), which is both necessary and sufficient for association with membranes. An additional phosphoinositide-binding mechanism has been proposed for this protein, through phosphoinositide binding to its PDZ2 domain (Wu et al. 2007). No other polarity protein is known to contain lipid-binding motifs, but Par4 is farne-

sylated at its carboxyl terminus, and this hydrophobic posttranslational modification could facilitate association with membranes.

3.2 Oligomerization

The amino-terminal conserved region 1 (CR1) is necessary for self-association of Par3 into higher-order complexes (Fig. 2) (Benton and Johnston 2003a; Mizuno et al. 2003; Feng et al. 2007). It is essential but not sufficient for membrane attachment. How the oligomerization of Par3 maintains the protein at the plasma membrane is unclear, but oligomerization might complement weak phosphoinositide binding by increasing avidity (Fig. 4).

3.3 Anchoring to Membrane Proteins

Enrichment and retention in a specific region of the cell cortex is frequently mediated by direct interactions with transmembrane proteins. Pals1 and Par6 can both bind directly to the carboxy-terminal sequence of Crumbs, via their PDZ domains. Par6 and Patj can also associate indirectly with Crumbs through Pals1. The recruitment of Par3 to tight junctions in mammalian epithelial cells is mediated,

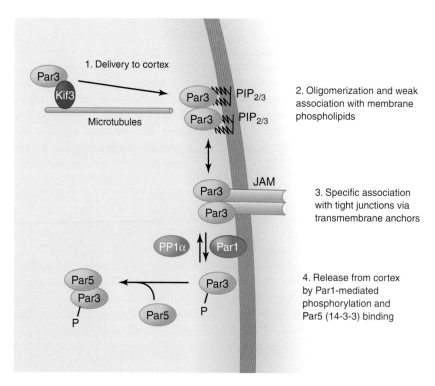

Figure 4. Mechanisms for the transport, cortical association, and anchoring of Par3 in mammalian cells. It is not yet known if all of these mechanisms operate in any one cell type, and additional processes, such as RNA localization, might play roles in certain circumstances. Junctional adhesion molecule (JAM) is shown as an example of a transmembrane protein to which Par3 can be anchored, but others exist, such as the neurotrophin receptor, p75NTR, in mammalian Schwann cells (Chan et al. 2006). PP1α is a phosphatase.

at least in part, through association between the first PDZ domain of Par3 and the carboxyl terminus of junction adhesion molecules (JAMs) (Fig. 4) (Ebnet et al. 2000; Itoh et al. 2001). Because Par6 also binds to the first PDZ domain of Par3, the associations with JAM and Par6 are mutually exclusive. A similar mechanism occurs in the mouse neuroepithelium, where Par3 is recruited to newly formed intercellular junctions through a direct interaction with nectin 1 or nectin 3 (Takekuni et al. 2003). In the *Drosophila* wing disc epithelium, Par3 is also recruited to junctions through its PDZ domains, by interaction with the adherens junction proteins β-catenin and Echinoid, an immunoglobulin-domain transmembrane protein (Wei et al. 2005). In embryonic epithelium, Par3 instead associates with E-cadherin (Harris and Peifer 2005). Thus, different tissues use distinct mechanisms to organize the spatial distribution of Par3 and other polarity proteins.

3.4 Localized mRNA Translation

RNA localization and local translation are important drivers of cell polarization in several situations and have been intensively studied in the *Drosophila* oocyte, in budding yeast, and in vertebrate neurons. The mRNAs for two epithelial-specific polarity proteins, Crumbs and Pals1 (also called Stardust), are enriched near the apical surface in *Drosophila* epithelial cells (Horne-Badovinac and Bilder 2008; Li et al. 2008). Why localize RNAs? One potential function is to regulate translation, perhaps in response to extracellular signals, or to cell density changes. As an example, the Par3 mRNA—but not that of Par6 or aPKC—is transported out along the axons of mammalian motor neurons, where it can be locally translated in response to stimulation of the neurons by nerve growth factor (NGF), a process that is essential for NGF-dependent axon outgrowth (Hengst et al. 2009).

3.5 Active Exclusion

None of the mechanisms described above stably anchors Par proteins at the plasma membrane. Cortical Par proteins are highly dynamic in the *C. elegans* zygote, rapidly exchanging with cytoplasmic pools and undergoing lateral diffusion along the plasma membrane (Goehring et al. 2011). Lgl protein is also highly dynamic in *Drosophila* embryonic cells (Mayer et al. 2005). Therefore, the targeting and retention mechanisms described above are insufficient to maintain their polarized distribution within the cell. Active mechanisms drive the segregation of polarity proteins into different cortical domains and maintain their asymmetry. As mentioned above, the Par3–Par6–aPKC complex is often localized in a complementary pattern to

that of Par1, and through mutual phosphorylation reactions, the two kinases, aPKC and Par1, exclude each other from their respective regions of the cortex. Par1 directly phosphorylates Par3 on S144 (equivalent to S151 in *Drosophila*) (Benton and Johnston 2003b; Hurd et al. 2003a) near CR1, the region necessary for oligomerization of Par3. The phosphorylated S144/151 residue acts as a docking site for Par5 (14-3-3), which might reduce oligomerization and thereby destabilize membrane association of Par3 (Fig. 4). This mechanism can therefore exclude Par3 from cortical regions that contain Par1. The phosphorylation and binding of Par5 (14-3-3) to Par3 can be reversed by protein phosphatase 1α (PP1α), allowing recycling of Par3 to appropriate cortical sites.

Conversely, aPKC can phosphorylate Par1, which both inhibits Par1 kinase activity and disassociates it from the plasma membrane (Fig. 5A). Two distinct mechanisms have been identified, one in which aPKC directly phosphorylates Par1 on T595 (Hurov et al. 2004) and an indirect mechanism by which aPKC activates protein kinase D (PKD), which then phosphorylates Par1 on S400 (Watkins et al. 2008). Phosphorylation of T595 reduces the kinase activity and displaces Par1 from the membranes. Furthermore, phosphorylation of S400 recruits Par5 (14-3-3), which displaces Par1 from the membrane.

The *C. elegans* aPKC can also phosphorylate Par2 and exclude it from the anterior cortical domain of the zygote, whereas Par2 in the posterior domain recruits Par1 to phosphorylate and exclude Par3 (Hao et al. 2006).

The spatial distributions of many downstream effectors of the Par signaling pathway are also controlled by phosphorylation. For example, the cell fate determinants Numb and Miranda, the polarity protein Lgl, and the spindle orientation factor Pins (called LGN in mammals), are all removed from the plasma membrane by aPKC-dependent phosphorylation (Fig. 5B) (Betschinger et al. 2003; Hao et al. 2006; Smith et al. 2007; Atwood and Prehoda 2009). Pins associates with the cell cortex in mitosis, by binding to Gαi subunits, where it functions to attach astral microtubules, which orient the mitotic spindle. In epithelial tissues, normal organization is often dictated by the ability of cells to divide in the plane of the epithelial sheet, but not perpendicular to the sheet. To this end, apical aPKC phosphorylates any Pins that diffuses into the apical region, resulting in the recruitment of Par5 (14-3-3), which disengages Pins from Gαi. In this way, Pins is excluded from the apical cortex, preventing astral microtubule attachment and perpendicular orientation of mitosis. Pins also orients the mitotic spindles in *Drosophila* stem cells and in some mammalian progenitors. Similar mechanisms, involving not only aPKC and Par1 but other protein kinases that target Par5 (14-3-3) consensus sites, probably also exist.

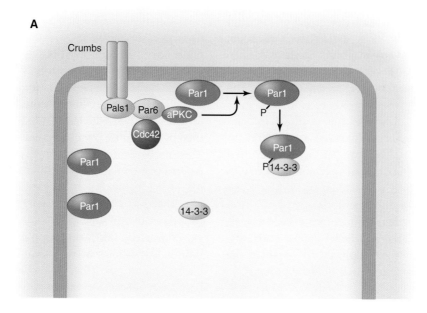

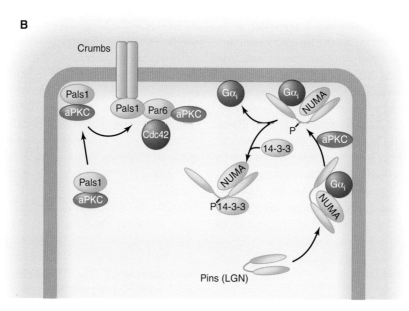

Figure 5. Active exclusion from cortical domains. A common mechanism for the establishment of discrete cortical regions of the cell occurs through the active removal of unwanted proteins from within a particular region. A kinase within the region phosphorylates the protein, resulting in the association of Par5 (14-3-3), which displaces the protein from the cell cortex. (*A*) Par1 is restricted to the lateral membranes in epithelial cells. Any Par1 that strays onto the apical surface is phosphorylated by aPKC, which recruits Par5 (14-3-3), causing its disassociation from the membrane. (*B*) aPKC at the apical surface of an epithelial cell phosphorylates any Pins (also known as LGN) protein that binds to Gαi within this domain. Association of the phosphorylated Pins with 14-3-3 triggers its dissociation from Gαi. Because phosphorylation does not occur at the basolateral membrane, Pins can remain associated with this region of the cortex. Pins is diffuse in the cytoplasm in interphase cells and can only associate with Gαi at the cell cortex after it has undergone a conformational switch triggered by the binding of NuMA, a nuclear protein that is released in mitosis. Also shown in this schematic is the recruitment by Par3 of aPKC to the apical surface, where the aPKC is disengaged and binds to the Crumbs–Pals1–Par6 complex. The apical aPKC is activated by the binding of Cdc42-GTP to Par6.

The mechanism underlying exclusion of Numb from the apical region of progenitor cells in *Drosophila* has been worked out in considerable detail. At the onset of mitosis, Aurora-A phosphorylates Par6 within the PB1 domain that binds to aPKC (Fig. 2), releasing Par6 and relieving its inhibition of aPKC (Wirtz-Peitz et al. 2008). The activated aPKC phosphorylates Lgl, which causes it to dissociate from the Par6–aPKC complex and allows the complex to interact with Par3. Par3 then acts to recruit Numb as a substrate for aPKC. Phosphorylation of Numb by aPKC causes it to be released from the cell cortex (Smith et al. 2007; Wirtz-Peitz et al. 2008). Because the Par complex is initially asymmetrically distributed, the loss of Numb occurs at only one side of the mitotic cell; thus, one daughter will inherit Numb while the other does not. Interestingly, Lgl seems here to function as a buffer, to suppress the phosphorylation of Numb until the appropriate time in the cell cycle. The recruitment of Numb by Par3 seems to be an evolutionarily conserved function, because in migrating mammalian fibroblasts, the same mechanism is used to release Numb from the cell cortex and regulates the internalization of integrins (Nishimura and Kaibuchi 2007).

4 POLARITY SIGNALING THROUGH PAR3–PAR6–APKC

4.1 Polarity Signaling through Small GTPases

Actin filaments and microtubules are the two major asymmetric components of cells. Both are vectorial polymers, and their organization is, therefore, of fundamental importance to cell polarization. This organization is dynamic and is highly regulated by signaling networks that respond to external and internal cues. Central to these signaling networks are the Rho family GTPases (Fig. 3), which regulate and are regulated by polarity proteins.

Cdc42 is a pivotal component of the polarity machinery in yeast and is conserved throughout the metazoa. Dominant-negative Cdc42 mutants disrupt polarized migration in mammalian fibroblasts (Nobes and Hall 1999), and Cdc42-GTP binds to Par6, providing a mechanism by which the GTPase can control cell polarization. Par6 contains a partial CRIB domain, a motif conserved among most Cdc42 effectors. This domain interacts with a region of the GTPase that undergoes a GTP-dependent switch in conformation (Garrard et al. 2003). Binding of Cdc42-GTP to Par6 relieves the inhibition of aPKC activity by Par6 (Yamanaka et al. 2001). In addition, the cytoplasmic tail of the receptor tyrosine kinase ephrin B1 competes with Cdc42 for binding to Par6, blocking tight junction formation (Lee et al. 2008). Tyrosine phosphorylation of ephrin B1 releases it from Par6. In this way, signaling through

tyrosine kinase receptors could affect aPKC activity and consequently cell polarity decisions.

In addition to counteracting the inhibition of aPKC by Par6, Cdc42 can also recruit the Par6–aPKC complex to specific regions of the cell cortex where Cdc42 is activated. For example, in *Drosophila* neuroblasts mutant for Cdc42, Par6–aPKC is mislocalized to the cytoplasm (Atwood et al. 2007), and depletion of Cdc42 from mammalian epithelial cells can partially mislocalize aPKC from the apical cortex (Martin-Belmonte et al. 2007). A positive-feedback loop probably reinforces the positioning of these proteins, because robust Cdc42 localization in the neuroblasts also requires Par6. In the *C. elegans* zygote, Cdc42 is not essential for the initial anterior enrichment of Par6, although the asymmetry is lost later, during the first cell division, which suggests that other factors set up the polarity (Aceto et al. 2006). In mammalian epithelial cells, Par6 localization to the apical surface probably requires its association with Pals1 and/or Crumbs, rather than Cdc42 (Gao et al. 2002; Hurd et al. 2003b).

Another small GTPase, Rho1, helps organize the polarity of the *C. elegans* zygote. The RhoGEF Ect1 is excluded from the posterior cortex, which restricts Rho-GTP production to the anterior end of the zygote (Motegi and Sugimoto 2006), where it stimulates myosin contractility. This contraction generates a cortical actin flow, translocating Par6, aPKC, Par3, and Cdc42 to the anterior end of the zygote. Cdc42-GTP then maintains this distribution of the Par proteins.

The Rho GTPase may play a distinct role in mammalian cells by controlling the association of Par3 with Par6–aPKC (Fig. 6). ROCK, a protein kinase downstream from RhoA, can phosphorylate Par3 on T833, adjacent to the aPKC-binding site in the carboxyl terminus of Par3, and this phosphorylation blocks the association with aPKC (Nakayama et al. 2008). However, aPKC can also phosphorylate ROCK, which suppresses its association with epithelial junctions. Because ROCK phosphorylates the myosin light chain and activates actomyosin contractility, this represents an additional polarity mechanism (Ishiuchi and Takeichi 2011).

4.2 Rho GTPases as Downstream Effectors of Par3–Par6–aPKC

The carboxy-terminal region of Par3 can bind to a Rac GEF called Tiam1 (Chen and Macara 2005; Mertens et al. 2005; Nishimura et al. 2005; Zhang and Macara 2006). Par3 sequesters Tiam1 to prevent inappropriate activation of Rac (Chen and Macara 2005; Zhang and Macara 2006). Loss of Par3 causes an increase in Rac-GTP levels, which results in a misorganization of actin filaments at the cell cortex. A

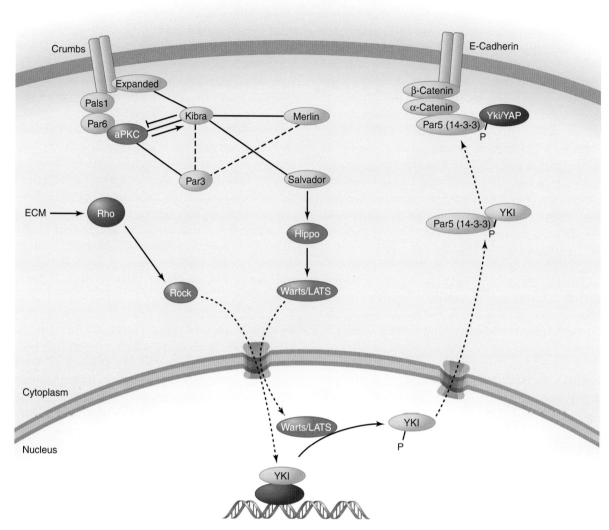

Figure 6. Interaction map showing potential links between the Par proteins and components of the Hippo pathway. A signaling cascade involving Salvador, Hippo, and Warts controls phosphorylation and nuclear localization of the transcription factors Yorkie (YKI) and TAZ to regulate epithelial growth. Epithelial integrity is monitored by cell-adhesion complexes (E-cadherin, α-catenin) and by the Par and Crumbs polarity complexes through the adaptor Kibra. The extracellular matrix (ECM) also has an impact on YKI nuclear localization through the Rho GTPase, independently of the Hippo pathway.

similar mechanism operates in the notum epithelium of *Drosophila*, where Par3 inhibits Sif, the fly homolog of Tiam1, to restrict filopodium formation to basal regions of the cell (Georgiou and Baum 2010). In other cell types, Par3 might recruit Tiam1 to sites within the cell where Rac needs to be activated. In such cases, loss of Par3 might reduce Rac activation at these sites (Pegtel et al. 2007).

Par6 can also regulate RhoA activity. By associating with aPKC, Par6 can activate a RhoGAP called p190, thereby reducing Rho-GTP levels (Zhang and Macara 2008). The link between Par6–aPKC and the p190 RhoGAP is unknown, but the pathway is important for controlling

synapse density in hippocampal neurons. Elevated expression of Par6 increases dendritic spine density, whereas silencing of Par6 reduces spine density.

A second, different mechanism by which Par6 can affect Rho is through association with Smad-ubiquitin regulatory factor 1 (Smurf1), an E3 ligase that down-regulates the TGFβ signaling pathway (Fig. 4). In addition to targeting Smads for degradation, Smurf1 can also ubiquitylate RhoA (Wang et al. 2003). TGFβ type II receptor can bind to and phosphorylate Par6 on a conserved carboxy-terminal residue (S345) (Ozdamar et al. 2005). This stimulates the association of Par6 (or aPKC) with Smurf1, which mediates the localized ubiquitylation and destruction of RhoA. In mammary NMuMG cells, loss of RhoA can cause the dissolution of tight junctions and an EMT (Wang et al. 2003). However, these effects might be cell type specific because, for example, in MDCK cells, dominant-negative RhoA expression had no effect on tight junctions (Bruewer et al. 2004).

4.3 Cross Talk between Wnt Signaling and Polarity Proteins

Planar cell polarity (PCP) signaling establishes directional asymmetry at the tissue level rather than at the cellular level (as occurs in apical/basal polarization or in the asymmetric divisions of some stem cells). PCP signaling establishes the directionality of wing hair orientation in *Drosophila*, for example, and the orientation of cilia in the outer hair cells of the mammalian cochlea. It is also essential for the oriented migration of cell sheets during gastrulation. PCP is stimulated by Wnt ligands that activate seven-transmembrane-span receptors of the Frizzled family (Simons and Mlodzik 2008). Other transmembrane proteins, including Vangl and Flamingo (also called Celsr), are also essential for PCP signaling. Downstream from these proteins, an adaptor called Dishevelled (Dvl) can recruit a GEF to activate RhoA (Tsuji et al. 2010), which stimulates the downstream kinase ROCK. As described above, ROCK phosphorylates Par3, in addition to the myosin light chain and other targets. Dvl is itself a target for phosphorylation by the Par1 polarity protein and can directly associate with aPKC (Sun et al. 2001; Ossipova et al. 2005). The association with aPKC stabilizes and activates the kinase and has been implicated in axonal differentiation and polarized migration of mammalian cells (Schlessinger et al. 2007; Zhang et al. 2007). Moreover, the *Drosophila* Frizzled receptor is phosphorylated and inhibited by aPKC, which is recruited to Frizzled through another polarity protein, Patj (Djiane et al. 2005). Thus, aPKC mediates multiple signaling connections between PCP components. Whether its binding partners Cdc42 and Par6 participate in PCP regulation remains to be determined.

Another apical/basal polarity protein, Scribble, has also been implicated in PCP. The mechanism remains obscure, but it interacts genetically and physically with Vangl, which, in turn, can bind to the extracellular domain of Frizzled, perhaps enabling non-cell-autonomous signaling between neighboring cells (Courbard et al. 2009). Signaling downstream from Scribble is poorly understood and requires the carboxyl terminus of the protein, including the third and fourth PDZ domains, which bind to Vangl. Control of apical/basal polarity by Scribble in *Drosophila*, in contrast, does not require any of the PDZ domains (Zeitler et al. 2004).

4.4 Connections between the Hippo Pathway and Polarity Signaling

The Hippo pathway was first discovered in *Drosophila* as a central regulator of organ size (Grusche et al. 2010). It controls the balance between proliferation and apoptosis in flies and mammals and involves a protein kinase cascade that ultimately phosphorylates and inactivates the transcriptional coactivators Yorkie (YKI, also known as Yap1) and TAZ (Fig. 6) (p. 133 [Harvey and Hariharan 2012]).

Phospho-Yap1 is sequestered at adherens junctions through association with α-catenin. This catenin can function as a tumor suppressor, and in the epidermis, loss of α-catenin results in the nuclear accumulation of YKI and increased cell proliferation (Schlegelmilch et al. 2011; Silvis et al. 2011). How cell density controls the α-catenin–YKI interaction remains unclear, but another Hippo pathway component, Merlin, also binds directly to α-catenin and is required for stable adherens junction formation (Gladden et al. 2010). Merlin couples α-catenin to Par3 during junction maturation in keratinocytes, which might be important for the apical positioning of tight junctions relative to the adherens junctions. In addition, Par5 (14-3-3) is required for the association of α-catenin with YKI. Merlin might in some way regulate the association of YKI–Par5 with α-catenin.

Remarkably, mechanical forces also regulate YKI (Fig. 6). Increasing the stiffness of the extracellular matrix activates Rho and its downstream kinase ROCK—through an unknown mechanism—which drives the nuclear accumulation and activation of YKI (Dupont et al. 2011). This pathway seems to be entirely separate from the canonical Hippo pathway but provides a direct mechanism through which the physical environment can reprogram gene expression. Whether α-catenin is involved in this pathway, through interactions with actin, remains to be determined.

Apical/basal polarity proteins also interact with components of the Hippo pathway (Fig. 6). For example, the *Drosophila* protein Expanded binds to the intracellular domain of Crumbs (Chen et al. 2010; Grzeschik et al. 2010; Ling et al. 2010; Robinson et al. 2010). Expanded is a FERM-domain protein distantly related to Merlin. Changes in the level of Crumbs expression cause a mislocalization of Expanded and are sufficient to induce hyperproliferation of epithelial cells in *Drosophila* (Lu and Bilder 2005). The vertebrate homolog of Expanded has not yet been unambiguously identified, however, and it remains unknown whether a similar signaling pathway operates in mammalian epithelial cells. The wiring might be somewhat different, because in mouse epithelial cell lines, Crumbs3-dependent apico–basal polarization leads to sequestration of SMADs in the cytoplasm through the Hippo pathway, which inhibits TGFβ signaling and EMT (Varelas et al. 2010), but there is no evidence of a link to TGF-β in *Drosophila*.

An important component of the Hippo pathway is Kibra, which binds to many pathway components, including Expanded, Merlin, Warts, Hippo, and Salvador, but how these interactions are regulated remains unclear. Kibra also associates with the Par complex and localizes to tight junctions and the apical membrane (Yoshihama et al. 2010). Kibra is a substrate for aPKC phosphorylation (Büther et al. 2004), but Kibra can also negatively regulate aPKC activity to control exocytosis of apical membrane components (Yoshihama et al. 2010). A similar mechanism might function in cell migration, during which an aPKC–Kibra–exocyst complex regulates focal adhesion dynamics at the leading edge by recruiting the MAP kinase ERK (Rosse et al. 2009). It seems likely that these connections could play important roles in the misorganization, overgrowth, and invasion of epithelial cancers.

4.5 Polarity Signaling and Cancer

The majority of human cancers arise from epithelial cells, and tumorigenesis is commonly believed to involve a loss of apical/basal polarity (Ch. 21 [Sever and Brugge 2014]). In particular, cells at the invasive fronts of carcinomas often express proteins associated with a mesenchymal phenotype, such as vimentin and Snail, whereas epithelial proteins such as E-cadherin and Crumbs are either lost or mislocalized away from the cell cortex. Some of these changes in expression are caused by reduced expression of micro-RNAs in the MiR-200 family, which suppress the expression of a transcription factor, ZEB1. ZEB1, in turn, suppresses the expression of epithelial genes, including E-cadherin and the polarity genes *Crb3* (a Crumbs homolog), *Lgl*, and *Patj* (Brabletz and Brabletz 2010). Because only the outermost layers of tumor cells in a carcinoma usually show mesenchymal characteristics, they are probably induced by interactions of the tumor cells with the surrounding stroma, but the mechanisms involved remain unclear (Vidal et al. 2010). Cells that have undergone an EMT are more motile, more invasive, and less adherent, and might contribute to dissemination of tumors throughout the body and establishment of metastases. However, it is important to note that there is very little evidence of a role for the polarity machinery in human cancer. Although several polarity proteins, including Scribble and Lgl, have tumor suppressor behavior in *Drosophila*, their role in human cancers remains ill defined (Wu et al. 2010). Moreover, it remains to be established whether loss of polarity is an essential aspect of malignancy. Epithelial cells could escape from their tissue of origin by mechanisms independent of classical EMT, including changes in spindle orientation during mitosis and collective migration of clusters of cells (Rorth 2009).

5 CONCLUSION

Understanding cell polarization is one of the major goals of cell biology and will inevitably have a broad impact on research into diseases such as cancer and neurological degeneration. A complicated web of signaling systems surrounds and intersects with the polarity machinery, yet we still understand very little about what the Par proteins do, how they are localized, how their various interactions are regulated, and which signaling components operate in which contexts. After all, the organization of a polarized cell is a formidably complicated process that involves cytoskeletal remodeling, membrane traffic, RNA localization, and protein complex assembly and disassembly, with feedback to gene expression and protein turnover. It is conceivable that the Par proteins participate in all of these processes, either directly or indirectly. It is worth remembering, however, that although the polarity machinery can work in a cell-autonomous fashion, the Par genes do not exist in any unicellular organism, which suggests that a key role for the Par proteins is to facilitate, mediate, or interpret cell–cell and cell–matrix interactions. The tissue context might, therefore, be expected to modulate Par protein functions. It will be interesting to determine whether regulation of these interactions controls morphogenesis and to what extent differential Par function contributes to phenotypic variation both between different cell types in one organism and between species.

REFERENCES

*Reference is in this book.

Aceto D, Beers M, Kemphues KJ. 2006. Interaction of PAR-6 with CDC-42 is required for maintenance but not establishment of PAR asymmetry in *C. elegans*. *Dev Biol* **299**: 386–397.

Atwood SX. Prehoda KE. 2009. aPKC phosphorylates Miranda to polarize fate determinants during neuroblast asymmetric cell division. *Curr Biol* **19**: 723–729.

Atwood SX, Chabu C, Penkert RR, Doe CQ, Prehoda KE. 2007. Cdc42 acts downstream of Bazooka to regulate neuroblast polarity through Par-6 aPKC. *J Cell Sci* **120**: 3200–3206.

Baas AF, Kuipers J, van der Wel NN, Batlle E, Koerten HK, Peters PJ, Clevers HC. 2004. Complete polarization of single intestinal epithelial cells upon activation of LKB1 by STRAD. *Cell* **116**: 457–466.

Beatty A, Morton D, Kemphues K. 2010. The *C. elegans* homolog of *Drosophila* Lethal giant larvae functions redundantly with PAR-2 to maintain polarity in the early embryo. *Development* **137**: 3995–4004.

Benton R, Johnston DS. 2003a. A conserved oligomerization domain in *Drosophila* Bazooka/PAR-3 is important for apical localization and epithelial polarity. *Curr Biol* **13**: 1330–1334.

Benton R, Johnston DS. 2003b. *Drosophila* PAR-1 and 14-3-3 inhibit Bazooka/PAR-3 to establish complementary cortical domains in polarized cells. *Cell* **115**: 691–704.

Cite this chapter as *Cold Spring Harb Perspect Biol* doi: 10.1101/cshperspect.a009654

Betschinger J, Mechtler K, Knoblich JA. 2003. The Par complex directs asymmetric cell division by phosphorylating the cytoskeletal protein Lgl. *Nature* **422**: 326–330.

Brabletz S, Brabletz T. 2010. The ZEB/miR-200 feedback loop—a motor of cellular plasticity in development and cancer? *EMBO Rep* **11**: 670–677.

Bruewer M, Hopkins AM, Hobert ME, Nusrat A, Madara JL. 2004. RhoA, Rac1, and Cdc42 exert distinct effects on epithelial barrier via selective structural and biochemical modulation of junctional proteins and F-actin. *Am J Physiol Cell Physiol* **287**: C327–C335.

Büther K, Plaas C, Barnekow A, Kremerskothen J. 2004. KIBRA is a novel substrate for protein kinase Cζ. *Biochem Biophys Res Commun* **317**: 703–707.

Butty AC, Perrinjaquet N, Petit A, Jaquenoud M, Segall JE, Hofmann K, Zwahlen C, Peter M. 2002. A positive feedback loop stabilizes the guanine-nucleotide exchange factor Cdc24 at sites of polarization. *EMBO J* **21**: 1565–1576.

Chan JR, Jolicoeur C, Yamauchi J, Elliott J, Fawcett JP, Ng BK, Cayouette M. 2006. The polarity protein Par-3 directly interacts with p75^{NTR} to regulate myelination. *Science* **314**: 832–836.

Charest PG, Firtel RA. 2006. Feedback signaling controls leading-edge formation during chemotaxis. *Curr Opin Genet Dev* **16**: 339–347.

Chen X, Macara IG. 2005. Par-3 controls tight junction assembly through the Rac exchange factor Tiam1. *Nat Cell Biol* **7**: 262–269.

Chen CL, Gajewski KM, Hamaratoglu F, Bossuyt W, Sansores-Garcia L, Tao C, Halder G. 2010. The apical–basal cell polarity determinant Crumbs regulates Hippo signaling in *Drosophila*. *Proc Natl Acad Sci* **107**: 15810–15815.

Courbard JR, Djiane A, Wu J, Mlodzik M. 2009. The apical/basal-polarity determinant Scribble cooperates with the PCP core factor Stbm/Vang and functions as one of its effectors. *Dev Biol* **333**: 67–77.

* Devreotes P, Horwitz AF. 2012. Cell migration and chemotaxis. *Cold Spring Harb Perspect Biol* **4**: a005959.

Djiane A, Yogev S, Mlodzik M. 2005. The apical determinants aPKC and dPatj regulate Frizzled-dependent planar cell polarity in the *Drosophila* eye. *Cell* **121**: 621–631.

Dupont S, Morsut L, Aragona M, Enzo E, Giulitti S, Cordenonsi M, Zanconato F, Le Digabel J, Forcato M, Bicciato S, et al. 2011. Role of YAP/TAZ in mechanotransduction. *Nature* **474**: 179–183.

Ebnet K, Schulz CU, Meyer Zu Brickwedde MK, Pendl GG, Vestweber D. 2000. Junctional adhesion molecule interacts with the PDZ domain-containing proteins AF-6 and ZO-1. *J Biol Chem* **275**: 27979–27988.

Etemad-Moghadam B, Guo S, Kemphues KJ. 1995. Asymmetrically distributed PAR-3 protein contributes to cell polarity and spindle alignment in early *C. elegans* embryos. *Cell* **83**: 743–752.

Feng W, Wu H, Chan LN, Zhang M. 2007. The Par-3 NTD adopts a PB1-like structure required for Par-3 oligomerization and membrane localization. *EMBO J* **26**: 2786–2796.

Gao L, Joberty G, Macara IG. 2002. Assembly of epithelial tight junctions is negatively regulated by Par6. *Curr Biol* **12**: 221–225.

Garrard SM, Capaldo CT, Gao L, Rosen MK, Macara IG, Tomchick DR. 2003. Structure of Cdc42 in a complex with the GTPase-binding domain of the cell polarity protein, Par6. *EMBO J* **22**: 1125–1133.

Georgiou M, Baum B. 2010. Polarity proteins and Rho GTPases cooperate to spatially organise epithelial actin-based protrusions. *J Cell Sci* **123**: 1089–1098.

Gladden AB, Hebert AM, Schneeberger EE, McClatchey AI. 2010. The NF2 tumor suppressor, Merlin, regulates epidermal development through the establishment of a junctional polarity complex. *Dev Cell* **19**: 727–739.

Goehring NW, Hoege C, Grill SW, Hyman AA. 2011. PAR proteins diffuse freely across the anterior–posterior boundary in polarized *C. elegans* embryos. *J Cell Biol* **193**: 583–594.

Goldstein B, Macara IG. 2007. The PAR proteins: Fundamental players in animal cell polarization. *Dev Cell* **13**: 609–622.

Gonczy P. 2008. Mechanisms of asymmetric cell division: Flies and worms pave the way. *Nat Rev Mol Cell Biol* **9**: 355–366.

Grusche FA, Richardson HE, Harvey KF. 2010. Upstream regulation of the Hippo size control pathway. *Curr Biol* **20**: R574–R582.

Grzeschik NA, Parsons LM, Allott ML, Harvey KF, Richardson HE. 2010. Lgl, aPKC, and Crumbs regulate the Salvador/Warts/Hippo pathway through two distinct mechanisms. *Curr Biol* **20**: 573–581.

Guo S, Kemphues KJ. 1995. *par-1*, a gene required for establishing polarity in *C. elegans* embryos, encodes a putative Ser/Thr kinase that is asymmetrically distributed. *Cell* **81**: 611–620.

Hao Y, Boyd L, Seydoux G. 2006. Stabilization of cell polarity by the *C. elegans* RING protein PAR-2. *Dev Cell* **10**: 199–208.

Harris TJ, Peifer M. 2005. The positioning and segregation of apical cues during epithelial polarity establishment in *Drosophila*. *J Cell Biol* **170**: 813–823.

* Harvey KF, Hariharan IK. 2012. The Hippo pathway. *Cold Spring Harb Perspect Biol* **4**: a011288.

Hengst U, Deglincerti A, Kim HJ, Jeon NL, Jaffrey SR. 2009. Axonal elongation triggered by stimulus-induced local translation of a polarity complex protein. *Nat Cell Biol* **11**: 1024–1030.

Hirano Y, Yoshinaga S, Takeya R, Suzuki NN, Horiuchi M, Kohjima M, Sumimoto H, Inagaki F. 2005. Structure of a cell polarity regulator, a complex between atypical PKC and Par6 PB1 domains. *J Biol Chem* **280**: 9653–9661.

Hoege C, Constantinescu AT, Schwager A, Goehring NW, Kumar P, Hyman AA. 2010. LGL can partition the cortex of one-cell *Caenorhabditis elegans* embryos into two domains. *Curr Biol* **20**: 1296–1303.

Horne-Badovinac S, Bilder D. 2008. Dynein regulates epithelial polarity and the apical localization of *stardust A* mRNA. *PLoS Genet* **4**: e8.

Hung TJ, Kemphues KJ. 1999. PAR-6 is a conserved PDZ domain-containing protein that colocalizes with PAR-3 in *Caenorhabditis elegans* embryos. *Development* **126**: 127–135.

Hurd TW, Fan S, Liu C-J, Kweon HK, Hakansson K, Margolis B. 2003a. Phosphorylation-dependent binding of 14-3-3 to the polarity protein Par3 regulates cell polarity in mammalian epithelia. *Curr Biol* **13**: 2082–2090.

Hurd TW, Gao L, Roh MH, Macara IG, Margolis B. 2003b. Direct interaction of two polarity complexes implicated in epithelial tight junction assembly. *Nat Cell Biol* **5**: 137–142.

Hurov JB, Watkins JL, Piwnica-Worms H. 2004. Atypical PKC phosphorylates PAR-1 kinases to regulate localization and activity. *Curr Biol* **14**: 736–741.

Ishiuchi T, Takeichi M. 2011. Willin and Par3 cooperatively regulate epithelial apical constriction through aPKC-mediated ROCK phosphorylation. *Nat Cell Biol* **13**: 860–866.

Itoh M, Sasaki H, Furuse M, Ozaki H, Kita T, Tsukita S. 2001. Junctional adhesion molecule (JAM) binds to PAR-3: A possible mechanism for the recruitment of PAR-3 to tight junctions. *J Cell Biol* **154**: 491–497.

Izumi Y, Hirose T, Tamai Y, Hirai S, Nagashima Y, Fujimoto T, Tabuse Y, Kemphues KJ, Ohno S. 1998. An atypical PKC directly associates and colocalizes at the epithelial tight junction with ASIP, a mammalian homologue of *Caenorhabditis elegans* polarity protein PAR-3. *J Cell Biol* **143**: 95–106.

Joberty G, Petersen C, Gao L, Macara IG. 2000. The cell-polarity protein Par6 links Par3 and atypical protein kinase C to Cdc42. *Nat Cell Biol* **2**: 531–539.

Kemphues KJ, Priess JR, Morton DG, Cheng NS. 1988. Identification of genes required for cytoplasmic localization in early *C. elegans* embryos. *Cell* **52**: 311–320.

Knoblich JA. 2008. Mechanisms of asymmetric stem cell division. *Cell* **132**: 583–597.

Krahn MP, Buckers J, Kastrup L, Wodarz A. 2010. Formation of a Bazooka–Stardust complex is essential for plasma membrane polarity in epithelia. *J Cell Biol* **190**: 751–760.

Lee HS, Nishanian TG, Mood K, Bong YS, Daar IO. 2008. EphrinB1 controls cell–cell junctions through the Par polarity complex. *Nat Cell Biol* **10**: 979–986.

Levitan DJ, Boyd L, Mello CC, Kemphues KJ, Stinchcomb DT. 1994. *par-2*, a gene required for blastomere asymmetry in *Caenorhabditis elegans*, encodes zinc-finger and ATP-binding motifs. *Proc Natl Acad Sci* **91**: 6108–6112.

Li Z, Wang L, Hays TS, Cai Y. 2008. Dynein-mediated apical localization of crumbs transcripts is required for Crumbs activity in epithelial polarity. *J Cell Biol* **180**: 31–38.

Lin D, Edwards AS, Fawcett JP, Mbamalu G, Scott JD, Pawson T. 2000. A mammalian PAR-3–PAR-6 complex implicated in Cdc42/Rac1 and aPKC signalling and cell polarity. *Nat Cell Biol* **2**: 540–547.

Ling C, Zheng Y, Yin F, Yu J, Huang J, Hong Y, Wu S, Pan D. 2010. The apical transmembrane protein Crumbs functions as a tumor suppressor that regulates Hippo signaling by binding to Expanded. *Proc Natl Acad Sci* **107**: 10532–10537.

Lu H, Bilder D. 2005. Endocytic control of epithelial polarity and proliferation in *Drosophila*. *Nat Cell Biol* **7**: 1232–1239.

Macara IG, Mili S. 2008. Polarity and differential inheritance—universal attributes of life? *Cell* **135**: 801–812.

Martin-Belmonte F, Mostov K. 2008. Regulation of cell polarity during epithelial morphogenesis. *Curr Opin Cell Biol* **20**: 227–234.

Martin-Belmonte F, Gassama A, Datta A, Yu W, Rescher U, Gerke V, Mostov K. 2007. PTEN-mediated apical segregation of phosphoinositides controls epithelial morphogenesis through Cdc42. *Cell* **128**: 383–397.

Mayer B, Emery G, Berdnik D, Wirtz-Peitz F, Knoblich JA. 2005. Quantitative analysis of protein dynamics during asymmetric cell division. *Curr Biol* **15**: 1847–1854.

Mertens AE, Rygiel TP, Olivo C, van der Kammen R, Collard JG. 2005. The Rac activator Tiam1 controls tight junction biogenesis in keratinocytes through binding to and activation of the Par polarity complex. *J Cell Biol* **170**: 1029–1037.

Mizuno K, Suzuki A, Hirose T, Kitamura K, Kutsuzawa K, Futaki M, Amano Y, Ohno S. 2003. Self-association of PAR-3-mediated by the conserved N-terminal domain contributes to the development of epithelial tight junctions. *J Biol Chem* **278**: 31240–31250.

Morais-de-Sa E, Mirouse V, St Johnston D. 2010. aPKC phosphorylation of Bazooka defines the apical/lateral border in *Drosophila* epithelial cells. *Cell* **141**: 509–523.

Morton DG, Shakes DC, Nugent S, Dichoso D, Wang W, Golden A, Kemphues KJ. 2002. The *Caenorhabditis elegans par-5* gene encodes a 14-3-3 protein required for cellular asymmetry in the early embryo. *Dev Biol* **241**: 47–58.

Motegi F, Sugimoto A. 2006. Sequential functioning of the ECT-2 RhoGEF, RHO-1 and CDC-42 establishes cell polarity in *Caenorhabditis elegans* embryos. *Nat Cell Biol* **8**: 978–985.

Munro EM. 2006. PAR proteins and the cytoskeleton: A marriage of equals. *Curr Opin Cell Biol* **18**: 86–94.

Nagai-Tamai Y, Mizuno K, Hirose T, Suzuki A, Ohno S. 2002. Regulated protein–protein interaction between aPKC and PAR-3 plays an essential role in the polarization of epithelial cells. *Genes Cells* **7**: 1161–1171.

Nakayama M, Goto TM, Sugimoto M, Nishimura T, Shinagawa T, Ohno S, Amano M, Kaibuchi K. 2008. Rho-kinase phosphorylates PAR-3 and disrupts PAR complex formation. *Dev Cell* **14**: 205–215.

Nishimura T, Kaibuchi K. 2007. Numb controls integrin endocytosis for directional cell migration with aPKC and PAR-3. *Dev Cell* **13**: 15–28.

Nishimura T, Yamaguchi T, Kato K, Yoshizawa M, Nabeshima YI, Ohno S, Hoshino M, Kaibuchi K. 2005. PAR-6–PAR-3 mediates Cdc42-induced Rac activation through the Rac GEFs STEF/Tiam1. *Nat Cell Biol* **7**: 270–277.

Nobes CD, Hall A. 1999. Rho GTPases control polarity, protrusion, and adhesion during cell movement. *J Cell Biol* **144**: 1235–1244.

Ossipova O, Dhawan S, Sokol S, Green JB. 2005. Distinct PAR-1 proteins function in different branches of Wnt signaling during vertebrate development. *Dev Cell* **8**: 829–841.

Ozdamar B, Bose R, Barrios-Rodiles M, Wang HR, Zhang Y, Wrana JL. 2005. Regulation of the polarity protein Par6 by TGFβ receptors controls epithelial cell plasticity. *Science* **307**: 1603–1609.

Pegtel DM, Ellenbroek SI, Mertens AE, van der Kammen RA, de Rooij J, Collard JG. 2007. The *par–tiam1* complex controls persistent migration by stabilizing microtubule-dependent front–rear polarity. *Curr Biol* **17**: 1623–1634.

Robinson BS, Huang J, Hong Y, Moberg KH. 2010. Crumbs regulates Salvador/Warts/Hippo signaling in *Drosophila* via the FERM-domain protein Expanded. *Curr Biol* **20**: 582–590.

Rorth P. 2009. Collective cell migration. *Annu Rev Cell Dev Biol* **25**: 407–429.

Rosse C, Formstecher E, Boeckeler K, Zhao Y, Kremerskothen J, White MD, Camonis JH, Parker PJ. 2009. An aPKC–exocyst complex controls paxillin phosphorylation and migration through localised JNK1 activation. *PLoS Biol* **7**: e1000235.

Schlegelmilch K, Mohseni M, Kirak O, Pruszak J, Rodriguez JR, Zhou D, Kreger BT, Vasioukhin V, Avruch J, Brummelkamp TR, et al. 2011. Yap1 acts downstream of α-catenin to control epidermal proliferation. *Cell* **144**: 782–795.

Schlessinger K, McManus EJ, Hall A. 2007. Cdc42 and noncanonical Wnt signal transduction pathways cooperate to promote cell polarity. *J Cell Biol* **178**: 355–361.

Schneider SQ, Bowerman B. 2003. Cell polarity and the cytoskeleton in the *Caenorhabditis elegans* zygote. *Annu Rev Genet* **37**: 221–249.

⋆ Sever R, Brugge JS. 2014. Signal transduction in cancer. *Cold Spring Harb Perspect Med* doi: 10.1101/cshperspect.a006098.

Silvis MR, Kreger BT, Lien WH, Klezovitch O, Rudakova GM, Camargo FD, Lantz DM, Seykora JT, Vasioukhin V. 2011. α-Catenin is a tumor suppressor that controls cell accumulation by regulating the localization and activity of the transcriptional coactivator Yap1. *Sci Signal* **4**: ra33.

Simons M, Mlodzik M. 2008. Planar cell polarity signaling: From fly development to human disease. *Annu Rev Genet* **42**: 517–540.

Slaughter BD, Smith SE, Li R. 2009. Symmetry breaking in the life cycle of the budding yeast. *Cold Spring Harb Perspect Biol* **1**: a003384.

Smith CA, Lau KM, Rahmani Z, Dho SE, Brothers G, She YM, Berry DM, Bonneil E, Thibault P, Schweisguth F, et al. 2007. aPKC-mediated phosphorylation regulates asymmetric membrane localization of the cell fate determinant Numb. *Embo J* **26**: 468–480.

St Johnston D. 2005. Moving messages: The intracellular localization of mRNAs. *Nat Rev Mol Cell Biol* **6**: 363–375.

Sun TQ, Lu B, Feng JJ, Reinhard C, Jan YN, Fantl WJ, Williams LT. 2001. PAR-1 is a Dishevelled-associated kinase and a positive regulator of Wnt signalling. *Nat Cell Biol* **3**: 628–636.

Tabuse Y, Izumi Y, Piano F, Kemphues KJ, Miwa J, Ohno S. 1998. Atypical protein kinase C cooperates with PAR-3 to establish embryonic polarity in *Caenorhabditis elegans*. *Development* **125**: 3607–3614.

Takekuni K, Ikeda W, Fujito T, Morimoto K, Takeuchi M, Monden M, Takai Y. 2003. Direct binding of cell polarity protein PAR-3 to cell–cell adhesion molecule nectin at neuroepithelial cells of developing mouse. *J Biol Chem* **278**: 5497–5500.

Tong Z, Gao XD, Howell AS, Bose I, Lew DJ, Bi E. 2007. Adjacent positioning of cellular structures enabled by a Cdc42 GTPase-activating protein-mediated zone of inhibition. *J Cell Biol* **179**: 1375–1384.

Tsuji T, Ohta Y, Kanno Y, Hirose K, Ohashi K, Mizuno K. 2010. Involvement of p114–RhoGEF and Lfc in Wnt-3a- and Dishevelled-induced RhoA activation and neurite retraction in N1E-115 mouse neuroblastoma cells. *Mol Biol Cell* **21**: 3590–3600.

Varelas X, Samavarchi-Tehrani P, Narimatsu M, Weiss A, Cockburn K, Larsen BG, Rossant J, Wrana JL. 2010. The Crumbs complex couples cell density sensing to Hippo-dependent control of the TGF-β–SMAD pathway. *Dev Cell* **19**: 831–844.

Vidal M, Salavaggione L, Ylagan L, Wilkins M, Watson M, Weilbaecher K, Cagan R. 2010. A role for the epithelial microenvironment at tumor boundaries: Evidence from *Drosophila* and human squamous cell carcinomas. *Am J Pathol* **176**: 3007–3014.

Wang HR, Zhang Y, Ozdamar B, Ogunjimi AA, Alexandrova E, Thomsen GH, Wrana JL. 2003. Regulation of cell polarity and protrusion formation by targeting RhoA for degradation. *Science* **302:** 1775–1779.

Watkins JL, Lewandowski KT, Meek SE, Storz P, Toker A, Piwnica-Worms H. 2008. Phosphorylation of the Par-1 polarity kinase by protein kinase D regulates 14-3-3 binding and membrane association. *Proc Natl Acad Sci* **105:** 18378–18383.

Watts JL, Morton DG, Bestman J, Kemphues KJ. 2000. The *C. elegans par-4* gene encodes a putative serine-threonine kinase required for establishing embryonic asymmetry. *Development* **127:** 1467–1475.

Wei SY, Escudero LM, Yu F, Chang LH, Chen LY, Ho YH, Lin CM, Chou CS, Chia W, Modolell J, et al. 2005. Echinoid is a component of adherens junctions that cooperates with DE-Cadherin to mediate cell adhesion. *Dev Cell* **8:** 493–504.

Wirtz-Peitz F, Nishimura T, Knoblich JA. 2008. Linking cell cycle to asymmetric division: Aurora-A phosphorylates the Par complex to regulate Numb localization. *Cell* **135:** 161–173.

Wodarz A, Ramrath A, Grimm A, Knust E. 2000. *Drosophila* atypical protein kinase C associates with Bazooka and controls polarity of epithelia and neuroblasts. *J Cell Biol* **150:** 1361–1374.

Wu H, Feng W, Chen J, Chan LN, Huang S, Zhang M. 2007. PDZ domains of Par-3 as potential phosphoinositide signaling integrators. *Mol Cell* **28:** 886–898.

Wu M, Pastor-Pareja JC, Xu T. 2010. Interaction between Ras(V12) and scribbled clones induces tumour growth and invasion. *Nature* **463:** 545–548.

Xiong Y, Huang CH, Iglesias PA, Devreotes PN. 2010. Cells navigate with a local-excitation, global-inhibition-biased excitable network. *Proc Natl Acad Sci* **107:** 17079–17086.

Yamanaka T, Horikoshi Y, Suzuki A, Sugiyama Y, Kitamura K, Maniwa R, Nagai Y, Yamashita A, Hirose T, Ishikawa H, et al. 2001. PAR-6 regulates aPKC activity in a novel way and mediates cell–cell contact-induced formation of the epithelial junctional complex. *Genes Cells* **6:** 721–731.

Yoshihama Y, Sasaki K, Horikoshi Y, Suzuki A, Ohtsuka T, Hakuno F, Takahashi S, Ohno S, Chida K. 2010. KIBRA suppresses apical exocytosis through inhibition of aPKC kinase activity in epithelial cells. *Curr Biol* **21:** 705–711.

Zeitler J, Hsu CP, Dionne H, Bilder D. 2004. Domains controlling cell polarity and proliferation in the *Drosophila* tumor suppressor Scribble. *J Cell Biol* **167:** 1137–1146.

Zhang H, Macara IG. 2006. The polarity protein PAR-3 and TIAM1 cooperate in dendritic spine morphogenesis. *Nat Cell Biol* **8:** 227–237.

Zhang H, Macara IG. 2008. The PAR-6 polarity protein regulates dendritic spine morphogenesis through p190 RhoGAP and the Rho GTPase. *Dev Cell* **14:** 216–226.

Zhang X, Zhu J, Yang GY, Wang QJ, Qian L, Chen YM, Chen F, Tao Y, Hu HS, Wang T, et al. 2007. Dishevelled promotes axon differentiation by regulating atypical protein kinase C. *Nat Cell Biol* **9:** 743–754.

CHAPTER **10**

Signaling Mechanisms Controlling Cell Fate and Embryonic Patterning

Norbert Perrimon[1,2], Chrysoula Pitsouli[1,3], and Ben-Zion Shilo[4]

[1]Department of Genetics, Harvard Medical School, Boston, Massachusetts 02115

[2]Howard Hughes Medical Institute, Boston, Massachusetts 02115

[3]Department of Biological Sciences, University of Cyprus, 1678 Nicosia, Cyprus

[4]Department of Molecular Genetics, Weizmann Institute of Science, Rehovot 76100, Israel

Correspondence: perrimon@receptor.med.harvard.edu

SUMMARY

During development, signaling pathways specify cell fates by activating transcriptional programs in response to extracellular signals. Extensive studies in the past 30 years have revealed that surprisingly few pathways exist to regulate developmental programs and that dysregulation of these can lead to human diseases, including cancer. Although these pathways use distinct signaling components and signaling strategies, a number of common themes have emerged regarding their organization and regulation in time and space. Examples from *Drosophila*, such as Notch, Hedgehog, Wingless/WNT, BMP (bone morphogenetic proteins), EGF (epidermal growth factor), and FGF (fibroblast growth factor) signaling, illustrate their abilities to act either at a short range or over a long distance, and in some instances to generate morphogen gradients that pattern fields of cells in a concentration-dependent manner. They also show how feedback loops and transcriptional cascades are part of the logic of developmental regulation.

Outline

Cite this chapter as *Cold Spring Harb Perspect Biol* doi: 10.1101/cshperspect.a005975

1 INTRODUCTION

Key to multicellularity is the coordinated interaction of the various cells that make up the body. Indeed, patterning of embryos, establishment of cell type diversity, and formation of tissues and organs all rely on cell-to-cell communication during development. Thus, arguably one of the most important principles of developmental biology involves "one group of cells changing the behavior of an adjacent set of cells, causing them to change their shape, mitotic rate, or fate" (Gilbert 2000).

Classically, the ability of one group of cells to affect the fate of another is called "induction." The cells that produce the signals are referred to as "inducing cells," whereas the receiving cells are termed "responders" (Spemann and Mangold 1924). The ability of cells to respond to the inducers, referred to as "competence" (Waddington 1940), usually reflects the presence of a receptor at the top of a pathway that regulates the expression of specific transcription factors in the responding cells. The responding cells, in turn, can become inductive and change the fate of their neighbors by producing new signals, thus generating sequential inductive events that increase cell-fate diversity in tissues.

Identification and characterization of the signaling pathways involved in development has led to the surprising realization that only a few exist (Gerhart 1999; Gilbert 2000; Barolo and Posakony 2002). These fall into 11 main classes, defined by the ligand or signal transducers involved: Notch, FGF, EGF, Wnt/Wingless (Wg), Hedgehog (Hh), transforming growth factor β (TGFβ)/BMPs, cytokine (nonreceptor tyrosine kinase JAK-STAT [signal transducers and activators of transcription] pathway), Hippo, Jun kinase (JNK), NF-κB, and retinoic acid receptor (RAR). These pathways involve either cell-to-cell contact via surface proteins (juxtacrine signaling), or secreted diffusible growth and differentiation factors (paracrine signaling). Among the pathways mentioned above, only two of them, Notch and Hippo, are juxtacrine, whereas the others are paracrine.

With the exception of those that release steroid hormones and retinoic acid, which cross the membrane and activate gene expression by binding directly to receptor proteins that act as transcriptional regulators, inducing cells generally produce secreted or transmembrane ligands, which in some cases require complex processing in the producing cells or the extracellular matrix. When these ligands bind to transmembrane receptors on target cells they activate a cascade of events that ultimately regulate the activity of a small number of transcription factors and/or cofactors, triggering gene-expression programs that drive the cellular changes. For example, Notch signaling (p. 109 [Kopan 2012]) regulates CSL (for CBF1, Suppressor of Hairless,

and Lag1) proteins that possess an integrase domain, receptor tyrosine kinases (RTKs) regulate ETS (erythroblast transformation-specific) transcription factors, Wnt ligands (p. 103 [Nusse 2012]) mostly regulate the high-mobility group (HMG) box-containing TCF (T-cell factor) transcription factor, Hh proteins (p. 107 [Ingham 2012]) regulate Gli (glioblastoma) transcription factors that have DNA-binding zinc-finger domains, and BMPs (p. 113 [Wrana 2013]) regulate Smads (Sma- and Mad-related proteins) transcription factors. Cytokine pathways (p. 117 [Harrison 2012]) regulate STATs, and Hippo (p. 133 [Harvey and Hariharan 2012]) regulates TAZ (for transcriptional coactivator with PDZ-binding motif) proteins that contain a WW domain and a carboxy-terminal PDZ-binding motif (Table 1). In addition, many pathways activate feedback loops that modulate or terminate the incoming signal (Perrimon and McMahon 1999; Freeman 2000).

The response to signaling-pathway activation is usually complex and involves the regulation of many processes, such as control of cell fate, apoptosis, cell proliferation, cytoskeletal reorganization, cell polarity, adhesion, and cell migration. Importantly, each pathway does not specifically regulate a single biological process but can elicit diverse effects, depending on the state of the cell at the time the pathway is activated. Furthermore, because few pathways exist, there are no unique signals for induction of each cell type. Instead, the response of a given cell to a signal depends on its amplitude, duration, interactions between pathways, and integration of transcription factor effectors at promoters and enhancers of target genes. It may also be predetermined by the set of transcription factors expressed in the cell at the time the signal is received.

Here, we use specific examples, mostly taken from *Drosophila*, to illustrate general principles and mechanisms by which signaling pathways operate in development to specify cell fates. Thus, this is not a comprehensive review of the structures and roles of all the pathways that have been implicated in developmental processes. A number of excellent reviews elsewhere describe in detail the roles of individual pathways in development (Notch [Artavanis-Tsakonas et al. 1999; Lai 2004; Fortini 2009], FGF [Ghabrial et al. 2003; Pownall and Isaacs 2010], EGF [Shilo 2005], Wnt/Wg [Logan and Nusse 2004; MacDonald et al. 2009], Hh [Ingham and McMahon 2001; Jiang and Hui 2008], TGFβ [Feng and Derynck 2005; Wu and Hill 2009], JAK/STAT [Hou et al. 2002; Arbouzova and Zeidler 2006], and Hippo [Pan 2007; Saucedo and Edgar 2007]).

Note also that a number of other pathways, such as those involving cadherins and integrins, are not discussed here as they are involved in permissive interactions whereby a tissue is made competent to respond and requires the proper environment to trigger the appropriate cellular changes

Cite this chapter as *Cold Spring Harb Perspect Biol* doi: 10.1101/cshperspect.a005975

Table 1. Key signaling pathways that orchestrate development—receptors, ligands, transcription factors, and outputs are shown for each

Signaling pathway	Receptor	Ligand	Transcriptional effector	Output
Wnt/Wg	Frizzled, dFrizzled2	Wg/Wnt	Armadillo/β-catenin with TCF/LEF	Patterning, growth, PCP (β-catenin independent)
Hh	Patched	Hh	Ci/Gli	Patterning, growth
TGFβ	Thickveins	Dpp/TGFβ	Smad (Mad/Medea)	Patterning, growth
RTK	EGFR	Spitz, Gurken, Keren, Vein	Pointed/Yan	Patterning, morphogenesis
	FGFR (Breathless, Heartless)	Branchless, Thisbe, Pyramus	Pointed/Yan	Patterning, morphogenesis, migration
	InR	dIlp1-dIlp7	Pointed/Yan, Foxo	Growth, metabolism, aging
	PDGF/VEGF receptor (PVR)	Pvf1-3	Pointed/Yan	Morphogenesis, migration
	Torso	Trunk, PTTH	Pointed/Yan	Patterning, metamorphosis
	dALK	Jelly belly	Pointed/Yan	Growth on starvation (CNS)
	Sevenless	Boss	Pointed/Yan	Patterning, cell-fate specification
Notch	Notch	Delta, Serrate	NICD with Su(H)	Patterning, lateral inhibition, cell-fate specification
Hippo	Fat	Dachsous	Yorkie with Scalloped	Growth, PCP
NF-κB	Toll	Spatzle	Dorsal/Dif	Patterning, innate immunity
JAK/STAT	Domeless	Unpaired1-3	STAT92E	Patterning, innate immunity
JNK	Eiger/TNF	Wengen	Jun and Fos	Migration, patterning, innate immunity
Nuclear receptors	EcRA, EcRB	Ecdysone	EcRA, EcRB with USP	Patterning, growth, metabolism

Abbreviations: TCF, T-cell factor; LEF, lymphoid enhancer-binding factor; PCP, planar cell polarity; TGF, transforming growth factor; RTK, receptor-tyrosine kinase; EGFR, epidermal growth factor receptor; FGFR, fibroblast growth factor receptor; PVDF, polyvinylidene difluoride; VEGFR, vascular endothelial growth factor; PTTH, prothoracicotropic hormone; CNS, central nervous system; NICD, Notch intracellular domain; STAT, signal transducer and activator of transcription; JNK, JUN kinase; TNF, tumor necrosis factor; USP, ubiquitin-specific protease.

(Gilbert 2000). We also do not discuss signaling pathways that control cellular behavior and cytoskeletal reorganization—for example, cell migration and axonal pathfinding.

2 EMBRYONIC PATTERNING: INTERPLAY BETWEEN TRANSCRIPTIONAL CASCADES AND SIGNALING

Following fertilization, as embryonic development proceeds, different cell types are formed progressively. With time, cells become more and more restricted in their developmental potential, and become determined to a specific fixed fate that represents a stable change in the internal state of the cell as a result of alterations in gene expression. The gradual increase in complexity occurring during determination and subsequent differentiation involves complex combinations of transcription factors. Some of these factors are common to many cell types, whereas others are present in only specific cell types. The changes in gene expression rely in part on the activation of signaling pathways by cell–cell communication. In the context of development, signaling pathways dictate developmental switches and as such are usually irreversible, pushing forward the developmental program in a ratchetlike mechanism by regulating the activity of transcription factors.

For example, patterning along the anteroposterior axis of the *Drosophila* embryo is initially set up by graded activity of the Bicoid transcription factor, which acts in a concentration-dependent manner to control the expression of gap genes (Fig. 1A) (St Johnston and Nusslein-Volhard 1992). These gap genes, in turn, coordinately define the domain of expression of pair-rule genes, which then define the expression of segment-polarity genes (Nusslein-Volhard and Wieschaus 1980). Although both gap and pair-rule genes encode diverse types of transcription factors, some of the segment-polarity genes, such as *hh* and *wg*, encode signaling molecules that activate pathways that operate in positive regulatory loops, to maintain each other's expression and the induced cell fates within the embryonic segmental unit (Heemskerk et al. 1991). In addition to the Bicoid patterning system, the Torso RTK pathway activates the Ras/ MAP kinase (MAPK) pathway to control the spatial expression of the Tailless and Huckebein transcription factors at the embryonic termini (Fig. 1C) (Duffy and Perrimon 1994). Finally, along the dorso–ventral axis of the embryo activation of the Toll receptor by the Spätzle ligand activates Dorsal (the fly homolog of NF-κB), which regulates the expression of Twist and Snail, two transcription factors that control mesoderm development, while repressing other genes, such as *rhomboid* and *sog* (Fig. 1B) (Levine 2008).

Hierarchies of transcription factor expression that progressively dictate distinct cell fates are common at later developmental stages in a variety of tissues. For example,

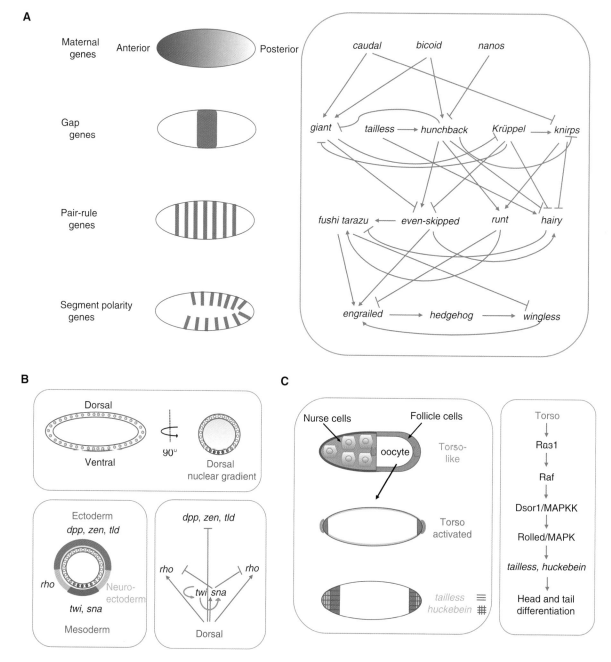

Figure 1. Patterning of the early *Drosophila* embryo. (*A*) Anterior–posterior patterning and segmentation of the embryo is initiated by maternally deposited gene products that regulate the expression of gap genes. Gap genes in turn control the expression of pair-rule genes, which themselves regulate segment-polarity genes. The gene hierarchy and activation/repression interactions between different transcription factors that coordinate patterning of the anterior–posterior axis of the early *Drosophila* embryo are shown to the right. (*B*) Dorsal–ventral patterning is initiated by a Dorsal nuclear gradient regulated by the Toll/NF-κB pathway. Graded nuclear localization of Dorsal subdivides the dorso–ventral axis into distinct domains expressing *twist* (*twi*) and *snail* (*sna*), *rhomboid* (*rho*) and *decapendaplegic* (*dpp*), *zerknullt* (*zen*) and *tolloid* (*tld*), which will form the prospective mesoderm, neurogenic ectoderm, and dorsal ectoderm, respectively. Dorsal activates the zygotic transcription program in the dorsoventral axis. (*C*) Terminal patterning is initiated in the germline by the localized expression of Torsolike in the space outside the poles of the embryo. Torsolike activates the Torso ligand (Trunk) locally and this is followed by Torso activation at the poles of the embryo, which will lead to induction of the terminal patterning genes *tailless* and *huckebein*. Torso is an RTK and its action is propagated through the MAPK pathway.

Cite this chapter as *Cold Spring Harb Perspect Biol* doi: 10.1101/cshperspect.a005975

in response to Dorsal signaling, Twist is activated to define the mesoderm and in turn activates MEF2 and Tinman in different cells to induce the skeletal muscle and cardiac muscle fates, respectively (Fig. 2A) (Sandmann et al. 2007). Another example of progressive specification owing to hierarchical expression of transcription factors and activity of signaling pathways is the specification of *Drosophila* blood cell types (Jung et al. 2005). In the *Drosophila* hemocyte (blood cell) lineage the blood cell precursors are specified in the embryo by expression of the transcription factors

Serpent (SRP) and Odd paired (ODD) and progressively express Hemese (HE) and activate the RTK PDGF/VEGF receptor (PVR), as well as the cytokine receptor Dome, to finally reach the prohemocyte stage. Then, cell-type-specific transcription factors are activated in response to signaling by the Notch, PVR or Notch and JAK/STAT pathways, which specify the different populations of mature hemocytes, namely, the plasmatocytes, the crystal cells, and lamellocytes, respectively, that are destined to perform specialized functions (Fig. 2B).

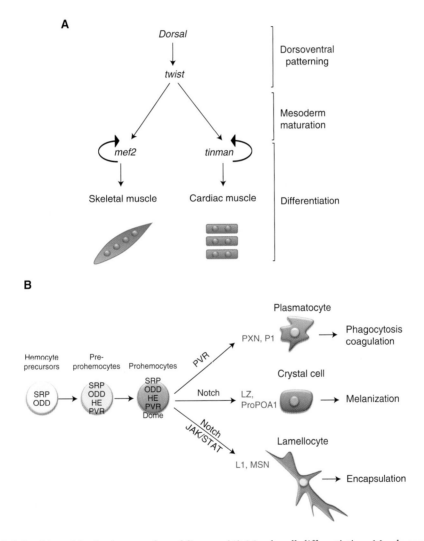

Figure 2. Cell-fate hierarchies in the mesodermal lineage. (*A*) Muscle cell differentiation. Muscle progenitors are specified in the embryonic mesoderm by Dorsal and activation of the transcription factor Twist. Further subdivision of Twist-positive cells to skeletal and cardiac muscle lineages depends on the expression of the transcription factors MEF2 and Tinman, respectively. (*B*) Hemocyte maturation in the *Drosophila* lymph gland. The earliest lymph gland cells, the hemocyte precursors, express SRP and ODD. As these cells transition into preprohemocyte fate, they initiate the expression of HE and PVR. Prohemocytes initiate Dome expression. Maturation to the various hemocyte fates requires down-regulation of Dome, up-regulation of different maturation markers, and the involvement of the indicated signaling pathways. Srp, Serpent; ODD, Odd Skipped; He, Hemese; PVR, PDGF/VEGF receptor; Dome, Domeless; PXN, Peroxidasin; P1, P1 antigen; LZ, Lozenge; ProPOA1, Prophenoloxidase A1; L1, L1 antigen; MSN, Misshapen.

3 JUXTACRINE SIGNALING: NOTCH AS AN EXAMPLE

The Notch signaling pathway is a highly conserved mechanism for cell communication between adjacent cells. Both the receptor Notch and its ligands, which belong to the Delta/Serrate/Lag2 (DSL) family, are transmembrane proteins. The requirement for direct cell–cell contact between the signal-sending and signal-receiving cells is necessitated by the membrane-anchored nature of the ligands. Interestingly, studies of the specification of sensory organs in the *Drosophila* thorax indicate that in some instances DSL ligands can activate Notch signaling beyond directly adjacent cells, because the signal-sending Delta cells extend filopodia that can reach cells a few cell diameters away (de Joussineau et al. 2003; Cohen et al. 2010).

Notch signaling is a simple linear pathway with no amplification step. Interaction of Notch receptors with DSL ligands presented by neighboring cells triggers two proteolytic cleavages within the receptor. The first one is extramembrane, executed by ADAM-family metalloproteases; this generates the substrate for the second cleavage, which is intramembrane, secretase-dependent (like amyloid generation), and releases the intracellular domain of Notch (NICD). NICD is subsequently transported to the nucleus and acts as a transcriptional coactivator that associates with a member of the CSL DNA-binding transcription factor family and turns on target gene expression. Among the targets of the NICD-CSL complex are the E(spl)/HES family genes, which are transcriptional repressors and account for many of the downstream effects of the pathway (reviews by Artavanis-Tsakonas et al. 1999; Bray 2006).

Notch signaling regulates a broad range of cellular processes in organisms ranging from sea urchins to humans, including cell-fate specification, formation of growth-organizing boundaries, stem cell maintenance, proliferation, apoptosis, and migration. Therefore, it is not surprising that its dysfunction has been implicated in many heritable developmental diseases, including Allagille and CADASIL syndromes, as well as cancer, where it promotes tumor growth in some contexts but can prevent it in others. How Notch signaling, especially considering the simplicity of the pathway, specifies so many different biological outcomes, depending on the cell context, is a major question in the field (reviews by Artavanis-Tsakonas et al. 1999; Lai 2004; Fortini 2009). Below we provide just a few examples of developmental processes regulated by different Notch modes of action: lateral inhibition, lineage decisions, and inductive signaling.

One of the best-characterized roles of Notch signaling is lateral inhibition, in which a specific cell fate is defined for a single cell within a group of equivalent cells (Fig. 3). For example, in the *Drosophila* embryonic neuroepithelium, equivalent ectodermal cells differentiate into either neuroblasts or epithelial cells through the action of Notch signaling. Initially, all neuroepithelial cells express low levels of both the Delta ligand and the Notch receptor. However, probably as the result of stochastic variations, some cells begin to express higher levels of Delta. These small differences are amplified through a positive-feedback loop that activates its transcription. Because the cells expressing high levels of the ligand cannot activate signaling because of *cis*-inhibitory interactions with the receptor (Heitzler and Simpson 1993), the system quickly resolves into Delta-expressing signal-sending cells and signal-receiving cells with low Delta levels that activate Notch signaling, which differentiate into neuronal and epithelial cells, respectively. This Notch-dependent lateral inhibition mechanism is used widely in development to pattern tissues containing initially identical cells. The same mechanism is used to select myoblast founder cells in the mesoderm (Bate and Rushton 1993; Rushton et al. 1995) and R8 photoreceptor fate from neural preclusters during eye development (review by Roignant and Treisman 2009). Another well-characterized example of lateral inhibition between two cells is the AC/VU (anchor cell/ventral uterine precursor cell) decision in *Caenorhabditis elegans* vulva, which is induced by activation of the Notch ortholog Lin12 that specifies the VU fate (reviews by Greenwald and Rubin 1992; Greenwald 1998).

Notch signaling also operates in control of lineage decisions and inductive signaling between nonequivalent cells. In these cases, the cells are initially distinct from each other either because they asymmetrically express regulators of the Notch pathway or because the ligand and receptor are differentially distributed in adjacent cells (review by Bray 2006). For instance, asymmetric segregation of Numb, which down-regulates Notch signaling through polarized receptor-mediated endocytosis, in the progeny of sensory organ precursors (SOPs) makes the Numb-positive cell Notch sending (Jan and Jan 1995). In contrast, during wing vein specification in *Drosophila*, expression of Delta in the vein regions induces Notch signaling in the intervein cells to inhibit vein fate, and patterning is established through a positive-feedback loop (Huppert et al. 1997). Often a combination of these mechanisms can account for the developmental outcomes. For example, in the *Drosophila* wing disc, both restricted expression of the glycosyltransferase Fringe, which increases the ability of Notch to bind to Delta, as well as restricted expression of ligands, lead to the specification of the wing margin (Panin et al. 1997).

As a rule of thumb, Notch represents a signaling modality that provides an on/off switch. How is this switch modulated and how is precise signaling ensured? First, multiple levels of regulation of both the receptor and

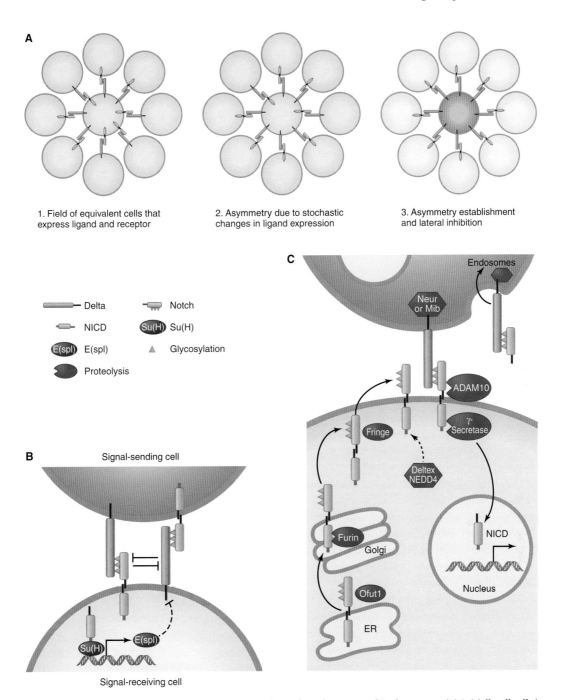

Figure 3. Lateral inhibition. (*A*) The process is progressive and can be separated in three steps: (1) Initially, all cells in the cluster express both Delta and Notch and are equivalent. (2) Stochastic changes in gene expression change the balance of ligand and receptor molecules, such that the cell in the middle expresses more Delta. (3) Asymmetry is established when Delta expression in the middle cell is stabilized through a positive-feedback loop, resulting in lateral inhibition whereby the Delta-expressing cell becomes the signal sender whereas its neighbors activate Notch signaling and adopt the receiving-cell fate. (*B*) During *Drosophila* neurogenesis the cell that activates Delta in the proneural cluster becomes a neuroblast, whereas its neighbors will be laterally inhibited and adopt an epidermal cell fate. The asymmetry between the neighboring cells is established by a negative-feedback loop that inhibits Delta in the signal-receiving cell mediated by repressors of the E(spl) complex. In addition, *cis*-inhibitory interactions between Notch and Delta exist and contribute to asymmetry generation and lateral inhibition. (*C*) The Notch receptor and its ligands are subject to a number of protein modifications, such as glycosylation (Ofut1 and Fringe), proteolysis (Furin, ADAM10, and γ-Secreatase), and ubiquitylation (Deltex plus NEDD₄). These events are critical for maturation of the receptor and its presentation on the cell membrane, for Notch activation on ligand binding, degradation, and trafficking of ligand-receptor complexes.

ligands are deployed. These include posttranslational modifications such as ubiquitylation that leads to proteasomal degradation, glycosylation, and phosphorylation, as well as trafficking into specific cellular compartments (Shilo and Schejter 2011). Second, when the pathway is used iteratively with a specific duration (e.g., *Drosophila* and vertebrate neurogenesis and vertebrate somitogenesis), then oscillatory activation/termination mechanisms are utilized. This is achieved not only because the NICD is a very short-lived transcription cofactor, but also because the pathway targets, the HES/E(spl) family, have very unstable messenger RNAs (mRNAs) and proteins, and exert autoinhibitory effects on their own transcription (reviewed by Fior and Henrique 2009).

4 PATTERNING BY SECRETED PARACRINE FACTORS

In the case of signaling pathways that are triggered by secreted ligands, a different set of rules applies. First and foremost, the range of signaling elicited by the ligand-producing cell can span tens of cell diameters. Different ligands have diverse distribution ranges, which are used in distinct contexts. The distribution of ligands over a distance of several cell diameters generates a graded signaling profile, which is used in many cases to generate several distinct responses, rather than a single on/off switch.

The observation that a single diffusible molecule could specify and pattern different cell fates in a concentration-dependent manner led to the concept of morphogen gradients (Turing 1952; Wolpert 1969; Meinhardt 1978). Morphogen gradients provide spatial information and generate different cell types in a distinct spatial order. The concentration gradient of the diffusing morphogen subdivides a field of cells by inducing or maintaining the expression of different target genes at distinct concentration thresholds. Accordingly, cells close to the morphogen source receive high levels of morphogen and express both low- and high-threshold target genes. Cells far from the source of the morphogen receive low levels of morphogen and express only low-threshold target genes. As a result, distinct cell types emerge.

The physical properties of the ligand, as well as its diffusion capacity, mode of transport, endocytosis, and interactions with heparan-sulfate proteoglycans (HSPGs), all affect the final distribution of the ligand and hence the resulting signaling profile. A clear hierarchy of ranges is evident in paracrine signaling in *Drosophila* tissues. In the case of RTK ligands, including Spitz (which activates the *Drosophila* EGF receptor), Branchless (which triggers the *Drosophila* FGF receptor Breathless), and the ligands that activate the *Drosophila* FGF receptor Heartless, the signaling range is restricted to a small number of cell diameters, typically two to eight. HH displays a similarly limited range. In contrast, the ligands for the BMP and Wnt/Wg pathways have a longer range, which can extend up to 30 cell diameters. Interestingly, as discussed below, lipid modifications of both HH and Wnts are important for distribution of these molecules in tissues. Below, we provide several examples illustrating the mechanisms underlying the regulation of ligand distribution for each of these major signaling pathways.

5 CONTROLLING THE SIGNALING RANGE OF SECRETED FACTORS

5.1 EGF and FGF

The range of EGFR signaling in *Drosophila* is regulated primarily by the amount of secreted ligand provided to the receiving cells. Three of the four EGFR ligands (Spitz, Keren, and Gurken) are produced as inactive transmembrane precursors that are sequestered in the endoplasmic reticulum (ER). The fourth ligand, Vein, is produced from the outset as a secreted molecule. Trafficking of ligand to a secretory compartment where processing takes place is facilitated by a transmembrane chaperone named Star (Lee et al. 2001; Tsruya et al. 2002). Within the secretory compartment, cleavage of the precursor is performed by intramembrane proteases of the Rhomboid family, and the cleaved extracellular ligand portion is subsequently secreted (Urban et al. 2001). The chaperone Star is also cleaved by Rhomboid proteins but this cleavage generates an inactive molecule (Tsruya et al. 2007). Some of the Rhomboid proteins localize not only to the secretory compartment but also to the ER. When Star encounters Rhomboid in the ER, it is inactivated before it can promote trafficking of the ligand precursors to the secretory compartment where ligand cleavage should take place (Yogev et al. 2008). Thus, only a fraction of the chaperone molecules escape inactivating cleavage in the ER, and hence the level of ligand precursor that is trafficked and secreted is significantly reduced. This leads to a corresponding reduction in the range of signaling. In tissues where a restricted range of EGFR activation is required, such as the eye disc or the germline, Rhomboid proteins are present in both the ER and secretory compartment.

Once ligand is secreted, another tier of regulation is used. High levels of EGFR activation induce the expression of the target gene *argos*, which encodes a secreted molecule that neutralizes the ligand (Golembo et al. 1996; Klein et al. 2004). Induction of Argos thus reduces the levels of active ligand that can diffuse from the source, and hence the range of signaling.

In responding cells, additional mechanisms restrict the signaling range, functioning in a cell-autonomous manner. Two inhibitor-encoding genes (*kekkon1* [Ghiglione et al.

Cite this chapter as *Cold Spring Harb Perspect Biol* doi: 10.1101/cshperspect.a005975

1999] and *sprouty* [Casci et al. 1999; Kramer et al. 1999; Reich et al. 1999]), in particular, are induced in a classical negative-feedback loop. In both cases, the induction of the inhibitors in a fairly broad range of the receiving cells results in productive signaling only in cells that are closer to the ligand source, and receive enough input to overcome the inhibitory effects. Kekkon1 encodes a transmembrane protein that generates inactive heterodimers with EGFR. Sprouty is an inhibitor of ERK/MAPK signaling whose mechanism of inhibition of RTK signaling remains incompletely understood. It interacts with several proteins impinging on signaling, including Grb2, Raf, Cbl, and PP2A, and undergoes phosphorylation that alters its binding properties and stability (Edwin et al. 2009; Reddi et al. 2010). Because Sprouty operates downstream from the receptor, by interacting with components common to multiple RTK pathways, it attenuates signaling by both FGF- and EGF-induced pathways (Hacohen et al. 1998). Both Kekkon1 and Sprouty are conserved in vertebrates. Sprouty, in particular, is an essential component that modulates RTK pathways in normal development and disease (Edwin et al. 2009).

The transcriptional output of RTK signaling is mediated by members of the ETS family of transcription factors. Most prominent is the ETS-domain protein Pointed, which has two isoforms generated by alternative splicing (Klambt 1993; O'Neill et al. 1994). ERK activates each of the two forms in a different manner. Phosphorylation of PointedP2 converts an inactive protein to the active form. The mechanistic basis for activation by phosphorylation is not known but may involve stabilization, nuclear translocation, and exposure of the transcriptional activation or DNA-binding domains. The second isoform, PointedP1, is constitutively active even in the absence of ERK signaling. However, its expression is dependent on ERK activity (Gabay et al. 1996). The transcription factor that responds to ERK activity to trigger PointedP1 expression is not known.

The YAN protein contains an ETS DNA-binding domain but is devoid of a transcriptional activation domain. YAN is also a target for ERK phosphorylation, but in this case phosphorylation leads to its inactivation by promoting nuclear exit and degradation (Rebay and Rubin 1995). The dual and opposite effects of ERK on the activators and inhibitor may make the induction of ETS-target genes more robust (Fig. 4).

An interesting variation occurs in the case of the Breathless FGF receptor. The receptor itself restricts diffusion of the ligand Branchless. The role of Breathless is to guide migration of tracheal cells toward the ligand source. To increase the sharpness of the attracting ligand gradient, expression of the receptor is induced by high levels of signaling, generating a trap that restricts the diffusion of the Branchless ligand (Oshiro et al. 2002).

5.2 Hedgehog

HH transmits information over several cell diameters, but its range is restricted. The distribution of HH has been studied most intensively in the wing imaginal disc, where it defines a zone of activation in the boundary between the posterior and anterior compartments of the disc. All posterior cells produce HH but do not respond to it, whereas the anterior cells do not produce it but can respond to it (review by Ingham and McMahon 2001). The range of HH diffusion from the posterior compartment determines the signaling range, and the region of the anterior compartment that receives HH subsequently becomes the domain that produces the BMP family ligand DPP, which directs long-range patterning of the wing.

HH is unusual as it undergoes dual lipid modification and autoproteolytic cleavage (Porter et al. 1996; Pepinsky et al. 1998; Chen et al. 2004). The cholesterol moiety that is added limits HH trafficking within and between cells and palmitoylation is required for the production of a soluble multimeric HH protein. Binding of HH to its receptor Patched (PTC) leads to its endocytosis and degradation. Because PTC functions by inhibiting the next step in the pathway (the transmembrane protein Smoothened [SMO]), this leads to pathway activation. Interestingly, *ptc* itself is a transcriptional target gene for HH signaling (reviewed in Wilson and Chuang 2010). As in the case of Branchless, this leads to more effective trapping of HH by the first rows of cells receiving the signal, and hence to a restriction of the signaling range.

Recently, studies in mammalian cells have shown that mammalian Hh signaling depends on the primary cilium, a small cellular projection found on most vertebrate cells (Goetz et al. 2009). In particular, Smo proteins participate in the transduction of Hh signals, moving into the cilium in response to Hh ligand. Interestingly, the absence of cilia in *Drosophila* suggests that a fundamental difference exists between the organization of the Hh pathway between invertebrates and mammals.

5.3 BMPs/TGFβ and Wnt/Wg Ligands

The BMP and Wnt/Wg family ligands act over a long range, especially in the wing disc, to pattern not only the cells close to the ligand source but also those positioned many cell diameters away. In these cases the regulation is more intricate as it involves shaping the distribution of the ligand over a long range, restricting signaling close to the source while facilitating signaling further away. This is important for maintaining the robustness of the resulting gradient to changes in the level of ligand produced (Eldar et al. 2003).

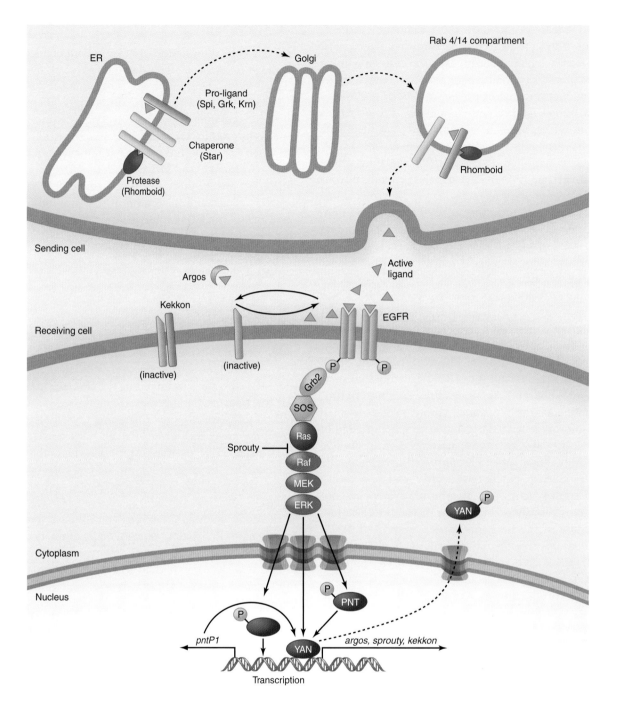

Figure 4. The EGFR pathway in *Drosophila*. Three membrane-anchored ligands—Spitz (SPI), Gurken (GRK), and Keren (KRN)—are retained in the ER, and are processed following trafficking by the chaperone protein Star, which is dedicated to these molecules, to the Rab4/14 compartment in the secretory pathway. In this compartment, the ligands encounter Rhomboid proteins, seven-transmembrane-span intramembrane serine proteases, which cleave the ligand precursors within the transmembrane domain, to release the active, secreted form. Rhomboids also reside in the ER cleave and inactivate Star, thus attenuating the level of ligand precursor that is trafficked to the Rab4/14 compartment. Within the receiving cells, the ligands encounter the EGF receptor, which on dimerization triggers the canonical SOS/Ras/Raf/MEK/MAPK pathway. The cardinal transcriptional output of the pathway is mediated by the ETS protein Pointed (PNT). In addition, the ETS protein YAN provides a constitutive repressor, which competes for Pointed binding sites, and can be removed from the nucleus and degraded upon phosphorylation by ERK. Several negative regulators keep the pathway in check. Especially important is a group of inducible repressive elements, which constitute a negative-feedback loop. Argos is a secreted molecule, which sequesters the ligand SPI, whereas Sprouty and Kekkon1 attenuate signaling within the receiving cell.

As in the case of Hh signaling, signaling by BMP regulates the expression level of the receptors; however, in this case the expression of the BMP receptor is inhibited by signaling (Lecuit et al. 1996). This generates a situation where less ligand trapping takes place close to the source, facilitating long-range diffusion. In addition, the elevated receptor levels further from the source make these cells more responsive to the low levels of ligand they encounter. Finally, the induction of inhibitors that block intracellular signaling, such as DAD (Tsuneizumi et al. 1997), which competes with the Smad proteins that transduce BMP signals, further restricts signaling close to the ligand source. The range of this response is dictated by the sensitivity of the promoter/enhancer of the inhibitory molecules to induction by signaling.

Another set of extracellular molecules that shape the distribution of BMPs and Wnts/Wg are HSPGs, which comprise a transmembrane protein core and long chains of sugars that emanate from this core (Perrimon and Bernfield 2000). The versatility of covalent links that can be formed between the sugar molecules has the potential to generate enormous complexity and hence specificity. Although the association between the ligands and HSPGs represents a low-affinity interaction, the sheer number of HSPGs may compensate for this low affinity. It is estimated that the number of HSPG molecules per cell is at least two orders of magnitude higher than that of the specific ligand receptors. Hence, the ligands travel in a "forest" of HSPGs, where they rarely encounter their specific receptors. HSPGs can have opposing effects on activation, and hence analysis of their function is complicated. They may facilitate signaling locally by trapping ligands and functioning as coreceptors that present the ligand to receptors such as FGFR. However, they may also facilitate signaling at a distance by either stabilizing the ligand or functioning as long-range carriers after cleavage of their extracellular protein stem. This latter activity also reduces the level of signaling close to the ligand source (reviewed in Yan and Lin 2009).

Wnt/Wg proteins, in addition to interacting with HSPGs, are modified by palmitoylation, a hydrophobic modification on a conserved cysteine residue that affects their distribution, as well as a second lipid modification by palmitoleic acid esterification of a serine residue. Studies in Drosophila and vertebrates have provided evidence that Wnt palmitoylation is controlled by Porcupine, predicted to be a membrane-bound O-acyl transferase, and that this modification is important for the generation of Wnt gradients. In Drosophila lack of palmitoylation in porcupine mutants affects WG secretion (Kadowaki et al. 1996). Similarly, in the chick neural tube porcupine-mediated lipid modification reduces the range of activity of Wnt1 and Wnt3a (Gali et al. 2007).

5.4 Long-Range Ligand Distribution

Studies of DPP and Wnt signaling in cell clones, in which the receptor is eliminated or a constitutively active receptor is expressed, have shown that the original signals are transmitted even to the most distant cells (Lecuit and Cohen 1996; Nellen and Basler 1996; Neumann and Cohen 1997), rather than being relayed by inducing secondary signals. Several models have been proposed for the long-range distribution of ligands, which may use multiple strategies. Although all of the proposed models are supported by several lines of compelling evidence, critical experiments directly eliminating one mode of trafficking and monitoring the outcome have not been performed for technical reasons. We therefore present the prevailing models below.

The simplest mechanism for ligand distribution is diffusion in the extracellular milieu. Reduction in ligand levels over a distance may be driven by endocytosis or extracellular degradation. HSPGs may enhance or reduce diffusion and keep the ligand in the plane of the epithelium by low-affinity interactions (Strigini and Cohen 2000). The association of some ligands with hydrophobic moieties, most notably Hh and cholesterol and palmitoylate, as well as Wnt/Wg proteins and palmitoylate, have raised the possibility of another mode of extracellular ligand trafficking, in which membrane fragments bearing these named argosomes are dispersed over large distances (Greco et al. 2001; Panáková et al. 2005). Exovesicles like these have been characterized in Drosophila imaginal discs. They contain Wnt/Wg proteins, are derived from basolateral membranes, and travel through tissues, where they are found predominantly in endosomes.

Another option is transcytosis of the ligand. In this scenario the ligand travels most of its journey in vesicles within cells. Its dilution over a distance is affected by the fraction of ligand that is endocytosed and by efficiency of ligand transfer between cells versus its intracellular degradation. Experiments have shown that elimination of the receptor in a group of cells adjacent to the ligand source reduces signaling in more distant cells that have a normal receptor. This suggests that the receptor is required in the proximal cells for incorporating the ligand into the cells and transferring it distally (reviewed in Wartlik and Gonzalez-Gaitan 2009).

One of the most provocative suggestions for the transfer of ligands over tens of cell diameters involves very thin cellular protrusions termed cytonemes. These have been proposed to serve as conduits in which morphogens move between producing and target cells (reviewed in Kornberg and Guha 2007). These structures have been identified in the epithelium of the wing disc and point toward cells that produce ligands. They contain the relevant ligand receptors, and their polarity is disrupted by uniform presentation of the respective ligand (Roy et al. 2011). Cytonemes have been proposed to serve to traffic the ligand to the cell body where signaling may take place. Interestingly, cytonemes are also found in vertebrate cells and thus may play a general role in long-range cell–cell communication.

6 THE LOGIC OF SIGNALING

Although pathways use distinct signaling components and signaling strategies, a number of common universal themes have emerged regarding their structures and regulation in time and space.

6.1 Linear Signaling Pathways

Developmental signaling elicited by ligand-receptor binding appears to be transmitted in a linear fashion within the cell, leading to induction of target genes. This is very different from typical signaling schemes in which multiple converging and diverging links are observed. The most compelling evidence for such linearity is that mutations in different components along a pathway give rise to very similar phenotypes (reviewed by Friedman and Perrimon 2007). This linearity of developmental signaling stems from the need to transmit a clear signal, in view of the irreversibility of the resulting decisions. This holds true both for cases where an on/off switch is induced and for situations where graded signaling elicits diverse responses, according to the level of signaling. Each pathway regulates the activity of one or more transcription factors, which bind to specific signaling pathway response elements in the enhancers and promoters of target genes (Barolo and Posakony 2002).

6.2 Negative-Feedback Switches

Tight regulation of signaling is essential for generation of reproducible patterns during development. In the case of pathways that function as switches that induce a particular cell fate within a zone of competent cells, this will determine the spatial boundaries of signaling. For ligands that function as morphogens to induce several distinct cell fates, this regulation will determine the overall spatial profile of resulting patterns. Another important consideration in signaling is the need to buffer against noise stemming from heterozygosity, unequal distribution of components between dividing cells and environmental fluctuations. Negative feedback provides a way of fine-tuning the signal over a range of signaling levels and sharpening boundaries between regions that respond differently.

Many examples of transcriptional induction of negative regulators exist. In some cases, these regulators function extracellularly to restrict the level or distribution of active ligand. For example, Noggin inhibits TGFβ signaling by binding to TGFβ family ligands and preventing them from binding to their receptors (Smith 1999). Similarly, the Argos molecule binds to EGFR ligands, thus effectively reducing their levels (Klein et al. 2004). In other cases, the inhibitor induced is acting only in the receiving cells. Examples of transmembrane molecules that compromise receptor activity include Kekkon1, which inhibits EGFR (Ghiglione et al. 1999). Inducible molecules that interfere with signaling intracellularly include Sprouty, which is a general repressor of ERK kinase signaling (Casci et al. 1999; Kramer et al. 1999; Reich et al. 1999), Dad, which competes with Smad proteins (Tsuneizumi et al. 1997), and axin, which negatively regulates Wnt/Wg signaling (Ikeda et al. 1998). For the cell-autonomous inhibitors, a relatively broad range of induction by signaling is required, such that only the cells that receive a signal above a certain threshold level experience productive signaling.

6.3 Generating a Threshold

In the case of Notch signaling, in which the ligand is membrane anchored, the boundaries of signaling are dictated by the contact zones between the sending and receiving cells. However, in cases where the ligand is diffusible, a graded signaling profile will be generated. Regardless of the range, this graded activation pattern is converted to sharp borders of induction of gene expression.

The underlying mechanisms for generating transcriptional thresholds are crucial for proper patterning. Several mechanisms have been identified. During early dorsoventral patterning in the *Drosophila* embryo, graded nuclear localization of the transcription factor Dorsal (an NF-κB homolog) is converted to sharp borders of zygotic gene expression (e.g., *twist* and *snail* in the ventralmost cells, which define the future mesoderm). The *snail* regulatory region contains multiple adjacent binding sites for Dorsal. Binding of one Dorsal molecule to DNA may facilitate the binding of additional molecules by protein–protein interactions, generating a sharper response (Rusch and Levine 1996).

In the case of BMP target genes in the wing disc, different stringencies of regulation may apply depending on the position of the responding cell relative to the ligand source. Of particular interest is the potential transcription factor Brinker (BRK), which negatively regulates DPP target genes in both the *Drosophila* wing disc and embryo. BRK antagonizes transcription of target genes, and forms a gradient that opposes the BMP activation gradient. Genes that are expressed closer to the ligand source require simultaneous suppression of *brk* expression and activation of transcription by Smads (along with binding of accessory transcription factors). For genes that are expressed in a broader pattern and hence require lower signaling levels for their induction, suppression of *brk* expression is sufficient (Affolter and Basler 2007).

Finally, a mechanism for generating transcriptional thresholds termed zero-order hypersensitivity has been proposed. In cases where a transcription factor, or transcriptional repressor, undergoes reversible phosphorylation and is in excess, even small differences in the rates of the reversible phosphorylation and dephosphorylation reactions will lead to the complete accumulation of the protein in one form or another. This generates a sharp threshold response (Melen et al. 2005).

7 INTEGRATING SIGNALING PATHWAYS

Many of the mechanisms underlying cell-type specification and formation of distinct tissues rely on interactions between signaling pathways. Often the activation of one pathway leads to activation of a second, the two pathways acting in a sequential or relay mode. In addition, two signaling pathways can act in parallel and converge to regulate the activity of the same target. Finally, cross talk can result from "pathway interference," in which one pathway modulates the activity of a canonical component of another.

A striking example of how the spatial and temporal interaction of signaling pathways can produce complex patterns during development is somitogenesis, the process that generates the spine through the periodic establishment of the embryonic segments from the paraxial mesoderm in vertebrates (Dequeant and Pourquie 2008). Somites are masses of mesoderm distributed along the two sides of the neural tube that give rise to the dermis, skeletal muscles, and vertebrae. During somitogenesis an oscillating mechanism, called the segmentation clock, drives pulses of expression of a limited number of genes repeatedly in the presomitic mesoderm (PSM) every time a new somite is formed. The first evidence that cyclic gene expression drives somitogenesis came from the observation that the HES1 (Hairy and Enhancer of split 1) mRNA is expressed in a dynamic cyclic pattern coinciding with the formation of each somite (Fig. 5A). Subsequently, several other genes with similar cyclic behavior were identified, the vast majority of which have been shown to be components of the Notch, FGFR, and Wnt signaling pathways (Fig. 5B). In particular, in the mouse, Notch-FGF-regulated genes oscillate out of phase with Wnt-regulated genes and their activation in the PSM is mutually exclusive. This suggests tight, coordinated regulation of signaling (Dequeant et al. 2006), which is achieved by a large number of negative-feedback loops (Fig. 5B) and the presence of a pacemaker that triggers the rhythmic coordinated activation of these signaling pathways.

FGF and Wnt signaling are regulated temporally and spatially. The ligands are expressed in gradients in the precursor tissue of the segments where they regulate the progressive maturation/differentiation of the tissue and define the domain of the clock activities. The Wnt pathway, for example, is activated in the PSM before segmentation, plays a role upstream of both the Wnt and Notch oscillations, and is thought to entrain the Notch feedback loop. As a result, the spatial and highly dynamic temporal regulations of these signaling activities guarantee the robust segmentation patterning of the vertebrate axis and are evolutionarily conserved in vertebrates (Dequeant and Pourquie 2008). In addition, experiments in zebrafish have indicated that the Notch pathway is required for synchrony of the oscillations at the cellular level and the coordinated expression of the correct targets within neighboring cells, because lack of Notch leads to a "salt and pepper" pattern of oscillations (Fig. 5A) (Dequeant and Pourquie 2008).

Determination of mesodermal progenitors in the *Drosophila* embryo (Carmena et al. 1998) exemplifies the complex interplay and integration of signaling pathways at the promoter level (Fig. 6). Using the regulation of the *even-skipped* (*eve*) promoter, Halfon et al. (2000) have illustrated how the synergistic integration of transcription factors, regulated by the Wnt/Wg, DPP/BMP, and EGF/FGF/ERK pathways, generates a specific developmental transcriptional response at a single defined enhancer. Because some of the pathways are activated earlier than others and in a broader domain, they determine the "competence group" of cells (expressing markers like Lethal of scute, L'sc) and lead to subsequent activation of additional pathways within a more restricted cell population (Fig. 6A). These later pathways are regarded as inductive, and it is the final integration of the transcriptional signals from all pathways, within a single enhancer, that induces the relevant target gene. In this system, the WG and DPP signals are orthogonal to each other and define the intersection zone as the competence group, through signal-responsive transcription factors (MAD and TCF) that induce two tissue-specific transcription factors (Tinman and Twist). In addition, TCF also contributes to the expression of essential elements for ERK signaling (i.e., Rhomboid, Heartless, and Heartbroken). Once activated, ERK provides the inductive signal, by activating the transcription factor Pointed and inactivating the YAN repressor (Fig. 6B). Finally, singling out of mesodermal Eve progenitors is achieved through the process of lateral inhibition mediated by Notch/Delta signaling (Carmena et al. 2002).

Such integration of signaling pathways at the promoter/enhancer level allows each gene to define its "rules" of regulation, according to the tissue setting in which it is activated. Two given pathways can act synergistically in one setting and antagonistically in another. Thus, whereas only a small number of signaling pathways are used during

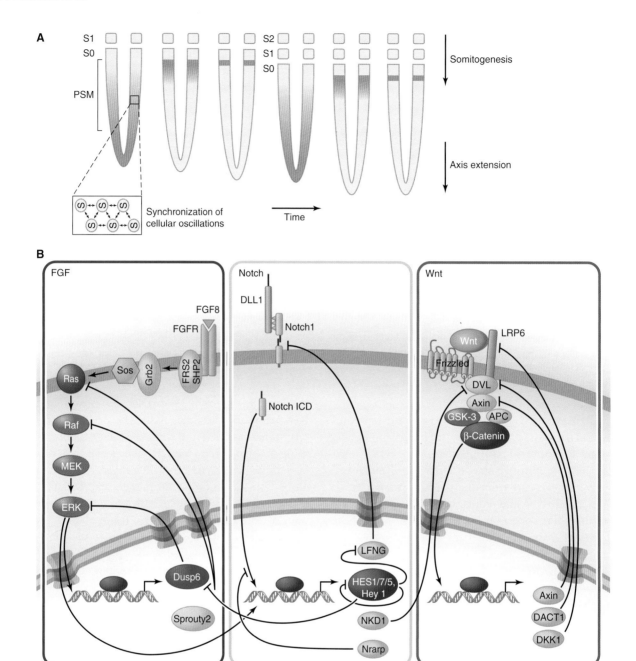

Figure 5. The segmentation clock oscillator. (*A*) Evidence of an oscillator underlying vertebrate segmentation comes from the transcriptional expression of the *hairy1* gene (dark green) in periodic waves in the presomitic mesoderm (PSM). These waves are associated with the timely formation of pairs of somites that are added sequentially. Experiments in zebrafish have shown that a remarkable property among neighboring PSM cells is that they undergo synchronized gene-expression oscillations (as shown in boxed area), which are coordinated by the Notch signaling pathway. (*B*) The FGF, Notch, and Wnt signaling pathways underlie the mouse oscillator. Cyclic genes belonging to the FGF (*left*) and Notch (*middle*) pathways oscillate in opposite phase to cyclic genes of the Wnt pathway (*right*). Several feedback loops are indicated. These are involved in reinforcing activity or shutting down a pathway. Some instances of pathway cross talk have also been observed. APC, adenomatous polyposis coli; DACT1, dapper homolog 1; DKK1, dickkopf homolog 1; FGFR, fibroblast growth factor receptor; Grb2, growth factor receptor bound protein; Dll1, Delta-like 1; DSH, dishevelled; DUSP6, dual specificity phosphatase 6; ERK, mitogen-activated protein kinase 1; GSK3, glycogen synthase kinase 3; HES, hairy enhancer of split-related; LFng, lunatic fringe; LPR6, low-density lipoprotein receptor-related protein 6; MEK, mitogen-activated protein kinase 1; NICD, Notch intracellular domain; NKD1, naked cuticle 1 homolog; Nrarp, Notch-regulated ankyrin repeat protein; SHP2, Src homology region 2-containing protein tyrosine phosphatase 2; SOS, son of sevenless.

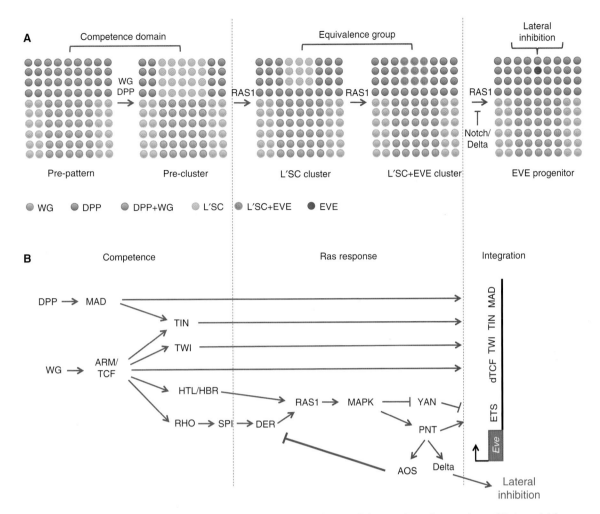

Figure 6. The WG/DPP/FGF interplay during specification of *Drosophila* mesodermal progenitors. (*A*) A model for patterning of the embryonic *Drosophila* mesoderm through the combinatorial actions of WG, DPP, and RAS/ERK signals. This model applies to both somatic muscle and pericardial muscle. The intersection between WG (red) and DPP (blue) delineates a prepattern (purple) in which Lethal of scute (L'SC) is initially activated in a precluster (orange). The entire L'SC precluster is competent to respond to RAS1. However, the spatially restricted activation of Heartless (HTL) and EGFR restricts L'SC to a subset of precluster cells that correspond to an equivalence group. RAS1 signaling activates EVE expression in all cells of the L'sc cluster (green) and subsequently a single EVE-expressing progenitor (red) is determined by lateral inhibition mediated by the Notch/Delta pathway. (*B*) WG, DPP, and RAS1 signal integration during specification of mesodermal EVE progenitors. WG and DPP provide developmental competence by regulating tissue-specific transcription factors (Tinman [TIN] and Twist [TWI]), signal-responsive transcription factors (MAD, TCF), and proximal components of the RTK/ERK pathway (FGFR/ HTL, Heartbroken [HBR]/DOF and Rhomboid [RHO]). The RAS pathway leads to activation of the ETS-binding transcription factor Pointed (PNT) and inactivation of the ETS-binding YAN repressor. The activities of all five transcriptional activators (TIN, TWI, MAD, TCF, and PNT) are integrated at the MHE (Muscle and Heart Enhancer) of *eve*, which is located 6 kb downstream from its transcription start site, and synergistically promote *eve* expression. In the absence of inductive RAS signaling, YAN represses *eve* by binding to ETS sites. In addition, RAS/ PNT signaling in the EVE progenitor promotes Delta and Argos (AOS) expression, which in combination activate Notch and shut down EGFR signaling in the nonprogenitor cells, to ensure lateral inhibition.

development, the combinatorial flexibility at the promoter level generates a vast array of possible responses. This would not be possible if more stringent and hardwired interactions between pathways operated more broadly at the cytoplasmic level.

Studies of cell-fate specification in the *Drosophila* eye have illustrated how two pathways, EGFR and Notch, can be utilized both sequentially and in parallel (Flores et al. 2000). Specifically, cone cell differentiation, visualized by the expression of the transcription factor PAX2, requires inputs from the EGFR and Notch pathways by neighboring R photoreceptor cells that produce both EGF and Delta ligands. In addition, the expression of the Delta ligand in R photoreceptor cells requires high levels of EGFR signaling (Flores et al. 2000; Tsuda et al. 2002; see reviews by Nagaraj and Banerjee 2004; Doroquez and Rebay 2006). Interestingly, differentiation of the R7 photoreceptor cell requires input from another RTK, Sevenless (SEV), activated by the transmembrane protein Bride of Sevenless (BOSS) (Perrimon and Perkins 1997).

Finally, a number of pathway interference mechanisms operating upstream of transcription have been documented. For example, studies in mammals and *Drosophila* have identified a mechanism by which the Hippo pathway coordinates Wnt/Wg morphogenetic signaling with growth control. Signaling via the Hippo pathway is critical for the precise control of organ size. Activation of the Hippo serine/threonine kinase leads to inhibition of the transcriptional co-activator TAZ and YAP (also known as Yorkie [YKI] in *Drosophila*) through phosphorylation and nuclear exclusion dependent on binding to 14-3-3 proteins. Varelas et al. (2010) showed that the Hippo pathway restricts Wnt/β-catenin signaling by promoting interactions between TAZ and Dishevelled, a cytoplasmic component of the canonical Wnt/Wg pathway. Similarly, Xia et al. (2010) have reported that the Fused (FU) serine/threonine kinase, a component of the canonical Hh pathway, functions together with the E3 ligase Smurf to regulate the ubiquitylation and subsequent degradation of Thick veins (TKV), a BMP receptor, during oogenesis. This mechanism ensures the generation of a steep gradient of BMP activity between the germline stem cell and its progeny. The degradation of TKV then permits expression of differentiation genes in the daughter cell.

Such examples may represent special cases rather than the norm in the context of developmental processes. Indeed, as stated above, during developmental signaling, cytoplasmic cross talk between pathways appears to be kept to a minimum to ensure that cells integrate signals quickly and effectively (Noselli and Perrimon 2000). An example of the simplicity more commonly seen is ERK regulation during *Drosophila* embryogenesis. When the active form of MAPK is monitored immunohistochemically, for every pattern that is observed, a single RTK has been shown to be responsible (Gabay et al. 1997). This reveals the striking absence of overlaps between RTK pathways in time and space.

8   CONCLUDING REMARKS: DEVELOPMENTAL VERSUS PHYSIOLOGICAL SIGNALING

In addition to their developmental roles, the signaling pathways discussed here play central roles in animal physiology. However, in contrast to their roles in development, where they act in ratchetlike mechanisms pushing forward developmental programs by regulating the activity of transcription factors, in physiological contexts the pathways are used to gauge the environment and fine-tune the physiological state of the cell, and as such are reversible.

ACKNOWLEDGMENTS

We thank Mary-Lee Dequeant for helpful comments. Work in the Perrimon laboratory is supported by the Howard Hughes Medical Institute and NIH, and work in the Shilo laboratory by a grant from the Minerva Foundation.

REFERENCES

*Reference is in this book.

Affolter M, Basler K. 2007. The decapentaplegic morphogen gradient: From pattern formation to growth regulation. *Nat Rev Genet* **8:** 663–674.

Arbouzova NI, Zeidler MP. 2006. JAK/STAT signalling in *Drosophila*: Insights into conserved regulatory and cellular functions. *Development* **133:** 2605–2616.

Artavanis-Tsakonas S, Rand MD, Lake RJ. 1999. Notch signaling: Cell fate control and signal integration in development. *Science* **284:** 770–776.

Barolo S, Posakony JW. 2002. Three habits of highly effective signaling pathways: Principles of transcriptional control by developmental cell signaling. *Genes Dev* **16:** 1167–1181.

Bate M, Rushton E. 1993. Myogenesis and muscle patterning in *Drosophila*. *C R Acad Sci III* **316:** 1047–1061.

Bray SJ. 2006. Notch signalling: A simple pathway becomes complex. *Nat Rev Mol Cell Biol* **7:** 678–689.

Carmena A, Gisselbrecht S, Harrison J, Jimenez F, Michelson AM. 1998. Combinatorial signaling codes for the progressive determination of cell fates in the *Drosophila* embryonic mesoderm. *Genes Dev* **12:** 3910–3922.

Carmena A, Buff E, Halfon MS, Gisselbrecht S, Jimenez F, Baylies MK, Michelson AM. 2002. Reciprocal regulatory interactions between the Notch and Ras signaling pathways in the *Drosophila* embryonic mesoderm. *Dev Biol* **244:** 226–242.

Casci T, Vinos J, Freeman M. 1999. Sprouty, an intracellular inhibitor of Ras signaling. *Cell* **96:** 655–665.

Chen MH, Li YJ, Kawakami T, Xu SM, Chuang PT. 2004. Palmitoylation is required for the production of a soluble multimeric Hedgehog protein complex and long-range signaling in vertebrates. *Genes Dev* **18:** 641–659.

Cohen M, Georgiou M, Stevenson NL, Miodownik M, Baum B. 2010. Dynamic filopodia transmit intermittent Delta-Notch signaling to drive pattern refinement during lateral inhibition. *Dev Cell* **19:** 78–89.

De Joussineau C, Soule J, Martin M, Anguille C, Montcourrier P, Alexandre D. 2003. Delta-promoted filopodia mediate long-range lateral inhibition in *Drosophila*. *Nature* **426:** 555–559.

Dequeant ML, Pourquie O. 2008. Segmental patterning of the vertebrate embryonic axis. *Nat Rev Genet* **9:** 370–382.

Dequeant ML, Glynn E, Gaudenz K, Wahl M, Chen J, Mushegian A, Pourquie O. 2006. A complex oscillating network of signaling genes underlies the mouse segmentation clock. *Science* **314:** 1595–1598.

Doroquez DB, Rebay I. 2006. Signal integration during development: Mechanisms of EGFR and Notch pathway function and cross-talk. *Crit Rev Biochem Mol Biol* **41:** 339–385.

Duffy JB, Perrimon N. 1994. The torso pathway in *Drosophila*: Lessons on receptor tyrosine kinase signaling and pattern formation. *Dev Biol* **166:** 380–395.

Edwin F, Anderson K, Ying C, Patel TB. 2009. Intermolecular interactions of Sprouty proteins and their implications in development and disease. *Mol Pharmacol* **76:** 679–691.

Eldar A, Rosin D, Shilo BZ, Barkai N. 2003. Self-enhanced ligand degradation underlies robustness of morphogen gradients. *Dev Cell* **5:** 635–646.

Feng XH, Derynck R. 2005. Specificity and versatility in tgf-β signaling through Smads. *Annu Rev Cell Dev Biol* **21:** 659–693.

Fior R, Henrique D. 2009. "Notch-Off": A perspective on the termination of Notch signalling. *Int J Dev Biol* **53:** 1379–1384.

Flores GV, Duan H, Yan H, Nagaraj R, Fu W, Zou Y, Noll M, Banerjee U. 2000. Combinatorial signaling in the specification of unique cell fates. *Cell* **103:** 75–85.

Fortini ME. 2009. Notch signaling: The core pathway and its posttranslational regulation. *Dev Cell* **16:** 633–647.

Freeman M. 2000. Feedback control of intercellular signalling in development. *Nature* **408:** 313–319.

Friedman A, Perrimon N. 2007. Genetic screening for signal transduction in the era of network biology. *Cell* **128:** 225–231.

Gabay L, Scholz H, Golembo M, Klaes A, Shilo BZ, Klambt C. 1996. EGF receptor signaling induces pointed P1 transcription and inactivates Yan protein in the *Drosophila* embryonic ventral ectoderm. *Development* **122:** 3355–3362.

Gabay L, Seger R, Shilo BZ. 1997. MAP kinase in situ activation atlas during *Drosophila* embryogenesis. *Development* **124:** 3535–3541.

Galli LM, Barnes TL, Secrest SS, Kadowaki T, Burrus LW. 2007. Porcupine-mediated lipid-modification regulates the activity and distribution of Wnt proteins in the chick neural tube. *Development* **134:** 3339–3348.

Gerhart J. 1999. 1998 Warkany lecture: Signaling pathways in development. *Teratology* **60:** 226–239.

Ghabrial A, Luschnig S, Metzstein MM, Krasnow MA. 2003. Branching morphogenesis of the *Drosophila* tracheal system. *Annu Rev Cell Dev Biol* **19:** 623–647.

Ghiglione C, Carraway KLIII, Amundadottir LT, Boswell RE, Perrimon N, Duffy JB. 1999. The transmembrane molecule kekkon 1 acts in a feedback loop to negatively regulate the activity of the *Drosophila* EGF receptor during oogenesis. *Cell* **96:** 847–856.

Gilbert SF. 2000. *Developmental biology*, 6th ed. Sinauer Associates, Sunderland, MA.

Goetz SC, Ocbina PJ, Anderson KV. 2009. The primary cilium as a Hedgehog signal transduction machine. *Methods Cell Biol* **94:** 199–222.

Golembo M, Raz E, Shilo BZ. 1996. The *Drosophila* embryonic midline is the site of Spitz processing, and induces activation of the EGF receptor in the ventral ectoderm. *Development* **122:** 3363–3370.

Greco V, Hannus M, Eaton S. 2001. Argosomes: A potential vehicle for the spread of morphogens through epithelia. *Cell* **106:** 633–645.

Greenwald I. 1998. LIN-12/Notch signaling: Lessons from worms and flies. *Genes Dev* **12:** 1751–1762.

Greenwald I, Rubin GM. 1992. Making a difference: The role of cell-cell interactions in establishing separate identities for equivalent cells. *Cell* **68:** 271–281.

Hacohen N, Kramer S, Sutherland D, Hiromi Y, Krasnow MA. 1998. Sprouty encodes a novel antagonist of FGF signaling that patterns apical branching of the *Drosophila* airways. *Cell* **92:** 253–263.

Halfon MS, Carmena A, Gisselbrecht S, Sackerson CM, Jimenez F, Baylies MK, Michelson AM. 2000. Ras pathway specificity is determined by the integration of multiple signal-activated and tissue-restricted transcription factors. *Cell* **103:** 63–74.

* Harrison DA. 2012. The JAK/STAT pathway. *Cold Spring Harb Perspect Biol* **4:** a011205.

* Harvey KF, Hariharan IK. 2012. The Hippo pathway. *Cold Spring Harb Perspect Biol* **4:** a011288.

Heemskerk J, DiNardo S, Kostriken R, O'Farrell PH. 1991. Multiple modes of engrailed regulation in the progression towards cell fate determination. *Nature* **352:** 404–410.

Heitzler P, Simpson P. 1993. Altered epidermal growth factor-like sequences provide evidence for a role of Notch as a receptor in cell fate decisions. *Development* **117:** 1113–1123.

Hou SX, Zheng Z, Chen X, Perrimon N. 2002. The Jak/STAT pathway in model organisms: Emerging roles in cell movement. *Dev Cell* **3:** 765–778.

Huppert SS, Jacobsen TL, Muskavitch MA. 1997. Feedback regulation is central to Δ-Notch signalling required for *Drosophila* wing vein morphogenesis. *Development* **124:** 3283–3291.

Ikeda S, Kishida S, Yamamoto H, Murai H, Koyama S, Kikuchi A. 1998. Axin, a negative regulator of the Wnt signaling pathway, forms a complex with GSK-3β and β-catenin and promotes GSK-3β-dependent phosphorylation of β-catenin. *EMBO J* **17:** 1371–1384.

* Ingham PW. 2012. Hedgehog signaling. *Cold Spring Harb Perspect Biol* **4:** a011221.

Ingham PW, McMahon AP. 2001. Hedgehog signaling in animal development: Paradigms and principles. *Genes Dev* **15:** 3059–3087.

Jan YN, Jan LY. 1995. Maggot's hair and bug's eye: Role of cell interactions and intrinsic factors in cell fate specification. *Neuron* **14:** 1–5.

Jiang J, Hui CC. 2008. Hedgehog signaling in development and cancer. *Dev Cell* **15:** 801–812.

Jung SH, Evans CJ, Uemura C, Banerjee U. 2005. The *Drosophila* lymph gland as a developmental model of hematopoiesis. *Development* **132:** 2521–2533.

Kadowaki T, Wilder E, Klingensmith J, Zachary K, Perrimon N. 1996. The segment polarity gene porcupine encodes a putative multitransmembrane protein involved in Wingless processing. *Genes Dev* **10:** 3116–3128.

Klambt C. 1993. The *Drosophila* gene pointed encodes two ETS-like proteins which are involved in the development of the midline glial cells. *Development* **117:** 163–176.

Klein DE, Nappi VM, Reeves GT, Shvartsman SY, Lemmon MA. 2004. Argos inhibits epidermal growth factor receptor signalling by ligand sequestration. *Nature* **430:** 1040–1044.

* Kopan R. 2012. Notch signaling. *Cold Spring Harb Perspect Biol* **4:** a011213.

Kornberg TB, Guha A. 2007. Understanding morphogen gradients: A problem of dispersion and containment. *Curr Opin Genet Dev* **17:** 264–271.

Kramer S, Okabe M, Hacohen N, Krasnow MA, Hiromi Y. 1999. Sprouty: A common antagonist of FGF and EGF signaling pathways in *Drosophila*. *Development* **126:** 2515–2525.

Lai EC. 2004. Notch signaling: Control of cell communication and cell fate. *Development* **131:** 965–973.

Lecuit T, Brook WJ, Ng M, Calleja M, Sun H, Cohen SM. 1996. Two distinct mechanisms for long-range patterning by Decapentaplegic in the *Drosophila* wing. *Nature* **381:** 387–393.

Lee JR, Urban S, Garvey CF, Freeman M. 2001. Regulated intracellular ligand transport and proteolysis control EGF signal activation in *Drosophila*. *Cell* **107:** 161–171.

Levine M. 2008. Dorsal-ventral patterning of the *Drosophila* embryo. In *The legacy of Drosophila genetics: From "defining the gene" to "analyzing genome function"* (ed. Bier E). Henry Stewart Talks, London.

Logan CY, Nusse R. 2004. The Wnt signaling pathway in development and disease. *Annu Rev Cell Dev Biol* **20:** 781–810.

MacDonald BT, Tamai K, He X. 2009. Wnt/β-catenin signaling: Components, mechanisms, and diseases. *Dev Cell* **17:** 9–26.

Meinhardt H. 1978. Space-dependent cell determination under the control of morphogen gradient. *J Theor Biol* **74:** 307–321.

Melen GJ, Levy S, Barkai N, Shilo BZ. 2005. Threshold responses to morphogen gradients by zero-order ultrasensitivity. *Mol Syst Biol* **1:** 2005.0028.

Nagaraj R, Banerjee U. 2004. The little R cell that could. *Int J Dev Biol* **48:** 755–760.

Nellen D, Burke R, Struhl G, Basler K. 1996. Direct and long-range action of a DPP morphogen gradient. *Cell* **85:** 357–368.

Neumann CJ, Cohen SM. 1997. Long-range action of Wingless organizes the dorsal-ventral axis of the *Drosophila* wing. *Development* **124:** 871–880.

Noselli S, Perrimon N. 2000. Signal transduction. Are there close encounters between signaling pathways? *Science* **290:** 68–69.

* Nusse R. 2012. Wnt signaling. *Cold Spring Harb Perspect Biol* **4:** a011163.

Nusslein-Volhard C, Wieschaus E. 1980. Mutations affecting segment number and polarity in *Drosophila*. *Nature* **287:** 795–801.

O'Neill EM, Rebay I, Tjian R, Rubin GM. 1994. The activities of two Ets-related transcription factors required for *Drosophila* eye development are modulated by the Ras/MAPK pathway. *Cell* **78:** 137–147.

Ohshiro T, Emori Y, Saigo K. 2002. Ligand-dependent activation of breathless FGF receptor gene in *Drosophila* developing trachea. *Mech Dev* **114:** 3–11.

Pan D. 2007. Hippo signaling in organ size control. *Genes Dev* **21:** 886–897.

Panakova D, Sprong H, Marois E, Thiele C, Eaton S. 2005. Lipoprotein particles are required for Hedgehog and Wingless signalling. *Nature* **435:** 58–65.

Panin VM, Papayannopoulos V, Wilson R, Irvine KD. 1997. Fringe modulates Notch-ligand interactions. *Nature* **387:** 908–912.

Pepinsky RB, Zeng C, Wen D, Rayhorn P, Baker DP, Williams KP, Bixler SA, Ambrose CM, Garber EA, Miatkowski K, et al. 1998. Identification of a palmitic acid-modified form of human Sonic hedgehog. *J Biol Chem* **273:** 14037–14045.

Perrimon N, Bernfield M. 2000. Specificities of heparan sulphate proteoglycans in developmental processes. *Nature* **404:** 725–728.

Perrimon N, McMahon AP. 1999. Negative feedback mechanisms and their roles during pattern formation. *Cell* **97:** 13–16.

Perrimon N, Perkins L. 1997. There must be 50 ways to rule the signal: The case of the *Drosophila* EGF receptor. *Cell* **89:** 13 16.

Porter JA, Ekker SC, Park WJ, von Kessler DP, Young KE, Chen CH, Ma Y, Woods AS, Cotter RJ, Koonin EV, et al. 1996. Hedgehog patterning activity: Role of a lipophilic modification mediated by the carboxy-terminal autoprocessing domain. *Cell* **86:** 21–34.

Pownall ME, Isaacs HV. 2010. *FGF signalling in vertebrate development.* Morgan & Claypool Life Sciences, San Rafael, CA.

Rebay I, Rubin GM. 1995. Yan functions as a general inhibitor of differentiation and is negatively regulated by activation of the Ras1/MAPK pathway. *Cell* **81:** 857–866.

Reddi HV, Madde P, Marlow LA, Copland JA, McIver B, Grebe SK, Eberhardt NL. 2010. Expression of the PAX8/PPARγ fusion protein is associated with decreased neovascularization in vivo: Impact on tumorigenesis and disease prognosis. *Genes Cancer* **1:** 480–492.

Reich A, Sapir A, Shilo B. 1999. Sprouty is a general inhibitor of receptor tyrosine kinase signaling. *Development* **126:** 4139–4147.

Roignant JY, Treisman JE. 2009. Pattern formation in the *Drosophila* eye disc. *Int J Dev Biol* **53:** 795–804.

Roy S, Hsiung F, Kornberg TB. 2011. Specificity of *Drosophila* cytonemes for distinct signaling pathways. *Science* **332:** 354–358.

Rusch J, Levine M. 1996. Threshold responses to the dorsal regulatory gradient and the subdivision of primary tissue territories in the *Drosophila* embryo. *Curr Opin Genet Dev* **6:** 416–423.

Rushton E, Drysdale R, Abmayr SM, Michelson AM, Bate M. 1995. Mutations in a novel gene, myoblast city, provide evidence in support of the founder cell hypothesis for *Drosophila* muscle development. *Development* **121:** 1979–1988.

Sandmann T, Girardot C, Brehme M, Tongprasit W, Stolc V, Furlong EE. 2007. A core transcriptional network for early mesoderm development in *Drosophila melanogaster*. *Genes Dev* **21:** 436–449.

Saucedo LJ, Edgar BA. 2007. Filling out the Hippo pathway. *Nat Rev Mol Cell Biol* **8:** 613–621.

Shilo BZ. 2005. Regulating the dynamics of EGF receptor signaling in space and time. *Development* **132:** 4017–4027.

Shilo BZ, Schejter ED. 2011. Regulation of developmental intercellular signalling by intracellular trafficking. *EMBO J* **30:** 3516–3526.

Smith WC. 1999. TGFβ inhibitors. New and unexpected requirements in vertebrate development. *Trends Genet* **15:** 3–5.

Spemann H, Mangold H. 1924. Über induktion von Embryonalagen durch Implantation Artfremder Organisatoren. *Roux' Arch Entw Mech* **100:** 599–638.

St Johnston D, Nusslein-Volhard C. 1992. The origin of pattern and polarity in the *Drosophila* embryo. *Cell* **68:** 201–219.

Strigini M, Cohen SM. 2000. Wingless gradient formation in the *Drosophila* wing. *Curr Biol* **10:** 293–300.

Tsruya R, Schlesinger A, Reich A, Gabay L, Sapir A, Shilo BZ. 2002. Intracellular trafficking by Star regulates cleavage of the *Drosophila* EGF receptor ligand Spitz. *Genes Dev* **16:** 222–234.

Tsruya R, Wojtalla A, Carmon S, Yogev S, Reich A, Bibi E, Merdes G, Schejter E, Shilo BZ. 2007. Rhomboid cleaves Star to regulate the levels of secreted Spitz. *EMBO J* **26:** 1211–1220.

Tsuda L, Nagaraj R, Zipursky SL, Banerjee U. 2002. An EGFR/Ebi/Sno pathway promotes δ expression by inactivating Su(H)/SMRTER repression during inductive notch signaling. *Cell* **110:** 625–637.

Tsuneizumi K, Nakayama T, Kamoshida Y, Kornberg TB, Christian JL, Tabata T. 1997. Daughters against dpp modulates dpp organizing activity in *Drosophila* wing development. *Nature* **389:** 627–631.

Turing AM. 1952. The chemical basis of morphogenesis. *Philos Trans R Soc London B* **237:** 37–72.

Urban S, Lee JR, Freeman M. 2001. *Drosophila* rhomboid-1 defines a family of putative intramembrane serine proteases. *Cell* **107:** 173–182.

Varelas X, Miller BW, Sopko R, Song S, Gregorieff A, Fellouse FA, Sakuma R, Pawson T, Hunziker W, McNeill H, et al. 2010. The Hippo pathway regulates Wnt/β-catenin signaling. *Dev Cell* **18:** 579–591.

Waddington CH. 1940. *Organizers and genes.* Cambridge University Press, Cambridge.

Wartlick O, Kicheva A, Gonzalez-Gaitan M. 2009. Morphogen gradient formation. *Cold Spring Harb Perspect Biol* **1:** a001255.

Wilson CW, Chuang PT. 2010. Mechanism and evolution of cytosolic Hedgehog signal transduction. *Development* **137:** 2079–2094.

Wolpert L. 1969. Positional information and the spatial pattern of cellular differentiation. *J Theor Biol* **25:** 1–47.

* Wrana JL. 2013. Signaling by the TGFβ superfamily. *Cold Spring Harb Perspect Biol* **5:** a011197.

Wu MY, Hill CS. 2009. Tgf-β superfamily signaling in embryonic development and homeostasis. *Dev Cell* **16:** 329–343.

Xia L, Jia S, Huang S, Wang H, Zhu Y, Mu Y, Kan L, Zheng W, Wu D, Li X, et al. 2010. The Fused/Smurf complex controls the fate of *Drosophila* germline stem cells by generating a gradient BMP response. *Cell* **143:** 978–990.

Yan D, Lin X. 2009. Shaping morphogen gradients by proteoglycans. *Cold Spring Harb Perspect Biol* **1:** a002493.

Yogev S, Schejter ED, Shilo BZ. 2008. *Drosophila* EGFR signalling is modulated by differential compartmentalization of Rhomboid intramembrane proteases. *EMBO J* **27:** 1219–1230.

CHAPTER 11

Signaling by Sensory Receptors

David Julius[1] and Jeremy Nathans[2]

[1]Department of Physiology, University of California School of Medicine, San Francisco, California 94158
[2]Department of Molecular Biology and Genetics, Johns Hopkins Medical School, Baltimore, Maryland 21205

Correspondence: David.Julius@ucsf.edu and jnathans@jhmi.edu

SUMMARY

Sensory systems detect small molecules, mechanical perturbations, or radiation via the activation of receptor proteins and downstream signaling cascades in specialized sensory cells. In vertebrates, the two principal categories of sensory receptors are ion channels, which mediate mechanosensation, thermosensation, and acid and salt taste; and G-protein-coupled receptors (GPCRs), which mediate vision, olfaction, and sweet, bitter, and umami tastes. GPCR-based signaling in rods and cones illustrates the fundamental principles of rapid activation and inactivation, signal amplification, and gain control. Channel-based sensory systems illustrate the integration of diverse modulatory signals at the receptor, as seen in the thermosensory/pain system, and the rapid response kinetics that are possible with direct mechanical gating of a channel. Comparisons of sensory receptor gene sequences reveal numerous examples in which gene duplication and sequence divergence have created novel sensory specificities. This is the evolutionary basis for the observed diversity in temperature- and ligand-dependent gating among thermosensory channels, spectral tuning among visual pigments, and odorant binding among olfactory receptors. The coding of complex external stimuli by a limited number of sensory receptor types has led to the evolution of modality-specific and species-specific patterns of retention or loss of sensory information, a filtering operation that selectively emphasizes features in the stimulus that enhance survival in a particular ecological niche. The many specialized anatomic structures, such as the eye and ear, that house primary sensory neurons further enhance the detection of relevant stimuli.

Outline

1 INTRODUCTION

An organism's perception of the world is filtered through its sensory systems. The properties of these systems dictate the types of stimuli that can be detected and constrain the ways in which these stimuli are reconstructed, integrated, and interpreted. Here we discuss how sensory signals are received and transduced, focusing on the first steps in the complex process of perceiving an external stimulus. A recurrent theme is the way in which the biochemical and biophysical properties of sensory receptor molecules, and the neurons in which they reside, have been sculpted by evolution to capture those signals that are most salient for the survival and reproduction of the organism. As a result, some classes of sensory receptors, such as the night vision receptor rhodopsin, show great conservation, whereas others, such as olfactory receptors, show great diversity.

Evolutionary comparisons are fascinating at many levels, not least of which is their power to highlight the logic of the stimulus–response relationship. For example, honeybees can see UV light, enabling them to locate sources of nectar and pollen based on the UV reflectance of flower petals (Kevan et al. 2001), whereas humans and Old World primates have excellent sensitivity and chromatic discrimination at longer wavelengths, permitting the identification of red, orange, and yellow fruit against a background of green foliage (Mollon 1989). Star-nosed moles use a specialized mechanoreceptive organ on their snout to locate meals and navigate through lightless subterranean tunnels (Catania 2005), and pit vipers have evolved thermoreceptive organs to detect the infrared radiation emitted by their warm-blooded prey (Campbell et al. 2002). In each of these cases, evolution has fine-tuned a sensory organ through anatomical and/or molecular changes to enhance the detection of relevant stimuli.

For simplicity, we focus on eukaryotic sensory systems, in which G-protein-coupled receptors (GPCRs) and ion channels predominate as sensory receptors. The one exception is mechanosensation, in which the molecular basis of membrane stretch detection has been beautifully delineated in bacteria but remains less clear in eukaryotes. Thus, our discussion of mechanosensation is focused largely on prokaryotic systems. We also describe the diverse cast of downstream transduction pathways and the manner in which receptors and transduction pathways are regulated to terminate signaling and set receptor sensitivity.

Before discussing individual receptors, it is worth noting that the physiological attributes of sensory systems are dictated not only by the molecular properties of receptor molecules and their associated signal transduction proteins, but also by the architecture of the sensory organs, cells, and subcellular structures in which they reside. A notable example is the retina, where visual pigments, together with other components of the phototransduction pathway, are localized within outer segments of rod and cone photoreceptor cells at near millimolar concentrations, thereby enhancing light sensitivity and transduction efficiency (Yau and Hardie 2009). Another example is seen in the cochlea, where sound pressure waves are transmitted through the middle ear to induce a localized and frequency-dependent distortion of the basilar membrane in the cochlea. Progressive changes in both the mechanical properties of the basilar membrane and the electrical properties of the auditory hair cells along the length of the cochlea generate a tonotopic map in which the amplitudes of different frequency components in a complex sound are reflected in the magnitudes of auditory receptor activation at different locations within the cochlea (Roberts et al. 1988).

2 RECEPTORS: DETECTION AND TRANSDUCTION

Both GPCRs and ion channels contribute to sensory transduction pathways by initiating or modulating stimulus-evoked responses (Fig. 1; Table 1). In vertebrates, GPCRs predominate as stimulus detectors in vertebrate visual and olfactory receptor cells. In contrast, recent studies suggest that fly olfactory neurons use ionotropic glutamate receptor-like channels to detect some classes of odorants (Benton et al. 2009), revealing a striking divergence of signal transduction mechanisms between insect and vertebrate chemosensory systems. Ion channels predominate in the detection of auditory and somatosensory stimuli, and both GPCRs and ion channels serve as stimulus detectors in the gustatory (taste) system.

Because GPCRs transduce information through multicomponent second-messenger-based "metabotropic" signaling pathways (Henrik-Heldin et al. 2012), they endow physiological systems with a tremendous capacity for signal amplification (Fig. 2). In vertebrate chemosensory and visual systems, GPCR-based signaling enables sensory cells to detect nanomolar concentrations of ligands or single photons, respectively. Such high sensitivity is possible because a single GPCR, during its active lifetime, can activate dozens to hundreds of G proteins, and each activated G protein in conjunction with its associated target enzyme can synthesize or destroy thousands of second-messenger molecules. This type of signaling cascade has been most thoroughly analyzed in vertebrate rod photoreceptors, where light-evoked activation of a single rhodopsin molecule leads to the activation of about 500 downstream effector proteins [the G protein transducin and its associated cyclic (c)GMP phosphodiesterase], with the consequent hydrolysis of about 100,000 molecules of cGMP per second (Fig. 1A) (Stryer 1986; Arshavsky et al. 2002).

Cite this chapter as *Cold Spring Harb Perspect Biol* doi: 10.1101/cshperspect.a005991

The reduction in cytosolic cGMP leads to the closure of cGMP-gated ion channels in the outer segment plasma membrane, thereby hyperpolarizing the cell and decreasing the release of the neurotransmitter glutamate onto bipolar cells, the second-order neurons within the retina.

A similar biochemical logic governs signaling in vertebrate olfactory sensory neurons, where activation of G-protein-coupled odorant receptors increases the synthesis of cAMP, which binds directly to and thereby opens cyclic-nucleotide-gated ion channels in the plasma membrane of olfactory cilia (Fig. 1B). The resulting depolarization of the plasma membrane initiates an action potential that is transmitted from the sensory neuron's body in the olfactory epithelium to its presynaptic terminal in the olfactory bulb. Thus, in both visual and olfactory sensory neurons, temporal control of signaling comes down to a balance between cyclic nucleotide synthesis and degradation. Interestingly, the detection of tastants and pheromones by GPCR-containing gustatory and vomeronasal sensory neurons, respectively, proceeds through a somewhat different signaling pathway involving G-protein-mediated activation of phospholipase C, which promotes hydrolysis of membrane phospholipids to generate second messengers (such as diacylglycerols, inositol phosphates, and polyunsaturated fatty acids) and thereby triggers calcium release from intracellular stores (Fig. 1C). These actions promote the opening of excitatory TRP ion channels, leading to depolarization and neurotransmitter release (Chandrashekar et al. 2006).

The same principles that underlie signal amplification—namely, the involvement of multiple sequential steps in a transduction pathway—endow metabotropic (i.e., GPCR) systems with a great capacity for adaptation and other forms of signal modulation. Here, again, the most detailed analysis has been performed in the visual system. The retina faces the daunting task of discriminating luminance changes on a timescale of tens of milliseconds and under conditions as disparate as sunny afternoons and moonless nights, in which background light intensity varies by more than six orders of magnitude. These challenges are met, in part, through negative-feedback mechanisms in the photoreceptor outer segment that terminate signaling and/or reset the baseline in the presence of a persistent stimulus. Although first delineated in studies of rod phototransduction, these feedback mechanisms are now known to be more-or-less generic to many GPCR signaling pathways (DeWire et al. 2007; Moore et al. 2007). The most receptor-proximal signal termination mechanism involves phosphorylation of activated receptors by specific serine/threonine kinases, such as rhodopsin kinase, a member of the G-protein receptor kinase (GRK) family, at several residues in the long carboxy-terminal tail of the receptor on a timescale of tens of milliseconds (Arshavsky et al. 2002). Once phosphorylated, the receptor is capped by the inhibitory protein arrestin, which blocks subsequent G-protein activation. In hormone and neurotransmitter receptor systems, arrestin binding also facilitates receptor endocytosis and recycling. In contrast, in the photoreceptor outer segment, rhodopsin remains stably localized to the disc membrane. Rhodopsin is recycled to its dark state by the combination of dephosphorylation and exchange of the photoisomerized all-*trans* retinal chromophore for a new molecule of 11-*cis* retinal, reactions that occur on a timescale of minutes (Arshavsky et al. 2002; Yau and Hardie 2009).

The time course of photoreceptor signal termination is also shaped by the accelerated hydrolysis of G-protein-bound GTP via an allosteric interaction between the regulator of G-protein signaling (RGS) family member RGS9 and the α-subunit of the photoreceptor G protein, transducin (Krispel et al. 2006). RGS-modulated signal termination plays an analogous role in primary olfactory sensory neurons in *Caenorhabditis elegans* (Ferkey et al. 2007). RGS-dependent enhancement of GTP hydrolysis is an ancient and evolutionarily conserved mechanism that accelerates signal termination across a wide variety of heterotrimeric and small G-protein signaling systems in organisms as diverse as yeast and man (Dohlman and Thorner 1997; Ross and Wilkie 2000; Netzel and Hepler 2006). In vertebrate photoreceptors, a third mechanism for activity-dependent feedback involves a light-dependent decline in intracellular calcium levels, which leads to the allosteric activation of guanylate cyclase, the enzyme that synthesizes cGMP, by small calcium-binding proteins termed guanylate-cyclase-activating proteins (GCAPs) (Yau 1991; Yau and Hardie 2009). In other GPCR signaling systems, calcium feedback acts on a wide variety of cellular effectors.

Ion channels are distinguished from multicomponent signaling cascades by the rapidity of their response, enabling ionotropic receptors to convert stimuli into neuronal depolarization on a millisecond timescale, as compared with tens or hundreds of milliseconds for most metabotropic systems (Fig. 2). This is especially relevant when the physiological timescale of stimulus presentation is rapid, as in acoustic signals. In this case, sound pressure waves with vibration frequencies on the order of several thousand cycles per second (or, for bats, up to 100,000 cycles per second) are detected by mechanically gated ion channels with open probabilities that are modulated by the rapid back-and-forth movements of a tightly interconnected set of microvilli, the stereociliary bundle (Fig. 1D). In this system, adaptation is effected by dynamic changes in the tension in the elastic elements that gate the channels.

Ion channels also have a great capacity for signal integration and gain control. This phenomenon is nicely

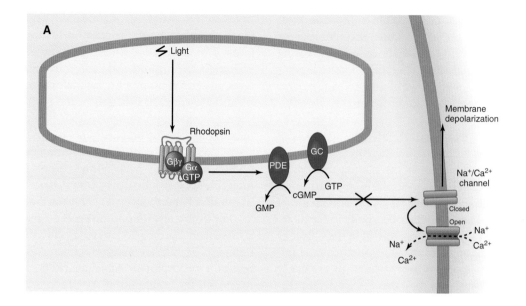

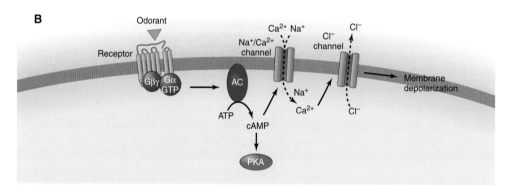

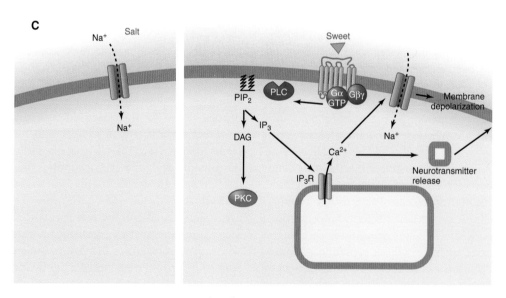

Figure 1. (*See facing page for legend.*)

D

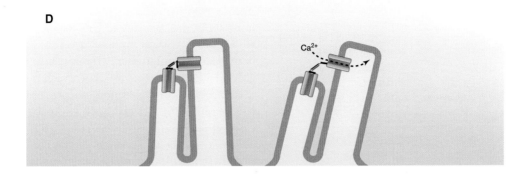

E

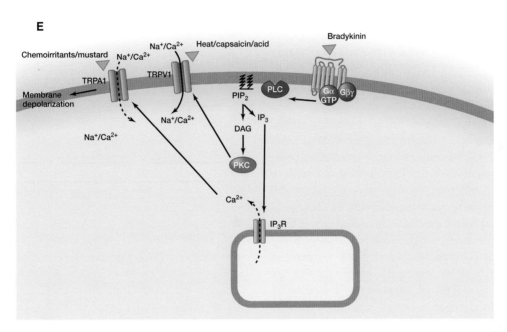

Figure 1. Comparison of sensory signaling systems for vision, olfaction, hearing and balance, taste, and pain/thermosensation. The events underlying signal transduction are shown schematically for (*A*) rod and cone photoreception; (*B*) olfaction in the main olfactory epithelium; (*C*) salt (*left*) and sweet (*right*) taste; (*D*) hearing and balance; and (*E*) pain/thermosensation. Schematics *A*–*E* refer to vertebrates. The final step in olfactory signaling consists of the calcium-dependent opening of anion channel TMEM16B, although recent work suggests that the resulting anion current plays only a minor role in olfactory signal transduction (Billig et al. 2011). For pain/thermosensation mediated by TRPV1, the figure shows inflammatory agents (extracellular protons, bioactive lipids, peptides, and neurotrophins) acting to enhance channel opening either as direct allosteric modulators of TRPV1 or through second-messenger signaling pathways. In auditory and vestibular hair cell bundles, it is not known whether transduction channels reside at both ends or at only one end of the extracellular elastic elements (tip links); in panel *D* the channels are shown at both ends.

illustrated in the somatosensory system, where the capsaicin- and heat-activated receptor TRPV1 is modulated by multiple components of the inflammatory milieu, including bioactive lipids, extracellular protons, neurotrophic factors, and inflammatory peptides (Fig. 1E). These agents enhance the sensitivity of TRPV1 to heat by functioning either directly as allosteric modulators of the channel or indirectly through metabotropic pathways that modulate TRPV1 channel function. These actions contribute to heightened pain sensitivity following tissue injury, such as sunburn, and constitute an important part of the pain pathway's protective function (Caterina and Julius 2001).

One outstanding question in sensory transduction concerns the molecular mechanisms used by cells to detect and

D. Julius and J. Nathans

Table 1. Classes of sensory receptors

Sensory modality[a]	Receptor(s)	Transduction mechanism
Vision	GPCRs	G-protein/cGMP phospho-diesterase/cGMP-gated ion channel
Olfaction[b]	GPCRs	G-protein/adenylyl cyclase/cAMP-gated ion channel
Hearing/balance	Nonselective cation channel	Direct gating by mechanical force
Taste (sweet and bitter)	GPCRs	G-protein/TRP channel
Taste (sour)	Ion channel	Direct sensing of ion flux
Mechanosensation (in bacteria)	Ion channel	Membrane stretch

[a]Data are for vertebrates unless otherwise noted.

[b]Data are for the main olfactory epithelium. A minority of olfactory sensory neurons appear to use transmembrane guanylate cyclases as receptors.

respond to mechanical stimuli (focal pressure, stretch, and osmotic challenge) (Gillespie and Walker 2001; Kung 2005). To date, this is best understood in prokaryotic systems, in which potentially lethal hypotonic shock leads to the activation of both small- and large-conductance cell surface mechanosensory channels, MscS and MscL, respectively, that equalize solute gradients by conducting anions, cations, and other small molecules relatively nonselectively. Reconstitution of purified MscS and MscL proteins in synthetic lipid bilayers has shown that these channels are intrinsically mechanosensitive, opening and closing in direct response to changes in lateral membrane pressure (Sukharev et al. 1994; Sukharev 2002; Vásquez et al. 2008; Kung et al. 2010).

A more complex model has been proposed for mechanosensory transduction in metazoan systems, based on genetic studies in *C. elegans*. Here, screens for touch-insensitive mutants have identified loci encoding microtubule-associated proteins as well as members of the amiloride-sensitive sodium channel family, arguing for the existence of a mechanosensory complex in which membrane stretch promotes channel opening via cytoskeletal changes (Chalfie 2009). In contrast to the single-component bacterial Msc system, this model has not yet been fully validated through functional reconstitution in heterologous systems. Genetic and physiologic studies in flies and mammals have identified several additional candidates for mechanotransducers, including members of the TRP and Piezo channel families. Whether these channels respond to mechanical stimuli directly, as in bacteria, or indirectly through membrane/cytoskeletal attachment, as in nematodes, remains to be determined.

3 STRUCTURAL BASIS OF SENSORY RECEPTOR ACTIVATION

Among eukaryotic sensory receptors, we currently know most about the structure of GPCRs and how they interact with ligands, G proteins, and arrestins. Much insight has been gleaned from studies of rhodopsin and the β-adrenergic and adenosine receptors, for which three-dimensional (3D) structures of active and inactive conformational states have recently been determined (Rosenbaum et al. 2009; Choe et al. 2011; Rasmussen et al. 2011; Standfuss et al. 2011; Xu et al. 2011). Crystallographic studies confirm that these receptors contain a membrane-embedded core of seven transmembrane α-helices with amino- and carboxy-terminal tails facing outside and inside the cell,

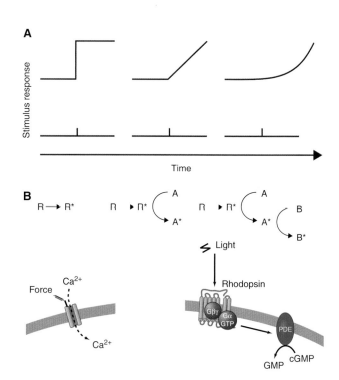

Figure 2. Schematic representation of temporal response profiles for single or multicomponent transduction systems. (A) Time course of stimulus-evoked responses for transduction systems consisting of zero, one, or two enzymatic stages of amplification. (B) For each system, the schematic diagram shows the flow of information in the transduction system, with R and R* representing the sensory receptor molecule in the inactive and active states, respectively. (*Left* panel) Here, there are no transduction components beyond the receptor itself, as in the case of a sensory ion channel, where activation is nearly instantaneous. (*Middle* panel) Here, the active receptor functions as a catalyst and converts inactive molecules of A to their active derivatives A*, which accumulate linearly with time following receptor activation. (*Right* panel) Here, the active derivative A* functions as a catalyst to convert inactive molecules of B to their active derivatives B*, resulting in second-order response kinetics. The *left* panel represents a mechanosensory system, and the *right* panel represents vertebrate phototransduction, as diagrammed below.

Cite this chapter as *Cold Spring Harb Perspect Biol* doi: 10.1101/cshperspect.a005991

respectively. We note that that in the case of rhodopsin, the amino terminus faces the lumen of the outer segment disc, which is topologically equivalent to the outside of the cell.

Binding of low-molecular-weight agonists and antagonists to the β-adrenergic and adenosine receptors, or light-induced isomerization of covalently bound retinal in rhodopsin, occurs within a pocket that is formed by four of the seven α-helices and is located below the surface of the plasma membrane (Fig. 3A). Subtle agonist-induced structural changes within this pocket promote rotational movements in hinge regions of helices 5 and 6 that sit at the midpoint of the lipid bilayer. This motion, in turn, exposes a hydrophobic G-protein-binding site formed by loops linking the cytoplasmic ends of transmembrane domains 5, 6, and 7.

Crystallographic and molecular dynamics simulations suggest that interactions with both ligand and G protein are required to fully stabilize the activated conformation of some GPCRs. This is consistent with classic pharmacological observations showing that agonist affinities decrease substantially when the G-protein is released. Interestingly, the conformational equilibrium between inactive and active states differs among GPCRs. Some receptors, such as the β2-adrenergic receptor, make occasional transitions to the activated state in the absence of ligand. In contrast, rhodopsin remains almost completely inactive in the absence of retinal isomerization, thereby maintaining the low level of dark noise that is characteristic of vertebrate rod phototransduction.

Much less is currently known about the tertiary structure of most ion channels that function as sensory receptors. Bacterial MscS and MscL mechanosensory channels are the exception: their crystal structures have been resolved to approximately 4 Å resolution (Perozo and Rees 2003; Kung et al. 2010). These stretch-activated channels assume two rather different overall configurations. MscS is a homoheptameric channel in which the seven subunits are arranged around a central axis. The third transmembrane helix of each subunit lines the ion-conducting pore, whereas the other two helices form a cone-shaped outer shell that interacts with membrane lipids. MscL is a homopentameric channel in which each subunit has two transmembrane helices, one facing the ion pore and the other facing the lipid bilayer. Despite their distinct stoichiometries and architectures, MscS and MscL perform the same function of opening and closing in response to changes in lateral pressure or surface tension exerted by the surrounding lipid bilayer as it undergoes deformation. Structural and molecular dynamics simulation studies suggest that in response to membrane stretching, both channels undergo rotational and kinking movements of their transmembrane helices to open a central ion-conducting pore, much as an iris opens in an old-fashioned camera (Fig. 3B).

At present, structural insights into metazoan sensory channel function are limited. For TRP channels, progress is currently limited to high-resolution structures of small soluble domains and low-resolution (~20 Å) structures of the intact channel derived from electron microscopic images (Malnic et al. 1999; Gaudet 2008; Li et al. 2011). How some TRP channels, such as TRPV1 and TRPM8, detect and respond to changes in ambient temperature remains an intriguing and unresolved question. Heterologous expression and in vitro proteoliposome reconstitution studies suggest that TRPV1 and TRPM8 are intrinsically temperature-sensitive, but whether gating in vivo involves channel–lipid interactions or association with other cellular factors remains uncertain. Mutational studies have implicated several channel domains as being important in temperature detection or specification of thermal activation thresholds, but the structural underpinnings of these processes have yet to be elucidated.

4 THE LOGIC OF SENSORY CODING

Sensory systems create an internal representation of the external world filtered through the molecular specificities of primary receptor proteins. Thus, the act of sensory coding can be conceptualized as an act of remapping: One multidimensional space (of sensory stimuli) is remapped onto another multidimensional space (of receptor cell responses). We focus here only on the representation of sensory stimuli at the level of the primary receptors, but note that the remapping process continues at each stage of sensory processing up to and including brain circuitry. For example, in the visual system, the distinct attributes of form, color, motion, and depth are extracted from the retinal image and processed in partially overlapping information streams in the visual cortex (Livingstone and Hubel 1988).

In bacteria, plants, and animals, light-sensing systems are based on the photoisomerization of a chromophore with a conjugated π-electron system: retinal or one of its derivatives in bacterial and animal rhodopsins (Fig. 4), and a tetrapyrrole in plant phytochromes. Although the energy difference between ground and excited electronic states is quantized, the receptor's absorbance spectrum (i.e., absorbance as a function of wavelength) appears as a relatively broad and approximately Gaussian curve. The breadth of the absorbance band arises from the large number of closely spaced vibrational energy states that are superimposed on the larger electronic energy gap (Abrahamson and Japar 1972). These broad absorbance bands permit a relatively small number of photoreceptors with partially overlapping spectral sensitivities to cover the biologically relevant region of the electromagnetic spectrum, the interval from near ultraviolet (~350 nm)

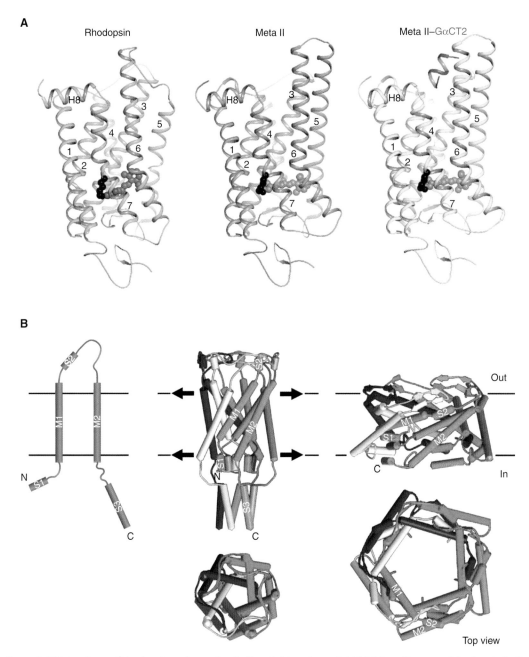

Figure 3. 3D structures of rhodopsin and a mechanically gated ion channel. (*A*) Ribbon diagram of rhodopsin in the inactive state with bound 11-*cis*-retinal (red spheres) compared with the light-activated Metarhodopsin II state containing all-*trans*-retinal (blue spheres) and Metarhodopsin II state in complex with a peptide (pink) corresponding to the receptor-binding site on the Gα subunit of transducin. Rotation and elongation of light-activated retinal lead to a slight rotational tilt of transmembrane helices 5 and 6, thereby enlarging a crevice at the cytoplasmic side of the receptor in which Gα can dock (Choe et al. 2011). (*B*) Bacterial mechanosensory ion channel, MscL, showing transmembrane topology of a monomer (*top left*) and the assembled pentameric channel complex (*top middle*). Membrane stretch and consequent changes in tension at the membrane–protein interface produce structural rearrangements that result in compression of the channel and expansion of the central ion permeation pore, as depicted in the *side* and *top* views (Kung 2005).

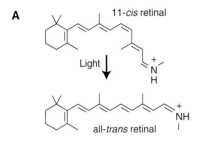

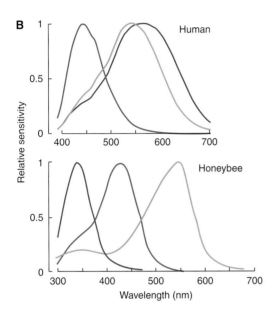

Figure 4. Color vision in humans and honeybees. (*A*) Photoisomerization of retinal from 11-*cis* to all-*trans*, the photochemical event that initiates receptor activation in vertebrate and invertebrate photoreceptors. (*B*) Spectral sensitivities of human cone photoreceptors and honeybee rhabdomeric photoreceptors. (Adapted from Osorio and Vorobyev 2008; reprinted, with permission, from Elsevier © 2008.)

to far red (650 nm). In humans, there are five classes of light receptors: cone photoreceptors have sensitivity maxima of ∼440 nm, ∼530 nm, and ∼560 nm; rod photoreceptors have sensitivity maxima of ∼500 nm; and intrinsically photosensitive retinal ganglion cells—which mediate pupil constriction and light entrainment of circadian rhythms—express a highly divergent photoreceptor protein, melanopsin, that confers a sensitivity maximum of ∼480 nm. The differential excitation of the three cone pigments provides information about the wavelength composition of a stimulus, a process that we recognize as color vision (Fig. 4).

The strategy of overlapping spectral sensitivities was first suggested by Thomas Young in a remarkably prescient passage in his 1801 Bakerian Lecture before the Royal Society:

> As it is almost impossible to conceive each sensitive point of the retina to contain an infinite number of particles (receptors), each capable of vibrating in perfect unison with every possible

undulation (frequency), it becomes necessary to suppose the number limited … and that each particle is capable of being put in motion more or less forcibly by undulations differing less or more from perfect unison (Young 1802).

Following Young's argument, we could measure light intensity as a function of wavelength from a particular location in a scene, sampling from 400 nm to 650 nm in 1-nm steps, and then represent the data by a single point in a 250-dimensional space. Young correctly surmised that the visual system remaps this high-dimensional stimulus space onto a lower-dimensional space of receptor activities.

One inevitable result of such remapping is that some pairs of points that reside at distinct locations in the higher-dimensional stimulus space will reside at indistinguishably close locations in the lower-dimensional receptor space. For color vision based on only three or four classes of receptors, surprisingly little information is lost relative to that which could be extracted by a larger ensemble of receptors. The reason for this is that the biological pigments that dominate natural scenes—like the chromophores of photoreceptors—have broad and relatively smooth bell-shaped absorbance curves. (Recall that the visual stimulus, i.e., the light reflected from an object, is proportional to the reciprocal of the absorbance spectrum.) In other words, the information content in the absorbance spectra of biomolecules is contained largely in their low-frequency components, with the word "frequency" referring not to a particular wavelength of light but to a Fourier decomposition of the curve of reflectance versus wavelength (Maloney 1986).

A simpler transformation between stimulus and response spaces is effected by taste receptors (Yarmolinsky et al. 2009). In mammals, sweet, umami (amino acid), and bitter tastants stimulate GPCRs, whereas sodium and hydrogen ions (i.e., salty and acidic modalities) are detected by members of the amiloride-sensitive and TRP ion channel families, respectively. Two classes of specialized taste cells express low-affinity GPCRs to detect common nutrients: T1R1-T1R3 heterodimers detect ʟ-amino acids, and T1R2-T1R3 heterodimers detect sugars. The relatively low ligand–receptor affinities (in the millimolar range) are appropriate given the organism's interest in identifying quantities of ligand sufficient for its nutritional needs. The umami and sweet taste receptors are of additional interest because, together with the gamma-aminobutyric acid (GABA)-B receptor, they represent the best-validated examples in which receptor dimerization is required for G-protein signaling (Milligan 2009).

A distinct class of taste receptors coexpress a mixture of about 30 high-affinity GPCRs (T2Rs) that recognize a broad array of bitter compounds, many of which are plant-derived toxins. The uniformly bitter sensation elicited by diverse T2R ligands attests to the compression of a multidimensional chemical stimulus space onto a single psychophysical dimension of bitterness—a logical feature given that the only behavioral output is aversion. This arrangement sacrifices discriminatory power, but it enables an organism to determine whether a substance is nutritionally beneficial (eliciting an attractive response) or potentially toxic (eliciting an aversive response).

Another apparently simple transformation between stimulus and response spaces is seen in mammalian thermosensation (Lumpkin and Caterina 2007). In this case, a scalar quantity, temperature, is detected by largely distinct sets of sensory neurons, each of which expresses one type of temperature-gated TRP channel. The critical features of this system are (1) the polarity of the response—channel opening that is either heat-activated (TRPV1) or cold-activated (TRPM8); (2) the temperature at the midpoint of the S-shaped stimulus–response curve; and (3) the steepness of the stimulus–response curve. The last of these features may reflect cooperative interactions among TRP channel subunits that line a central pore. For each class of thermosensory neurons, the monotonic mapping of stimulus temperature to response shows systematic compressions and expansions along the temperature axis, with maximal sensitivity to small changes in temperature occurring, as one would expect, in the steepest region of the receptor's stimulus–response curve. For mammals, the intervals of maximal sensitivity flank normal peripheral body temperatures: $\sim 15^\circ$C to $\sim 25^\circ$C for cold receptors and $\sim 35^\circ$C to $\sim 50^\circ$C for heat receptors (Iggo 1982). In keeping with this pattern, the temperature of the half-maximal response of TRPM8 orthologs differs among species and correlates with core body temperature (Myers et al. 2009).

In reality, the situation is more complex than the preceding paragraph indicates because, as noted above, thermosensory neurons integrate additional stimulus dimensions by virtue of the sensitivity of various TRP channels to chemical ligands and by the sensitizing action of inflammatory mediators at the cellular level (Basbaum et al. 2009). Many TRP channel activators are plant-derived compounds that provoke a sensation of irritation or burning pain. Primary afferent neurons coexpressing capsaicin (TRPV1) and wasabi (TRPA1) receptors are dually sensitive to heat and electrophilic irritants, a mixed-modality arrangement wherein chemical irritants and inflammatory agents excite heat-sensitive fibers to produce thermal hyperalgesia (increased sensitivity to pain). This phenomenon of cross-modality signaling appears as a distinctive feature of somatosensation, presumably reflecting the protective function of the pain pathway. Thus, thermosensory neurons provide the organism with a sensory space in which noxious chemical and thermal stimuli are integrated with tissue injury and inflammation to yield gradations of two relatively simple sensations, pain and temperature.

Olfactory systems illustrate the most complex stimulus–response relationships. This complexity reflects the extraordinary diversity of chemical space and the large number of distinct odorant receptors (~ 100 to ~ 1000) expressed in vertebrates and invertebrates (Su et al. 2009). The most naive mapping would assign each odorant an independent axis in stimulus space and each receptor an independent axis in response space. Even if the two spaces are compressed by combining the attributes of related compounds and related receptors into a smaller number of axes—a method referred to as principal component analysis—the dimensionality of each space remains extremely large.

Systematic analyses in which a transgenic receptor is expressed in a single neuron from which the endogenous receptor has been eliminated (the "empty neuron" technique) have delineated odorant responses for the entire olfactory repertoire in fruit flies and other insects (Hallem and Carlson 2006; Carey et al. 2010). Similarly, taking advantage of the one-receptor–one-neuron relationship in the main olfactory epithelium of mammals, researchers have used calcium imaging during odorant exposure followed by single-cell PCR to delineate basic patterns of odorant receptor specificities (Malnic et al. 1999). Several fundamental observations have emerged from these studies (Malnic et al. 1999; Hallem et al. 2004; Hallem and Carlson 2006; Carey et al. 2010; Wang et al. 2010): First, individual odorants typically activate more than one receptor, and, conversely, individual receptors are typically activated by more than one odorant. Second, receptors differ in the breadth of their odorant responses: Some receptors are activated by a small number of odorants, whereas others are activated by a larger number of odorants, often with related chemical structures. Third, both the level of receptor activation and the number of classes of activated receptors increase with increasing odorant concentration. Organisms face the further challenge of interpreting mixtures of odorants, and for this task, additional identifying information can be obtained based on antagonistic interactions between pairs of odorants and on odorant-specific temporal dynamics of receptor responses (Hallem et al. 2004; Oka et al. 2004, 2009; DasGupta and Wadell 2008; Su et al. 2011).

5 EVOLUTION AND VARIATION

The strongest selective pressure for optimal performance in sensory systems occurs when the stimulus affects behaviors most directly related to Darwinian selection: feeding, mating, and avoiding death from predation, poisoning, and so on. Conversely, sensory systems that gather information that is of little or no utility will disappear over time, a process recognizable by the loss or inactivation of their associated gene sequences. As the following examples illustrate, these considerations inform any comparisons among species of the performance characteristics of sensory systems.

Variations among different types of photoreceptors, both within and between species, beautifully illustrate the evolution of sensory system performance to fit diverse ecological needs (Yau and Hardie 2009). For example, different signal-to-noise ratios are apparent in rod-mediated night vision (where the ratio must be high) compared with cone-mediated daytime vision (where the ratio can be lower). The rod visual pigment, rhodopsin, has a half-life for spontaneous activation at 37°C of ~400 yr, corresponding to an energy barrier of ~22 kcal/mol for thermal isomerization of the 11-*cis* retinal chromophore (Baylor et al. 1980). In mammalian rods, which have 4×10^7 rhodopsins per cell, this works out to only approximately one spontaneous activation event per minute per rod. This very low level of receptor noise represents a critical performance feature that sets the absolute sensitivity of dim light vision (Hecht et al. 1942; Baylor et al. 1980). In contrast, mammalian cones operate at light levels that are several orders of magnitude higher than the rod operating range, and cones are correspondingly several orders of magnitude noisier than rods (Schnapf et al. 1990; Schneeweis and Schnapf 1999).

Visual pigment spectral sensitivity is one of the most intensively studied systems in which sensory receptor evolution has been explored at the molecular, organismal, and ecological levels. In the ocean, chlorophyll and other biomolecules in photosynthetic microorganisms selectively deplete longer-wavelength sunlight, and, as a result, there is a corresponding blueshift in the visual pigment spectral sensitivities of fish that live at greater ocean depths (Lythgoe 1979). Similarly, dolphin rhodopsin and the dolphin long-wavelength cone pigment are blueshifted relative to their terrestrial counterparts as an adaptation to the aquatic environment (Fasick et al. 1998; Fasick and Robinson 2000). At extreme depths, where little sunlight penetrates, visual pigment spectral sensitivities are under entirely different selective pressures: Bioluminescence permits communication among organisms of the same species, and visual pigments are tuned to the emitting wavelengths, in some cases at the far-red end of the spectrum (Douglas et al. 1998).

Visual pigment spectral tuning has also been studied in relation to discrimination among natural objects that are behaviorally relevant (Osorio and Vorobyev 2008). Comparing the color space defined by honeybee rhabdomeric photoreceptors, which have sensitivity maxima at ~350, ~425, and ~550 nm, with that defined by human cone

photoreceptors, which have sensitivity maxima of ~440, ~530, and ~560 nm (representative of Old World primate cone sensitivities) (Fig. 4), reveals a wider dispersion of floral hues in honeybee color space compared with primate color space (Fig. 5). This pattern supports the general idea that color vision in pollinator species such as honeybees, butterflies, and hummingbirds coevolved with floral pigments to enhance discrimination among floral species, an arrangement that enhances both feeding and pollination. Similarly, the surface hues of fruits that are consumed by primates occupy a wider swath of primate color space compared with honeybee color space (Fig. 5), a pattern that suggests that primate trichromacy may have coevolved with fruit coloration to enhance both fruit consumption and seed dispersal (Regan et al. 2001). In particular, yellow, orange, or red fruit is readily detected against a background of dappled green foliage if an animal compares the extents of excitation of a pair of visual pigments in the 500–600-nm region of the spectrum, as nearly all Old World and some New World primates can (Mollon 1989; Regan et al. 2001). Such discrimination is difficult with the single longer-wavelength-sensitive pigment typical of non-primate mammals. A chromatic discrimination task similar to this one is the basis of the Ishihara test for color vision deficiency in humans.

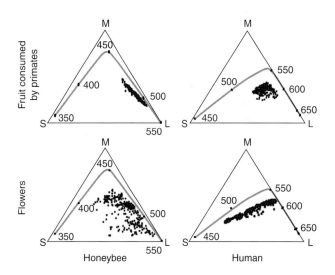

Figure 5. Chromaticity diagrams for humans and honeybees. The triangles represent a plane within a 3D receptor space, with each vertex corresponding to the point at which the plane intersects an axis representing the degree of excitation of one receptor type. (S) Short-wavelength receptor axis; (M) medium-wavelength receptor axis; (L) long-wavelength receptor axis. The small black circles within the triangles represent the chromaticities of a set of fruits that are consumed by primates (*upper* panels) or a set of flower petals (*lower* panels). The line within each chromaticity diagrams represents the locus of spectrally pure lights, with black circles and the adjacent numbers marking steps of 50 nm. (Adapted from Osorio and Vorobyev 2008; reprinted, with permission, from Elsevier © 2008.)

Fruit flies and mosquitoes provide an analogous instance of species-specific differences in the patterning of salient odorants in olfactory receptor space. *Drosophila melanogaster* feeds on ripe or rotting fruit and appears to be especially good at discriminating among esters, the dominant volatiles emitted by fruit (Su et al. 2009; Carey et al. 2010). In contrast, *Anopheles gambiae* feeds on human blood and is better at discriminating aromatics, including several that are characteristic of humans (Carey et al. 2010). High salience is also seen in the heterodimeric carbon dioxide receptors of *Drosophila* and *Anopheles*. In *Drosophila*, this odorant represents a stress signal, whereas in *Anopheles*, it is one of the chemotactic signals used to identify a human host (Jones et al. 2007; Lu et al. 2007). Finally, receptors specific for within-species olfactory communication, most especially in the context of pheromones, have been identified in both vertebrates and invertebrates (Touhara and Vosshall 2009). In *Drosophila*, *cis*-vaccenyl acetate (cVA) is produced by males, increases female receptivity, and then—upon transfer to the female partner during mating—decreases female attractiveness to other males. Interestingly, the responses to cVA are mediated through at least two odorant receptors (Or67d and Or65a), a two-transmembrane-domain coreceptor (SNMP), and a soluble odorant-binding protein (LUSH) (Vosshall 2008).

As noted in the discussion above on primate color vision and fruit consumption, an immobile plant has much to gain if it can encourage animal foragers to help spread its seeds. Avian foragers present the best opportunity for widespread seed dispersal, and, therefore, one might expect that plants and the birds that disperse their seeds would have coevolved a strategy that discourages competition from less desirable foragers such as mammals. A striking example of this phenomenon is seen in the insensitivity of the avian vanilloid thermosensory channel (TRPV1) to activation by capsaicin, which is the source of the painfully "hot" sensation elicited by chili peppers (Jordt and Julius 2002). Capsaicin thus appears to selectively repel mammals by activating their temperature/pain receptors, thereby preserving the pepper seeds for consumption and transport by birds. The avoidance or non-avoidance of food constitutes a major point of intersection between animal and plant ecologies. In this respect, it is interesting that the vertebrate bitter receptor (T2R) gene repertoire—which represents a sum of all undesirable edible compounds, many of which are plant-derived—is evolving rapidly (Dong et al. 2009).

The most extreme change in sensory signaling is a complete loss of receptor function. In the genomes of the bushbaby (*Otolemur crassicaudatus*) and owl monkey (*Aotus trivirgatus*)—primates with predominantly nocturnal lifestyles—the shortwave-sensitive cone pigment gene has decayed into a pseudogene, leaving only a single class of longer-wavelength cones to mediate daytime vision, with no possibility of color vision (Jacobs et al. 1996; Kawamura and Kubotera 2004). Similarly, in humans, all of the sequences coding for vomeronasal receptors of the V2R class, which mediate pheromone sensing in rodents, are pseudogenes (Touhara and Vosshall 2009). Among mammalian genes that encode the principal receptors of the main olfactory epithelium, the fraction that are pseudogenes ranges from ~15% to ~80%, depending on the species (Niimura and Nei 2007; Niimura 2009; Touhara and Vosshall 2009). In dolphins, all genes encoding class II receptors of the main olfactory epithelium, which are mostly specialized for volatile hydrophobic ligands, appear to be inactivated by mutation (Freitag et al. 1998; Niimura 2009).

Any discussion of sensory signaling would be incomplete without considering the evolution of specialized anatomic structures that facilitate sensory function. The frequency-dependent conversion of sound pressure waves to localized basilar membrane distortions in the cochlea represents a dramatic example of these auxiliary structures in action. This micro-mechanical system is remarkably sensitive: the threshold for detecting basilar membrane displacement is <1 nm (Johnstone et al. 1986). A larger-scale anatomic specialization is the increase in head width in hammerhead sharks. The great distance between left and right nasal openings, together with the evolution of olfactory receptor neurons that have sub-nanomolar affinities for amino acids, allows the hammerhead to sense shallow concentration gradients of dilute amino acids by comparing the excitation of left and right olfactory sensory epithelia (Tricas et al. 2009; Gardiner and Atema 2010). This spatial differencing strategy is the olfactory analog of binaural sound localization in mammals.

A particularly striking instance of anatomic specialization is seen in the pit organs of vipers such as rattlesnakes. Stretched across the pit is an extremely thin tissue that is densely innervated by thermosensory fibers. The fibers express a TRP channel that senses the minute increase in temperature evoked by infrared radiation from a nearby warm-blooded animal (Gracheva et al. 2010). The low body temperature of the viper relative to its prey and the small heat capacity of the sensory tissue are important for optimal pit organ function. Recent calculations suggest that the sensory fibers of the pit organ can respond to temperature changes as small as 0.001°C (Bakken and Krochmal 2007).

6 CONCLUDING REMARKS

The sensory receptors and signaling systems described here represent only a small sampling of the many that have been investigated. Because of space limitations, we have not discussed plant sensory systems, and we have only briefly touched on microbial systems. Even with this limited sampling, the diversity of sensory systems is striking, and it is evident that each lifestyle and ecological niche is accompanied by a distinctive set of adaptations in sensory system structure and function.

Over the past 30 years, the identities and primary structures of most of the major classes of vertebrate sensory receptor proteins have been defined. The one exception is the mechanosensory channels in the auditory and vestibular system, which remain enigmatic. The signaling cascades downstream from GPCR-type sensory receptors have also been largely defined. In vertebrate photoreception, the best studied of all GPCR signaling cascades, it is likely that all of the components involved in signaling and adaptation are now known. In less experimentally accessible systems, such as the taste and vomeronasal systems, additional signaling components remain to be identified, and the interactions between signal activation pathways and feedback loops are still incompletely understood.

A full molecular understanding of sensory receptor function requires the 3D structures of receptors and signaling components in their various active and inactive conformations, and, in some cases, in complex with each other. This has been achieved for rhodopsin, transducin, and the bacterial MscL mechanosensory channel, and it is an area of active investigation for other classes of sensory receptors and their downstream effectors. Structural studies can be expected to play a critical role in elucidating the molecular basis of receptor–ligand specificity in chemosensory systems.

A further challenge in the field of sensory biology comes from the need to diagnose and treat diseases that affect sensory signaling, including those associated with chronic pain or the loss of vision or hearing. Understanding sensory signaling at the molecular and cellular levels will inform these clinical investigations and will continue to be one of nature's grand scientific challenges for biologists, chemists, physicists, and engineers.

REFERENCES

Abrahamson EW, Japar SM. 1972. Principles of the interaction of light and matter. In *Handbook of sensory physiology. VII/1: Photochemistry of vision* (ed. Dartnall HJA), pp. 1–32. Springer-Verlag, Berlin.

Arshavsky VY, Lamb TD, Pugh EN. 2002. G proteins and phototransduction. *Annu Rev Physiol* **64**: 153–187.

Bakken GS, Krochmal AR. 2007. The imaging properties and sensitivity of the facial pit of pitvipers as determined by optical and heat-transfer analysis. *J Exp Biol* **210**: 2801–2810.

Basbaum AI, Bautista DM, Scherrer G, Julius D. 2009. Cellular and molecular mechanisms of pain. *Cell* **139**: 267–284.

Baylor DA, Matthews G, Yau KW. 1980. Two components of electrical dark noise in toad retinal rod outer segments. *J Physiol* **309**: 591–621.

Benton R, Vannice KS, Gomez-Diaz C, Vosshall LB. 2009. Variant ionotropic glutamate receptors as chemosensory receptors in *Drosophila*. *Cell* **136**: 149–162.

Billig GM, Pál B, Fidzinski P, Jentsch TJ. 2011. Ca^{2+}-activated Cl^- currents are dispensable for olfaction. *Nat Neurosci* **14**: 763–769.

Campbell AL, Naik RR, Sowards L, Stone MO. 2002. Biological infrared imaging and sensing. *Micron* **33**: 211–225.

Carey AF, Wang G, Su CY, Zwiebel LJ, Carlson JR. 2010. Odorant reception in the malaria mosquito *Anopheles gambiae*. *Nature* **464**: 66–71.

Catania KC. 2005. Star-nosed moles. *Curr Biol* **15**: R863–R864.

Caterina MJ, Julius D. 2001. The vanilloid receptor: A molecular gateway to the pain pathway. *Annu Rev Neurosci* **24**: 487–517.

Chalfie M. 2009. Neurosensory mechanotransduction. *Nat Rev Mol Cell Biol* **10**: 44–52.

Chandrashekar J, Hoon MA, Ryba NJ, Zuker CS. 2006. The receptors and cells for mammalian taste. *Nature* **444**: 288–294.

Choe HW, Kim YJ, Park JH, Morizumi T, Pai EF, Krauss N, Hofmann KP, Scheerer P, Ernst OP. 2011. Crystal structure of metarhodopsin. II. *Nature* **471**: 651–655.

DasGupta S, Wadell S. 2008. Learned odor discrimination in *Drosophila* without combinatorial odor maps in the antennal lobe. *Curr Biol* **18**: 1668–1674.

DeWire SM, Ahn S, Lefkowitz RJ, Shenoy SK. 2007. β-Arrestins and cell signaling. *Annu Rev Physiol* **69**: 483–510.

Dohlman HG, Thorner J. 1997. RGS proteins and signaling by heterotrimeric G proteins. *J Biol Chem* **272**: 3871–3874.

Dong D, Jones G, Zhang S. 2009. Dynamic evolution of bitter taste receptor genes in vertebrates. *BMC Evol Biol* **9**: 12.

Douglas RH, Partridge JC, Marshall NJ. 1998. The eyes of deep-sea fish. I: Lens pigmentation, tapeta and visual pigments. *Prog Retin Eye Res* **17**: 597–636.

Fasick JI, Cronin TW, Hunt DM, Robinson PR. 1998. The visual pigments of the bottlenose dolphin (*Tursiops truncatus*). *Vis Neurosci* **15**: 643–651.

Fasick JI, Robinson PR. 2000. Spectral-tuning mechanisms of marine mammal rhodopsins and correlations with foraging depth. *Vis Neurosci* **17**: 781–788.

Ferkey DM, Hyde R, Haspel G, Dionne HM, Hess HA, Suzuki H, Schafer WR, Koelle MR, Hart AC. 2007. *C. elegans* G protein regulator RGS-3 controls sensitivity to sensory stimuli. *Neuron* **53**: 39–52.

Freitag J, Ludwig G, Andreini I, Rössler P, Breer H. 1998. Olfactory receptors in aquatic and terrestrial vertebrates. *J Comp Physiol A* **183**: 635–650.

Gardiner JM, Atema J. 2010. The function of bilateral odor arrival time differences in olfactory orientation of sharks. *Curr Biol* **20**: 1187–1191.

Gaudet R. 2008. TRP channels entering the structural era. *J Physiol* **586**: 3565–3575.

Gillespie PG, Walker RG. 2001. Molecular basis of mechanosensory transduction. *Nature* **413**: 194–202.

Gracheva EO, Ingolia NT, Kelly YM, Cordero-Morales JF, Hollopeter G, Chesler AT, Sánchez EE, Perez JC, Weissman JS, Julius D. 2010. Molecular basis of infrared detection by snakes. *Nature* **464**: 1006–1011.

Hallem EA, Carlson JR. 2006. Coding of odors by a receptor repertoire. *Cell* **125**: 143–160.

Hallem EA, Ho MG, Carlson JR. 2004. The molecular basis of odor coding in the *Drosophila* antenna. *Cell* **117**: 965–979.

Hecht S, Schlaer S, Pirenne MH. 1942. Energy, quanta, and vision. *J Gen Physiol* **25**: 819–840.

Henrik-Heldin C, Evans R, Gutkind S. 2012. *Cold Spring Harb Perspect Biol* (in press).

Iggo A. 1982. Cutaneous sensory mechanisms. In *The senses* (ed. Barlow HB, Mollon JD), pp. 369–408. Cambridge University Press, Cambridge.

Jacobs GH, Neitz M, Neitz J. 1996. Mutations in S-cone pigment genes and the absence of colour vision in two species of nocturnal primate. *Proc Biol Sci* **263**: 705–710.

Johnstone BM, Patuzzi R, Yates GK. 1986. Basilar membrane measurements and the travelling wave. *Hear Res* **22:** 147–153.

Jones WD, Cayirlioglu P, Kadow IG, Vosshall LB. 2007. Two chemosensory receptors together mediate carbon dioxide detection in *Drosophila*. *Nature* **445:** 86–90.

Jordt SE, Julius D. 2002. Molecular basis for species-specific sensitivity to "hot" chili peppers. *Cell* **108:** 421–430.

Kawamura S, Kubotera N. 2004. Ancestral loss of short wave-sensitive cone visual pigment in lorisiform prosimians, contrasting with its strict conservation in other prosimians. *J Mol Evol* **58:** 314–321.

Kevan PG, Chittka L, Dyer AG. 2001. Limits to the salience of ultraviolet: Lessons from colour vision in bees and birds. *J Exp Biol* **204:** 2571–2580.

Krispel CM, Chen D, Melling N, Chen YJ, Martemyanov KA, Quillinan N, Arshavsky VY, Wensel TG, Chen CK, Burns ME. 2006. RGS expression rate-limits recovery of rod photoresponses. *Neuron* **51:** 409–416.

Kung C. 2005. A possible unifying principle for mechanosensation. *Nature* **436:** 647–654.

Kung C, Martinac B, Sukharev S. 2010. Mechanosensitive channels in microbes. *Annu Rev Microbiol* **64:** 313–329.

Li M, Yu Y, Yang J. 2011. Structural biology of TRP channels. *Adv Exp Med Biol* **704:** 1–23.

Livingstone M, Hubel D. 1988. Segregation of form, color, movement, and depth: Anatomy, physiology, and perception. *Science* **240:** 740–749.

Lu T, Qiu YT, Wang G, Kwon JY, Rutzler M, Kwon HW, Pitts RJ, van Loon JJ, Takken W, Carlson JR, et al. 2007. Odor coding in the maxillary palp of the malaria vector mosquito *Anopheles gambiae*. *Curr Biol* **17:** 1533–1544.

Lumpkin EA, Caterina MJ. 2007. Mechanisms of sensory transduction in the skin. *Nature* **445:** 858–865.

Lythgoe JN. 1979. *The ecology of vision*. Clarendon Press, Oxford.

Malnic B, Hirono J, Sato T, Buck LB. 1999. Combinatorial receptor codes for odors. *Cell* **96:** 713–723.

Maloney LT. 1986. Evaluation of linear models of surface spectral reflectance with a small number of parameters. *J Opt Soc Am A* **3:** 1673–1683.

Milligan G. 2009. G protein-coupled receptor hetero-dimerization: Contribution to pharmacology and function. *Br J Pharmacol* **58:** 5–14.

Mollon JD. 1989. "Tho she kneel'd in that Place where they grew. . .": The uses and origins of primate colour vision. *J Exp Biol* **146:** 21–38.

Moore CA, Milano SK, Benovic JL. 2007. Regulation of receptor trafficking by GRKs and arrestins. *Annu Rev Physiol* **69:** 451–482.

Myers BR, Sigal YM, Julius D. 2009. Evolution of thermal response properties in a cold-activated TRP channel. *PLoS One* **4:** e5741.

Neitzel KL, Hepler JR. 2006. Cellular mechanisms that determine selective RGS protein regulation of G protein-coupled receptor signaling. *Semin Cell Dev Biol* **17:** 383–389.

Niimura Y. 2009. On the origin and evolution of vertebrate olfactory receptor genes: Comparative genome analysis among 23 chordate species. *Genome Biol Evol* **1:** 34–44.

Niimura Y, Nei M. 2007. Extensive gains and losses of olfactory receptor genes in mammalian evolution. *PLoS One* **2:** e708.

Oka Y, Omura M, Kataoka H, Touhara K. 2004. Olfactory receptor antagonism between odorants. *EMBO J* **23:** 120–126.

Oka Y, Takai Y, Touhara K. 2009. Nasal airflow rate affects the sensitivity and pattern of glomerular odorant responses in the mouse olfactory bulb. *J Neurosci* **29:** 12070–12078.

Osorio D, Vorobyev M. 2008. A review of the evolution of animal colour vision and visual communication signals. *Vis Res* **48:** 2042–2051.

Perozo E, Rees DC. 2003. Structure and mechanism in prokaryotic mechanosensitive channels. *Curr Opin Struct Biol* **13:** 432–442.

Rasmussen SG, DeVree BT, Zou Y, Kruse AC, Chung KY, Kobilka TS, Thian FS, Chae PS, Pardon E, Calinski D, et al. 2011. Crystal structure of the β2 adrenergic receptor-Gs protein complex. *Nature* **477:** 549–555.

Regan BC, Julliot C, Simmen B, Viénot F, Charles-Dominique P, Mollon JD. 2001. Fruits, foliage and the evolution of primate colour vision. *Philos Trans R Soc Lond B Biol Sci* **356:** 229–283.

Roberts WM, Howard J, Hudspeth AJ. 1988. Hair cells: Transduction, tuning, and transmission in the inner ear. *Annu Rev Cell Biol* **4:** 63–92.

Rosenbaum DM, Rasmussen SG, Kobilka BK. 2009. The structure and function of G-protein-coupled receptors. *Nature* **459:** 356–363.

Ross EM, Wilkie TM. 2000. GTPase-activating proteins for heterotrimeric G proteins: Regulators of G protein signaling (RGS) and RGS-like proteins. *Annu Rev Biochem* **69:** 795–827.

Schnapf JL, Nunn BJ, Meister M, Baylor DA. 1990. Visual transduction in cones of the monkey *Macaca fascicularis*. *J Physiol* **427:** 681–713.

Schneeweis DM, Schnapf JL. 1999. The photovoltage of macaque cone photoreceptors: Adaptation, noise, and kinetics. *J Neurosci* **19:** 1203–1216.

Standfuss J, Edwards PC, D'Antona A, Fransen M, Xie G, Oprian DD, Schertler GF. 2011. The structural basis of agonist-induced activation in constitutively active rhodopsin. *Nature* **471:** 656–660.

Stryer L. 1986. Cyclic GMP cascade of vision. *Annu Rev Neurosci* **9:** 87–119.

Su CY, Menuz K, Carlson JR. 2009. Olfactory perception: Receptors, cells, and circuits. *Cell* **139:** 45–59.

Su CY, Martelli C, Emonet T, Carlson JR. 2011. Temporal coding of odor mixtures in an olfactory receptor neuron. *Proc Natl Acad Sci* **108:** 5075–5080.

Sukharev S. 2002. Purification of the small mechanosensitive channel of *Escherichia coli* (MscS): The subunit structure, conduction, and gating characteristics in liposomes. *Biophys J* **83:** 290–298.

Sukharev SI, Blount P, Martinac B, Blattner FR, Kung C. 1994. A large-conductance mechanosensitive channel in *E. coli* encoded by *mscL* alone. *Nature* **368:** 265–268.

Touhara K, Vosshall LB. 2009. Sensing odorants and pheromones with chemosensory receptors. *Annu Rev Physiol* **71:** 307–332.

Tricas TC, Kajiura SM, Summers AP. 2009. Response of the hammerhead shark olfactory epithelium to amino acid stimuli. *J Comp Physiol A Neuroethol Sens Neural Behav Physiol* **195:** 947–954.

Vásquez V, Sotomayor M, Cordero-Morales J, Schulten K, Perozo E. 2008. A structural mechanism for MscS gating in lipid bilayers. *Science* **321:** 1210–1214.

Vosshall LB. 2008. Scent of a fly. *Neuron* **59:** 685–689.

Wang G, Carey AF, Carlson JR, Zwiebel LJ. 2010. Molecular basis of odor coding in the malaria vector mosquito *Anopheles gambiae*. *Proc Natl Acad Sci* **107:** 4418–4423.

Xu F, Wu H, Katritch V, Han GW, Jacobson KA, Gao ZG, Cherezov V, Stevens RC. 2011. Structure of an agonist-bound human A2A adenosine receptor. *Science* **332:** 322–327.

Yarmolinsky DA, Zuker CS, Ryba NJP. 2009. Common sense about taste: From mammals to insects. *Cell* **139:** 234–244.

Yau KW. 1991. Calcium and light adaptation in retinal photoreceptors. *Curr Opin Neurobiol* **1:** 252–257.

Yau KW, Hardie RC. 2009. Phototransduction motifs and variations. *Cell* **139:** 246–264.

Young T. 1802. The Bakerian lecture. On the theory of light and colours. *Phil Trans Roy Soc* **92:** 12–48.

CHAPTER 12

Synaptic Signaling in Learning and Memory

Mary B. Kennedy

Division of Biology and Biological Engineering, California Institute of Technology, Pasadena, California 91125

Correspondence: kennedym@its.caltech.edu

SUMMARY

Learning and memory require the formation of new neural networks in the brain. A key mechanism underlying this process is synaptic plasticity at excitatory synapses, which connect neurons into networks. Excitatory synaptic transmission happens when glutamate, the excitatory neurotransmitter, activates receptors on the postsynaptic neuron. Synaptic plasticity is a higher-level process in which the strength of excitatory synapses is altered in response to the pattern of activity at the synapse. It is initiated in the postsynaptic compartment, where the precise pattern of influx of calcium through activated glutamate receptors leads either to the addition of new receptors and enlargement of the synapse (long-term potentiation) or the removal of receptors and shrinkage of the synapse (long-term depression). Calcium/calmodulin-regulated enzymes and small GTPases collaborate to control this highly tuned mechanism.

Outline

1 INTRODUCTION

The function of the brain is to process and store information about the environment and direct behavior in response to that information. Three major cell types in the brain—excitatory neurons that use glutamate as their transmitter, inhibitory neurons that use γ-aminobutyric acid (GABA) as their transmitter, and glial cells—work together to respond to the environment while maintaining the overall connectivity among neurons within an acceptable homeostatic range. Excitatory neurons, the most numerous in the brain, each receive thousands of synaptic inputs and, in turn, make thousands of synaptic connections onto other neurons. A human brain contains, on average, 86 billion neurons (Herculano-Houzel 2009) that in toto make trillions of synaptic connections.

A typical cortical excitatory neuron (Fig. 1) comprises a neuronal soma (cell body), several branched dendrites, and a single axon that can extend for many millimeters and often branches to make thousands of individual synaptic connections. The soma is the site of the nucleus and most of the neuron's protein synthetic machinery. Most inhibitory synaptic contacts occur on the somal plasma membrane. In contrast, the highly branched dendrites receive most of the excitatory synaptic contacts, which are made onto small membrane protuberances called dendritic spines (Fig. 1). When the neuronal membrane becomes depolarized to a threshold level, an action potential is initiated at the base of the axon near the soma; this wave of depolarization travels unabated to each of the thousands of presynaptic endings along the axon. Depolarization causes membranous synaptic vesicles within the presynaptic terminals to fuse with the plasma membrane at the "active zone" opposite the postsynaptic site and flood the synaptic cleft with neurotransmitters. Some of the transmitter molecules bind to specific receptors, which are ligand-gated channels in the postsynaptic membrane. Sodium and potassium ions flow through the channels of the activated receptors, decreasing the gradient in their concentration across the membrane and thus producing a localized depolarization called an excitatory postsynaptic potential (EPSP). If the synapse fires repeatedly, or if several different synapses on a neuron fire at the same time, the EPSPs can sum to produce a depolarization that extends to the soma and initiates an action potential.

Information is stored when individual synapses that connect a particular group of neurons become more able (or less able) to generate an action potential in the postsynaptic neuron in response to environmental signals. Memories are stored initially in the hippocampus, where synapses among excitatory neurons begin to form new circuits within seconds of the events to be remembered. An

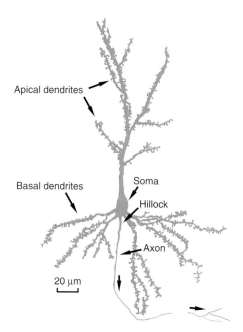

Figure 1. Features of excitatory neurons in the brain. The cell body (soma) of a typical excitatory pyramidal neuron is ∼10 μm in diameter and is located in one of several sheets of tightly packed somas that define the layers of the neocortex and hippocampus. Apical and basal dendrites extend from the soma, reaching into adjacent areas that are referred to as neuropil. Postsynaptic structures are located in tiny membrane protuberances called spines that can be seen along the dendrites. Each soma gives rise to one axon, which has a smaller diameter than the numerous dendrites. The axon can extend for millimeters from the soma and branches to form thousands of presynaptic terminals where transmitter is released onto the postsynaptic sites of other neurons. The axon hillock is located at the base of the axon. Action potentials are usually initiated at this site; they travel along the axon (arrows) to presynaptic terminals keeping a uniform amplitude of depolarization. Back-propagating action potentials travel in the opposite direction through the soma and into the dendrites. The size of their depolarization decreases as they travel and is regulated by the composition of dendritic ion channels. (From Peters and Kaiserman-Abramof 1970; modified, with permission, © Wiley.)

increase in the strength of a relatively small number of synapses can bind connected neurons into a circuit that stores a new memory. A deceptively simple principle guides the direction and amplitude of this synaptic plasticity: neurons that fire together, wire together. When release of transmitter at a synapse is repeatedly correlated with firing of action potentials in the postsynaptic neuron, they become stronger. In contrast, when release of transmitter at a synapse repeatedly fails to correlate with postsynaptic firing, because the resulting EPSPs do not sum to produce the required threshold depolarization, they become gradually weaker and may disappear altogether. Essentially all of the glutamatergic synapses between excitatory neurons in the hippocampus and cortex of the mammalian brain display this behavior. Such synapses are referred to as Hebbian

Cite this chapter as *Cold Spring Harb Perspect Biol* doi: 10.1101/cshperspect.a016824

synapses, after Donald Hebb, who first suggested a similar principle in 1949. Nonglutamatergic synapses do not display this Hebbian behavior. The unique plasticity of excitatory glutamatergic synapses is an essential mechanism of memory formation. Glutamatergic synaptic plasticity, and thus memory formation, can be modulated by release of the inhibitory transmitter GABA and by the influence of acetylcholine, biogenic amines, small peptides, and larger protein hormones released from neurons and glial cells; but synapses that release these other transmitters do not display Hebbian behavior.

The ongoing pattern of electrical activity through Hebbian synapses influences cellular processes in the postsynaptic neuron at many time scales. In a few seconds, changes can be triggered in the structure of the postsynapse itself. Over minutes, the summation of synaptic activity can result in increased levels of the classical second messengers cAMP and calcium and activation of the mTOR and MAPK pathways (Kennedy et al. 2005), leading to up-regulation of translation of mRNAs stored in the dendritic shaft near active synapses (Ho et al. 2011). Local translation is believed to provide proteins needed for remodeling of synapses and dendrites in response to high synaptic activity. Over a few hours, activity-dependent nuclear transcription factors stored near the synapse can become activated and travel from the synapse into the nucleus (Flavell and Greenberg 2008; Ch'ng and Martin 2011). These changes in dendritic protein synthesis and in nuclear transcription can influence the structure of the neuron and its role in neuronal networks for hours, days, or a lifetime. For example, new ion channels may be transcribed and inserted into the membrane to change the intrinsic electrical firing pattern of the neuron, or the overall production of excitatory receptors may be dampened to maintain homeostatic balance.

Here I focus on the specialized machinery that underlies Hebbian behavior of synapses between excitatory neurons. At this time, signal transduction involved in memory formation is best understood, although still incompletely, for the early phases of plasticity.

2 SPINE SYNAPSES

In the brain, most synapses between excitatory neurons are located on spines, tiny compartments that protrude from the neuron's highly branched dendrites (Fig. 2). A typical excitatory pyramidal neuron in the hippocampus or cortex has $\sim$10,000 such synapses, most spines hosting just a single synapse. Spines vary in size from $\sim$0.5–2 μm in length, from $\sim$0.25–1 μm in width, and from $\sim$10–100 attoliters in volume. The synaptic contact itself usually occurs at the tip of the spine. It comprises the presynaptic active zone, the synaptic cleft, and the postsynaptic receptor cluster and

varies in diameter from $\sim$0.1 to 0.8 μm. Spines also vary in shape from stubby to thin to "mushroom-shaped." In general, the larger, mushroom-shaped spines contain stronger synapses. Functionally, a stronger synapse is defined as one that contributes more depolarization to the neuronal membrane upon activation than a weaker one; thus, its activation is more likely to generate an action potential in the postsynaptic neuron.

The postsynaptic membranes of spine synapses contain two distinct types of ligand-gated channels that are receptors for the neurotransmitter glutamate but are distinguishable by their ability to respond to pharmacological agents. For 2-amino-3-(3-hydroxy-5-methyl-isoxazol-4-yl) propanoic acid (AMPA)-type glutamate receptors (AMPARs), binding of glutamate triggers a small, relatively rapid EPSP, resulting from an influx of potassium and sodium ions. AMPA-type receptors produce an EPSP each time glutamate is released at a synapse. The other class, N-methyl-D-aspartate (NMDA)-type glutamate receptors (NMDARs), are evolutionarily related to AMPARs but are more complex; they are responsible for the Hebbian behavior of spine synapses (Mayer et al. 1984; Nowak et al. 1984). As I discuss in detail below, the channel in NMDARs allows passage of calcium ions, as well as potassium and sodium, and opens only if two conditions are met: glutamate is bound and the synaptic membrane is strongly depolarized, as occurs when release of transmitter is correlated with firing of action potentials in the postsynaptic neuron.

Synaptic plasticity is most often studied by recording electrical responses from synapses of the Schaffer-collateral pathway, which connects two different sets of excitatory neurons in the hippocampus (Lüscher and Malenka 2012). Repeated activation of this pathway, and thus its spine synapses, at a frequency between $\sim$10 and 100 Hz for a few seconds usually initiates a process referred to as long-term potentiation (LTP), in which the activated synapses increase in size and more effectively depolarize the postsynaptic membrane. In contrast, activation of these synapses at a lower frequency, between $\sim$1 and 5 Hz, for several minutes usually initiates a process called long-term depression (LTD), in which the activated synapses decrease in size and less effectively depolarize the postsynaptic membrane. The change in strength of the synapses can last for hours to a lifetime, depending on how often the stimulation is repeated.

Intriguingly, although these processes have opposite effects on the strength of the synapse, both are controlled by the influx of calcium caused by activation of NMDARs. Differences in the timing and amount of calcium entry into the spine through NMDARs account for the opposite outcomes after stimulation of synapses in the two different

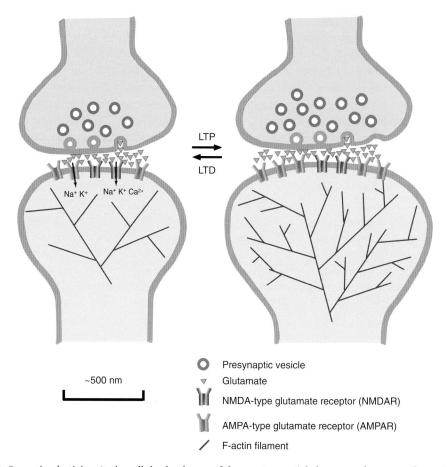

~500 nm

○ Presynaptic vesicle

▽ Glutamate

NMDA-type glutamate receptor (NMDAR)

AMPA-type glutamate receptor (AMPAR)

/ F-actin filament

Figure 2. Synaptic plasticity. At the cellular level, one of the most essential elements of memory formation is the adjustment in synaptic strength of excitatory synapses between neurons. AMPA-type glutamate receptors (yellow) allow passage of sodium and potassium through their channel. Their principal function is to depolarize the membrane, producing an excitatory postsynaptic potential (EPSP). NMDA-type glutamate receptors (blue) also depolarize the membrane, but in addition to sodium and potassium, calcium flows through their channel and can initiate synaptic plasticity. A long-lasting increase in synaptic strength is referred to as long-term potentiation (LTP). LTP involves the addition of new synaptic AMPA-type glutamate receptors (AMPARs) and an increase in the size of the head of the postsynaptic spine, supported by an increase in the size and branching of the actin cytoskeleton. Long-term depression (LTD) is a long-lasting decrease in synaptic strength that involves a decrease in the number of synaptic AMPARs and shrinkage of the spine head. LTP is induced when repeated firing of an action potential in the presynaptic terminal and the resulting release of glutamate cause firing of action potentials in the postsynaptic neuron. LTD is induced when repeated firing of an action potential in the presynaptic terminal does not cause firing of action potentials in the postsynaptic neuron.

frequency ranges (Franks and Sejnowski 2002; Sjostrom and Nelson 2002). The higher frequency, larger amplitude influx of calcium leads to two important molecular changes in the spine synapse over the next 15–60 sec (Fig. 2). More AMPARs are inserted into the postsynaptic membrane, increasing the size of the EPSP in the spine upon subsequent synaptic activation. At the same time, the actin cytoskeleton, which gives the spine its shape, is remodeled, producing a larger and more branched cytoskeleton that supports a larger spine head. In contrast, lower frequency, lower amplitude, but more prolonged influx of calcium produces the opposite effect on the spine. The number of

AMPARs in the postsynaptic membrane is reduced, resulting in a smaller EPSP upon subsequent activation, and the actin cytoskeleton shrinks, leaving a smaller spine head. We do not yet know precisely how this delicate differential regulation by calcium influx is achieved. However, much has been learned about the molecular machinery that is responsible.

Although NMDARs initiate synaptic plasticity, their numbers are not altered by the processes that alter AMPA receptor numbers during LTP or LTD. A variety of modulatory agents in the brain can adjust the flux of calcium through NMDARs (Salter and Kalia 2004) or the frequency range at which the switch between LTP and LTD occurs, a process called metaplasticity (Abraham and Bear 1996; Yang et al. 2012). Homeostatic regulation can lead to slow changes in the steady-state levels of NMDARs and AMPARs; however, these changes are not directly related to storage of memories.

3 SIGNALING MOLECULES IN SPINE SYNAPSES

3.1 The NMDA-Type Glutamate Receptor and the Hebbian Response

Both AMPARs and NMDARs are homologous tetramers arranged such that a channel is formed by the intersection of their intramembrane domains (Mayer 2006). Binding of glutamate to two sites on the extracellular portion of the receptor opens the channel. However, the similarity between AMPARs and NMDARs ends there. Opening of the AMPAR channel produces the predictable rapid EPSP (Seeburg 1993). In contrast, opening of the NMDAR channel alone is not sufficient to produce depolarization because the mouth of the channel is blocked by a bound magnesium ion, which acts like a cork in a bottle. Ions can flow through the open channel of the NMDAR only if the membrane is sufficiently depolarized to loosen the binding of the magnesium ion and relieve the "magnesium block" (Ascher and Nowak 1988).

One event that can depolarize the spine membrane sufficiently is a "back-propagating action potential," a type of dendritic depolarization that was discovered relatively recently (Spruston et al. 1995; Magee and Johnston 1997; Stuart et al. 1997; Magee et al. 1998). Axonal action potentials are initiated near the neuronal cell body at the base of the axon, where the threshold for triggering an action potential is lowest. They then propagate to synapses at the end of the axon in a nondecremental fashion; that is, the size of the depolarizing wave does not decrease as it moves along the axon. Dendritic action potentials are also believed to begin at the base of the axon, but they are decremental, decreasing in size as they back-propagate from the base of the axon through the cell body and into the dendrites. The size of the back-propagating potential and the length that it travels depend on the configuration of potassium and sodium ion channels in each dendrite. However, the depolarization produced can be sufficient to relieve the magnesium block at spines along the dendrite. Therefore, the coincidence of glutamate binding to receptors at a spine and the arrival of a back-propagating action potential will allow the NMDAR channels to open. Electrophysiologists still debate whether there are additional circumstances associated with firing of the postsynaptic neuron that result in strong local depolarization of dendrites during synaptic activity. Nonetheless, it is clear that when a synapse repeatedly contributes to the triggering of postsynaptic action potentials, NMDARs in that synapse will be powerfully activated.

A second important difference between NMDARs and AMPARs is the mixture of ions that flow through their channels. Most AMPARs only allow passage of sodium and potassium ions, which produces depolarization. In contrast, calcium passes through NMDAR channels along with sodium and potassium (MacDermott et al. 1986) and acts as a second messenger in the spine. NMDARs stay activated for several tens of milliseconds, during which the channel flickers open in short bursts, partly because magnesium bounces in and out of the mouth of the channel and partly because opening and closing of any protein channel is stochastic in nature. As a result, calcium flows into the spine in irregular bursts, for tens of milliseconds, and is rapidly pumped out by calcium-ATPases and sodium/calcium exchangers (see p. 95 [Bootman 2012]). Proteins in the spine cytosol that are sensitive to calcium are thus subjected to rapidly fluctuating levels of the ion that may never reach a stable equilibrium as long as the channel is open. In contrast, AMPARs are active for just a few milliseconds and produce only a brief, transient depolarization.

Finally, NMDARs contain very long carboxy-terminal "tails" ($\sim$600 residues in length) that extend into the cytosol and help to organize the postsynaptic signaling machinery. In adults, the tetrameric NMDAR is assembled from mixtures of five receptor subunit isoforms: GluN1, GluN2A, GluN2B, GluN2C, and GluN2D (Kutsuwada et al. 1992; Monyer et al. 1992). Each individual receptor contains two GluN1 subunits, which are necessary for formation of the channel, and a pair of GluN2 subunits (Furukawa et al. 2005; Mayer 2006), which contribute the long cytosolic tails. The carboxy-terminal tail of GluN1 is shorter ($\sim$100–120 residues) compared with those of the GluN2 subunits. In the hippocampus and cortex, the predominant GluN2 subunits are GluN2A and GluN2B (Monyer et al. 1994). Their 600-residue tails associate with distinct but overlapping sets of signaling enzymes and scaffold proteins (Foster et al. 2010). Both of them can associate with the primary postsynaptic density scaffold protein PSD95 and with calcium-/calmodulin-dependent protein kinase II (CaMKII); however, the affinity of CaMKII for GluN2B is considerably higher than for GluN2A (Gardoni et al. 2001).

3.2 Calcium-Regulated Signaling Enzymes in the Postsynaptic Density

The postsynaptic density (PSD) is the name that was given by electron microscopists to a densely staining plaque of proteinaceous material attached to the cytosolic face of the postsynaptic membrane opposite presynaptic vesicle release sites at excitatory synapses. Subcellular fractionation and biochemical analyses (see Kennedy 1997) revealed that it contains specialized scaffold proteins that physically link NMDARs, AMPARs, and signaling enzymes responsible for synaptic plasticity (Kennedy 2000; Sheng and Kim 2011).

About eight calcium-sensitive enzymes reside in significant numbers in or near the PSD, although some appear to have a more central role in synaptic plasticity than others (Table 1). CaMKII and the phosphoprotein phosphatase calcineurin (also known as PP2B) are required for NMDAR-dependent induction of LTP and LTD, respectively. The others participate in signaling pathways that can regulate synaptic function in response to modulatory agents, such as acetylcholine, biogenic amines, or neuropeptides.

3.2.1 CaMKII

CaMKII makes up ∼1% of total protein in the forebrain and ∼2% in the hippocampus (Bennett et al. 1983; Erondu and Kennedy 1985). These high levels of expression, at least 10 times higher than those of other signaling enzymes, are a specialization of excitatory neurons, which represent most of the mass of the forebrain (Sik et al. 1998). CaMKII is present throughout the cytosol of somas, axons, and dendrites, including spines in which it is present both in the cytosol and the PSD (Kennedy et al. 1983; Chen et al. 2005; Khan et al. 2012). Activation of synaptic NMDARs increases association of CaMKII with spines and the PSD; however, the role and mechanism of this translocation are still incompletely understood (Khan et al. 2012).

CaMKII is a complex holoenzyme, the structure of which has interesting consequences for the dynamics of its activation by calcium/CaM (Fig. 3). Each holoenzyme comprises 12 catalytic subunits held together by their carboxy-terminal association domains (Bennett et al. 1983; Kolb et al. 1998; Rosenberg et al. 2005, 2006). Mammalian

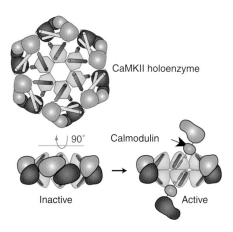

Figure 3. CaMKII. CaMKII is a ring of six dimers of calcium-/calmodulin (CaM)-activated catalytic subunits. The subunits are bound together by a central "hub" structure (light orange) formed from the carboxy-terminal association domains of each subunit. The inactive dimers (light and dark blue) are docked against the central hub by interactions among helices in the association domains (red) and residues in their inhibitory domains (light yellow). Binding of activated calmodulin to the subunit dimers is cooperative because binding to one subunit dissociates the dimer and makes the other subunit more available for calmodulin binding. The activated subunits are mobile and, in addition to phosphorylating other synaptic proteins, they can autophosphorylate each other at a critical threonine residue that locks the subunit in an active state until it is dephosphorylated by phosphatase 1 or phosphatase 2A. (From Rosenberg et al. 2005; modified, with permission, © Elsevier.)

genomes encode four highly similar CaMKII subunits: α, β, γ, and δ (Gaertner et al. 2004). They can form stable homo-oligomers or hetero-oligomers that contain differing numbers of each isoform; the numbers depend on their relative rates of synthesis. Their major sequence differences occur in the linker region between the catalytic and association domains. Only the α and β subunits are highly expressed in brain; and the α subunit is only expressed in neurons. In forebrain, the α:β ratio is ∼3:1 (Bennett et al. 1983). Thus, the unusually high levels of CaMKII in forebrain are primarily a result of the level of expression of the α subunit.

Atomic structures of holoenzymes from *Caenorhabditis elegans* (Rosenberg et al. 2005) and human (Rellos et al. 2010; Chao et al. 2011) revealed that the 12 subunits are arranged in two closely apposed rings (Fig. 3). The association domains are located at the center of the ring and the catalytic domains are around the outside. When the enzyme is inactive, each catalytic domain in the upper ring forms a dimer with a corresponding subunit in the lower ring. Binding of calcium-CaM to one of the subunits in a dimer activates that subunit and frees its partner, increasing its availability to bind calcium-CaM. The magnitude of the resulting cooperativity for binding of calcium-CaM depends on the nature of the subunit dimer interface and

TABLE 1. Calcium-sensitive enzymes in or near the PSD

Calcium/CaM regulated
Calcium-/calmodulin-dependent protein kinase II (CaMKII)
Calcineurin (also known as protein phosphatase 2B)
Calcium/CaM-stimulated adenylyl cyclase (AC1)
Neuronal nitric oxide synthase (nNOS)
Calcium/CaM-dependent phosphodiesterase (PDE1)
Ras-GRF1/2

Calcium regulated
Protein kinase C (PKC)
Calpain protease

Cite this chapter as *Cold Spring Harb Perspect Biol* doi: 10.1101/cshperspect.a016824

the length of the linker region between the catalytic and association domains (Chao et al. 2010, 2011). Thus, the sensitivity of CaMKII to activation by calcium is highly tuned, differing among different species and among individual isoforms in the same species.

When CaMKII is inactive, an inhibitory domain blocks the substrate-binding site. Upon activation by calcium-CaM, CaMKII subunits become autophosphorylated. The first autophosphorylation occurs on a threonine residue (T286) in the inhibitory domain, preventing its binding to the substrate-binding site. The phosphothreonine keeps the active site open even after the calcium concentration returns to baseline and calcium-CaM dissociates (Miller and Kennedy 1986; Miller et al. 1988; Schworer et al. 1988). This first autophosphorylation event is intersubunit; occurring when one activated subunit in the holoenzyme binds to and phosphorylates the inhibitory domain of a neighboring subunit (Hanson et al. 1994). The target subunit must also have bound calcium-CaM so that its inhibitory domain is exposed (Fig. 3) (Hanson et al. 1994; Rellos et al. 2010). Thus, conversion of CaMKII to a calcium-independent state depends on the square of the active CaM concentration. This combined cooperativity of CaM binding and CaM-induced autophosphorylation contributes to the dependence of CaMKII activity on the frequency of calcium influx into the dendritic spine (Chao et al. 2010).

A second autophosphorylation occurs on threonine residues located in the CaM-binding domain (T305 or T306) and blocks binding of calcium-CaM to the kinase (Patton et al. 1990). This event desensitizes CaMKII to subsequent activation by calcium/CaM. Reversal of calcium-independent activity and alleviation of desensitization requires dephosphorylation of the respective sites by protein phosphatase PP1 or PP2A (Shields et al. 1985). Thus, the duration of activation of CaMKII at the postsynaptic site by synaptic activity is regulated reversibly by the magnitude of the transient formation of calcium-CaM during synaptic activity and by local regulation of protein phosphatase activity, which is incompletely understood.

The earliest evidence for a role of CaMKII in synaptic plasticity came from pharmacological experiments in which inhibitors of protein kinases injected postsynaptically blocked induction of LTP (e.g., Malinow et al. 1989). However, definitive evidence for its involvement was obtained when disruptions in learning behavior and synaptic plasticity were observed in a series of mouse mutants. Deletion of the α subunit of CaMKII, for example, results in a deficiency in LTP and impaired spatial learning (Silva et al. 1992a,b). Remarkably, mutation of T286 to alanine in the α subunit abolishes LTP and spatial learning altogether, establishing that autophosphorylation of T286 in CaMKII plays a central role in induction of LTP (Giese et al. 1998).

The concentration of CaMKII in spines and the PSD is highly variable, and regulated by synaptic activity. Estimates of the number of holoenzymes in an average spine (64 attoliters in volume) vary from ~200 to ~1000 (Bennett et al. 1983; Erondu and Kennedy 1985; Lee et al. 2009). Estimates of the number of CaMKII holoenzymes in an average PSD are more variable, ranging from ~10 to ~100 (Chen et al. 2005; Ding et al. 2013). Synaptic activity has been observed to recruit CaMKII to postsynaptic sites, although the mechanism of this movement is not clear (Shen and Meyer 1999). Once in the spine, CaMKII can attach to several docking sites: F-actin filaments, which bind a site on the β subunit (Shen et al. 1998); the cytosolic tails of NR2 subunits of NMDARs (Leonard et al. 2002), which bind a site on both the α and β subunits; and a scaffold protein termed densin (Walikonis et al. 2001; Carlisle et al. 2011), which binds specifically to α subunits. The significance and precise dynamics of these regulated movements of the CaMKII holoenzyme remain to be determined.

The spine and PSD contain many potential targets for phosphorylation by CaMKII, including AMPAR subunit GluA1, the neuronal GTPase-activating protein (GAP) synGAP, and AMPAR-associated transmembrane AMPAR regulatory proteins (TARPs) (see below), all of which are important for orchestrating the early stages of synaptic plasticity. However, the precise timing and coordination of phosphorylation events following influx of calcium are unknown.

3.2.2 Calcineurin

Calcineurin is a calcium/CaM-activated protein phosphatase found in many cell types. Injection of inhibitors of calcineurin into postsynaptic neurons in hippocampal slices first suggested that its activity is required for induction of LTD (Mulkey et al. 1994). Both brain isoforms of its catalytic subunit (CNAα and CNAβ) bind calcium-CaM and form a heterodimer with the regulatory subunit CNB1, which also binds calcium (Kuno et al. 1992). The requirement of calcineurin for induction of LTD was confirmed in hippocampal slices from mice with a forebrain-specific deletion of CNB1. The magnitude of LTD was reduced in these slices and the frequency threshold for the transition from induction of LTD to induction of LTP was shifted to a lower value (Zeng et al. 2001). These findings led to the hypothesis that the relative activation of CaMKII and calcineurin determines whether LTP or LTD will be induced—a hypothesis that remains to be proven. An implication of this hypothesis is that the steady-state number of AMPARs at a synapse and the steady-state size of the actin cytoskeleton are maintained by a balance of CaMKII and calcineurin activities. More recent work reveals that the mechanisms controlling the size and strength of the post-synapse involve the action of several signaling enzymes. For example, the broader-specificity protein phosphatases PP1 and PP2A also appear to be necessary for induction of LTP (Mulkey et al. 1993). It is not yet clear how synaptic activity regulates these two phosphatases. Furthermore, protein kinases PKA and PKC, which mediate the action of other major second messenger pathways, can regulate synaptic plasticity when activated by any of several modulatory neurotransmitters, including acetylcholine, biogenic amines, and neuropeptides (Abeliovich et al. 1993; Blitzer et al. 1995; Kennedy et al. 2005).

A possible link between PKA, calcineurin, and PP1 involves a cycle similar to that found in glycogen metabolism in which a small protein inhibitor of PP1 (inhibitor 1) is activated by phosphorylation by PKA and inactivated by dephosphorylation by calcineurin (Huang and Glinsmann 1976; Lisman 1989). However, deletion of inhibitor 1 in the mouse has no effect on LTP in the Schaffer-collateral pathway or in the medial perforant pathway that arises in the entorhinal cortex, but reduces LTP in synapses from the lateral perforant path (Allen et al. 2000). Importantly, the mutation has no effect on performance of spatial learning tasks. Paralogs of inhibitor 1 that regulate PP1 have been identified in the basal ganglia and cerebellum, but none has been found in the hippocampus or cortex. The divergent effects of inhibitor 1 deletion on different synaptic pathways in the hippocampus show that distinct, heterogeneous molecular mechanisms underlie synaptic plasticity both in different dendritic subregions and in different neuronal subtypes. Two other PP1 regulatory proteins, spinophilin and neurabin, can regulate PP1 activity and its association with the actin cytoskeleton. Deletion of spinophilin eliminates LTD induced by low-frequency stimulation of the Schaffer-collateral pathway (Feng et al. 2000). This is consistent with a requirement for PP1 activity but does not shed light on the relationship of calcium influx to PP1 activity during induction of LTD.

Calcineurin is a more selective phosphatase than PP1 or PP2A, requiring upstream motifs, such as LxVP, in its substrates (Grigoriu et al. 2013). Although there is a small amount of overlap in the target sites for CaMKII and calcineurin, many of their target sites do not overlap. Therefore, the shift in phosphorylation status of individual proteins in the spine during and after an influx of calcium is difficult to predict and is still the subject of study.

3.2.3 Modulatory Calcium-Sensitive Enzymes

The calcium/CaM-stimulated adenylyl cyclase isoform AC1 (Wang and Storm 2003), the calcium/CaM-activated cyclic nucleotide phosphodiesterase PDE1 (Sharma et al. 2006), the calcium/CaM-regulated neuronal nitric oxide synthase isoform (nNOS) (Salerno et al. 2013), members of a family of calcium-sensitive guanine nucleotide exchange factors (GEFs) called RasGRF1 and RasGRF2 (Feig 2011), calcium-sensitive PKC isoforms (Lipp and Reither 2011), and the calcium-dependent protease calpain (Croall and DeMartino 1991) are all present at low and varying levels in spines and can modulate the sensitivity of a synapse to induction of synaptic plasticity or the magnitude and duration of plastic changes. These modulatory mechanisms are outside our scope here, however.

3.3 Scaffold Proteins in the PSD

Signaling enzymes are organized within the PSD by three major classes of scaffold proteins: the PSD95 family (also called MAGUKS [for membrane-associated guanylate kinases]); the SHANK family (for SH3 domain and ankyrin repeat domain proteins, also called ProSAPs [for proline-rich-synapse-associated proteins]); and the Homer family. Additional scaffold proteins, including A-kinase-anchoring protein (AKAP) 79 (see Ch. 3 [Newton et al. 2014]), neurabin, and spinophilin, serve to position enzymes such as PKA, PKC, calcineurin, and PP1 in the PSD. TARPs, also called stargazins, associate tightly with AMPARs and help to control their trafficking into the synapse (Fig. 4) (Hastie et al. 2013).

PSD95 and its three relatives SAP97, SAP102, and PSD93 are centered ~12–20 nm from the postsynaptic membrane (Valtschanoff and Weinberg 2001; Chen et al. 2008) and link glutamate receptors to the PSD structure and to proximal signaling enzymes. Each contains three PDZ domains, an SH3 domain, and a carboxy-terminal degenerate guanylate kinase (GuK) domain, all of which act as protein-docking sites (Kornau et al. 1997; Sheng and Kim 2011). The first two PDZ domains of the PSD95 family can bind to several synaptic membrane proteins via PDZ-binding motifs in their carboxyl termini, including the GluN subunits of NMDARs, TARPs, and neuroligin, a transmembrane adhesion protein. In adult mammals, PSD95 is the most abundant of the family members and associates preferentially with GluN2A, whereas SAP102 predominates in synapses during the first few weeks of development and associates preferentially with GluN2B (Sans et al. 2000). These two scaffold proteins facilitate the developmental shift from NMDARs that contain predominantly GluN2B to the adult form that contains a mixture of GluN2A and GluN2B (Sans et al. 2000; Elias et al. 2006). The transition is important for synaptic signaling because complexes formed by PSD95 and SAP102 contain different sets of signaling enzymes.

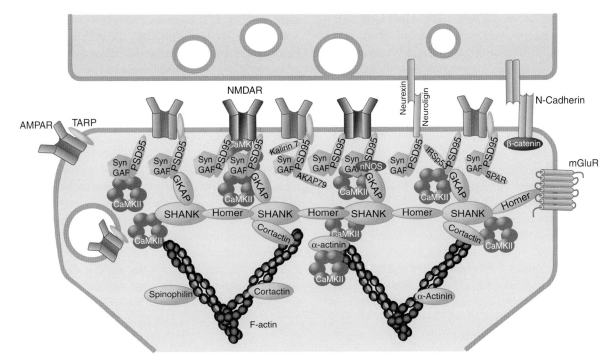

Figure 4. Schematic diagram of the postsynaptic density (PSD) scaffold. NMDARs are immobilized in the post-synaptic membrane by association with PSD95. AMPARs associate with TARPs that bind PSD95. Signaling enzymes are positioned near the receptors by association with PSD95 or other scaffolds. For example, PKA, PKC, and calcineurin bind to AKAP79, which binds to PSD95 and actin in the spine. AMPARs are recruited to the PSD during induction of LTP by diffusional trapping at new docking sites. This diffusional trapping may involve phosphorylation of TARPs by CaMKII, which increases their binding affinity for PSD95. AMPARs are added to or removed from the extrasynaptic plasma membrane pool by exocytosis or endocytosis, respectively. (Modified from Sheng and Kim 2011.)

The PDZ domains can also bind to cytosolic signaling enzymes, including nNOS and synGAP. The carboxy-terminal SH3 domains of the PSD95 family have an unusual split structure created by the insertion of the GuK domain between two α helices (McGee et al. 2001; Tavares et al. 2001). This structure may help regulate oligomerization at the postsynaptic site. The SH3 domain can also bind directly to an AKAP79/150 scaffold protein that localizes PKA, PKC, and calcineurin to the PSD (Gold et al. 2011). Finally, the terminal GuK domain forms a docking site for linker proteins of the GKAP family (for guanylate kinase-associating protein), providing a critical bridge between the PSD95 family and the next layer of postsynaptic scaffold proteins, the SHANK family (Kim et al. 1997; Takeuchi et al. 1997; Naisbitt et al. 1999).

SHANK proteins act as a "scaffold of scaffolds" within the PSD (Fig. 4) (see Sheng and Kim 2011). They form an interacting network centered ~25 nm from the postsynaptic membrane and contain multiple, differentially spliced protein-interaction domains (Sheng and Kim 2000; Boeckers et al. 2002). GKAP binds to a PDZ domain in SHANK, linking it to the PSD95 scaffold (Fig. 4). A pro-

line-rich domain in SHANK links it to the actin cytoskeleton via the SH3 domain of cortactin. Cortactin is an F-actin-binding protein that also enhances binding of actin to the ARP2/3 complex, facilitating branching of actin filaments.

A second proline-rich domain in SHANK interacts with the Homer family of scaffold proteins (Fig. 4). Homer proteins form tetramers linked by their carboxy-terminal coiled-coil domains. Each tetramer contains four identical amino-terminal EVH domains that bind to proline-rich sites on three classes of synaptic proteins: SHANK, the cytosolic tails of metabotropic glutamate receptors (mGluRs), and inositol 1,4,5-trisphosphate (IP$_3$) receptors (IP$_3$Rs). Thus, the Homer proteins can form a bridge between mGluRs and internal calcium stores located in vesicles of smooth endoplasmic reticulum that contain IP$_3$Rs. Not all spines contain such stores, but when they are present calcium is released into the cytoplasm when IP$_3$ binds to the IP$_3$Rs (p. 95 [Bootman 2012]). Finally, the EVH domains link all of the Homer complexes directly to SHANK.

The dense protein network formed by these three families of scaffold proteins is highly dynamic. For example,

PSD95 exchanges between neighboring spines with a median retention time of ~30–100 min (Gray et al. 2006). Retention time decreases during sensory deprivation (Gray et al. 2006) and also after synaptic activity (Steiner et al. 2008). The composition and size of PSDs thus appear to be regulated in an ongoing way by neural activity. This dynamic scaffold provides the underlying spatial organization for the signaling events that regulate the strength of the synapse (Fig. 4).

Genome-wide association (GWAS) studies have identified mutations or copy-number variants in PSD scaffolds as risk factors for autism spectrum disorders. In humans and in mice, deletion of SHANK scaffold proteins has been repeatedly linked to autistic behaviors (Durand et al. 2007; Herbert 2011; Peca et al. 2011). Similarly, deletion of PSD95 in mice results in behaviors associated with autism, and two human single-nucleotide polymorphisms in the gene encoding PSD95 are associated with characteristics of William's syndrome, a genetic disorder that includes a highly social personality and cognitive difficulties (Feyder et al. 2010).

4 REGULATION OF THE NUMBER OF AMPARs IN A SYNAPSE

One of the two central processes associated with LTP and LTD is regulation of the number of synaptic AMPARs. What is the link between activation of calcium-dependent enzymes and control of the number of AMPARs at a synapse? AMPAR number appears to be regulated at several levels, and we are beginning to unravel some of the signaling pathways involved.

AMPARs are tetramers formed from mixtures of four subunits: GluA1, GluA2, GluA3, and GluA4. In adult forebrain neurons, the most prominent subunit combinations are GluA1-GluA2 (so-called GluA1-2 receptors) and GluA2-GluA3 (so-called GluA2-3 receptors) (Lu et al. 2009). Each subunit can be alternatively spliced to form additional isomers. The tetramer combinations are formed during synthesis in the soma and dendrites, and vesicles containing the receptors are then carried to synaptic sites by constitutive membrane trafficking. One important functional distinction among the isomers is the sequence of the cytoplasmic tail (Malinow and Malenka 2002). GluA1, GluA2L, and GluA4 each have a cytoplasmic tail of ~100 residues, whereas GluA2 and GluA3 have shorter tails (~50–70 residues). The two types of tails each contain distinct phosphorylation sites and distinct PDZ-domain ligands that govern trafficking of the receptor and movement into the synaptic membrane. GluA1-2 receptors are less abundant in PSDs, but they are the receptors that are added to synapses during induction of LTP (Shi et al. 2001).

They are also the receptors that are removed when recently potentiated synapses undergo activity-dependent LTD. In contrast, GluA2-3 receptors cycle constitutively into and out of the PSD, independently of synaptic activity. The GluA2-3 tetramers gradually replace GluA1-2 tetramers during periods of low activity, while maintaining the total steady-state number of AMPA receptors (Shi et al. 2001).

Experiments in which individually tagged surface AMPARs are tracked in real time have provided evidence for a three-step mechanism of activity-induced movement of GluA1-2 tetramers to the synapse. The steps include: (1) exocytosis of AMPARs at extrasynaptic and perisynaptic sites, (2) lateral diffusion into synapses, and (3) rate-limiting diffusional trapping in the PSD (Opazo and Choquet 2011). In this model, activation of NMDARs initiates two parallel signaling cascades: one that facilitates diffusional trapping of AMPARs in the PSD, and a second that increases AMPAR exocytosis perisynaptically (Fig. 5) (Makino and Malinow 2009; Petrini et al. 2009). Thus, a relatively rapid increase in the number of AMPARs at the synapse (less than a minute) is accomplished by diffusional trapping (Opazo et al. 2010); concurrently, the mobile population of AMPARs in the dendritic and perisynaptic region is replenished by exocytotic delivery of new AMPARs over a period of minutes (Makino and Malinow 2009).

Recent work has clarified the mechanism of diffusional trapping. Opazo et al. (2010) found that, in the absence of recent NMDAR activation, AMPARs on the dendritic surface are highly mobile and exchange rapidly between extrasynaptic and synaptic sites. Under these basal conditions, CaMKII is not highly enriched in the PSD (Ding et al. 2013). Activation of NMDARs causes influx of calcium into the spine, which activates CaMKII and causes it to translocate into the PSD, where it phosphorylates several sites on the cytoplasmic carboxyl terminus of the AMPAR-associated TARPs (Tomita et al. 2005). This phosphorylation facilitates binding of the TARPs to PDZ domains on PSD95, and thus restricts diffusion of AMPARs, trapping them in the PSD (Opazo et al. 2010). This diffusional trapping mechanism (see Fig. 5A) may result in the addition of new placeholders (slots) for AMPARs, as suggested by Shi et al. (2001).

New AMPARs are added by exocytosis to the plasma membrane of spines and adjoining dendrites during a period lasting several seconds to a few minutes following induction of LTP (Patterson et al. 2010). The biochemical mechanisms by which activation of NMDARs and subsequent activation of CaMKII lead to increased exocytosis of AMPARs are unknown. One clue is that inhibition of the Ras-ERK1/2 pathway partially blocks the activity-induced increase in exocytosis rate (Patterson et al. 2010).

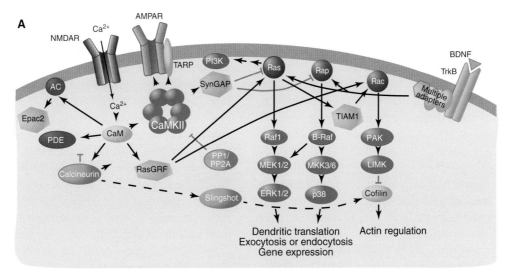

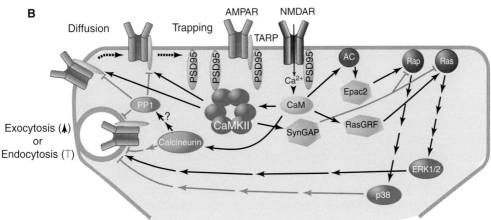

Figure 5. Signaling pathways in the spine. (A) Schematic diagram of major signaling enzymes that mediate changes in synaptic plasticity driven by calcium influx through activated NMDARs. Critical early targets of CaMKII include AMPARs, TARPs, and SynGAP. RasGRF can activate both Ras and Rac. The critical targets of calcineurin are not known. One possibility is the cofilin phosphatase Slingshot, which dephosphorylates and activates the F-actin regulator cofilin. Calcium/CaM-dependent AC and cAMP-phosphodiesterase (PDE) are both present in spines; their responses to calcium could generate a transient spike in cAMP, activating Epac. Ras and Rap regulate trafficking of AMPARs, but their downstream targets, beyond the MAPKs, are not yet known. RasGRF and Ras, acting through the GTPase exchange factor TIAM1 can activate Rac, which regulates pathways that control polymerization of actin. TrkB receptors that respond to the central nervous system (CNS) hormone brain-derived neural factor (BDNF) can provide tonic activation of Ras and Rap. (B) Synaptic regulation of AMPAR trafficking. A critical step in the induction of LTP is the trapping of additional AMPARs in the PSD scaffold through association of TARP and PSD95, which is increased by phosphorylation of residues on TARP by CaMKII. Dephosphorylation of these residues by PP1 can produce loss of AMPARs and depotentiation. Calcineurin, a calcium/CaM-dependent phosphatase, regulates PP1 and endocytosis. Addition of AMPARs to the dendritic plasma membrane by exocytosis, and their removal by endocytosis, occurs at perisynaptic sites in the spine and along the dendritic shaft. Changes in the activity of Ras and Rap are regulated by downstream targets of calcium/CaM, including RasGRF, SynGAP, and adenylyl cyclase (AC). Active Ras and Rap, in turn, activate the MAPKs ERK1/2 and p38. Pathways downstream from ERK facilitate exocytosis, whereas those downstream from p38 facilitate endocytosis. Many of the intermediate steps in the processes downstream from MAPKs are unknown.

5 THE ROLE OF SMALL GTPases IN SYNAPTIC PLASTICITY

The Ras and Rho families of small GTPases play important roles in synaptic plasticity, regulating both insertion and removal of AMPARs at the membrane, and growth and shrinkage of the spine actin cytoskeleton (Fig. 5). Their activity influences both homeostatic maintenance of synaptic structure and alterations in structure that occur during changes in synaptic strength. However, most of what we know about their roles is still phenomenological; many mechanistic questions remain about how they are activated in the spine by synaptic activity and which downstream targets drive changes in synaptic structure.

Ras and the downstream MAPKK (MEK), which activates the MAPK ERK1/2, appear to be important for activity-driven increases in the number of synaptic AMPARs during induction of LTP (Fig. 5A). Inhibition of MEK before electrophysiological induction of LTP blocks the normal increase in the number of synaptic AMPARs (Zhu et al. 2002). Furthermore, transfections of primary cultures of hippocampal neurons with wild-type, constitutively active, or dominant-negative forms of Ras for 15 h cause changes in AMPAR-mediated synaptic currents, which is consistent with the notion that endogenous Ras activity contributes to an increase in the number of synaptic AMPARs. Similar experiments with the GTPase Rap suggest that endogenous Rap contributes to a decrease in synaptic AMPAR levels and that inhibition of the downstream MAPK p38 blocks the normal decrease in the number of synaptic AMPARs produced after LTD induction. Because of the relatively long timescale of these experimental manipulations of the Ras and Rap pathways, it has been difficult to separate their effects on homeostatic maintenance of synaptic structure from their roles in acute changes in synaptic strength.

Ras and Rap are activated by several mechanisms in spines (Kennedy et al. 2005). Ras can be activated directly by the calcium/CaM-dependent RasGRF (Feig 2011). A variety of hormones in the brain stimulate receptor tyrosine kinases to activate Ras and Rap. A prominent example in spines is TrkB, which responds to the peptide hormone brain-derived neurotrophic factor (BDNF) and regulates synaptic function in a variety of ways during development and in the adult (Minichiello 2009; Yoshii and Constantine-Paton 2010). Active, phosphorylated TrkB can activate Ras or Rap1, depending on which adapter proteins it binds (York et al. 1998; Stork 2003). In addition, Rap is activated by Epac2, a cAMP-activated Rap GEF (RapGEF), which responds to cAMP formed in the spine by either calcium/CaM-dependent or G-protein-activated adenylyl cyclases (Grewal et al. 1999; Penzes et al. 2011; p. 99 [Sassone-Corsi 2012]).

Downstream Ras effectors in spines include Raf1, which initiates the phosphorylation cascade leading to activation of ERK1/2; phosphoinositide 3-kinase (PI3K), an enzyme that forms 3′ phosphorylated phosphatidylinositol lipids such as PIP_3, an activator of several membrane-bound signaling proteins (p. 87 [Hemmings and Restuccia 2012]); and TIAM1, a membrane-associated RacGEF (Tolias et al. 2005) that links activation of Ras to activation of Rac. The principal Rap effector in spines is B-Raf, a form of Raf that, in neurons, can activate the p38 MAPK cascade as well as the ERK1/2 cascade (Shi et al. 2005).

A second line of evidence for critical roles of Ras and Rap in regulation of synaptic plasticity comes from studies of synGAP (Chen et al. 1998; Kim et al. 1998). SynGAP is abundant in the PSD and its GAP activity against both Ras and Rap is increased by phosphorylation by CaMKII (Oh et al. 2002, 2004; Krapivinsky et al. 2004). Thus, activation of CaMKII by NMDARs produces a decrease in levels of active Ras. However, this occurs with a delay, and because of the presence of calcium/CaM-sensitive RasGRF in the spine, there is first a transient spike in the level of active Ras (Carlisle et al. 2008). Mice lacking synGAP entirely die shortly after birth (Kim et al. 2003; Vazquez et al. 2004). However, mice heterozygous for synGAP survive and exhibit defects in dendritic spine development, a reduced amplitude of hippocampal LTP, and defective learning behavior (Komiyama et al. 2002; Kim et al. 2003; Vazquez et al. 2004; Carlisle et al. 2008). Thus, synGAP is unusual among PSD proteins in that it exhibits a gene-dosage effect; simply reducing the amount of synGAP by half produces a striking phenotype. The importance of synGAP for cognition is underscored by reports that copy-number variants or single point mutations in the gene encoding it are found in ~5% of human patients with nonsyndromic intellectual disability (NSID) and none are found in controls (Hamdan et al. 2009, 2011).

Phenotypes of cultured $synGAP^{-/-}$ neurons and $synGAP^{+/-}$ heterozygous mice highlight the role of Ras and Rap in modification of the spine actin cytoskeleton. Cultured $synGAP^{-/-}$ hippocampal neurons show accelerated spine development, and at maturity their spines are significantly larger than those of wild type (Vazquez et al. 2004). Neurons in intact brains of $synGAP^{+/-}$ heterozygotes also have larger spines, display markedly increased activation of Ras and Rac, and increased phosphorylation of cofilin, a regulator of actin polymerization (Carlisle et al. 2008). In neurons cultured from wild-type mice, activation of NMDARs causes transient dephosphorylation of cofilin, most likely by the specific phosphatase Slingshot, which is activated by calcineurin (Carlisle et al. 2008). The dephosphorylation activates cofilin, leading to depolymerization of actin. Subsequent activation of the kinase PAK leads to phosphorylation and activation of LIM kinase, which then phosphorylates and inactivates cofilin, so that the actin cytoskeleton can repolymerize in a new form. In $synGAP^{-/-}$ neurons, the basal level of active PAK is elevated, and the initial transient dephosphorylation of cofilin is blunted (Carlisle et al. 2008), apparently tipping the balance toward increasing actin polymerization and larger spines.

$SynGAP^{-/-}$ neurons also have increased levels of synaptic AMPARs (Vazquez et al. 2004; Rumbaugh et al. 2006) and alterations in the regulation of ERK1/2 and p38 (Krapivinsky et al. 2004; Rumbaugh et al. 2006; Carlisle et al.

2008). Thus, it appears that synGAP exerts a carefully balanced restrictive effect both on spine size and on synaptic strength through its regulation of Ras and Rap activity.

The precise timing of activation of Ras and Rap by their various effectors during induction of LTP or LTD has not been studied with precision; thus, we do not yet know the exact mechanisms by which their activation is coordinated to control transient changes during synaptic plasticity versus homeostatic regulation of neuronal excitability.

6 CONCLUDING REMARKS

Synapses in the brain release a number of different neurotransmitters including GABA, acetylcholine, the biogenic amines serotonin, dopamine, and norepinephrine, and a wide variety of peptide neurohormones. However, as far as we now know, it is only excitatory glutamatergic synapses that display the Hebbian form of regulation discussed here. The sculpting of excitatory connections in response to input from the environment is the principal mechanism of memory formation in the brain. As excitatory connections are altered by the Hebbian mechanism, new neural networks are formed, and others are weakened or strengthened. All of the other types of synapses contribute to regulation of Hebbian plasticity and help to determine the conditions under which specific memories are formed, as well as how long the memories will last. We know much less about regulation of the size of the signal and response in these other synaptic types, which are fewer in number and are dispersed among the more abundant glutamatergic synapses, making them less accessible to molecular manipulation or measurement.

Another obstacle to our full understanding of synaptic regulation is the subtle variation in mechanisms of synaptic plasticity in spines of different excitatory neuronal types and among neurons in different brain regions. These differences effectively obscure our vision because most experimental methods either sample blindly from the mixture of synapses in a preparation or record average changes from a poorly understood mixture of synaptic types. As we learn which receptors and enzymes play critical roles in modulating synaptic plasticity, new anatomical techniques such as array tomography (Micheva et al. 2010) and superresolution confocal microscopy (Dani et al. 2010) will help to sort out distinct synaptic types.

A final experimental frontier concerns the delicate timing of synaptic regulation required for healthy brain function. To paraphrase Marc Kirschner describing regulation of embryonic development, "*In the regulation of the brain*, as in the theater, timing is everything. Imagine if, one night, the actors in a play were to miss every single cue, delivering each line perfectly, but always too early or too late. The

evening would be a disaster. The same is true in *brain function*. Starting at the moment when *the environment stimulates sensory endings, neurons in the brain send signals to each other to coordinate sensory perception, emotional and motor responses, and the laying down of memories*. Not only do the signals have to be correct, they also must be perfectly timed. Otherwise, disasters like *mental illness* can result" (paraphrased words in italics) (see kirschner.med.harvard.edu).

A challenge arises from the fact that the biochemical reactions that initiate and sculpt changes in spine structure underlying activity-dependent synaptic plasticity occur in a tiny compartment that contains tens to several hundred copies of the requisite enzymes and effectors. Some of these are immobilized by scaffold proteins that hold them in close proximity to the most important downstream targets. Additional complexity arises from the fact that the initiating calcium signal is always fluctuating, driven by the stuttering kinetics of the NMDAR channel and active calcium pumps in the spine membrane. Thus, time-resolved, high-resolution mass spectroscopy and engineered biochemical real-time sensors, in concert with modeling methods such as those afforded by the spatially accurate, stochastic modeling program MCell (e.g., see Kennedy et al. 2005), will be needed to help resolve rapid, transient molecular events involved in memory formation from those underlying homeostatic mechanisms.

REFERENCES

*Reference is in this book.

Abeliovich A, Chen C, Goda Y, Silva AJ, Stevens CF, Tonegawa S. 1993. Modified hippocampal long-term potentiation in PKCγ-mutant mice. *Cell* **75**: 1253–1262.

Abraham WC, Bear MF. 1996. Metaplasticity: The plasticity of synaptic plasticity. *Trends Neurosci* **19**: 126–130.

Allen PB, Hvalby O, Jensen V, Errington ML, Ramsay M, Chaudhry FA, Bliss TV, Storm-Mathisen J, Morris RG, Andersen P, et al. 2000. Protein phosphatase 1 regulation in the induction of long-term potentiation: Heterogeneous molecular mechanisms. *J Neurosci* **20**: 3537–3543.

Ascher P, Nowak L. 1988. The role of divalent cations in the N-methyl-D-aspartate responses of mouse central neurones in culture. *J Physiol* **399**: 247–266.

Bennett MK, Erondu NE, Kennedy MB. 1983. Purification and characterization of a calmodulin-dependent protein kinase that is highly concentrated in brain. *J Biol Chem* **258**: 12735–12744.

Blitzer RD, Wong T, Nouranifar R, Iyengar R, Landau EM. 1995. Postsynaptic cAMP pathway gates early LTP in hippocampal CA1 region. *Neuron* **15**: 1403–1414.

Boeckers TM, Bockmann J, Kreutz MR, Gundelfinger ED. 2002. ProSAP/Shank proteins—a family of higher order organizing molecules of the postsynaptic density with an emerging role in human neurological disease. *J Neurochem* **81**: 903–910.

* Bootman MD. 2012. Calcium signaling. *Cold Spring Harb Perspect Biol* **4**: a011171.

Carlisle HJ, Manzerra P, Marcora E, Kennedy MB. 2008. SynGAP regulates steady-state and activity-dependent phosphorylation of cofilin. *J Neurosci* **28**: 13673–13683.

Carlisle HJ, Luong TN, Medina-Marino A, Schenker LT, Khorosheva EM, Indersmitten T, Gunapala KM, Steele AD, O'Dell TJ, Patterson PH, et al. 2011. Deletion of densin-180 results in abnormal behaviors associated with mental illness and reduces mGluR5 and DISC1 in the postsynaptic density fraction. *J Neurosci* **31**: 16194–16207.

Chao LH, Pellicena P, Deindl S, Barclay LA, Schulman H, Kuriyan J. 2010. Intersubunit capture of regulatory segments is a component of cooperative CaMKII activation. *Nat Struct Mol Biol* **17**: 264–272.

Chao LH, Stratton MM, Lee IH, Rosenberg OS, Levitz J, Mandell DJ, Kortemme T, Groves JT, Schulman H, Kuriyan J. 2011. A mechanism for tunable autoinhibition in the structure of a human Ca^{2+}/calmodulin-dependent kinase II holoenzyme. *Cell* **146**: 732–745.

Chen H-J, Rojas-Soto M, Oguni A, Kennedy MB. 1998. A synaptic Ras-GTPase activating protein (p135 SynGAP) inhibited by CaM Kinase II. *Neuron* **20**: 895–904.

Chen X, Vinade L, Leapman RD, Petersen JD, Nakagawa T, Phillips TM, Sheng M, Reese TS. 2005. Mass of the postsynaptic density and enumeration of three key molecules. *Proc Natl Acad Sci* **102**: 11551–11556.

Chen X, Winters C, Azzam R, Li X, Galbraith JA, Leapman RD, Reese TS. 2008. Organization of the core structure of the postsynaptic density. *Proc Natl Acad Sci* **105**: 4453–4458.

Ch'ng TH, Martin KC. 2011. Synapse-to-nucleus signaling. *Curr Opin Neurobiol* **21**: 345–352.

Croall DE, DeMartino GN. 1991. Calcium-activated neutral protease (calpain) system: Structure, function, and regulation. *Physiol Rev* **71**: 813–847.

Dani A, Huang B, Bergan J, Dulac C, Zhuang X. 2010. Superresolution imaging of chemical synapses in the brain. *Neuron* **68**: 843–856.

Ding J-D, Kennedy MB, Weinberg R. 2013. Subcellular organization of CaMKII in rat hippocampal pyramidal neurons. *J Comp Neurol* **521**: 3570–3583.

Durand CM, Betancur C, Boeckers TM, Bockmann J, Chaste P, Fauchereau F, Nygren G, Rastam M, Gillberg IC, Anckarsater H, et al. 2007. Mutations in the gene encoding the synaptic scaffolding protein SHANK3 are associated with autism spectrum disorders. *Nat Genet* **39**: 25–27.

Elias GM, Funke L, Stein V, Grant SG, Bredt DS, Nicoll RA. 2006. Synapse-specific and developmentally regulated targeting of AMPA receptors by a family of MAGUK scaffolding proteins. *Neuron* **52**: 307–320.

Erondu NE, Kennedy MB. 1985. Regional distribution of type II Ca^{2+}/calmodulin-dependent protein kinase in rat brain. *J Neurosci* **5**: 3270–3277.

Feig LA. 2011. Regulation of neuronal function by Ras-GRF exchange factors. *Genes Cancer* **2**: 306–319.

Feng J, Yan Z, Ferreira A, Tomizawa K, Liauw JA, Zhuo M, Allen PB, Ouimet CC, Greengard P. 2000. Spinophilin regulates the formation and function of dendritic spines. *Proc Natl Acad Sci* **97**: 9287–9292.

Feyder M, Karlsson RM, Mathur P, Lyman M, Bock R, Momenan R, Munasinghe J, Scattoni ML, Ihne J, Camp M, et al. 2010. Association of mouse Dlg4 (PSD-95) gene deletion and human DLG4 gene variation with phenotypes relevant to autism spectrum disorders and Williams' syndrome. *Am J Psychiatry* **167**: 1508–1517.

Flavell SW, Greenberg ME. 2008. Signaling mechanisms linking neuronal activity to gene expression and plasticity of the nervous system. *Annu Rev Neurosci* **31**: 563–590.

Foster KA, McLaughlin N, Edbauer D, Phillips M, Bolton A, Constantine-Paton M, Sheng M. 2010. Distinct roles of NR2A and NR2B cytoplasmic tails in long-term potentiation. *J Neurosci* **30**: 2676–2685.

Franks KM, Sejnowski TJ. 2002. Complexity of calcium signaling in synaptic spines. *Bioessays* **24**: 1130–1144.

Furukawa H, Singh SK, Mancusso R, Gouaux E. 2005. Subunit arrangement and function in NMDA receptors. *Nature* **438**: 185–192.

Gaertner TR, Kolodziej SJ, Wang D, Kobayashi R, Koomen JM, Stoops JK, Waxham MN. 2004. Comparative analyses of the three-dimensional structures and enzymatic properties of the α, β, γ and δ isoforms of Ca^{2+}-calmodulin-dependent protein kinase II. *J Biol Chem* **279**: 12484–12494.

Gardoni F, Schrama LH, Kamal A, Gispen WH, Cattabeni F, Di Luca M. 2001. Hippocampal synaptic plasticity involves competition between Ca^{2+}/calmodulin-dependent protein kinase II and postsynaptic density 95 for binding to the NR2A subunit of the NMDA receptor. *J Neurosci* **21**: 1501–1509.

Giese KP, Fedorov NB, Filipkowski RK, Silva AJ. 1998. Autophosphorylation at Thr286 of the α calcium-calmodulin kinase II in LTP and learning. *Science* **279**: 870–873.

Gold MG, Stengel F, Nygren PJ, Weisbrod CR, Bruce JE, Robinson CV, Barford D, Scott JD. 2011. Architecture and dynamics of an A-kinase anchoring protein 79 (AKAP79) signaling complex. *Proc Natl Acad Sci* **108**: 6426–6431.

Gray NW, Weimer RM, Bureau I, Svoboda K. 2006. Rapid redistribution of synaptic PSD-95 in the neocortex in vivo. *PLoS Biol* **4**: e370.

Grewal SS, York RD, Stork PJ. 1999. Extracellular-signal-regulated kinase signalling in neurons. *Curr Opin Neurobiol* **9**: 544–553.

Grigoriu S, Bond R, Cossio P, Chen JA, Ly N, Hummer G, Page R, Cyert MS, Peti W. 2013. The molecular mechanism of substrate engagement and immunosuppressant inhibition of calcineurin. *PLoS Biol* **11**: e1001492.

Hamdan FF, Gauthier J, Spiegelman D, Noreau A, Yang Y, Pellerin S, Dobrzeniecka S, Cote M, Perreau-Linck E, Carmant L, et al. 2009. Mutations in SYNGAP1 in autosomal nonsyndromic mental retardation. *N Engl J Med* **360**: 599–605.

Hamdan FF, Daoud H, Piton A, Gauthier J, Dobrzeniecka S, Krebs MO, Joober R, Lacaille JC, Nadeau A, Milunsky JM, et al. 2011. De novo SYNGAP1 mutations in nonsyndromic intellectual disability and autism. *Biol Psychiatry* **69**: 898–901.

Hanson PI, Meyer T, Stryer L, Schulman H. 1994. Dual role of calmodulin in autophosphorylation of multifunctional CaM kinase may underlie decoding of calcium signals. *Neuron* **12**: 943–956.

Hastie P, Ulbrich MH, Wang HL, Arant RJ, Lau AG, Zhang Z, Isacoff EY, Chen L. 2013. AMPA receptor/TARP stoichiometry visualized by single-molecule subunit counting. *Proc Natl Acad Sci* **110**: 5163–5168.

* Hemmings BA, Restuccia DF. 2012. PI3K-PKB/Akt pathway. *Cold Spring Harb Perspect Biol* **4**: a011189.

Herbert MR. 2011. SHANK3, the synapse, and autism. *N Engl J Med* **365**: 173–175.

Herculano-Houzel S. 2009. The human brain in numbers: A linearly scaled-up primate brain. *Front Hum Neurosci* **3**: 31.

Ho VM, Lee JA, Martin KC. 2011. The cell biology of synaptic plasticity. *Science* **334**: 623–628.

Huang FL, Glinsmann WH. 1976. Separation and characterization of two phosphorylase phosphatase inhibitors from rabbit skeletal muscle. *Eur J Biochem* **70**: 419–426.

Kennedy MB. 1997. The postsynaptic density at glutamatergic synapses. *Trends Neurosci* **20**: 264–268.

Kennedy MB. 2000. Signal-processing machines at the postsynaptic density. *Science* **290**: 750–754.

Kennedy MB, Bennett MK, Erondu NE. 1983. Biochemical and immunochemical evidence that the "major postsynaptic density protein" is a subunit of a calmodulin-dependent protein kinase. *Proc Natl Acad Sci* **80**: 7357–7361.

Kennedy MB, Beale HC, Carlisle HJ, Washburn LR. 2005. Integration of biochemical signalling in spines. *Nat Rev Neurosci* **6**: 423–434.

Khan S, Reese TS, Rajpoot N, Shabbir A. 2012. Spatiotemporal maps of CaMKII in dendritic spines. *J Comput Neurosci* **33**: 123–139.

Kim E, Naisbitt S, Hsueh YP, Rao A, Rothschild A, Craig AM, Sheng M. 1997. GKAP, a novel synaptic protein that interacts with the guanylate kinase-like domain of the PSD-95/SAP90 family of channel clustering molecules. *J Cell Biol* **136**: 669–678.

Cite this chapter as *Cold Spring Harb Perspect Biol* doi: 10.1101/cshperspect.a016824

Kim JH, Liao D, Lau L-F, Huganir RL. 1998. SynGAP: A synaptic RasGAP that associates with the PSD-95/SAP90 protein family. *Neuron* **20**: 683–691.

Kim JH, Lee HK, Takamiya K, Huganir RL. 2003. The role of synaptic GTPase-activating protein in neuronal development and synaptic plasticity. *J Neurosci* **23**: 1119–1124.

Kolb SJ, Hudmon A, Ginsberg TR, Waxham MN. 1998. Identification of domains essential for the assembly of calcium/calmodulin-dependent protein kinase II holoenzymes. *J Biol Chem* **273**: 31555–31564.

Komiyama NH, Watabe AM, Carlisle HJ, Porter K, Charlesworth P, Monti J, Strathdee DJ, O'Carroll CM, Martin SJ, Morris RG, et al. 2002. SynGAP regulates ERK/MAPK signaling, synaptic plasticity, and learning in the complex with postsynaptic density 95 and NMDA receptor. *J Neurosci* **22**: 9721–9732.

Kornau H-C, Seeburg PH, Kennedy MB. 1997. Interaction of ion channels and receptors with PDZ domain proteins. *Curr Opin Neurobiol* **7**: 368–373.

Krapivinsky G, Medina I, Krapivinsky L, Gapon S, Clapham DE. 2004. SynGAP-MUPP1-CaMKII synaptic complexes regulate p38 MAP kinase activity and NMDA receptor-dependent synaptic AMPA receptor potentiation. *Neuron* **43**: 563–574.

Kuno T, Mukai H, Ito A, Chang CD, Kishima K, Saito N, Tanaka C. 1992. Distinct cellular expression of calcineurin Aα and Aβ in rat brain. *J Neurochem* **58**: 1643–1651.

Kutsuwada T, Kashiwabuchi N, Mori H, Sakimura K, Kushiya E, Kazuaki A, Meguro H, Masaki H, Kumanishi T, Arakawa M, et al. 1992. Molecular diversity of the NMDA receptor channel. *Nature* **358**: 36–41.

Lee SJ, Escobedo-Lozoya Y, Szatmari EM, Yasuda R. 2009. Activation of CaMKII in single dendritic spines during long-term potentiation. *Nature* **458**: 299–304.

Leonard AS, Bayer KU, Merrill MA, Lim IA, Shea MA, Schulman H, Hell JW. 2002. Regulation of calcium/calmodulin-dependent protein kinase II docking to N-methyl-D-aspartate receptors by calcium/calmodulin and α-actinin. *J Biol Chem* **277**: 48441–48448.

Lipp P, Reither G. 2011. Protein kinase C: The "masters" of calcium and lipid. *Cold Spring Harb Perspect Biol* **3**: a004556.

Lisman J. 1989. A mechanism for the Hebb and the anti-Hebb processes underlying learning and memory. *Proc Natl Acad Sci* **86**: 9574–9578.

Lu W, Shi Y, Jackson AC, Bjorgan K, During MJ, Sprengel R, Seeburg PH, Nicoll RA. 2009. Subunit composition of synaptic AMPA receptors revealed by a single-cell genetic approach. *Neuron* **62**: 254–268.

Lüscher C, Malenka RC. 2012. NMDA receptor-dependent long-term potentiation and long-term depression (LTP/LTD). *Cold Spring Harb Perspect Biol* **4**: a005710.

MacDermott AB, Mayer ML, Westbrook GL, Smith SJ, Barker JL. 1986. NMDA-receptor activation increases cytoplasmic calcium concentration in cultured spinal cord neurones. *Nature* **321**: 519–522.

Magee JC, Johnston D. 1997. A synaptically controlled, associative signal for Hebbian plasticity in hippocampal neurons. *Science* **275**: 209–213.

Magee J, Hoffman D, Colbert C, Johnston D. 1998. Electrical and calcium signaling in dendrites of hippocampal pyramidal neurons. *Annu Rev Physiol* **60**: 327–346.

Makino H, Malinow R. 2009. AMPA receptor incorporation into synapses during LTP: The role of lateral movement and exocytosis. *Neuron* **64**: 381–390.

Malinow R, Malenka RC. 2002. AMPA receptor trafficking and synaptic plasticity. *Annu Rev Neurosci* **25**: 103–126.

Malinow R, Schulman H, Tsien RW. 1989. Inhibition of post-synaptic PKC or CaMKII blocks induction but not expression of LTP. *Science* **245**: 862–866.

Mayer ML. 2006. Glutamate receptors at atomic resolution. *Nature* **440**: 456–462.

Mayer ML, Westbrook GL, Guthrie PB. 1984. Voltage-dependent block by Mg^{2+} of NMDA responses in spinal cord neurones. *Nature* **309**: 261–263.

McGee AW, Dakoji SR, Olsen O, Bredt DS, Lim WA, Prehoda KE. 2001. Structure of the SH3-guanylate kinase module from PSD-95 suggests a mechanism for regulated assembly of MAGUK scaffolding proteins. *Mol Cell* **8**: 1291–1301.

Micheva KD, Busse B, Weiler NC, O'Rourke N, Smith SJ. 2010. Single-synapse analysis of a diverse synapse population: Proteomic imaging methods and markers. *Neuron* **68**: 639–653.

Miller SG, Kennedy MB. 1986. Regulation of brain type II Ca^{2+}/calmodulin-dependent protein kinase by autophosphorylation: A Ca^{2+}-triggered molecular switch. *Cell* **44**: 861–870.

Miller SG, Patton BL, Kennedy MB. 1988. Sequences of autophosphorylation sites in neuronal type II CaM kinase that control Ca^{2+}-independent activity. *Neuron* **1**: 593–604.

Minichiello L. 2009. TrkB signalling pathways in LTP and learning. *Nat Rev Neurosci* **10**: 850–860.

Monyer H, Sprengel R, Schoepfer R, Herb A, Higuchi M, Lomeli H, Burnashev N, Sakmann B, Seeburg PH. 1992. Heteromeric NMDA receptors: Molecular and functional distinction of subtypes. *Science* **256**: 1217–1221.

Monyer H, Burnashev N, Laurie DJ, Sakmann B, Seeburg PH. 1994. Developmental and regional expression in the rat brain and functional properties of four NMDA receptors. *Neuron* **12**: 529–540.

Mulkey RM, Herron CE, Malenka RC. 1993. An essential role for protein phosphatases in hippocampal long-term depression. *Science* **261**: 1051–1055.

Mulkey RM, Endo S, Shenolikar S, Malenka RC. 1994. Involvement of a calcineurin/inhibitor-1 phosphatase cascade in hippocampal long-term depression. *Nature* **369**: 486–488.

Naisbitt S, Kim E, Tu JC, Xiao B, Sala C, Valtschanoff J, Weinberg RJ, Worley PF, Sheng M. 1999. Shank, a novel family of postsynaptic density proteins that binds to the NMDA receptor/PSD-95/GKAP complex and cortactin. *Neuron* **23**: 569–582.

* Newton AC, Bootman MD, Scott JD. 2014. Second messengers. *Cold Spring Harb Perspect Biol* doi: 10.1101/cshperspect.a005926.

Nowak L, Bregestovski P, Ascher P, Herbet A, Prochiantz A. 1984. Magnesium gates glutamate-activated channels in mouse central neurones. *Nature* **307**: 462–465.

Oh JS, Chen H-J, Rojas-Soto M, Oguni A, Kennedy MB. 2002. Erratum. *Neuron* **33**: 151.

Oh JS, Manzerra P, Kennedy MB. 2004. Regulation of the neuron-specific Ras GTPase activating protein, synGAP, by Ca^{2+}/calmodulin-dependent protein kinase II. *J Biol Chem* **279**: 17980–17988.

Opazo P, Choquet D. 2011. A three-step model for the synaptic recruitment of AMPA receptors. *Mol Cell Neurosci* **46**: 1–8.

Opazo P, Labrecque S, Tigaret CM, Frouin A, Wiseman PW, De Koninck P, Choquet D. 2010. CaMKII triggers the diffusional trapping of surface AMPARs through phosphorylation of stargazin. *Neuron* **67**: 239–252.

Patterson MA, Szatmari EM, Yasuda R. 2010. AMPA receptors are exocytosed in stimulated spines and adjacent dendrites in a Ras-ERK-dependent manner during long-term potentiation. *Proc Natl Acad Sci* **107**: 15951–15956.

Patton BL, Miller SG, Kennedy MB. 1990. Activation of type II calcium/calmodulin-dependent protein kinase by Ca^{2+}/calmodulin is inhibited by autophosphorylation of threonine within the calmodulin-binding domain. *J Biol Chem* **265**: 11204–11212.

Peca J, Feliciano C, Ting JT, Wang W, Wells MF, Venkatraman TN, Lascola CD, Fu Z, Feng G. 2011. Shank3 mutant mice display autistic-like behaviours and striatal dysfunction. *Nature* **472**: 437–442.

Penzes P, Woolfrey KM, Srivastava DP. 2011. Epac2-mediated dendritic spine remodeling: Implications for disease. *Mol Cell Neurosci* **46**: 368–380.

Peters A, Kaiserman-Abramof IR. 1970. The small pyramidal neurons of the cerebral cortex. The perikaryon, dendrites and spines. *Am J Anat* **127**: 321–356.

Petrini EM, Lu J, Cognet L, Lounis B, Ehlers MD, Choquet D. 2009. Endocytic trafficking and recycling maintain a pool of mobile surface

AMPA receptors required for synaptic potentiation. *Neuron* **63:** 92–105.

Rellos P, Pike ACW, Niesen FH, Salah E, Lee WH, von Delft F, Knapp S. 2010. Structure of the CaMKIIδ/calmodulin complex reveals the molecular mechanism of CaMKII kinase activation. *PLoS Biol* **8:** e1000426.

Rosenberg OS, Deindl S, Sung RJ, Nairn AC, Kuriyan J. 2005. Structure of the autoinhibited kinase domain of CaMKII and SAXS analysis of the holoenzyme. *Cell* **123:** 849–860.

Rosenberg OS, Deindl S, Comolli LR, Hoelz A, Downing KH, Nairn AC, Kuriyan J. 2006. Oligomerization states of the association domain and the holoenzyme of Ca/CaM kinase II. *FEBS J* **273:** 682–694.

Rumbaugh G, Adams JP, Kim JH, Huganir RL. 2006. SynGAP regulates synaptic strength and mitogen-activated protein kinases in cultured neurons. *Proc Natl Acad Sci* **103:** 4344–4351.

Salerno JC, Ray K, Poulos T, Li H, Ghosh DK. 2013. Calmodulin activates neuronal nitric oxide synthase by enabling transitions between conformational states. *FEBS Lett* **587:** 44–47.

Salter MW, Kalia LV. 2004. Src kinases: A hub for NMDA receptor regulation. *Nat Rev Neurosci* **5:** 317–328.

Sans N, Petralia RS, Wang YX, Blahos J, Hell JW, Wenthold RJ. 2000. A developmental change in NMDA receptor-associated proteins at hippocampal synapses. *J Neurosci* **20:** 1260–1271.

★ Sassone-Corsi P. 2012. The cyclic AMP pathway. *Cold Spring Harb Perspect Biol* **4:** a011148.

Schworer CM, Colbran RJ, Keefer JR, Soderling TR. 1988. Ca²⁺/calmodulin-dependent protein kinase II. Identification of a regulatory autophosphorylation site adjacent to the inhibitory and calmodulin-binding domains. *J Biol Chem* **263:** 13486–13489.

Seeburg PH. 1993. The molecular biology of mammalian glutamate receptor channels. *Trends Neurosci* **16:** 359–365.

Sharma RK, Das SB, Lakshmikuttyamma A, Selvakumar P, Shrivastav A. 2006. Regulation of calmodulin-stimulated cyclic nucleotide phosphodiesterase (PDE1): Review. *Int J Mol Med* **18:** 95–105.

Shen K, Meyer T. 1999. Dynamic control of CaMKII translocation and localization in hippocampal neurons by NMDA receptor stimulation. *Science* **284:** 162–166.

Shen K, Teruel MN, Subramanian K, Meyer T. 1998. CaMKIIβ functions as an F-actin targeting module that localizes CaMKIIα/β hetero-oligomers to dendritic spines. *Neuron* **21:** 593–606.

Sheng M, Kim E. 2000. The Shank family of scaffold proteins. *J Cell Sci* **113:** 1851–1856.

Sheng M, Kim E. 2011. The postsynaptic organization of synapses. *Cold Spring Harb Perspect Biol* **3:** a005678.

Shi S, Hayashi Y, Esteban JA, Malinow R. 2001. Subunit-specific rules governing AMPA receptor trafficking to synapses in hippocampal pyramidal neurons. *Cell* **105:** 331–343.

Shi G-X, Han J, Andres DA. 2005. Rin GTPase couples nerve growth factor signaling to p38 and b-Raf/ERK pathways to promote neuronal differentiation. *J Biol Chem* **280:** 37599–37609.

Shields SM, Ingebritsen TS, Kelly PT. 1985. Identification of protein phosphatase-1 in synaptic junctions: Dephosphorylaton of endogenous calmodulin-dependent kinase II and synapse-enriched phosphoproteins. *J Neurosci* **5:** 3414–3422.

Sik A, Hajos N, Gulacsi A, Mody I, Freund TF. 1998. The absence of a major Ca²⁺ signaling pathway in GABAergic neurons of the hippocampus. *Proc Natl Acad Sci* **95:** 3245–3250.

Silva AJ, Paylor R, Wehner JM, Tonegawa S. 1992a. Impaired spatial learning in α-calcium-calmodulin kinase II mutant mice. *Science* **257:** 206–211.

Silva AJ, Stevens CF, Tonegawa S, Wang Y. 1992b. Deficient hippocampal long-term potentiation in α-calcium-calmodulin kinase II mutant mice. *Science* **257:** 201–206.

Sjostrom PJ, Nelson SB. 2002. Spike timing, calcium signals and synaptic plasticity. *Curr Opin Neurobiol* **12:** 305–314.

Spruston N, Schiller Y, Stuart G, Sakmann B. 1995. Activity-dependent action potential invasion and calcium influx into hippocampal CA1 dendrites. *Science* **268:** 297–300.

Steiner P, Higley MJ, Xu W, Czervionke BL, Malenka RC, Sabatini BL. 2008. Destabilization of the postsynaptic density by PSD-95 serine 73 phosphorylation inhibits spine growth and synaptic plasticity. *Neuron* **60:** 788–802.

Stork PJ. 2003. Does Rap1 deserve a bad Rap? *Trends Biochem Sci* **28:** 267–275.

Stuart G, Spruston N, Sakmann B, Häusser M. 1997. Action potential initiation and backpropagation in neurons of the mammalian CNS. *Trends Neurosci* **20:** 125–131.

Takeuchi M, Hata Y, Hirao K, Toyoda A, Irie M, Takai Y. 1997. SAPAPs. A family of PSD-95/SAP90-associated proteins localized at postsynaptic density. *J Biol Chem* **272:** 11943–11951.

Tavares GA, Panepucci EH, Brunger AT. 2001. Structural characterization of the intramolecular interaction between the SH3 and guanylate kinase domains of PSD-95. *Mol Cell* **8:** 1313–1325.

Tolias KF, Bikoff JB, Burette A, Paradis S, Harrar D, Tavazoie S, Weinberg RJ, Greenberg ME. 2005. The Rac1-GEF Tiam1 couples the NMDA receptor to the activity-dependent development of dendritic arbors and spines. *Neuron* **45:** 525–538.

Tomita S, Stein V, Stocker TJ, Nicoll RA, Bredt DS. 2005. Bidirectional synaptic plasticity regulated by phosphorylation of stargazin-like TARPs. *Neuron* **45:** 269–277.

Valtschanoff JG, Weinberg RJ. 2001. Laminar organization of the NMDA receptor complex within the postsynaptic density. *J Neurosci* **21:** 1211–1217.

Vazquez LE, Chen HJ, Sokolova I, Knuesel I, Kennedy MB. 2004. SynGAP regulates spine formation. *J Neurosci* **24:** 8862–8872.

Walikonis RS, Oguni A, Khorosheva EM, Jeng C-J, Asuncion FJ, Kennedy MB. 2001. Densin-180 forms a ternary complex with the α-subunit of CaMKII and α-actinin. *J Neurosci* **21:** 423–433.

Wang H, Storm DR. 2003. Calmodulin-regulated adenylyl cyclases: Cross-talk and plasticity in the central nervous system. *Mol Pharmacol* **63:** 463–468.

Yang K, Trepanier C, Sidhu B, Xie YF, Li H, Lei G, Salter MW, Orser BA, Nakazawa T, Yamamoto T, et al. 2012. Metaplasticity gated through differential regulation of GluN2A versus GluN2B receptors by Src family kinases. *EMBO J* **31:** 805–816.

York RD, Yao H, Dillon T, Ellig CL, Eckert SP, McCleskey EW, Stork PJ. 1998. Rap1 mediates sustained MAP kinase activation induced by nerve growth factor. *Nature* **392:** 622–626.

Yoshii A, Constantine-Paton M. 2010. Postsynaptic BDNF-TrkB signaling in synapse maturation, plasticity, and disease. *Dev Neurobiol* **70:** 304–322.

Zeng H, Chattarji S, Barbarosie M, Rondi-Reig L, Philpot BD, Miyakawa T, Bear MF, Tonegawa S. 2001. Forebrain-specific calcineurin knockout selectively impairs bidirectional synaptic plasticity and working/episodic-like memory. *Cell* **107:** 617–629.

Zhu JJ, Qin Y, Zhao M, Van Aelst L, Malinow R. 2002. Ras and Rap control AMPA receptor trafficking during synaptic plasticity. *Cell* **110:** 443–455.

CHAPTER 13

Signaling in Muscle Contraction

Ivana Y. Kuo[1] and Barbara E. Ehrlich[1,2]

[1]Department of Pharmacology, School of Medicine, Yale University, New Haven, Connecticut 06520

[2]Department of Cellular and Molecular Physiology, School of Medicine, Yale University, New Haven, Connecticut 06520

Correspondence: barbara.ehrlich@yale.edu

SUMMARY

Signaling pathways regulate contraction of striated (skeletal and cardiac) and smooth muscle. Although these are similar, there are striking differences in the pathways that can be attributed to the distinct functional roles of the different muscle types. Muscles contract in response to depolarization, activation of G-protein-coupled receptors, and other stimuli. The actomyosin fibers responsible for contraction require an increase in the cytosolic levels of calcium, which signaling pathways induce by promoting influx from extracellular sources or release from intracellular stores. Increases in cytosolic calcium stimulate numerous downstream calcium-dependent signaling pathways, which can also regulate contraction. Alterations to the signaling pathways that initiate and sustain contraction and relaxation occur as a consequence of exercise and pathophysiological conditions.

Outline

1 INTRODUCTION

Muscle can be subdivided into two general categories: striated muscle, which includes skeletal and cardiac muscles; and nonstriated muscle, which includes smooth muscle such as vascular, respiratory, uterine, and gastrointestinal muscles. In all muscle types, the contractile apparatus consists of two main proteins: actin and myosin. Striated muscle is so called because the regular arrangement of alternating actomyosin fibers gives it a striped appearance. This arrangement allows coordinated contraction of the whole muscle in response to neuronal stimulation through a voltage- and calcium-dependent process known as excitation–contraction coupling. The coupling enables the rapid and coordinated contraction required of skeletal muscles and the heart. Smooth muscle does not contain regular striations or undergo the same type of excitation–contraction coupling. Instead, it typically uses second messenger signaling to open intracellular channels that release the calcium ions that control the contractile apparatus. These processes, in contrast to excitation–contraction coupling, are slow and thus suitable for the slower and more sustained contractions required of smooth muscle. The actomyosin contractile apparatus is both calcium- and phosphorylation-dependent, and restoration of basal calcium levels or its phosphorylation status returns an actively contracting muscle to a noncontractile state. Muscle-specific signals modulate these processes, depending on the type of muscle, its function, and the amount of force required.

In all muscle cells, contraction thus depends on an increase in cytosolic calcium concentration (Fig. 1). Calcium has an extracellular concentration of $2-4$ mM and a resting cytosolic concentration of ~100 nM. It is also stored inside cells within the sarcoplasmic (SR, referring to skeletal and cardiac muscle) and endoplasmic reticulum (ER, referring to smooth muscle) at a concentration of ~0.4 mM (p. 95 [Bootman 2012]). In striated muscle, the increase in calcium levels is due to its release from the SR stores via ryanodine receptor (RyRs). Neurotransmitters such as acetylcholine bind to receptors on the muscle surface and elicit a depolarization by causing sodium/calcium ions to enter through associated channels. This shifts the resting membrane potential to a more positive value, which in turn activates voltage-gated channels, resulting in an action potential (the "excitation" part). The action potential stimulates L-type calcium channels (also known as dihydropyridine receptors). In skeletal muscle, these are mechanically

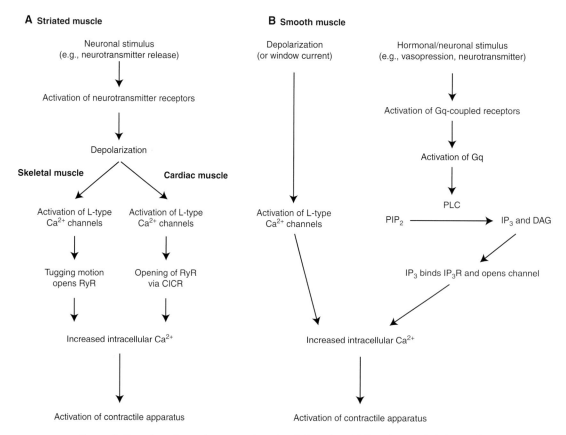

Figure 1. Overview of muscle contraction signals in striated (*A*) and smooth (*B*) muscle.

Cite this chapter as *Cold Spring Harb Perspect Biol* doi: 10.1101/cshperspect.a006023

coupled to the SR RyRs and open them directly. In cardiac muscle, calcium influx through the L-type channels opens RyRs via calcium-induced calcium release (CICR) (p. 95 [Bootman 2012]). The RyR is a large tetrameric six-transmembrane-span calcium-release channel. Of the three RyR subtypes, RyR1 is predominantly found in skeletal muscle (see review by Klein et al. 1996), and RyR2 is predominantly found in cardiac muscle (Cheng et al. 1993).

Smooth muscle also contains voltage-gated calcium channels and RyRs responsible for increases in intracellular calcium concentration (see below). Depolarization causes L-type calcium channels to open, enabling calcium to enter down its concentration gradient into the cell (Fig. 1B). Opening of RyRs is usually associated with CICR. As the intracellular calcium concentration rises, calcium binds to RyRs, whose consequent opening further enhances the increase in cytoplasmic calcium concentration. Another major mechanism controlling contraction in these cells, however, involves a different tetrameric six-transmembrane-span calcium channel: the inositol 1,4,5-trisphosphate (IP_3) receptor (IP_3R). Circulating hormones (e.g., vasopressin and bradykinin) and neurotransmitters released by sympathetic nerves (e.g., endothelin and norepinephrine) act through G-protein-coupled receptors (GPCRs) to generate the second messenger IP_3 via activation of phospholipase C (PLC). IP_3 binds to and opens IP_3Rs on the ER/SR, causing the calcium release that drives contraction. IP_3Rs are present in both skeletal and cardiac muscle; however, they do not contribute significantly to the excitation–contraction coupling in striated muscle. Note that both RyRs and IP_3Rs are stimulated by low concentrations of cytoplasmic calcium but close when the concentration gets higher, showing bell-shaped response curves (Bezprozvanny et al. 1991; Finch et al. 1991).

Once intracellular calcium levels are raised, calcium binds to either troponin C on actin filaments (in striated muscle) or calmodulin (CaM), which regulates myosin filaments (in smooth muscle). In striated muscle, calcium causes a shift in the position of the troponin complex on actin filaments, which exposes myosin-binding sites (Fig. 2A). Myosin bound by ADP and inorganic phosphate (Pi) can then form cross-bridges with actin, and the release of ADP and Pi produces the power stroke that drives contraction. This force causes the thin actin filament to slide past the thick myosin filament and shortens the muscle. Binding of ATP to myosin then releases myosin from actin, and myosin hydrolyzes ATP to repeat the process (Fig. 2B).

In smooth muscle, by contrast, calcium binds to CaM, which then interacts with myosin light-chain kinase (MLCK), causing it to phosphorylate the myosin light-chain (MLC) at S19 or Y18. The phosphorylated MLC then forms cross-bridges with actin, producing phosphorylated actomyosin, which leads to contraction (Fig. 2C). Note that striated muscle contraction can also be regulated by calcium-bound CaM and MLCK; however, this is not the dominant mechanism. Finally, calcium and calcium-CaM also bind to various other proteins in muscle cells, including the phosphatase calcineurin and protein kinases such as CaMKIV, respectively. These regulate other cellular targets, including transcription factors such as NFAT and CREB, which control gene expression programs that can have longer-term effects on muscle physiology.

These different calcium-release mechanisms all also stimulate the pumping of calcium from the cytoplasm back into intracellular stores via the SR/ER calcium ATPase (SERCA) pump. The plasma membrane calcium ATPase (PMCA) pump and the sodium/calcium exchanger (NCX), both of which reside on the plasma membrane, can also remove calcium from the cytosol. Calcium dissociates from troponin C or calmodulin as the cytosolic calcium concentration decreases as a consequence, which terminates the contraction process.

The main pathways promoting muscle relaxation involve the second messengers cAMP and cyclic guanosine monophosphate (cGMP). cAMP is generated by adenylyl cyclases, downstream from the β-adrenergic G_S-coupled receptor, which is activated by noradrenaline. Note that the cAMP pathway generally promotes contraction in cardiac muscle; however, in smooth muscle, activation of cAMP causes relaxation. The cGMP pathway can be activated either by nitric oxide (NO) or natriuretic peptides (NPs). In the case of blood vessels and other smooth muscles, NO produced by endothelial NO synthase (eNOS) diffuses across the muscle cell membrane to activate soluble guanylyl cyclase (sGC), which in turn increases levels of cGMP. NPs, such as atrial (ANP, released by the heart atria under high blood pressure), brain (BNP, primarily released by the heart ventricle), and c-type (CNP, mainly involved in pathological conditions, and released by the vascular and central nervous system), instead bind to transmembrane guanylyl cyclase, whose intracellular domain possesses the enzymatic activity (Nishikimi et al. 2011). The cAMP and cGMP generated act via the protein kinase PKA and PKG on the contractile process in multiple ways: (1) their phosphorylation of calcium pumps leads to increased activity; (2) activation of MLC phosphatase (MLCP) by PKG antagonizes MLCK; and (3) both PKA and PKG cause a reduction in the sensitivity of the contractile machinery by inhibiting the GTPase RhoA (this increases MLCP activity and causes MLC dephosphorylation and muscle relaxation). The levels of cAMP and cGMP are in turn regulated through their degradation by phosphodiesterases to yield the inactive metabolites 5′-AMP and 5′-GMP.

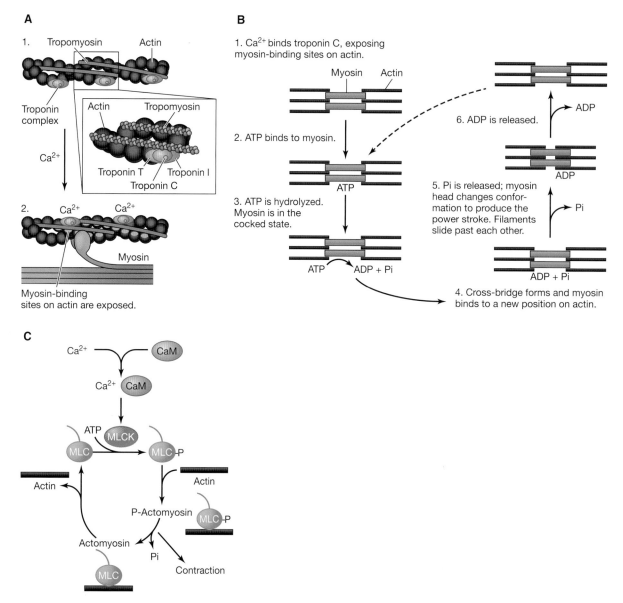

Figure 2. Calcium triggers contraction in striated muscle. (*A*) Actomyosin in striated muscle. (*1*) Striated muscle in the relaxed state has tropomyosin covering myosin-binding sites on actin. (*2*) Calcium binds to troponin C, which induces a conformational change in the troponin complex. This causes tropomyosin to move deeper into the actin groove, revealing the myosin-binding sites. (*B*) Cross-bridge cycle in striated muscle. (*1*) Calcium binds to troponin C, causing the conformational shift in tropomyosin that reveals myosin-binding sites on actin. (*2*) ATP then binds to myosin. (*3*) ATP is then hydrolyzed. (*4*) A cross-bridge forms and myosin binds to a new position on actin. (*5*) Pi is released and myosin changes conformation, resulting in the power stroke that causes the filaments to slide past each other. (*6*) ADP is then released. (*C*) Contraction in smooth muscle. In smooth muscle, calcium binds to calmodulin and causes the activation of myosin light chain (MLC) kinase (MLCK). This phosphorylates MLC, which then binds to actin to form phosphorylated actomyosin, enabling the cross-bridge cycle to start.

Below, we examine the key differences between the signaling mechanisms controlling contraction of skeletal, cardiac, and smooth muscle, and how these relate to their differing functions. In addition, we discuss the changes to the signaling pathways that occur as a consequence of exercise and pathological situations.

2 SKELETAL MUSCLE CONTRACTION

Skeletal muscles comprise multiple individual muscle fibers that are stimulated by motor neurons stemming from the spinal cord. They are grouped together to form "motor units" and more than one type of muscle fiber can be

Cite this chapter as *Cold Spring Harb Perspect Biol* doi: 10.1101/cshperspect.a006023

present within each motor unit. Muscle fibers can be divided into fast- and slow-twitch muscles. Fast-twitch muscles use glycolytic metabolism and are recruited for phasic activity (an active contraction). Slow-twitch muscles (also known as red muscles) are rich in myoglobin, mitochondria, and oxidative enzymes and specialized for sustained or tonic activity. See Schiaffino and Reggiani (2011) for a more complete discussion of skeletal muscle types and the types of myosin isoforms that make up fast- and slow-twitch muscles.

The neuromuscular junction (NMJ) that connects skeletal muscle with the nerves that innervate them consists of three distinct parts: the distal motor nerve ending, the synaptic cleft, and the postsynaptic region, located on the muscle membrane. Motor neurons branch into multiple termini, which are juxtaposed to motor endplates, specialized regions of muscle where neurotransmitter receptors

are concentrated (Fig. 3A). The transfer of information between the nerve and muscle is mediated by the release of acetylcholine from the motor neuron, which diffuses across the synaptic cleft, and binds to and activates the ligand-gated, nicotinic acetylcholine receptors (nAChRs) on the endplate. Activation of the nAChR leads to an influx of cations (sodium and calcium) that causes depolarization of the muscle cell membrane. This depolarization in turn activates a high density of voltage-gated sodium channels on the muscle membrane, eliciting an action potential.

The action potential runs along the top of the muscle and invades the T-tubules (specialized invaginations of the membrane containing numerous ion channels). The opening of voltage-gated sodium channels activates L-type voltage-gated calcium channels lining the T-tubule. A conformational change in these enables release of calcium on the closely apposed SR via activation of RyR1.

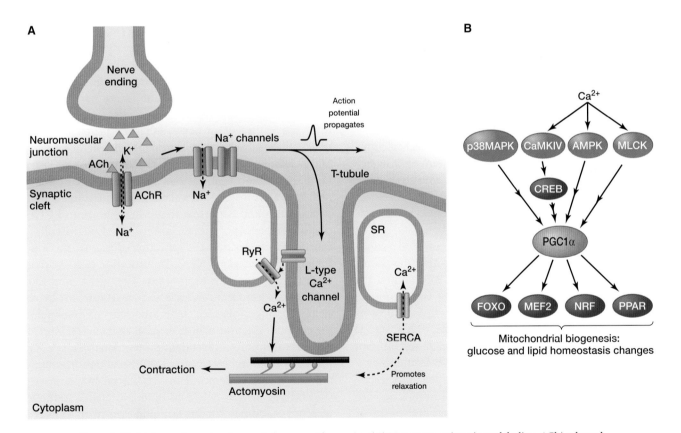

Figure 3. Skeletal muscle contraction and changes with exercise. (A) Neurotransmitter (acetylcholine, ACh) released from nerve endings binds to receptors (AChRs) on the muscle surface. The ensuing depolarization causes sodium channels to open, which elicits an action potential that propagates along the cell. The action potential invades T-tubules and causes the L-type calcium channels to open, which in turn causes ryanodine receptors (RyRs) in the SR to open and release calcium, which stimulates contraction. Calcium is pumped back into the SR by (SR/ER calcium ATPase SERCA) pumps. The decreasing cytosolic calcium levels cause calcium to disassociate from troponin C and, consequently, tropomyosin reverts to a conformation that covers the myosin-binding sites. (B) Signaling in exercised skeletal muscle. Both calcium and calcium-independent signals stimulate the transcriptional coactivator PGC1α. This activates a number of transcription factors that regulate genes associated with mitochondrial biogenesis, glucose, and lipid homeostasis.

Calcium then binds to troponin as described above, initiating the contraction process. Calcium-bound CaM also activates MLCK, whose phosphorylation of the MLC changes cross-bridge properties. This modulates the troponin-dependent contraction, although there is no effect on the ATPase activity of MLC. MLC phosphorylation instead enhances force development at submaximal saturating calcium concentrations (see below). The phosphate group is subsequently removed by protein phosphatase 1 (PP1).

3 SKELETAL MUSCLE FIBER TYPES AND EXERCISE

Skeletal muscle is plastic. Exercise can lead to pronounced changes in its metabolic properties and, sometimes, a change in the fiber type. Physical differences between fast- and slow-twitch muscles underlie the functional roles of these fibers, including the type of myosin used and differing resting calcium levels. The free calcium level is twofold higher in slow-twitch muscle, even though the SR volume is greater. The level of MLC phosphorylation is higher in fast-twitch muscle, however, because of higher levels of expression of MLCK (Bozzo et al. 2005). The force enhancement produced by MLC phosphorylation, under submaximal saturating calcium concentrations, counteracts the reduction in force caused by fatigue in fast-twitch muscle fibers (Schiaffino and Reggiani 2011).

Fast- and slow-twitch fibers also have different calcium-sequestering and -buffering systems. Different SERCA isoforms are present: SERCA2A is the main isoform in slow-twitch muscle fibers, whereas SERCA1A is expressed in fast-twitch muscle fibers. Similarly, different cytosolic calcium buffers are expressed. Calsequestrin (CSQ) is the main SR-luminal calcium-buffering protein. It is a high-capacity, low-affinity calcium-binding protein that binds calcium cooperatively (Campbell et al. 1983). When the muscle is at rest, the SR is primed to release large amounts of calcium, because CSQ is polymerized, which reduces its ability to bind calcium. In cardiac muscle, only CSQ2 is expressed. In skeletal muscle, CSQ1 and CSQ2 are found in slow-twitch muscle fibers, but only CSQ1 is found in fast-twitch muscle fibers. The two isoforms differ in their carboxy-terminal tail; functionally, CSQ1 reduces the activity of RyR1, whereas CSQ2 increases the open probability of RyR1 and RyR2 (Wei et al. 2009).

Other differences between the muscle fiber types include posttranslational modifications such as phosphorylation of RyR by PKA, and interactions between RyR and other proteins, such as CaM and FK506-binding protein (FKBP) 12 and FKBP12.6. Phosphorylation of RyR by either PKA or CaMKII fully activates the channel. PKA and CaMKII can also phosphorylate phospholamban, a protein that inhibits SERCA; phosphorylation causes

phospholamban to dissociate from SERCA. The FKBPs are immunophilins that bind to immunosuppressants such as rapamycin and FK506. FKBP12 and FKBP12.6 have differing expression levels in muscle tissue, but both bind all three forms of the RyR and stabilize its closed state. Collectively, these calcium-dependent differences between fast- and slow-twitch muscle fibers, in addition to differences in the myosin isoform used and the number of mitochondria, account for the different functional outputs of the two muscle fiber types.

Long-term exercise causes a general shift in muscle fiber type from slow twitch to fast twitch. It induces a number of changes, including altered expression and activity of membrane transporters and mitochondrial metabolic enzymes, together with increased blood supply to skeletal muscle (Ch. 14 [Hardie 2012]). These, in turn, enhance the oxidative capacity and increase expression of enzymes preventing damage by reactive oxygen species (ROS). One major signaling pathway is through the peroxisome-proliferator-activated receptor (PPAR) γ coactivator (PGC) 1α (Fig. 3B). PGC1α coactivates a number of transcription factors that regulate genes important for muscle function. These include PPARs (which regulate glucose and lipid homeostasis, proliferation, and differentiation), nuclear respiratory factors (NRFs, which regulate metabolism and mitochondrial biosynthesis), myocyte enhancer factor 2 (MEF2, which is involved in development and hematopoesis), and Forkhead box O (FoxO) family transcription factors (which counter oxidative stress and promote cell-cycle arrest and apoptosis) (Handschin and Spiegelman 2006; Ronnebaum and Patterson 2010). In addition to PGC1α, calcium-dependent processes are also involved. Rises in cytosolic calcium result in the activation of calcineurin, which then dephosphorylates NFAT. Translocation of NFAT to the nucleus results in activation of slow-fiber gene expression. Rises in nuclear calcium levels also cause calcium-dependent signaling molecules to become active. These include the phosphorylation of histone deacetylases (HDACs) by nuclear calmodulin-dependent protein kinase. HDACs repress transcription by causing DNA to be tightly wrapped around histones. Removal of HDACs enables transcription factors such as MEF2 to bind and enable induction of genes encoding proteins found in slow fibers (Liu et al. 2005).

As mentioned above, exercise induces an increase in the levels of mitochondrial metabolic enzymes to compensate for the increased metabolic demand on skeletal muscle. Unsurprisingly, PGC1α is a potent stimulator of mitochondrial biogenesis (see review by Olesen et al. 2010). This was shown elegantly by experiments in which overexpression of PGC1α in white, glycolytic skeletal muscle could turn it into red, oxidative muscle by increasing the levels and activity of a number of mitochondrial proteins (Lin et al. 2002; Wenz et al. 2009). These proteins include most components of the mitochondrial respiratory chain and ATP synthase, as well as several enzymes in the Krebs cycle and enzymes involved in fatty acid oxidation.

4 MALIGNANT HYPERTHERMIA IN SKELETAL MUSCLE

Mutations in RyR and CSQ isoforms cause malignant hyperthermia, demonstrating the importance of proteins involved in calcium signaling in skeletal muscle. The mutations in RyR1 appear to increase its open probability when levels of luminal calcium are low and account for the majority of malignant hyperthermia cases (80%); the remainder are caused by mutations to CSQ1.

In the case of RyR1 mutations, volatile anesthetics (inhaled anesthetics such as isoflurane or halothane) lead to a rapid opening of RyR1 and an uncontrolled release of calcium from the SR, which in turn leads to sustained skeletal muscle contraction (Robinson et al. 2006). In response to the elevated calcium levels, there is activation of SERCA to pump calcium, using ATP, back into the SR. However, the continual activation of SERCA consumes excessive ATP, leading to hypermetabolism. This then leads to a drop in ATP levels, acidosis, tachycardia, and an abnormal increase in body temperature. These symptoms can be treated with dantrolene, an inhibitor of the RyR signaling pathway. The mutations in RyR1 associated with malignant hyperthermia are clustered in three hot spots on the 500 kDa protein (Lanner et al. 2010). The first cluster is near the amino terminus and the second cluster is in the middle of the protein. The third cluster lies in the carboxy-terminal region surrounding the channel-forming domains. How mutations in all three regions exert similar effects is yet to be determined.

Mutations in CSQ can also result in malignant hyperthermia. A lack of buffering causes uncontrolled calcium transients that lead to lethal malignant hyperthermia in response to heat stress and volatile anesthetics (Dainese et al. 2009).

5 CARDIAC MUSCLE CONTRACTION

In cardiac muscle, depolarization starts in the pacemaker cells (modified cardiac myocytes that set the heart rate and are rich in signaling molecules) in the sinoatrial node, which is innervated by both parasympathetic and sympathetic nerves. The external stimuli modulate the activity of the pacemaker cells—they undergo spontaneous self-depolarization to produce action potentials. This is achieved by a slow leak of potassium ions and a concurrent influx of sodium and calcium ions. The action potential then traverses to the cardiac myocytes, where it invades the T-tubule. However, unlike skeletal muscle, where L-type calcium channels are directly coupled to RyRs, in cardiomyocytes the influx of calcium across the plasma membrane elicits calcium release from the SR via RyRs by CICR (Fig. 4B). The predominant isoform in the heart is RyR2. As in skeletal muscle, contraction is controlled by phosphorylation of troponin but can also be modulated by calcium-CaM and MLCK. Mice with a nonphosphorylatable MLC in ventricular myocytes display depressed contractile function and develop atrial hypertrophy and dilatation (Sanbe et al. 1999).

Catecholamines, such as adrenaline and noradrenaline, act on β-adrenergic receptors (metabotropic GPCRs) to release cAMP that in turn activates PKA. PKA can be viewed as a primary regulator of the contractile pathway, as it phosphorylates a number of targets, including L-type calcium channels and RyRs. In most cases, phosphorylation of these proteins increases calcium release (for example, phosphorylation of RyR increases its open probability), and thus the outcome is to stimulate contraction (Ibrahim et al. 2011). Another target of PKA is phospholamban (an inhibitor of SERCA), which, when phosphorylated, loses its inhibitory effect on SERCA.

6 EXERCISE HYPERTROPHY IN CARDIAC MUSCLE

Cardiac hypertrophy is an abnormal enlargement of the heart that occurs because of increases in cell size and proliferation of nonmuscle cells. These changes can either be beneficial (e.g., exercised hearts), in which changes are correlated with increased contractility, or detrimental, in which changes lead to decreased contractility and subsequent heart failure.

Exercised hearts develop a form of mild cardiac hypertrophy that does not lead to cardiac failure. The main structural changes include a thickening of the ventricle wall, which leads to increased contractility and thus a greater ability to pump blood. Within myocytes, expression of the α myosin heavy chain increases, which leads to high ATPase activity and increased contractility (Fig. 4B). Various signals are involved, including growth factors such as insulin-like growth factor (IGF1), vascular endothelial growth factor (VEGF), and hepatocyte growth factor (HGF) (Fig. 4B) (p. 87 [Hemmings and Restuccia 2012]). There is increased signaling through the phosphoinositide 3 kinase (PI3K)/ Akt pathway, which leads to proliferation and growth of cardiomyocytes (Matsui et al. 2003). Transcription factors up-regulated in exercised hearts include GATA4, which regulates genes involved in myocardial differentiation. Other pathways, such as the calcineurin/NFAT pathway are down-

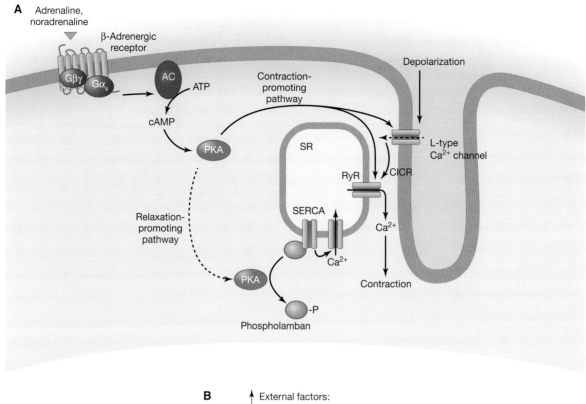

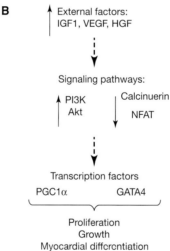

Figure 4. Cardiac muscle contraction and changes with exercise. (*A*) Cardiac muscle contraction can occur as a consequence of calcium entry through L-type calcium channels, which activate ryanodine receptor (RyR) channels in the SR. Alternatively, β-adrenergic receptors on the cell membrane lead to activation of adenylyl cyclase (AC), which stimulates PKA. This can promote contraction by phosphorylating RyR and L-type calcium channels or relaxation by phosphorylating the SERCA pump inhibitor phospholamban. (*B*) Changes with exercise lead to an activation of the PI3K/Akt pathway, and a down-regulation of NFAT and calcinurin.

regulated (Oliveira et al. 2009). Cardiac muscle, like skeletal muscle, consumes tremendous amounts of ATP. Thus, PGC1α is also up-regulated in exercised hearts, facilitating transcription of metabolic and oxidative genes (Ventura-Clapier et al. 2007; Watson et al. 2007).

7 PATHOPHYSIOLOGICAL CARDIAC HYPERTROPHY

The main pathways driving pathological cardiac hypertrophy are overstimulation of the sympathetic nervous system, increased oxidative stress, and inflammatory signaling (Balakumar and Jagadeesh 2010). These collectively lead to induction of fetal isoforms of heart proteins and a corresponding decrease in adult forms (Chien 1999), including the myosin heavy chain (see below). The signals responsible include the GPCR agonist endothelin 1, peptide growth factors such as platelet-derived growth factor (PDGF), epidermal growth factor (EGF), and cytokines such as cardiotrophin and leukemia inhibitory factor (LIF). Mechanical stress can also induce hypertrophy. In each case, activation of the ERK mitogen-activated protein kinase (MAPK) pathway (p. 81 [Morrison 2012]) is often observed in hypertrophy and leads to regulation of transcription factors that alter expression of the myosin heavy chain, IP_3R2, and other proteins (see below).

In hypertrophy, paracrine and autocrine neurohormonal factors that activate the heterotrimeric G protein Gq, and consequently PLCβ, are released. This results in an increase in cytosolic calcium levels and activation of PKC by diacylglycerol (DAG) as well as activation of CaMKII (Mishra et al. 2010). The importance of the Gq pathway in hypertrophy has been shown in studies of transgenic mice: mice overexpressing Gq have heart failure (D'Angelo et al. 1997), whereas mice with reduced Gq levels are protected against hypertrophy (Wettschureck et al. 2001). There is also a switch from the α form of the myosin heavy chain to the fetal β isoform (Miyata et al. 2000). This has a lower ATPase activity and a lower rate of contraction. Other changes include increased SERCA2A activity (Hasenfuss et al. 1994; Meyer et al. 1995), up-regulation of IP_3R2 (Harzheim et al. 2010), and changes to a neuronal calcium sensor (NCS1). NCS1 is a calcium-binding protein that also interacts with IP_3R (Schlecker et al. 2006).

8 HEART FAILURE

The structural organization of the T-tubules breaks down in heart failure. This breakdown, caused by myocardial insults (such as myocardial infarction causing ischemia) among other factors, leads to impaired contractility owing to reduced, asynchronous, and chaotic calcium release. Several signaling pathways are compromised in heart failure. Initially, there can be reorganization of the β-adrenergic system. Activation of the β2-adrenergic receptor is normally limited to the T-tubule, whereas in heart failure, there is a redistribution of the receptor across the entire plasma membrane (Nikolaev et al. 2010). With chronic adrenergic activation, the hyperphosphorylation of RyRs results in leaky RyR channels, leading to a reduction of SR calcium and, thus, weaker contractions.

Other alterations to RyR2 that are observed include increased nitrosylation and loss of the regulatory protein FKBP12.6 (Andersson and Marks 2010). Both of these changes result in increased RyR activity. Moreover, mutations in RyR2 that result in leaky channels have been linked to catecholaminergic polymorphic ventricular tachycardia (CPVT) and arrhythmogenic right ventricular dysplasia type 2.

Mutations in CSQ2 cause CPVT (Postma et al. 2002) by lowering buffering capacity within the SR, which results in premature calcium release and thus arrhythmias. Another important modulator of RyRs is junctin. Its levels are reduced in heart failure, which may be a compensatory mechanism to increase contractility (Pritchard and Kranias 2009).

Heart failure also leads to up-regulation of molecules that may have a protective function. One pathway is through cGMP, which promotes relaxation. The cGMP pathway is regulated by cGMP-targeted phosphodiesterases, of which one, PDE5A, looks to be a promising target for protective therapy against hypertrophy.

9 SMOOTH MUSCLE TYPES

Smooth muscle is found lining the walls of various organs and tubular structures in the body, including the intestine, bladder, airway, uterus, blood vessels, and stomach. It receives neural innervation from the autonomic nervous system, and its contractile state is also controlled by hormonal and autocrine/paracrine stimuli. Smooth muscle can be divided into two types: unitary and multiunit smooth muscles. In unitary smooth muscle, individual smooth muscle cells are coupled to neighboring cells by gap junctions. These gap junctions permit cell-to-cell passage of small molecules such as ATP and ions. These include those mobilized in response to electrical signals causing depolarization, which enable the whole area (known as a syncytium) to coordinate activity. In contrast, in multiunit smooth muscle, cells are not coupled to each other and are intermingled with connective tissue. Smooth muscle can undergo tonic (sustained) or phasic contractions. In the case of vascular smooth muscle, a sustained contraction is required to provide vessel tone. This enables the regulation of blood flow. Blood vessels are divided into the larger diameter conduit vessels (e.g., the thoracic aorta) and the smaller diameter resistance vessels. The resistance blood vessels display a myogenic response, in which increasing pressures over the physiological range (~70–100 mmHg) result in a sustained contractile state. However, overconstriction of the vessels leads to hypertension (see below). Other smooth muscle, such as that found in the gut, including the stomach, small intestine, or gall bladder, shows variable tone and rhythmic contractions known as slow waves.

10 THE CONTRACTILE PROCESS IN SMOOTH MUSCLE

The important distinction between striated muscle and smooth muscle is that calcium mediates contraction by regulating the availability of actin filaments in striated muscle, whereas in smooth muscle MLC is the target (Fig. 5). The source of the increase in cytosolic calcium levels can be extracellular or intracellular, or a combination of the two. In the case of tonic constriction of blood vessels, a constant supply of calcium comes from influx via the L-type calcium channels. The resting membrane potential of smooth muscle (between -50 and -40 mV) is such that it lies in an overlap (the window current) between the activation and inactivation curves of the L-type channel. Thus a small population of the L-type channels is always open. An alternatively spliced high-voltage-gated form of T-type channels may also contribute to calcium influx (Kuo et al. 2011), along with stretch-activated channels residing on the plasma membrane, such as TRPC6.

In stomach muscle, the rhythmic contractions are due to the activity of pacemaker cells, but activation of voltage-gated calcium channels can trigger calcium entry and contraction. Sympathetic nerves run along the vascular smooth muscle and can release stimuli such as acetylcholine, norepinephrine, angiotensin, and endothelin. Moreover, circulating blood factors such as cytokines and diffusible factors such as nitric oxide can also act on receptors in the plasma membrane or cross the plasma membrane, respectively, to regulate pathways controlling intracellular calcium levels.

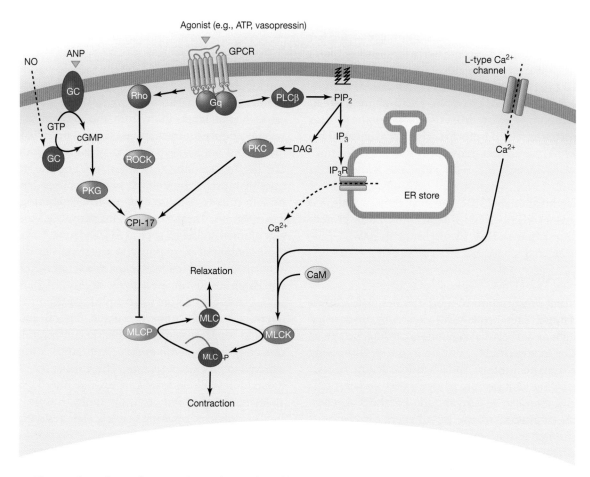

Figure 5. Smooth muscle contraction. Calcium released by L-type calcium channels or IP3Rs downstream from Gq-coupled cell-surface receptors causes smooth muscle contraction. It binds to calmodulin (CaM) and the resulting complex stimulates myosin light-chain (MLC) kinase (MLCK). This phosphorylates MLC to promote contraction. A RhoA/ROCK pathway and a diacylglycerol (DAG) pathway contribute to calcium sensitization by altering the phosphorylation status of myosin light-chain phosphatase (MLCP). Relaxation is mediated through the cGMP/PKG pathway downstream from nitric oxide (NO) and agonists such as atrial natriuretic peptide (ANP).

Cite this chapter as *Cold Spring Harb Perspect Biol* doi: 10.1101/cshperspect.a006023

The activation of receptor-operated channels (ROCs) also causes calcium influx, which enables additional calcium release from intracellular stores. GPCRs activate PLCβ to generate IP$_3$, which releases calcium via IP3Rs. In vascular smooth muscle and the circular smooth muscle of the gut, the main isoform is IP$_3$R1. Note, however, that there is some heterogeneity. In longitudinal smooth muscle of the gut, RyRs, rather than IP$_3$Rs, are expressed. Agonists such as cholecystokinin bind to the GPCR cholecystokinin A receptor (CCK-AR), which activates phospholipase A2, which in turn produces arachidonic acid. Arachidonic acid (AA) can also be generated through the cleavage of DAG. AA activates chloride channels, which depolarize the cell membrane, enabling the opening of voltage-gated calcium channels and an initial influx of calcium. This calcium can either act directly on the RyR causing CICR or enable the release of cyclic ADP ribose, which interacts with RyRs to enhance CICR.

In all smooth muscle, calcium-bound CaM then binds to MLCK, stimulating phosphorylation of MLC, which leads to muscle contraction. The necessity for MLCK has been shown in MLCK-knockout mice, in which smooth muscle MLC cannot be phosphorylated by other kinases (He et al. 2008; Zhang et al. 2010). The dephosphorylation of MLC is catalyzed by MLCP and a complex of the myosin-targeting protein MYPT1 and the phosphatase PP1 and results in relaxation.

11 CALCIUM SENSITIZATION

Calcium sensitization is an essentially calcium-independent process that enables the amount of constriction in smooth muscle to be tuned by an alteration in the sensitivity of MLC to calcium (Fig. 5). This process enables the muscle to sustain a contraction once the initial calcium transient has dissipated. There are two mechanisms for calcium sensitization: a DAG-PLC-PKC pathway and a RhoA pathway (Lincoln 2007).

Diacylglycerol (DAG) is produced by PLCβ downstream from certain GPCRs and activates the conventional and novel protein kinase C (cPKC and nPKC), but not atypical PKC (aPKC) (Steinberg 2008). PKC has a variety of downstream targets, such as MLCK and C-kinase potentiated protein phosphatase 1 inhibitor, molecular mass 17 kDa (CPI-17), both of which enhance constriction. CPI-17 is a smooth-muscle-specific inhibitor of MLCP that binds to its catalytic subunit and inhibits phosphatase activity, allowing contraction to persist.

Several agonists, including angiotensin II, norepinephrine, and endothelin, activate the small G protein RhoA. RhoA in turn activates Rho kinase (ROCK), which can mediate calcium sensitization through two main pathways. First, ROCK stimulates phosphorylation of MYPT1 (Feng et al. 1999). This can be direct, at T695 or T853, with a preference for T853. Alternatively, it can phosphorylate another kinase, zipper-interacting protein kinase (ZIPK, also known as DAPK3), which phosphorylates MYPT1 primarily at T695 (Kiss et al. 2002). ZIPK also phosphorylates MLC at T18/S19. Phosphorylation of MYPT1 interferes with binding of MLCP to MLC, and thus is believed to decrease phosphatase activity. ROCK can also phosphorylate CPI-17 (MacDonald et al. 2001).

The preference for the MYPT1 or CPI-17 pathway depends on the type of smooth muscle. Whereas MYPT1 is ubiquitously expressed in smooth muscle, CPI-17 is differentially expressed. Moreover, RhoA and associated proteins are expressed at lower levels in phasic smooth muscle compared with tonic smooth muscle (Patel and Rattan 2006). Note that PKC can also phosphorylate CPI-17 to prevent MLCP activity. Within resistance arteries, an increase in vascular pressure also activates the RhoA pathway; however, the signaling intermediates linking the change in vascular pressure and the activation of RhoA remain unknown (Cole and Welsh 2011).

12 VASCULAR SMOOTH MUSCLE IN DISEASE

Smooth muscle cells are remarkably plastic, altering their phenotype in response to conditions such as vascular injury, altered blood flow conditions, or disease states. The changes in phenotype that can occur include cell proliferation, apoptosis, and cell migration and are induced by many factors, including cytokines and growth factors, mechanical forces, neuronal stimuli, and genetic factors. Here we limit our discussion to hypertension.

In hypertension, there is often a change in the sympathetic nervous system and the renin–angiotensin system that leads to increased blood pressure. Angiotensinogen is converted to angiotensin I by renin, which in turn is converted to angiotensin II by angiotensin-converting enzyme (ACE). Increased circulating angiotensin II acts on the angiotensin receptors (AT1 and AT2), which, when activated, cause increased peripheral resistance. The consequence for smooth muscle cells is they become hypercontractile. Treatments include ACE inhibitors (which inhibit the conversion of angiotensin I to angiotensin II), α1-adrenergic antagonists (which block the AT1 and AT2 GPCRs), and calcium channel blockers (such as dihydropyridines, which inhibit the voltage-gated calcium channels). All of these treatments aim to reduce the contractility of smooth muscle. Interfering with downstream targets such as RhoA signaling in hypertensive animals has also been shown to be effective (Uehata et al. 1997; Seko et al. 2003; Moriki et al. 2004).

The sustained contractile state of vascular smooth muscle is associated with the activation of calcium-dependent transcription factors. These include SRF, FOS, NFAT, and CREB. SRF, which is activated by the RhoA pathway, promotes the expression of genes encoding components of the contractile apparatus. Calcium-stimulated CaMKII activates and causes the translocation of CaMKIV to the nucleus, where it can activate CREB, which promotes transcription of components of the contractile apparatus and other targets. However, CaMKII can also activate a phosphatase that dephosphorylates and thus inactivates CREB (Matchkov et al. 2012). NFAT is activated on dephosphorylation by calcium-activated calcineurin, which induces genes associated with proliferation and migration.

NO produced by eNOS in endothelial cells protects against the changes observed in hypertension: the cGMP pathway inhibits DNA synthesis, mitogenesis, and cell proliferation (Forstermann and Sessa 2012). However, endothelial dysfunction is a hallmark of vascular disease, including hypertension. In many types of vascular diseases, eNOS is up-regulated but owing to reduced oxygen availability it is converted to a dysfunctional enzyme that produces superoxides, which contribute to vascular oxidative stress (Forstermann and Sessa 2012).

In some disease states, smooth muscle cells adopt a noncontractile phenotype. Although these cells still have signaling machinery that increases intracellular calcium levels, they have significantly reduced calcium influx through voltage-gated calcium channels. Thus, there is a shift to intracellular-store-operated calcium release, similar to the changes observed in cardiac hypertrophy. Concomitant with decreases in the levels of SERCA, RyR2, PMCA1, and the sodium/calcium exchanger, the levels of STIM, ORAI (proteins associated with refilling of intracellular calcium stores; see p. 95 [Bootman 2012]), SERCA2B, and IP$_3$R increase and there is a change in RyR receptor subtypes from RyR2 to RyR3 (Lipskaia and Lompre 2004; Berra-Romani et al. 2008; Baryshnikov et al. 2009; Matchkov et al. 2012). These changes collectively reflect a less contractile phenotype.

13 CONCLUDING REMARKS

Signal transduction is essential for the function of contractile cells. The stimulatory signal results in an increase in cytosolic calcium levels, which activates muscle contraction. We now know the main contributors to the various types of muscle contraction, and have a better appreciation of the changes that occur to the contractile apparatus under exercise and pathophysiological conditions. For example, the identification of PGC1α as a master regulator of transcription factors up-regulated in both

exercised and pathological striated muscle provides new avenues to modulate muscles in a therapeutic setting. It is also apparent that many signaling proteins in both smooth and striated muscles are activated by changes in cytosolic calcium levels, and these signaling pathways often lead to alterations in gene expression. Because we now have a better appreciation of the changes that occur to the contractile apparatus under pathophysiological conditions, this knowledge can be harnessed to allow us to treat disease strategically.

ACKNOWLEDGMENTS

Research in the Ehrlich Laboratory is supported by National Institutes of Health funds. I.Y.K. is an American Heart Association postdoctoral fellow.

REFERENCES

*Reference is in this book.

Andersson DC, Marks AR. 2010. Fixing ryanodine receptor Ca leak—A novel therapeutic strategy for contractile failure in heart and skeletal muscle. *Drug Discov Today* 7: e151–e157.
Balakumar P, Jagadeesh G. 2010. Multifarious molecular signaling cascades of cardiac hypertrophy: Can the muddy waters be cleared? *Pharmacol Res* 62: 365–383.
Baryshnikov SG, Pulina MV, Zulian A, Linde CI, Golovina VA. 2009. Orai1, a critical component of store-operated Ca^{2+} entry, is functionally associated with Na$^+$/Ca^{2+} exchanger and plasma membrane Ca^{2+} pump in proliferating human arterial myocytes. *Am J Physiol Cell Physiol* 297: C1103–C1112.
Berra-Romani R, Mazzocco-Spezzia A, Pulina MV, Golovina VA. 2008. Ca^{2+} handling is altered when arterial myocytes progress from a contractile to a proliferative phenotype in culture. *Am J Physiol Cell Physiol* 295: C779–C790.
Bezprozvanny I, Watras J, Ehrlich BE. 1991. Bell-shaped calcium-response curves of Ins(1,4,5)P3- and calcium-gated channels from endoplasmic reticulum of cerebellum. *Nature* 351: 751–754.
*Bootman MD. 2012. Calcium signaling. *Cold Spring Harb Perpsect Biol* 4: a011171.
Bozzo C, Spolaore B, Toniolo L, Stevens L, Bastide B, Cieniewski-Bernard C, Fontana A, Mounier Y, Reggiani C. 2005. Nerve influence on myosin light chain phosphorylation in slow and fast skeletal muscles. *FEBS J* 272: 5771–5785.
Campbell KP, MacLennan DH, Jorgensen AO, Mintzer MC. 1983. Purification and characterization of calsequestrin from canine cardiac sarcoplasmic reticulum and identification of the 53,000 dalton glycoprotein. *J Biol Chem* 258: 1197–1204.
Cheng H, Lederer WJ, Cannell MB. 1993. Calcium sparks: Elementary events underlying excitation-contraction coupling in heart muscle. *Science* 262: 740–744.
Chien KR. 1999. Stress pathways and heart failure. *Cell* 98: 555–558.
Cole WC, Welsh DG. 2011. Role of myosin light chain kinase and myosin light chain phosphatase in the resistance arterial myogenic response to intravascular pressure. *Arch Biochem Biophys* 510: 160–173.
Dainese M, Quarta M, Lyfenko AD, Paolini C, Canato M, Reggiani C, Dirksen RT, Protasi F. 2009. Anesthetic- and heat-induced sudden death in calsequestrin-1-knockout mice. *FASEB J* 23: 1710–1720.
D'Angelo DD, Sakata Y, Lorenz JN, Boivin GP, Walsh RA, Liggett SB, Dorn GW 2nd. 1997. Transgenic Gαq overexpression induces cardiac contractile failure in mice. *Proc Natl Acad Sci* 94: 8121–8126.

Cite this chapter as *Cold Spring Harb Perspect Biol* doi: 10.1101/cshperspect.a006023

Feng J, Ito M, Ichikawa K, Isaka N, Nishikawa M, Hartshorne DJ, Nakano T. 1999. Inhibitory phosphorylation site for Rho-associated kinase on smooth muscle myosin phosphatase. *J Biol Chem* **274:** 37385–37390.

Finch EA, Turner TJ, Goldin SM. 1991. Calcium as a coagonist of inositol 1,4,5-trisphosphate-induced calcium release. *Science* **252:** 443–446.

Forstermann U, Sessa WC. 2012. Nitric oxide synthases: Regulation and function. *Eur Heart J* **33:** 829–837, 837a–837d.

Handschin C, Spiegelman BM. 2006. Peroxisome proliferator-activated receptor γ coactivator 1 coactivators, energy homeostasis, and metabolism. *Endocr Rev* **27:** 728–735.

* Hardie DG. 2012. Organismal carbohydrate and lipid homeostasis. *Cold Spring Harb Perspect Biol* **4:** a006031.

Harzheim D, Talasila A, Movassagh M, Foo RS, Figg N, Bootman MD, Roderick HL. 2010. Elevated InsP3R expression underlies enhanced calcium fluxes and spontaneous extra-systolic calcium release events in hypertrophic cardiac myocytes. *Channels (Austin)* **4:** 67–71.

Hasenfuss G, Reinecke H, Studer R, Meyer M, Pieske B, Holtz J, Holubarsch C, Posival H, Just H, Drexler H. 1994. Relation between myocardial function and expression of sarcoplasmic reticulum Ca^{2+}-ATPase in failing and nonfailing human myocardium. *Circ Res* **75:** 434–442.

He WQ, Peng YJ, Zhang WC, Lv N, Tang J, Chen C, Zhang CH, Gao S, Chen HQ, Zhi G, et al. 2008. Myosin light chain kinase is central to smooth muscle contraction and required for gastrointestinal motility in mice. *Gastroenterology* **135:** 610–620.

* Hemmings BA, Restuccia DF. 2012. The PI3K-PKB/Akt pathway. *Cold Spring Harb Perspect Biol* **4:** a011189.

Ibrahim M, Gorelik J, Yacoub MH, Terracciano CM. 2011. The structure and function of cardiac t-tubules in health and disease. *Proc Biol Sci* **278:** 2714–2723.

Kiss E, Muranyi A, Csortos C, Gergely P, Ito M, Hartshorne DJ, Erdodi F. 2002. Integrin-linked kinase phosphorylates the myosin phosphatase target subunit at the inhibitory site in platelet cytoskeleton. *Biochem J* **365:** 79–87.

Klein MG, Cheng H, Santana LF, Jiang YH, Lederer WJ, Schneider MF. 1996. Two mechanisms of quantized calcium release in skeletal muscle. *Nature* **379:** 455–458.

Kuo IY, Wolfle SE, Hill CE. 2011. T-type calcium channels and vascular function: The new kid on the block? *J Physiol* **589:** 783–795.

Lanner JT, Georgiou DK, Joshi AD, Hamilton SL. 2010. Ryanodine receptors: Structure, expression, molecular details, and function in calcium release. *Cold Spring Harb Perspect Biol* **2:** a003996.

Lin J, Wu H, Tarr PT, Zhang CY, Wu Z, Boss O, Michael LF, Puigserver P, Isotani E, Olson EN, et al. 2002. Transcriptional co-activator PGC-1α drives the formation of slow-twitch muscle fibres. *Nature* **418:** 797–801.

Lincoln TM. 2007. Myosin phosphatase regulatory pathways: Different functions or redundant functions? *Circ Res* **100:** 10–12.

Lipskaia L, Lompre AM. 2004. Alteration in temporal kinetics of Ca^{2+} signaling and control of growth and proliferation. *Biol Cell* **96:** 55–68.

Liu Y, Shen T, Randall WR, Schneider MF. 2005. Signaling pathways in activity-dependent fiber type plasticity in adult skeletal muscle. *J Muscle Res Cell Motil* **26:** 13–21.

MacDonald JA, Eto M, Borman MA, Brautigan DL, Haystead TA. 2001. Dual Ser and Thr phosphorylation of CPI-17, an inhibitor of myosin phosphatase, by MYPT-associated kinase. *FEBS Lett* **493:** 91–94.

Matchkov VV, Kudryavtseva O, Aalkjaer C. 2012. Intracellular Ca^{2+} signalling and phenotype of vascular smooth muscle cells. *Basic Clin Pharmacol Toxicol* **110:** 42–48.

Matsui T, Nagoshi T, Rosenzweig A. 2003. Akt and PI 3-kinase signaling in cardiomyocyte hypertrophy and survival. *Cell Cycle* **2:** 220–223.

Meyer M, Schillinger W, Pieske B, Holubarsch C, Heilmann C, Posival H, Kuwajima G, Mikoshiba K, Just H, Hasenfuss G, et al. 1995. Alterations of sarcoplasmic reticulum proteins in failing human dilated cardiomyopathy. *Circulation* **92:** 778–784.

Mishra S, Ling H, Grimm M, Zhang T, Bers DM, Brown JH. 2010. Cardiac hypertrophy and heart failure development through Gq and CaM kinase II signaling. *J Cardiovasc Pharmacol* **56:** 598–603.

Miyata S, Minobe W, Bristow MR, Leinwand LA. 2000. Myosin heavy chain isoform expression in the failing and nonfailing human heart. *Circ Res* **86:** 386–390.

Moriki N, Ito M, Seko T, Kureishi Y, Okamoto R, Nakakuki T, Kongo M, Isaka N, Kaibuchi K, Nakano T. 2004. RhoA activation in vascular smooth muscle cells from stroke-prone spontaneously hypertensive rats. *Hypertens Res* **27:** 263–270.

* Morrison DK. 2012. MAP kinase pathways. *Cold Spring Harb Perspect Biol* **4:** a0011254.

Nikolaev VO, Moshkov A, Lyon AR, Miragoli M, Novak P, Paur H, Lohse MJ, Korchev YE, Harding SE, Gorelik J. 2010. β2-Adrenergic receptor redistribution in heart failure changes cAMP compartmentation. *Science* **327:** 1653–1657.

Nishikimi T, Kuwahara K, Nakao K. 2011. Current biochemistry, molecular biology, and clinical relevance of natriuretic peptides. *J Cardiol* **57:** 131–140.

Olesen J, Kiilerich K, Pilegaard H. 2010. PGC-1α-mediated adaptations in skeletal muscle. *Pflugers Archiv Eur J Phys* **460:** 153–162.

Oliveira RS, Ferreira JC, Gomes ER, Paixao NA, Rolim NP, Medeiros A, Guatimosim S, Brum PC. 2009. Cardiac anti-remodelling effect of aerobic training is associated with a reduction in the calcineurin/NFAT signalling pathway in heart failure mice. *J Physiol* **587:** 3899–3910.

Patel CA, Rattan S. 2006. Spontaneously tonic smooth muscle has characteristically higher levels of RhoA/ROK compared with the phasic smooth muscle. *Am J Physiol Gastrointest Liver Physiol* **291:** G830–G837.

Postma AV, Denjoy I, Hoorntje TM, Lupoglazoff JM, Da Costa A, Sebillon P, Mannens MM, Wilde AA, Guicheney P. 2002. Absence of calsequestrin 2 causes severe forms of catecholaminergic polymorphic ventricular tachycardia. *Circ Res* **91:** e21–e26.

Pritchard TJ, Kranias EG. 2009. Junctin and the histidine-rich Ca^{2+} binding protein: Potential roles in heart failure and arrhythmogenesis. *J Physiol* **587:** 3125–3133.

Robinson R, Carpenter D, Shaw MA, Halsall J, Hopkins P. 2006. Mutations in RYR1 in malignant hyperthermia and central core disease. *Hum Mutat* **27:** 977–989.

Ronnebaum SM, Patterson C. 2010. The FoxO family in cardiac function and dysfunction. *Ann Rev Physiol* **72:** 81–94.

Sanbe A, Fewell JG, Gulick J, Osinska H, Lorenz J, Hall DG, Murray LA, Kimball TR, Witt SA, Robbins J. 1999. Abnormal cardiac structure and function in mice expressing nonphosphorylatable cardiac regulatory myosin light chain 2. *J Biol Chem* **274:** 21085–21094.

Schiaffino S, Reggiani C. 2011. Fiber types in mammalian skeletal muscles. *Physiol Rev* **91:** 1447–1531.

Schlecker C, Boehmerle W, Jeromin A, DeGray B, Varshney A, Sharma Y, Szigeti-Buck K, Ehrlich BE. 2006. Neuronal calcium sensor-1 enhancement of InsP3 receptor activity is inhibited by therapeutic levels of lithium. *J Clin Invest* **116:** 1668–1674.

Seko T, Ito M, Kureishi Y, Okamoto R, Moriki N, Onishi K, Isaka N, Hartshorne DJ, Nakano T. 2003. Activation of RhoA and inhibition of myosin phosphatase as important components in hypertension in vascular smooth muscle. *Circ Res* **92:** 411–418.

Steinberg SF. 2008. Structural basis of protein kinase C isoform function. *Physiol Rev* **88:** 1341–1378.

Uehata M, Ishizaki T, Satoh H, Ono T, Kawahara T, Morishita T, Tamakawa H, Yamagami K, Inui J, Maekawa M, et al. 1997. Calcium sensitization of smooth muscle mediated by a ρ-associated protein kinase in hypertension. *Nature* **389:** 990–994.

Ventura-Clapier R, Mettauer B, Bigard X. 2007. Beneficial effects of endurance training on cardiac and skeletal muscle energy metabolism in heart failure. *Cardiovasc Res* **73:** 10–18.

Watson PA, Reusch JE, McCune SA, Leinwand LA, Luckey SW, Konhilas JP, Brown DA, Chicco AJ, Sparagna GC, Long CS, et al. 2007. Restoration of CREB function is linked to completion and stabilization of adaptive cardiac hypertrophy in response to exercise. *Am J Physiol Heart Circ Physiol* **293:** H246–H259.

Wei L, Hanna AD, Beard NA, Dulhunty AF. 2009. Unique isoform-specific properties of calsequestrin in the heart and skeletal muscle. *Cell Calcium* **45:** 474–484.

Wenz T, Rossi SG, Rotundo RL, Spiegelman BM, Moraes CT. 2009. Increased muscle PGC-1α expression protects from sarcopenia and metabolic disease during aging. *Proc Natl Acad Sci* **106:** 20405–20410.

Wettschureck N, Rutten H, Zywietz A, Gehring D, Wilkie TM, Chen J, Chien KR, Offermanns S. 2001. Absence of pressure overload induced myocardial hypertrophy after conditional inactivation of Gαq/Gα11 in cardiomyocytes. *Nat Med* **7:** 1236–1240.

Zhang WC, Peng YJ, Zhang GS, He WQ, Qiao YN, Dong YY, Gao YQ, Chen C, Zhang CH, Li W, et al. 2010. Myosin light chain kinase is necessary for tonic airway smooth muscle contraction. *J Biol Chem* **285:** 5522–5531.

Cite this chapter as *Cold Spring Harb Perspect Biol* doi: 10.1101/cshperspect.a006023

CHAPTER 14

Organismal Carbohydrate and Lipid Homeostasis

D. Grahame Hardie

College of Life Sciences, University of Dundee, Dundee DD1 5EH, Scotland, United Kingdom

Correspondence: d.g.hardie@dundee.ac.uk

SUMMARY

All living organisms maintain a high ATP:ADP ratio to drive energy-requiring processes. They therefore need mechanisms to maintain energy balance at the cellular level. In addition, multi-cellular eukaryotes have assigned the task of storing energy to specialized cells such as adipo-cytes, and therefore also need a means of intercellular communication to signal the needs of individual tissues and to maintain overall energy balance at the whole body level. Such signal-ing allows animals to survive periods of fasting or starvation when food is not available and is mainly achieved by hormonal and nervous communication. Insulin, adipokines, epinephrine, and other agonists thus stimulate pathways that regulate the activities of key enzymes involved in control of metabolism to integrate organismal carbohydrate and lipid metabolism. Overnu-trition can dysregulate these pathways and have damaging consequences, causing insulin re-sistance and type 2 diabetes.

Outline

1 INTRODUCTION

Heterotrophic organisms, including mammals, gain energy from the ingestion and breakdown (catabolism) of reduced carbon compounds, mainly carbohydrates, fats, and proteins. A large proportion of the energy released, rather than appearing simply as heat, is used to convert ADP and inorganic phosphate (Pi) into ATP. The high intracellular ratio of ATP to ADP thus created is analogous to the fully charged state of a rechargeable battery, representing a store of energy that can be used to drive energy-requiring processes, including the anabolic pathways required for cell maintenance and growth. Individual cells must constantly adjust their rates of nutrient uptake and catabolism to balance their rate of ATP consumption, so that they can maintain a constant high ratio of ATP to ADP. The main control mechanism used to achieve this energy homeostasis is AMP-activated protein kinase (AMPK) (see Box 1) (Hardie 2011).

One potential problem for heterotrophic organisms is that there may be prolonged periods when food is not available. They must therefore store molecules like glucose and fatty acids during "times of plenty" to act as reserves for use during periods of fasting or starvation. In multicellular organisms, much of this energy storage function has been devolved to specialized cells. For example, although all mammalian cells (with the possible exception of neurons) can store some glucose in the form of glycogen, large quantities are stored only in muscle and the liver. Similarly, most mammalian cells can store fatty acids in the form of triglyceride droplets in the cytoplasm, but storing large amounts appears to be harmful to many cells (see Sec. 12). Adipocytes in white adipose tissue have therefore become specialized for triglyceride storage, releasing fatty acids when other tissues need them. In this chapter, we examine the signaling pathways that coordinate carbohydrate and lipid metabolism between energy-utilizing tissues such as muscle, energy-storing tissues such as adipose tissue, and the liver (an organ that coordinates whole body metabolism).

2 MAINTAINING ENERGY HOMEOSTASIS — HORMONES AND ADIPOKINES

The development of multicellular organisms during eukaryotic evolution required the acquisition of systems of hormonal and neuronal signaling that allowed tissues to communicate their needs to each other. In this section, the critical endocrine glands and nerve centers involved in regulation of whole body energy homeostasis, and the hormones and cytokines they produce, are briefly discussed; they are also summarized in Fig. 1. Subsequently,

our main focus will be the signaling pathways by which target cells respond to these agents.

2.1 The Hypothalamus and Pituitary Gland

The hypothalamus is a small region at the base of the brain that controls critical functions such as body temperature, thirst, hunger, and circadian rhythms. It modulates feeding behavior by producing neuropeptides that promote or repress appetite. The hypothalamus also produces "releasing hormones" (e.g., corticotrophin-releasing hormone [CRH] and thyrotropin-releasing hormone [TRH]) that travel the short distance to the neighboring pituitary gland. Here they trigger release of peptide hormones (e.g., adrenocorticotrophic hormone [ACTH] and thyroid-stimulating hormone [TSH], released in response to CRH and TRH, respectively).

2.2 The Adrenal Gland

The adrenal gland contains two distinct regions: a central medulla and an outer cortex. The medulla contains modified sympathetic neurons that release epinephrine (a catecholamine also known as adrenaline, formed by oxidation and deamination of the amino acid tyrosine) into the bloodstream. The hypothalamus has projections that connect with sympathetic nerves and can thus trigger epinephrine release. For example, this occurs when blood glucose levels drop, a response that appears to involve activation of AMPK in glucose-sensitive neurons within the hypothalamus (McCrimmon et al. 2008). Release of epinephrine also occurs during exercise and other stressful situations and triggers the "fight or flight response." As well as altering blood flow via effects on the heart and vasculature, epinephrine has many metabolic effects, acting via G-protein-coupled receptors (GPCRs) to increase intracellular cyclic AMP or calcium (Ch. 1 [Heldin et al. 2014]). In muscle, it stimulates glycogen and triglyceride breakdown, providing fuels for accelerated ATP production. In the liver, it promotes release of glucose from glycogen into the bloodstream. In adipose tissue, it mobilizes triglyceride stores, the fatty acids derived either being used as catabolic fuels or, in brown adipose tissue, to generate heat.

The adrenal cortex releases the glucocorticoid cortisol, which (like other steroid hormones) is synthesized from cholesterol. This also occurs in response to starvation or stressful conditions, but in this case is triggered by release of ACTH from the pituitary gland. Acting via nuclear receptors that directly regulate transcription, glucocorticoids promote gluconeogenesis in the liver, protein breakdown in muscle, and triglyceride breakdown in adipose tissue, while reducing insulin-stimulated glucose uptake by muscle.

Cite this chapter as *Cold Spring Harb Perspect Biol* doi: 10.1101/cshperspect.a006031

BOX 1. THE ENERGY CHARGE HYPOTHESIS AND ENERGY SENSING BY AMP-ACTIVATED PROTEIN KINASE

Catabolism generates ATP from ADP, whereas anabolism and most other cellular processes, such as the action of motor proteins or membrane pumps, require energy and is usually driven by hydrolysis of ATP to ADP. There is no reason a priori why these opposing processes should automatically remain in balance, and the fact that cellular ATP:ADP ratios are usually rather constant indicates that there are systems inside cells that maintain energy balance. Daniel Atkinson proposed in his energy charge hypothesis that the major signals that regulate cellular energy homeostasis would be ATP, ADP, and AMP (Ramaiah et al. 1964). AMP, like ADP, is a good indicator of energy stress, because its concentration increases as the ADP:ATP ratio increases, owing to displacement of the near-equilibrium reaction catalyzed by adenylate kinase ($2ADP \leftrightarrow ATP + AMP$). Atkinson's hypothesis was based on findings that two enzymes, glycogen phosphorylase and phosphofructokinase, which catalyze key control points in glycogen breakdown and glycolysis (see main text), are activated by AMP and inhibited by ATP. The discovery of the AMP-activated protein kinase (AMPK) revitalized this concept. Regulation of AMPK by adenine nucleotides is surprisingly complex (Hardie et al. 2011), but provides great flexibility.

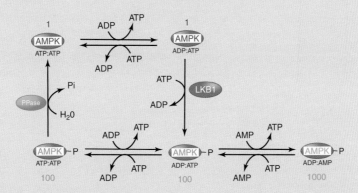

In the diagram above, the numerals below each form refer to its activity relative to the basal state (top left). AMPK is activated >100-fold by phosphorylation of a threonine residue (172) within the activation loop of the kinase domain. This is mainly catalyzed by the upstream kinase LKB1, which has a high basal activity. One of the regulatory subunits of AMPK contains two sites that competitively bind the adenine nucleotides AMP, ADP, or ATP. Binding of ADP or AMP (but not ATP) to the first site (*top center*) causes a conformational change that promotes net phosphorylation of 172 (Oakhill et al. 2011; Xiao et al. 2011), thus causing a switch to the active, phosphorylated form (*bottom center*; during a mild metabolic stress, the concentration of AMP is much lower than that of ADP, so binding of ADP may be the key event responsible for this change). As stress becomes more severe, binding of AMP (but not ADP or ATP) at the second site causes a further 10-fold allosteric activation (bottom right), the combination of the two effects yielding >1000-fold activation overall. As metabolic stress subsides, AMP and ADP concentrations will decrease and they will be replaced in the two regulatory sites by ATP (moving from right to left on the bottom row). This initially causes a loss of the allosteric activation, then a conformational change that promotes dephosphorylation, so that the kinase returns to the basal state (top left). This complex mechanism allows AMPK to be activated in a sensitive but dynamic manner over a wide range of ADP:ATP and AMP:ATP ratios, phosphorylating more and more downstream targets as stress becomes more severe.

2.3 The Thyroid Gland

The thyroid gland synthesizes the thyroid hormone thyroxine (also known as T4 because it contains four iodine atoms), via iodination and subsequent combination of two molecules of the amino acid tyrosine present within a large precursor protein called thyroglobulin. Proteolytic breakdown of thyroglobulin releases T4, but the more potent hormone tri-iodothyronine (T3) is produced by removal of one iodine atom from T4 in other tissues, including the liver. T3 acts, like glucocorticoids, by binding to nuclear receptors that regulate transcription, and has effects on almost all cells. However, a major effect of T3 on whole body energy homeostasis involves inhibition of AMPK in the hypothalamus

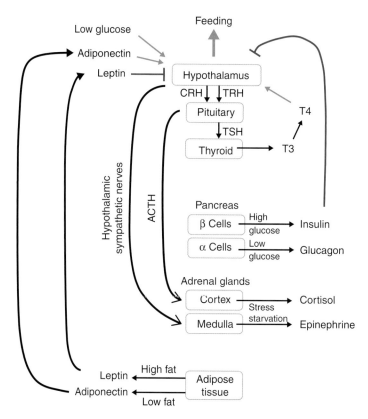

Figure 1. Summary of endocrine systems that regulate energy balance by modulating carbohydrate and lipid metabolism. Neurons in the hypothalamus promote feeding, and the successive release of corticotrophin-releasing hormone (CRH), adenocorticotrophin-releasing hormone (ACTH), and cortisol from the hypothalamus, pituitary, and adrenal cortex, respectively, and release of epinephrine from the adrenal medulla. Hypothalamic neurones are activated by low glucose, the thyroid hormone T3, and adiponectin, while being inhibited by leptin and insulin. The α and β cells in the pancreas monitor blood glucose independently, releasing glucagon and insulin, respectively. Insulin and leptin are hormones that represent nutrient surplus, whereas cortisol, epinephrine, and adiponectin are hormones that represent either deprivation of nutrients (e.g., starvation) or demand for energy (e.g., during exercise).

(Lopez et al. 2010). This promotes firing of sympathetic nerves that trigger release of epinephrine from the adrenal medulla. This in turn increases energy expenditure by stimulating white adipose tissue to release fatty acids (which are oxidized in other tissues) and by promoting fatty acid oxidation and heat production in brown adipose tissue.

2.4 Pancreatic Islets

The pancreas contains small islands of endocrine tissue called the islets of Langerhans, which are responsible for the secretion of the hormones insulin and glucagon. The β cells synthesize insulin, a peptide hormone with two disulfide-linked chains formed by cleavage of a single precursor polypeptide, proinsulin. They release insulin in response to high concentrations of glucose and amino acids that are derived from the gut after feeding, i.e., during "times of plenty." Almost all cells in the body express the insulin receptor, which (like the IGF1 receptor) has two disulfide-linked

polypeptide chains, one with a cytoplasmic tyrosine kinase domain. Binding of insulin to this receptor activates phosphoinositide (PI) 3-kinase, which synthesizes phosphatidylinositol 3,4,5-trisphosphate (PIP$_3$) from phosphatidylinositol 4,5-bisphosphate (PIP$_2$). The membrane lipid PIP$_3$ is a second messenger that activates the Akt signaling pathway (p. 87 [Hemmings and Restuccia 2012]), promoting the cellular uptake of glucose and amino acids, as well as synthesis of fatty acids, and the conversion of these components to forms in which they are stored, i.e., glycogen, proteins, and triglycerides.

The α cells within the islets release another peptide hormone, glucagon, in response to low blood glucose during fasting or starvation. Glucagon binds to Gα-linked receptors that increase cyclic AMP levels in the liver (see below), causing a switch from glycolysis to glucose production by gluconeogenesis. It also mobilizes triglyceride stores in adipose tissue, releasing fatty acids into the bloodstream by the process of lipolysis.

2.5 Adipose Tissue

The main function of adipose tissue is to store triglycerides, but it also acts as an endocrine organ, releasing peptide hormones termed adipokines that play key roles in regulating whole body energy balance. One is leptin, a small globular protein whose concentration is elevated in the blood of obese individuals, which suggests that it represents a signal that fat reserves are adequate (Friedman and Halaas 1998). The leptin receptor is a member of the cytokine receptor family, a single-pass membrane protein that lacks intrinsic kinase activity but is coupled via Janus kinases (JAKs) to transcription factors of the STAT family (p. 177 [Harrison 2012]). Leptin receptors are expressed in the hypothalamus, and their activation represses synthesis of neuropeptides that promote feelings of hunger, while enhancing synthesis of those that promote satiety, thus reducing appetite. Unfortunately, many obese people appear to have become resistant to the appetite-suppressing effects of leptin.

Another adipokine, adiponectin, is an unusual peptide hormone that has a globular domain linked to a collagen-like sequence that causes it to form disulphide-linked trimers as well as higher-order oligomers (Shetty et al. 2009). Although secreted by adipocytes, adiponectin paradoxically displays high plasma concentrations in lean individuals and low levels in obese individuals, which suggests that it is a signal indicating that fat stores are low. Adiponectin binds to two receptors (AdipoRI and AdipoRII) that are predicted to have seven transmembrane helices, although they differ from GPCRs in that their amino termini are intracellular (Kadowaki and Yamauchi 2005). In general, adiponectin has catabolic and anti-anabolic effects: binding to AdipoRI activates AMPK (via a mechanism that remains unclear), promoting fat oxidation in the liver and muscle, and inhibiting glucose production by the liver. Adiponectin also increases appetite by activating AMPK in the hypothalamus, thus opposing the effects of leptin (Kubota et al. 2007).

3 MUSCLE—ACUTE ACTIVATION OF GLYCOGEN BREAKDOWN

The mechanisms by which target cells respond to the hormones and cytokines described in the previous section are discussed below. Skeletal muscle represents the major site of glycogen storage within the body, although because it lacks glucose-6-phosphatase it cannot release glucose back into the bloodstream. Muscle glycogen breakdown is therefore used entirely to meet the energy demands of the muscle itself, and is especially important during periods of intense exercise. The enzyme phosphorylase uses phosphate to split the terminal glycosidic linkages of the outer chains of glycogen, releasing glucose 1-phosphate, which immediately

enters glycolysis to generate ATP. 5′-AMP allosterically activates phosphorylase, with ATP antagonizing this effect. Thus, phosphorylase should be activated by an increase in the cellular AMP:ATP ratio, a signal that the energy status of the cell is compromised (see Box 1). One of the key glycolytic enzymes in muscle, phosphofructokinase (PFK1), is also allosterically activated by AMP and inhibited by ATP, so glycolysis should be activated at the same time (Fig. 2).

Phosphorylase occurs not only as the form activated by AMP (called phosphorylase *b*), but also as a second form (phosphorylase *a*) that is phosphorylated at a serine residue near the amino terminus and active even in the absence of AMP. The enzyme that catalyzes the *b*-to-*a* transition, phosphorylase kinase, is activated by calcium. When a muscle is stimulated to contract, the neurotransmitter acetylcholine is released at the specialized synapse between the motor nerve and the muscle (the neuromuscular junction). Activation of nicotinic acetylcholine receptors on the muscle cell then causes firing of action potentials that pass down the transverse tubules, triggering opening of voltage-gated calcium channels. This in turn causes opening of calcium-activated calcium channels (ryanodine receptors) on the sarcoplasmic reticulum, leading to a sudden release of calcium from there into the cytoplasm (Ch. 13 [Kuo and Ehrlich 2014]). This calcium influx triggers muscle contraction (creating a massive demand for ATP), while at the same time activating phosphorylase kinase. Thus, contraction is synchronized with glycogen breakdown, which helps to satisfy the demand for ATP. Phosphorylase kinase was the first of >500 mammalian protein kinases to be identified (Fischer and Krebs 1989). It is a large multisubunit complex $(\alpha_4\beta_4\,\gamma_4\delta_4)$, with the γ subunit carrying the kinase activity and the δ subunit being a tightly bound molecule of the calcium-binding protein calmodulin, responsible for activation by calcium. Cyclic-AMP-dependent protein kinase (PKA), which is activated by increases in cyclic AMP when epinephrine acts on muscle, phosphorylates both the α and β subunits. This greatly increases the kinase activity of the γ subunit, while also making the complex more sensitive to calcium. Note that the PKA $\rightarrow$ phosphorylase kinase pathway was the first protein kinase cascade to be described.

Why does phosphorylase need three tiers of regulation, mediated by AMP, calcium, and cyclic AMP (Fig. 2)? Imagine a mouse suddenly encountering a cat: if the mouse had no phosphorylase kinase (and therefore only had the AMP-activated *b* form of phosphorylase), it could start to run, but glycogen breakdown would not occur until some ATP had been used up and AMP had increased. The delay involved in this feedback mechanism might be fatal. Indeed, humans with muscle phosphorylase kinase deficiency have been described, and they experience muscle weakness or pain during exercise. If, however, the mouse

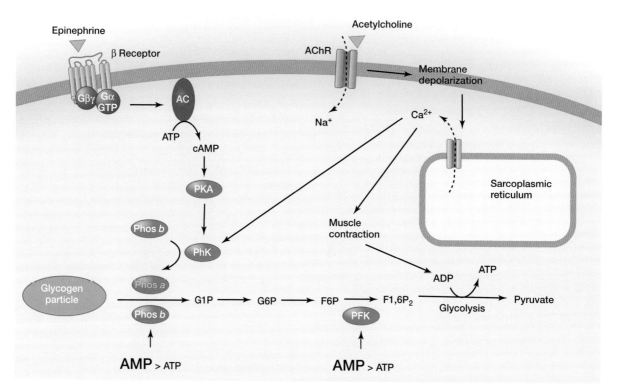

Figure 2. Regulation of glycogen breakdown in skeletal muscle. Muscle contraction increases ADP and AMP and decreases ATP, activating both phosphorylase b (phos b) and phosphofructokinase (PFK) through allosteric regulation, thus promoting glycogen breakdown and glycolysis to generate ATP. However, contraction (initiated by firing of motor nerves that release acetyl choline) is triggered by release of calcium from channels in the sarcoplasmic reticulum membrane, also activating phosphorylase kinase. The latter phosphorylates phosphorylase and converts it to the a form (phos a), which no longer requires AMP for activity. Increases in cyclic AMP levels, triggered by binding of epinephrine to receptors on the plasma membrane, activate cyclic-AMP-dependent protein kinase (PKA). PKA phosphorylates phosphorylase kinase and amplifies its activation by the calcium-dependent mechanism.

had phosphorylase kinase, calcium-dependent phosphorylation of phosphorylase would now occur, overriding the allosteric mechanism, and the onset of glycogen breakdown would be synchronized with the onset of muscle contraction. This feed-forward effect of calcium would anticipate the demand for ATP and remove the delay implicit in the allosteric mechanism, thus increasing the chances of escape. Finally, if the mouse is foraging in a place where it might expect trouble, it will be nervous and have high levels of circulating epinephrine. This causes phosphorylation of phosphorylase kinase by PKA, so that when calcium goes up in response to contraction, the conversion of phosphorylase to the more active a form is even more rapid, maximizing the chances of escape.

4 MUSCLE—ACUTE REGULATION OF GLUCOSE UPTAKE AND GLYCOGEN SYNTHESIS

Muscle stores of glycogen are finite and cannot maintain rapid ATP production for long periods. Prolonged exercise

therefore requires increased uptake of glucose from the bloodstream. In addition, glycogen must be replenished when exercise terminates, and this also requires increased glucose uptake. Muscle expresses the "insulin-sensitive" glucose transporter GLUT4. In the fasted state, it is mainly present in intracellular GLUT4 storage vesicles (GSVs) but insulin released after a meal causes these to fuse with the plasma membrane, increasing the number of transporters at the membrane and hence the rate of glucose uptake (Fig. 3). Fusion of GSVs with the membrane is promoted by small proteins of the Rab family. Under basal conditions, these are maintained in their inactive GDP-bound form by proteins with Rab-GTPase activator protein (Rab-GAP) domains that bind to GSVs. One of these is TBC1D4 (also known as AS160). Binding of insulin to its receptor causes activation of PI 3-kinase (p. 87 [Hemmings and Restuccia 2012]), causing the formation of PIP$_3$ and thus activation of Akt. Akt phosphorylates TBC1D4 at multiple sites, causing it to interact with 14-3-3 proteins and dissociate from GSVs. No longer restrained by the Rab-GAP activity of

Cite this chapter as *Cold Spring Harb Perspect Biol* doi: 10.1101/cshperspect.a006031

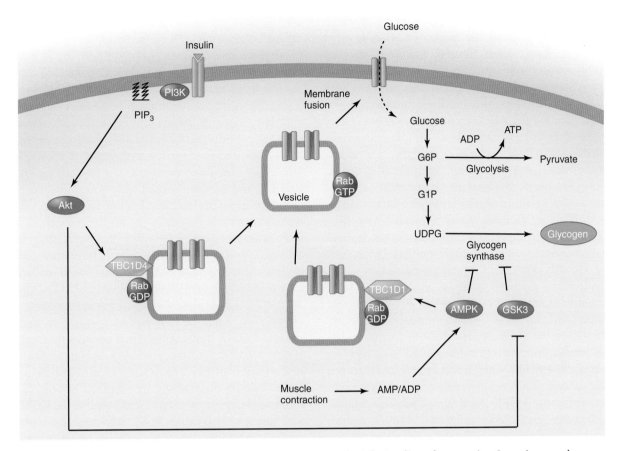

Figure 3. Regulation of muscle glucose uptake and glycogen synthesis by insulin and contraction. In resting muscle in the fed state, insulin binding to its receptor activates the PI-3-kinase → Akt pathway. Akt phosphorylates the Rab-GAP protein TBC1D4 (AS160) attached to GLUT4 storage vesicles (GSVs), causing its dissociation and promoting Rab:GTP-mediated fusion of GSVs with the plasma membrane and increased glucose uptake. This causes accumulation of glucose 6-phosphate (G6P), which activates glycogen synthase; the latter is also dephosphorylated and activated following inactivation of GSK3 by Akt. Muscle contraction, by contrast, causes increases in ADP and AMP levels that activate AMPK. AMPK phosphorylates TBC1D1, causing GSVs to fuse with the membrane. In this case, G6P does not accumulate because of the demand for ATP, and AMPK also inactivates glycogen synthase. This drives flux from increased glucose uptake into ATP production rather than glycogen synthesis.

TBC1D4, Rab proteins are converted to their active, GTP-bound forms, stimulating fusion of GSVs with the plasma membrane and increasing glucose uptake (Chen et al. 2011).

Glucose uptake triggered by muscle contraction uses a similar mechanism, but is switched on by different signaling pathways. In this case, members of the AMPK family (Box 1) are the key mediators. Exercise consumes ATP and thus increases muscle ADP:ATP and AMP:ATP ratios, activating AMPK. The use of pharmacological activators of AMPK in perfused muscle shows that activating AMPK is sufficient for increased glucose uptake (Merrill et al. 1997), whereas the use of muscle-specific-AMPK-knockout mice shows that AMPK is necessary for a full response, although a small response does remain in its absence (O'Neill et al. 2011). Because a muscle-specific knockout of the upstream kinase for AMPK, LKB1, does appear to abolish the response (Sakamoto et al.

2005), it is possible that another kinase downstream from LKB1 compensates when AMPK is absent. One candidate is the SNF1- and AMPK-related kinase (SNARK, also known as NUAK2) (Koh et al. 2010), although how it is activated during muscle contraction remains unclear.

AMPK stimulates glucose uptake, at least in part, by phosphorylating TBC1D1 (Sakamoto and Holman 2008). As with phosphorylation of its close relative TBC1D4 (also known as AS160) by Akt, this causes association with 14-3-3 proteins and dissociation from GSVs, promoting their fusion with the plasma membrane. Because the kinase domains of AMPK and SNARK/NUAK2 have closely related sequences, SNARK/NUAK2 might also phosphorylate TBC1D1. Thus, insulin and contraction increase glucose uptake via parallel signaling pathways that converge on the activation of Rab proteins.

The effects of insulin on muscle are anabolic, and the increased flux through GLUT4 is mainly directed into glycogen synthesis. By contrast, the effects of AMPK are catabolic and the increased flux through GLUT4 is directed into glycolysis and glucose oxidation instead. How are these different metabolic fates determined? Insulin stimulates glucose uptake in resting muscle, when there would not be a large demand for ATP. Glucose 6-phosphate (G6P) would therefore not be metabolized rapidly and would accumulate and activate glycogen synthase, for which G6P is a critical allosteric activator (Bouskila et al. 2010). In addition, glycogen synthase is inactivated by phosphorylation at carboxy-terminal sites by glycogen synthase kinase 3 (GSK3). However, in the presence of insulin, Akt phosphorylates and inactivates GSK3, thus causing a net dephosphorylation and activation of glycogen synthase (McManus et al. 2005)—this represents a second mechanism by which insulin stimulates glycogen synthesis. By contrast, AMPK is activated in contracting muscle, when glycolysis would be activated and G6P would not accumulate. In addition, AMPK itself inactivates glycogen synthase by phosphorylating sites distinct from those phosphorylated by GSK3 (Jorgensen et al. 2004). Thus, whereas insulin activates glycogen synthase, driving flux from increased glucose uptake into glycogen synthesis, AMPK inactivates glycogen synthase, driving flux from increased glucose uptake into glycolysis and glucose oxidation instead (Fig. 3).

5 MUSCLE—ACUTE REGULATION OF FATTY ACID OXIDATION

During prolonged, low-intensity exercise, ATP is partly generated by the mitochondrial oxidation of fatty acids. Muscle uptake of plasma fatty acids is catalyzed by transporters such as CD36 that translocate from intracellular vesicles to the membrane. Like GLUT4 translocation, this process is stimulated by AMPK (Bonen et al. 2007), although the mechanism remains unclear in this case. Fatty acids then enter the mitochondrion for oxidation in the form of acyl-carnitine esters, which requires a carnitine: palmitoyl transferase, CPT1, on the outer mitochondrial membrane. CPT1 is inhibited by malonyl-CoA, a metabolic intermediate produced in muscle by the ACC2 isoform of acetyl-CoA carboxylase. ACC2 is phosphorylated and inactivated by AMPK, which lowers malonyl-CoA levels and thus stimulates fatty acid oxidation during exercise (Merrill et al. 1997).

6 MUSCLE—LONG-TERM ADAPTATION TO EXERCISE

Athletes who train for endurance events have elevated mitochondrial content, allowing them to produce ATP more

rapidly by glucose and fatty acid oxidation. Regular endurance exercise increases mitochondrial biogenesis in part through effects of AMPK on the transcriptional coactivator PPAR-γ coactivator-1α (PGC-1α). PGC-1α is recruited to DNA by transcription factors that bind to promoters of nuclear genes encoding mitochondrial proteins (Lin et al. 2005). These include the nuclear respiratory factors NRF1 and NRF2, which switch on expression of mitochondrial transcription factor A (TFAM), a mitochondrial matrix protein required for the replication of mitochondrial DNA. PGC-1α is also recruited to promoters by PPAR-α, PPAR-δ, and estrogen-related receptor α (ERR-α), which switch on genes involved in mitochondrial fatty acid oxidation. One of these encodes pyruvate dehydrogenase kinase 4 (PDK4) (Wende et al. 2005). By phosphorylating and inactivating pyruvate dehydrogenase, PDK4 reduces entry of pyruvate into the TCA cycle and thus favors oxidation of fatty acids rather than carbohydrates.

AMPK may stimulate PGC-1α in part via direct phosphorylation, which is proposed to promote its ability to activate its own transcription, in a positive-feedback loop (Jager et al. 2007). However, AMPK activation also causes deacetylation of PGC-1α (Canto et al. 2010). Up to 13 lysine residues on PGC-1α are modified by acetylation, a reaction catalyzed by acetyltransferases of the GCN5 and SRC families. This causes PGC-1α to relocalize within the nucleus, inhibiting its transcriptional activity. The acetyl groups on PGC-1α are removed by the NAD$^+$-dependent deacetylase SIRT1, which reactivates PGC-1α. AMPK may increase the activity of SIRT1 by increasing the concentration of cytoplasmic NAD$^+$, although the exact mechanism remains uncertain. The "nutraceutical" resveratrol, which is produced by plants in response to fungal infection and is present in small amounts in red wine, has garnered much interest because it extends lifespan in nematode worms and in mice fed a high-fat diet (Baur et al. 2006). It was originally thought to be a direct activator of SIRT1, but it now appears that it may activate SIRT1 indirectly by inhibiting mitochondrial ATP synthesis and thus activating AMPK (Hawley et al. 2010; Um et al. 2010).

7 LIVER—ACUTE REGULATION OF CARBOHYDRATE METABOLISM

During starvation, blood glucose levels must be maintained to provide fuel for catabolism, particularly in neurons, which cannot use fatty acids. During short-term fasting, liver glycogen breakdown is the major source of glucose. Epinephrine (acting via Gq-linked, α1 adrenergic receptors coupled to release of inositol 1,4,5-trisphosphate [IP$_3$] by phospholipase C) increases intracellular calcium levels, whereas glucagon (acting via a Gα-linked receptor)

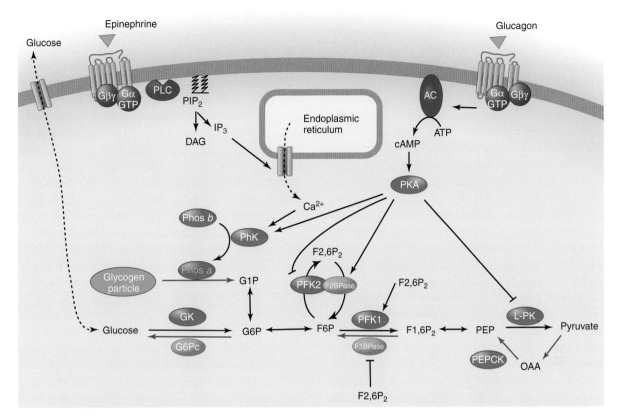

Figure 4. Acute regulation of glycolysis and gluconeogenesis in the liver. Reaction steps unique to glucose release via gluconeogenesis or glycogenolysis are shown in blue. The steps opposing liver pyruvate kinase (L-PK) in gluconeogenesis are not shown in detail. During fasting or starvation, epinephrine and glucagon increase calcium and cyclic AMP (cAMP) levels, activating phosphorylase kinase and cAMP-dependent protein kinase (PKA), which act together to promote glycogen breakdown. During starvation, PKA also phosphorylates L-PK and 6-phosphofructo-2-kinase/fructose 2,6-bisphosphatase (PFK2/F2BPase), inactivating the former, and inhibiting the kinase and activating the phosphatase activity of the latter. This causes a drop in fructose 2,6-bisphosphate levels, which triggers a net switch from glycolysis to gluconeogenesis.

increases the levels of cyclic AMP. Calcium and cyclic AMP then trigger glycogen breakdown in the liver via essentially the same mechanisms as those described above for muscle (Fig. 4). One important difference is that liver cells express glucose-6-phosphatase (and an associated transporter that carries G6P into the ER lumen), which are required for the release of glucose derived from glycogen breakdown into the bloodstream (van Schaftingen and Gerin 2002). The liver also expresses a different isoform of phosphorylase that is not sensitive to AMP. This is consistent with the view that liver glycogen is a store of glucose for use by other tissues, rather than for internal use during periods of energy deficit, as in muscle.

In the liver, glycolysis is most active in the fed state and is an anabolic pathway, because it provides precursors for lipid biosynthesis. The liver is also the major site of gluconeogenesis, the synthesis of glucose from noncarbohydrate

precursors, which is essentially a reversal of glycolysis except for three irreversible steps in which different reactions are used (blue arrows in Fig. 4). Gluconeogenesis becomes particularly important as a source of glucose during starvation, particularly for the brain, which cannot use fatty acids. The liver therefore must have mechanisms to trigger a switch from glycolysis to gluconeogenesis during the transition from the fed to the starved state. A key mediator of this switch is a metabolite that has a purely regulatory role, fructose 2-6-bisphosphate, which is synthesized and broken down to fructose 6-phosphate by distinct domains of a single bienzyme polypeptide termed 6-phosphofructo-2-kinase/fructose-2,6-bisphosphatase (PFK2/FBPase) (Fig. 4).

A key step in glycolysis is the conversion of fructose 6-phosphate to fructose 1,6-bisphosphate, catalyzed by 6-phosphofructo-1-kinase (PFK1). PFK1 is allosterically activated by fructose 2, 6-bisphosphate, which also inhibits the opposing reaction in gluconeogenesis, catalyzed by fructose-1,6-bisphosphatase. On the transition from the fed to the starved state, glucagon is released, increasing cyclic AMP levels in the liver and activating PKA. PKA phosphorylates the liver isoform of PFK2/FBPase, inhibiting its kinase and activating its phosphatase activity (Rider et al. 2004). The consequent drop in fructose 2,6-bisphosphate levels both reduces PFK1 activation and relieves inhibition of fructose-1,6-bisphosphatase, causing a net switch from glycolysis to gluconeogenesis. In addition, PKA phosphorylates and inactivates the liver isoform of pyruvate kinase (L-PK) (Riou et al. 1978), causing additional inhibition of glycolysis at a later step (Fig. 4).

On returning to the fed state again, blood glucose increases, glucagon levels decrease, and the effects just described are reversed. Some of the increased flux of glucose into the liver caused by the high blood glucose levels also enters the pentose phosphate pathway, generating the intermediate xylulose 5-phosphate. Xylulose 5-phosphate has been found to activate a protein phosphatase that dephosphorylates PFK2/FBPase, thus switching it back to the state that favors fructose 2,6-bisphosphate synthesis (Nishimura et al. 1994).

In muscle, where gluconeogenesis is absent and glycolysis has a purely catabolic role, it would not make sense for hormones that increase cyclic AMP (such as epinephrine) to inhibit glycolysis. Indeed, muscle expresses different isoforms of pyruvate kinase and PFK2/FBPase, which lack the PKA sites.

8 LIVER—LONG-TERM REGULATION OF GLUCONEOGENESIS VIA EFFECTS ON GENE EXPRESSION

Another important tier of regulation of glycolysis and gluconeogenesis occurs at the level of transcription. Although expression of most genes involved in these pathways is regulated, research has particularly focused on the genes encoding the catalytic subunit of glucose-6-phosphatase (G6Pc), and phosphoenolpyruvate carboxykinase (PEPCK) (Yabaluri and Bashyam 2010). Although often referred to as "gluconeogenic genes," in fact neither is involved exclusively with that pathway. Thus, glucose-6-phosphatase releases into the bloodstream glucose derived from glycogen breakdown as well as gluconeogenesis (Fig. 4), whereas phosphoenolpyruvate produced by PEPCK is used as a precursor for biosynthesis of products other than glucose, including glycerol 3-phosphate used in triglyceride synthesis.

Three important hormonal regulators of transcription of these genes are glucocorticoids and glucagon (which are released during fasting or starvation and increase transcription) and insulin (which is released after carbohydrate feeding and represses transcription). The promoters for these genes contain hormone response units that bind the critical transcription factors and are most well defined in the case of the *G6Pc* promoter (Fig. 5).

The Glucocorticoid Response Unit. The promoter contains three glucocorticoid response elements (GREs) that

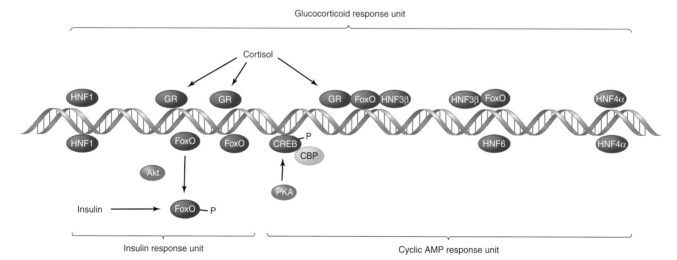

Figure 5. Regulation of the *G6Pc* promoter, showing the approximate location of elements binding the key transcription factors. Glucocorticoids such as cortisol, in complex with the glucocorticoid receptor (GR), bind to three sites within the glucocorticoid response unit, enhancing transcription. Cyclic AMP-dependent protein kinase (PKA) phosphorylates cyclic AMP response element binding protein (CREB), recruiting CREB-binding protein (CBP), and activating transcription. Finally, Akt phosphorylates FoxO at multiple sites, triggering the binding of 14-3-3 proteins and their nuclear exclusion, thus inhibiting transcription. Not shown are the roles of coactivators other than CBP described in the text, i.e., PGC-1α and CRTC2.

bind the glucocorticoid-receptor complex, which activates transcription. However, accessory elements that bind additional transcription factors, including hepatocyte nuclear factors (HNF1/3β/4α/6) and FOXO1, are necessary for a full response.

The Cyclic AMP Response Unit. The promoter contains two cyclic-AMP-response elements (CREs), sequence elements that bind the transcription factor CRE-binding protein (CREB). Glucagon activates PKA, which phosphorylates CREB at 133, promoting binding of the CBP transactivator to activate transcription.

The Insulin Response Unit. The promoter contains two insulin-response elements that bind FoxO1, a transcription factor whose loss leads to decreased expression of both *PEPCK* and *G6Pc* on fasting. FoxO1, along with other members of the forkhead box family, is phosphorylated at multiple conserved sites by Akt, causing its relocalization from the nucleus to the cytoplasm owing to binding of 14-3-3 proteins (Brunet et al. 1999). Insulin also induces expression of COP1, an E3 ubiquitin ligase that promotes FoxO1 degradation by the proteasome (Kato et al. 2008).

Transcription of these genes is also regulated by modulation of various coactivators not shown in Fig. 5, including PGC-1α. The *PGC-1α* promoter contains CREs, and in the liver *PGC-1α* is a cyclic-AMP-induced gene activated by glucagon. PGC-1α promotes PEPCK expression in part by interacting with the glucocorticoid receptor. The NAD$^+$-dependent deacetylase SIRT1 is also induced in the liver on fasting and, as described above, it deacetylates PGC-1α, increasing transcription of target genes. SIRT1 also deacetylates FOXO1, thus enhancing transcription of *G6Pc*.

Interestingly, some of these signaling events appear to have arisen during early eukaryotic evolution. In the nematode worm *Caenorhabditis elegans*, restricting the diet in early life switches development to the long-lived Dauer larval form, and this can be mimicked by mutations in genes encoding orthologues of the insulin/IGF1 receptor (*daf-2*) or PI-3-kinase (*age-1*), or overexpression of orthologues of SIRT1 (Sir-2.1) and FoxO1 (Daf-16) (Tissenbaum and Guarente 2001). Thus, dietary restriction extends life span in part by preventing activation of the insulin-like receptor → PI-3-kinase → Akt pathway, which in turn inhibits the transcription factor FoxO1 (Daf-16) by triggering its phosphorylation and acetylation. Dietary restriction also activates the *C. elegans* orthologue of AMPK, which activates FoxO1 by phosphorylation at sites different from those targeted by Akt (Greer et al. 2007), and perhaps also by deacetylation. All of these pathways are conserved in mammals, and there is currently intense interest as to whether they regulate mammalian life span in the same manner.

Another transcriptional coactivator that regulates *PEPCK* expression is CREB-regulated transcription coactivator 2 (CRTC2, formerly called TORC2), which is recruited to the *PEPCK* promoter by CREB. CRTC2 is phosphorylated by the protein kinase salt-inducible kinase 1 (SIK1), which triggers binding of 14-3-3 proteins and its relocation from the nucleus to the cytoplasm. However, PKA phosphorylates SIK1, causing its relocalization from the nucleus to the cytoplasm, thus preventing CRTC2 phosphorylation and promoting PEPCK expression. SIK1 is a member of the AMPK-related kinase family and AMPK also phosphorylates CRTC2 at the same site, explaining how AMPK activation switches off *PEPCK* expression. Interestingly, phosphorylation of the CRTC2 orthologue by AMPK is conserved in *C. elegans*, and is required for extension of life span by AMPK (Mair et al. 2011).

When would AMPK switch off gluconeogenesis? Adiponectin activates AMPK in the liver via the AdipoRI receptor (see above). This explains how it inhibits hepatic glucose production, and why low adiponectin levels in obese humans correlate with elevated liver glucose production. In addition, AMPK is activated by the antidiabetic drug metformin (see below), which lowers blood glucose levels mainly by repressing gluconeogenesis.

9 LIVER—REGULATION OF FATTY ACID, TRIGLYCERIDE, AND CHOLESTEROL METABOLISM

Liver cells carry out both the synthesis and oxidation of fatty acids, and express both isoforms of acetyl-CoA carboxylase (ACC1 and ACC2). By phosphorylating ACC1 and ACC2 at conserved sites to cause their inactivation, AMPK switches off fatty acid synthesis to conserve energy, while switching on fatty acid oxidation to generate more ATP (Hardie 2007). This occurs, for example, when AMPK is activated in the liver by adiponectin. AMPK activation also causes inactivation of glycerol phosphate acyl transferase (GPAT), the first enzyme in the pathway of triglyceride and phospholipid synthesis, although it has not yet been shown to be a direct target for AMPK. Finally, AMPK inhibits cholesterol synthesis by phosphorylation of HMG-CoA reductase (ACC1 and HMG-CoA reductase were, in fact, the first AMPK targets to be identified (Munday et al. 1988; Clarke and Hardie 1990).

In addition to these acute effects, fatty acid and triglyceride synthesis are regulated in the longer term at the level of gene expression. The genes targeted include those whose products are involved directly in these pathways (ACC1, fatty acid synthase, stearoyl-CoA desaturase and GPAT), in the glycolytic pathway that provides the precursors for lipid synthesis (glucokinase, PFK1, aldolase, and L-PK), and in pathways that provide NADPH for the reductive steps in lipid synthesis (glucose-6-phosphate dehydrogenase and malic enzyme). These are referred to collectively as lipogenic enzymes, and their expression is up-regulated by carbohydrate feeding, so that excess dietary carbohydrates are converted to triglycerides. The latter are then exported from the liver as very-low-density lipoproteins (VLDL) and carried to adipose tissue for long-term storage.

Carbohydrate feeding causes an increase in blood glucose that also triggers insulin release. Studies with cultured liver cells suggest that increases in both glucose and insulin are necessary for the increased transcription of most lipogenic genes. A transcription factor involved in the effects of insulin is sterol response element-binding protein 1c (SREBP1c). SREBP1a and SREBP1c are derived from the same gene by use of alternate transcription start sites and are closely related to the product of another gene, SREBP2 (Raghow et al. 2008). All three have amino-terminal transcription factor domains (TFDs) linked to carboxy-terminal regulatory domains by two transmembrane α helices that anchor them within the endoplasmic reticulum membrane (Fig. 6). The TFDs are released from the membrane by regulated proteolytic processing; the TFD from SREBP1a/1c targets mainly lipogenic genes, whereas that from SREBP2 targets genes involved in cholesterol biosynthesis and uptake (including HMG-CoA reductase).

The carboxy-terminal domains of SREBPs bind to ER membrane proteins called SREB cleavage activator protein (SCAP) and Insigs (Insig1/Insig2). When membrane sterol levels are high, they bind to SCAP and Insig1, causing SREBP2 to be retained within the ER. Conversely, when membrane sterol levels are low, SCAP dissociates from Insig2 and the SCAP–SREBP2 complex moves to the Golgi apparatus, where proteinases release the TFD (Yang et al. 2002). Whereas SREBP2 appears to be regulated by sterols mainly at the proteolytic processing step (i.e., sterols inhibit processing), insulin appears to regulate SREBP1c at additional levels, enhancing its transcription and reducing its degradation, and enhancing degradation of Insig2.

Mice lacking SREBP1c show reduced expression of most lipogenic genes, but their response to carbohydrate feeding is not entirely eliminated. This suggests that other transcription factors also play a role. It has been proposed that the response of lipogenic genes to glucagon and fatty acids (which down-regulate their expression), and high glucose (which induces expression), involves the carbohydrate response element-binding protein (ChREBP) (Uyeda and Repa 2006), a transcription factor with a DNA-binding domain related to that of SREBP-1. ChREBP is phosphorylated at two sites by PKA in response to glucagon, which

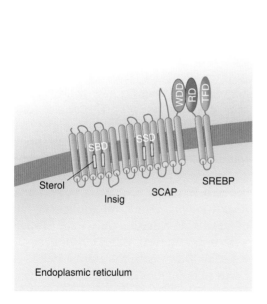
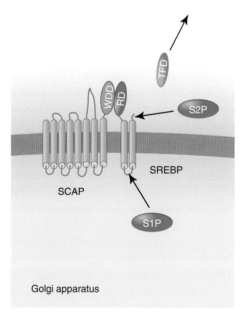

Figure 6. Regulation of processing of SREBPs. The precursor forms of SREBPs bind to the membrane protein SCAP through interactions between their carboxy-terminal regulatory domain (RD) and the WD repeat domain (WDD) of SCAP. The SREBP-SCAP complex is retained in the endoplasmic reticulum by interaction with Insigs. Reduced binding of sterols to the sterol-binding domain (SBD) of Insig1 and the sterol sensor domain (SSD) of SCAP causes their dissociation, and the SCAP-SREBP2 complex then translocates to the Golgi, where the site 1 and site 2 proteases (S1P and S2P) cleave SREBP2, releasing the transcription factor domain (TFD) that translocates to the nucleus. Regulation of SREB1c is similar, except that there appears to be multiple mechanisms that trigger its release from the ER, including insulin-induced degradation of Insig2.

prevents its nuclear import and inhibits promoter binding. Phosphorylation of a third site by AMPK also inhibits promoter binding and may account for down-regulation of lipogenic genes by adiponectin. It has also been proposed that generation of AMP during conversion of fatty acids to their CoA esters could activate AMPK and explain the effects of fatty acids on lipogenic gene expression. One hypothesis to explain the effect of high glucose is that it is metabolized by the pentose phosphate pathway to xylulose 5-phosphate, which activates the phosphatase that dephosphorylates PFK2/FBPase (see above). This phosphatase is also thought to dephosphorylate the PKA/AMPK sites on ChREBP, promoting its nuclear localization and promoter binding (Kabashima et al. 2003).

10 ADIPOCYTES—REGULATION OF FATTY ACID METABOLISM

Although subcutaneous fat provides thermal insulation, the main metabolic function of adipocytes is to store fatty acids as triglycerides, neutral lipids that are very insoluble in water and are deposited in lipid droplets. White adipocytes, unlike other cells, have a single central triglyceride droplet that occupies almost the entire volume of the cell. The phospholipid monolayer that forms its cytoplasmic face is lined with a protein called peripilin1 (Brasaemle 2007). To release fatty acids back into the circulation, the triglycerides in the lipid droplet must be hydrolyzed back to free fatty acids (lipolysis). Three lipases are involved, which remove the first, second, and third fatty acids: (1) adipose tissue triglyceride lipase (ATGL), (2) hormone-sensitive lipase (HSL), and (3) monacylglycerol lipase. Lipolysis is greatly enhanced during fasting by glucagon and/or epinephrine, acting via increases in cyclic AMP; insulin opposes this because Akt phosphorylates and activates the cyclic AMP phosphodiesterase PDE3B, thus lowering cyclic AMP levels (Berggreen et al. 2009). HSL is directly phosphorylated and activated by PKA, although the effect on activity is modest (about twofold) compared with the effects on lipolysis (at least 100-fold). Phosphorylation of HSL also triggers its translocation from the cytoplasm to the lipid droplet, thus increasing its accessibility to substrate (Clifford et al. 2000). However, some hormone-stimulated release of fatty acids still occurs even in adipocytes from HSL-deficient mice (Haemmerle et al. 2002). This may be because perilipin1, which seems to regulate access of lipolytic enzymes to the surface of the lipid droplet, is also phosphorylated by PKA (Clifford et al. 2000). Adipocytes from perilipin1 knockout mice have a high basal lipolytic rate that is only marginally stimulated by cyclic-AMP-elevating agents (Tansey et al. 2001). The crucial effect of PKA may therefore be to phosphorylate

perilipin1, which alters the accessibility of triglycerides within the lipid droplet to both ATGL and HSL.

Activation of AMPK opposes the effects of epinephrine and glucagon on lipolysis, in part because it phosphorylates HSL at sites close to the PKA sites, antagonizing the activation and translocation induced by PKA (Daval et al. 2005). Whether AMPK also antagonizes the effect of phosphorylation of perilipin1 by PKA remains unclear. It might appear paradoxical that AMPK inhibits lipolysis, because fatty acids are an excellent fuel for oxidative catabolism. However, the fatty acids produced by lipolysis are not usually oxidized within the adipocyte, but are released for use elsewhere. If the fatty acids generated by lipolysis are not rapidly removed from adipocytes either by export or by oxidative metabolism, they are recycled into triglycerides, an energy-intensive process in which two molecules of ATP are consumed per fatty acid. Thus, inhibition of lipolysis by AMPK may ensure that the rate of lipolysis does not exceed the rate at which the fatty acids can be removed from the system.

11 BROWN ADIPOCYTES—REGULATION OF FATTY ACID OXIDATION AND HEAT PRODUCTION

Brown adipocytes are so called because, unlike white fat cells, they have abundant mitochondria containing cytochromes that produce their characteristic color. Increases in cyclic AMP levels induced by epinephrine trigger lipolysis as in white fat cells, but brown adipocytes differ in that the fatty acids are not released but are oxidized within their own mitochondria. Another unique feature of brown adipocytes is that they express uncoupling protein1 (UCP1), which dissipates the electrochemical gradient produced by pumping of protons across the inner mitochondrial membrane by the respiratory chain. The energy expended in brown fat cells therefore mainly appears in the form of heat, rather than ATP. This heat-generating system is particularly important in neonatal animals (including humans), but also occurs in adult rodents exposed to cold environments. Once thought to be absent in adult humans, improved methodology has shown that brown fat does indeed occur (van Marken Lichtenbelt et al. 2009). These findings have rekindled interest in the idea that regulation of energy expenditure by brown fat might be a way of controlling obesity.

12 INSULIN RESISTANCE AND TYPE 2 DIABETES—A RESPONSE TO OVERNUTRITION?

Insulin resistance is a condition in which tissues become resistant to the effects of the hormone, and individuals with type 2 diabetes have a high fasting blood glucose level caused primarily by insulin resistance (rather than lack of insulin as in type 1 diabetes). In insulin-resistant individuals, muscle takes up less glucose in response to insulin, and the hormone is also less effective at suppressing glucose production by the liver. Blood glucose therefore remains elevated for longer periods after a carbohydrate meal, a condition known as glucose intolerance. Many individuals who are insulin resistant compensate by secreting more insulin so that, although they may show glucose intolerance, in the fasting state their blood glucose is within the normal range such that they are not classed as diabetic. However, this compensation mechanism may eventually fail, and insulin-resistant individuals often become diabetic.

As the world has become more developed and urbanized, there has been an alarming increase in the prevalence of type 2 diabetes. By 2025 it is predicted that the number with the disorder will increase to >300 million (nearly 4% of the world population). The increase is particularly evident in countries where economic development has been very rapid, like China, where in 2010 >90 million adults were estimated to have diabetes (Yang et al. 2010).

What is the reason for this dramatic increase? Type 2 diabetes is strongly associated with obesity: In one large study, females who were obese (body mass index [BMI] $> 30 \text{ kg/m}^2$) had a 20-fold higher risk of developing diabetes compared with those who were lean (BMI < 23 kg/m^2) (Hu et al. 2001). This is a serious problem, because over one-quarter of the U.S. population now have a BMI > 30. Obesity is caused by excessive energy intake (overnutrition) and/or reduced energy expenditure (less physical activity). At the cellular level, insulin resistance can be regarded as a response to excessive storage of nutrients, especially lipids, and it can be reproduced in vitro by incubating cells with high concentrations of glucose or fatty acids. Excessive storage of triglycerides appears to be a particular culprit (Samuel et al. 2010). This is dramatically illustrated by lipodystrophy, a lack of white adipose tissue that can be caused either by rare genetic disorders, or as a side effect of antiviral drugs used to treat AIDS patients (Garg 2004). In both cases, excessive amounts of triglyceride are stored in the liver and muscle instead, and these tissues become profoundly insulin resistant. It appears that large amounts of triglyceride can be stored with relative safety in adipocytes, but that, if stored elsewhere, they become very damaging. Indeed, the tendency to insulin resistance in obese people may be because their adipose tissue stores are already full, leading to increased lipid storage (steatosis) in the liver and muscle.

The current frontline drug used to treat type 2 diabetes is metformin, prescribed to more than 100 million people in 2010. Derived from an ancient herbal remedy, it activates AMPK, which it does either by inhibiting complex I of the mitochondrial respiratory chain (Owen et al. 2000) or by inhibiting AMP deaminase (Ouyang et al. 2011), both of which increase the cellular AMP:ATP and ADP:ATP ratios (Hawley et al. 2010). By mechanisms already discussed, AMPK activation inhibits synthesis and storage of fat and promotes its oxidation instead, promotes glucose uptake and oxidation by muscle, and inhibits gluconeogenesis in the liver. All of these effects would be beneficial in individuals with insulin resistance and/or type 2 diabetes. There remains some controversy as to whether AMPK activation explains all therapeutic effects of metformin (Foretz et al. 2010). Nevertheless, it is intriguing that a drug used to treat diabetes activates a signaling pathway switched on by exercise. There is much evidence to show that regular exercise protects against development of insulin resistance, and conversely that physical inactivity is a factor contributing to its occurrence. Thus, metformin may at least partly be acting as an exercise mimetic.

There is also much debate about the underlying causes of insulin resistance at the molecular level. Disturbances in lipid metabolites, inflammatory mediators, and adipokines have all been proposed. Whatever the underlying mechanisms, insulin resistance is likely to be a feedback mechanism that has evolved to limit nutrient uptake under conditions when cellular stores of nutrients are already replete. The key to understanding insulin resistance may lie in working out how cells detect when their nutrient stores are full. Although leptin and adiponectin appear to be released from adipocytes when their triglyceride stores are high and low, respectively, we currently know very little at the molecular level about how the level of intracellular stores of lipids (or other nutrients, like glycogen) are monitored.

13 CONCLUDING REMARKS

Energy balance in multicellular organisms involves a complex interplay between energy-utilizing tissues such as muscle, energy-storing tissues such as adipose tissue, and organs involved in metabolic coordination such as the liver. These tissues signal to each other via hormones and cytokines that are either secreted by specialized endocrine cells (e.g., the hypothalamus, islets of Langerhans, pituitary, thyroid, and adrenal glands) or by the tissues themselves (e.g., adipokines, released by adipocytes). These hormones either act at receptors that switch on protein kinase signaling cascades triggered by second messengers such as PIP_3 (insulin), calcium (epinephrine acting at α1 receptors) or cyclic AMP (glucagon), or bind to nuclear receptors that are transcription factors (cortisol and T3). These signaling cascades interact with other signaling pathways involved in regulating energy balance at the cell-autonomous level (e.g., AMPK). The net effect is modulation of carbohydrate and lipid metabolism, both by direct phosphorylation of metabolic enzymes and by effects of gene expression or protein turnover. One important remaining challenge is to understand how cells monitor their levels of energy reserves such as triglyceride, a process likely to be important in understanding disorders such as obesity and type 2 diabetes.

REFERENCES

*Reference is in this book.

Baur JA, Pearson KJ, Price NL, Jamieson HA, Lerin C, Kalra A, Prabhu VV, Allard JS, Lopez-Lluch G, Lewis K, et al. 2006. Resveratrol improves health and survival of mice on a high-calorie diet. Nature 444: 337–342.

Berggreen C, Gormand A, Omar B, Degerman E, Goransson O. 2009. Protein kinase B activity is required for the effects of insulin on lipid metabolism in adipocytes. Am J Physiol Endocrinol Metab 296: E635–E646.

Bonen A, Han XX, Habets DD, Febbraio M, Glatz JF, Luiken JJ. 2007. A null mutation in skeletal muscle FAT/CD36 reveals its essential role in insulin- and AICAR-stimulated fatty acid metabolism. Am J Physiol Endocrinol Metab 292: E1740–E1749.

Bouskila M, Hunter RW, Ibrahim AF, Delattre L, Peggie M, van Diepen JA, Voshol PJ, Jensen J, Sakamoto K. 2010. Allosteric regulation of glycogen synthase controls glycogen synthesis in muscle. Cell Metab 12: 456–466.

Brasaemle DL. 2007. Thematic review series: Adipocyte biology. The perilipin family of structural lipid droplet proteins: Stabilization of lipid droplets and control of lipolysis. J Lipid Res 48: 2547–2559.

Brunet A, Bonni A, Zigmond MJ, Lin MZ, Juo P, Hu LS, Anderson MJ, Arden KC, Blenis J, Greenberg ME. 1999. Akt promotes cell survival by phosphorylating and inhibiting a Forkhead transcription factor. Cell 96: 857–868.

Canto C, Jiang LQ, Deshmukh AS, Mataki C, Coste A, Lagouge M, Zierath JR, Auwerx J. 2010. Interdependence of AMPK and SIRT1 for metabolic adaptation to fasting and exercise in skeletal muscle. Cell Metab 11: 213–219.

Chen S, Wasserman DH, MacKintosh C, Sakamoto K. 2011. Mice with AS160/TBC1D4-Thr649Ala knockin mutation are glucose intolerant with reduced insulin sensitivity and altered GLUT4 trafficking. Cell Metab 13: 68–79.

Clarke PR, Hardie DG. 1990. Regulation of HMG-CoA reductase: Identification of the site phosphorylated by the AMP-activated protein kinase in vitro and in intact rat liver. EMBO J 9: 2439–2446.

Clifford GM, Londos C, Kracmcr FB, Vernon RG, Yeaman SJ. 2000. Translocation of hormone-sensitive lipase and perilipin upon lipolytic stimulation of rat adipocytes. J Biol Chem 275: 5011–5015.

Daval M, Diot-Dupuy F, Bazin R, Hainault I, Viollet B, Vaulont S, Hajduch E, Ferre P, Foufelle F. 2005. Anti-lipolytic action of AMP-activated protein kinase in rodent adipocytes. J Biol Chem 280: 25250–25257.

Fischer EH, Krebs EG. 1989. Commentary on "The phosphorylase b to a converting enzyme of rabbit skeletal muscle". Biochim Biophys Acta 1000: 297–301.

Foretz M, Hebrard S, Leclerc J, Zarrinpashneh E, Soty M, Mithieux G, Sakamoto K, Andreelli F, Viollet B. 2010. Metformin inhibits hepatic gluconeogenesis in mice independently of the LKB1/AMPK pathway via a decrease in hepatic energy state. J Clin Invest 120: 2355–2369.

Friedman JM, Halaas JL. 1998. Leptin and the regulation of body weight in mammals. Nature 395: 763–770.

Garg A. 2004. Acquired and inherited lipodystrophies. New Engl J Med 350: 1220–1234.

Greer EL, Dowlatshahi D, Banko MR, Villen J, Hoang K, Blanchard D, Gygi SP, Brunet A. 2007. An AMPK-FOXO pathway mediates longevity induced by a novel method of dietary restriction in C. elegans. Curr Biol 17: 1646–1656.

Haemmerle G, Zimmermann R, Hayn M, Theussl C, Waeg G, Wagner E, Sattler W, Magin TM, Wagner EF, Zechner R. 2002. Hormone-sensitive lipase deficiency in mice causes diglyceride accumulation in adipose tissue, muscle, and testis. J Biol Chem 277: 4806–4815.

Hardie DG. 2007. AMP-activated/SNF1 protein kinases: Conserved guardians of cellular energy. Nat Rev Mol Cell Biol 8: 774–785.

Hardie DG. 2011. AMP-activated protein kinase—An energy sensor that regulates all aspects of cell function. Genes Dev 25: 1895–1908.

Hardie DG, Carling D, Gamblin SJ. 2011. AMP-activated protein kinase: Also regulated by ADP? Trends Biochem Sci 36: 470–477.

* Harrison DA. 2012. The JAK/STAT pathway. Cold Spring Harb Perspect Biol 4: a011205.

Hawley SA, Ross FA, Chevtzoff C, Green KA, Evans A, Fogarty S, Towler MC, Brown LJ, Ogunbayo OA, Evans AM, et al. 2010. Use of cells expressing γ subunit variants to identify diverse mechanisms of AMPK activation. Cell Metab 11: 554–565.

* Heldin CH, Lu B, Evans R, Gutkind JS. 2014. Signals and receptors. Cold Spring Harb Perspect Biol doi: 10.1101/cshperspect.a005900.

* Hemmings BA, Restuccia DF. 2012. The PI3K-PKB/Akt pathway. Cold Spring Harb Perspect Biol 4: a011189.

Hu FB, Manson JE, Stampfer MJ, Colditz G, Liu S, Solomon CG, Willett WC. 2001. Diet, lifestyle, and the risk of type 2 diabetes mellitus in women. New Engl J Med 345: 790–797.

Jager S, Handschin C, St-Pierre J, Spiegelman BM. 2007. AMP-activated protein kinase (AMPK) action in skeletal muscle via direct phosphorylation of PGC-1α. Proc Natl Acad Sci 104: 12017–12022.

Jorgensen SB, Nielsen JN, Birk JB, Olsen GS, Viollet B, Andreelli F, Schjerling P, Vaulont S, Hardie DG, Hansen BF, et al. 2004. The α2-5'AMP-activated protein kinase is a site 2 glycogen synthase kinase in skeletal muscle and is responsive to glucose loading. Diabetes 53: 3074–3081.

Kabashima T, Kawaguchi T, Wadzinski BE, Uyeda K. 2003. Xylulose 5-phosphate mediates glucose-induced lipogenesis by xylulose 5-phosphate-activated protein phosphatase in rat liver. Proc Natl Acad Sci 100: 5107–5112.

Kadowaki T, Yamauchi T. 2005. Adiponectin and adiponectin receptors. Endocr Rev 26: 439–451.

Kato S, Ding J, Pisck E, Jhala US, Du K. 2008. COP1 functions as a FoxO1 ubiquitin E3 ligase to regulate FoxO1-mediated gene expression. J Biol Chem 283: 35464–35473.

Koh HJ, Toyoda T, Fujii N, Jung MM, Rathod A, Middelbeek RJ, Lessard SJ, Treebak JT, Tsuchihara K, Esumi H, et al. 2010. Sucrose nonfermenting AMPK-related kinase (SNARK) mediates contraction-stimulated glucose transport in mouse skeletal muscle. Proc Natl Acad Sci 107: 15541 15546.

Kubota N, Yano W, Kubota T, Yamauchi T, Itoh S, Kumagai H, Kozono H, Takamoto I, Okamoto S, Shiuchi T, et al. 2007. Adiponectin stimulates AMP-activated protein kinase in the hypothalamus and increases food intake. Cell Metab 6: 55–68.

* Kuo IY, Ehrlich BE. 2014. Signaling in muscle contraction. Cold Spring Harb Perspect Biol doi: 10.1101/cshperspect.a006023.

Lin J, Handschin C, Spiegelman BM. 2005. Metabolic control through the PGC-1 family of transcription coactivators. Cell Metab 1: 361–370.

Lopez M, Varela L, Vazquez MJ, Rodriguez-Cuenca S, Gonzalez CR, Velagapudi VR, Morgan DA, Schoenmakers E, Agassandian K, Lage R, et al. 2010. Hypothalamic AMPK and fatty acid metabolism mediate thyroid regulation of energy balance. Nat Med 16: 1001–1008.

Mair W, Morantte I, Rodrigues AP, Manning G, Montminy M, Shaw RJ, Dillin A. 2011. Lifespan extension induced by AMPK and calcineurin is mediated by CRTC-1 and CREB. Nature 470: 404–408.

McCrimmon RJ, Shaw M, Fan X, Cheng H, Ding Y, Vella MC, Zhou L, McNay EC, Sherwin RS. 2008. Key role for AMP-activated protein kinase in the ventromedial hypothalamus in regulating counterregulatory hormone responses to acute hypoglycemia. *Diabetes* **57**: 444–450.

McManus EJ, Sakamoto K, Armit LJ, Ronaldson L, Shpiro N, Marquez R, Alessi DR. 2005. Role that phosphorylation of GSK3 plays in insulin and Wnt signalling defined by knockin analysis. *EMBO J* **24**: 1571–1583.

Merrill GM, Kurth E, Hardie DG, Winder WW. 1997. AICAR decreases malonyl-CoA and increases fatty acid oxidation in skeletal muscle of the rat. *Am J Physiol* **273**: E1107–E1112.

Munday MR, Campbell DG, Carling D, Hardie DG. 1988. Identification by amino acid sequencing of three major regulatory phosphorylation sites on rat acetyl-CoA carboxylase. *Eur J Biochem* **175**: 331–338.

Nishimura M, Fedorov S, Uyeda K. 1994. Glucose-stimulated synthesis of fructose 2,6-bisphosphate in rat liver. Dephosphorylation of fructose 6-phosphate, 2-kinase:fructose 2,6-bisphosphatase and activation by a sugar phosphate. *J Biol Chem* **269**: 26100–26106.

O'Neill HM, Maarbjerg SJ, Crane JD, Jeppesen J, Jorgensen SB, Schertzer JD, Shyroka O, Kiens B, van Denderen BJ, Tarnopolsky MA, et al. 2011. AMP-activated protein kinase (AMPK) β1β2 muscle null mice reveal an essential role for AMPK in maintaining mitochondrial content and glucose uptake during exercise. *Proc Natl Acad Sci* **108**: 16092–16097.

Oakhill JS, Steel R, Chen ZP, Scott JW, Ling N, Tam S, Kemp BE. 2011. AMPK is a direct adenylate charge-regulated protein kinase. *Science* **332**: 1433–1435.

Ouyang J, Parakhia RA, Ochs RS. 2011. Metformin activates AMP kinase through inhibition of AMP deaminase. *J Biol Chem* **286**: 1–11.

Owen MR, Doran E, Halestrap AP. 2000. Evidence that metformin exerts its anti-diabetic effects through inhibition of complex 1 of the mitochondrial respiratory chain. *Biochem J* **348**: 607–614.

Raghow R, Yellaturu C, Deng X, Park EA, Elam MB. 2008. SREBPs: The crossroads of physiological and pathological lipid homeostasis. *Trends Endocrinol Metab* **19**: 65–73.

Ramaiah A, Hathaway JA, Atkinson DE. 1964. Adenylate as a metabolic regulator. Effect on yeast phosphofructokinase kinetics. *J Biol Chem* **239**: 3619–3622.

Rider MH, Bertrand L, Vertommen D, Michels PA, Rousseau GG, Hue L. 2004. 6-phosphofructo-2-kinase/fructose-2,6-bisphosphatase: Head-to-head with a bifunctional enzyme that controls glycolysis. *Biochem J* **381**: 561–579.

Riou JP, Claus TH, Pilkis SJ. 1978. Stimulation of glucagon of in vivo phosphorylation of rat hepatic pyruvate kinase. *J Biol Chem* **253**: 656–659.

Sakamoto K, Holman GD. 2008. Emerging role for AS160/TBC1D4 and TBC1D1 in the regulation of GLUT4 traffic. *Am J Physiol Endocrinol Metab* **295**: E29–E37.

Sakamoto K, McCarthy A, Smith D, Green KA, Hardie DG, Ashworth A, Alessi DR. 2005. Deficiency of LKB1 in skeletal muscle prevents AMPK activation and glucose uptake during contraction. *EMBO J* **24**: 1810–1820.

Samuel VT, Petersen KF, Shulman GI. 2010. Lipid-induced insulin resistance: Unravelling the mechanism. *Lancet* **375**: 2267–2277.

Shetty S, Kusminski CM, Scherer PE. 2009. Adiponectin in health and disease: Evaluation of adiponectin-targeted drug development strategies. *Trends Pharmacol Sci* **30**: 234–239.

Tansey JT, Sztalryd C, Gruia-Gray J, Roush DL, Zee JV, Gavrilova O, Reitman ML, Deng CX, Li C, Kimmel AR, et al. 2001. Perilipin ablation results in a lean mouse with aberrant adipocyte lipolysis, enhanced leptin production, and resistance to diet-induced obesity. *Proc Natl Acad Sci* **98**: 6494–6499.

Tissenbaum HA, Guarente L. 2001. Increased dosage of a sir-2 gene extends lifespan in *Caenorhabditis elegans*. *Nature* **410**: 227–230.

Um JH, Park SJ, Kang H, Yang S, Foretz M, McBurney MW, Kim MK, Viollet B, Chung JH. 2010. AMP-activated protein kinase-deficient mice are resistant to the metabolic effects of resveratrol. *Diabetes* **59**: 554–563.

Uyeda K, Repa JJ. 2006. Carbohydrate response element binding protein, ChREBP, a transcription factor coupling hepatic glucose utilization and lipid synthesis. *Cell Metab* **4**: 107–110.

van Marken Lichtenbelt WD, Vanhommerig JW, Smulders NM, Drossaerts JM, Kemerink GJ, Bouvy ND, Schrauwen P, Teule GJ. 2009. Cold-activated brown adipose tissue in healthy men. *New Engl J Med* **360**: 1500–1508.

van Schaftingen E, Gerin I. 2002. The glucose-6-phosphatase system. *Biochem J* **362**: 513–532.

Wende AR, Huss JM, Schaeffer PJ, Giguere V, Kelly DP. 2005. PGC-1α coactivates PDK4 gene expression via the orphan nuclear receptor ERRα: A mechanism for transcriptional control of muscle glucose metabolism. *Mol Cell Biol* **25**: 10684–10694.

Xiao B, Sanders MJ, Underwood E, Heath R, Mayer FV, Carmena D, Jing C, Walker PA, Eccleston JF, Haire LF, et al. 2011. Structure of mammalian AMPK and its regulation by ADP. *Nature* **472**: 230–233.

Yabaluri N, Bashyam MD. 2010. Hormonal regulation of gluconeogenic gene transcription in the liver. *J Biosci* **35**: 473–484.

Yang T, Espenshade PJ, Wright ME, Yabe D, Gong Y, Aebersold R, Goldstein JL, Brown MS. 2002. Crucial step in cholesterol homeostasis: Sterols promote binding of SCAP to INSIG-1, a membrane protein that facilitates retention of SREBPs in ER. *Cell* **110**: 489–500.

Yang W, Lu J, Weng J, Jia W, Ji L, Xiao J, Shan Z, Liu J, Tian H, Ji Q, et al. 2010. Prevalence of diabetes among men and women in China. *New Engl J Med* **362**: 1090–1101.

CHAPTER 15

Signaling in Innate Immunity and Inflammation

Kim Newton and Vishva M. Dixit

Department of Physiological Chemistry, Genentech, Inc., South San Francisco, California 94080

Correspondence: dixit@gene.com

SUMMARY

Inflammation is triggered when innate immune cells detect infection or tissue injury. Surveillance mechanisms involve pattern recognition receptors (PRRs) on the cell surface and in the cytoplasm. Most PRRs respond to pathogen-associated molecular patterns (PAMPs) or host-derived damage-associated molecular patterns (DAMPs) by triggering activation of NF-κB, AP1, CREB, c/EBP, and IRF transcription factors. Induction of genes encoding enzymes, chemokines, cytokines, adhesion molecules, and regulators of the extracellular matrix promotes the recruitment and activation of leukocytes, which are critical for eliminating foreign particles and host debris. A subset of PRRs activates the protease caspase 1, which causes maturation of the cytokines IL1β and IL18. Cell adhesion molecules and chemokines facilitate leukocyte extravasation from the circulation to the affected site, the chemokines stimulating G-protein-coupled receptors (GPCRs). Binding initiates signals that regulate leukocyte motility and effector functions. Other triggers of inflammation include allergens, which form antibody complexes that stimulate Fc receptors on mast cells. Although the role of inflammation is to resolve infection and injury, increasing evidence indicates that chronic inflammation is a risk factor for cancer.

Outline

K. Newton and V.M. Dixit

1 INTRODUCTION

The role of the inflammatory response is to combat infection and tissue injury. Innate immune cells residing in tissues, such as macrophages, fibroblasts, mast cells, and dendritic cells, as well as circulating leukocytes, including monocytes and neutrophils, recognize pathogen invasion or cell damage with intracellular or surface-expressed pattern recognition receptors (PRRs). These receptors detect, either directly or indirectly, pathogen-associated molecular patterns (PAMPs), such as microbial nucleic acids, lipoproteins, and carbohydrates, or damage-associated molecular patterns (DAMPs) released from injured cells. Activated PRRs then oligomerize and assemble large multisubunit complexes that initiate signaling cascades that trigger the release of factors that promote recruitment of leukocytes to the region.

Vascular alterations play an important role in the inflammatory response (Fig. 1). Histamine, prostaglandins, and nitric oxide act on vascular smooth muscle to cause vasodilation, which increases blood flow and brings in circulating leukocytes, whereas inflammatory mediators including histamine and leukotrienes act on endothelial cells to increase vascular permeability and allow plasma proteins and leukocytes to exit the circulation. Cytokines such as

tumor necrosis factor (TNF) and interleukin 1 (IL1) promote leukocyte extravasation by increasing the levels of leukocyte adhesion molecules on endothelial cells. Activated innate immune cells at the site of infection or injury, including dendritic cells, macrophages, and neutrophils, remove foreign particles and host debris by phagocytosis, plus they also secrete cytokines that shape the slower, lymphocyte-mediated adaptive immune response.

Below we examine how PRRs signal recognition of infection and injury. We then describe how the ensuing inflammatory response is amplified by the cytokines TNF and IL1β. Next, we discuss the mechanisms that get leukocytes to where they are needed. Finally, we consider inflammatory signaling pathways triggered during allergic or hypersensitivity reactions and the possibility that chronic inflammation promotes tumor development.

2 DAMPs AND PAMPs TRIGGER THE INNATE IMMUNE RESPONSE

DAMPs are endogenous molecules normally found in cells that get released during necrosis and contribute to sterile inflammation. They include ATP, the cytokine IL1α, uric acid, the calcium-binding, cytoplasmic proteins S100A8 and S100A9, and the DNA-binding nuclear protein HMGB1.

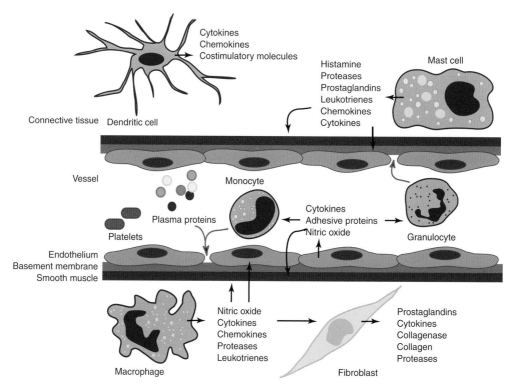

Figure 1. Cells and mediators of the inflammatory response. Molecules derived from plasma proteins and cells in response to tissue damage or pathogens mediate inflammation by stimulating vascular changes, plus leukocyte migration and activation. Granulocytes include neutrophils, basophils, and eosinophils.

Cite this chapter as *Cold Spring Harb Perspect Biol* doi: 10.1101/cshperspect.a006049

Amyloid β fibrils associated with Alzheimer's disease have also been shown to be proinflammatory. PAMPs, in contrast, are pathogen-derived, often essential for microbe survival, and, like DAMPs, structurally diverse. PAMPs include bacterial and viral nucleic acids, fungal β-glucan and α-mannan cell wall components, the bacterial protein flagellin, components of the peptidoglycan bacterial cell wall, and lipopolysaccharide (LPS) from Gram-negative bacteria.

3 TOLL-LIKE RECEPTORS (TLRs)

Members of the TLR family are major PRRs in cells. They are type I transmembrane proteins containing leucine-rich repeats (LRRs) that recognize bacterial and viral PAMPs in the extracellular environment (TLR1, TLR2, TLR4, TLR5, TLR6, and TLR11) or endolysosomes (TLR3, TLR7, TLR8, TLR9, and TLR10). The ligands identified for the different TLRs are listed in Table 1. Signal transduction by TLRs relies on a cytoplasmic Toll/IL1 receptor (TIR) domain that serves as the docking site for TIR-containing cytoplasmic adaptor proteins (Fig. 2). TIR domains in the receptor for the proinflammatory cytokine IL1 function in a similar fashion.

3.1 MyD88-Dependent TAK1 and IKK Activation by TLRs

With the exception of TLR3, all known TLRs engage the adaptor MyD88 either directly (TLR5, TLR7, TLR8, TLR9, TLR10, and TLR11, heterodimeric TLR1-TLR2 and TLR2-TLR6, and the IL1Rs) or in combination with the adaptor TIRAP/Mal (TLR1-TLR2, TLR2-TLR6, and TLR4). MyD88 contains a death domain (DD) in addition to a TIR domain, and this mediates interactions with the DD of the serine/threonine kinase IRAK4 (Lin et al. 2010). Clustering

Table 1. Agonists of mouse and human Toll-like receptors

TLR	Ligand
1/2	Triacyl lipopeptides
2/6	Diacyl lipopeptides
3	dsRNA
4	Lipopolysaccharide
5	Flagellin
7	ssRNA
8	ssRNA in humans; unclear in mice
9	CpG DNA, malarial hemozoin
10[a]	Unknown
11[b]	Uropathogenic bacteria, *Toxoplasma gondii* profilin-like protein
12[b]	Unknown
13[b]	Unknown

[a]Expressed only in humans.
[b]Expressed only in mice.

of IRAK4 within the receptor complex probably results in its autophosphorylation. The kinase activity of IRAK4 is required for its DD to bind the DD of the related kinases IRAK1 and IRAK2. MyD88 and the IRAKs nucleate a larger complex, which in the case of TLR4 includes E3 ubiquitin ligases (TRAF6, cIAP1, and cIAP2) and the E2 ubiquitin-conjugating enzyme Ubc13 (Tseng et al. 2010). TRAF6 and Ubc13 catalyze the formation of polyubiquitin chains in which the carboxyl terminus of one ubiquitin forms an isopeptide bond with the ε-amino group of K63 of an adjacent ubiquitin. TRAF6 can build polyubiquitin chains on lysines within itself and IRAK1, and it also promotes K63-linked polyubiquitylation of cIAPs (Skaug et al. 2009; Tseng et al. 2010). These K63-linked chains are thought to recruit the adaptor proteins TAB2 and TAB3, which exist in a complex with the kinase TAK1. The regulatory subunit of the IκB kinase (IKK) complex, known as IKKγ or NEMO, also binds K63-linked polyubiquitin and is recruited to the TLR4 signaling complex (Laplantine et al. 2009; Tseng et al. 2010).

Analyses of knockout mice indicate that both IKK and TAK1 are important for activation of NF-κB transcription factors, whereas only TAK1 is required for activation of the mitogen-activated protein kinases (MAPKs) p38α and JNK (Sato et al. 2005; Shim et al. 2005; Israel 2010). How TAK1 and IKK are activated, however, requires further clarification. Unanchored K63-linked polyubiquitin chains synthesized by TRAF6 and Ubc13 were proposed to activate TAK1 by inducing its autophosphorylation (Xia et al. 2009). Although TAK1 can phosphorylate the activation loops in the kinases IKKβ and MKK6 in vitro, the latter a MAPK kinase upstream of p38α (Skaug et al. 2009), it is not clear whether TAK1 phosphorylates IKKβ in cells. A recent study suggested that TLR4-induced TAK1 autophosphorylation and activation, but not IKK activation, require translocation of the MyD88–TRAF6–Ubc13–cIAP–TAK1–IKKγ signaling complex from TLR4 into the cytosol (Tseng et al. 2010). This translocation depends on TRAF6 and the cIAPs. Studies of gene-targeted mice expressing a kinase-dead version of TAK1 would clarify whether TAK1 requires its kinase activity to activate IKK.

IKK activation, like TAK1 activation, has been linked to polyubiquitin binding, but the chain linkages and the key E2/E3 ubiquitin enzyme combinations appear to differ. Both unanchored polyubiquitin formed by TRAF6 with the E2 UbcH5c (Xia et al. 2009) and linear polyubiquitin conjugated to IKKγ (Rahighi et al. 2009; Tokunaga et al. 2009) have been shown to induce IKK activation. The linear ubiquitin chain assembly complex (LUBAC) joins ubiquitins in a head-to-tail fashion with the carboxy-terminal glycine residue of one ubiquitin linked to the amino terminus of another ubiquitin. E2 enzymes that support LUBAC activity in vitro include UbcH5c and UbcH7

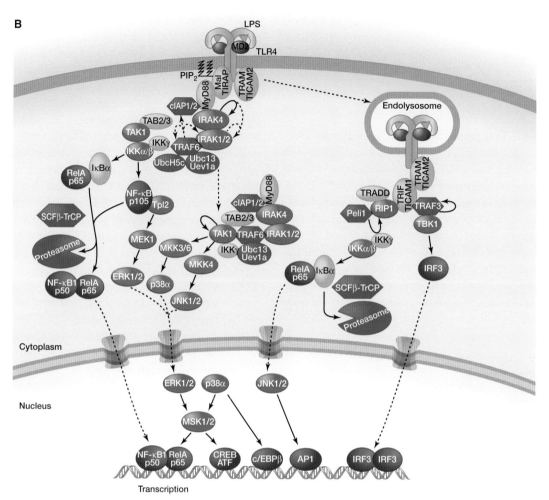

C

Output	Genes
Leukocyte recruitment	*Ccl2, Ccl3, Ccl4, Ccl5, Cxcl1, Cxcl2, Cxcl5, Cxcl10, Ccrl2*
Cell adhesion	*Icam1, Vcam1*
Cell survival	*Bcl2a1, Cflar*
Remodeling of extracellular matrix	*Mmp13*
Vascular effects	*Edn1*
Synthesis of inflammatory mediators	*Hdc, Nos2, Ptges, Ptgs2*
Inflammatory cytokines	*Il1a, Il1b, Il6, Il18, Tnf*
Antiviral response	*Ifnb*
Intracellular signaling (positive)	*Birc2, Birc3, Casp4, Mefv, Nfkbiz*
Intracellular signaling (negative)	*Bcl3, Dusp1, Nfkbia, Socs3, Tnfaip3, Zc3h12a*
PRRs	*Fpr1, Nlrp3*
Regulators of adaptive immune response	*Ch25h, Icosl, Il10, Il12a, Il12b, Il15, Tnfsf9*

Figure 2. Signaling by TLR4. (*A*) Domain structure of human TLR4. (*B*) Binding of LPS to TLR4 and the coreceptor MD2 triggers interactions between the cytoplasmic TIR domain of TLR4 and TIR-containing adaptor proteins (Mal, MyD88, and TRAM). MyD88 binds IRAK4, which requires its kinase activity to bind the kinases IRAK1 and IRAK2 sequentially. The MyD88–IRAK complex also engages the ubiquitin ligase TRAF6 to make polyubiquitin chains that activate the IKK complex for NF-κB- and ERK-dependent gene transcription. Ubiquitin ligases cIAP1 and cIAP2 recruited to the TLR4 signaling complex regulate translocation of a subset of signaling components to the cytoplasm, where TAK1 activation initiates a MAPK cascade that stimulates gene expression. TLR4 activated at the plasma membrane is endocytosed but can signal within the endosomal compartment via the adaptors TRAM and TRIF. The kinase and ubiquitin ligase combination of RIP1 and Peli1 interacts with TRIF to signal NF-κB activation, whereas TBK1 and TRAF3 stimulate IRF3-dependent transcription. (*C*) Functional outputs of some of the genes up-regulated by TLR4 signaling.

Cite this chapter as *Cold Spring Harb Perspect Biol* doi: 10.1101/cshperspect.a006049

(Gerlach et al. 2011; Ikeda et al. 2011; Tokunaga et al. 2011). Binding of the carboxy-terminal UBAN domain in IKKγ to linear ubiquitin chains may produce conformational changes necessary for IKK activation. The importance of IKKγ binding to polyubiquitin is supported by the identification of mutations in the UBAN domain that impair NF-κB activation and cause immunodeficiency in humans (Table 2) (Döffinger et al. 2001). These studies may explain why NF-κB activation is largely normal in the absence of Ubc13 while MAPK activation is compromised (Yamamoto et al. 2006).

The MAPKs JNK and p38α, like TAK1, are activated downstream from TLR2 and TLR4 in a cIAP-dependent manner. The TAK1-containing signaling complex translocates into the cytosol and recruits the kinase MKK4 to phosphorylate and activate JNK (Tseng et al. 2010). It probably also recruits MKK3 and MKK6 to activate p38α because both kinases associate with TRAF6 in response to LPS (Wan et al. 2009). p38α is required for activation of the transcription factors CREB and c/EBPβ, and it contributes to the induction of several genes, including those encoding chemokines (*Cxcl1*, *Cxcl2*), cytokines (*IL10*, *IL12b*, *IL1a*, and *IL1b*), and regulators of extracellular matrix remodeling (*Mmp13*) and cell adhesion (*Vcam1*) (Kang et al. 2008; Kim et al. 2008). JNK regulates the activity of the AP1 transcription factor and stimulates expression of proinflammatory mediators such as TNF (Das et al. 2009).

Activation of the IKK complex is required for NF-κB-dependent transcription as well as transcriptional responses downstream from the MAPK ERK. IKKβ substrates include p105, the precursor of the p50 NF-κB1 transcription factor, as well as the IκB proteins that sequester NF-κB transcription factors in the cytosol. Phosphorylation by IKKβ targets these substrates for K48-linked polyubiquitylation by the E3 ubiquitin ligase SCFβ-TrCP and subsequent proteasomal degradation (Kanarek et al. 2010). Degradation of p105, which exists in a complex with the kinase Tpl2, activates a Tpl2–MEK1–ERK kinase cascade that leads to the induction of genes such as *Ptgs2*

by the CREB/ATF family of transcription factors (Banerjee and Gerondakis 2007). The cyclooxygenase 2 (COX2) enzyme encoded by *Ptgs2* is involved in the synthesis of prostaglandins, which are important mediators of pain, inflammation, and fever. IκB degradation allows dimeric NF-κB transcription factors composed largely of RelA (p65) and NF-κB1 (p50) subunits to accumulate in the nucleus and drive expression of a large number of proinflammatory genes (Table 3 lists a subset of these genes).

NF-κB also induces genes that limit the duration and magnitude of the inflammatory response, such as *Tnfaip3* and *Nfkbia* (the latter encodes IκBα and thus forms a negative-feedback loop). These genes prevent the inflammatory response from causing more tissue damage than the initial injury. For example, mice lacking the A20 deubiquitylating enzyme encoded by *Tnfaip3* die from unchecked inflammation. A20 is thought to switch off TLR signaling by countering ubiquitylation by TRAF6 or cIAPs (Newton et al. 2008; Skaug et al. 2009; Shembade et al. 2010). Somatic mutation of *TNFAIP3* occurs frequently in some human B-cell lymphomas, which suggests that A20 may function as a tumor suppressor, and polymorphisms within the *TNFAIP3* locus have been associated with autoimmune disorders, including systemic lupus erythematosus, rheumatoid arthritis, Crohn's inflammatory bowel disease, and psoriasis (Table 2) (Vereecke et al. 2011). The tumor suppressor gene *CYLD*, which is mutated in familial cylindromatosis, also encodes a deubiquitylating enzyme that limits NF-κB signaling. CYLD cleaves both linear and K63-linked polyubiquitin efficiently in vitro (Komander et al. 2009), but mice lacking CYLD do not develop the multiorgan inflammation seen in A20-deficient mice, perhaps because CYLD normally is phosphorylated and inactivated by IKK (Reiley et al. 2005). Another NF-κB target gene that suppresses TLR signaling is *Cd200*, which encodes the membrane glycoprotein ligand for CD200R1 (Mukhopadhyay et al. 2010). TAM (Tyro3, Axl, and Mer) receptor tyrosine kinases are further examples of receptor systems that negatively regulate TLR signaling (Rothlin et al. 2007).

Table 2. Genes regulating inflammation that are mutated in human disease

Gene	Protein	Disease
CIAS1	NLRP3	Familial cold autoinflammatory syndrome; Muckle–Wells syndrome; neonatal-onset multisystem inflammatory disease
IKBKG	IKKγ	Anhidrotic ectodermal dysplasia with immunodeficiency
NOD2	NOD2	Crohn's inflammatory bowel disease; Blau syndrome characterized by arthritis and uveitis
TNFAIP3	A20	B-cell lymphomas

3.2 MyD88-Dependent Type I Interferon (IFN) Induction by TLRs 7 and 9

The MyD88–IRAK4–IRAK1–TRAF6 signaling complex required for proinflammatory cytokine production by TLR7 and TLR9 also stimulates synthesis of interferon α (IFNα) and IFNβ in plasmacytoid dendritic cells (pDCs). The induction of type I IFNs and *Ifn*-related genes is critical to the antiviral response and also requires TRAF3, IKKα, and the transcription factor IRF7. Phosphorylation of IRF7 by IKKα promotes its dimerization and translocation

Table 3. Genes induced by the canonical IKKβ/NF-κB signaling pathway

Gene[a]	Protein	Function
Ager	RAGE	PRR belonging to the immunoglobulin superfamily that recognizes multiple ligands including HMGB1
Birc3	cIAP2	Ubiquitin ligase that regulates NF-κB activation
Casp4	Caspase 11	Aspartate-specific cysteine protease implicated in inflammation
Ccl2	MCP1	Chemokine for monocyte recruitment
Ccl3	MIP1α	Chemokine for leukocyte recruitment
Ccl5	RANTES	Chemokine for monocyte and T-cell recruitment
Cd200	CD200	Binds CD200R1 and inhibits macrophage activation
Cfb	Complement factor B	Serine protease in the alternative complement activation pathway
Cflar	c-FLIP	Inhibitor of death receptor-induced apoptosis
Csf2	GM-CSF	Growth factor that promotes differentiation and activation of DCs, macrophages, and neutrophils
Cxcl1	KC	Chemokine for neutrophil recruitment
Cxcl2	MIP2	Chemokine for neutrophil recruitment
F3	Tissue factor	Coagulation factor
Icam1	ICAM1	Cell adhesion molecule that interacts with β2 integrins
Ifnb1	IFNβ	Suppressor of virus replication
Il1b	IL1β	Cytokine that amplifies the inflammatory response
Il6	IL6	Pleiotropic cytokine that stimulates fever, production of hepatocyte acute phase proteins, and lymphocyte differentiation
Il12b	IL12 p40	Component of heterodimeric IL12 and IL23, which modulates NK cells and lymphocyte effector functions
Mmp9	MMP9	Metalloproteinase that degrades extracellular matrix
Nfkbia	IκBα	Inhibitor of NF-κB signaling
Nfkbib	IκBβ	NF-κB transcriptional coactivator
Nos2	iNOS	Enzyme that makes antimicrobial nitric oxide
Sele	E-selectin	Cell adhesion molecule
Selp	P-selectin	Cell adhesion molecule
Sod2	MnSOD	Enzyme that converts superoxide to hydrogen peroxide
Tnf	TNF	Cytokine that amplifies the inflammatory response
Tnfaip3	A20	Inhibitor of NF-κB signaling by TLRs and TNFR1
Vcam1	VCAM1	Cell adhesion molecule that interacts with β1 integrin VLA-4

[a]Mouse gene nomenclature used.

into the nucleus, where it up-regulates expression of type I *Ifn* genes. Differences have been noted in myeloid DCs[1]; IRF1 is more important than IRF7, and there does not seem to be a requirement for TRAF3 and IRAK1 (Hoshino et al. 2010; Takeuchi and Akira 2010).

3.3 TRIF-Dependent Signaling by TLRs 3 and 4

TLR4 is endocytosed following ligand binding and, like TLR3 stimulated with dsRNA, transduces signals from within the endosomal compartment. Both receptors recruit the TIR-containing cytoplasmic adaptor TRIF (also known as TICAM1), which in the case of endocytosed TLR4 occurs via the bridging adaptor TRAM (also known as TICAM2). TRIF can activate NF-κB using its RIP homotypic interaction motif (RHIM) to recruit the RHIM-containing kinase RIP1 (Cusson-Hermance et al. 2005), which is, in turn, bound and ubiquitylated by the E3 ubiquitin ligase Peli1 (Chang et al. 2009). K63-linked polyubiquitylation of RIP1 by Peli1 has not been shown, but could be a mechanism for the recruitment of IKKγ and TAK1. Interaction of the DD in RIP1 with the DD in the cytoplasmic adaptor TRADD is important for TRIF-dependent NF-κB and MAPK activation in some cell types (Chen et al. 2008; Ermolaeva et al. 2008; Pobezinskaya et al. 2008). The TRIF-containing complex also induces type I IFN, and this is dependent on TRAF3 plus the kinase TBK1, the latter phosphorylating and activating the transcription factor IRF3 (Takeuchi and Akira 2010). TRAF3 can modify itself with K63-linked polyubiquitin, and this may be important for its interaction with TBK1 and subsequent IRF3 activation.

4 RIG-I-LIKE RECEPTORS (RLRs)

The RLR family of PRRs, which comprises RIG-I, MDA5, and LGP2, signals the production of proinflammatory cytokines and type I IFN in response to viral and bacterial nucleic acids in the cytoplasm (Fig. 3). These cytosolic proteins have a central DExD/H-box helicase domain and a carboxy-terminal regulatory domain (CTD), the latter binding RNA. RIG-I and MDA5 also have two amino-terminal caspase activation and recruitment domains (CARDs), which function as protein–protein interaction motifs. RIG-I recognizes double-stranded RNAs (dsRNAs) that have a 5′ triphosphate and are either viral in origin or generated by RNA polymerase III from microbial DNA templates (Lu et al. 2010; Wang et al. 2010b). MDA5 and

---

[1]DCs are very heterogeneous. pDCs acquire DC morphology and secrete large amounts of IFN during virus infections. They can be distinguished from other DC subsets by their cell surface markers. The myeloid DCs referenced were derived in vitro from bone marrow cells with granulocyte/macrophage colony-stimulating factor (GM-CSF).

 Cite this chapter as *Cold Spring Harb Perspect Biol* doi: 10.1101/cshperspect.a006049

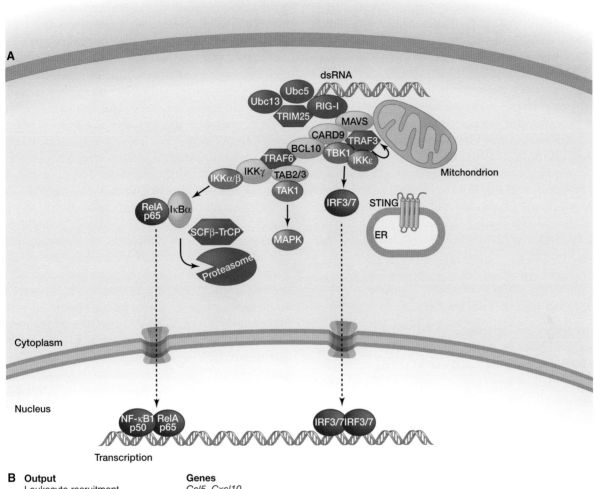

Figure 3. Signaling by RIG-I. (*A*) RIG-I binding to dsRNA that has a 5′ triphosphate and polyubiquitin, the latter generated by the ubiquitin ligase TRIM25 and E2 ubiquitin-conjugating enzymes Ubc5 and Ubc13, promotes RIG-I binding to mitochondrial MAVS. Subsequently, a larger complex containing the adaptor proteins CARD9 and BCL10 is assembled for MAPK and NF-κB activation. TRAF3, the kinases TBK1 and IKKε, and ER-resident protein STING are required for activation of transcription factors IRF3 and IRF7. (*B*) Functional outputs of some of the genes up-regulated by MAVS signaling.

LGP2 also bind RNA, but MDA5 and RIG-I have non-redundant functions in sensing certain viruses. LGP2 appears, in most contexts, to act as a positive regulator of signaling by MDA5 and RIG-I (Venkataraman et al. 2007; Satoh et al. 2010).

Binding of RIG-I to RNA causes a conformational change (Jiang et al. 2011) that enables its CARDs to bind K63-linked polyubiquitin generated by the E3 ubiquitin ligase TRIM25. In an ill-defined manner, this interaction causes RIG-I to engage the mitochondrial CARD-containing protein MAVS (also called IPS-1, VISA, and Cardif)

(Zeng et al. 2010). Subsequently, MAVS forms large aggregates (Hou et al. 2011) that stimulate MAPK activation and transcription induced by IRF3, IRF7, and NF-κB. Many of the components found downstream from TRIF in TLR signaling also are engaged by MAVS. For example, TRAF3 plus the kinases TBK1 and IKKε mediate IRF3/7 activation and induction of type I *Ifn* genes (Takeuchi and Akira 2010). RIG-I, but not MDA5, also requires the transmembrane protein STING to induce IFN (Ishikawa and Barber 2008). STING is located in the endoplasmic reticulum (ER), but its precise role in IFN induction by dsRNA requires further study. Experiments with fibro-blasts from gene-targeted mice also implicate TRAF6, IKKγ, and the DD-containing proteins TRADD, RIP1, and FADD in IFN induction (Balachandran et al. 2004; Zhao et al. 2007; Michallet et al. 2008; Yoshida et al. 2008). Note that loss of TRADD, RIP1, or FADD produces a defect less severe than does MAVS or TBK1 deficiency. The extent of IRF3 phosphorylation and dimerization has not been determined in cells lacking TRADD, RIP1, or FADD; it remains possible that these proteins, like TRAF6, contribute to type I *Ifn* gene expression by activating NF-κB (Wang et al. 2010a). In DCs, MAVS engages the CARD-containing adaptors CARD9 and BCL10 to activate NF-κB (Poeck et al. 2010). BCL10 engages TRAF6 in lymphocytes to activate NF-κB (Sun et al. 2004b), and a similar pathway may operate downstream from MAVS. A role for FADD, TRADD, and RIP1 in MAVS signaling by DCs has not been examined.

5 NOD-LIKE RECEPTORS (NLRs)

Members of the Nod-like receptor (NLR) family of cytosolic PRRs are best known for their ability to signal NF-κB activation (NOD1 and NOD2) or secretion of the proinflammatory cytokines IL1β and IL18 (NLRP1/NALP1, NLRP3/ NALP3/cryopyrin, and NLRC4/Ipaf) (Fig. 4). These proteins typically contain a CARD or pyrin domain at the amino terminus, a central nucleotide-binding oligomerization NACHT domain, and carboxy-terminal LRRs. NOD2 and NLRP3 have received considerable attention because their mutation is linked to inflammatory disease. *NOD2* mutations are associated with Crohn's inflammatory bowel disease and Blau syndrome, whereas mutations in the *CIAS1* gene encoding NLRP3 are associated with familial cold autoinflammatory syndrome, Muckle–Wells syndrome, and neonatal-onset multisystem inflammatory disease (Table 2).

NOD1 and NOD2 are sensors of different bacterial peptidoglycan components, but they both interact with the CARD-containing kinase RIP2 to activate MAPK and NF-κB signaling (Park et al. 2007). cIAPs are proposed to bind and ubiquitylate RIP2, and K63-linked polyubiquitylation of RIP2 recruits TAK1 for IKK and MAPK activation (Yang et al. 2007; Hitotsumatsu et al. 2008; Bertrand et al. 2009). NOD1 and NOD2 also have been shown to stimulate autophagy[2] independently of RIP2 (Travassos et al. 2010).

Distinct PAMPS and DAMPs trigger NLRP1, NLRP3, and NLRC4 to nucleate signaling complexes termed "inflammasomes" (Table 4). The CARD- and PYRIN-domain-containing adaptor ASC is a critical inflammasome component, binding the CARD in the zymogen form of the aspartate-specific cysteine protease caspase 1 (Mariathasan et al. 2004). The proximity of caspase-1 zymogens within the inflammasome complex is believed to facilitate their autocatalytic activation. Caspase-1 substrates include pro-IL18 and pro-IL1β, the latter being up-regulated transcriptionally by MyD88-dependent TLR signaling. Inflammasome activation also results in an extremely rapid form of cell death termed "pyroptosis." The suicide of infected macrophages by pyroptosis is important for bacterial clearance (Miao et al. 2010), but the critical substrates of caspase 1 in this process still have to be determined.

Intriguingly, immunofluorescence microscopy of endogenous inflammasome components in mouse macrophages infected with *Salmonella typhimurium* suggests that inflammasome assembly occurs at a single focus within a cell (Broz et al. 2010). One question that continues to vex the field is how NLRP1, NLRP3, and NLRC4 sense PAMPS and DAMPs, because direct binding has not been shown. Phosphorylation of NLRC4 is critical for inflammasome activation (Qu et al. 2012) and the diversity of entities that trigger NLRP3-dependent caspase-1 activation suggests that NLRP3 might respond to a particular stress-activated signaling pathway. Both potassium efflux and the generation of reactive oxygen species (ROS) have been proposed as critical events upstream of NLRP3 activation, but the precise nature of NLR activation remains obscure.

ASC-dependent caspase-1 activation is also triggered in response to cytoplasmic DNA that appears during an infection or after tissue injury. The responsible PRR is not an NLR but the IFN-induced protein AIM2, which has a HIN200 domain for DNA binding and a pyrin domain to engage ASC. Cytoplasmic DNA also triggers type I IFN production, but this requires neither AIM2 nor TLRs. Instead, DNA binding to the enzyme cGAS (cyclic guanosine monophosphate-adenosine monophosphate synthase) promotes production of the second messenger cGAMP, which binds to STING and thereby stimulates activation of IRF3 (Sun et al. 2013; Wu et al. 2013a).

Caspase-1 activation also occurs in a caspase-11-dependent, TLR4-independent manner in response to intracellular LPS (Kayagaki et al. 2013). The nature of the intracellular LPS sensor remains to be elucidated.

6 THE PROINFLAMMATORY CYTOKINE TUMOR NECROSIS FACTOR (TNF)

Induction of the cytokines IL1β and TNF by PRRs serves to amplify the inflammatory response because they too promote NF-κB and MAPK activation. Binding of IL1β to

---

[2]Autophagy is the process by which cytoplasmic components, including organelles and invading bacteria, are sequestered inside double-membrane vesicles and then delivered to the lysosome for degradation.

Cite this chapter as *Cold Spring Harb Perspect Biol* doi: 10.1101/cshperspect.a006049

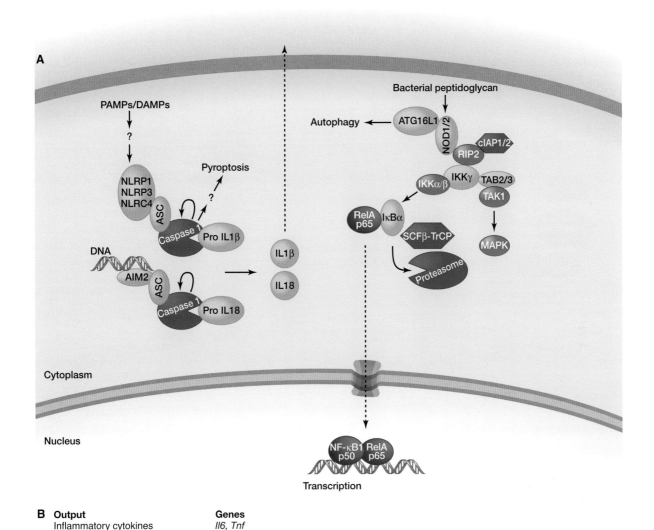

Figure 4. Signaling by NLRs. (*A*) NLRP1, NLRP3, and NLRC4 respond to diverse PAMPS and DAMPs by engaging the caspase-1 adaptor protein ASC, whereas AIM2 binds ASC in response to cytoplasmic dsDNA. Activation of caspase 1 within each inflammasome complex results in processing of pro-IL1β and pro-IL18 and secretion of their biologically active forms. Caspase-1 activation also triggers a rapid form of cell death termed pyroptosis. NOD1 and NOD2 sense different components of bacterial peptidoglycan and stimulate either the autophagy machinery or gene transcription via NF-κB and MAPK activation. The latter outcome requires interaction of NOD1 or NOD2 with the kinase RIP2, which may be ubiquitylated by cIAPs in order to recruit TAB2/3 and IKKγ for TAK1 and IKK activation. (*B*) Functional outputs of some of the genes up-regulated by NOD1 or NOD2 signaling.

IL1R triggers MyD88-dependent signaling (Muzio et al. 1997), whereas TNF mediates most of its proinflammatory effects by binding to TNF receptor I (TNFRI) (Peschon et al. 1998).

6.1 NF-κB and MAPK Activation by TNF

TNFRI (also called TNFRSF1A) is a type I transmembrane protein that has cysteine-rich extracellular domains (CRDs) for TNF binding. Its cytoplasmic tail contains a DD that

recruits the DD-containing adaptor TRADD and kinase RIP1 (Fig. 5). TRADD facilitates binding of RIP1 to TNF-R1 and recruits TRAF2, which is an adaptor for the ubiquitin ligases cIAP1 and cIAP2 (Chen et al. 2008; Ermolaeva et al. 2008; Pobezinskaya et al. 2008). Analyses of cells lacking cIAPs, RIP1, or TRAF2 indicate that all three contribute to NF-κB and MAPK activation, but the details of how they do so continue to be unraveled. Ubiquitylation of RIP1 by the cIAPs and E2 UbcH5 is believed to be important for recruitment of NEMO and TAK1, and subsequently for IKK

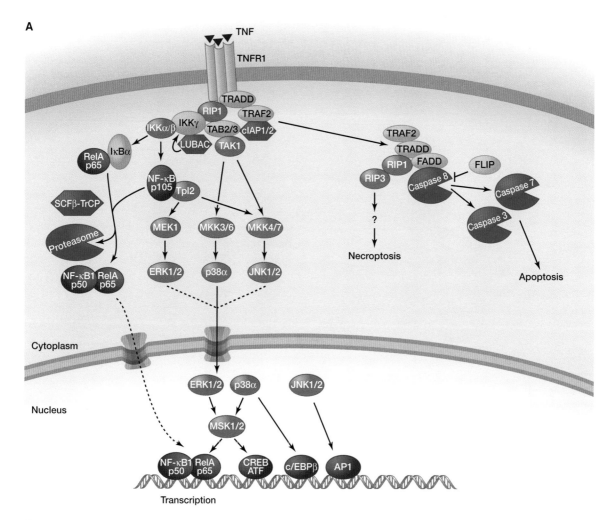

Figure 5. Signaling by TNFR1. (*A*) Binding of TNF to TNFR1 causes the cytoplasmic death domain (DD) in TNFR1 to bind the DD-containing proteins TRADD and RIP1. TRADD also binds TRAF2, which serves as an adaptor for the ubiquitin ligases cIAP1 and cIAP2. Ubiquitylation of RIP1, and potentially other components of the complex, recruits IKKγ and TAK1 for NF-κB and MAPK activation. Recruitment of LUBAC for linear ubiquitylation of IKKγ may stabilize the signaling complex. Translocation of TRADD, TRAF2, and RIP1 to the cytoplasm nucleates a second complex that contains the adaptor protein FADD and caspase 8. If c-FLIP levels are low, activation of caspase 8 and its substrates caspase 3 and caspase 7 causes apoptotic cell death. Inhibition of protein synthesis and caspases, as might occur in a virus-infected cell, promotes necroptotic cell death that is dependent on the kinase activities of RIP1 and RIP3. (*B*) Functional outputs of some of the genes up-regulated by TNF signaling.

Table 4. Stimuli detected by inflammasome sensors that activate caspase 1

Sensor	Known stimuli
AIM2	Cytoplasmic DNA
NLRP1	*Bacillus anthracis* lethal toxin
NLRP3	ATP
	Ionophore nigericin
	Marine toxin maitotoxin
	Crystals such as monosodium urate, calcium pyrophosphate dihydrate, silica, and asbestos
	Fungi such as *Candida albicans*
	Bacteria such as *Staphylococcus aureus, Neisseria gonorrhoeae, Salmonella typhimurium,* and *Listeria monocytogenes*
	Viruses such as sendai, adenovirus, and influenza
NLRC4	Bacterial flagellin
	Certain bacterial type III secretion systems

activation (Varfolomeev et al. 2008; Xu et al. 2009), although TRAF2 ubiquitin ligase activity has been invoked recently as well (Alvarez et al. 2010). In addition, in some cell types, ubiquitylation of TRAF2 or cIAPs rather than RIP1 may support IKK activation (Li et al. 2009; Wong et al. 2010). Recruitment of LUBAC components sharpin, HOIL-1, and HOIP to the TNFRI complex is linked to cIAP ubiquitin ligase activity, and linear ubiquitylation of RIP1 and NEMO may stabilize the TNFRI signaling complex (Haas et al. 2009; Gerlach et al. 2011; Ikeda et al. 2011; Tokunaga et al. 2011). Underscoring the importance of sharpin in TNF signaling, mutation of *Sharpin* in mice causes chronic proliferative dermatitis that is rescued by TNF deficiency (Gerlach et al. 2011).

IKKβ activation by TNF triggers not only NF-κB transcription but also the Tpl2–MEK1–ERK kinase cascade activated by TLRs (see above). In fibroblasts, but not macrophages or B cells, Tpl2 activation by TNF has been linked to activation of the MKK4–JNK pathway as well. In addition, activation of the kinase MSK1 by ERK may enhance NF-κB transcriptional activity through phosphorylation of RelA (Banerjee and Gerondakis 2007). Genetic studies indicate that TNF-induced JNK activation is mediated largely by upstream kinases TAK1 and MKK7, whereas p38 activation requires TAK1 and MKK3/MKK6 (Brancho et al. 2003; Sato et al. 2005; Shim et al. 2005).

6.2 Apoptosis and Necroptosis Induction by TNF

The TNFRI-associated signaling complex formed in response to TNF is transient as TRADD, RIP1, and TRAF2 shift to the cytoplasm to form what is referred to as complex II (Micheau and Tschopp 2003). Here, the DD in TRADD binds the DD in the adaptor FADD, whereas the death effector domain (DED) in FADD interacts with the DEDs

in the prodomain of caspase 8 or its catalytically inactive homolog FLIP. If FLIP levels are low, then stable active dimers of caspase 8 are formed. Caspase-8 substrates include caspase 3 and caspase 7, which are activated by proteolytic processing and cleave vital cellular proteins to cause apoptotic cell death (Ch. 19 [Green and Llambi 2014]). Because expression of FLIP is driven by NF-κB (Micheau et al. 2001) and many viruses have evolved strategies to inhibit NF-κB activation (Shisler and Jin 2004; Taylor et al. 2009), apoptosis in the absence of c-FLIP constitutes a host defense mechanism against infection. Certain viruses encode their own FLIPs, but the infected host cell may yet prevail in its attempt to die because TNF activates a form of cell death called necroptosis when protein synthesis and caspases are inhibited. This death is dependent on the kinase activity of RIP1 and the related kinase RIP3 (Cho et al. 2009; He et al. 2009; Zhang et al. 2009). These kinases interact through their RHIMs with RIP3, then phosphorylate the pseudokinase mixed-lineage kinase domain-like (MLKL) (Sun et al. 2012; Wu et al. 2013b). It is not clear how phosphorylation of MLKL leads to cell death.

7 SELECTINS AND INTEGRINS

One important output of LPS, TNF, and IL1β signaling in endothelial cells is increased surface expression of transmembrane proteins involved in cell adhesion, such as P-selectin, E-selectin, ICAM1, and VCAM1. All four are induced by NF-κB in mice, whereas up-regulation of human P-selectin relies on fusion of secretory granules with the plasma membrane. Glycoproteins including CD44, P-selectin glycoprotein ligand 1 (PSGL1), and E-selectin ligand (ESL1) on leukocytes interact with the selectins to mediate leukocyte rolling along vessel walls. Engagement of either PSGL1 or CD44 triggers a signaling pathway that causes leukocyte β2 integrins LFA1 and MAC1 expressed on the cell surface to adopt a more extended conformation that increases their affinity for endothelial ICAM1. Interactions between the β2 integrins and ICAM1 then slow leukocyte rolling further. β2 integrin activation via this "inside-out" signaling (Ch. 8 [Devreotes and Horwitz 2014]) requires the membrane-associated Src family kinases Fgr, Hck, and Lyn. These tyrosine kinases probably phosphorylate cytoplasmic immunoreceptor tyrosine-based activation motifs (ITAMs) in the transmembrane adaptors DAP12 and FcRγ to create docking sites for the tandem SH2 domains in the tyrosine kinase Syk (Ch. 16 [Cantrell 2014]), but note that they can also phosphorylate integrin β subunits. Studies with neutrophils from gene-targeted mice indicate that sequential activation of Syk and the kinase Btk is essential for slow rolling on E-selectin and ICAM1 (Zarbock et al. 2008; Yago et al. 2010).

Further β2 integrin activation needed for leukocyte arrest before migration across the endothelial cell barrier is mediated by chemokines and chemoattractants immobilized on the endothelial cell surface. These factors engage G-protein-coupled receptors (GPCRs) on the leukocyte surface (see below). Note that integrin engagement also elicits "outside-in" signaling in leukocytes, and, similarly to Fc receptor signaling (see below), this activates leukocyte effector functions (Lowell 2011). Clustering of ICAM1 on endothelial cells also triggers signals that facilitate leukocyte migration across the endothelium into the surrounding tissue. Phosphorylation of VE-cadherin appears to loosen adherens junctions, whereas activation of myosin light chain kinase (MLCK) mediates endothelial cell contraction (Muller 2011).

8 G-PROTEIN-COUPLED RECEPTORS (GPCRs)

Lipid-based inflammatory mediators such as prostaglandins, leukotrienes, and platelet-activating factor; vasoactive amines such as histamine and serotonin; complement fragments C3b, C3a, and C5a; chemokines; proteases; and bacterial or mitochondrial formylated peptides all activate signaling by GPCRs linked to heterotrimeric G-proteins

composed of α, β, and γ subunits. Following ligand binding, or cleavage in the case of protease-activated receptors (PARs), Gα and Gβγ interact with ion channels or enzymes such as adenylyl cyclase, phospholipase C (PLC), and phosphoinositide 3-kinase (PI3K). In addition, active GPCRs are phosphorylated by GPCR kinases (GRKs) to stimulate binding of arrestins, adaptors that stimulate GPCR endocytosis as well as MAPK activation. To highlight some of the pathways engaged by GPCRs during inflammation, below we focus on signaling by the chemoattractants C5a and the prototypical formylated peptide formyl-Met-Leu-Phe (fMLP) (Fig. 6).

8.1 C5a Receptor (C5aR) and Formyl Peptide Receptors (FPRs)

Complement protein C5a is produced by complement plasma proteases activated by IgM- and IgG-containing antibody complexes (the classical pathway), pathogens coated with host mannose-binding lectin or C-reactive protein (the lectin pathway), or pathogens in isolation (the alternative pathway). C5a can also be generated by noncomplement proteases such as thrombin and kallikrein, which are components of the clotting system activated

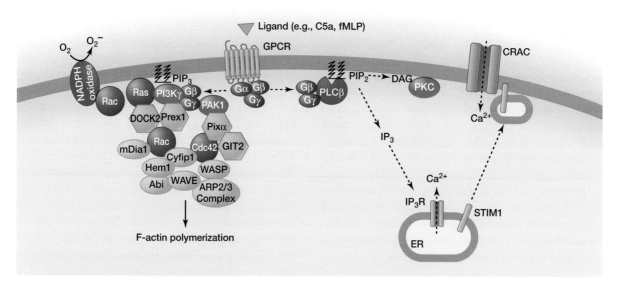

Figure 6. Signaling by GPCRs activated by chemoattractants C5a and fMLP. C5a or fMLP binding to their respective GPCRs triggers dissociation of $G_{\alpha i}$-GTP from $G_{\beta \gamma}$, the latter interacting with PLCβ, the p101 regulatory subunit of PI3Kγ, and PAK1. Activated PLCβ generates the second messengers IP_3 and DAG to elevate intracellular calcium and activate PKC, respectively. These outcomes regulate JNK activation, vesicle exocytosis, and superoxide production by the NADPH oxidase. PIP_3 generated by PI3Kγ, whose activation also involves the GTPase Ras, stimulates GEFs (DOCK2 and Prex1) that activate Rac GTPases. PAK1 interacts with the GEF PIXα for activation of another Rho family GTPase called Cdc42. Rac1, Rac2, and Cdc42 together regulate chemotaxis by coordinating alterations to the actin cytoskeleton via mDia1 and the ARP2/3 complex. Rac2 is also an essential component of the NADPH oxidase. The signaling components regulating gene transcription are less defined.

Cite this chapter as *Cold Spring Harb Perspect Biol* doi: 10.1101/cshperspect.a006049

in response to endothelial cell injury. C5a and formyl peptides, the latter of bacterial or mitochondrial origin, stimulate leukocyte chemotaxis, degranulation, superoxide production for microbe killing, and, as mentioned above, activation of integrins for cell adhesion. Similarly to TNF, C5a stimulates endothelial cells to increase expression of cytokines, chemokines, and cell adhesion molecules such as E-selectin, ICAM1, and VCAM1 (Albrecht et al. 2004).

C5a and fMLP activate predominantly pertussis toxin-sensitive G_i proteins. The $G\beta\gamma$ dimer that is released activates several enzymes, including PLCβ (Camps et al. 1992). Hydrolysis of phosphatidylinositol 4,5-bisphosphate in the plasma membrane by PLC yields the second messengers inositol 1,4,5-trisphosphate (IP$_3$) and diacylglycerol (DAG) (p. 95 [Bootman 2012]). Binding of IP$_3$ to its receptor causes depletion of calcium stores within the ER and relocation of the calcium-binding type-I transmembrane protein STIM1 from the ER to structures near the plasma membrane (Brechard et al. 2009). STIM1 then activates the plasma membrane calcium-release-activated calcium channel (CRAC) to cause an influx of calcium into the cell. Elevated intracellular calcium together with protein kinase C (PKC) activation by DAG is important for vesicle exocytosis, superoxide production by the NADPH oxidase, and JNK activation (Li et al. 2000). Calcium stimulates lysosome exocytosis by activating the synaptotagmin regulator of vesicle fusion SYT7 (Colvin et al. 2010).

GTP-bound Ras, activated by a mechanism that is unclear, and $G\beta\gamma$ dimers activate PI3Kγ by binding to its p101 regulatory subunit and p110γ catalytic subunit, respectively. Phosphatidylinositol 3,4,5-trisphosphate (PIP$_3$) produced by PI3Kγ activates Rac guanine-nucleotide exchange factors (GEFs), such as Prex1 and DOCK2 (Welch et al. 2002; Kunisaki et al. 2006), and contributes to superoxide production and chemokinesis[3] (Suire et al. 2006; Ferguson et al. 2007; Nishio et al. 2007). The GTPase RhoG also has a role in superoxide production but is dispensable for neutrophil migration (Condliffe et al. 2006). The GTPase Rac2 appears to be essential for the assembly of filamentous actin (F-actin) and the NADPH oxidase, whereas Rac1 localizes F-actin to the leading edge of the cell facing the chemoattractant (Sun et al. 2004a). This asymmetrical polymerization of F-actin drives membrane protusions in the direction of migration. Active Rac is thought to exert its effect on the actin cytoskeleton by interacting with the adaptor Cyfip1 (also called Sra1), which, in combination with several proteins, stimulates the ARP2/3 actin nucleation complex (Ch. 8 [Devreotes and Horwitz 2014]).

The GEF PIXα also is required for F-actin assembly at the leading edge in C5a-stimulated neutrophils. It is recruited to $G\beta\gamma$ via the kinase PAK1 and appears to function by interacting with the GTPase Cdc42 and the GTPase-activating (GAP) protein GIT2 (Li et al. 2003; Mazaki et al. 2006). Activated Cdc42 interacts with the adaptor WASP to engage the ARP2/3 actin nucleation complex. WASP appears to work in concert with the actin-nucleating protein mDia1, because neutrophils lacking both WASP and mDia1 show a profound defect in chemotaxis (Shi et al. 2009).

9 Fc RECEPTORS

Repeated exposure to a polyvalent foreign substance can elicit an inflammatory response called a hypersensitivity reaction if the host makes antibodies against the substance. Immune complexes containing the antigen and IgG or IgM antibodies activate complement proteases, culminating in the generation of C3a and C5a, which signal leukocyte recruitment and activation (see above); the opsonin C3b, which coats and promotes phagocytosis of bacteria; and the membrane attack complex for bacterial cell lysis (C5b-9). In addition, complexes containing IgG or IgE antibodies engage Fc receptors on leukocytes. Members of the Fc receptor family are type I transmembrane proteins (with the exception of human GPI-anchored FcγRIIIB) that produce activating (human FcγRI, FcγRIIA, FcγRIIC, FcγRIIIA, FcγRIIIB, and FcϵRI) or inhibitory (human FcγRIIB) signals. Mast cells expressing the high-affinity receptor for IgE, FcϵRI, play a central role in allergic reactions. FcϵRI engagement causes intracellular granules to fuse with the plasma membrane such that preformed inflammatory mediators including histamine, serotonin, and proteases are released into the extracellular environment. Activated mast cells also secrete proinflammatory prostaglandins, leukotrienes, and cytokines, but these are synthesized de novo.

9.1 FcϵRI

FcϵRI is an $\alpha\beta\gamma_2$ heterotetramer. Its α-chain contains extracellular Ig-like domains for binding the heavy-chain constant region of IgE, whereas the β-chain and a γ-chain homodimer transduce signals via cytoplasmic ITAMs (Fig. 7) (Ch. 16 [Cantrell 2014]). IgE-induced clustering of FcϵRI promotes activation of Src family kinases Lyn and Fyn. Lyn substrates include both positive and negative regulators of mast cell activation, which fine-tune the magnitude and duration of the response. Lyn stimulates activation by phosphorylating the FcRγ ITAM, which recruits the SH2 domains in Syk. Subsequent Syk-dependent phosphorylation of the transmembrane adaptors LAT1 and

---

[3]Chemokinesis refers to random cell migration, whereas chemotaxis is directed cell migration along a chemical gradient.

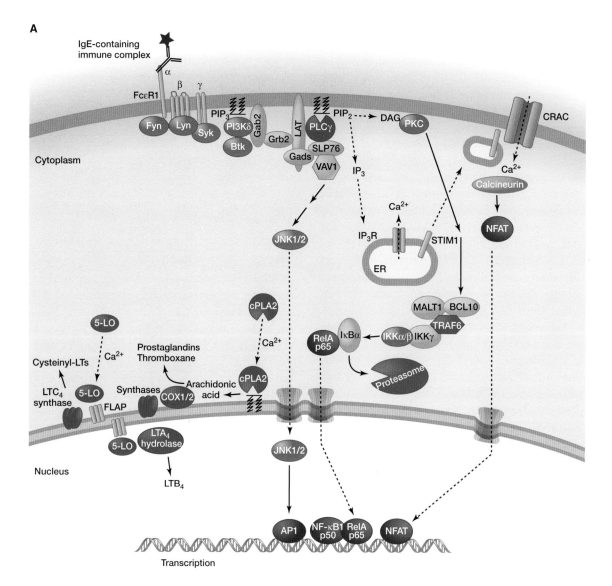

B **Output**

Output	Genes
Leukocyte recruitment	*Ccl1, Ccl2, Ccl3, Ccl4*
Leukocyte activation	*Csf2, Il3*
Cell survival	*Bcl2a1*
Inflammatory cytokines	*Il1a, Il1b, Il6, Lif, Tnf*
Cell adhesion	*Itga2, Icam1*
Regulators of adaptive immune response	*Il11, Il25, Slamf1, Tnfrsf9*

Figure 7. Signaling by FcεRI. (*A*) Binding of the Fc region of antigen-bound IgE to FcεRI activates the Src family kinases Lyn and Fyn. Tyrosine phosphorylation of the FcRγ ITAM recruits the tyrosine kinase Syk, which is required for phosphorylation of LAT transmembrane adaptor proteins. Phosphorylated LAT1 binds PLCγ and the adaptors Gads and Grb2. Gads recruits the adaptor SLP76, which regulates activation of PLCγ and the GEF Vav1. Grb2 binds Gab2, which is phosphorylated by Fyn and binds the p85 regulatory subunit of PI3Kδ. PIP$_3$ generated by PI3Kδ retains signaling components such as Gab2, PLCγ, and Btk at the plasma membrane. IP$_3$ generated by PLCγ depletes ER calcium stores, which causes a STIM1-dependent influx of calcium that promotes mast cell degranulation. Elevated intracellular calcium also activates the phosphatase calcineurin, stimulates NFAT-dependent gene expression, and triggers the translocation of cPLA2 and 5-lipoxygenase (5-LO) to the nuclear envelope, cytoplasmic lipid bodies, or ER. cPLA2 releases arachidonic acid from membrane phospholipids. COX enzymes and downstream synthases metabolize arachidonic acid into prostaglandins and thromboxane, whereas leukotriene (LT) synthesis from arachidonic acid involves five-lipoxygenase-activating protein (FLAP), 5-LO, and downstream LTC$_4$ synthase or LTA$_4$ hydrolase. DAG generated by PLCγ activates PKC, which is important for IKK activation via MALT1, BCl10, and TRAF6, as well as subsequent NF-κB-dependent gene transcription. IKKβ has also been implicated in mast cell degranulation independent of NF-κB activation. (*B*) Functional outputs of some of the genes up-regulated by FcεRI signaling.

LAT2 recruits additional SH2-containing signaling components, such as PLCγ and the adaptors Grb2 and Gads. SH3 domains in Grb2 and Gads bind proline-rich regions in additional proteins such as the adaptors SLP76 and Gab2. SLP76 interacts with the Rho/Rac GEF VAV1, which contributes to PLCγ and JNK activation. Fyn-dependent phosphorylation of Gab2 recruits the SH2-containing p85 regulatory subunit of PI3Kδ. PIP$_3$ produced by PI3Kδ retains proteins containing plextrin homology (PH) domains at the plasma membrane, such as PLCγ, Gab2, Akt, and Btk. The kinase Btk phosphorylates and enhances the activity of PLCγ (Alvarez-Errico et al. 2009).

PLC-γ signaling triggers STIM1-dependent calcium influx (p. 95 [Bootman 2012]), which is essential for normal mast cell degranulation, leukotriene synthesis, and activation of NFAT transcription factors via the calcium-dependent phosphatase calcineurin (Baba et al. 2008; Vig et al. 2008). NFAT promotes expression of the cytokines TNF and IL13. Eicosanoids including leukotrienes and prostaglandins are derived from arachidonic acid, which is liberated from phospholipids by cytosolic phospholipase A$_2$ in response to elevated intracellular calcium and MAPK activation (Fujishima et al. 1999).

PKC activation by DAG is required for degranulation and activation of NF-κB, the latter contributing to the induction of TNF and IL6. BCL10, MALT1 (also called paracaspase), and TRAF6 regulate NF-κB activation but are dispensable for degranulation (Klemm et al. 2006; Chen et al. 2007; Yang et al. 2008), whereas IKKβ is required for both functions (Suzuki and Verma 2008). The mechanism by which IKKβ is activated for degranulation remains unclear, but, once activated, IKKβ appears to promote exocytosis by phosphorylating the SNARE receptor SNAP23.

9.2 Fcγ Receptors

Activating Fcγ receptors, in common with FcεRI, contain cytoplasmic ITAMs and stimulate Src family kinases plus Syk. The downstream signaling events that promote phagocytosis, degranulation, cytokine production, and superoxide production are less well defined but probably involve many of the components engaged by FcεRI. Superoxide production by the NADPH oxidase that assembles on phagosomal membranes requires Vav-mediated activation of Rac GTPases, the putative Rac adaptor CAPRI, and PLCγ. Depending on the cell type and context, CAPRI, VAV, and Rac also contribute to remodeling of the actin cytoskeleton for phagocytosis, along with PI3K and its adaptor Gab2 (Gu et al. 2003; Zhang et al. 2005; Hall et al. 2006; Utomo et al. 2006; Jakus et al. 2009). FcγRIIB is unique among Fc receptors in that it suppresses ITAM signaling through an immunoreceptor tyrosine-based inhibitory motif (ITIM). Lyn-dependent ITIM phosphorylation recruits the SH2-containing inositol 5′-phosphatase (SHIP), which hydrolyzes PIP$_3$ and thereby limits recruitment of PH-domain-containing proteins such as PLC-γ, Btk, and VAV (Lowell 2011).

10 INFLAMMATION AS A RISK FACTOR FOR CANCER

Bacteria and viruses that establish persistent infections are linked to certain cancers. For example, *Helicobacter pylori* bacteria increase the risk of gastric cancer and MALT lymphoma, whereas Hepatitis B and Hepatitis C viruses increase the risk of hepatocellular carcinoma. Similarly, pancreatitis is a risk factor for pancreatic ductal adenocarcinoma, and this has been modeled successfully in mice expressing oncogenic K-Ras in adult pancreatic acinar cells (Guerra et al. 2011). Studies with gene-targeted mice have confirmed that inflammatory signaling pathways promote tumor development. For example, deletion of FcRγ suppresses squamous cell carcinoma driven by keratinocyte-specific expression of the human papilloma virus oncogene HPV16 (Andreu et al. 2010). Similarly, IL6 deficiency in hematopoietic cells suppresses colon tumors that develop in response to procarcinogen azoxymethane plus colitis-inducing dextran sulfate sodium (DSS) (Grivennikov et al. 2009). IL6 enhances proliferation and survival of intestinal epithelial cells through activation of the transcription factor STAT3.

In another mouse model, the ability of tobacco smoke to promote lung tumors driven by oncogenic K-Ras is reduced by IKKβ deletion in myeloid cells (Takahashi et al. 2010). Finally, IL6 or TNFR1 deficiency protects obese mice from hepatocellular carcinomas that form in response to the procarcinogen diethylnitrosamine by limiting lipid accumulation and inflammatory infiltrates in the liver (Park et al. 2010). Note, however, that in all of these tumor models, inflammation alone is insufficient for tumor development, which implies that carcinogen-induced mutations are needed.

11 CONCLUDING REMARKS

Many of the major players in inflammatory signaling have been identified, but the importance and complexity of posttranslational modifications such as ubiquitylation in these pathways continue to be unraveled. Binding of TLRs, TNF and IL1 receptors, GPCRs, integrins, selectins, and Fc receptors to their ligands triggers the formation of multisubunit signaling complexes, but it remains to be seen how diverse inflammatory stimuli can activate intracellular

PRRs such as NLRP3 and NLRC4. An attractive hypothesis is that posttranslational modifications to NLR family members are key to their activation. Once activated, both surface and intracellular PRRs stimulate transcription of inflammatory genes; TLRs, RLRs, and some NLRs (e.g., NOD1 and NOD2) engage common downstream signaling pathways to stimulate transcription factors such as NF-κB, AP1, CREB, and c/EBPβ, whereas caspase-1-activating PRRs (e.g., AIM2, NLRP3, and NLRC4) stimulate similar pathways indirectly via the secretion of IL1β and IL18. Going forward, it will be important to understand how innate immune cells exposed to multiple inflammatory mediators and stimuli in vivo integrate signals from diverse receptors, because this will offer insight into what critical components might be targeted for therapeutic benefit in inflammatory disorders.

REFERENCES

*Reference is in this book.

Albrecht EA, Chinnaiyan AM, Varambally S, Kumar-Sinha C, Barrette TR, Sarma JV, Ward PA. 2004. C5a-induced gene expression in human umbilical vein endothelial cells. *Am J Pathol* **164:** 849–859.

Alvarez SE, Harikumar KB, Hait NC, Allegood J, Strub GM, Kim EY, Maceyka M, Jiang H, Luo C, Kordula T, et al. 2010. Sphingosine-1-phosphate is a missing cofactor for the E3 ubiquitin ligase TRAF2. *Nature* **465:** 1084–1088.

Alvarez-Errico D, Lessmann E, Rivera J. 2009. Adapters in the organization of mast cell signaling. *Immunol Rev* **232:** 195–217.

Andreu P, Johansson M, Affara NI, Pucci F, Tan T, Junankar S, Korets L, Lam J, Tawfik D, DeNardo DG, et al. 2010. FcRγ activation regulates inflammation-associated squamous carcinogenesis. *Cancer Cell* **17:** 121–134.

Baba Y, Nishida K, Fujii Y, Hirano T, Hikida M, Kurosaki T. 2008. Essential function for the calcium sensor STIM1 in mast cell activation and anaphylactic responses. *Nat Immunol* **9:** 81–88.

Balachandran S, Thomas E, Barber GN. 2004. A FADD-dependent innate immune mechanism in mammalian cells. *Nature* **432:** 401–405.

Banerjee A, Gerondakis S. 2007. Coordinating TLR-activated signaling pathways in cells of the immune system. *Immunol Cell Biol* **85:** 420–424.

Bertrand MJ, Doiron K, Labbe K, Korneluk RG, Barker PA, Saleh M. 2009. Cellular inhibitors of apoptosis cIAP1 and cIAP2 are required for innate immunity signaling by the pattern recognition receptors NOD1 and NOD2. *Immunity* **30:** 789–801.

*Bootman MD. 2012. Calcium signaling. *Cold Spring Harb Perspect Biol* **4:** a011171.

Brancho D, Tanaka N, Jaeschke A, Ventura JJ, Kelkar N, Tanaka Y, Kyuuma M, Takeshita T, Flavell RA, Davis RJ. 2003. Mechanism of p38 MAP kinase activation in vivo. *Genes Dev* **17:** 1969–1978.

Brechard S, Plancon S, Melchior C, Tschirhart EJ. 2009. STIM1 but not STIM2 is an essential regulator of Ca^{2+} influx-mediated NADPH oxidase activity in neutrophil-like HL-60 cells. *Biochem Pharmacol* **78:** 504–513.

Broz P, Newton K, Lamkanfi M, Mariathasan S, Dixit VM, Monack DM. 2010. Redundant roles for inflammasome receptors NLRP3 and NLRC4 in host defense against *Salmonella*. *J Exp Med* **207:** 1745–1755.

Camps M, Carozzi A, Schnabel P, Scheer A, Parker PJ, Gierschik P. 1992. Isozyme-selective stimulation of phospholipase C-β2 by G protein β γ-subunits. *Nature* **360:** 684–686.

*Cantrell D. 2014. Signaling in lymphocyte activation. *Cold Spring Harb Perspect Biol* doi: 10.1101/cshperspect.a018788.

Chang M, Jin W, Sun SC. 2009. Peli1 facilitates TRIF-dependent Toll-like receptor signaling and proinflammatory cytokine production. *Nat Immunol* **10:** 1089–1095.

Chen Y, Pappu BP, Zeng H, Xue L, Morris SW, Lin X, Wen R, Wang D. 2007. B cell lymphoma 10 is essential for FcεR-mediated degranulation and IL-6 production in mast cells. *J Immunol* **178:** 49–57.

Chen NJ, Chio II, Lin WJ, Duncan G, Chau H, Katz D, Huang HL, Pike KA, Hao Z, Su YW, et al. 2008. Beyond tumor necrosis factor receptor: TRADD signaling in toll-like receptors. *Proc Natl Acad Sci* **105:** 12429–12434.

Cho YS, Challa S, Moquin D, Genga R, Ray TD, Guildford M, Chan FK. 2009. Phosphorylation-driven assembly of the RIP1–RIP3 complex regulates programmed necrosis and virus-induced inflammation. *Cell* **137:** 1112–1123.

Colvin RA, Means TK, Diefenbach TJ, Moita LF, Friday RP, Sever S, Campanella GS, Abrazinski T, Manice LA, Moita C, et al. 2010. Synaptotagmin-mediated vesicle fusion regulates cell migration. *Nat Immunol* **11:** 495–502.

Condliffe AM, Webb LM, Ferguson GJ, Davidson K, Turner M, Vigorito E, Manifava M, Chilvers ER, Stephens LR, Hawkins PT. 2006. RhoG regulates the neutrophil NADPH oxidase. *J Immunol* **176:** 5314–5320.

Cusson-Hermance N, Khurana S, Lee TH, Fitzgerald KA, Kelliher MA. 2005. Rip1 mediates the Trif-dependent toll-like receptor 3- and 4-induced NF-κB activation but does not contribute to interferon regulatory factor 3 activation. *J Biol Chem* **280:** 36560–36566.

Das M, Sabio G, Jiang F, Rincon M, Flavell RA, Davis RJ. 2009. Induction of hepatitis by JNK-mediated expression of TNF-α. *Cell* **136:** 249–260.

*Devreotes P, Horwitz AR. 2014. Signaling networks that regulate cell migration. *Cold Spring Harb Perspect Biol* doi: 10.1101/cshperspect.a005959.

Döffinger R, Smahi A, Bessia C, Geissmann F, Feinberg J, Durandy A, Bodemer C, Kenwrick S, Dupuis-Girod S, Blanche S, et al. 2001. X-linked anhidrotic ectodermal dysplasia with immunodeficiency is caused by impaired NF-κB signaling. *Nat Genet* **27:** 277–285.

Ermolaeva MA, Michallet MC, Papadopoulou N, Utermohlen O, Kranidioti K, Kollias G, Tschopp J, Pasparakis M. 2008. Function of TRADD in tumor necrosis factor receptor 1 signaling and in TRIF-dependent inflammatory responses. *Nat Immunol* **9:** 1037–1046.

Ferguson GJ, Milne L, Kulkarni S, Sasaki T, Walker S, Andrews S, Crabbe T, Finan P, Jones G, Jackson S, et al. 2007. PI(3)Kγ has an important context-dependent role in neutrophil chemokinesis. *Nat Cell Biol* **9:** 86–91.

Fujishima H, Sanchez Mejia RO, Bingham CO III, Lam BK, Sapirstein A, Bonventre JV, Austen KF, Arm JP. 1999. Cytosolic phospholipase A2 is essential for both the immediate and the delayed phases of eicosanoid generation in mouse bone marrow-derived mast cells. *Proc Natl Acad Sci* **96:** 4803–4807.

Gerlach B, Cordier SM, Schmukle AC, Emmerich CH, Rieser E, Haas TL, Webb AI, Rickard JA, Anderton H, Wong WW, et al. 2011. Linear ubiquitination prevents inflammation and regulates immune signalling. *Nature* **471:** 591–596.

*Green DR, Llambi F. 2014. Cell death signaling. *Cold Spring Harb Perspect Biol* doi: 10.1101/cshperspect.a006080.

Grivennikov S, Karin E, Terzic J, Mucida D, Yu GY, Vallabhapurapu S, Scheller J, Rose-John S, Cheroutre H, Eckmann L, et al. 2009. IL-6 and Stat3 are required for survival of intestinal epithelial cells and development of colitis-associated cancer. *Cancer Cell* **15:** 103–113.

Gu H, Botelho RJ, Yu M, Grinstein S, Neel BG. 2003. Critical role for scaffolding adapter Gab2 in FcγR-mediated phagocytosis. *J Cell Biol* **161:** 1151–1161.

Guerra C, Collado M, Navas C, Schuhmacher AJ, Hernández-Porras I, Cañamero M, Rodriguez-Justo M, Serrano M, Barbacid M. 2011. Pancreatitis-induced inflammation contributes to pancreatic cancer by inhibiting oncogene-induced senescence. *Cancer Cell* **19:** 728–739.

Haas TL, Emmerich CH, Gerlach B, Schmukle AC, Cordier SM, Rieser E, Feltham R, Vince J, Warnken U, Wenger T, et al. 2009. Recruitment of the linear ubiquitin chain assembly complex stabilizes the TNF-R1 signaling complex and is required for TNF-mediated gene induction. *Mol Cell* **36:** 831–844.

Hall AB, Gakidis MA, Glogauer M, Wilsbacher JL, Gao S, Swat W, Brugge JS. 2006. Requirements for Vav guanine nucleotide exchange factors and Rho GTPases in FcγR- and complement-mediated phagocytosis. *Immunity* **24:** 305–316.

He S, Wang L, Miao L, Wang T, Du F, Zhao L, Wang X. 2009. Receptor interacting protein kinase-3 determines cellular necrotic response to TNF-α. *Cell* **137:** 1100–1111.

Hitotsumatsu O, Ahmad RC, Tavares R, Wang M, Philpott D, Turer EE, Lee BL, Advincula R, Malynn BA, Werts C, et al. 2008. The ubiquitin-editing enzyme A20 restricts nucleotide-binding oligomerization domain containing 2-triggered signals. *Immunity* **28:** 381–390.

Hornung V, Latz E. 2010. Intracellular DNA recognition. *Nat Rev Immunol* **10:** 123–130.

Hoshino K, Sasaki I, Sugiyama T, Yano T, Yamazaki C, Yasui T, Kikutani H, Kaisho T. 2010. Critical role of IκB kinase α in TLR7/9-induced type I IFN production by conventional dendritic cells. *J Immunol* **184:** 3341–3345.

Hou F, Sun L, Zheng H, Skaug B, Jiang Q, Chen ZJ. 2011. MAVS forms functional prion-like aggregates to activate and propagate antiviral innate immune response. *Cell* **146:** 448–461.

Ikeda F, Deribe YL, Skånland SS, Stieglitz B, Grabbe C, Franz-Wachtel M, van Wijk SJ, Goswami P, Nagy V, Terzic J, et al. 2011. SHARPIN forms a linear ubiquitin ligase complex regulating NF-κB activity and apoptosis. *Nature* **471:** 637–641.

Ishikawa H, Barber GN. 2008. STING is an endoplasmic reticulum adaptor that facilitates innate immune signalling. *Nature* **455:** 674–678.

Israel A. 2010. The IKK complex, a central regulator of NF-κB activation. *Cold Spring Harb Perspect Biol* **2:** a000158.

Jakus Z, Simon E, Frommhold D, Sperandio M, Mocsai A. 2009. Critical role of phospholipase Cγ2 in integrin and Fc receptor-mediated neutrophil functions and the effector phase of autoimmune arthritis. *J Exp Med* **206:** 577–593.

Jiang F, Ramanathan A, Miller MT, Tang CQ, Gale M, Patel SS, Marcotrigiano J. 2011. Structural basis of RNA recognition and activation by innate immune receptor RIG-I. *Nature* **479:** 423–427.

Kanarek N, London N, Schueler-Furman O, Ben-Neriah Y. 2010. Ubiquitination and degradation of the inhibitors of NF-κB. *Cold Spring Harb Perspect Biol* **2:** a000166.

Kang YJ, Chen J, Otsuka M, Mols J, Ren S, Wang Y, Han J. 2008. Macrophage deletion of p38α partially impairs lipopolysaccharide-induced cellular activation. *J Immunol* **180:** 5075–5082.

Kayagaki N, Wong MT, Stowe IB, Ramani SR, Gonzalez LC, Akashi-Takamura S, Miyake K, Zhang J, Lee WP, Muszynski A, et al. 2013. Noncanonical inflammasome activation by intracellular LPS independent of TLR4. *Science* doi: 10.1126/science.1240248.

Kim C, Sano Y, Todorova K, Carlson BA, Arpa L, Celada A, Lawrence T, Otsu K, Brissette JL, Arthur JS, et al. 2008. The kinase p38α serves cell type-specific inflammatory functions in skin injury and coordinates pro- and anti-inflammatory gene expression. *Nat Immunol* **9:** 1019–1027.

Klemm S, Gutermuth J, Hultner L, Sparwasser T, Behrendt H, Peschel C, Mak TW, Jakob T, Ruland J. 2006. The Bcl10–Malt1 complex segregates FcεRI-mediated nuclear factor κB activation and cytokine production from mast cell degranulation. *J Exp Med* **203:** 337–347.

Komander D, Reyes-Turcu F, Licchesi JD, Odenwaelder P, Wilkinson KD, Barford D. 2009. Molecular discrimination of structurally equivalent Lys 63-linked and linear polyubiquitin chains. *EMBO Rep* **10:** 466–473.

Kunisaki Y, Nishikimi A, Tanaka Y, Takii R, Noda M, Inayoshi A, Watanabe K, Sanematsu F, Sasazuki T, Sasaki T, et al. 2006. DOCK2 is a Rac activator that regulates motility and polarity during neutrophil chemotaxis. *J Cell Biol* **174:** 647–652.

Laplantine E, Fontan E, Chiaravalli J, Lopez T, Lakisic G, Veron M, Agou F, Israel A. 2009. NEMO specifically recognizes K63-linked polyubiquitin chains through a new bipartite ubiquitin-binding domain. *EMBO J* **28:** 2885–2895.

Li Z, Jiang H, Xie W, Zhang Z, Smrcka AV, Wu D. 2000. Roles of PLC-β2 and -β3 and PI3Kγ in chemoattractant-mediated signal transduction. *Science* **287:** 1046–1049.

Li Z, Hannigan M, Mo Z, Liu B, Lu W, Wu Y, Smrcka AV, Wu G, Li L, Liu M, Huang CK, Wu D. 2003. Directional sensing requires Gβγ-mediated PAK1 and PIX α-dependent activation of Cdc42. *Cell* **114:** 215–227.

Li S, Wang L, Dorf ME. 2009. PKC phosphorylation of TRAF2 mediates IKKα/β recruitment and K63-linked polyubiquitination. *Mol Cell* **33:** 30–42.

Lin SC, Lo YC, Wu H. 2010. Helical assembly in the MyD88–IRAK4–IRAK2 complex in TLR/IL-1R signalling. *Nature* **465:** 885–890.

Lowell CA. 2011. Src-family and Syk kinases in activating and inhibitory pathways in innate immune cells: Signaling cross talk. *Cold Spring Harb Perspect Biol* **3:** a002352.

Lu C, Xu H, Ranjith-Kumar CT, Brooks MT, Hou TY, Hu F, Herr AB, Strong RK, Kao CC, Li P. 2010. The structural basis of 5′ triphosphate double-stranded RNA recognition by RIG-I C-terminal domain. *Structure* **18:** 1032–1043.

Mariathasan S, Newton K, Monack DM, Vucic D, French DM, Lee WP, Roose-Girma M, Erickson S, Dixit VM. 2004. Differential activation of the inflammasome by caspase-1 adaptors ASC and Ipaf. *Nature* **430:** 213–218.

Mazaki Y, Hashimoto S, Tsujimura T, Morishige M, Hashimoto A, Aritake K, Yamada A, Nam JM, Kiyonari H, Nakao K, et al. 2006. Neutrophil direction sensing and superoxide production linked by the GTPase-activating protein GIT2. *Nat Immunol* **7:** 724–731.

Miao EA, Leaf IA, Treuting PM, Mao DP, Dors M, Sarkar A, Warren SE, Wewers MD, Aderem A. 2010. Caspase-1-induced pyroptosis is an innate immune effector mechanism against intracellular bacteria. *Nat Immunol* **11:** 1136–1142.

Michallet MC, Meylan E, Ermolaeva MA, Vazquez J, Rebsamen M, Curran J, Poeck H, Bscheider M, Hartmann G, Konig M, et al. 2008. TRADD protein is an essential component of the RIG-like helicase antiviral pathway. *Immunity* **28:** 651–661.

Micheau O, Tschopp J. 2003. Induction of TNF receptor I-mediated apoptosis via two sequential signaling complexes. *Cell* **114:** 181–190.

Micheau O, Lens S, Gaide O, Alevizopoulos K, Tschopp J. 2001. NF-κB signals induce the expression of c-FLIP. *Mol Cell Biol* **21:** 5299–5305.

Mukhopadhyay S, Plüddemann A, Hoe JC, Williams KJ, Varin A, Makepeace K, Aknin M, Bowdish DME, Smale ST, Barclay AN, et al. 2010. Immune inhibitory ligand CD200 induction by TLRs and NLRs limits macrophage activation to protect the host from meningococcal septicemia. *Cell Host Microbe* **16:** 236–247.

Muller WA. 2011. Mechanisms of leukocyte transendothelial migration. *Annu Rev Pathol* **28:** 323–344.

Muzio M, Ni J, Feng P, Dixit VM. 1997. IRAK (Pelle) family member IRAK-2 and MyD88 as proximal mediators of IL-1 signaling. *Science* **278:** 1612–1615.

Newton K, Matsumoto ML, Wertz IE, Kirkpatrick DS, Lill JR, Tan J, Dugger D, Gordon N, Sidhu SS, Fellouse FA, et al. 2008. Ubiquitin chain editing revealed by polyubiquitin linkage-specific antibodies. *Cell* **134:** 668–678.

Nishio M, Watanabe K, Sasaki J, Taya C, Takasuga S, Iizuka R, Balla T, Yamazaki M, Watanabe H, Itoh R, et al. 2007. Control of cell polarity and motility by the PtdIns(3,4,5)P3 phosphatase SHIP1. *Nat Cell Biol* **9:** 36–44.

Park JH, Kim YG, McDonald C, Kanneganti TD, Hasegawa M, Body-Malapel M, Inohara N, Nunez G. 2007. RICK/RIP2 mediates innate immune responses induced through Nod1 and Nod2 but not TLRs. *J Immunol* **178:** 2380–2386.

Park EJ, Lee JH, Yu GY, He G, Ali SR, Holzer RG, Osterreicher CH, Takahashi H, Karin M. 2010. Dietary and genetic obesity promote liver inflammation and tumorigenesis by enhancing IL-6 and TNF expression. *Cell* **140:** 197–208.

Peschon JJ, Torrance DS, Stocking KL, Glaccum MB, Otten C, Willis CR, Charrier K, Morrissey PJ, Ware CB, Mohler KM. 1998. TNF receptor-deficient mice reveal divergent roles for p55 and p75 in several models of inflammation. *J Immunol* **160:** 943–952.

Pobezinskaya YL, Kim YS, Choksi S, Morgan MJ, Li T, Liu C, Liu Z. 2008. The function of TRADD in signaling through tumor necrosis factor receptor 1 and TRIF-dependent Toll-like receptors. *Nat Immunol* **9:** 1047–1054.

Poeck H, Bscheider M, Gross O, Finger K, Roth S, Rebsamen M, Hannesschlager N, Schlee M, Rothenfusser S, Barchet W, et al. 2010. Recognition of RNA virus by RIG-I results in activation of CARD9 and inflammasome signaling for interleukin 1β production. *Nat Immunol* **11:** 63–69.

Qu Y, Misaghi S, Izrael-Tomasevic A, Newton K, Gilmour LL, Lamkanfi M, Louie S, Kayagaki N, Liu J, Komuves L, et al. 2012. Phosphorylation of NLRC4 is critical for inflammasome activation. *Nature* **490:** 539–542.

Rahighi S, Ikeda F, Kawasaki M, Akutsu M, Suzuki N, Kato R, Kensche T, Uejima T, Bloor S, Komander D, et al. 2009. Specific recognition of linear ubiquitin chains by NEMO is important for NF-κB activation. *Cell* **136:** 1098–1109.

Reiley W, Zhang M, Wu X, Granger E, Sun SC. 2005. Regulation of the deubiquitinating enzyme CYLD by IκB kinase γ-dependent phosphorylation. *Mol Cell Biol* **25:** 3886–3895.

Rothlin CV, Ghosh S, Zuniga EI, Oldstone MB, Lemke G. 2007. TAM receptors are pleiotropic inhibitors of the innate immune response. *Cell* **131:** 1124–1136.

Sato S, Sanjo H, Takeda K, Ninomiya-Tsuji J, Yamamoto M, Kawai T, Matsumoto K, Takeuchi O, Akira S. 2005. Essential function for the kinase TAK1 in innate and adaptive immune responses. *Nat Immunol* **6:** 1087–1095.

Satoh T, Kato H, Kumagai Y, Yoneyama M, Sato S, Matsushita K, Tsujimura T, Fujita T, Akira S, Takeuchi O. 2010. LGP2 is a positive regulator of RIG-I- and MDA5-mediated antiviral responses. *Proc Natl Acad Sci* **107:** 1512–1517.

Shembade N, Ma A, Harhaj EW. 2010. Inhibition of NF-κB signaling by A20 through disruption of ubiquitin enzyme complexes. *Science* **327:** 1135–1139.

Shi Y, Zhang J, Mullin M, Dong B, Alberts AS, Siminovitch KA. 2009. The mDia1 formin is required for neutrophil polarization, migration, and activation of the LARG/RhoA/ROCK signaling axis during chemotaxis. *J Immunol* **182:** 3837–3845.

Shim JH, Xiao C, Paschal AE, Bailey ST, Rao P, Hayden MS, Lee KY, Bussey C, Steckel M, Tanaka N, et al. 2005. TAK1, but not TAB1 or TAB2, plays an essential role in multiple signaling pathways in vivo. *Genes Dev* **19:** 2668–2681.

Shisler JL, Jin XL. 2004. The vaccinia virus K1L gene product inhibits host NF-κB activation by preventing IκBα degradation. *J Virol* **78:** 3553–3560.

Skaug B, Jiang X, Chen ZJ. 2009. The role of ubiquitin in NF-κB regulatory pathways. *Annu Rev Biochem* **78:** 769–796.

Suire S, Condliffe AM, Ferguson GJ, Ellson CD, Guillou H, Davidson K, Welch H, Coadwell J, Turner M, Chilvers ER, et al. 2006. Gβγs and the Ras binding domain of p110γ are both important regulators of PI(3)Kγ signalling in neutrophils. *Nat Cell Biol* **8:** 1303–1309.

Sun CX, Downey GP, Zhu F, Koh AL, Thang H, Glogauer M. 2004a. Rac1 is the small GTPase responsible for regulating the neutrophil chemotaxis compass. *Blood* **104:** 3758–3765.

Sun L, Deng L, Ea CK, Xia ZP, Chen ZJ. 2004b. The TRAF6 ubiquitin ligase and TAK1 kinase mediate IKK activation by BCL10 and MALT1 in T lymphocytes. *Mol Cell* **14:** 289–301.

Sun L, Wang H, Wang Z, He S, Chen S, Liao D, Wang L, Yan J, Liu W, Lei X, et al. 2012. Mixed lineage kinase domain-like protein mediates necrosis signaling downstream of RIP3 kinase. *Cell* **148:** 213–227.

Sun L, Wu J, Du F, Chen X, Chen ZJ. 2013. Cyclic GMP-AMP synthase is a cytosolic DNA sensor that activates the type I interferon pathway. *Science* **339:** 786–791.

Suzuki K, Verma IM. 2008. Phosphorylation of SNAP-23 by IκB kinase 2 regulates mast cell degranulation. *Cell* **134:** 485–495.

Takahashi H, Ogata H, Nishigaki R, Broide DH, Karin M. 2010. Tobacco smoke promotes lung tumorigenesis by triggering IKKβ- and JNK1-dependent inflammation. *Cancer Cell* **17:** 89–97.

Takeuchi O, Akira S. 2010. Pattern recognition receptors and inflammation. *Cell* **140:** 805–820.

Taylor SL, Frias-Staheli N, Garcia-Sastre A, Schmaljohn CS. 2009. Hantaan virus nucleocapsid protein binds to importin α proteins and inhibits tumor necrosis factor α-induced activation of nuclear factor κB. *J Virol* **83:** 1271–1279.

Tokunaga F, Sakata S, Saeki Y, Satomi Y, Kirisako T, Kamei K, Nakagawa T, Kato M, Murata S, Yamaoka S, et al. 2009. Involvement of linear polyubiquitylation of NEMO in NF-κB activation. *Nat Cell Biol* **11:** 123–132.

Tokunaga F, Nakagawa T, Nakahara M, Saeki Y, Taniguchi M, Sakata S, Tanaka K, Nakano H, Iwai K. 2011. SHARPIN is a component of the NF-κB-activating linear ubiquitin chain assembly complex. *Nature* **471:** 633–636.

Travassos LH, Carneiro LA, Ramjeet M, Hussey S, Kim YG, Magalhaes JG, Yuan L, Soares F, Chea E, Le Bourhis L, et al. 2010. Nod1 and Nod2 direct autophagy by recruiting ATG16L1 to the plasma membrane at the site of bacterial entry. *Nat Immunol* **11:** 55–62.

Tseng PH, Matsuzawa A, Zhang W, Mino T, Vignali DA, Karin M. 2010. Different modes of ubiquitination of the adaptor TRAF3 selectively activate the expression of type I interferons and proinflammatory cytokines. *Nat Immunol* **11:** 70–75.

Utomo A, Cullere X, Glogauer M, Swat W, Mayadas TN. 2006. Vav proteins in neutrophils are required for FcγR-mediated signaling to Rac GTPases and nicotinamide adenine dinucleotide phosphate oxidase component p40(phox). *J Immunol* **177:** 6388–6397.

Varfolomeev E, Goncharov T, Fedorova AV, Dynek JN, Zobel K, Deshayes K, Fairbrother WJ, Vucic D. 2008. c-IAP1 and c-IAP2 are critical mediators of tumor necrosis factor α (TNFα)-induced NF-κB activation. *J Biol Chem* **283:** 24295–24299.

Venkataraman T, Valdes M, Elsby R, Kakuta S, Caceres G, Saijo S, Iwakura Y, Barber GN. 2007. Loss of DExD/H box RNA helicase LGP2 manifests disparate antiviral responses. *J Immunol* **178:** 6444–6455.

Vereecke L, Beyaert R, van Loo G. 2011. Genetic relationships between A20/TNFAIP3, chronic inflammation, and autoimmune disease. *Biochem Soc Trans* **39:** 1086–1091.

Vig M, DeHaven WI, Bird GS, Billingsley JM, Wang H, Rao PE, Hutchings AB, Jouvin MH, Putney JW, Kinet JP. 2008. Defective mast cell effector functions in mice lacking the CRACM1 pore subunit of store-operated calcium release-activated calcium channels. *Nat Immunol* **9:** 89–96.

Wan Y, Xiao H, Affolter J, Kim TW, Bulek K, Chaudhuri S, Carlson D, Hamilton T, Mazumder B, Stark GR, et al. 2009. Interleukin-1 receptor-associated kinase 2 is critical for lipopolysaccharide-mediated post-transcriptional control. *J Biol Chem* **284:** 10367–10375.

Wang J, Basagoudanavar SH, Wang X, Hopewell E, Albrecht R, Garcia-Sastre A, Balachandran S, Beg AA. 2010a. NF-κB RelA subunit is crucial for early IFN-β expression and resistance to RNA virus replication. *J Immunol* **185:** 1720–1729.

Wang Y, Ludwig J, Schuberth C, Goldeck M, Schlee M, Li H, Juranek S, Sheng G, Micura R, Tuschl T, et al. 2010b. Structural and functional insights into 5′-ppp RNA pattern recognition by the innate immune receptor RIG-I. *Nat Struct Mol Biol* **17:** 781–787.

Welch HC, Coadwell WJ, Ellson CD, Ferguson GJ, Andrews SR, Erdjument-Bromage H, Tempst P, Hawkins PT, Stephens LR. 2002. P-Rex1, a PtdIns(3,4,5)P3- and Gβγ-regulated guanine-nucleotide exchange factor for Rac. *Cell* **108:** 809–821.

Wong WW, Gentle IE, Nachbur U, Anderton H, Vaux DL, Silke J. 2010. RIPK1 is not essential for TNFR1-induced activation of NF-κB. *Cell Death Differ* **17**: 482–487.

Wu J, Sun L, Chen X, Du F, Shi H, Chen C, Chen ZJ. 2013a. Cyclic GMP-AMP is an endogenous second messenger in innate immune signaling by cytosolic DNA. *Science* **339**: 826–830.

Wu J, Huang Z, Ren J, Zhang Z, He P, Li Y, Ma J, Chen W, Zhang Y, Zhou X, et al. 2013b. Mlkl knockout mice demonstrate the indispensable role of Mlkl in necroptosis. *Cell Res* **23**: 994–1006.

Xia ZP, Sun L, Chen X, Pineda G, Jiang X, Adhikari A, Zeng W, Chen ZJ. 2009. Direct activation of protein kinases by unanchored polyubiquitin chains. *Nature* **461**: 114–119.

Xu M, Skaug B, Zeng W, Chen ZJ. 2009. A ubiquitin replacement strategy in human cells reveals distinct mechanisms of IKK activation by TNFα and IL-1β. *Mol Cell* **36**: 302–314.

Yago T, Shao B, Miner JJ, Yao L, Klopocki AG, Maeda K, Coggeshall KM, McEver RP. 2010. E-selectin engages PSGL-1 and CD44 through a common signaling pathway to induce integrin αLβ2-mediated slow leukocyte rolling. *Blood* **116**: 485–494.

Yamamoto M, Okamoto T, Takeda K, Sato S, Sanjo H, Uematsu S, Saitoh T, Yamamoto N, Sakurai H, Ishii KJ, et al. 2006. Key function for the Ubc13 E2 ubiquitin-conjugating enzyme in immune receptor signaling. *Nat Immunol* **7**: 962–970.

Yang Y, Yin C, Pandey A, Abbott D, Sassetti C, Kelliher MA. 2007. NOD2 pathway activation by MDP or *Mycobacterium tuberculosis* infection involves the stable polyubiquitination of Rip2. *J Biol Chem* **282**: 36223–36229.

Yang YJ, Chen W, Carrigan SO, Chen WM, Roth K, Akiyama T, Inoue J, Marshall JS, Berman JN, Lin TJ. 2008. TRAF6 specifically contributes to FcεRI-mediated cytokine production but not mast cell degranulation. *J Biol Chem* **283**: 32110–32118.

Yoshida R, Takaesu G, Yoshida H, Okamoto F, Yoshioka T, Choi Y, Akira S, Kawai T, Yoshimura A, Kobayashi T. 2008. TRAF6 and MEKK1 play a pivotal role in the RIG-I-like helicase antiviral pathway. *J Biol Chem* **283**: 36211–36220.

Zarbock A, Abram CL, Hundt M, Altman A, Lowell CA, Ley K. 2008. PSGL-1 engagement by E-selectin signals through Src kinase Fgr and ITAM adapters DAP12 and FcRγ to induce slow leukocyte rolling. *J Exp Med* **205**: 2339–2347.

Zeng W, Sun L, Jiang X, Chen X, Hou F, Adhikari A, Xu M, Chen ZJ. 2010. Reconstitution of the RIG-I pathway reveals a signaling role of unanchored polyubiquitin chains in innate immunity. *Cell* **141**: 315–330.

Zhang J, Guo J, Dzhagalov I, He YW. 2005. An essential function for the calcium-promoted Ras inactivator in Fcγ receptor-mediated phagocytosis. *Nat Immunol* **6**: 911–919.

Zhang DW, Shao J, Lin J, Zhang N, Lu BJ, Lin SC, Dong MQ, Han J. 2009. RIP3, an energy metabolism regulator that switches TNF-induced cell death from apoptosis to necrosis. *Science* **325**: 332–336.

Zhao T, Yang L, Sun Q, Arguello M, Ballard DW, Hiscott J, Lin R. 2007. The NEMO adaptor bridges the nuclear factor-κB and interferon regulatory factor signaling pathways. *Nat Immunol* **8**: 592–600.

CHAPTER 16

Signaling in Lymphocyte Activation

Doreen Cantrell

College of Life Sciences, Wellcome Trust Biocentre, University of Dundee, Dundee DD1 5EH, Scotland, United Kingdom

Correspondence: d.a.cantrell@dundee.ac.uk

SUMMARY

The fate of T and B lymphocytes, the key cells that direct the adaptive immune response, is regulated by a diverse network of signal transduction pathways. The T- and B-cell antigen receptors are coupled to intracellular tyrosine kinases and adaptor molecules to control the metabolism of inositol phospholipids and calcium release. The production of inositol polyphosphates and lipid second messengers directs the activity of downstream guanine-nucleotide-binding proteins and protein and lipid kinases/phosphatases that control lymphocyte transcriptional and metabolic programs. Lymphocyte activation is modulated by costimulatory molecules and cytokines that elicit intracellular signaling that is integrated with the antigen-receptor-controlled pathways.

Outline

1 INTRODUCTION

The adaptive immune response is directed by B and T lymphocytes. These cells express specific receptors that recognize pathogen-derived antigens: the B-cell antigen receptor (BCR) and the T-cell antigen receptor (TCR), respectively. B lymphocytes have two principal roles: to produce and secrete specific antibodies/immunoglobulins, and to function as antigen-presenting cells (APCs). T cells have multiple roles in adaptive immune responses. In this context, peripheral T cells can be subdivided on the basis of whether they express CD8 or CD4, receptors that recognize class I and class II major histocompatibility complex (MHC) molecules, respectively. CD8$^+$ T cells differentiate to cytolytic effectors that directly kill virus- or bacteria-infected cells. CD4$^+$ T cells are referred to as "helper" T cells because they produce regulatory cytokines and chemokines that mediate autocrine or paracrine control of T-cell differentiation and/or regulate the differentiation of B cells and/or direct the activity of macrophages and neutrophils (O'Shea and Paul 2010). At least five major subpopulations of mature CD4$^+$ cells exist with distinct functions that are tailored to deal with different pathogens. Th1 cells, characterized by interferon (IFN)γ production; Th2 cells, characterized by interleukin 4 (IL4) and IL13 production; Th17 cells, which produce the proinflammatory cytokines IL17 and IL22; regulatory T (Treg) cells that function to restrain autoimmunity and strong inflammatory responses; and follicular helper T (Tfh) cells, a class of effector CD4$^+$ T cells that regulate the development of antigen-specific B-cell immunity.

The paradigm of the adaptive immune response is that a primary response to an antigen causes clonal expansion of antigen-reactive T or B cells and produces a large number of effector lymphocytes that cause clearance of the pathogen. Once the pathogen is cleared there is a contraction phase of the immune response characterized by loss of effector lymphocytes and the emergence of long-lived memory cells capable of mounting rapid secondary responses to reinfection with the original pathogen.

The proliferation and differentiation of mature lymphocytes in adaptive immune responses are directed by antigen receptors, costimulatory molecules, adhesion molecules, cytokines, and chemokines. These extrinsic stimuli are coupled to a diverse network of signal transduction pathways that control the transcriptional and metabolic programs that determine lymphocyte function. At the core of lymphocyte signal transduction is the regulated metabolism of inositol phospholipids and the resultant production of inositol polyphosphates and lipids such as polyunsaturated diacylglycerols (DAGs). These second messengers direct the activity of protein and lipid kinases

and guanine-nucleotide-binding proteins that control lymphocyte proliferation, differentiation, and effector function. Below, I outline both the unique and the conserved aspects of signaling in lymphocytes, focusing on signaling pathways controlled by antigen receptors and how these responses are subsequently shaped and modulated by cytokines and chemokines.

2 ANTIGEN-RECEPTOR STRUCTURE AND FUNCTION

The TCR and BCR are multiprotein complexes comprising subunits containing highly variable antigen-binding regions linked noncovalently to invariant signal transduction subunits. In both cases, rearrangements of the DNA sequences that encode the antigen-binding region create a diversity in antigen-receptor structures. A key feature of T- and B-cell populations is that each individual lymphocyte will express multiple copies of a unique antigen receptor with a single antigen specificity (defined by three complementarity-determining regions [CDRs]). It is the selectivity of antigen receptors that underpins immune specificity by ensuring that only those lymphocytes that recognize a specific pathogen are activated by it.

The BCR is composed of a highly variable membrane-bound immunoglobulin of either the IgM or IgD subclass in a complex with the invariant also known as Igα and Igβ (CD79a and CD79b) heterodimer (Tolar et al. 2009). Immunoglobulin subunits are highly variable because the genes that encode these proteins undergo rearrangements and somatic hypermutation during B-cell development, which produces a high degree of protein diversity ($\geq 10^{11}$ different receptors) (Schatz and Ji 2011).

The TCR is also characterized by highly variable antigen-binding subunits, either an αβ or a γδ dimer (Davis 2004; Krogsgaard and Davis 2005; Xiong and Raulet 2007). These are coupled to the invariant CD3 subunits γε, δε, and ζζ, which are essential for trafficking and stability of the γδ and αβ subunits at the plasma membrane. CD3 antigens also transmit signals into the cell across the plasma membrane. Like the BCR immunoglobulin sequences, the TCR-αβ or γδ dimers are highly variable because the genes that encode them undergo rearrangements (but not hypermutation) during their development. Indeed, there is potential for the production of $\sim 10^{18}$ different TCR-αβ receptor complexes. This is compared with to a minimal estimate of 10^{11} BCR complexes. The salient feature is that each T cell only expresses an αβ or a γδ receptor complex with a single specificity.

T cells that express TCR-γδ complexes are found predominantly at epithelial barriers (e.g., in the skin and gut epithelia). The ligands for TCR-γδ complexes are not well defined but can be bacterial phosphoantigens, alkylamines, and aminobisphosphonates (Hayday 2009). T cells that express TCR-αβ complexes typically recirculate between the blood, secondary lymphoid organs (spleen and lymph nodes), and the lymphatic system. The ligands for TCR-αβ complexes are not antigens per se but rather pathogen- (or transplantation-antigen)-derived peptides bound to MHC molecules, a group of molecules that display the short, approximately nine-residue peptides on the surface of APCs. TCR-αβ-expressing T cells are thus not triggered by soluble pathogen-derived peptides but only by peptide-MHC complexes on the surface of dendritic cells, B cells, and other cells that can function as APCs (Krogsgaard and Davis 2005).

3 IMMUNORECEPTOR TYROSINE-BASED ACTIVATION MOTIFS

The antigen-receptor subunits that mediate signal transduction are the invariant chains CD3γ, δ, ε, ζ in T cells, Igα and Igβ in B lymphocytes, and the FcRγ chain in mast cells (see below). These signaling subunits have no intrinsic signaling capacity, but all contain a $YxxL/I-X_{6-8}-YxxL/I$ motif referred to as an immunoreceptor tyrosine-based activation motif (ITAM) (Abram and Lowell 2007; Love and Hayes 2010). The CD3γ, δ, and ε subunits each contain a single ITAM, and there are three ITAMs in the CD3ζ chain. The minimal TCR complex thus has 10 ITAMs. These couple the TCR to intracellular tyrosine kinases (see below). ITAM motifs are a defining feature of antigen-receptor complexes. Igα and Igβ, the signaling subunits of the BCR, both have a single ITAM.

ITAM motifs are not restricted to the TCR and BCR. For example, mast cells comprise an important group of lymphocytes whose fate is determined by antigen-specific immunoglobulin. These cells respond to antigen because they express a high-affinity receptor for IgE. This receptor, termed FcεR1, binds to the immunoglobulin IgE with high affinity. When FcεR1-IgE complexes are cross-linked by polyvalent antigen they can trigger mast cell degranulation and the release of cytokines and allergic mediators. The FcεR1 is assembled from three subunits: the α subunit that binds to the Fc region of IgE, a β subunit that provides important accessory signaling, and the FcRγ chain, which is a signaling subunit that contains a single ITAM (Beaven and Metzger 1993; Abram and Lowell 2007; p. 125 [Samelson 2011]).

TCR/BCR/FcεR1 signaling is initiated by the tyrosine phosphorylation of ITAMs by Src-family tyrosine kinases such as Lck and Fyn in T cells, Lyn in B cells, and Fyn in mast cells (Salmond et al. 2009). When both tyrosine residues are phosphorylated, the ITAM forms a high-affinity binding site for Syk-family tyrosine kinases; generally in T cells this is Zap-70 (Wang et al. 2010), whereas in B cells and

mast cells Syk is recruited (Chu et al. 1998). Zap-70 and Syk contain tandem SH2 domains that bind with high affinity to the doubly phosphorylated ITAM (Chu et al. 1998). The activation of Zap-70 or Syk is initiated by binding to phosphorylated ITAMs. This is proposed to release Syk/Zap-70 from an autoinhibited conformation and expose regulatory tyrosine residues for phosphorylation by Src-family kinases (Au-Yeung et al. 2009). The phosphorylation of tyrosine residues in the activation loop in the Zap-70/Syk catalytic domain, as well as two residues in the adjacent linker region, then further stimulates their catalytic activity. Antigen-receptor control of Syk-family tyrosine kinases is fundamental for lymphocyte activation and underpins the ability of antigen receptors to transduce signals from pathogen-derived antigens to the interior of lymphocytes (Mocsai et al. 2010; Wang et al. 2010).

How the Src-family kinases such as Lck are regulated is central to antigen-receptor signal transduction (Salmond et al. 2009). The activity of Lck is regulated by phosphorylation and dephosphorylation of a carboxy-terminal tyrosine (Y505) by the ubiquitously expressed kinase carboxy-terminal Src kinase (CSK), as well as autophosphorylation of the activation loop tyrosine residue, Y394. Phosphorylated Y505 forms an intramolecular binding site for the Lck SH2 domain, thereby locking the kinase into an autoinhibited state. The key to initiating the activation of Lck and its relatives is to dephosphorylate the carboxy-terminal tyrosine and relieve autoinhibition of the kinase. This is mediated by transmembrane-receptor-like tyrosine phosphatases, such as CD45 and CD148 (Hermiston et al. 2009; Zikherman et al. 2010). Hence in T cells, the Lck activation threshold is set by the balanced activity of the kinase-phosphatase pair CSK, which phosphorylates Y505, and CD45, which dephosphorylates this residue (Zikherman et al. 2010).

It is frequently assumed that triggering antigen receptors stimulates Src kinase family activity, and antigen receptors are often depicted as molecular switches that are either on or off. In reality, antigen receptors are always signaling and it is the intensity of the signal that changes. The assembly of antigen receptors at the plasma membrane is thus proposed to mediate low-level signaling and the engagement with high-affinity ligands (antigen or antigen–MHC) increases the intensity. Indeed Src-family kinases such as Lck are constitutively active before antigen-receptor engagement and cause low-level ITAM phosphorylation (Nika et al. 2010). The levels of ITAM phosphorylation are limited by tyrosine phosphatases, and the increases in ITAM phosphorylation that follow antigen-receptor engagement probably result from spatial constraints on the ITAM-phosphatase interaction (van der Merwe and Dushek 2011).

How are these spatial constraints regulated to explain how ligand occupancy triggers TCR signaling? Surprisingly, we do not know, although there is no shortage of theories. Current models range from the ligand-induced conformational change to the idea that the TCR is a mechanosensor that converts the mechanical energy generated by antigen binding into a biochemical signal (Kim et al. 2009). One other idea well supported by experimental data is that binding of the TCR to peptide-MHC complexes on the surface of APCs causes spatial segregation of TCR complexes (which have small ectodomains) away from receptor tyrosine phosphatases such as CD45 and CD148 (which have very large ectodomains). This might locally perturb the kinase–phosphatase balance sufficiently to favor ITAM phosphorylation and Zap-70 recruitment (van der Merwe and Davis 2003; van der Merwe and Dushek 2011). Note that the MHC-binding coreceptors CD4 and CD8 are also thought to play a role in perturbing the kinase-phosphatase balance in localized areas of the T-cell membrane. CD4 and CD8 can thus promote TCR signaling by stabilizing interactions between the TCR and peptide-MHC ligands. However, the cytoplasmic domains of CD4 and CD8 constitutively bind Lck and hence facilitate the recruitment of this kinase to ligand-engaged TCR complexes (Artyomov et al. 2010).

What about the BCR and FcɛR1? In quiescent B cells, the BCR may exist in an oligomeric autoinhibited state, and ligand occupancy could drive the dissociation of these oligomers into monomers that interact more effectively with downstream tyrosine kinases (Yang and Reth 2010a,b). For the FcɛR1, the opposite is probably the case. This receptor binds IgE but is only effectively triggered when antigen oligomerizes the receptor (Beaven and Metzger 1993).

4 ADAPTOR MOLECULES FOR ANTIGEN RECEPTORS

The immediate substrates for tyrosine kinases activated by TCRs/BCRs/FcɛR1s are specialized adaptor proteins that coordinate the localization and activation of key effector enzymes. In T cells and mast cells, the adaptors LAT and SLP76 are substrates for Zap-70 and Syk, respectively (Jordan and Koretzky 2010; p. 125 [Samelson 2011]). In B cells, the adaptor coupling Syk to effector enzymes is B-cell linker protein (BLNK), also known as SLP65 (Koretzky et al. 2006).

LAT is an integral membrane protein with a cytoplasmic tail containing nine tyrosine residues. When phosphorylated, these act as docking sites for effector enzymes containing SH2 domains. For example, phosphorylated Y132 of LAT recruits phospholipase Cγ (PLCγ), a critical molecule for lymphocyte activation. The subsequent tyrosine phosphorylation of PLCγ activates the enzyme, resulting in the hydrolysis of its substrate phosphatidylinositol 4,5-bisphosphate (PIP$_2$). LAT not only recruits PLCγ but also

plays a complex role as a scaffold that ensures PLCγ activation. Phosphorylated Y171, Y191, and Y226 in LAT can thus bind to the SH2 domain of Grb2 family members such as Gads, which recruits SLP76 to the LAT complex.

SLP-76 contains three key tyrosine residues, a central SH3-binding proline-rich domain and a carboxy-terminal SH2 domain. The SLP76 proline-rich domain binds to the SH3 domain of Gads; the SLP76-Gads complex is then recruited to LAT via binding of the Gads SH2 domain binding to tyrosine-phosphorylated LAT.

Tyrosine-phosphorylated SLP76 can recruit a number of effector molecules into the LAT complex, notably the Tec-family tyrosine kinase Itk, which phosphorylates PLCγ, leading to its activation. The SH2 domain of SLP76 is also important because it binds to the cytosolic adaptor ADAP, which links SLP76 to the regulation of integrin-mediated cell adhesion. The LAT-SLP76 complex thus nucleates and organizes multiple TCR-dependent signaling pathways in T cells. Indeed, LAT and SLP76 are essential for TCR function: there are multiple defects in thymus T-cell development and peripheral T-cell function in the absence of these adaptors.

LAT and SLP-76 are equally important for mast cell function, coupling Syk to signaling pathways downstream from the FcɛR1 (Alvarez-Errico et al. 2009; Kambayashi et al. 2009). However, neither LAT nor SLP76 is expressed in B cells; there, the predominant adaptor molecule is BLNK (Kurosaki and Hikida 2009). BLNK is a Syk substrate and contains nine tyrosine residues that are rapidly phosphorylated following BCR triggering. Its recruitment to the plasma membrane requires association with CIN85, and the BLNK-CIN85 complex coordinates recruitment of effectors such as PLCγ and Grb2-family adaptors (Oellerich et al. 2011). BLNK is essential for normal B-cell development and for peripheral B-cell function (see Fig. 1).

5 CALCIUM AND DIACYLGLYCEROL SIGNALING

A major function for antigen-receptor-coupled tyrosine kinases and adaptors is to regulate intracellular calcium levels and control DAG-mediated signaling (Oh-hora and Rao 2008; Matthews and Cantrell 2009). Inositol 1,4,5-trisphosphate (IP$_3$) produced by PLCγ binds to IP$_3$ receptors on endoplasmic reticulum (ER) membranes, initiating release of calcium from stores and an increase in cytosolic calcium concentration (p. 95 [Bootman 2012]). This in turn triggers calcium entry across the plasma membrane via activation of highly selective store-operated calcium-release-activated calcium (CRAC) channels. Stromal interaction molecules 1 and 2 (STIM1 and STIM2) sense depletion of the ER stores and relocate to ER–plasma-membrane junctions. There they bind to the CRAC channel protein Orai1, which

Cite this chapter as *Cold Spring Harb Perspect Biol* doi: 10.1101/cshperspect.a018788

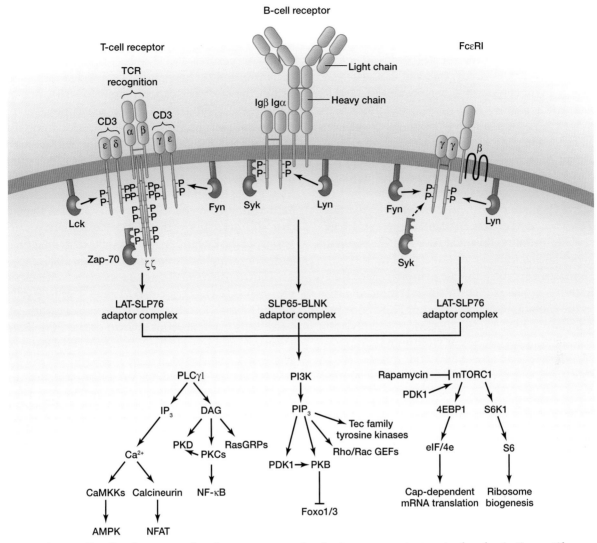

Figure 1. Signaling downstream from immune receptors bearing immunoreceptor tyrosine-based activation motifs (ITAMs; yellow rectangles). T-cell receptors, B-cell receptors, and FcεR1s all contained ITAMs that can be tyrosine phosphorylated (red circles) by Src-family kinases such as Fyn and Lck. This creates docking sites for the recruitment and activation of the tyrosine kinases Zap-70 and Syk. These in turn phosphorylate adaptor complexes that recruit numerous additional signaling molecules that control phospholipid, calcium, small G protein, and kinase signaling.

activates the channels to allow entry of extracellular calcium to promote a sustained increase in intracellular calcium levels. This coupling of antigen receptors to CRAC channels allows lymphocytes to sustain high levels of intracellular calcium concentrations during an immune response (Hogan et al. 2010).

6 DOWNSTREAM FROM CALCIUM SIGNALING IN LYMPHOCYTES

Increases in intracellular calcium concentration in lymphocytes initiate signaling by the calcium/calmodulin-dependent protein kinase kinases (CaMKKs) (Matthews and Cantrell 2009). The best-studied role for calcium signaling

in both B and T lymphocytes, however, is control of calcineurin (also known as protein phosphatase 2B, PP2B), a protein phosphatase that controls the intracellular localization of members of the NFAT (nuclear factor of activated T cells) family of transcription factors (Im and Rao 2004; Muller and Rao 2010). These are key regulators of cytokine gene expression in B and T lymphocytes, in which they control expression of IL2, IL4, TNF, and IFNγ. In quiescent lymphocytes, before antigen-receptor engagement, NFATs are constitutively phosphorylated via the actions of NFAT kinases that include CK1 and GSK3. This phosphorylation of NFATs causes their nuclear exclusion as a result of binding to 14-3-3 proteins, thus maintaining them inactive in the cytosol. NFATs remain inactive until triggering of antigen receptors raises intracellular free calcium levels, which activates calcineurin, which then dephosphorylates NFATs, allowing their translocation to the nucleus.

In the nucleus, NFATs form complexes with other transcription factors, bind to target genes and modulate gene transcription. In the context of *IL2* expression, NFAT–AP1 complexes act as positive regulators of IL2 production, whereas complexes containing NFAT with the Foxp3 transcription factor appear to repress cytokine gene expression (Im and Rao 2004; Muller and Rao 2010). The impact of NFAT translocation to the nucleus on the T-cell transcriptional program thus depends on cellular context and the available NFAT-binding partners. Nevertheless, the rate-limiting step for NFAT activation is antigen-receptor-regulated increases in intracellular calcium and the resultant activation of calcineurin.

The importance of calcium/calcineurin signaling for T-cell activation is emphasized by the clinical efficacy of drugs based on the compound cyclosporin A or FK506 that prevent calcineurin activation and NFAT dephosphorylation (Gallo et al. 2006). These are potent T-cell immunosuppressants used for the prevention of organ transplant rejection and for the treatment of chronic T-cell-mediated autoimmune diseases, such as ectopic eczema.

7 DIACYLGLYCEROL SIGNALING IN LYMPHOCYTES

Multiple species of DAG are produced as intermediates in phospholipid resynthesis pathways. Consequently, quiescent lymphocytes have high levels of DAG before immune activation. However, antigen-receptor stimulation induces further production of polyunsaturated DAG by triggering PLCγ-mediated hydrolysis of PIP_2; in particular, triggering localized increases in DAG levels in membrane microdomains (Spitaler et al. 2006; Quann et al. 2009). DAG binds with high affinity to proteins that contain a conserved cysteine-rich domain (CRD) ($H-X_{12}-C-X_2-C-X_{13/14}-C-X_2-C-X_4-H-X_2-C-X_7-C$). In lymphocytes, these proteins include the Ras/Rap guanyl-releasing protein (GRP) family of guanine nucleotide exchange factors (GEFs), which activate Ras and Rap GTPases, and the serine/threonine kinases protein kinase C (PKC) and protein kinase D (PKD).

8 PKC AND LYMPHOCYTES

Lymphocytes express multiple PKC isoforms, including α, βI, βII, δ, ε, η, and θ, and these have key roles in lymphocyte activation (Matthews and Cantrell 2009). They are important regulators of lymphocyte transcriptional programs and, in particular, control expression of genes encoding cytokines and cytokine receptors. DAG/PKC signaling also

plays a key role in controlling integrin-mediated cell adhesion and lymphocyte polarity. The direct substrates for PKCs include PKDs (Matthews et al. 2010). Lymphocytes predominantly express PKD2 and activation of this kinase requires *trans*-phosphorylation of conserved serine residues within the enzyme's catalytic domain (S701 and S711). These sites are substrates for both conventional and novel PKCs, and their phosphorylation is essential for efficient TCR-induced cytokine production and for optimal antibody production by B lymphocytes. Other PKC substrates include scaffolding proteins such as Carma1 and GEFs for the GTPases Ras and Rap1 (Matthews and Cantrell 2009). In particular, PKC-mediated phosphorylation of RapGEF2 is critical for activation of the GTPase Rap1, which controls the activity of the integrin LFA1 (also known as integrin $\alpha_L\beta_2$) and hence lymphocyte adhesion (Kinashi 2005).

The coordination of integrin-mediated cell adhesion by PKC and GTPases is essential to allow T cells, B cells, and natural killer (NK) cells to form tight contacts with APCs or target cells via a structure known as the immunological synapse (Dustin et al. 2010; Springer and Dustin 2011). These are formed between naïve T cells and APCs or effector cytolytic T cells and pathogen-infected target cells. B cells can also form immunological synapses with APCs in a process that potentiates antigen binding and processing of even membrane-tethered antigens (Harwood and Batista 2011). Immunological synapses are highly ordered structures characterized by the segregation of receptors and signaling molecules into distinct areas known as supramolecular activation clusters (SMACs). Stable immune synapses are arranged in concentric zones: antigen receptors accumulate in the center (cSMAC), whereas integrins segregate to the periphery (pSMAC). One common misconception is that the immune synapse is involved in the initiation of antigen-receptor signaling. The reality is that immune synapses are formed as a downstream consequence of antigen-receptor engagement. Immunological synapses provide a focus for DAG signaling following antigen-receptor engagement (Spitaler et al. 2006). Moreover, formation of immunological synapses is associated with the polarization of the microtubule-organizing center (MTOC) toward the target cell. This MTOC polarization is coordinated by calcium and DAG signaling pathways, with PKC family members playing a crucial role. The reorientation of the MTOC controls the ability of lymphocytes to direct cytokine secretion and to direct the exocytosis of secretory or lytic granules. For example, in cytotoxic T cells the immunological synapse directs the secretion of the granules that contain cytolytic effector molecules such as perforin and granzymes toward the target cell (Jenkins and Griffiths 2010).

One of the best characterized roles for PKCs in lymphocytes is the control of gene expression via the transcription factor NF-κB1 (also known as p50) (Oeckinghaus et al. 2011; Gerondakis and Siebenlist 2012). In quiescent lymphocytes, NF-κB1 is sequestered in the cytosol in a complex with inhibitor of NF-κB (IκB). The activation of PKC results in the assembly of a complex comprising the scaffolding protein Carma1, Bcl10, and MALT1 (Blonska and Lin 2009). This PKC-induced Carma1-Bcl10-MALT1 complex subsequently binds to and activates the IκB kinase (IKK) complex, which then phosphorylates IκB, triggering its rapid ubiquitylation by the E3 ligase SCF-βTrCP and degradation by the proteasome. The removal of IκB unmasks the nuclear localization sequence of NF-κB1 and permits its translocation to the nucleus, where it stimulates the transcription of target genes. This mechanism is common to all lymphocytes but there is redundancy between PKC isoforms: in B lymphocytes, PKCβ isoforms are involved, whereas in T cells PKCε and θ are essential.

9 Ras SIGNALING AND LYMPHOCYTES

In quiescent lymphocytes Ras GTPases are predominantly inactive. Engagement of antigen receptors stimulates Ras proteins to accumulate in a GTP-bound state. This allows Ras to bind to the serine/threonine kinase Raf1, which in turn activates the kinase MEK1 that phosphorylates and activates the MAP kinases (MAPKs) ERK1 and ERK2 (p. 81 [Morrison 2012]). Two major classes of GEFs couple antigen receptors to Ras activation: the Ras GRPs and SOS. Ras GRPs are activated by DAG and PKC-mediated phosphorylation. RasGRP1 acts downstream from antigen receptors in T cells, whereas RasGRP1 and RasGRP3 function in B cells, and RasGRP4 functions in mast cells. SOS is activated independently of DAG/PKC via a tyrosine-kinase-dependent pathway. It thus binds constitutively to the SH3 domains of the adaptor Grb2 and is recruited to the plasma membrane when the SH2 domain of Grb2 binds to tyrosine-phosphorylated adaptors such as LAT in T cells or Shc in B cells. Note that Ras is also activated by members of the common cytokine-receptor γ chain (γ_c) family of cytokines (see below), such as IL2. Receptors for these cytokines recruit SOS to the plasma membrane via Grb2 and the adaptor Shc (p. 117 [Harrison 2012]).

The prototypical role for Ras in lymphocytes is to control gene transcription via ERK1 and ERK2 (Matthews and Cantrell 2009). These phosphorylate and regulate a number of key substrates, including the ternary complex factor (TCF) subfamily of ETS-domain transcription factors. They also control the activity of the RSK serine/threonine kinases that are known to have important functions in lymphocyte development and peripheral lymphocyte function. The initiating step for RSK activation is thus ERK1/2-mediated phosphorylation of S369, T365, and T577 in the carboxy-terminal catalytic domain of the kinase. The activated carboxy-terminal catalytic domain of RSK then phosphorylates S386 intramolecularly to create a docking site for the kinase PDK1, which then phosphorylates S227 in the amino-terminal RSK kinase domain, thereby activating the enzyme (Finlay and Cantrell 2011a).

A full list of ERK1/2 substrates in lymphocytes is beyond our scope here but there have been some unexpected insights into the complexity of ERK signaling pathways in lymphocytes that warrant discussion. Flow cytometric-based assays that assess ERK activity at the single-cell level have shown that, when lymphocytes respond to an increasing strength of antigen-receptor stimulus, ERK activation is a digital (all or nothing) rather than an analog response (Chakraborty et al. 2009; Das et al. 2009). In this digital response, the frequency of cells within a population that activate ERK changes, each cell activating it to an equivalent level. This means, in practice, that even a strong antigen-receptor stimulus can only trigger a proportion of lymphocytes to activate ERKs at any one time. The digital nature of this ERK response creates signaling heterogeneity within the responding lymphocyte population.

10 COSTIMULATORY MOLECULES, CYTOKINES, AND LYMPHOCYTE ACTIVATION

Lymphocyte responses both prior and subsequent to antigen-receptor engagement are modulated by multiple costimulatory and coinhibitory receptors. Signaling via Toll-like receptors (TLRs) is also a major factor influencing the fate of lymphocytes during an immune response. Because T and B lymphocytes respond to antigens presented to them by APCs, lymphocyte activation can be regulated by the adhesion molecules and costimulatory molecules expressed by the APC. Note also that many of the cytokines that control lymphocyte fate are produced in response to TLR-mediated activation of dendritic cells and macrophages (Ch. 15 [Newton and Dixit 2012]). Hence, the nature of the pathogen challenge to the innate immune system, and the resultant cytokine milieu modulate the adaptive immune response.

For T cells, key coreceptor molecules include the MHC receptors CD4 and CD8, and proteins such as CD28 (a positive coregulator) and CTLA4 and PD-1 (negative coregulators) (Artyomov et al. 2010; Francisco et al. 2010; Bour-Jordan et al. 2011; Walker and Sansom 2011). In B cells, molecules such as CD19 and the CD21 receptor for complement component C3d are essential (Carter and Fearon 1992; Depoil et al. 2008; Elgueta et al. 2009; Mackay et al. 2010) as are the TNF receptor family members CD40 and receptor for B-cell-activating factor (BAFFR) (Watts 2005; Elgueta et al. 2009; Karin and Gallagher 2009).

A full review of lymphocyte regulation by costimulatory factors is beyond our scope here but there are some general themes. Costimulatory molecules frequently work as adaptors to recruit signaling molecules to the plasma membrane and hence amplify antigen-receptor-mediated signaling. For example, CD4 and CD8 in T cells recruit Lck to the plasma membrane. Similarly, CD28 in T cells and CD19 in B cells both have cytoplasmic domains that can be tyrosine phosphorylated and thus can act as docking sites for SH2-domain-containing adaptors and enzymes. The CD19 cytoplasmic tail contains nine tyrosine residues with the potential to be phosphorylated and interact with signaling molecules including lipid kinases, Vav-family GEFs, and adaptor proteins such as Grb2. Other important examples of molecules that recruit key adaptor molecules to the plasma membrane are the lymphocytic activation molecule (SLAM) family of receptors and associated intracellular adaptors of the SLAM-associated protein (SAP) family (Veillette 2010).

The engagement of CD40 by its ligand (CD40L) leads to signals via adaptor proteins known as TNFR-associated factors (TRAFs), which activate signaling pathways, including MAPKs and NF-κB (p. 121 [Lim and Staudt 2012]).

The plethora of costimulatory molecules that can contribute to lymphocyte activation can be confusing, particularly because all seem to activate similar signal transduction pathways. The key message is that these receptors function at different times and in different contexts. For example, CD28 binds to the B7 family members CD80 and CD86, which are mainly expressed on APCs responding to TLR signaling. The ligand for CD40 is produced transiently by antigen-activated T cells and plays a key role in promoting specific T cell "help" to B cells by ensuring integration of signals between CD40-expressing B cells and antigen-primed T cells. In contrast, BAFF is mainly produced by neutrophils, monocytes, and macrophages and hence allows cross talk between B cells and these cells of the innate immune system.

11 CYTOKINE SIGNALING IN LYMPHOCYTES

Cytokines that signal via the Janus tyrosine kinases (JAKs) (p. 117 [Harrison 2012]), such as the γ_c family of cytokines, IFNs, and cytokines such as IL12 and IL23, are particularly important to the adaptive immune system (Rochman et al. 2009). For example, CD4-expressing $\alpha\beta$ T cells differentiate during immune responses to produce distinct effector subpopulations (O'Shea and Paul 2010) and the specification of these CD4[+] T-cell subsets is controlled by cytokines that direct the combinatorial action of multiple chromatin regulators and key lineage-specifying transcription factors. For example, IL12 drives Th1 T-cell differentiation and IL6,

IL21, and IL23 drive Th17 cell differentiation. Moreover, cytokines have pleotropic roles. IL2 is important for the differentiation of antigen-primed CD8[+] T cells to effector cytotoxic T cells (CTLs) but is also required for optimal Th1 T-cell differentiation and for the development of Treg cells.

One striking feature of lymphocyte biology is that the ability of cells to respond to cytokines (i.e., to express particular cytokine receptors) can be shaped by antigen-receptor triggering. Cytokine production by cells of the immune system is, in turn, controlled by triggering of antigen receptors in T and B cells or by receptors of the innate immune system. A prototypical example is IL2, which is only produced by antigen-receptor-activated T cells and B cells or pathogen-triggered dendritic cells. Moreover, expression of the IL2 receptor (IL2R) is tightly controlled by immune activation. The ILR2 receptor complex consists of a γ_c, a β subunit (CD122), and an α subunit (CD25). The expression of CD25 is rate limiting as it determines the ability of the receptor to bind IL2 with high affinity. Importantly, CD25 is not expressed on naïve CD4 and CD8 T cells but only on activated T cells. In addition, the expression of CD25 is transient and its sustained expression requires constant immune stimulation. IL2 responsiveness is thus tightly linked to antigen-receptor triggering to ensure the tight control of T cells by IL2. IL12 receptors are similar: these are only expressed on activated T cells. Furthermore, IL12 receptor expression needs to be sustained by IL2 and there is tight control of IL12 secretion by pathogen-activated dendritic cells and macrophages. Such dynamic regulation of cytokine and cytokine-receptor expression during immune activation ensures the immune specificity of cytokine action (i.e., only lymphocytes that have been primed by antigen-receptor triggering can respond to IL12). Note the production of cytokines is also limited to either pathogen-activated innate immune cells or antigen-activated lymphocytes (Fig. 2).

Cytokines that activate JAKs regulate the function of SH2-domain-containing transcription factors known as STATs (signal transducers and activators of transcription) (Ghoreschi et al. 2009; p. 117 [Harrison 2012]). There are four JAKs (JAK1, JAK2, JAK3, and Tyk2) and 7 STATs (STAT1, STAT2, STAT3, STAT4, STAT5a, STAT5b, and STAT6). A single JAK, or combination of JAKs, associates selectively with the cytoplasmic domains of the cytokine receptors. The model for JAK activation is that ligand occupancy of cytokine-receptor dimers results in JAK transphosphorylation and activation. The type I IFN receptors signal via JAK1 and Tyk2; IL12 and IL23 receptors signal via JAK2 and Tyk2. The IFNγ receptor activates JAK1 and JAK2, whereas γ_c-containing receptors, which include the receptors for IL2, IL4, IL7, IL9, IL15, and IL21, use JAK1 and JAK3. JAK activation results in phosphorylation of tyrosine

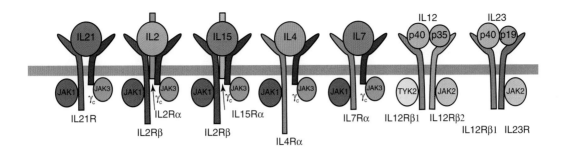

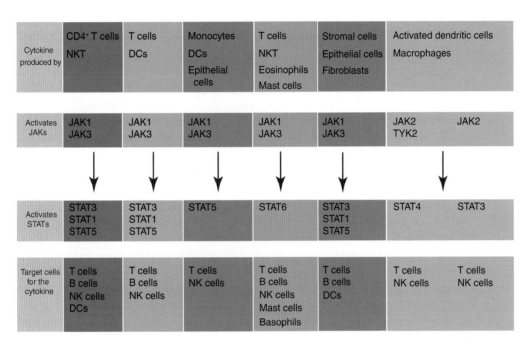

Figure 2. Signaling by interleukin (IL) receptors. Many cytokines signal via receptors linked to Janus tyrosine kinases (JAKs), which regulate the SH2-domain-containing transcription factors STATs. The different ILs produced by different cell types activate receptors coupled to different combinations of JAKs and STATs.

residues within the cytoplasmic tails of cytokine-receptor subunits that act as docking sites for the SH2 domains of the STATs. The recruitment of STATs leads to their phosphorylation by the JAKs. The STATs then form homodimers via SH2 domain interactions and translocate to the nucleus to bind STAT-response elements in DNA. STATs control lymphocyte transcriptional programs by working as transcriptional activators but they can also function as gene repressors (O'Shea and Paul 2010).

The specificity of STAT activation is determined by the selectivity of STAT SH2 domains for the STAT-recruitment motifs in the different cytokine-receptor subunits. For example, IL2 predominantly activates STAT5, because tyrosine-phosphorylated IL2Rβ subunits contain a high-affinity binding site for STAT5. The IL4 receptor, which comprises γc and a unique IL4 receptor α chain, activates STAT6 because tyrosine-phosphorylated IL4 receptors selectively bind STAT6. Figure 2 summarizes current information

about the JAK/STAT signaling combinations that function downstream from the major cytokine receptors.

It should be stressed that although the activation of STATs is pivotal for cytokine actions it is usually not sufficient to mimic the effects of cytokines. Indeed, cytokines can regulate other signal transduction pathways, some of which are shared with other receptors (e.g., IL2 and IL15 also activate Ras/ERK signaling) (Cantrell 2003). Moreover, many cytokines induce accumulation of phosphatidylinositol 3,4,5-trisphosphate (PIP_3), a product of phosphoinositide 3-kinases (PI3Ks) (Okkenhaug and Fruman 2010; Finlay and Cantrell 2011).

12 PI3K-MEDIATED SIGNALING IN LYMPHOCYTES

PI3K signaling is important for lymphocyte activation and integrates multiple receptor inputs. For example, in naïve T cells, low basal levels of PIP_3 are maintained by IL7 signaling; these increase strikingly in response to triggering of the antigen-receptor complex and are then sustained by stimuli from costimulatory molecules such as CD28. Cytokines such as IL2 and IL15 can then further sustain intracellular concentrations of PIP₃. Similarly, in B cells, cytokines such as BAFF and low-level signaling by non-antigen-engaged BCRs maintain a low level of PIP₃ (Srinivasan et al. 2009). The levels of PIP₃ increase following BCR activation, and costimulatory molecules such as CD19 and cytokines such as IL4 can also sustain levels of this lipid.

Antigen receptor and cytokines control PIP₃ metabolism in lymphocytes via class I PI3Ks, which typically exist in a complex comprising a p110 catalytic subunit and an 85-kDa SH2-domain-containing regulatory/adaptor subunit. Four p110 isoforms exist (α, β, γ, and δ) and two p85 subunits (α and β) exist. These different isoforms function in distinct pathways in lymphocytes, and expression of p110δ is restricted to hematopoietic cells. p110δ produces the PIP₃ that is generated in response to many antigen receptors and cytokines, whereas p110γ, which heterodimerizes with the p101 regulatory subunit rather than a p85-type subunit, is involved in chemokine receptor signaling (Okkenhaug and Fruman 2010).

The production of PIP₃ requires recruitment of PI3K to the plasma membrane. There are two possible mechanisms: binding of the SH2 domain of p85 to phosphorylated tyrosine residues in receptor cytoplasmic domains or membrane-localized adaptors; and direct recruitment of p110 by Ras. In the case of the BCR, CD19 recruits PI3K to the plasma membrane via binding of p85 to its tyrosine-phosphorylated cytoplasmic domain. Tyrosine-phosphorylated cytokine receptors similarly recruit PI3K by binding p85. Surprisingly, how TCR and CD28 signaling induces PIP₃ accumulation is not known, but direct recruitment to tyrosine-phosphorylated CD28 does not occur, and it is more likely that adaptors such as LAT or SLP76 are important.

PIP₃ binds to pleckstrin homology (PH) domains in other signaling proteins to control their activity and subcellular localization. In lymphocytes, these include Tec-family tyrosine kinases such as Itk and Btk, GEFs for Rho family GTPases, and the kinases PDK1 and Akt (also known as PKB) (p. 87 [Hemmings and Restuccia 2012]). Akt is activated by PDK1-mediated phosphorylation of T308 within its catalytic domain. This is PIP₃ dependent probably because the binding of PIP₃ to the Akt PH domain causes a conformational change that allows PDK1 to phosphorylate T308. PDK1 also has a PIP₃-binding PH domain, but this promotes translocation of the enzyme to the plasma membrane (where it can colocalize with Akt) rather than enzyme activation (Finlay and Cantrell 2011).

Once activated, Akt phosphorylates a number of critical signaling molecules. For example, it phosphorylates and

inactivates the Rheb GAP TSC2, causing accumulation of Rheb-GTP complexes, which play a role in activating the mTORC1 complex (mammalian target of rapamycin complex 1) (p. 91 [Laplante and Sabatini 2012]). Akt also phosphorylates the transcription factors Foxo1/3 and Fox4A. These Foxo family transcription factors are nuclear and active in quiescent cells but, when phosphorylated, they exit the nucleus and form a complex with 14-3-3 proteins in the cytosol, which terminates their transcriptional activity.

Akt is fundamentally important in many cells because it controls nutrient uptake and cellular metabolism. In particular, activated lymphocytes up-regulate glucose, amino acid and iron uptake, and switch their metabolism to glycolysis (see Ch. 7 [Ward and Thompson 2012]). This increases cellular energy production and nutrient uptake to support the increased biosynthetic demands of rapid cell proliferation. Note, however, that it is difficult to ascribe a universal function for Akt that holds for all lymphocyte subpopulations. For example, Akt is important for metabolism and cell survival in peripheral B lymphocytes (Srinivasan et al. 2009) and in T lymphocyte progenitors in the thymus, but is not essential for metabolism or for the survival of peripheral or effector cytotoxic T cells (Finlay and Cantrell 2011). Moreover, the Akt/Foxo pathway has a critical role controlling expression of the recombinase genes responsible for antigen-receptor diversity in B cells (Kuo and Schlissel 2009) but there is no evidence for such a role in T cells. The molecular basis for these differences is not understood but probably reflects redundancies with other kinases that have similar substrate specificities (e.g., SGK1).

Akt/Foxo signaling is also uniquely linked to the regulation of the expression of key cytokine and chemokine receptors and adhesion molecules in lymphocytes (Hedrick 2009; Lorenz 2009; Macintyre et al. 2011). Hence, when Akt is inactive in quiescent lymphocytes, nonphosphorylated Foxo1, Foxo3, Foxo3A, and Foxo4 are found in the nucleus, where they drive transcription of genes encoding the receptor for IL7, an essential homeostatic cytokine for lymphocytes. Moreover, Foxo transcription factors also drive expression of the transcription factor KLF2; this directly regulates transcription of adhesion molecules and chemokine receptors that together control lymphocyte entry and egress from secondary lymphoid tissues and lymphocyte positioning in lymphoid tissue. The activation of Akt thus causes lymphocytes to change their trafficking program around the body. Akt activation also changes the cytokine-receptor profile of T cells and hence the ability of cytokines to determine T-cell fate.

In many cells, a key role for Akt is to control the activity of the mammalian target of rapamycin complex 1 (mTORC1) (p. 91 [Laplante and Sabatini 2012]). Rapamycin is a powerful immunosuppressant that is used in the clinic to prevent rejection of organ transplants. mTORC1 coordinates inputs from nutrients and antigen and cytokine receptors to control T-cell differentiation (Powell and Delgoffe 2010). The molecular mechanisms used by mTORC1 to control T-cell differentiation are not fully understood; neither are the signaling processes that activate mTORC1. There is, however, evidence that mTORC1 controls expression of genes encoding effector cytokines and cytolytic molecules. Moreover, mTORC1 directs the tissue-homing properties of T cells by regulating the expression of chemokine and adhesion receptors (Sinclair et al. 2008).

13 INHIBITORY SIGNALS AND LYMPHOCYTE ACTIVATION

Signals from APCs and other immune cells can also deliver inhibitory signals to lymphocytes to ensure immune homeostasis. Indeed these are vital for a balanced immune response because a failure to limit immune responses results in excessive inflammation and potentially autoimmunity. Examples of signaling molecules that mediate key negative-feedback pathways in lymphocytes include SHIP, a lipid phosphatase with specificity for the $5'$ position of PIP_3 (Parry et al. 2010). SHIP is recruited to the plasma membrane by the binding of its SH2 domain to a tyrosine-phosphorylated immune cell tyrosine-based inhibitory motif (ITIM) located in the cytosolic domain of cell-surface receptors and dampens production of PIP_3. A prototypical example of this feedback process occurs in B cells when coligation of the BCR with the $Fc\gamma RIIB$ by antigen-antibody complexes results in tyrosine phosphorylation of the ITIM in $Fc\gamma RIIB$ (Daëron and Lesourne 2006). SHIP binds to the phosphorylated ITIM, thereby recruiting this inositol $5'$ phosphatase into the BCR-$Fc\gamma RIIB$ complex. SHIP dephosphorylates PIP_3 to produce $PI(3,4)P_2$ and, accordingly, diminishes the BCR-dependent elevation of intracellular PIP_3 levels. There are many other examples of ITIM-containing receptors that play an important role in immune homeostasis. For example, an extensive family of sialic-acid-binding immunoglobulin-like lectins, siglecs, responds to sialylated glycans to regulate lymphocyte function (Nitschke 2009; Cao and Crocker 2011). Siglecs are key regulators of B cell, NK cell, and macrophage biology.

In T cells, transmembrane receptors such CTLA4 and PD1 are critical for limiting T-cell function during immunity and tolerance (Veillette et al. 2002; Francisco et al. 2010; Bour-Jordan et al. 2011). The purpose of PD1 signaling is to limit the expansion of effector T cells during an immune response and hence to limit the pathology and tissue damage associated with effector $CD8^+$ T-cell-mediated tissue destruction. However, the failure to control chronic viral infections such as HIV results from inhibitory-receptor-driven exhaustion of antigen-specific T cells, demonstrating how the balancing of positive- and negative-feedback signaling needs to be finely tuned to ensure a favorable outcome. The impact of any imbalance of these pathways on human health is enormous: a failure of feedback control leads to autoimmunity; too much feedback control can limit the ability of the immune system to clear the pathogen.

B lymphocytes express the siglec family member CD22 (also known as siglec2), which inhibits B-cell signaling and B-cell-mediated autoimmunity by recruiting SHP1 (Lorenz 2009; Nitschke 2009). CD22 interacts with ligands carrying $\alpha 2-6$-linked sialic acids both in *cis* and in *trans* to modulate the BCR signaling threshold. The importance of SHP1 is strikingly illustrated by the phenotype of the moth-eaten (me/me) mouse, which lacks SHP1 tyrosine phosphatase activity and displays a variety of hematopoietic and immune disorders that result in death two or three weeks after birth (Lorenz 2009).

There are additional key negative regulator receptors in which there is either no classical ITIM or controversy as to the importance of the recruitment of phosphatases. CTLA4 is an example. It is an essential negative regulator of T-cell-mediated immune responses: CTLA4-deficient mice show a fatal lymphoproliferative disorder. CTLA4 binds the same two ligands (CD80 and CD86) the costimulatory molecule CD28 binds. The engagement of CD28 by CD80 or CD86 results in T-cell costimulation, whereas CTLA4 engagement results in inhibition of T-cell activation. CTLA4 might deliver a negative signal to the T cell by recruiting tyrosine phosphatases to the plasma membrane. However, two other models exist. One proposes that CTLA4 activates T cells to increase their motility and that this prevents T cells from making stable contacts with APCs (Rudd 2008). The other proposes that CTLA4 competes with CD28 for ligand but binds to CD80/86 with higher avidity than does CD28. Indeed CTLA4 has now been shown to capture its ligands CD80 and CD86 by *trans*-endocytosis (Qureshi et al. 2011). It could thus inhibit CD28 costimulation by depleting CD28 ligands. These models are not necessarily mutually exclusive, and how CTLA4 and the other inhibitory molecules exert essential feedback control is still the subject of much debate.

14 CONCLUDING REMARKS

In lymphocytes, signal inputs generated by specific pathogens regulate the activity of evolutionarily conserved signaling pathways. Antigen receptors direct the immune response but lymphocyte signaling is also controlled by cytokines and chemokines that are not antigen specific. These antigen-specific and -nonspecific elements of lymphocyte signal transduction are tightly coupled because antigen-receptor signaling controls the repertoire of cytokine and chemokine receptors and adhesion molecules expressed by lymphocytes. Antigen receptors also direct lymphocyte trafficking between the blood, peripheral tissues, and secondary lymphoid organs and hence control the cytokine milieu available to these cells. This coordination of antigen receptor and cytokine signaling ensures the immune specificity of lymphocyte activation and is fundamental for adaptive immune responses.

REFERENCES

*Reference is in this book.

Abram CL, Lowell CA. 2007. The expanding role for ITAM-based signaling pathways in immune cells. *Sci STKE* **2007:** re2.

Alarcón B, Swamy M, van Santen HM, Schamel WW. 2006. T-cell antigen-receptor stoichiometry: Pre-clustering for sensitivity. *EMBO Rep* **7:** 490–495.

Alvarez-Errico D, Lessmann E, Rivera J. 2009. Adapters in the organization of mast cell signaling. *Immunol Rev* **232:** 195–217.

Artyomov MN, Lis M, Devadas S, Davis MM, Chakraborty AK. 2010. CD4 and CD8 binding to MHC molecules primarily acts to enhance Lck delivery. *Proc Natl Acad Sci* **107:** 16916–16921.

Au-Yeung BB, Deindl S, Hsu LY, Palacios EH, Levin SE, Kuriyan J, Weiss A. 2009. The structure, regulation, and function of ZAP-70. *Immunol Rev* **228:** 41–57.

Beaven MA, Metzger H. 1993. Signal transduction by Fc receptors: The FcεRI case. *Immunol Today* **14:** 222–226.

Blonska M, Lin X. 2009. CARMA1-mediated NF-κB and JNK activation in lymphocytes. *Immunol Rev* **228:** 199–211.

* Bootman MD. 2012. Calcium signaling. *Cold Spring Harb Perspect Biol* **4:** a011171.

Bour-Jordan H, Esensten JH, Martinez-Llordella M, Penaranda C, Stumpf M, Bluestone JA. 2011. Intrinsic and extrinsic control of peripheral T-cell tolerance by costimulatory molecules of the CD28/B7 family. *Immunol Rev* **241:** 180–205.

Cantrell DA. 2003. GTPases and T cell activation. *Immunol Rev* **192:** 122–130.

Cao H, Crocker PR. 2011. Evolution of CD33-related siglecs: Regulating host immune functions and escaping pathogen exploitation? *Immunology* **132:** 18–26.

Carter RH, Fearon DT. 1992. CD19: Lowering the threshold for antigen receptor stimulation of B lymphocytes. *Science* **256:** 105–107.

Chakraborty AK, Das J, Zikherman J, Yang M, Govern CC, Ho M, Weiss A, Roose J. 2009. Molecular origin and functional consequences of digital signaling and hysteresis during Ras activation in lymphocytes. *Sci Signal* **2:** t2.

Chow LM, Veillette A. 1995. The Src and Csk families of tyrosine protein kinases in hemopoietic cells. *Semin Immunol* **7:** 207–226.

Chu DH, Morita CT, Weiss A. 1998. The Syk family of protein tyrosine kinases in T-cell activation and development. *Immunol Rev* **165:** 167–180.

Daëron M, Lesourne R. 2006. Negative signaling in Fc receptor complexes. *Adv Immunol* **89:** 39–86.

Das J, Ho M, Zikherman J, Govern C, Yang M, Weiss A, Chakraborty AK, Roose JP. 2009. Digital signaling and hysteresis characterize ras activation in lymphoid cells. *Cell* **136:** 337–351.

Davis MM. 2004. The evolutionary and structural "logic" of antigen receptor diversity. *Semin Immunol* **16:** 239–243.

Depoil D, Fleire S, Treanor BL, Weber M, Harwood NE, Marchbank KL, Tybulewicz VL, Batista FD. 2008. CD19 is essential for B cell activation by promoting B cell receptor-antigen microcluster formation in response to membrane-bound ligand. *Nat Immunol* **9:** 63–72.

Dustin ML, Chakraborty AK, Shaw AS. 2010. Understanding the structure and function of the immunological synapse. *Cold Spring Harb Perspect Biol* **2:** a002311.

Elgueta R, Benson MJ, de Vries VC, Wasiuk A, Guo Y, Noelle RJ. 2009. Molecular mechanism and function of CD40/CD40L engagement in the immune system. *Immunol Rev* **229:** 152–172.

Finlay D, Cantrell D. 2011a. The coordination of T-cell function by serine/threonine kinases. *Cold Spring Harb Perspect Biol* **3:** a002261.

Finlay D, Cantrell D. 2011b. Metabolism, migration and memory in cytotoxic T cells. *Nat Rev Immunol* **11:** 109–117.

Francisco LM, Sage PT, Sharpe AH. 2010. The PD-1 pathway in tolerance and autoimmunity. *Immunol Rev* **236:** 219–242.

Gallo EM, Canté-Barrett K, Crabtree GR. 2006. Lymphocyte calcium signaling from membrane to nucleus. *Nat Immunol* **7:** 25–32.

Gerondakis S, Siebenlist U. 2012. Roles of the NF-κB pathway in lymphocyte development and function. *Cold Spring Harb Perspect Biol* **2:** a000182.

Ghoreschi K, Laurence A, O'Shea JJ. 2009. Janus kinases in immune cell signaling. *Immunol Rev* **228:** 273–287.

Harwood NE, Batista FD. 2011. The cytoskeleton coordinates the early events of B-cell activation. *Cold Spring Harb Perspect Biol* **3:** a002360.

* Harrison DA. 2012. The JAK/STAT pathway. *Cold Spring Harb Perspect Biol* **4:** a011205.

Hayday AC. 2009. γδ T cells and the lymphoid stress-surveillance response. *Immunity* **31:** 184–196.

Hedrick SM. 2009. The cunning little vixen: Foxo and the cycle of life and death. *Nat Immunol* **10:** 1057–1063.

* Hemmings BA, Restuccia DF. 2012. PI3K-PKB/Akt pathway. *Cold Spring Harb Perspect Biol* **4:** a011189.

Hermiston ML, Zikherman J, Zhu JW. 2009. CD45, CD148, and Lyp/Pep: Critical phosphatases regulating Src family kinase signaling networks in immune cells. *Immunol Rev* **228:** 288–311.

Hogan PG, Lewis RS, Rao A. 2010. Molecular basis of calcium signaling in lymphocytes: STIM and ORAI. *Annu Rev Immunol* **28:** 491–533.

Im SH, Rao A. 2004. Activation and deactivation of gene expression by Ca^{2+}/calcineurin-NFAT-mediated signaling. *Mol Cells* **18:** 1–9.

Jenkins MR, Griffiths GM. 2010. The synapse and cytolytic machinery of cytotoxic T cells. *Curr Opin Immunol* **22:** 308–313.

Jordan MS, Koretzky GA. 2010. Coordination of receptor signaling in multiple hematopoietic cell lineages by the adaptor protein SLP-76. *Cold Spring Harb Perspect Biol* **2:** a002501.

Kambayashi T, Larosa DF, Silverman MA, Koretzky GA. 2009. Cooperation of adapter molecules in proximal signaling cascades during allergic inflammation. *Immunol Rev* **232:** 99–114.

Karin M, Gallagher E. 2009. TNFR signaling: Ubiquitin-conjugated TRAFfic signals control stop-and-go for MAPK signaling complexes. *Immunol Rev* **228:** 225–240.

Kim ST, Takeuchi K, Sun ZY, Touma M, Castro CE, Fahmy A, Lang MJ, Wagner G, Reinherz EL. 2009. The αβ T cell receptor is an anisotropic mechanosensor. *J Biol Chem* **284:** 31028–31037.

Kinashi T. 2005. Intracellular signalling controlling integrin activation in lymphocytes. *Nat Rev Immunol* **5:** 546–559.

Koretzky GA, Abtahian F, Silverman MA. 2006. SLP76 and SLP65: Complex regulation of signalling in lymphocytes and beyond. *Nat Rev Immunol* **6:** 67–78.

Krogsgaard M, Davis MM. 2005. How T cells "see" antigen. *Nat Immunol* **6:** 239–245.

Kuo T, Schlissel MS. 2009. Mechanisms controlling expression of the RAG locus during lymphocyte development. *Curr Opin Immunol* **21:** 173–178.

Kurosaki T, Hikida M. 2009. Tyrosine kinases and their substrates in B lymphocytes. *Immunol Rev* **228:** 132–148.

* Laplante M, Sabatini DM. 2012. mTOR signaling. *Cold Spring Harb Perspect Biol* **4:** a011593.

⋆ Lim K-H, Staudt LM. 2013. Toll-like receptor signaling. *Cold Spring Harb Perspect Biol* **5:** a011247.

Lorenz U. 2009. SHP-1 and SHP-2 in T cells: Two phosphatases functioning at many levels. *Immunol Rev* **228:** 342–359.

Love PE, Hayes SM. 2010. ITAM-mediated signaling by the T-cell antigen receptor. *Cold Spring Harb Perspect Biol* **2:** a002485.

Macintyre AN, Finlay D, Preston G, Sinclair LV, Waugh CM, Tamas P, Feijoo C, Okkenhaug K, Cantrell DA. 2011. Protein kinase B controls transcriptional programs that direct cytotoxic T cell fate but is dispensable for T cell metabolism. *Immunity* **34:** 224–236.

Mackay F, Figgett WA, Saulep D, Lepage M, Hibbs ML. 2010. B-cell stage and context-dependent requirements for survival signals from BAFF and the B-cell receptor. *Immunol Rev* **237:** 205–225.

Matthews SA, Cantrell DA. 2009. New insights into the regulation and function of serine/threonine kinases in T lymphocytes. *Immunol Rev* **228:** 241–252.

Matthews SA, Navarro MN, Sinclair LV, Emslie E, Feijoo-Carnero C, Cantrell DA. 2010. Unique functions for protein kinase D1 and protein kinase D2 in mammalian cells. *Biochem J* **432:** 153–163.

Mocsai A, Ruland J, Tybulewicz VL. 2010. The SYK tyrosine kinase: A crucial player in diverse biological functions. *Nat Rev Immunol* **10:** 387–402.

⋆ Morrison DK. 2012. MAP kinase pathways. *Cold Spring Harb Perspect Biol* **4:** a011254.

Muller MR, Rao A. 2010. NFAT, immunity and cancer: A transcription factor comes of age. *Nat Rev Immunol* **10:** 645–656.

⋆ Newton K, Dixit VM. 2012. Signaling in innate immunity and inflammation. *Cold Spring Harb Perspect Biol* **4:** a006049.

Nika K, Soldani C, Salek M, Paster W, Gray A, Etzensperger R, Fugger L, Polzella P, Cerundolo V, Dushek O, et al. 2010. Constitutively active Lck kinase in T cells drives antigen receptor signal transduction. *Immunity* **32:** 766–777.

Nitschke L. 2009. CD22 and Siglec-G: B-cell inhibitory receptors with distinct functions. *Immunol Rev* **230:** 128–143.

Oeckinghaus A, MS Hayden, Ghosh S. 2011. Crosstalk in NF-κB signaling pathways. *Nat Immunol* **12:** 695–708.

Oellerich T, Bremes V, Neumann K, Bohnenberger H, Dittmann K, Hsiao HH, Engelke M, Schnyder T, Batista FD, Urlaub H, et al. 2011. The B-cell antigen receptor signals through a preformed transducer module of SLP65 and CIN85. *EMBO J* **30:** 3620–3634.

Oh-hora M, Rao A. 2008. Calcium signaling in lymphocytes. *Curr Opin Immunol* **20:** 250–258.

Okkenhaug K, Fruman DA. 2010. PI3Ks in lymphocyte signaling and development. *Curr Top Microbiol Immunol* **346:** 57–85.

O'Shea JJ, Paul WE. 2010. Mechanisms underlying lineage commitment and plasticity of helper CD4+ T cells. *Science* **327:** 1098–1102.

Parry RV, Harris SJ, Ward SG. 2010. Fine tuning T lymphocytes: A role for the lipid phosphatase SHIP-1. *Biochim Biophys Acta* **1804:** 592–597.

Powell JD, Delgoffe GM. 2010. The mammalian target of rapamycin: Linking T cell differentiation, function, and metabolism. *Immunity* **33:** 301–311.

Quann EJ, Merino E, Furuta T, Huse M. 2009. Localized diacylglycerol drives the polarization of the microtubule-organizing center in T cells. *Nat Immunol* **6:** 627–635.

Qureshi OS, Zheng Y, Nakamura K, Attridge K, Manzotti C, Schmidt EM, Baker J, Jeffery LE, Kaur S, Briggs Z, et al. 2011. *Trans*-endocytosis of CD80 and CD86: A molecular basis for the cell-extrinsic function of CTLA-4. *Science* **332:** 600–603.

Rochman Y, Spolski R, Leonard WJ. 2009. New insights into the regulation of T cells by γ$_c$ family cytokines. *Nat Rev Immunol* **9:** 480–490.

Rudd CE. 2008. The reverse stop-signal model for CTLA4 function. *Nat Rev Immunol* **8:** 153–160.

Salmond RJ, Filby A, Qureshi I, Caserta S, Zamoyska R. 2009. T-cell receptor proximal signaling via the Src-family kinases, Lck and Fyn, influences T-cell activation, differentiation, and tolerance. *Immunol Rev* **228:** 9–22.

⋆ Samelson LE. 2011. Immunoreceptor signaling. *Cold Spring Harb Perspect Biol* **3:** a011510.

Schatz DG, Ji Y. 2011. Recombination centres and the orchestration of V(D)J recombination. *Nat Rev Immunol* **11:** 251–263.

Sinclair LV, Finlay D, Feijoo C, Cornish GH, Gray A, Ager A, Okkenhaug K, Hagenbeek TJ, Spits H, Cantrell DA. 2008. Phosphatidylinositol-3-OH kinase and nutrient-sensing mTOR pathways control T lymphocyte trafficking. *Nat Immunol* **9:** 513–521.

Spitaler M, Emslie E, Wood CD, Cantrell D. 2006. Diacylglycerol and protein kinase D localization during T lymphocyte activation. *Immunity* **24:** 535–546.

Springer TA, Dustin ML. 2011. Integrin inside-out signaling and the immunological synapse. *Curr Opin Cell Biol* **24:** 107–115.

Srinivasan L, Sasaki Y, Calado DP, Zhang B, Paik JH, DePinho RA, Kutok JL, Kearney JF, Otipoby KL, Rajewsky K. 2009. PI3 kinase signals BCR-dependent mature B cell survival. *Cell* **139:** 573–586.

Tolar P, Sohn HW, Liu W, Pierce SK. 2009. The molecular assembly and organization of signaling active B-cell receptor oligomers. *Immunol Rev* **232:** 34–41.

van der Merwe PA, Davis SJ. 2003. Molecular interactions mediating T cell antigen recognition. *Annu Rev Immunol* **21:** 659–684.

van der Merwe PA, Dushek O. 2011. Mechanisms for T cell receptor triggering. *Nat Rev Immunol* **11:** 47–55.

Veillette A. 2010. SLAM-family receptors: Immune regulators with or without SAP-family adaptors. *Cold Spring Harb Perspect Biol* **2:** a002469.

Veillette A, Latour S, Davidson D. 2002. Negative regulation of immunoreceptor signaling. *Annu Rev Immunol* **20:** 669–707.

Walker LS, Sansom DM. 2011. The emerging role of CTLA4 as a cell-extrinsic regulator of T cell responses. *Nat Rev Immunol* **11:** 852–863.

Wang H, Kadlecek TA, Au-Yeung BB, Goodfellow HE, Hsu LY, Freedman TS, Weiss A. 2010. ZAP-70: An essential kinase in T-cell signaling. *Cold Spring Harb Perspect Biol* **2:** a002279.

⋆ Ward PS, Thompson CB. 2012. Signaling in control of cell growth and metabolism. *Cold Spring Harb Perspect Biol* **4:** a006783.

Watts TH. 2005. TNF/TNFR family members in costimulation of T cell responses. *Annu Rev Immunol* **23:** 23–68.

Xiong N, Raulet DH. 2007. Development and selection of γδ T cells. *Immunol Rev* **215:** 15–31.

Yang J, Reth M. 2010a. The dissociation activation model of B cell antigen receptor triggering. *FEBS Lett* **584:** 4872–4877.

Yang J, Reth M. 2010b. Oligomeric organization of the B-cell antigen receptor on resting cells. *Nature* **467:** 465–469.

Zikherman J, Jenne C, Watson S, Doan K, Raschke W, Goodnow CC, Weiss A. 2010. CD45-Csk phosphatase-kinase titration uncouples basal and inducible T cell receptor signaling during thymic development. *Immunity* **32:** 342–354.

CHAPTER 17

Vertebrate Reproduction

Sally Kornbluth[1] and Rafael Fissore[2]

[1]Duke University School of Medicine, Durham, North Carolina 27710

[2]University of Massachusetts, Amherst, Veterinary and Animal Sciences, Amherst, Massachusetts 01003

Correspondence: sally.kornbluth@duke.edu

SUMMARY

Vertebrate reproduction requires a myriad of precisely orchestrated events—in particular, the maternal production of oocytes, the paternal production of sperm, successful fertilization, and initiation of early embryonic cell divisions. These processes are governed by a host of signaling pathways. Protein kinase and phosphatase signaling pathways involving Mos, CDK1, RSK, and PP2A regulate meiosis during maturation of the oocyte. Steroid signals—specifically testosterone—regulate spermatogenesis, as does signaling by G-protein-coupled hormone receptors. Finally, calcium signaling is essential for both sperm motility and fertilization. Altogether, this signaling symphony ensures the production of viable offspring, offering a chance of genetic immortality.

Outline

Cite this chapter as *Cold Spring Harb Perspect Biol* doi: 10.1101/cshperspect.a006064

1 INTRODUCTION

Mammalian reproduction depends on the proper development and maturation of both the female egg and the male sperm. These gametes fuse through a complex series of events, known as fertilization, that ensure the highest quality of offspring. Both gamete development and fertilization depend on numerous connected signaling pathways, a flaw in any of which can lead to infertility or birth defects.

The egg and sperm are haploid germ cells that, upon fertilization, reconstitute a diploid cell—the embryo. Production of haploid gametes from diploid precursors requires a modified cell cycle known as meiosis. Before meiosis, the full complement of parental chromosomes is first duplicated in S phase, to produce so-called sister chromatids (i.e., four copies of each chromosome per cell) and paternal and maternal chromosomes pair up. The homologous chromosomes from each chromosome pair are then separated in the first meiotic M phase (meiosis I, also known as MI). Subsequently, without further replication, the cells reenter M phase (meiosis II, also known as MII) to divide the sister chromatids equally into four haploid daughter cells. For male gametes, four mature sperm are generated, whereas in the female, a single final gamete (the egg) is produced together with three polar bodies. Movement through these stages of meiosis is carefully controlled by kinases, phosphatases, ubiquitin-dependent degradation of key regulators, and calcium flux.

Before acquiring the capacity to fertilize eggs, a sperm must reside in the female reproductive tract and undergo physiological changes that render it fertilization competent (Bedford 1970). The acquisition of fertilization competence and the biochemical, membrane, and enzymatic changes that underlie it are collectively known as capacitation (Austin 1951; Chang 1951). As with gamete development and maturation, capacitation and fertilization depend on careful regulation through signaling pathways. These include pathways involving gonadotropins, G-protein-coupled receptors (GPCRs), kinases, and calcium signaling (Salicioni et al. 2007).

2 OOCYTE MATURATION

Oocyte maturation has been most extensively studied in the frog *Xenopus laevis*, because its very large oocytes allow both physical manipulation of the cell (microinjection of proteins, RNAs, and antisense oligonucleotides) and observation of the progression through meiosis with the naked eye. Although some notable differences have been observed, genetic studies in mammals (primarily mouse) have revealed similar overall regulation of meiotic progression (Fig. 1).

Oocytes and sperm both begin life as primordial germ cells (PGCs) that migrate to the nascent gonads (ovaries in females, testes in males) in early embryonic development. Under the influence of a variety of cytokines and growth factors, PGCs that will become oocytes continue dividing mitotically within cell clusters. In oogenesis, the premeiotic S phase is followed by a prolonged arrest in prophase I of meiosis until sexual maturity. During this phase, the oocyte is maintained in a G_2-phase-arrested state through G-protein-coupled signaling (see below). When mitosis ceases, these oocytes each become surrounded by somatic granulosa and theca cells, which form the primordial follicles that serve as repositories of dormant oocytes for later ovulation. Oocytes nestled within the follicles grow and stockpile nutrients until they become competent to undergo maturation; upon receipt of appropriate hormonal signals, one follicle from the larger pool will mature fully during each menstrual cycle in the mammal. Stimulated by pituitary hormones (gonadotropins) and as a consequence of maturation-inducing steroid hormones (e.g., progesterone) synthesized by the ovarian follicle cells, the oocyte exits prophase arrest, and progresses through MI, transitioning promptly to MII without any intervening DNA replication. At MII, the oocyte arrests again awaiting fertilization.

The end product of oocyte maturation is a haploid egg capable of being fertilized. A strong MII arrest helps to prevent parthenogenesis, which is the aberrant entry of the haploid egg into the mitotic cell cycle in the absence of fertilization. In an effort to define the factors responsible for MII arrest, Masui and colleagues injected extract prepared from a mature M-phase-arrested frog egg into blastomeres formed after the first embryonic cell division (Masui and Markert 1971); injected cells remained arrested in M phase, whereas the uninjected cells continued to divide. These experiments helped to identify both maturation promoting factor (MPF), which drives entry into both MI and MII during oocyte development, and the cytostatic factor (CSF), which maintains MII arrest. We now know that MPF is equivalent to the complex of cyclin B and cyclin-dependent kinase 1 (cyclin-B–CDK1) that drives entry into mitosis in the somatic cell cycle (Dunphy et al. 1988; Labbe et al. 1989; Ch. 6 [Rhind and Russell 2012]). CSF was shown to be the kinase Mos, the cellular counterpart of the viral oncoprotein v-Mos, which is expressed primarily in germ cells (Propst et al. 1987; Sagata et al. 1989). Proper maturation from MI entry through MII arrest depends on tightly controlled temporal regulation of both cyclin-B–CDK1 and Mos activity.

2.1 Meiosis I

During MI, oocytes are maintained in the G_2-arrested state by high levels of cytosolic cAMP. Constitutive signaling by

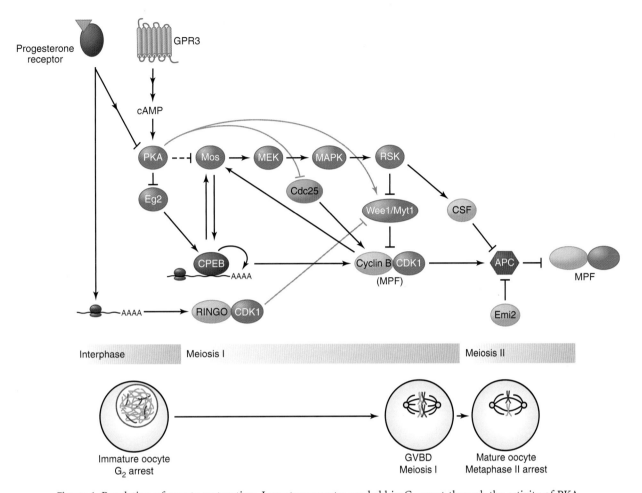

Figure 1. Regulation of oocyte maturation. Immature oocytes are held in G₂ arrest through the activity of PKA, which is stimulated by GPR3-dependent production of cAMP. Progesterone signaling leads to a loss of PKA activity, leading to disinhibition of the CDK1 activator Cdc25 and stimulation of the CDK1 inhibitor Wee1. Additionally, progesterone stimulates translation of both Mos and cyclin B proteins (the former also stimulating translation of the latter) through induction of Eg2, which phosphorylates and activates CPEB to unmask the messages and promote their polyadenylation. In parallel, progesterone stimulates translation of RINGO, which can activate CDK1 independently of cyclin and make it phosphorylate and inhibit the CDK1 inhibitor Myt1. All of these events result in activation of MPF, which drives the oocyte into MI. Maturation continues via the Mos-MEK-ERK pathway as shown. Blockade of APC (anaphase-promoting complex) activity by CSF and Emi2 holds the mature oocyte at a second arrest until the time of fertilization.

the GPCR GPR3 probably stimulates adenylyl cyclase to keep cAMP levels high. Indeed, overexpression of GPR3 in frog oocytes makes them resistant to the maturation effects of progesterone, and in mice lacking GPR3, oocytes mature in the absence of additional stimuli (Freudzon et al. 2005; Hinckley et al. 2005; Mehlmann 2005; Deng et al. 2008). Although sphingosine 1-phosphate and sphingosyl-phosphorylcholine have been proposed as GPR3 ligands, unliganded GPR3 appears to be able to stimulate adenylyl cyclase and thus the role of GPR3 in maintenance of G₂ arrest is not entirely clear (Eggerickx et al. 1995; Uhlen-brock et al. 2002; Hinckley et al. 2005). Progesterone stimulation seems to antagonize the GPR3 signal, triggering

a decrease in cAMP levels that is at least partly mediated by stimulation of phosphodiesterases that degrade cAMP (primarily PDE3 in the oocytes) (Tsafriri et al. 1996). This leads to a diminution of protein kinase A (PKA) activity. If PDE3 is artificially inhibited or cAMP synthesis is artificially stimulated, progesterone-induced maturation can be blocked. Conversely, injection of oocytes with the PKA inhibitor PKI can promote resumption of meiosis even without progesterone stimulation (Stanford et al. 2003).

As in the mitotic cell cycle, cyclin-B–CDK1 activity is controlled by phosphorylation of CDK1 on Y15 by Wee1-family kinases, which is opposed by the Cdc25 phosphatase (Watanabe et al. 1995; Berry and Gould 1996). The Wee1 relative Myt1 contributes to suppressing CDK1 phosphorylation, and the Cdc25c isoform mediates its subsequent dephosphorylation. In mice, this process is mediated by oocyte-specific isoforms: WEE1B and CDC25B. Loss of WEE1B irreversibly arrests oocytes in prophase (Han et al. 2005). In the G_2-arrested oocyte, PKA directly phosphorylates Cdc25 on S287 (*Xenopus* numbering), promoting the binding of the small acidic protein 14-3-3 (Duckworth et al. 2002). 14-3-3 interferes with the ability of Cdc25 to interact with and dephosphorylate cyclin-B–CDK1 and prevents its translocation into the nucleus, where it would promote rapid cyclin-B–CDK1 activation (Kumagai and Dunphy 1999; Lopez-Girona et al. 1999; Yang et al. 1999). Thus, the drop in PKA activity required for maturation promotes Cdc25 activation. PKA also phosphorylates and activates Wee1/Myt1 (Stanford and Ruderman 2005); so the drop in PKA also promotes CDK1 activation by alleviating its suppression by these kinases.

The formation and activity of MPF is also regulated by translation of cyclin B, whose mRNA is translationally dormant before the induction of oocyte maturation owing to its very short poly-A tail. At the time of oocyte maturation, *cis*-acting sequences within the 3′ UTR of the cyclin B mRNA promote cytoplasmic polyadenylation, elongating the tail more than 100 nucleotides. These cytoplasmic polyadenylation elements (CPEs) within the 3′ UTR of the mRNA are bound by CPE-binding protein (CPEB) (Hake and Richter, 1994). Through a process that is not entirely clear, the drop in PKA activity that heralds the onset of oocyte maturation also induces activation of a kinase, Eg2, which phosphorylates CPEB, activating it to both unmask the mRNA and recruit a poly(A) polymerase to elongate the poly(A) tail (Andresson and Ruderman 1998; Frank-Vaillant et al. 2000; Hodgman et al. 2001).

Note that in every species there is at least some preformed cyclin-B–CDK1 complex (known as pre-MPF) whose activity is suppressed by phosphorylation of CDK1 at T14 and Y15. Indeed, the initial discovery of MPF relied on the ability of the injected MPF to mobilize the pre-MPF pool through autoamplification (Masui and Markert 1971; Drury and Schorderet-Slatkine 1975; Wasserman and Masui 1975). Phosphorylation by cyclin-B–CDK1 suppresses Wee1/Myt1 and activates Cdc25, which promotes more conversion of pre-MPF to MPF. In species in which most of the CDK1 is bound to cyclin B in pre-MPF complexes, new cyclin B translation is not absolutely required for induction of oocyte maturation; however, in those species that have low amounts of pre-MPF and high levels of free CDK1, cyclin B synthesis is an obligate step in maturation (Jagiello 1969; Fulka et al. 1986; Moor and Crosby 1986; Hunter and Moor 1987; Gautier and Maller 1991; Mattioli et al. 1991).

In mammals, the signal emanating from a loss of PKA activity may be conveyed directly to CDK1 via Wee1B, as this appears to be a direct PKA target. For Myt1, there is evidence for indirect pathways of inhibition. First, soon after progesterone treatment, a non-cyclin alternative activator of CDK1 known as RINGO is translated (Ferby et al. 1999). This protein can bind to and activate CDK1, causing it to phosphorylate and suppress Myt1 (Ruiz et al. 2008). This, in turn, leads to activation of cyclin-B–CDK1 complexes. A second pathway is an oocyte-specific MAP kinase (MAPK) cascade involving the MAPKKK Mos, the MAPKK MEK, and the MAPK ERK. In frogs, the terminal effector in this pathway is RSK, which can phosphorylate and inhibit Myt1 (Palmer et al. 1998). RSK can also phosphorylate Cdc25, contributing to its activation. Accordingly, injection of activated RSK into *Xenopus* oocytes can induce meiotic maturation and RSK inhibition interferes with progesterone-induced maturation. In mice, alternative pathways must operate (e.g., the direct inhibition of WEE1B by PKA, as described above), because mice lacking all known isoforms of RSK do not show defects in oocyte maturation (Dumont et al. 2005).

In some species, including *Xenopus*, activation of the Mos-ERK pathway precedes completion of MI and breakdown of the nuclear envelope (known as germinal vesicle breakdown [GVBD] in oocytes). In others, it occurs after GVBD (because cyclin-B–CDK1 can actually activate ERK). Whether the Mos-MEK-ERK-RSK pathway is involved at MI depends on the organism, and Mos accumulation is controlled by multiple mechanisms (reviewed in Fan and Sun 2004). The 3′ end of the Mos mRNA in the immature oocyte has a short poly(A) tail whose elongation (necessary for efficient translation) is masked through binding of CPEB. In addition to the Eg2-induced phosphorylation of CPEB, which unmasks and enhances the translation of Mos (Mendez et al. 2000), the stability of Mos protein is greatly enhanced by phosphorylation at S3 as both dephosphorylation of this residue and the presence of a proline at residue 2 are required for recognition by the ubiquitin-proteasome degradation system (Nishizawa et al. 1993). This site is phosphorylated in a positive-feedback loop by ERK, which stabilizes Mos. Where Mos accumulates only after GVBD, it is phosphorylated at the same site by cyclin-B–CDK1. Together, increased translation and stabilization promote Mos accumulation during maturation. Note that, in *Xenopus*, redundant pathways allow MI progression even when Mos is ablated; yet, the kinetics are delayed, indicating that Mos normally enhances meiotic progression in this species. Indeed, when either cyclin B synthesis or Mos synthesis is impaired, progesterone-induced GVBD can proceed, but ablation of both abolishes this.

2.2 MI-MII Transition

Mos appears to be more widely important for MII. For example, unlike RSK-knockout mice, Mos-knockout mice are sterile and oocytes fail to mature properly (Colledge et al. 1994; Hashimoto et al. 1994). To understand the role of Mos, we must first consider cyclin B dynamics in meiosis. At the time of exit from MI, cyclin B must be degraded by the anaphase-promoting complex (APC), a multisubunit E3 ubiquitin ligase also known as the cyclosome. However, because cyclin-B–CDK1 inhibits formation of prereplicative complexes necessary for DNA replication, complete loss of cyclin-B–CDK1 kinase activity (as occurs in a somatic mitosis) would result in reinitiation of S phase. Thus, cyclin B translation is ramped up immediately after GVBD. Moreover, there must be sufficiently rapid reaccumulation of cyclin B to drive MII. Mos participates in two ways: it helps to drive cyclin B synthesis, and it helps control cyclin B degradation. This allows partial but not complete loss of cyclin B at MI exit—a decrease sufficient to exit MI but not initiate S phase—followed by unimpeded cyclin B accumulation to drive MII.

A key effector of this pathway is an inhibitor of the APC known as Emi2 (reviewed in Wu and Kornbluth 2008). Emi2 binds directly to the APC, inhibiting its ability to ubiquitylate substrates and so cause their proteasomal degradation (e.g., cyclin B). Emi2 is also regulated at the level of protein stability and is a substrate of another multisubunit E3 ubiquitin ligase, SCFβ–TrCP (Liu and Maller 2005; Rauh et al. 2005; Hansen et al. 2006). Recognition by this E3 ligase, which requires a phosphodegron in its targets, depends on phosphorylation of Emi2 at critical sites within the amino-terminal half of the protein, catalyzed by cyclin-B–CDK1. It is this feedback phosphorylation of Emi2 that helps to precisely control cyclin B levels. As cyclin B is synthesized, it binds to and activates its partner CDK1. Active cyclin-B–CDK1 complexes phosphorylate Emi2, promoting its degradation, thereby alleviating suppression of the APC to allow cyclin B degradation and MI exit. If Emi2 is artificially stabilized at this transition, it causes MI arrest by preventing cyclin B degradation (Fig. 2).

If cyclin-B–CDK1 phosphorylation of Emi2 were unopposed, then ultimately accumulation of cyclin B would eradicate Emi2, allowing complete, rather than the required partial, cyclin B degradation. However, the Mos-ERK pathway interferes at this point. Emi2 is phosphorylated directly by RSK, the kinase downstream from ERK. RSK-mediated phosphorylation of Emi2 promotes docking of the protein phosphatase PP2A on Emi2 (Wu et al. 2007b). PP2A, in turn, dephosphorylates the sites on Emi2 that are phosphorylated by cyclin-B–CDK1, thereby stabilizing Emi2 and restoring its inhibitory binding to the APC (Tang et al. 2008).

This allows the accumulation of cyclin B necessary for blocking S phase and, ultimately, for entry into MII. Mos may also promote inhibition of Myt1 and consequently T14 and Y15 dephosphorylation and activation of CDK1. This could help to activate cyclin-B–CDK1 as cyclin B accumulates. As MI is completed, cyclin B synthesis is markedly enhanced; eventually, this exceeds the ability of the APC to keep pace (even without Emi2 inhibition), cyclin B is degraded, and MII ensues. Although the general role of Emi2 appears to be conserved in mammals, the pathway appears to be somewhat different from that in frogs; because RSK is dispensable, another downstream target of Mos, MEK or ERK, probably phosphorylates Emi2 and recruits PP2A.

2.3 MII Arrest

Once MII is initiated, the oocyte arrests again, this time as a mature egg awaiting fertilization. Mos and Emi2 are also central to the activity that maintains this arrest (Kanki and Donoghue 1991; Hashimoto et al. 1994; Dupre et al. 2002; Madgwick et al. 2006; Ohe et al. 2007). The critical nature of Mos is clear because removal of Mos from egg extracts by immunodepletion destroys CSF activity (Daar et al. 1991), and eggs from mice or frogs lacking Mos fail to arrest in MII (Colledge et al. 1994; Hashimoto et al. 1994; Araki et al. 1996). The target of the Mos-MAPK cascade that produces the MII arrest is again the APC. Indeed, when radiolabeled cyclin B is injected into *Xenopus* eggs, treatment with the MEK inhibitor UO126 promotes cyclin B degradation because the APC inhibition mediated by the ERK MAPK pathway is lifted (Gross et al. 2000). Inhibition of the APC in a CSF-arrested egg requires Emi2, which is targeted by the ERK MAPK pathway during MI and MII (Tung et al. 2005). Loss of Emi2 in either frog or mouse eggs prevents MII arrest and allows parthenogenetic divisions.

Accumulation of cyclin B must be carefully regulated to maintain MII arrest. Cyclin B is synthesized continuously. If unopposed, this would make it difficult to achieve the rapid degradation of cyclin B required for a sharp cell-cycle transition upon fertilization. Again, tight regulation of Emi2 occurs through a negative-feedback loop involving its phosphorylation by cyclin-B–CDK1 and RSK. The carboxy-terminal Emi2 phosphorylations impede association of Emi2 with the APC; although the precise manner in which Emi2 inhibits the APC is not yet clear, the physical association of Emi2 with the APC is critical for APC inhibition and the amino-terminal phosphorylations dissociate Emi2 from the APC, allowing cyclin B degradation (Wu et al. 2007a). Thus, the Mos-ERK-RSK pathway maintains the CSF arrest, restricting cyclin B levels within a narrow limit. Although the precise sites of phosphorylation do not appear to be conserved from *Xenopus* to mammals,

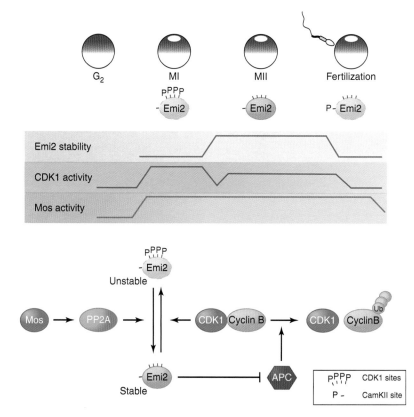

Figure 2. Regulation of Emi2 and the APC during the MI–MII transition. Phosphorylation controls Emi2 stability during oocyte maturation. At MI, CDK1 phosphorylates four amino-terminal sites (S213, T239, T252, and T267) on Emi2; this triggers Emi2 degradation, required for MI exit. At MI anaphase, cyclin B is degraded, leading to a drop in CDK1 activity. Emi2 is stabilized by dephosphorylation triggered by Mos signaling. Emi accumulates, resulting in APC inhibition, critical for S phase block and MII entry. CDK1 activity is low in MII relative to MI. Emi2 is stable in MII, as required for CSF arrest. At fertilization, Emi2 is quickly degraded through a CaMKII-mediated pathway, allowing activation of the APC and exit from MII. At the onset of MI anaphase, APC-mediated cyclin B degradation results in decreased CDK1 activity. With the Mos-PP2A pathway predominant, dephosphorylated and stabilized Emi2 protein prevents complete ubiquitylation of cyclin B by the APC. This is essential for the inhibition of S phase between MI and MII. (From Tang et al. 2008; adapted, with permission.)

the overall mode of regulation may be conserved through phosphorylation of alternative sites.

3 SPERM MATURATION

Spermatogenesis occurs over the course of several weeks and encompasses three successive phases (Sharpe 1994): proliferation, meiosis, and differentiation. During proliferation, spermatogonial stem cells (SSCs) differentiate into spermatogonia. These undergo several mitotic divisions, giving rise to spermatocytes. After two meiotic divisions, spermatocytes form haploid spermatids. The final transformation of spermatids into mature sperm entails a major physical and structural reorganization of the cell that is known as spermiogenesis (Fig. 3).

As oocyte maturation is supported by follicular cells, so is spermatogenesis supported by nongermline cells in and around the seminiferous tubules. These include Sertoli, Leydig, and myoid cells (Hermo et al. 2010). The Sertoli cells perform a plethora of functions to sustain germ cells through all stages of development. They express receptors for growth factors and hormones and secrete many essential regulatory factors required for spermatogenesis. Lying outside the tubules in the interstitial space and in close proximity to blood vessels, Leydig cells are responsible for the regulated production of androgens in the testis. Lastly, in the peritubular tissue, along with extracellular matrix, myoid cells support the seminiferous tubules (Yoshida et al. 2007; de Rooij 2009). These cells create a microenvironment that allows SSCs to self-renew and/or differentiate, as is required for continual fertility (Nalam and Matzuk 2010).

Complete spermatogenesis requires the coordinated action of peptide and steroid hormones, which are important

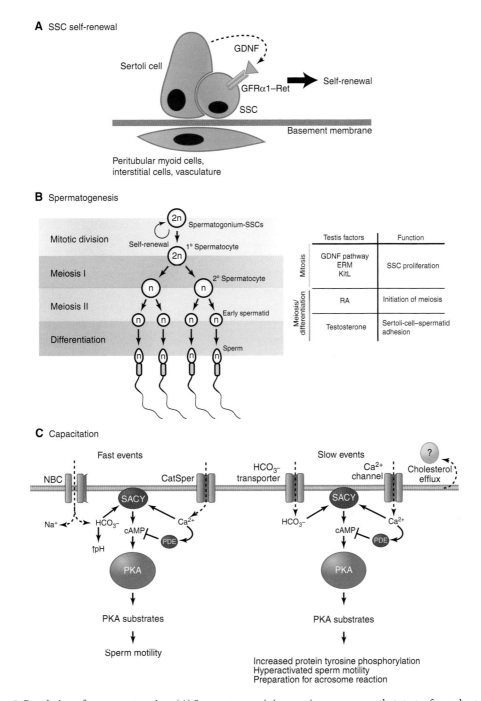

Figure 3. Regulation of sperm maturation. (*A*) Spermatogenesis is a continuous process that starts after puberty. This is possible because a subset of spermatogonium, spermatogonial stem cells (SSCs), are capable of self-renewal. Glial-cell-line-derived neurotrophic factor (GDNF) and the receptor complex composed of Ret protooncoprotein and the GDNF family receptor α1 (GFRα1) are key signaling events for self-renewal. (From Hofmann 2008; adapted, with permission.) (*B*) Spermatocytes undergo a series of maturation steps before differentiating into sperm. These are tightly regulated by interaction with Sertoli cells and testis factors such as GDNF, ERM, KitL, and retinoic acid (RA). (*C*) Sperm must additionally undergo capacitation before they are capable of fertilization. This includes both fast and slow events. The fast events stimulate flagellar activity through a calcium flux that is controlled by the adenylyl cyclase SACY and mediated by the CatSper and sodium/bicarbonate (NBC) channels. Slow events increase motility and introduce changes that prepare sperm for fertilization. These include increases in both intracellular calcium levels and tyrosine phosphorylation of PKA substrates. cAMP levels are also controlled by endogenous phosphodiesterase (PDE). Unknown cholesterol acceptors present in uterine/oviduct fluids mediate cholesterol efflux during capacitation. (From Visconti 2009; adapted, with permission.)

regulators of seminiferous tubule function (McLachlan et al. 2002). Only somatic cells in the testis express hormone receptors; therefore these cells (i.e., Sertoli cells) are the exclusive mediators of hormone activity in spermatogenesis. Of the two gonadotropins, follicle-stimulating hormone (FSH) and luteinizing hormone (LH), which also play important roles during oogenesis (Richards and Pangas 2010), FSH has broader involvement in spermatogenesis whereas LH functions primarily in the regulation of testosterone production by Leydig cells (Ruwanpura et al. 2010). However, both gonadotropins exert their biological effects by activating cognate GPCRs. In mice, following formation of the seminiferous cords, gonocytes, the precursors of SSCs, proliferate until day E15–16, when they become quiescent, which is coincident with changes in the expression of several cell-cycle proteins (van den Ham et al. 2003). Formation and proliferation of SSCs is observed during the first postnatal week and continues with the synchronous first wave of spermatogenesis that extends in the prepubertal testis over the first 35 days after birth (Itman et al. 2006).

3.1 Stem Cell Proliferation and Maintenance

The predominant role of FSH is regulating Sertoli cell proliferation during prepubertal development (Holdcraft and Braun 2004). Stimulation of the FSH receptor on Sertoli cells activates several downstream signaling cascades, including those involving cAMP, calcium release, ERK, PI3K, and phospholipases A2 and C. These activate the cAMP-responsive transcription factor CREB, leading to stimulation of gene expression (Ruwanpura et al. 2010). Testosterone produced in response to LH mostly functions via the nuclear androgen receptor (AR) and its subsequent stimulation of gene expression (Wang et al. 2009; p. 129 [Sever and Glass 2013]). It may also act via alternative mechanisms, which also result in phosphorylation of CREB. These "nongenomic" pathways are thought to involve the activation of several kinases, including Src (Cheng et al. 2007). Importantly for proliferation of both Sertoli cells and SSCs, FSH stimulation increases expression and secretion of glial-cell-line-derived neurotrophic factor (GDNF) (Hu et al. 1999), a distant member of the transforming growth factor β (TGFβ) family superfamily that regulates the proliferation of uncommitted SSCs (Meng et al. 2000; Hermo et al. 2010).

GDNF promotes the activation and expression of many signaling molecules, including the transcription factors Fos and BCL6. GDNF also stimulates Ras, Akt, and Src-family kinases; Akt activation, specifically, is important for preventing apoptosis of SSCs (Sariola and Saarma 2003; Hermo et al. 2010; Oatley and Brinster 2012). These signals are

initiated following its engagement of a receptor complex composed of the Ret proto-oncoprotein (McGuinness et al. 1996) and the GDNF family receptor α1 (GFRα1) protein (Meng et al. 2000). GFRA1 appears to be expressed only by a subpopulation of SSCs, possibly a marker of true "stemness." This is consistent with the effects of its downregulation, which causes widespread inhibition of SSC proliferation and differentiation (reviewed in Nalam and Matzuk 2010). SSCs lacking GFRα1 display reduced phosphorylation of Ret at Y1062, a known binding site for several of the downstream targets of this pathway. Additionally, loss of either member of the receptor complex leads to defective SSC proliferation and differentiation (Naughton et al. 2006). The GDNF pathway is thus essential for the maintenance of uncommitted SSCs.

SSC renewal requires the transcription factor Ets variant gene 5 (ERM) a product of the Sertoli cells. Deletion of ERM in mice results in inhibition of spermatogenesis following the first wave (Chen et al. 2005). Thus, although GDNF plays a critical role in spermatogenesis during the perinatal period, ERM does so at puberty. Expression of ERM and GDNF is increased in Sertoli cells by addition of the fibroblast growth factor (FGF) 2 and activation of the FGF2 receptor, which stimulates ERK and PI3K signaling. Therefore, FGFs may also play an important role in regulating the functions of Sertoli cells that control the establishment of the SSC pool (Simon et al. 2007).

Another factor regulating proliferation and differentiation of SSCs is the protooncoprotein Kit, a receptor tyrosine kinase expressed at high levels in differentiating spermatogonia and early spermatocytes (Sorrentino et al. 1991). In undifferentiated spermatogonia, Kit expression is suppressed by the transcription factor Plzf. Plzf is necessary for the maintenance of undifferentiated SSCs. Plzf-null mice display accumulation of Kit-positive cells and depletion of germ cells (Buaas et al. 2004; Costoya et al. 2004), and dominant white spotting (W) mutants affecting the Kit locus inhibit spermatogonia differentiation without affecting either mitosis of undifferentiated SSCs or initiation of meiosis in spermatocytes (Yoshinaga et al. 1991). Furthermore, in the postnatal testis, expression of the Kit ligand (KitL, also known as stem cell factor [SCF]), in Sertoli cells is critical for spermatogenesis (Flanagan et al. 1991), and mutations in KitL phenocopy the spermatogenic defects observed in W mutant mice (Bedell and Mahakali Zama 2004). The binding of KitL to Kit causes receptor dimerization and phosphorylation of cytoplasmic tyrosine residues. These phosphorylated residues become anchoring sites for SH2-domain containing proteins such as phospholipase Cγ (PLCγ), Src, and PI3K. Consequent induction of calcium release, phosphorylation of target proteins such as p70S6K (Feng et al. 2000), and gene expression leads to proliferation and differentiation of spermatogonia (Sette et al. 2000). Note that spermatogonia from Kit-null testes can nevertheless undergo spermatogenesis, which indicates other parallel pathways must also operate (Nalam and Matzuk 2010).

3.2 Spermatocyte Meiosis and Release

After differentiation, spermatogonia undergo a few mitotic divisions before entering meiosis, which is delayed until puberty. Retinoic acid (RA) (see p. 129 [Sever and Glass 2013]) is a key signaling molecule in meiotic initiation (Vernet et al. 2006), and Sertoli cells are believed to be the main source of RA in the postnatal testes. Importantly, although RA is generated in Sertoli cells by the enzyme aldehyde dehydrogenase family 1, subfamily A1 (ALDH1A1), male gonads also express an enzyme responsible for RA degradation, cytochrome P450, family 26, subfamily b polypeptide 1 (CYP26B1) (Bowles et al. 2006). This limits the availability of RA and postpones meiosis in the male.

Hormonal signaling plays an important role in control of meiosis and production of functional spermatids. Suppression of FSH production and signaling leads to reduced numbers of pachytene spermatocytes (Matthiesson et al. 2006) and compromises the release of sperm, which suggests that FSH regulates adhesion between Sertoli cells and spermatids (Saito et al. 2000; Ruwanpura et al. 2010). Spermatocyte meiosis is arguably even more dependent on testosterone signaling. In mice lacking AR in Sertoli cells spermatogenesis is arrested at the late spermatocyte stage and spermiation, which is the release of the mature spermatids into the lumen of the seminiferous tubules, fails (Chang et al. 2004; De Gendt et al. 2004). Moreover, in adult mice that lack both gonadotropins owing to a defect in the *GnRH* gene, testosterone supplementation alone can restore full spermatogenesis (Haywood et al. 2003). The processes modulated by testosterone during spermiogenesis include adhesion between Sertoli cells and different stage spermatids. Testosterone is thought to regulate expression and/or function of proteins required for the assembly and/or disassembly of adhesion junctions, including α and β integrins, focal adhesion kinases, and Src kinases (Ruwanpura et al. 2008, 2010; Shupe et al. 2011).

3.3 Sperm Capacitation and Calcium Channels

After entering the oviduct, most mammalian sperm associate with epithelial cells in the isthmus creating a sperm reservoir. Sperm are sporadically released from this reservoir and migrate to the ampulla, which is the site of fertilization (Suarez 2008b). The mechanisms that facilitate this release are unclear but they involve the loss of BSPs (originally isolated from the cow, so named bovine seminal plasma proteins) and other extrinsic proteins from the sperm; these events are seemingly precipitated by the progression of capacitation events, including hyperactivation (a change in sperm motility) (Suarez 2008a). Unfolding over several hours, sperm capacitation begins as sperm come in contact with the female reproductive tract and involves a series of sequential and simultaneous processes. Sperm acquire motility immediately after their release from the cauda epididymis (for reviews see Yanagimachi 1994; Salicioni et al. 2007). These subsequent changes involve both modification of the plasma membrane phospholipid composition and increases in intracellular calcium concentration. Early in capacitation, activation of soluble adenylyl cyclase (SACY) (Chen et al. 2000) causes an increase in cAMP levels, which activates PKA containing a unique catalytic subunit (Cα2 or Cs). These events are initiated by an increase in the intracellular concentration of bicarbonate, which activates the enzyme by promoting closure of the catalytic active site and metal recruitment (Steegborn et al. 2005). This is itself triggered by the high concentrations of bicarbonate in the seminal fluid, which enters via the Na^+/HCO_3^- cotransporter. PKA activation coincides with phosphorylation of a number of proteins, although its targets remain mostly unknown (Signorelli et al. 2012) except for FSCB, a 270-kDa protein involved in the biogenesis of the fibrous sheath (Li et al. 2007), a cytoskeletal structure in the mammalian sperm flagellum. The ongoing alkalinization, which activates the downstream sperm-specific plasma membrane calcium channel CatSper (Ren et al. 2001), also increases intracellular calcium levels (Carlson et al. 2003). Together, this increase and the PKA-regulated phosphorylation of proteins on tyrosine residues, including A-kinase-anchoring proteins (AKAPs) (Ficarro et al. 2003), drive the activation of flagellar motility that is necessary for both sperm migration and residence in the female reproductive tract.

Additional aspects of capacitation that are more protracted are required for sperm to fertilize the egg. These occur closer to the site of the ovulated egg(s) and include a general increase in protein phosphorylation, loss of cholesterol from the plasma membrane, hyperactivation (acquisition of an asymmetrical, nonlinear motility pattern), and increased susceptibility to stimuli that promote the acrosome reaction. These slow processes are also regulated by bicarbonate; however, they additionally require the transfer of cholesterol from the plasma membrane to unknown cholesterol acceptors present in the uterine/oviduct fluids. Also, despite the dominant role of PKA in sperm capacitation, research suggests that other kinases, such as Src-family kinases and PKC, and phosphatases such as PP1α, PP1γ2, PP2A, and PP2B may play a role regulating the rapid and slow events of capacitation (reviewed in Visconti et al. 2011; Signorelli et al. 2012).

Hyperactivated sperm show increased flagellar bend amplitudes (Yanagimachi 1970), which is most often observed in the oviduct. This amplification generates enough power for the sperm to make its way through the oviductal mucus, cumulus matrix, and the zona pellucida (ZP) surrounding the egg (Chang and Suarez 2010; Hung and Suarez 2010). CATSPER channels and calcium are critical for hyperactivation; either the absence of external calcium or the presence of mutations in any of the four CATSPER subunits results in a failure to hyperactivate (Carlson et al. 2003), and CATSPER-defective sperm cannot leave the oviduct sperm reservoir and are incapable of penetrating the ZP. The mechanism of CATSPER regulation in the oviduct remains unclear. Progesterone is a possible regulator (Lishko et al. 2011; Strunker et al. 2011) produced by cumulus cells and is present in the follicular fluid (Sun et al. 2005). It may act as the long-sought chemoattractant. In this regard, hyperactivation could promote sperm chemotaxis (Chang and Suarez 2010). Indeed, progesterone can induce increases in intracellular calcium near the base of the sperm head—the site where the calcium stores are located (Fukami et al. 2001). Note that in marine species such as the sea urchin *Arbacia punctulata*, in which sperm migrate toward the oocyte along a chemoattractant gradient, motility is regulated by incremental increases in intracellular calcium concentration caused by the binding of the chemoattractant resact to the sperm (Bohmer et al. 2005). Progesterone could act similarly in mammals.

4 FERTILIZATION

4.1 The Acrosome Reaction in Sperm

Before fusing with the egg's plasma membrane, sperm must undergo the acrosome reaction (Yanagimachi and Mahi 1976). The acrosome is a Golgi-derived organelle that lies above the tip of the sperm head. Its contents are released following fusion between the outer acrosomal membrane and the plasma membrane (Kim et al. 2011), and this reaction is critical for interaction with the ZP of the egg. Only capacitated sperm are capable of undergoing the acrosome reaction. Progesterone produced by the cumulus cells surrounding the oocyte has been proposed as the possible inducer of this reaction (Osman et al. 1989; Jin et al. 2011). Upon breaching the egg's plasma membrane, the sperm induces the initiation of embryonic development by evoking an increase in the intracellular concentration of free calcium, a signaling mechanism that regulates numerous cellular processes (Fig. 4) (Berridge et al. 2000b; p. 95 [Bootman 2012]).

The signaling mechanisms leading to the acrosome reaction are well characterized (Arnoult et al. 1996; Evans and Florman 2002). Activation of PLCδ4 by a calcium influx and the consequent generation of inositol 1,4,5-trisphosphate (IP_3) (Distelhorst and Bootman 2011) causes a second influx of calcium (Fukami et al. 2001). This promotes exocytosis by stimulating the reorganization of a SNARE protein in the sperm membranes (Mayorga et al. 2007). This reorganization is facilitated by the NSF and α-SNAP proteins and the calcium sensor synaptotagmin. Release of proteolytic enzymes and hyaluronidase from the acrosome aids the penetration of the cumulus cells and ZP by the sperm (Kim et al. 2008)

How the sperm interacts with ZP has been a point of some dispute. Early studies implicated several sperm-surface enzymes as critical for the interaction of the sperm with the egg coat (Lu and Shur 1997; Ikawa et al. 2010), but more recent work supports the involvement of a disintegrin and metalloproteinase family member, ADAM3 (Shamsadin et al. 1999). On the egg side, recent studies point to the amino-terminal domain of ZP2, which is cleaved after fertilization by the egg cortical granule component ovastacin, as the ZP ligand (reviewed in Avella et al. 2013). After binding, sperm must penetrate the ZP, probably using sperm-associated serine proteases such as acrosin and testisin (also known as PRSS21 or TESP5) (Baba et al. 1994; Yamashita et al. 2008).

4.2 Fusion and Egg Activation Leading to Fertilization

Soon after negotiating the ZP, the sperm fuses with the egg plasma membrane (Evans and Florman 2002). Early studies implicated the proteases ADAM1b and ADAM2, together known as fertilin, but gene knockout studies showed that fertilin is not required for fusion (Inoue et al. 2007). Subsequent work identified IZUMO, a member of the immunoglobulin superfamily, as the protein that appears to mediate fusion (Inoue et al. 2005; Sosnik et al. 2009). In the egg, the tetraspanin protein CD9 has been shown to be critical for fusion (Le Naour et al. 2000; Miyado et al. 2000). Whether CD9 and IZUMO directly interact is not clear.

Fusion of the gametes produces a calcium signal in the egg. A variety of calcium signals are required for egg activation; these reflect both the plasticity of the signaling machinery as well as the distinct requirements for egg activation in different species. Species typically either display a single increase in calcium concentration (e.g., in sea urchins, starfish, frogs, and fish) or show calcium oscillations (e.g., in nemertian worms, ascidians, and mammals) (Stricker 1999; Stricker and Whitaker 1999; Miyazaki and Ito 2006). The mechanism(s) that mediate calcium influx remains poorly characterized in mammalian eggs. Cells use several calcium influx mechanisms, including

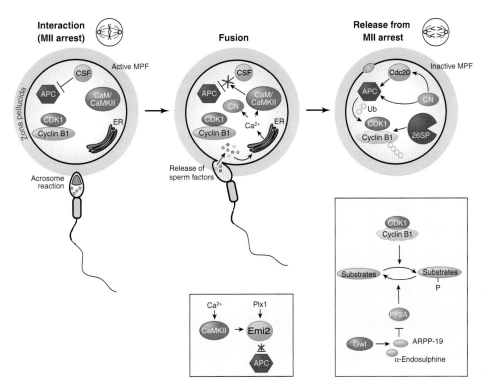

Figure 4. Fertilization events. Both egg and sperm must undergo complex changes before fertilization can occur. Upon interaction with the zona pelucida, sperm undergo a calcium-flux-driven reorganization of SNARE fusion proteins, called the acrosome reaction. Fusion with the egg membrane releases factors from the sperm into the cytoplasm. These factors stimulate calcium release from the egg ER and subsequent activation of CaMKII and calcineurin (CN). The consequences of CaMKII activation include phosphorylation of Emi2, which promotes its Plx1 and SCFβ-TrCP-dependent degradation and consequent reactivation of the APC. CN activation promotes dephosphorylation of Cdc20 and also the Cdc27 subunit of the APC. At the same time, dephosphorylation of M phase phosphoproteins is promoted by inactivation of the Greatwall kinase (Gwl), thereby alleviating inhibition of PP2A, allowing it to dephosphorylate M phase CDK1 substrates.

receptor-operated channels (ROCs) and voltage-operated calcium channels (VOCCs) (Berridge et al. 2000a; Tosti and Boni 2004; Smyth et al. 2006). Calcium influx may be attained, at least in part, by store-operated calcium entry (SOCE), a mechanism regulated by ER calcium levels and driven by store-operated calcium channels (SOCs) (Park et al. 2009). Stromal interaction molecule (STIM1) in the ER acts as a calcium sensor (Liou et al. 2005; Roos et al. 2005) and causes opening of Orai1, a channel partner protein in the plasma membrane to replenish stores (Feske et al. 2006; Vig et al. 2006).

The IP$_3$ receptor (IP$_3$R) on the ER is the main intracellular calcium-release channel in many mammalian cell types (reviewed in Berridge et al. 2000b; Bootman et al. 2001; p. 95 [Bootman 2012]). Although mammalian oocytes, eggs, and the surrounding cells express all three IP$_3$R isoforms (reviewed in Fissore et al. 1999a; Berridge et al. 2000b; Díaz-Muñoz et al. 2008), oocytes and eggs overwhelmingly express the type I IP$_3$R isoform (Kume et al. 1997; Fissore et al. 1999b; Jellerette et al. 2000; Tokmakov

et al. 2002), which requires binding by both calcium and IP$_3$ for activation and is stimulated at low calcium levels and inhibited at high calcium levels (Iino 1990a,b; Finch et al. 1991). This makes it especially suited to support long-lasting oscillations. The importance of IP$_3$R1 in fertilization is supported by studies showing injection of an anti-IP$_3$R1 antibody blocks sperm-initiated calcium oscillations (Miyazaki et al. 1992) and egg activation in mice (Xu et al. 1994) and works in the same manner in other vertebrates (Parys et al. 1994; Thomas et al. 1998; Yoshida et al. 1998; Runft et al. 1999; Goud et al. 2002; Iwasaki et al. 2002).

The signals that initiate the calcium increase at fertilization have proven elusive (Whitaker 2006; Parrington et al. 2007). Nonetheless, in some species, especially those with a single calcium transient at fertilization, Src-family kinases and PLCγ have been shown to lead to production of IP$_3$ during fertilization (Giusti et al. 1999; Sato et al. 2000). The receptor responsible for recruiting and activating the kinases remains to be identified (Mahbub Hasan et al. 2005). Similarly, it is unclear how sperm induce calcium oscillations in mammals. Calcium responses can be initiated in eggs from various species, including mammals by the same agonists that cause calcium release in somatic cells (Katayama et al. 1993; Miyazaki and Ito 2006). However, these fail to replicate the pattern of calcium oscillations associated with fertilization; other mechanism(s) might, therefore, be at play. Studies of sea urchin, ascidian eggs, and later, mammalian eggs showed that injection of sperm extracts or whole sperm can replicate fertilization-like responses in these species (Stice and Robl 1990; Swann 1990; Tesarik and Testart 1994; Nakano et al. 1997; Stricker 1997; Swann and Lai 1997; Wu et al. 1997; Kurokawa and Fissore 2003; Malcuit et al. 2006; Swann et al. 2006) strengthening the notion of a sperm cytosolic factor (SF). The mechanism may involve the release of SF into the ooplasm after fusion of the gametes. Importantly, the SF is not IP_3 or calcium but an uncharacterized protein moiety (Swann 1990; Wu et al. 1997; Kyozuka et al. 1998; Harada et al. 2007). Once initiated, calcium oscillations trigger all events of egg activation (Schultz and Kopf 1995). The presence of oscillations ensures the persistent degradation of Emi2 and the stepwise progression of activation, as events such as cortical granule exocytosis exit require fewer calcium increases than exit from MII or recruitment of maternal RNAs (Ducibella et al. 2002).

Cytosolic preparations from mammalian sperm possess high PLC activity (Parrington et al. 1999; Jones et al. 2000; Rice et al. 2000), and this is highly sensitive to calcium and could therefore be (Rice et al. 2000) the oscillation initiator. Indeed, a novel sperm-specific PLCζ (Saunders et al. 2002) has been identified (Fujimoto et al. 2004; Kouchi et al. 2004) and can evoke appropriate oscillations in mouse (Saunders et al. 2002), rat (Ito et al. 2008), human (Rogers et al. 2004), bovine (Malcuit et al. 2005; Ross et al. 2008), porcine (Yoneda et al. 2006), and equine (Bedford-Guaus et al. 2008) eggs. The findings that aberrant PLCζ expression is associated with infertility, and sperm from patients with repeated fertilization failure after intracytoplasmic sperm injection (ICSI), which also fail to initiate calcium oscillations (Yoon et al. 2008; Heytens et al. 2009), supports the idea that PLCζ released from the sperm triggers oscillations in fertilized eggs. Nevertheless, questions remain regarding its expression during spermatogenesis and storage, the means by which release into the ooplasm occurs, and how activation occurs upon egg entry.

5 FROM EGG TO ZYGOTE

Once fertilization has occurred, interphase nuclei form separately around the paternal (sperm) and maternal (Eggerickx et al. 1995) genomes in structures known as pronuclei. The paternal genome undergoes a series of

transformations that result in both exchange of spermatic chromatin packaging proteins (protamines) for maternal histones and demethylation of many repressed paternal genes. Upon completion of DNA replication, the pronuclear envelopes break down and the maternal and paternal chromosomes comingle on a common mitotic spindle in the zygote. In this transition to the zygote, maternal mRNA transcripts are largely degraded; many are replaced by zygote-specific transcripts both to take over the function of maternal housekeeping messages and to allow expression of new proteins that will build the early embryo.

At fertilization, the egg exits its MII arrest, in part as a consequence of events set in motion by the calcium response, which inactivates CSF through degradation of Mos and reactivates the APC, leading to cyclin degradation. Reactivation of the APC appears to be mediated in part by the calcium-dependent phosphatase calcineurin (also known as PP2B). Inhibition of calcineurin by cyclosporine A halts destruction of cyclin B. Not only has a core component of the APC, Cdc27, been shown to serve as a calcineurin substrate, but the APC-activating subunit Cdc20 is also a substrate of calcineurin. To exit meiosis, a host of CDK1–cyclin-B substrates must be dephosphorylated. This is accomplished largely by PP2A (as described above). Premature inactivation of CDK1–cyclin-B substrates would prohibit entry into or maintenance of MII. Inappropriate PP2A activation is therefore prevented through binding of a small peptide called Arpp19 and α endosulfine. Phosphorylation of Arpp19 by the conserved mitotic kinase Greatwall, which is activated by CDK1 at M phase entry, allows it to inhibit the PP2A isoform B55δ. At M phase exit, Greatwall is inactivated, which alleviates inhibition of PP2A and consequently allows the requisite dephosphorylation of M phase substrates. Additionally, rapid destruction of Emi2 alleviates inhibition of the APC, leading to loss of CDK1 activity (through degradation of cyclin B). In *Xenopus* eggs, this is triggered by calcium via a calcium-calmodulin-dependent kinase (CaMKII). CaMKII phosphorylates Emi2 protein on T195. This creates a docking site on Emi2 for another kinase, Plx1 in *Xenopus*, which then phosphorylates Emi2 within a sequence (the phosphodegron) required for recognition by the SCFβ-TrCP ubiquitin ligase. This series of events leads to ubiquitylation and degradation of Emi2, liberation of the APC, and degradation of substrates required for M phase exit. Note the Plx1-phosphorylation site does not appear to be conserved in mammals, and it is not known whether the ability of CaMKII to trigger Emi2 degradation in mammalian eggs is also triggered by recruitment Plk1 (the mammalian Plk1 ortholog). The culmination of these events allows the fertilized egg to exit arrest and proceed with cell division.

Cite this chapter as *Cold Spring Harb Perspect Biol* doi: 10.1101/cshperspect.a006064

6 CONCLUDING REMARKS

In gametes of both sexes, development, maturation, and fertilization require the careful coordination of complex processes. From hormone-initiated and kinase/phosphatase-controlled maturation, to calcium-induced capacitation and fertilization, regulatory mechanisms ensure that reproduction occurs only under conditions in which they are best poised for success. For males, the signals that regulate spermatogenesis are at first contained within the testis, but following spermiation and ejaculation, proper sperm function depends on factors outside of the male reproductive tract: for external fertilizers, factors in the outside environment; and for internal fertilizers, the milieu of the female reproductive tract. Taking into account these potentially harsh environments, male reproduction relies on the production of large quantities of sperm, with progenitor cells retaining mitotic capacity into adulthood. For female internal fertilizers, the production of eggs is more contained, restricted to the follicle until ovulation and to the female ducts until fertilization and beyond, relying on the production of few gametes; in contrast, externally fertilizing females, like their male counterparts, produce large numbers of gametes. Even for these organisms, though, oocytes and eggs are self-contained developmental units capable of sustaining the early stages of development whose progenitors have generally entered meiosis and have lost the capacity to regenerate.

In every species, regardless of the site of fertilization, cascades of protein modifications regulate cell-cycle transitions to guarantee that oocytes can be fertilized only at an appropriate time. Hormonal regulation and calcium signaling promote capacitation of sperm only in the correct environment for fertilization. Furthermore, multiple overlapping pathways provide the checks and balances that are necessary to prevent defective reproduction.

Although great progress has been made over the last 50 years in detailing the molecular events underlying most aspects of vertebrate fertilization, there are still aspects of gamete development and fertilization whose precise regulation by cell signaling events remain to be determined. Elucidation of these events is likely to have important implications for the continued development of reproductive technologies and for maximizing the health of gametes, and thus of progeny.

REFERENCES

*Reference is in this book.

Andresson T, Ruderman JV. 1998. The kinase Eg2 is a component of the *Xenopus* oocyte progesterone-activated signaling pathway. *EMBO J* **17:** 5627–5637.

Araki K, Naito K, Haraguchi S, Suzuki R, Yokoyama M, Inoue M, Aizawa S, Toyoda Y, Sato E. 1996. Meiotic abnormalities of c-mos knockout mouse oocytes: Activation after first meiosis or entrance into third meiotic metaphase. *Biol Reprod* **55:** 1315–1324.

Arnoult C, Cardullo RA, Lemos JR, Florman HM. 1996. Activation of mouse sperm T-type Ca^{2+} channels by adhesion to the egg zona pellucida. *Proc Natl Acad Sci* **93:** 13004–13009.

Austin CR. 1951. Observations on the penetration of the sperm in the mammalian egg. *Aust J Sci Res B* **4:** 581–596.

Avella MA, Xiong B, Dean J. 2013. The molecular basis of gamete recognition in mice and humans. *Mol Hum Reprod* **19:** 279–289.

Baba T, Azuma S, Kashiwabara S, Toyoda Y. 1994. Sperm from mice carrying a targeted mutation of the acrosin gene can penetrate the oocyte zona pellucida and effect fertilization. *J Biol Chem* **269:** 31845–31849.

Bedell MA, Mahakali Zama A. 2004. Genetic analysis of Kit ligand functions during mouse spermatogenesis. *J Androl* **25:** 188–199.

Bedford JM. 1970. Sperm capacitation and fertilization in mammals. *Biol Reprod* **2:** 128–158.

Bedford-Guaus SJ, Yoon SY, Fissore RA, Choi YH, Hinrichs K. 2008. Microinjection of mouse phospholipase Cζ complementary RNA into mare oocytes induces long-lasting intracellular calcium oscillations and embryonic development. *Reprod Fertil Dev* **20:** 875–883.

Berridge MJ, Lipp P, Bootman MD. 2000a. Signal transduction. The calcium entry pas de deux. *Science* **287:** 1604–1605.

Berridge MJ, Lipp P, Bootman MD. 2000b. The versatility and universality of calcium signalling. *Nat Rev Mol Cell Biol* **1:** 11–21.

Berry LD, Gould KL. 1996. Regulation of Cdc2 activity by phosphorylation at T14/Y15. *Prog Cell Cycle Res* **2:** 99–105.

Bohmer M, Van Q, Weyand I, Hagen V, Beyermann M, Matsumoto M, Hoshi M, Hildebrand E, Kaupp UB. 2005. Ca^{2+} spikes in the flagellum control chemotactic behavior of sperm. *EMBO J* **24:** 2741–2752.

* Bootman MD. 2012. Calcium signaling. *Cold Spring Harb Perspect Biol* **4:** a011171.

Bootman MD, Collins TJ, Peppiatt CM, Prothero LS, MacKenzie L, De Smet P, Travers M, Tovey SC, Seo JT, Berridge MJ, et al. 2001. Calcium signalling—An overview. *Sem Cell Dev Biol* **12:** 3–10.

Bowles J, Knight D, Smith C, Wilhelm D, Richman J, Mamiya S, Yashiro K, Chawengsaksophak K, Wilson MJ, Rossant J, et al. 2006. Retinoid signaling determines germ cell fate in mice. *Science* **312:** 596–600.

Buaas FW, Kirsh AL, Sharma M, McLean DJ, Morris JL, Griswold MD, de Rooij DG, Braun RE. 2004. Plzf is required in adult male germ cells for stem cell self-renewal. *Nat Genet* **36:** 647–652.

Carlson AE, Westenbroek RE, Quill T, Ren D, Clapham DE, Hille B, Garbers DL, Babcock DF. 2003. CatSper1 required for evoked Ca^{2+} entry and control of flagellar function in sperm. *Proc Natl Acad Sci* **100:** 14864–14868.

Chang MC. 1951. Fertilizing capacity of spermatozoa deposited into the fallopian tubes. *Nature* **168:** 697–698.

Chang H, Suarez SS. 2010. Rethinking the relationship between hyperactivation and chemotaxis in mammalian sperm. *Biol Reprod* **83:** 507–513.

Chang C, Chen YT, Yeh SD, Xu Q, Wang RS, Guillou F, Lardy H, Yeh S. 2004. Infertility with defective spermatogenesis and hypotestosteronemia in male mice lacking the androgen receptor in Sertoli cells. *Proc Natl Acad Sci* **101:** 6876–6881.

Chen Y, Cann MJ, Litvin TN, Iourgenko V, Sinclair ML, Levin LR, Buck J. 2000. Soluble adenylyl cyclase as an evolutionarily conserved bicarbonate sensor. *Science* **289:** 625–628.

Chen C, Ouyang W, Grigura V, Zhou Q, Carnes K, Lim H, Zhao GQ, Arber S, Kurpios N, Murphy TL, et al. 2005. ERM is required for transcriptional control of the spermatogonial stem cell niche. *Nature* **436:** 1030–1034.

Cheng J, Watkins SC, Walker WH. 2007. Testosterone activates mitogen-activated protein kinase via Src kinase and the epidermal growth factor receptor in sertoli cells. *Endocrinology* **148:** 2066–2074.

Colledge WH, Carlton MB, Udy GB, Evans MJ. 1994. Disruption of c-mos causes parthenogenetic development of unfertilized mouse eggs. *Nature* **370:** 65–68.

Costoya JA, Hobbs RM, Barna M, Cattoretti G, Manova K, Sukhwani M, Orwig KE, Wolgemuth DJ, Pandolfi PP. 2004. Essential role of Plzf in maintenance of spermatogonial stem cells. *Nat Genet* **36:** 653–659.

Daar I, Paules RS, Vande Woude GF. 1991. A characterization of cytostatic factor activity from *Xenopus* eggs and c-mos-transformed cells. *J Cell Biol* **114:** 329–335.

De Gendt K, Swinnen JV, Saunders PT, Schoonjans L, Dewerchin M, Devos A, Tan K, Atanassova N, Claessens F, Lecureuil C, et al. 2004. A Sertoli cell-selective knockout of the androgen receptor causes spermatogenic arrest in meiosis. *Proc Natl Acad Sci* **101:** 1327–1332.

Deng J, Lang S, Wylie C, Hammes SR. 2008. The *Xenopus laevis* isoform of G protein-coupled receptor 3 (GPR3) is a constitutively active cell surface receptor that participates in maintaining meiotic arrest in *X. laevis* oocytes. *Mol Endocrinol* **22:** 1853–1865.

de Rooij DG. 2009. The spermatogonial stem cell niche. *Microsc Res Tech* **72:** 580–585.

Díaz-Muñoz M, de la Rosa Santander P, Juárez-Espinosa AB, Arellano RO, Morales-Tlalpan V. 2008. Granulosa cells express three inositol 1,4,5-trisphosphate receptor isoforms: Cytoplasmic and nuclear Ca^{2+} mobilization. *Reprod Biol Endocrinol* **6:** 60.

Distelhorst CW, Bootman MD. 2011. Bcl-2 interaction with the inositol 1,4,5-trisphosphate receptor: Role in Ca^{2+} signaling and disease. *Cell Calcium* **50:** 234–241.

Drury KC, Schorderet-Slatkine S. 1975. Effects of cycloheximide on the "autocatalytic" nature of the maturation promoting factor (MPF) in oocytes of *Xenopus laevis*. *Cell* **4:** 269–274.

Ducibella T, Huneau D, Angelichio E, Xu Z, Schultz RM, Kopf GS, Fissore R, Madoux S, Ozil JP. 2002. Egg-to-embryo transition is driven by differential responses to Ca^{2+} oscillation number. *Dev Biol* **250:** 280–291.

Duckworth BC, Weaver JS, Ruderman JV. 2002. G_2 arrest in *Xenopus* oocytes depends on phosphorylation of cdc25 by protein kinase A. *Proc Natl Acad Sci* **99:** 16794–16799.

Dumont J, Umbhauer M, Rassinier P, Hanauer A, Verlhac MH. 2005. p90RSK is not involved in cytostatic factor arrest in mouse oocytes. *J Cell Biol* **169:** 227–231.

Dunphy WG, Brizuela L, Beach D, Newport J. 1988. The *Xenopus* cdc2 protein is a component of MPF, a cytoplasmic regulator of mitosis. *Cell* **54:** 423–431.

Dupre A, Jessus C, Ozon R, Haccard O. 2002. Mos is not required for the initiation of meiotic maturation in *Xenopus* oocytes. *EMBO J* **21:** 4026–4036.

Eggerickx D, Denef JF, Labbe O, Hayashi Y, Refetoff S, Vassart G, Parmentier M, Libert F. 1995. Molecular cloning of an orphan G-protein-coupled receptor that constitutively activates adenylate cyclase. *Biochem J* **309:** 837–843.

Evans JP, Florman HM. 2002. The state of the union: The cell biology of fertilization. *Nat Cell Biol* **4:** S57–S63.

Fan HY, Sun QY. 2004. Involvement of mitogen-activated protein kinase cascade during oocyte maturation and fertilization in mammals. *Biol Reprod* **70:** 535–547.

Feng LX, Ravindranath N, Dym M. 2000. Stem cell factor/c-kit up-regulates cyclin D3 and promotes cell cycle progression via the phosphoinositide 3-kinase/p70 S6 kinase pathway in spermatogonia. *J Biol Chem* **275:** 25572–25576.

Ferby I, Blazquez M, Palmer A, Eritja R, Nebreda AR. 1999. A novel p34cdc2-binding and activating protein that is necessary and sufficient to trigger G_2/M progression in *Xenopus* oocytes. *Genes Dev* **13:** 2177–2189.

Feske S, Gwack Y, Prakriya M, Srikanth S, Puppel SH, Tanasa B, Hogan PG, Lewis RS, Daly M, Rao A. 2006. A mutation in Orai1 causes immune deficiency by abrogating CRAC channel function. *Nature* **441:** 179–185.

Ficarro S, Chertihin O, Westbrook VA, White F, Jayes F, Kalab P, Marto JA, Shabanowitz J, Herr JC, Hunt DF, et al. 2003. Phosphoproteome analysis of capacitated human sperm. Evidence of tyrosine phosphorylation of a kinase-anchoring protein 3 and valosin-containing protein/p97 during capacitation. *J Biol Chem* **278:** 11579–11589.

Finch EA, Turner TJ, Goldin SM. 1991. Subsecond kinetics of inositol 1,4,5-trisphosphate-induced calcium release reveal rapid potentiation and subsequent inactivation by calcium. *Ann NY Acad Sci* **635:** 400–403.

Fissore RA, Long CR, Duncan RP, Robl JM. 1999a. Initiation and organization of events during the first cell cycle in mammals: Applications in cloning. *Cloning* **1:** 89–100.

Fissore RA, Longo FJ, Anderson E, Parys JB, Ducibella T. 1999b. Differential distribution of inositol trisphosphate receptor isoforms in mouse oocytes. *Biol Reprod* **60:** 49–57.

Flanagan JG, Chan DC, Leder P. 1991. Transmembrane form of the kit ligand growth factor is determined by alternative splicing and is missing in the Sl^d mutant. *Cell* **64:** 1025–1035.

Frank-Vaillant M, Haccard O, Thibier C, Ozon R, Arlot-Bonnemains Y, Prigent C, Jessus C. 2000. Progesterone regulates the accumulation and the activation of Eg2 kinase in *Xenopus* oocytes. *J Cell Sci* **113:** 1127–1138.

Freudzon L, Norris RP, Hand AR, Tanaka S, Saeki Y, Jones TL, Rasenick MM, Berlot CH, Mehlmann LM, Jaffe LA. 2005. Regulation of meiotic prophase arrest in mouse oocytes by GPR3, a constitutive activator of the Gs G protein. *J Cell Biol* **171:** 255–265.

Fujimoto S, Yoshida N, Fukui T, Amanai M, Isobe T, Itagaki C, Izumi T, Perry AC. 2004. Mammalian phospholipase Cζ induces oocyte activation from the sperm perinuclear matrix. *Dev Biol* **274:** 370–383.

Fukami K, Nakao K, Inoue T, Kataoka Y, Kurokawa M, Fissore RA, Nakamura K, Katsuki M, Mikoshiba K, Yoshida N, et al. 2001. Requirement of phospholipase Cδ4 for the zona pellucida-induced acrosome reaction. *Science* **292:** 920–923.

Fulka J Jr, Motlik J, Fulka J, Crozet N. 1986. Activity of maturation promoting factor in mammalian oocytes after its dilution by single and multiple fusions. *Dev Biol* **118:** 176–181.

Gautier J, Maller JL. 1991. Cyclin B in *Xenopus* oocytes: Implications for the mechanism of pre-MPF activation. *EMBO J* **10:** 177–182.

Giusti AF, Carroll DJ, Abassi YA, Foltz KR. 1999. Evidence that a starfish egg Src family tyrosine kinase associates with PLC-γ1 SH2 domains at fertilization. *Dev Biol* **208:** 189–199.

Goud PT, Goud AP, Leybaert L, Van Oostveldt P, Mikoshiba K, Diamond MP, Dhont M. 2002. Inositol 1,4,5-trisphosphate receptor function in human oocytes: Calcium responses and oocyte activation-related phenomena induced by photolytic release of InsP$_3$ are blocked by a specific antibody to the type I receptor. *Mol Hum Reprod* **8:** 912–918.

Gross SD, Schwab MS, Taieb FE, Lewellyn AL, Qian YW, Maller JL. 2000. The critical role of the MAP kinase pathway in meiosis II in *Xenopus* oocytes is mediated by p90RSK. *Curr Biol* **10:** 430–438.

Hake LE, Richter JD. 1994. CPEB is a specificity factor that mediates cytoplasmic polyadenylation during *Xenopus* oocyte maturation. *Cell* **79:** 617–627.

Han SJ, Chen R, Paronetto MP, Conti M. 2005. WEE1B is an oocyte-specific kinase involved in the control of meiotic arrest in the mouse. *Curr Biol* **15:** 1670–1676.

Hansen DV, Tung JJ, Jackson PK. 2006. CaMKII and polo-like kinase 1 sequentially phosphorylate the cytostatic factor Emi2/XErp1 to trigger its destruction and meiotic exit. *Proc Natl Acad Sci* **103:** 608–613.

Harada Y, Matsumoto T, Hirahara S, Nakashima A, Ueno S, Oda S, Miyazaki S, Iwao Y. 2007. Characterization of a sperm factor for egg activation at fertilization of the newt Cynops pyrrhogaster. *Dev Biol* **306:** 797–808.

Hashimoto N, Watanabe N, Furuta Y, Tamemoto H, Sagata N, Yokoyama M, Okazaki K, Nagayoshi M, Takeda N, Ikawa Y, et al. 1994. Parthenogenetic activation of oocytes in c-mos-deficient mice. *Nature* **370:** 68–71.

Cite this chapter as *Cold Spring Harb Perspect Biol* doi: 10.1101/cshperspect.a006064

Haywood M, Spaliviero J, Jimemez M, King NJ, Handelsman DJ, Allan CM. 2003. Sertoli and germ cell development in hypogonadal (hpg) mice expressing transgenic follicle-stimulating hormone alone or in combination with testosterone. *Endocrinology* **144:** 509–517.

Hermo L, Pelletier RM, Cyr DG, Smith CE. 2010. Surfing the wave, cycle, life history, and genes/proteins expressed by testicular germ cells. Part 1: Background to spermatogenesis, spermatogonia, and spermatocytes. *Microsc Res Tech* **73:** 241–278.

Heytens E, Parrington J, Coward K, Young C, Lambrecht S, Yoon SY, Fissore RA, Hamer R, Deane CM, Ruas M, et al. 2009. Reduced amounts and abnormal forms of phospholipase Cζ (PLCζ) in spermatozoa from infertile men. *Hum Reprod* **24:** 2417–2428.

Hinckley M, Vaccari S, Horner K, Chen R, Conti M. 2005. The G-protein-coupled receptors GPR3 and GPR12 are involved in cAMP signaling and maintenance of meiotic arrest in rodent oocytes. *Dev Biol* **287:** 249–261.

Hodgman R, Tay J, Mendez R, Richter JD. 2001. CPEB phosphorylation and cytoplasmic polyadenylation are catalyzed by the kinase IAK1/Eg2 in maturing mouse oocytes. *Development* **128:** 2815–2822.

Hofmann MC. 2008. Gdnf signaling pathways within the mammalian spermatogonial stem cell niche. *Mol Cell Endocrinol* **288:** 95–103.

Holdcraft RW, Braun RE. 2004. Hormonal regulation of spermatogenesis. *Int J Androl* **27:** 335–342.

Hu J, Shima H, Nakagawa H. 1999. Glial cell line-derived neurotropic factor stimulates sertoli cell proliferation in the early postnatal period of rat testis development. *Endocrinology* **140:** 3416–3421.

Hung PH, Suarez SS. 2010. Regulation of sperm storage and movement in the ruminant oviduct. *Soc Reprod Fertil Suppl* **67:** 257–266.

Hunter AG, Moor RM. 1987. Stage-dependent effects of inhibiting ribonucleic acids and protein synthesis on meiotic maturation of bovine oocytes in vitro. *J Dairy Sci* **70:** 1646–1651.

Iino M. 1990a. Biphasic Ca^{2+} dependence of inositol 1,4,5-trisphosphate-induced Ca release in smooth muscle cells of the guinea pig taenia caeci. *J Gen Physiol* **95:** 1103–1122.

Iino M. 1990b. Calcium release mechanisms in smooth muscle. *Jpn J Pharmacol* **54:** 345–354.

Ikawa M, Inoue N, Benham AM, Okabe M. 2010. Fertilization: A sperm's journey to and interaction with the oocyte. *J Clin Invest* **120:** 984–994.

Inoue N, Ikawa M, Isotani A, Okabe M. 2005. The immunoglobulin superfamily protein Izumo is required for sperm to fuse with eggs. *Nature* **434:** 234–238.

Inoue N, Yamaguchi R, Ikawa M, Okabe M. 2007. Sperm-egg interaction and gene manipulated animals. *Soc Reprod Fertil Suppl* **65:** 363–371.

Itman C, Mendis S, Barakat B, Loveland KL. 2006. All in the family: TGF-β family action in testis development. *Reproduction* **132:** 233–246.

Ito M, Shikano T, Oda S, Horiguchi T, Tanimoto S, Awaji T, Mitani H, Miyazaki S. 2008. Difference in Ca^{2+} oscillation-inducing activity and nuclear translocation ability of PLCZ1, an egg-activating sperm factor candidate, between mouse, rat, human, and medaka fish. *Biol Reprod* **78:** 1081–1090.

Iwasaki H, Chiba K, Uchiyama T, Yoshikawa F, Suzuki F, Ikeda M, Furuichi T, Mikoshiba K. 2002. Molecular characterization of the starfish inositol 1,4,5-trisphosphate receptor and its role during oocyte maturation and fertilization. *J Biol Chem* **277:** 2763–2772.

Jagiello GM. 1969. Meiosis and inhibition of ovulation in mouse eggs treated with actinomycin D. *J Cell Biol* **42:** 571–574.

Jellerette T, He CL, Wu H, Parys JB, Fissore RA. 2000. Down-regulation of the inositol 1,4,5-trisphosphate receptor in mouse eggs following fertilization or parthenogenetic activation. *Dev Biol* **223:** 238–250.

Jin M, Fujiwara E, Kakiuchi Y, Okabe M, Satouh Y, Baba SA, Chiba K, Hirohashi N. 2011. Most fertilizing mouse spermatozoa begin their acrosome reaction before contact with the zona pellucida during in vitro fertilization. *Proc Natl Acad Sci* **108:** 4892–4896.

Jones KT, Matsuda M, Parrington J, Katan M, Swann K. 2000. Different Ca^{2+}-releasing abilities of sperm extracts compared with tissue extracts and phospholipase C isoforms in sea urchin egg homogenate and mouse eggs. *Biochem J* **346:** 743–749.

Kanki JP, Donoghue DJ. 1991. Progression from meiosis I to meiosis II in *Xenopus* oocytes requires de novo translation of the mosxe protooncogene. *Proc Natl Acad Sci* **88:** 5794–5798.

Katayama Y, Miyazaki S, Oshimi Y, Oshimi K. 1993. Ca^{2+} response in single human T cells induced by stimulation of CD4 or CD8 and interference with CD3 stimulation. *J Immunol Meth* **166:** 145–153.

Kim E, Yamashita M, Kimura M, Honda A, Kashiwabara SI, Baba T. 2008. Sperm penetration through cumulus mass and zona pellucida. *Int J Dev Biol* **52:** 677–682.

Kim KS, Foster JA, Kvasnicka KW, Gerton GL. 2011. Transitional states of acrosomal exocytosis and proteolytic processing of the acrosomal matrix in guinea pig sperm. *Mol Reprod Dev* **78:** 930–941.

Kouchi Z, Fukami K, Shikano T, Oda S, Nakamura Y, Takenawa T, Miyazaki S. 2004. Recombinant phospholipase Cζ has high Ca^{2+} sensitivity and induces Ca^{2+} oscillations in mouse eggs. *J Biol Chem* **279:** 10408–10412.

Kumagai A, Dunphy WG. 1999. Binding of 14-3-3 proteins and nuclear export control the intracellular localization of the mitotic inducer Cdc25. *Genes Dev* **13:** 1067–1072.

Kume S, Yamamoto A, Inoue T, Muto A, Okano H, Mikoshiba K. 1997. Developmental expression of the inositol 1,4,5-trisphosphate receptor and structural changes in the endoplasmic reticulum during oogenesis and meiotic maturation of *Xenopus laevis*. *Dev Biol* **182:** 228–239.

Kurokawa M, Fissore RA. 2003. ICSI-generated mouse zygotes exhibit altered calcium oscillations, inositol 1,4,5-trisphosphate receptor-1 down-regulation, and embryo development. *Mol Hum Reprod* **9:** 523–533.

Kyozuka K, Deguchi R, Mohri T, Miyazaki S. 1998. Injection of sperm extract mimics spatiotemporal dynamics of Ca^{2+} responses and progression of meiosis at fertilization of ascidian oocytes. *Development* **125:** 4099–4105.

Labbe JC, Picard A, Peaucellier G, Cavadore JC, Nurse P, Doree M. 1989. Purification of MPF from starfish: Identification as the H1 histone kinase p34cdc2 and a possible mechanism for its periodic activation. *Cell* **57:** 253–263.

Le Naour F, Rubinstein E, Jasmin C, Prenant M, Boucheix C. 2000. Severely reduced female fertility in CD9-deficient mice. *Science* **287:** 319–321.

Li YF, He W, Jha KN, Klotz K, Kim YH, Mandal A, Pulido S, Digilio L, Flickinger CJ, Herr JC. 2007. FSCB, a novel protein kinase A-phosphorylated calcium-binding protein, is a CABYR-binding partner involved in late steps of fibrous sheath biogenesis. *J Biol Chem* **282:** 34104–34119.

Liou J, Kim ML, Heo WD, Jones JT, Myers JW, Ferrell JE Jr, Meyer T. 2005. STIM is a Ca^{2+} sensor essential for Ca^{2+}-store-depletion-triggered Ca^{2+} influx. *Curr Biol* **15:** 1235–1241.

Lishko PV, Botchkina IL, Kirichok Y. 2011. Progesterone activates the principal Ca^{2+} channel of human sperm. *Nature* **471:** 387–391.

Liu J, Maller JL. 2005. Calcium elevation at fertilization coordinates phosphorylation of XErp1/Emi2 by Plx1 and CaMK II to release metaphase arrest by cytostatic factor. *Curr Biol* **15:** 1458–1468.

Lopez-Girona A, Furnari B, Mondesert O, Russell P. 1999. Nuclear localization of Cdc25 is regulated by DNA damage and a 14-3-3 protein. *Nature* **397:** 172–175.

Lu Q, Shur BD. 1997. Sperm from β1,4-galactosyltransferase-null mice are refractory to ZP3-induced acrosome reactions and penetrate the zona pellucida poorly. *Development* **124:** 4121–4131.

Madgwick S, Hansen DV, Levasseur M, Jackson PK, Jones KT. 2006. Mouse Emi2 is required to enter meiosis II by reestablishing cyclin B1 during interkinesis. *J Cell Biol* **174:** 791–801.

Mahbub Hasan AK, Sato K, Sakakibara K, Ou Z, Iwasaki T, Ueda Y, Fukami Y. 2005. Uroplakin III, a novel Src substrate in *Xenopus* egg rafts, is a target for sperm protease essential for fertilization. *Dev Biol* **286:** 483–492.

Malcuit C, Knott JG, He C, Wainwright T, Parys JB, Robl JM, Fissore RA. 2005. Fertilization and inositol 1,4,5-trisphosphate (IP$_3$)-induced calcium release in type-1 inositol 1,4,5-trisphosphate receptor down-regulated bovine eggs. *Biol Reprod* **73**: 2–13.

Malcuit C, Maserati M, Takahashi Y, Page R, Fissore RA. 2006. Intra-cytoplasmic sperm injection in the bovine induces abnormal $[Ca^{2+}]_i$ responses and oocyte activation. *Reprod Fertil Dev* **18**: 39–51.

Masui Y, Markert CL. 1971. Cytoplasmic control of nuclear behavior during meiotic maturation of frog oocytes. *J Exp Zool* **177**: 129–145.

Matthiesson KL, McLachlan RI, O'Donnell L, Frydenberg M, Robertson DM, Stanton PG, Meachem SJ. 2006. The relative roles of follicle-stimulating hormone and luteinizing hormone in maintaining spermatogonial maturation and spermiation in normal men. *J Clin Endocrinol Metab* **91**: 3962–3969.

Mattioli M, Galeati G, Bacci ML, Barboni B. 1991. Changes in maturation-promoting activity in the cytoplasm of pig oocytes throughout maturation. *Mol Reprod Dev* **30**: 119–125.

Mayorga LS, Tomes CN, Belmonte SA. 2007. Acrosomal exocytosis, a special type of regulated secretion. *IUBMB Life* **59**: 286–292.

McGuinness OM, Moreton RB, Johnson MH, Berridge MJ. 1996. A direct measurement of increased divalent cation influx in fertilised mouse oocytes. *Development* **122**: 2199–2206.

McLachlan RI, O'Donnell L, Meachem SJ, Stanton PG, de Kretser DM, Pratis K, Robertson DM. 2002. Identification of specific sites of hormonal regulation in spermatogenesis in rats, monkeys, and man. *Recent Prog Horm Res* **57**: 149–179.

Mehlmann LM. 2005. Oocyte-specific expression of Gpr3 is required for the maintenance of meiotic arrest in mouse oocytes. *Dev Biol* **288**: 397–404.

Mendez R, Hake LE, Andresson T, Littlepage LE, Ruderman JV, Richter JD. 2000. Phosphorylation of CPE binding factor by Eg2 regulates translation of c-mos mRNA. *Nature* **404**: 302–307.

Meng X, Lindahl M, Hyvonen ME, Parvinen M, de Rooij DG, Hess MW, Raatikainen-Ahokas A, Sainio K, Rauvala H, Lakso M, et al. 2000. Regulation of cell fate decision of undifferentiated spermatogonia by GDNF. *Science* **287**: 1489–1493.

Miyado K, Yamada G, Yamada S, Hasuwa H, Nakamura Y, Ryu F, Suzuki K, Kosai K, Inoue K, Ogura A, et al. 2000. Requirement of CD9 on the egg plasma membrane for fertilization. *Science* **287**: 321–324.

Miyazaki S, Ito M. 2006. Calcium signals for egg activation in mammals. *J Pharmacol Sci* **100**: 545–552.

Miyazaki S, Yuzaki M, Nakada K, Shirakawa H, Nakanishi S, Nakade S, Mikoshiba K. 1992. Block of Ca^{2+} wave and Ca^{2+} oscillation by anti-body to the inositol 1,4,5-trisphosphate receptor in fertilized hamster eggs. *Science* **257**: 251–255.

Moor RM, Crosby IM. 1986. Protein requirements for germinal vesicle breakdown in ovine oocytes. *J Embryol Exp Morphol* **94**: 207–220.

Nakano Y, Shirakawa H, Mitsuhashi N, Kuwabara Y, Miyazaki S. 1997. Spatiotemporal dynamics of intracellular calcium in the mouse egg injected with a spermatozoon. *Mol Hum Reprod* **3**: 1087–1093.

Nalam RL, Matzuk MM. 2010. Local signalling environments and human male infertility: What we can learn from mouse models. *Expert Rev Mol Med* **12**: e15.

Naughton CK, Jain S, Strickland AM, Gupta A, Milbrandt J. 2006. Glial cell-line derived neurotrophic factor-mediated RET signaling regulates spermatogonial stem cell fate. *Biol Reprod* **74**: 314–321.

Nishizawa M, Furuno N, Okazaki K, Tanaka H, Ogawa Y, Sagata N. 1993. Degradation of Mos by the N-terminal proline (Pro2)-dependent ubiquitin pathway on fertilization of *Xenopus* eggs: Possible significance of natural selection for Pro2 in Mos. *EMBO J* **12**: 4021–4027.

Oatley JM, Brinster RL. 2012. The germline stem cell niche unit in mammalian testes. *Physiol Rev* **92**: 577–595.

Ohe M, Inoue D, Kanemori Y, Sagata N. 2007. Erp1/Emi2 is essential for the meiosis I to meiosis II transition in *Xenopus* oocytes. *Dev Biol* **303**: 157–164.

Osman RA, Andria ML, Jones AD, Meizel S. 1989. Steroid induced exocytosis: The human sperm acrosome reaction. *Biochem Biophys Res Commun* **160**: 828–833.

Palmer A, Gavin AC, Nebreda AR. 1998. A link between MAP kinase and $p34^{cdc2}$/cyclin B during oocyte maturation: $p90^{RSK}$ phosphorylates and inactivates the $p34^{cdc2}$ inhibitory kinase Myt1. *EMBO J* **17**: 5037–5047.

Park CY, Hoover PJ, Mullins FM, Bachhawat P, Covington ED, Raunser S, Walz T, Garcia KC, Dolmetsch RE, Lewis RS. 2009. STIM1 clusters and activates CRAC channels via direct binding of a cytosolic domain to Orai1. *Cell* **136**: 876–890.

Parrington J, Jones KT, Lai A, Swann K. 1999. The soluble sperm factor that causes Ca^{2+} release from sea-urchin (*Lytechinus pictus*) egg homogenates also triggers Ca^{2+} oscillations after injection into mouse eggs. *Biochem J* **341**: 1–4.

Parrington J, Davis LC, Galione A, Wessel G. 2007. Flipping the switch: How a sperm activates the egg at fertilization. *Dev Dyn* **236**: 2027–2038.

Parys JB, McPherson SM, Mathews L, Campbell KP, Longo FJ. 1994. Presence of inositol 1,4,5-trisphosphate receptor, calreticulin, and cal-sequestrin in eggs of sea urchins and *Xenopus laevis*. *Dev Biol* **161**: 466–476.

Propst F, Rosenberg MP, Iyer A, Kaul K, Vande Woude GF. 1987. c-mos proto-oncogene RNA transcripts in mouse tissues: Structural features, developmental regulation, and localization in specific cell types. *Mol Cell Biol* **7**: 1629–1637.

Rauh NR, Schmidt A, Bormann J, Nigg EA, Mayer TU. 2005. Calcium triggers exit from meiosis II by targeting the APC/C inhibitor XErp1 for degradation. *Nature* **437**: 1048–1052.

Ren D, Navarro B, Perez G, Jackson AC, Hsu S, Shi Q, Tilly JL, Clapham DE. 2001. A sperm ion channel required for sperm motility and male fertility. *Nature* **413**: 603–609.

* Rhind N, Russell P. 2012. Signaling pathways that regulate cell division. *Cold Spring Harb Perspect Biol* **4**: a005942.

Rice A, Parrington J, Jones KT, Swann K. 2000. Mammalian sperm contain a Ca^{2+}-sensitive phospholipase C activity that can generate $InsP_3$ from PIP_2 associated with intracellular organelles. *Dev Biol* **228**: 125–135.

Richards JS, Pangas SA. 2010. The ovary: Basic biology and clinical implications. *J Clin Invest* **120**: 963–972.

Rogers NT, Hobson E, Pickering S, Lai FA, Braude P, Swann K. 2004. Phospholipase Czeta causes Ca^{2+} oscillations and parthenogenetic activation of human oocytes. *Reproduction* **128**: 697–702.

Roos J, DiGregorio PJ, Yeromin AV, Ohlsen K, Lioudyno M, Zhang S, Safrina O, Kozak JA, Wagner SL, Cahalan MD, et al. 2005. STIM1, an essential and conserved component of store-operated Ca^{2+} channel function. *J Cell Biol* **169**: 435–445.

Ross PJ, Beyhan Z, Iager AE, Yoon SY, Malcuit C, Schellander K, Fissore RA, Cibelli JB. 2008. Parthenogenetic activation of bovine oocytes using bovine and murine phospholipase Cζ. *BMC Dev Biol* **8**: 16.

Ruiz EJ, Hunt T, Nebreda AR. 2008. Meiotic inactivation of *Xenopus* Myt1 by CDK/XRINGO, but not CDK/cyclin, via site-specific phosphorylation. *Mol Cell* **32**: 210–220.

Runft LL, Watras J, Jaffe LA. 1999. Calcium release at fertilization of *Xenopus* eggs requires type I IP_3 receptors, but not SH2 domain-mediated activation of PLCγ or G_q-mediated activation of PLCβ. *Dev Biol* **214**: 399–411.

Ruwanpura SM, McLachlan RI, Stanton PG, Meachem SJ. 2008. Follicle-stimulating hormone affects spermatogonial survival by regulating the intrinsic apoptotic pathway in adult rats. *Biol Reprod* **78**: 705–713.

Ruwanpura SM, McLachlan RI, Meachem SJ. 2010. Hormonal regulation of male germ cell development. *J Endocrinol* **205**: 117–131.

Sagata N, Watanabe N, Vande Woude GF, Ikawa Y. 1989. The c-mos proto-oncogene product is a cytostatic factor responsible for meiotic arrest in vertebrate eggs. *Nature* **342**: 512–518.

Saito K, O'Donnell L, McLachlan RI, Robertson DM. 2000. Spermiation failure is a major contributor to early spermatogenic suppression caused by hormone withdrawal in adult rats. *Endocrinology* **141**: 2779–2785.

Salicioni AM, Platt MD, Wertheimer EV, Arcelay E, Allaire A, Sosnik J, Visconti PE. 2007. Signalling pathways involved in sperm capacitation. *Soc Reprod Fertil Suppl* **65:** 245–259.

Sariola H, Saarma M. 2003. Novel functions and signalling pathways for GDNF. *J Cell Sci* **116:** 3855–3862.

Sato K, Tokmakov AA, Iwasaki T, Fukami Y. 2000. Tyrosine kinase-dependent activation of phospholipase Cγ is required for calcium transient in *Xenopus* egg fertilization. *Dev Biol* **224:** 453–469.

Saunders CM, Larman MG, Parrington J, Cox LJ, Royse J, Blayney LM, Swann K, Lai FA. 2002. PLCζ: A sperm-specific trigger of Ca²⁺ oscillations in eggs and embryo development. *Development* **129:** 3533–3544.

Schultz RM, Kopf GS. 1995. Molecular-basis of mammalian egg activation. *Curr Top Dev Biol* **30:** 21–62.

Sette C, Dolci S, Geremia R, Rossi P. 2000. The role of stem cell factor and of alternative c-kit gene products in the establishment, maintenance and function of germ cells. *Int J Dev Biol* **44:** 599–608.

★ Sever R, Glass CK. 2013. Signaling by nuclear receptors. *Cold Spring Harb Perspect Biol* **5:** a016709.

Shamsadin R, Adham IM, Nayernia K, Heinlein UA, Oberwinkler H, Engel W. 1999. Male mice deficient for germ-cell cyritestin are infertile. *Biol Reprod* **61:** 1445–1451.

Sharpe RM. 1994. *Regulation of spermatogenesis.* Raven Press, New York.

Shupe J, Cheng J, Puri P, Kostereva N, Walker WH. 2011. Regulation of Sertoli-germ cell adhesion and sperm release by FSH and nonclassical testosterone signaling. *Mol Endocrinol* **25:** 238–252.

Signorelli J, Diaz ES, Morales P. 2012. Kinases, phosphatases and proteases during sperm capacitation. *Cell Tissue Res* **349:** 765–782.

Simon L, Ekman GC, Tyagi G, Hess RA, Murphy KM, Cooke PS. 2007. Common and distinct factors regulate expression of mRNA for ETV5 and GDNF, Sertoli cell proteins essential for spermatogonial stem cell maintenance. *Exp Cell Res* **313:** 3090–3099.

Smyth JT, Dehaven WI, Jones BF, Mercer JC, Trebak M, Vazquez G, Putney JW Jr. 2006. Emerging perspectives in store-operated Ca²⁺ entry: Roles of Orai, Stim and TRP. *Biochim Biophys Acta* **1763:** 1147–1160.

Sorrentino V, Giorgi M, Geremia R, Besmer P, Rossi P. 1991. Expression of the c-kit proto-oncogene in the murine male germ cells. *Oncogene* **6:** 149–151.

Sosnik J, Miranda PV, Spiridonov NA, Yoon SY, Fissore RA, Johnson GR, Visconti PE. 2009. Tssk6 is required for Izumo relocalization and gamete fusion in the mouse. *J Cell Sci* **122:** 2741–2749.

Stanford JS, Ruderman JV. 2005. Changes in regulatory phosphorylation of Cdc25C Ser287 and WEE1 Ser549 during normal cell cycle progression and checkpoint arrests. *Mol Biol Cell* **16:** 5749–5760.

Stanford JS, Lieberman SL, Wong VL, Ruderman JV. 2003. Regulation of the G₂/M transition in oocytes of *Xenopus tropicalis*. *Dev Biol* **260:** 438–448.

Steegborn C, Litvin TN, Levin LR, Buck J, Wu H. 2005. Bicarbonate activation of adenylyl cyclase via promotion of catalytic active site closure and metal recruitment. *Nat Struct Mol Biol* **12:** 32–37.

Stice SL, Robl JM. 1990. Activation of mammalian oocytes by a factor obtained from rabbit sperm. *Mol Reprod Dev* **25:** 272–280.

Stricker SA. 1997. Intracellular injections of a soluble sperm factor trigger calcium oscillations and meiotic maturation in unfertilized oocytes of a marine worm. *Dev Biol* **186:** 185–201.

Stricker SA. 1999. Comparative biology of calcium signaling during fertilization and egg activation in animals. *Dev Biol* **211:** 157–176.

Stricker SA, Whitaker M. 1999. Confocal laser scanning microscopy of calcium dynamics in living cells. *Microsc Res Tech* **46:** 356–369.

Strunker T, Goodwin N, Brenker C, Kashikar ND, Weyand I, Seifert R, Kaupp UB. 2011. The CatSper channel mediates progesterone-induced Ca²⁺ influx in human sperm. *Nature* **471:** 382–386.

Suarez SS. 2008a. Control of hyperactivation in sperm. *Hum Reprod Update* **14:** 647–657.

Suarez SS. 2008b. Regulation of sperm storage and movement in the mammalian oviduct. *Int J Dev Biol* **52:** 455–462.

Sun F, Bahat A, Gakamsky A, Girsh E, Katz N, Giojalas LC, Tur-Kaspa I, Eisenbach M. 2005. Human sperm chemotaxis: Both the oocyte and its surrounding cumulus cells secrete sperm chemoattractants. *Hum Reprod* **20:** 761–767.

Swann K. 1990. A cytosolic sperm factor stimulates repetitive calcium increases and mimics fertilization in hamster eggs. *Development* **110:** 1295–1302.

Swann K, Lai FA. 1997. A novel signalling mechanism for generating Ca²⁺ oscillations at fertilization in mammals. *BioEssays* **19:** 371–378.

Swann K, Saunders CM, Rogers NT, Lai FA. 2006. PLCζ: A sperm protein that triggers Ca²⁺ oscillations and egg activation in mammals. *Sem Cell Dev Biol* **17:** 264–273.

Tang W, Wu JQ, Guo Y, Hansen DV, Perry JA, Freel CD, Nutt L, Jackson PK, Kornbluth S. 2008. Cdc2 and Mos regulate Emi2 stability to promote the meiosis I-meiosis II transition. *Mol Biol Cell* **19:** 3536–3543.

Tesarik J, Testart J. 1994. Treatment of sperm-injected human oocytes with Ca²⁺ ionophore supports the development of Ca²⁺ oscillations. *Biol Reprod* **51:** 385–391.

Thomas TW, Eckberg WR, Dube F, Galione A. 1998. Mechanisms of calcium release and sequestration in eggs of *Chaetopterus pergamentaceus*. *Cell Calcium* **24:** 285–292.

Tokmakov AA, Sato KI, Iwasaki T, Fukami Y. 2002. Src kinase induces calcium release in *Xenopus* egg extracts via PLCγ and IP₃-dependent mechanism. *Cell Calcium* **32:** 11–20.

Tosti E, Boni R. 2004. Electrical events during gamete maturation and fertilization in animals and humans. *Hum Reprod Update* **10:** 53–65.

Tsafriri A, Chun SY, Zhang R, Hsueh AJ, Conti M. 1996. Oocyte maturation involves compartmentalization and opposing changes of cAMP levels in follicular somatic and germ cells: Studies using selective phosphodiesterase inhibitors. *Dev Biol* **178:** 393–402.

Tung JJ, Hansen DV, Ban KH, Loktev AV, Summers MK, Adler JR 3rd, Jackson PK. 2005. A role for the anaphase-promoting complex inhibitor Emi2/XErp1, a homolog of early mitotic inhibitor 1, in cytostatic factor arrest of *Xenopus* eggs. *Proc Natl Acad Sci* **102:** 4318–4323.

Uhlenbrock K, Gassenhuber H, Kostenis E. 2002. Sphingosine 1-phosphate is a ligand of the human gpr3, gpr6 and gpr12 family of constitutively active G protein-coupled receptors. *Cell Signal* **14:** 941–953.

van den Ham R, van Dissel-Emiliani FM, van Pelt AM. 2003. Expression of the scaffolding subunit A of protein phosphatase 2A during rat testicular development. *Biol Reprod* **68:** 1369–1375.

Vernet N, Dennefeld C, Rochette-Egly C, Oulad-Abdelghani M, Chambon P, Ghyselinck NB, Mark M. 2006. Retinoic acid metabolism and signaling pathways in the adult and developing mouse testis. *Endocrinology* **147:** 96–110.

Vig M, Beck A, Billingsley JM, Lis A, Parvez S, Peinelt C, Koomoa DL, Soboloff J, Gill DL, Fleig A, et al. 2006. CRACM1 multimers form the ion-selective pore of the CRAC channel. *Curr Biol* **16:** 2073–2079.

Visconti PE. 2009. Understanding the molecular basis of sperm capacitation through kinase design. *Proc Natl Acad Sci* **106:** 667–668.

Visconti PE, Krapf D, de la Vega-Beltran JL, Acevedo JJ, Darszon A. 2011. Ion channels, phosphorylation and mammalian sperm capacitation. *Asian J Androl* **13:** 395–405.

Wang RS, Yeh S, Tzeng CR, Chang C. 2009. Androgen receptor roles in spermatogenesis and fertility: Lessons from testicular cell-specific androgen receptor knockout mice. *Endocr Rev* **30:** 119–132.

Wasserman WJ, Masui Y. 1975. Effects of cyclohexamide on a cytoplasmic factor initiating meiotic maturation in *Xenopus* oocytes. *Exp Cell Res* **91:** 381–388.

Watanabe N, Broome M, Hunter T. 1995. Regulation of the human WEE1Hu CDK tyrosine 15-kinase during the cell cycle. *EMBO J* **14:** 1878–1891.

Whitaker M. 2006. Calcium at fertilization and in early development. *Physiol Rev* **86:** 25–88.

Wu JQ, Kornbluth S. 2008. Across the meiotic divide—CSF activity in the post-Emi2/XErp1 era. *J Cell Sci* **121:** 3509–3514.

Wu H, He CL, Fissore RA. 1997. Injection of a porcine sperm factor triggers calcium oscillations in mouse oocytes and bovine eggs. *Mol Reprod Dev* **46:** 176–189.

Wu JQ, Hansen DV, Guo Y, Wang MZ, Tang W, Freel CD, Tung JJ, Jackson PK, Kornbluth S. 2007a. Control of Emi2 activity and stability through Mos-mediated recruitment of PP2A. *Proc Natl Acad Sci* **104:** 16564–16569.

Wu Q, Guo Y, Yamada A, Perry JA, Wang MZ, Araki M, Freel CD, Tung JJ, Tang W, Margolis SS, et al. 2007b. A role for Cdc2- and PP2A-mediated regulation of Emi2 in the maintenance of CSF arrest. *Curr Biol* **17:** 213–224.

Xu Z, Kopf GS, Schultz RM. 1994. Involvement of inositol 1,4,5-trisphosphate-mediated Ca^{2+} release in early and late events of mouse egg activation. *Development* **120:** 1851–1859.

Yamashita M, Honda A, Ogura A, Kashiwabara S, Fukami K, Baba T. 2008. Reduced fertility of mouse epididymal sperm lacking Prss21/Tesp5 is rescued by sperm exposure to uterine microenvironment. *Genes Cells* **13:** 1001–1013.

Yanagimachi R. 1970. The movement of golden hamster spermatozoa before and after capacitation. *J Reprod Fertil* **23:** 193–196.

Yanagimachi R. 1994. Mammalian fertilization. In *The physiology of reproduction*, 2nd ed. (ed. Knobil E, Neill JD), Vol. 1, pp. 189–317. Raven, New York.

Yanagimachi R, Mahi CA. 1976. The sperm acrosome reaction and fertilization in the guinea-pig: A study in vivo. *J Reprod Fertil* **46:** 49–54.

Yang J, Winkler K, Yoshida M, Kornbluth S. 1999. Maintenance of G$_2$ arrest in the *Xenopus* oocyte: A role for 14-3-3-mediated inhibition of Cdc25 nuclear import. *EMBO J* **18:** 2174–2183.

Yoneda A, Kashima M, Yoshida S, Terada K, Nakagawa S, Sakamoto A, Hayakawa K, Suzuki K, Ueda J, Watanabe T. 2006. Molecular cloning, testicular postnatal expression, and oocyte-activating potential of porcine phospholipase Cζ. *Reproduction* **132:** 393–401.

Yoon SY, Jellerette T, Salicioni AM, Lee HC, Yoo MS, Coward K, Parrington J, Grow D, Cibelli JB, Visconti PE, et al. 2008. Human sperm devoid of PLC, zeta 1 fail to induce Ca^{2+} release and are unable to initiate the first step of embryo development. *J Clin Invest* **118:** 3671–3681.

Yoshida M, Sensui N, Inoue T, Morisawa M, Mikoshiba K. 1998. Role of two series of Ca^{2+} oscillations in activation of ascidian eggs. *Dev Biol* **203:** 122–133.

Yoshida S, Sukeno M, Nabeshima Y. 2007. A vasculature-associated niche for undifferentiated spermatogonia in the mouse testis. *Science* **317:** 1722–1726.

Yoshinaga K, Nishikawa S, Ogawa M, Hayashi S, Kunisada T, Fujimoto T, Nishikawa S. 1991. Role of c-kit in mouse spermatogenesis: Identification of spermatogonia as a specific site of c-kit expression and function. *Development* **113:** 689–699.

Cite this chapter as *Cold Spring Harb Perspect Biol* doi: 10.1101/cshperspect.a006064

CHAPTER 18

Cell Signaling and Stress Responses

Gökhan S. Hotamisligil[1] and Roger J. Davis[2]

[1]Department of Genetics and Complex Diseases, Broad Institute of Harvard-MIT, Harvard School of Public Health, Boston, Massachusetts 02115

[2]Howard Hughes Medical Institute and Program in Molecular Medicine, University of Massachusetts Medical School, Worcester, Massachusetts 01605

Correspondence: roger.davis@umassmed.edu

SUMMARY

Stress-signaling pathways are evolutionarily conserved and play an important role in the maintenance of homeostasis. These pathways are also critical for adaptation to new cellular environments. The endoplasmic reticulum (ER) unfolded protein response (UPR) is activated by biosynthetic stress and leads to a compensatory increase in ER function. The JNK and p38 MAPK signaling pathways control adaptive responses to intracellular and extracellular stresses, including environmental changes such as UV light, heat, and hyperosmotic conditions, and exposure to inflammatory cytokines. Metabolic stress caused by a high-fat diet represents an example of a stimulus that coordinately activates both the UPR and JNK/p38 signaling pathways. Chronic activation of these stress-response pathways ultimately causes metabolic changes associated with obesity and altered insulin sensitivity. Stress-signaling pathways, therefore, represent potential targets for therapeutic intervention in the metabolic stress response and other disease processes.

Outline

1 INTRODUCTION

An important aspect of cellular physiology is the maintenance of homeostasis. Evolutionarily conserved biochemical mechanisms play a key role in this process. Thus,
exposure to extra- or intracellular stress disrupts cellular
homeostasis and causes the engagement of signaling pathways that serve to rebalance biochemical processes within
the cell. One example is the AMP-activated protein kinase
signaling pathway, which responds to increased AMP and
ADP concentrations within the cell by dampening anabolic pathways and promoting catabolic pathways that replenish the ATP supply (Ch. 14 [Hardie 2012]). A second
example is the cellular response to DNA damage that engages the ataxia telangiectasia mutated (ATM) stress-signaling pathway to induce growth arrest mediated by the
p53 tumor suppressor protein and promote DNA repair
before reentry into the cell cycle (Ch. 6 [Rhind and Russell
2012]). A third example is the regulation of receptor ligand
sensitivity to control the amplitude of signal transduction.
This type of stress response maintains, for example, the
dynamic range of vision following exposure to high- and
low-intensity light sources (see Ch. 11 [Julius and Nathans
2012]). Similarly, signaling by the metabolic hormones
leptin and insulin is dynamically regulated by stress-signaling pathways to control feeding behavior and biosynthetic
processes (Ch. 14 [Hardie 2012]). Such pathways are critical for normal cellular homeostasis and adaptive changes
in cell physiology that benefit the organism.

In addition to their contributions to normal physiology, stress-activated signaling pathways play roles in establishing dysfunctional states associated with stress exposure
and the development of disease. Here, we focus on two
different mammalian stress-activated-signaling pathways
to illustrate these concepts: the unfolded protein response
(UPR) and stress-activated MAP kinase (MAPK) pathways.
The UPR is engaged within the endoplasmic reticulum
(ER) during biosynthetic stress and leads to a coordinated
inhibition of general protein translation and specific up-
regulation of ER functional capacity. The pathways involved therefore serve to maintain cellular homeostasis.
Stress-activated MAPK pathways, in contrast, are regulated
by a diverse array of intra- and extracellular stresses, including environmental physical/chemical changes and exposure to inflammatory cytokines. These stress pathways
cause phosphorylation of nuclear and cytoplasmic substrates, leading to a network response and adaptation to
the new cellular environment.

The UPR and MAPK pathways can function separately or cooperatively. For example, increased saturated fatty
acid levels, caused by a high-fat diet, induce both the
UPR and stress-activated MAPK pathways. Together, these
pathways cause adaptation to the new diet by regulating
insulin signaling, blood glucose concentration, and obesity
(Fig. 1).

2 THE UNFOLDED PROTEIN RESPONSE

The ER is a multifunctional organelle that comprises retic-
ular and tubular structures spanning the cell and has numerous fundamental functions in all cells. It has two major
domains: smooth ER, which is devoid of ribosomes; and
rough ER, which is studded by attached ribosomes (Palade
and Porter 1954; Lynes and Simmen 2011). The rough ER
is responsible for the synthesis and trafficking of secreted
and integral membrane proteins (Gething and Sambrook
1992; Ellgaard and Helenius 2003), whereas smooth ER
is associated with lipid synthesis and metabolism, and calcium storage. The functional specialization within the ER
is more complex and various subregions support distinct
pathways involved in homeostasis and survival. For example, rough ER is involved in quality control and protein
degradation and harbors oxidoreductases (Gething and
Sambrook 1992; Ellgaard and Helenius 2003; Kostova
and Wolf 2003; Rutkowski and Kaufman 2004; Meusser
et al. 2005) and there are additional ER domains devoted
to specialized functions including the mitochondria-as-
sociated ER membrane (MAM), the nuclear envelope, per-
oxisomal components, Russell bodies, and lipid droplets

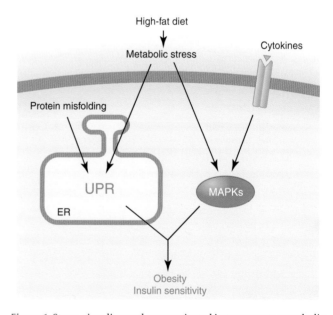

Figure 1. Stress-signaling pathways activated in response to metabolic
stress. Feeding mice a high-fat diet causes metabolic stress that leads
to the UPR and activation of stress-activated MAP kinases. These
signaling pathways result in an adaptive response associated with
obesity and altered insulin sensitivity.

Cite this chapter as *Cold Spring Harb Perspect Biol* doi: 10.1101/cshperspect.a006072

(Lynes and Simmen 2011). Similarly, the smooth ER has specialized domains, such as the plasma membrane-associated ER and regions that can also form MAM, autophagosomes, and lipid droplets (English et al. 2009; Hayashi et al. 2009).

Adaptation of the ER to a variety of metabolic and stress conditions is critical for cell function and survival, as well as organismal health (Walter and Ron 2011). Given the demands fluctuating conditions place on the functional capacity of the ER, a potent adaptive response, the UPR (Mori 2000; Marciniak and Ron 2006; Zhao and Ackerman 2006), has evolved to maintain the functional integrity of this organelle (Fig. 2). In eukaryotic cells, UPR signaling is initiated by three ER-membrane-associated proteins: PERK (PKR-like eukaryotic initiation factor 2α [eIF2α] kinase), IRE1 (inositol-requiring enzyme 1), and ATF6 (activating transcription factor 6). Acting in concert, signaling through these branches controls protein synthesis, facilitates protein degradation, and produces the molecules necessary for the ER to restore equilibrium (Mori 2000; Marciniak and Ron 2006; Zhao and Ackerman 2006).

Early studies of adaptive ER responses showed that the levels of two ER-localized chaperones, 78- and 94-kDa glucose-regulated proteins (GRP78 and GRP94), are increased on glucose starvation (Shiu et al. 1977) and protein N-glycosylation inhibitors and calcium ionophores enhance their expression (Welch et al. 1983; Resendez et al. 1985; Kim et al 1987). Sambrook and colleagues subsequently observed increased levels of ER chaperones on overexpression of mutant influenza virus hemagglutinin and were the first to propose that malfolded proteins in the lumen of the ER are detected and invoke a response (Kozutsumi et al. 1988). Subsequently, genetic screens in yeast identified IRE1 as an ER-localized receptor-like kinase and ribonuclease required for the ER-to-nucleus signaling that activates chaperone expression under ER stress conditions (Nikawa and Yamashita 1992; Cox et al. 1993; Mori et al. 1993). Hac1 was shown to be the leucine-zipper transcription factor that functions downstream from IRE1 to induce transcription by binding to a defined DNA sequence, the UPR element (UPRE) (Cox and Walter 1996; Nikawa et al. 1996) in promoters turned on by the UPR in yeast. The equivalent of Hac1 in multicellular organisms is XBP1 (Shen et al. 2001; Calfon et al. 2002).

An additional *cis*-acting element (ER stress element, ERSE) was later identified in promoter regions of mammalian ER chaperones and led to the description of another ER-resident leucine-zipper transcription factor, ATF6 (Yoshida et al. 1998). Together, ATF6 and XBP1 stimulate the expression of a broad array of genes involved in protein folding, secretion, and degradation to clear mis-folded proteins from the ER (Walter and Ron 2011). Finally, the third molecule activated during the UPR was identified as PERK, one of the four known eIF2α kinases (Baird and Wek 2012; Donnelly et al. 2013). It is involved in translational attenuation, temporarily halting arrival of new proteins in the ER (Harding et al. 1999). These three branches are now considered as mediators of the canonical UPR (Fig. 2).

In the canonical model, the intraluminal domains of these initiators (i.e., the amino termini of IRE1 and PERK, and the carboxyl terminus of ATF6) are bound by the chaperone Grp78 (also called BiP) in the absence of stress and rendered inactive (Bertolotti et al. 2000; Shen et al. 2002). Accumulation of improperly folded proteins in the ER lumen results in the recruitment of BiP away from these UPR sensors. Stripping off BiP allows for oligomerization and activation of PERK and IRE1, and translocation of ATF6 to the Golgi cisternae, which lead to a cascade of downstream signaling events (Shamu and Walter 1996; Bertolotti et al. 2000). Recent studies support the view that more complex luminal events underlie mounting of the UPR. For example, IRE1 can form higher-order oligomers on activation in vitro (Li et al. 2010) and directly interact with unfolded proteins (Gardner and Walter 2011). Dynamic regulation of IRE1 by BiP may thus adjust the magnitude of activation as opposed to providing an "on-or-off" switch (Pincus et al. 2010). Hence, it is likely that stress responses emanating from the ER are more complex than the canonical UPR model.

Activation of the ATF6 branch of the UPR requires translocation of ATF6 to the Golgi body and processing by the serine protease site-1 protease and the metalloprotease site-2 protease to release an active transcription factor (Chen et al. 2002). This branch also responds to signals other than BiP sequestration—for example, the redox status of the ER. Active ATF6 moves to the nucleus to stimulate the expression of genes containing the ERSE1, ERSE2, UPRE, and cAMP-response elements (p. 99 [Sassone Corsi 2012]) in their promoters (Yoshida et al. 1998). Genes required for ER-associated degradation and the gene encoding the ER degradation-enhancing α-mannosidase-like protein (EDEM) contain UPREs and, when induced, facilitate clearance and degradation of misfolded proteins from the ER lumen (Yoshida et al. 1998; Friedlander et al. 2000; Kokame et al. 2001).

The oldest branch of the UPR is mediated by IRE1, which is conserved from yeast to humans (Patil and Walter 2001; Calfon et al. 2002). IRE1 has two known isoforms, α and β, the latter being restricted primarily to the intestine (Wang et al. 1998). IRE1 harbors two distinct catalytic activities: a serine/threonine kinase for which the only known substrate is IRE1 itself, and an endoribonuclease

A

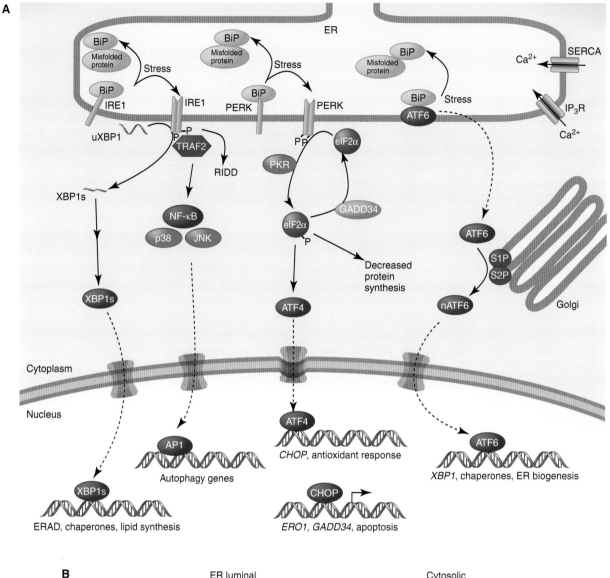

B

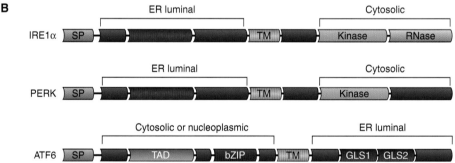

Figure 2. The canonical UPR. (*A*) In the canonical model of the UPR, unfolded or misfolded proteins activate the three major sensing molecules (IRE1, PERK, and ATF6) at the ER membrane by recruiting the ER chaperone BiP away from the lumenal domains of these proteins. IRE1 is a kinase and ribonuclease that on autophosphorylation activates splicing and produces the active transcription factor XBP1, which induces the expression of ER chaperones, degradation components, and lipid synthesis enzymes. PERK is a kinase that is also activated through dimerization and autophosphorylation and phosphorylates eIF2α to attenuate general protein synthesis. ATF6 is a transcription factor that once released from the ER will move to the Golgi. After processing at this site, it translocates to the nucleus to activate the transcription of chaperone genes. Together, these pathways reduce entry of proteins into the ER, facilitate disposal of the misfolding proteins, and produce the components for the ER to adapt its folding capacity to reach equilibrium. When these pathways fail to reach homeostasis, they can also trigger death. Under severe stress conditions, the synthesis of ATF4 is enhanced in an eIF2α-phosphorylation-independent manner that promotes apoptosis. (*B*) Domain structure of the ER stress sensors IRE1, PERK, and ATF6. SP, signal peptide; TM, transmembrane domain; TAD, transcriptional activation domain; bZIP, basic leucine zipper; GLS1 and GLS2, Golgi localization sequences 1 and 2. Dark gray bars represent regions of limited sequence similarity between IRE1 and PERK.

activity (Sidrauski and Walter 1997). The endoribonuclease activity is activated on dimerization and autophosphorylation and cleaves a 26-nucleotide intron from the XBP1 messenger RNA (mRNA), generating an mRNA whose translation produces functional XBP1 (so-called XBP1s) (Shamu and Walter 1996; Sidrauski and Walter 1997). XBP1s, alone or in conjunction with ATF6α, launches a transcriptional program that induces many ER chaperones (including BiP), proteins involved in ER biogenesis, and secretion (for example, EDEM, ERdj4, protein disulfide isomerase [PDI], and other ER proteins) (Yoshida et al. 2001, 2003; Lee et al. 2003). The endonuclease activity of IRE1 can also degrade other mRNAs, preventing their translation and thereby providing an additional way to reduce the translational burden and thus relieve ER stress (Hollien and Weissman 2006). This mechanism has been termed regulated IRE1-dependent degradation.

GTP-bound eIF2 is essential for loading of the initiator Met-tRNA (tRNA) onto an mRNA-charged 40S ribosomal subunit for translation initiation (Hinnebusch and Lorsch 2012). Phosphorylation of its GTP-binding subunit, eIF2α, at S51 by PERK is another important aspect of the UPR. This converts eIF2α into a competitive inhibitor of eIF2B (the GTP exchange factor for eIF2α). This sequesters eIF2B and reduces the rate of regeneration of the eIF2-GTP-tRNA$_i^{Met}$ ternary complex, which, in turn, results in lower rates of global protein synthesis, thereby reducing the ER workload (Shi et al. 1998; Harding et al. 1999). At least three other kinases can phosphorylate eIF2α at S51: double-stranded RNA-dependent kinase (PKR), general control nonderepressible 2, and heme-regulated inhibitor kinase (Baird and Wek 2012; Donnelly et al. 2013). The PERK branch of the UPR is also linked to transcriptional regulation through several distinct mechanisms, which increase the level of and/or activate the transcription factors ATF2, ATF4, C/EBP (Harding et al. 2000; Ma et al. 2002; Ron and Walter 2007), NRF2 (Cullinan et al. 2003), and NF-κB (Jiang et al. 2003; Deng et al. 2004). Generation of the protein products of the induced transcripts is achieved in the context of general translational attenuation through features in the mRNAs that permit their preferential translation. For example, the 5′-end of the ATF4 transcript has two upstream open reading frames (uORFs) that prevent translation under normal circumstances (Somers et al. 2013). However, under stressed conditions, ribosome capacitation is delayed, the uORFs are skipped, and functional ribosome complexes are assembled at the bona fide start codon (Harding et al. 2000). The synthesis of functional ATF4 consequently activates the expression of genes involved in apoptosis, ER redox control, glucose metabolism, and the relief of eIF2α inhibition (Harding et al. 2000; Ma et al. 2002; Jiang et al. 2004).

2.1 Noncanonical Aspects of the UPR and Other Stress Signals

The ER has developed additional strategies to ensure its proper function under stress conditions. For example, the Golgi reassembly stacking protein 1 facilitates the exit of mutant or misfolded proteins from the ER lumen through the activation of an unconventional secretory pathway (Gee et al. 2011). In addition, chaperone-mediated autophagy assists the disposal of misfolded proteins in the ER, and ER-phagy (selective autophagy of the ER) promotes the turnover of damaged ER in bulk (Klionsky 2010; Arias and Cuervo 2011). Moreover, during cytokinesis, alternative surveillance mechanisms other than the UPR are also activated to monitor the "fitness" of the ER and ensure its proper transmission into daughter cells (Babour et al. 2010). ER stress also generates oxidative stress that needs to be alleviated during recovery (Cullinan and Diehl 2006). Reactive oxygen species (ROS) are produced in the ER owing to UPR-stimulated up-regulation of protein chaperones involved in disulfide bond formation as well as oxidative phosphorylation in the mitochondria (Sevier et al. 2001). During disulfide bond formation in the ER, electrons are passed through a series of thiol-disulfide exchange reactions from the thiols of the substrate protein to PDI, then to ERO1, and finally to molecular oxygen (Tu and Weissman 2002). As a byproduct of these reactions, ROS (hydrogen peroxide and other peroxides, superoxide radical, hydroperoxyl radical, and hydroxyl radical) accumulate during the UPR-increased protein folding and can produce sufficiently high levels of ROS to be toxic to the cell (Harding et al. 2003; Sevier and Kaiser 2008). Finally, there are also numerous metabolic adaptations integrated into the UPR (see below).

2.2 Physiological Roles of the UPR

In addition to the pathways discussed above, the UPR also engages many other processes that are critical for normal cellular and organismal adaptations (Hotamisligil 2010). Three specific examples are discussed below: (1) survival pathways, apoptosis, and autophagy; (2) immune responses and inflammation; and (3) nutrient sensing and metabolic regulation. In each case, components of the UPR interact with other signaling pathways to control processes beyond simply protein folding and ER stress, which can have effects both within and beyond the cell in which they are activated. A challenging aspect of ER biology is to understand the mechanisms leading to adaptive versus maladaptive/apoptotic responses in the face of stress. It is likely that defective activity or disproportionate (prolonged or imbalanced) engagement of distinct signaling networks stimulated by each branch of the UPR is a critical determinant of detrimental outcomes (Hotamisligil 2010).

2.2.1 Cell Survival and Death Responses of the ER

Under ER stress conditions, activation of the UPR reduces unfolded protein load through several prosurvival mechanisms, including the expansion of the ER membrane, selective synthesis of key components of the protein folding and quality control machinery, and attenuation of the influx of proteins into the ER. In conditions in which ER homeostasis cannot be established owing to severe or prolonged stress, or unusual challenges (such as those presented by energy or nutrient overload and inflammation), the responses triggered by ER stress result in a maladaptive set of events leading to various cellular or systemic pathologies, including death by apoptosis (Rao et al. 2004; Holcik and Sonenberg 2005; Szegezdi et al. 2006; Scull and Tabas 2011).

When cells are subject to irreparable ER stress, the UPR drives proapoptotic signals to eliminate the damaged material (Ch. 19 [Green and Llambi 2014]). Both PERK and ATF6 induce expression of the transcription factor C/EBP homologous protein (CHOP), which in turn leads to reduced expression of the antiapoptotic gene *Bcl2* and increased expression of a number of proapoptotic genes (McCullough et al. 2001; Ma et al. 2002; Marciniak et al. 2004). The proteins p58[IPK], GADD34, and TRB3 are also involved in the PERK-mediated apoptotic pathway. These targets have individually been linked to the promotion of apoptosis; GADD34 and p58[IPK] both negatively regulate eIF2α signaling and downstream adaptive responses, and TRBP inhibits signaling by the kinase Akt (Novoa et al. 2001; Ladiges et al. 2005; Bromati et al. 2011). IRE1α activation is linked to apoptosis through its ability to activate the JNK MAPK (Urano et al. 2000) and subsequent downstream phosphorylation of the Bcl2 family members Bim and Bmf, which in turn cause Bax/Bak-dependent apoptosis (Lei and Davis 2003). IRE1 may also regulate the activation of ER-localized caspase-12 through the modulation of a TRAF2–caspase-12 complex (Yoneda et al. 2001; Walter and Ron 2011). Autophagosome formation is accelerated in cells under ER stress (Kroemer et al. 2010), and disturbance of autophagy can render them vulnerable to ER stress and, consequently, lead to death (Ogata et al. 2006). IRE1α–JNK signaling can trigger autophagy by leading to phosphorylation of Bcl2, which disrupts Bcl2–beclin-1 binding. This results in the activation of beclin 1, an essential autophagy regulator (see Wei et al. 2008; Ch. 19 [Green and Llambi 2014]).

In several different contexts, nutrient-sensing pathways are coupled to ER function, in particular, to the IRE1 axis. This is best illustrated by the coordinated regulation of a major cellular nutrient sensor, the mTORC1 complex (see p. 91 [Laplante and Sabatini 2012]), and the UPR. Loss of the mTOR inhibitors TSC1 or TSC2 in cell lines and mouse or human tumors hyperactivates mTORC1 and its downstream network (Yecies and Manning 2011). A key function of mTORC1 is to stimulate overall translational initiation (Proud 2009), which causes ER stress (Ozcan et al. 2008). This can promote mTORC1-mediated negative-feedback inhibition of insulin action and further increase the vulnerability of cells to apoptosis (Ozcan et al. 2008). The UPR is sensitive to the nutritional status of the cell, just like the mTORC1 complex, responding to glucose deprivation, exposure to excess fatty acids, hypoxia, and growth stimuli (Appenzeller-Herzog and Hall 2012). In pancreatic β cells, for example, glucose can regulate IRE1 activation by promoting the assembly of an IRE1α-RACK1-PP2A complex. RACK1 is a β-propeller protein that binds to the 40S ribosomal subunit (Coyle et al. 2009; Sharma et al. 2013), linking cell regulation and translation. In response to an acute increase in glucose levels, RACK1 directs PP2A to IRE1α, promoting its dephosphorylation (Qiu et al. 2010). Conversely, ER stress or prolonged exposure to high glucose levels causes RACK1 to dissociate from PP2A, resulting in disruption of this tripartite regulatory module (Qiu et al. 2010). In this scenario, RACK1-associated phosphorylated IRE1α may have altered functional outputs. The balance in IRE1 signaling is, thus, vital to the survival of these cells and insulin biosynthesis.

2.2.2 The UPR and Inflammation

The UPR and other stress-signaling networks are highly integrated with immune signaling (Gregor and Hotamisligil 2011). For example, all three main arms of the UPR regulate NF-κB signaling during ER stress through distinct mechanisms (Jiang et al. 2003; Kaneko et al. 2003; Deng et al. 2004; Hu et al. 2006; Yamazaki et al. 2009). Moreover, signaling through Toll-like receptors (TLRs) (on p. 121 [Lim and Staudt 2013]) can activate IRE1, via NOX2-mediated production of ROS, resulting in production of inflammatory cytokines (Martinon et al. 2010). TLRs engage IRE1α, but not the other branches of the UPR, to promote cytosolic splicing and activation of XBP1, which occur in the absence of an ER stress response and do not seem to contribute to the induction of ER-stress-induced genes. Instead, activation of XBP1 by IRE1 promotes sustained production of inflammatory mediators, including interleukin (IL) 6, in certain contexts (Martinon et al. 2010). It is important to emphasize that these responses may not necessarily be related to ER stress per se, despite using IRE1. This distinction and the duration and context of the signaling are critical to determining signaling outcome.

Links between the immune response and ER stress can be even more complex and involve both innate and adaptive immunity. This is exemplified by the secretory cells of the gut (McGuckin et al. 2011). In intestinal goblets cells, a mutation in mucin 2 (a component of mucus) causes a disease similar to ulcerative colitis in humans with a complex pattern of inflammation involving many immune mediators (Heazlewood et al. 2008). In addition, mutation of mucin 2 results in ER vacuolization and activation of GRP78 and XBP1. Similarly, experimentally triggered ER stress in cultured cells can cause increased expression of many inflammatory molecules, such as IL8, IL6, MCP1, and tumor necrosis factor (TNF) (Li et al. 2005). As mentioned above, ER stress and autophagy are linked, and autophagy is a critical regulator of innate immune responses (Levine et al. 2011). Importantly, cytokines and the pathways they activate influence the function of ER (Zhang et al. 2008; Jiao et al. 2011). Activation of certain pathways, such as those involving JNK and IκB kinase (IKK) (Deng et al. 2004; Hu et al. 2006), and production of certain mediators, such as ROS, all have negative effects on the function of ER (Cullinan and Diehl 2006; Gotoh and Mori 2006; Uehara et al. 2006). The extent of oxidative stress and the levels, duration, and magnitude of ROS and/or NO production can tip the balance in ER responses toward a maladaptive profile (Chan et al. 2011). The metabolic status of the ER (or the metabolic environment within which it has to operate) is thus a key determinant of the balance between adaptive and maladaptive responses.

The kinase PKR, although not known to be physically associated with the ER, is also activated during ER stress (Shimazawa et al. 2007; Nakamura et al. 2010). Once activated, it brings together several key stress and inflammatory signaling molecules, including JNK and insulin-signaling components such as IRS1, with eIF2α. PKR interacts with and directly phosphorylates IRS1 through which it links ER stress to suppression of insulin action. Inflammatory cytokines and toxic lipids, such as palmitate, induce phosphorylation of IRS by PKR, leading to inhibition of insulin signaling. In the absence of PKR activity, neither inflammatory cytokines nor toxic lipids can interfere with insulin action, and deletion of PKR in mice results in significantly improved glucose metabolism.

PKR also contributes to the activation of the NLRP3 inflammasome and HMGB1 (high-mobility group protein B1) production (Lu et al. 2012), which are responsible for activation of inflammatory processes (Lamkanfi and Dixit 2012) in response to stimuli such as double-stranded RNA, adjuvant alum and *Escherichia coli* (Ch. 15 [Newton and Dixit 2013]). PKR also physically interacts with NLRP3, and activates the inflammasome in a cell-free system with recombinant NLRP3, ASC, and procaspase-1 (Lu et al.

2012). Recent reports also suggest direct regulation of the inflammasome by IRE1 itself via the thioredoxin-interacting protein (TXNIP) (Lerner et al. 2012; Oslowski et al. 2012). Under irremediable ER stress conditions, activation of IRE1α results in elevated expression of TXNIP, leading to a terminal UPR featuring activation of the inflammasome. Hence, the UPR, through PKR and IRE1, appears to act at the interface of ER stress signaling, inflammatory responses, and metabolic regulation. However, other studies indicate that TXNIP is involved in the endocytosis of the glucose transporter GLUT1 (Wu et al. 2013); so, the connection to inflammasome function may be indirect.

2.2.3 Metabolic Responses Emanating from the ER

The ER is critical for regulation of metabolic homeostasis and its dysfunction plays an important role in the emergence of metabolic disease. The ER is sensitive to the metabolic status of cells, responding to feeding–fasting cycles, acute nutrient exposure, and circadian rhythms, and is equipped with direct and indirect means to alter metabolic responses (Cretenet et al. 2010; Hotamisligil 2010; Pfaffenbach et al. 2010; Boden et al. 2011; Hatori et al. 2011). These aspects of ER function are critical in a number of metabolic diseases, such as obesity and type 2 diabetes. Many of the signals generated through the UPR are directly linked to metabolic responses and also engage inflammatory and stress-signaling pathways (Fig. 3), which are also linked to metabolic regulation. Integration of these pathways is critical for glucose metabolism, insulin secretion, insulin action, and the metabolic activities of the ER (Marciniak and Ron 2006; Hotamisligil 2010; Fu et al. 2012).

Insulin is the master regulator of glucose metabolism in mammals and defective insulin action and/or production results in diabetes (Ch. 7 [Ward and Thompson 2012] and Ch. 14 [Hardie 2012]). Obesity and diabetes cause ER dysfunction and/or stress in animal models (Ozcan et al. 2004; Nakatani et al. 2005) and humans (Boden et al. 2008; Sharma et al. 2008; Gregor et al. 2009), and chemicals that reduce ER stress and improve ER function improve glucose metabolism in animal models (Ozawa et al. 2005; Ozcan et al. 2006; Kammoun et al. 2009) and in humans (Kars et al. 2010; Xiao et al. 2011). ER stress responses intersect with insulin action at several points, including the stimulation of JNK by PKR (Nakamura et al. 2010), CaM kinase II (Li et al. 2009; Ozcan and Tabas 2012; Ozcan et al. 2012), and IRE1-dependent interaction between TRAF2 and ASK1 (a MAP kinase kinase kinase, MAPKKK) (Urano et al. 2000). This inhibits insulin signaling by uncoupling the insulin receptor from the substrate, IRS1, that mediates its metabolic actions (Hirosumi et al. 2002; Ozcan et al. 2004; Sabio and Davis 2010; Ozcan et al. 2012).

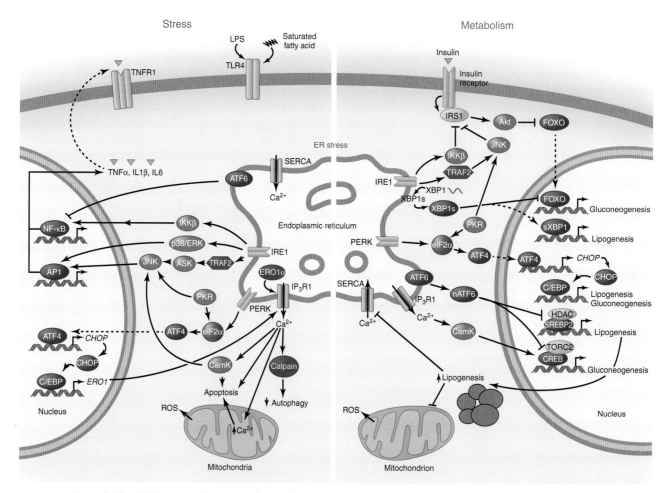

Figure 3. The UPR in stress signaling, inflammation, and metabolism. The UPR contributes to both inflammatory/stress signaling and metabolic regulation, as exemplified in chronic metabolic diseases. It activates several stress-related kinases including ERK, p38, JNK, and IKK. The resulting signals can impair signaling by insulin or other endocrine hormones and disrupt metabolism. The UPR can also modulate glucose and lipid metabolism directly. Nuclear ATF6 inhibits gluconeogenesis and lipogenesis by directly binding to TORC2 and SREBP2 proteins, respectively. This mechanism is defective in metabolic disease. The spliced form of XBP1 (XBP1s) can directly or indirectly (through SREBP1) activate the lipogenesis program while inhibiting gluconeogenesis. eIF2α phosphorylation leads to the synthesis of CHOP and the activation of lipogenesis programs via CEBPα/β. (Note that the opposing effects of the UPR on gluconeogenesis and lipogenesis are context dependent.) Activation of lipogenesis alters the membrane lipid composition of the ER, inhibits SERCA calcium pumps, and propagates ER stress, which in turn disrupts metabolism further. Hence, inflammatory, stress, and metabolic responses generate a vicious cycle if the ER dysfunction cannot be remedied.

ER stress and inflammation in the central nervous system and gut can also indirectly regulate glucose metabolism in peripheral tissues, particularly the liver (Caricilli et al. 2011; Purkayastha et al. 2011; Milanski et al. 2012). Chronic ER stress affects hypothalamic neuroendocrine pathways that regulate satiety, body weight, and metabolism. Acute ER stress in the brain can induce glucose intolerance and systemic and hepatic insulin resistance, and blocking brain TLR4 or TNF action in obese individuals improves insulin sensitivity and glucose homeostasis in the liver (Mighiu et al. 2012; Milanski et al. 2012). In the gut, alterations in the composition of the gut microbiota seen in TLR2-defi-

ciency are accompanied by ER stress, JNK activation, and impaired insulin signaling in the liver (Caricilli et al. 2011). Hence, ER function is part of the inter-organ communication network that ensures metabolic homeostasis.

Insulin is produced by pancreatic β cells, which, as professional secretory cells, are heavily dependent on a healthy, functioning ER. Mutations in the PERK branch (Wolcott-Rallison syndrome) of the UPR (Delepine et al. 2000; Harding et al. 2001; Zhang et al. 2002) or the WFS1 gene (for Wolfram syndrome 1), which encodes wolframin, result in death and dysfunction of β cells and cause diabetes (Ishihara et al. 2004; Fonseca et al. 2005; Yamada et al. 2006).

Wolframin targets ATF6 for proteasomal degradation, and its loss results in ER stress signaling mediated by increased levels of ATF6α (Fonseca et al. 2010). XBP1 deficiency in mice also compromises β-cell function and survival (Lee et al. 2011a). Moreover, in type 2 diabetes, pancreatic β cells also suffer from a vicious cycle of inflammatory alterations, which can compromise the folding capacity of the ER and normal functioning of β cells, which progressively lose their ability to produce insulin (Eguchi et al. 2012). This is a challenging problem for therapeutic intervention because both deficiency and hyperactivity of each one of the three UPR branches can be detrimental for β cells (Nozaki et al. 2004; Seo et al. 2008; Trusina et al. 2008) and potentially for other cells.

The UPR is thus a critical determinant of metabolic homeostasis. Lipid, carbohydrate, and protein metabolism are all regulated by the UPR, and both the synthesis of and sensitivity to key metabolic hormones, such as insulin, are regulated by these pathways. The survival and function of key cell types critical to metabolism, including pancreatic β cells, gut epithelium, and hypothalamic neurons, are controlled by the UPR. Furthermore, the link between inflammation and metabolism is intimately related to ER function.

3   STRESS SIGNALING BY MAP KINASES

MAPK pathways are universally conserved eukaryotic-signaling modules that transduce extracellular and intracellular signals to regulatory networks within the cell by phosphorylation of key protein targets (p. 81 [Morrison 2012]). MAPKs are activated by dual phosphorylation of tyrosine and threonine residues in a partially unstructured segment (activation loop) between the amino and carboxy-terminal lobes of the catalytic domain (Payne et al. 1991). Phosphorylation on the threonine residue improves the geometry of the active site catalytic residues, leading to formation of hydrogen bonds on the surface of the MAPK that connect the amino-terminal domain to the activation loop and promote closure of the active site cleft. Phosphorylation on the tyrosine residue leads to the creation of new hydrogen bonds and refolding of the activation loop to form a structure that contributes to the substrate-interaction sites (Canagarajah et al. 1997; Rodriguez Limardo et al. 2011). The ERK family of MAPKs are primarily activated by exposure of cells to cytokines and growth factors (Robinson and Cobb 1997). In contrast, the p38 and JNK MAPK families respond primarily to the exposure of cells to extracellular and intracellular stress (Davis 2000; Cuadrado and Nebreda 2010). Consequently, the JNK and p38 MAP kinases are often termed stress-activated MAPKs. They respond to inflammatory cytokines and many changes in the physical/chemical environment, as well as DNA damage and redox imbalance (Fig. 4).

The JNKs are encoded by three genes (*JNK1*, *JNK2*, and *JNK3*), which are alternatively spliced to yield 10 different isoforms (Gupta et al. 1996). JNK1 and JNK2 are ubiquitously expressed, but JNK3 is expressed primarily in the brain (Davis 2000). Gene disruption studies in mice show that these JNK isoforms can mediate different biological responses (Davis 2000). The p38 MAPKs are encoded by four genes that can be divided into two subgroups (*p38α/β* and *p38γ/δ*). These p38 isoforms show nonredundant functions and different sensitivities to small molecule inhibitors (Cuenda and Rousseau 2007; Cuadrado and Nebreda 2010).

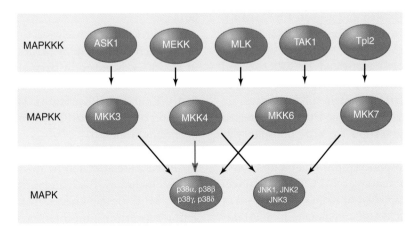

Figure 4. Stress-activated MAPK-signaling pathways. The p38 MAP kinases are primarily activated by the MAPKK isoforms MKK3 and MKK6, but a minor contribution of MKK4 can be detected. All p38 MAPK isoforms are activated by MKK3 and MKK6, although p38δ is activated by MKK3 significantly more potently than MKK6. The JNK group of MAPKs is activated by the MAPKK isoforms MKK4 and MKK7.

The minimal consensus sequence for target protein phosphorylation by MAPKs is -S/T-P-. However, the presence of this consensus motif is not sufficient for phosphorylation by a MAPK, which frequently requires a docking interaction between the MAPK and another region of the substrate (Enslen and Davis 2001; Tanoue and Nishida 2003; Akella et al. 2008). Two types of MAPK docking motifs have been identified in substrates: (1) the FXFP motif, and (2) the D domain, comprising a hydrophobic motif (LXL) plus a basic region (Bardwell and Thorner 1996; Jacobs et al. 1999; Enslen and Davis 2001). Structural analysis of proteins docked to p38 MAPK (Chang et al. 2002) and JNK3 (Heo et al. 2004) show extensive interactions between the docked proteins and regions of the MAPK outside the active site. The sites of D domain and FXFP interaction on MAPKs are different (Akella et al. 2008). In addition to these conserved interactions, the carboxy-terminal sequence of p38γ MAPK can dock directly to proteins that have PDZ domains (e.g., α1 syntrophin, PSD95 [also known as SAP90], and DLG [also known as SAP97]) to direct their phosphorylation (Hasegawa et al. 1999; Hou et al. 2010).

Bioinformatic analyses and protein interaction screens (e.g., two-hybrid assays) have identified many MAPK substrates. More recently, chemical genetic methods and mass spectroscopy have enabled a more comprehensive analysis of MAPK substrates in specific tissues (Allen et al. 2007; Carlson et al. 2011). These include membrane, cytosolic, and nuclear proteins that participate in many biological processes, and especially many transcription factors, such as ATF2 (activated by JNK1/2/3 and p38α/β MAPK), Jun (activated by JNK1/2/3), MEF2C (activated by p38α/β MAPK and ERK5), and Elk1 (activated by JNK1/2/3, p38α/β MAPK, and ERK1/2) (Whitmarsh and Davis 2000). Other MAPK targets include protein kinases phosphorylated and activated by MAPKs, including eEF2K (activated by p38γ/δ MAPK), MK2/3 (activated by p38α/β MAPK), MK5 (activated by ERK3/4), MNK1/2 and MSK1/2 (activated by ERK1/2 and p38α/β MAPK), and RSK1/2/3 (activated by ERK1/2) (Cargnello and Roux 2011). Protein phosphatases are also MAPK targets, including nuclear DUSP1 (protected against proteasomal degradation by ERK phosphorylation) and cytoplasmic DUSP6 (proteasomal degradation is promoted by ERK phosphorylation) (Caunt and Keyse 2013). A complete understanding of MAPK function will require a systems-level approach to define the network of interactions that mediate MAPK signaling (Ch. 4 [Azeloglu and Iyengar 2014]). Nevertheless, we already have a good understanding of the roles of these kinases in stress signaling from biochemical and genetic experiments in multiple organisms.

3.1 MAPK Activation by Stress Signals

Canonical phosphorylation-dependent activation of a MAPK is mediated by a MAPKK. Different MAPKK isoforms selectively activate particular MAPKs: MKK1 and MKK2 activate ERK1 and ERK2, MKK3 activates p38, MKK4 activates JNK and p38, MKK6 activates p38, and MKK7 activates JNK (Fig. 4). Gene disruption studies in mice have shown that MKK3 and MKK6 are the main MAPKKs responsible for p38 activation in vivo, although a minor role of MKK4 can be detected (Brancho et al. 2003). This role is supported by biochemical studies that show phosphorylation of p38 on threonine and tyrosine by both MKK3 and MKK6. Similar gene disruption studies show that MKK4 and MKK7 collaborate in JNK activation (Fleming et al. 2000; Tournier et al. 2001). JNK is preferentially phosphorylated on tyrosine by MKK4 and threonine by MKK7.

Although MAPK activation by MAPKK isoforms represents the major mechanism of MAPK regulation in vivo, noncanonical mechanisms of MAPK activation can be detected under specific circumstances. In yeast, association of the MAPK Fus3 with the scaffold protein Ste5 promotes autophosphorylation of the tyrosine in its activation loop (Bhattacharyya et al. 2006; Good et al. 2009) and autophosphorylation on the activation loop tyrosine of the Smk1 MAPK is driven by its association with the meiosis-specific protein Ssp2 (Whinston et al. 2013). In T cells, the tyrosine kinase Zap70 phosphorylates p38α on tyrosine and causes subsequent autophosphorylation of the dual phosphorylation motif in the activation loop (Salvador et al. 2005). Similarly, ERK7 is activated by autophosphorylation (Abe et al. 2001). Another noncanonical pathway is represented by the atypical MAPK isoforms ERK3 and ERK4, which can be activated by the protein kinase PAK1 (Deleris et al. 2011).

MAPKKs are themselves activated by phosphorylation of their activation loop by MAPKKKs. Selective activation of MAPKKK isoforms by stress stimuli creates parallel signaling cascades involving sequential phosphorylation of MAPKKs and MAPKs (Fig. 4). These MAPKKKs therefore function to trigger MAPK pathway activation in response to specific stimuli.

The mixed-lineage protein kinase (MLK) group of MAPKKKs are activated by members of the Rho GTPase family, including Rac1 and Cdc42 (Gallo and Johnson 2002). MLK protein kinases are autoinhibited by an interaction between an amino-terminal SH3 domain and a proline motif in the carboxy-terminal region. Binding of GTP-bound Rac1 or Cdc42 to MLK protein kinases, mediated by a CRIB motif, causes displacement of SH3-mediated autoinhibition and subsequent protein kinase activation that leads to activation of the MAPKK and, subsequently, the MAPK. This mechanism contributes to stress signaling mediated by inflammatory cytokines (Kant et al. 2011). The response to activated Rho proteins may also be mediated by members of the MEKK group of MAPKKKs, including MEKK1 and MEKK4 (Fanger et al. 1997).

The MAPKKK isoform MEKK2 is activated by receptor tyrosine kinases that bind epidermal growth factor, fibroblast growth factor 2 (FGF2), and stem cell factor (SCF, also known as Kit ligand) by a mechanism that has not been defined (Garrington et al. 2000; Kesavan et al. 2004). Activated MEKK2 causes coordinate activation of the ERK5 and JNK pathways (Kesavan et al. 2004).

The stress-induced immediate early gene response can also cause activation of the JNK and p38 MAPK pathways. GADD45 α, β, and γ proteins are expressed in cells exposed to inflammatory cytokines or environmental stress. These are small (18 kDa), acidic, mainly nuclear proteins belonging to the L7Ae/L30e/S12e RNA-binding protein superfamily. They lack obvious enzymatic activity and exert their pleiotropic function by interaction with multiple effectors that influence processes as diverse as cell cycle progression, apoptosis, and DNA repair and demethylation (Niehrs and Schäfer 2012). All GADD45 isoforms can bind to and activate the MAPKKK MEKK4 (Takekawa and Saito 1998). Indeed, mice lacking GADD45 or MEKK4 show similar defects in MAPK signaling (Chi et al. 2004). The binding of GADD45 to MEKK4 disrupts autoinhibition by the MEKK4 amino-terminal domain and causes dimerization, transphosphorylation, and activation of MEKK4 (Miyake et al. 2007).

Redox stress can activate the MAPKKK isoform ASK1 by causing the release of the inhibitor thioredoxin (Matsukawa et al. 2004). Consequently, knockout mice lacking ASK1 expression show severe defects in MAPK signaling during the response to oxidative stress (Matsuzawa et al. 2002).

Stress-activated MAPKs are activated during the DNA damage response by the ATM protein kinase (Ch. 6 [Rhind and Russell 2012]), which phosphorylates and activates members of the TAO group of atypical MAPKKKs (Raman et al. 2007). This mechanism appears to be selective for the p38 branch of stress-activated MAPK signaling.

Exposure of cells to the inflammatory cytokines TNF and IL1 causes activation of stress-activated MAPKs (p. 121 [Lim and Staudt 2013]; Ch. 15 [Newton and Dixit 2013]). This pathway requires the MAPKKK isoform TAK1 and the ubiquitin-binding accessory proteins TAB2 and TAB3. These TAB proteins activate TAK1 when bound to K63-linked polyubiquitin chains formed by the inflammatory receptor-signaling complex (Chen 2012). The TNF receptor causes K63-linked ubiquitylation of RIP1 and TRAF2/5 by the E3 ligases cIAP1/2, whereas the IL1 receptor causes K63-linked autoubiquitylation of the E3 ligase TRAF6. A similar ubiquitin-mediated TAK1 activation mechanism is engaged when TLRs bind pathogen-associated molecular patterns to stimulate stress-activated MAPK-signaling pathways (p. 121 [Lim and Staudt 2013]; Ch. 15 [Newton and Dixit 2013]). These roles of TAK1 in MAPK activation are coordinated with TAK1-mediated activation of the NF-κB-signaling pathway (p. 121 [Lim and Staudt 2013; Ch. 15 [Newton and Dixit 2013]).

3.2 Inactivation of Stress-Activated MAPK-Signaling Pathways

The major mechanism for inactivation of MAPK-signaling pathways is to reverse phosphorylation-mediated activation. Dual-specificity phosphatases (DUSPs, also known as MAPK phosphatases) play a key role in MAPK inactivation (Owens and Keyse 2007; Caunt and Keyse 2013). In addition, MAPKs can be inactivated by the serine/threonine protein phosphatases PP2A and PP2C, tyrosine phosphatases, and the death-effector-domain protein PEA15. The PP2C family member WIP1 plays a major role in switching off the response of p38 to DNA damage (Le Guezennec and Bulavin 2010) and the tyrosine phosphatase HePTP regulates p38 MAPK in B cells following stimulation by adrenalin (McAlees and Sanders 2009). This regulation can display complex dynamics because of phosphorylation-induced regulation of DUSP activity and signal-induced DUSP expression (Keyse 2008). Note that stress-activated MAPK pathways can also be inactivated by bacterial pathogens (Ch. 20 [Alto and Orth 2012]).

3.3 The Role of Scaffold Proteins

The protein kinase cascade that is initiated by an activated MAPKKK and includes subsequent MAPKK and MAPK activation represents a defined signaling module. The interactions among the constituent proteins can involve a series of binary docking interactions. For example, as outlined above, a D domain located in the amino-terminal region of a MAPKK can interact with a docking site on MAPKs (Bardwell and Thorner 1996; Enslen and Davis 2001). Cleavage by anthrax lethal factor protease that removes this D domain accounts for the prevention of MAPKK-mediated MAPK activation by this pathogenic bacterium (Duesbery et al. 1998). Similarly, docking interactions between a MAPKKK and MAPKK have been described, including the Phox/Bem1p domain-mediated interactions between MEKK2/3 and MKK5 (Nakamura and Johnson 2007).

MAPK-signaling modules can be assembled by scaffolding proteins (Morrison and Davis 2003; Good et al.

2011; Witzel et al. 2012). The best-characterized scaffold is Ste5, a component of the yeast-mating response MAPK pathway (Good et al. 2011). Other proteins that regulate mammalian MAPK signaling include scaffolds for the ERK pathway (KSR1), the p38 pathway (OSM, JIP2, and JIP4), and the JNK pathway (JIP1 and JIP3) (Morrison and Davis 2003). Gene disruption studies have confirmed that JIP1 plays a key role in JNK activation caused by the neurotransmitter glutamate (Whitmarsh et al. 2001) and metabolic stress caused by a high-fat diet (Jaeschke et al. 2004; Morel et al. 2010). JIP1 binds to JNK, the MAPKK isoform MKK7, and members of the MLK group of MAPKKKs to form a functional JNK-signaling module. The formation of this signaling complex is regulated by phosphorylation (Morel et al. 2010). Moreover, JIP1 interacts with kinesin light chain and is dynamically localized within the cell by microtubule-mediated protein trafficking. This scaffold protein allows regulatory control of localization, module activation, and substrate access to direct stress-activated MAPK signaling to specific targets within the cell (Morrison and Davis 2003).

3.4 Physiological Role of Stress-Activated MAPKs

Like the UPR, stress-activated MAPKs contribute to many aspects of normal cellular physiology and pathology (Davis 2000; Cuadrado and Nebreda 2010). Below, we discuss their roles in the context of three specific examples: cell death versus survival signaling, inflammatory stress, and metabolic stress.

3.4.1 Stress-Activated MAPKs and Cell Death

The initial response of cells to stress exposure is to mount a survival response, but prolonged exposure to stress may lead to cell death (Ch. 19 [Green and Llambi 2014]). MAPKs are implicated in these processes (Xia et al. 1995). Studies of TNF-stimulated cell death show that JNK initially promotes a survival response and that prolonged JNK activation is required for cell death (Lamb et al. 2003; Ventura et al. 2006). Sustained activation of JNK and p38 can contribute to both necrotic and apoptotic cell death by regulating the expression of cytotoxic ligands (e.g., FasL and TNF) (Das et al. 2009) and synthesis of ROS (Ventura et al. 2004). In addition, stress-activated MAPKs can regulate the intrinsic apoptosis pathway, mediated by mitochondria, by regulating members of the Bcl2 family (Tournier et al. 2000). Thus, JNK can phosphorylate and inhibit prosurvival signaling by Mcl1 by inducing ubiquitin-dependent proteasomal degradation (Morel et al. 2009) and JNK can cause induction of the proapoptotic BH3-only protein Bim (Wong et al. 2005; Perier et al.

2007). Bim is also subject to posttranslational regulation by MAPKs (Puthalakath and Strasser 2002). ERK phosphorylation on multiple serine residues sites triggers ubiquitin-independent proteasomal degradation of Bim (Wiggins et al. 2011). In contrast, JNK phosphorylation of Bim on a threonine residue disrupts the interaction of Bim with the dynein microtubule motor complex that can sequester Bim on the microtubule cytoskeleton (Lei and Davis 2003; Hubner et al. 2008). The Bim-related protein Bmf is similarly regulated by JNK-mediated phosphorylation, which disrupts the interaction of Bmf with the myosin V motor complex that sequesters Bmf on the actin cytoskeleton (Lei and Davis 2003; Hubner et al. 2010). JNK therefore promotes the intrinsic apoptotic pathway by degrading the antiapoptotic protein Mcl1 and up-regulating the pro-apoptotic protein Bim by multiple mechanisms, including increased transcription and sequestration of Bim by the cytoskeleton. The loss of the antiapoptotic protein Mcl1 combined with the increase in Bim activity causes a change in the balance of pro- and antiapoptotic pathways that leads to cell death. This balance of cell death and survival regulated by stress-activated MAPKs is also influenced by the activation state of survival signaling pathways, including those involving Akt, ERK, and NF-κB.

In vivo studies have shown that stress-activated MAPKs play a critical role in neurodegenerative disorders, including stroke (Kuan et al. 2003), seizure (Yang et al. 1997), Alzheimer's disease (Mazzitelli et al. 2011), and Parkinson's disease (Perier et al. 2007). Cell death regulated by stress-activated MAPKs may also play an important role during tumor development (Davis 2000).

3.4.2 Stress-Activated MAPKs and Inflammation

MAPKs contribute to both innate and adaptive immune responses (Dong et al. 2002). In particular, stress-activated MAPKs play an important role in the regulation of inflammatory cytokine expression. The p38 and JNK pathways target transcription factors (e.g., ATF, Jun, and MEF2 family members) and chromatin-remodeling enzymes that can regulate expression of chemokines (e.g., CCL2 and CCL5) and cytokines (e.g., IL1, IL6, IL12p40, and TNF) during inflammatory responses (Davis 2000; Cuadrado and Nebreda 2010). In addition, p38α increases cytokine mRNA stability controlled by the AU-rich element-binding proteins HuR (phosphorylated by p38) and TTP (phosphorylated by p38α-MAPK-activated MK2) (Clark et al. 2009). p38α can also stimulate cytokine mRNA translation by phosphorylation and activation of eIF4E/MAPK-interacting kinases (MNK1 and MNK2) (Noubade et al. 2011). Moreover, p38γ and p38δ increase cytokine translation by phosphorylation and activation of eEF2K, which increases protein synthesis (Gonzalez-Teran et al. 2013). Thus, JNK and p38 coordinately regulate transcription, mRNA stability, and mRNA translation to promote cytokine/chemokine expression.

The immune response elicited during inflammation depends on the involvement of specific innate and adaptive immune cells. These types of cells are influenced by stress-activated MAPK-signaling pathways. Macrophages represent an important part of the innate immune response. Classical activation of macrophages by interferon γ or endotoxin causes polarization to the M1 phenotype that is associated with expression of inflammatory cytokines (e.g., TNF) and inflammation. In contrast, alternatively activated macrophages polarized to the M2 phenotype can express anti-inflammatory cytokines (e.g., IL10) and are implicated in resolution of the immune response and tissue remodeling. The M2 phenotype is complex and represents a mixture of different cell types, including cells exposed to IL4 or IL13 (M2a), immune complexes (M2b), and IL10 or TGFβ (M2c). Stress-activated MAPKs can control macrophage polarization. For example, JNK is required for M1, but not M2, macrophage development (Han et al. 2013). Consequently, JNK can promote inflammation by polarizing macrophages to the M1 inflammatory phenotype. In contrast, JNK inhibition can suppress inflammation by increasing polarization to the M2 anti-inflammatory phenotype.

Stress-activated MAPKs can also regulate the formation of specific T-cell subsets. For example, JNK is required for CD8$^+$ cell proliferation and IL2 secretion (Conze et al. 2002). In contrast, it is not required for CD4$^+$ T-cell activation or IL2 secretion, but JNK is required for naïve CD4$^+$ T-cell differentiation into Th1 or Th2 effector cells (Dong et al. 2000). Moreover, p38α is required for differentiation of CD4$^+$ T cells into Th17 cells (Noubade et al. 2011). A consequence of these roles in T cells is that JNK is required for efficient viral clearance (Arbour et al. 2002) and p38 is required for suppression of autoimmunity (Noubade et al. 2011).

Collectively, these mechanisms lead to inflammation in tissues with increased stress-activated MAPK activity, which can promote metabolic dysfunction (Sabio and Davis 2010) and carcinogenesis (Das et al. 2011).

3.4.3 Stress-Activated MAPKs and Metabolism

Feeding a high-fat diet causes metabolic stress and activates stress-activated MAPK-signaling pathways. This is mediated by increased amounts of saturated free fatty acids that activate the MLK group of MAPKKKs (Jaeschke and Davis 2007; Kant et al. 2013). The mechanism of fatty acid signaling may require G-protein-coupled receptors (Talukdar

et al. 2011) or nonreceptor mechanisms (Holzer et al. 2011) that activate the tyrosine kinase Src in lipid raft domains of the plasma membrane (Holzer et al. 2011). p38α and p38δ appear to have roles in insulin resistance, oxidative stress-induced β-cell failure, and hepatic gluconeogenesis (Sumara et al. 2009; Lee et al. 2011b). Moreover, the JNK pathway is required for high-fat-diet-induced obesity and insulin resistance (Hirosumi et al. 2002).

The effects of JNK on obesity are due to a requirement for JNK in negative-feedback regulation of energy expenditure by the hypothalamus-pituitary-thyroid axis (Sabio and Davis 2010). The activation of JNK inhibits the expression of hypothalamic thyrotropin-releasing hormone and pituitary gland expression of thyroid-stimulating hormone. The consequence of JNK-mediated suppression of the hypothalamic-pituitary-thyroid axis is that levels of circulating thyroid hormone, thyroid-hormone-dependent target gene expression, and oxidative metabolism are all decreased, which reduces energy expenditure, and obesity is increased.

The effects of JNK on insulin resistance depend on roles of JNK in peripheral tissues that are independent of obesity (Sabio and Davis 2010). Activated JNK causes inhibition of signaling by the insulin receptor. Initial studies indicated that this might be mediated by inhibitory phosphorylation of IRS1 (Aguirre et al. 2000), but this conclusion was not supported by later studies (Copps and White 2012). The mechanism of inhibition of insulin receptor signaling by JNK remains to be elucidated. However, one potential mechanism that may contribute to insulin resistance is the requirement of JNK for macrophage polarization to the M1 phenotype because tissue infiltration by inflammatory macrophages is a key determinant of the development of insulin resistance (Han et al. 2013). It is clear that loss of JNK function in murine models protects mice against the development of insulin resistance when fed a high-fat diet (Sabio and Davis 2010). This observation is important because compensation of insulin resistance by hyperinsulinemia can lead to β-cell failure and diabetes. Because stress-activated MAPKs play a key role in energy homeostasis and normal glycemia, they represent potential targets for therapies aimed at treatment of metabolic syndrome and prediabetes (Sabio and Davis 2010).

4 CONCLUSIONS

Stress-signaling pathways are evolutionarily conserved and play an important role in the maintenance of homeostasis and adaptation to new cellular microenvironments (Fig. 5). Key areas for future research include structural studies of stress-response signaling proteins and integrated analysis of the signaling network's response to stress.

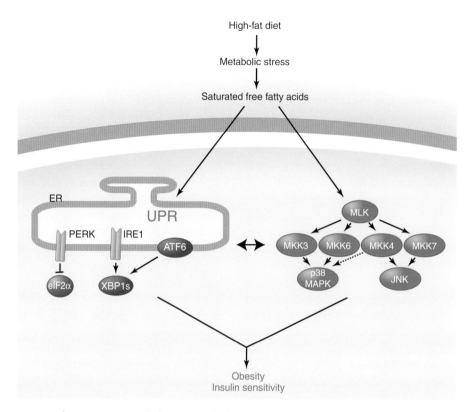

Figure 5. Integrated response to metabolic stress. A high-fat diet causes metabolic stress associated with increased amounts of saturated free fatty acids, which engage the UPR and stress-activated MAPKs. The UPR includes three different signaling pathways that are initiated by IRE1, PERK, and ATF6. The stress-activated MAPK response is initiated by the MLK group of MAPKKKs and leads to the activation of the JNK and p38 MAPKs. Cross talk between the UPR and stress-activated MAPK signaling leads to an integrated adaptive response.

Finally, studies of stress-signaling pathways that translate to the clinic are needed if we are to realize their potential for therapeutic intervention in disease processes. Effective and specific means to target both MAPK signaling and the UPR are required. An important area of research will be the development of molecular tools that allow us to specifically modulate the function of the key components of MAPK pathways. Similarly, we must define new targets within the UPR and develop new tools to modulate its components.

ACKNOWLEDGMENTS

Studies in the Hotamisligil laboratory are currently supported by grants from the National Institutes of Health (NIH), the Juvenile Diabetes Research Foundation, the American Diabetes Association, Servier and UCB Pharmaceuticals, and the Simmons Fund. Special thanks to Scott Widenmaier, Ling Yang, and Takahisa Nakamura for critical discussions, thoughtful comments, and help in preparing the manuscript, Ana Paula Arruda and Suneng Fu for help in illustrations, and Megan Washack and Claudia Garcia Wagner for editorial assistance. Studies in the Davis laboratory are currently supported by grants from the NIH, the American Diabetes Association, and the Howard Hughes Medical Institute. Kathy Gemme provided expert editorial assistance. We thank the students and fellows who contributed to the studies in our groups over the years and to our collaborators. We regret the inadvertent omission of references to important work by our colleagues because of space limitations.

REFERENCES

*Reference is in this book.

Abe MK, Kahle KT, Saelzler MP, Orth K, Dixon JE, Rosner MR. 2001. ERK7 is an autoactivated member of the MAPK family. *J Biol Chem* **276**: 21272–21279.

Aguirre V, Uchida T, Yenush L, Davis R, White MF. 2000. The c-Jun NH$_2$-terminal kinase promotes insulin resistance during association with insulin receptor substrate-1 and phosphorylation of Ser[307]. *J Biol Chem* **275**: 9047–9054.

Akella R, Moon TM, Goldsmith EJ. 2008. Unique MAP kinase binding sites. *Biochim Biophys Acta* **1784**: 48–55.

Allen JJ, Li M, Brinkworth CS, Paulson JL, Wang D, Hübner A, Chou W-H, Davis RJ, Burlingame AL, Messing RO, et al. 2007. A semi-synthetic epitope for kinase substrates. *Nat Methods* **4:** 511–516.

* Alto NM, Orth K. 2012. Subversion of cell signaling by pathogens. *Cold Spring Harb Perspect Biol* **4:** a006114.

Appenzeller-Herzog C, Hall MN. 2012. Bidirectional crosstalk between endoplasmic reticulum stress and mTOR signaling. *Trends Cell Biol* **22:** 274–282.

Arbour N, Naniche D, Homann D, Davis RJ, Flavell RA, Oldstone MB. 2002. c-Jun NH$_2$-terminal kinase (JNK)1 and JNK2 signaling pathways have divergent roles in CD8$^+$ T cell-mediated antiviral immunity. *J Exp Med* **195:** 801–810.

Arias E, Cuervo A. 2011. Chaperone-mediated autophagy in protein quality control. *Curr Opin Cell Biol* **23:** 184–189.

* Azeloglu EU, Iyengar R. 2014. Signaling networks: Information flow, computation, and decision making. *Cold Spring Harb Perspect Biol* doi: 10.1101/cshperspect.a005934.

Babour A, Bicknell A, Tourtellotte J, Niwa M. 2010. A surveillance pathway monitors the fitness of the endoplasmic reticulum to control its inheritance. *Cell* **142:** 256–269.

Baird TD, Wek RC. 2012. Eukaryotic initiation factor 2 phosphorylation and translational control in metabolism. *Adv Nutr* **3:** 307–321.

Bardwell L, Thorner J. 1996. A conserved motif at the amino termini of MEKs might mediate high-affinity interaction with the cognate MAPKs. *Trends Biochem Sci* **21:** 373–374.

Bertolotti A, Zhang Y, Hendershot LM, Harding HP, Ron D. 2000. Dynamic interaction of BiP and ER stress transducers in the unfolded-protein response. *Nat Cell Biol* **2:** 326–332.

Bhattacharyya RP, Reményi A, Good MC, Bashor CJ, Falick AM, Lim WA. 2006. The Ste5 scaffold allosterically modulates signaling output of the yeast mating pathway. *Science* **311:** 822–826.

Boden G, Duan X, Homko C, Molina EJ, Song W, Perez O, Cheung P, Merali S. 2008. Increase in endoplasmic reticulum stress-related proteins and genes in adipose tissue of obese, insulin-resistant individuals. *Diabetes* **57:** 2438–2444.

Boden G, Song W, Duan X, Cheung P, Kresge K, Barrero C, Merali S. 2011. Infusion of glucose and lipids at physiological rates causes acute endoplasmic reticulum stress in rat liver. *Obesity* **19:** 1366–1373.

Brancho D, Tanaka N, Jaeschke A, Ventura JJ, Kelkar N, Tanaka Y, Kyuuma M, Takeshita T, Flavell RA, Davis RJ. 2003. Mechanism of p38 MAP kinase activation in vivo. *Genes Dev* **17:** 1969–1978.

Bromati CR, Lellis-Santos C, Yamanaka TS, Nogueira TCA, Leonelli M, Caperuto LC, Gorjao R, Leite AR, Anhe GF, Bordin S. 2011. UPR induces transient burst of apoptosis in islets of early lactating rats through reduced AKT phosphorylation via ATF4/CHOP stimulation of TRB3 expression. *Am J Physiol* **300:** R92–R100.

Calfon M, Zeng H, Urano F, Till JH, Hubbard SR, Harding HP, Clark SG, Ron D. 2002. IRE1 couples endoplasmic reticulum load to secretory capacity by processing the XBP-1 mRNA. *Nature* **415:** 92–96.

Canagarajah BJ, Khokhlatchev A, Cobb MH, Goldsmith EJ. 1997. Activation mechanism of the MAP kinase ERK2 by dual phosphorylation. *Cell* **90:** 859–869.

Cargnello M, Roux PP. 2011. Activation and function of the MAPKs and their substrates, the MAPK-activated protein kinases. *Microbiol Mol Biol Rev* **75:** 50–83.

Caricilli A, Picardi P, de Abreu L, Ueno M, Prada P, Ropelle E, Hirabara S, Vieira P, Camara N, Curi R, et al. 2011. Gut microbiota is a key modulator of insulin resistance in TLR 2 knockout mice. *PLoS Biol* **9:** e1001212.

Carlson SM, Chouinard CR, Labadorf A, Lam CJ, Schmelzle K, Fraenkel E, White FM. 2011. Large-scale discovery of ERK2 substrates identifies ERK-mediated transcriptional regulation by ETV3. *Sci Signal* **4:** rs11.

Caunt CJ, Keyse SM. 2013. Dual-specificity MAP kinase phosphatases (MKPs): Shaping the outcome of MAP kinase signalling. *FEBS J* **280:** 489–504.

Chan J, Cooney G, Biden T, Laybutt D. 2011. Differential regulation of adaptive and apoptotic unfolded protein response signalling by cyto-kine-induced nitric oxide production in mouse pancreatic β cells. *Diabetologia* **54:** 1766–1776.

Chang CI, Xu BE, Akella R, Cobb MH, Goldsmith EJ. 2002. Crystal structures of MAP kinase p38 complexed to the docking sites on its nuclear substrate MEF2A and activator MKK3b. *Mol Cell* **9:** 1241–1249.

Chen ZJ. 2012. Ubiquitination in signaling to and activation of IKK. *Immunol Rev* **246:** 95–106.

Chen X, Shen J, Prywes R. 2002. The luminal domain of ATF6 senses endoplasmic reticulum (ER) stress and causes translocation of ATF6 from the ER to the Golgi. *J Biol Chem* **277:** 13045–13052.

Chi H, Lu B, Takekawa M, Davis RJ, Flavell RA. 2004. GADD45β/GADD45γ and MEKK4 comprise a genetic pathway mediating STAT4-independent IFNγ production in T cells. *EMBO J* **23:** 1576–1586.

Clark A, Dean J, Tudor C, Saklatvala J. 2009. Post-transcriptional gene regulation by MAP kinases via AU-rich elements. *Front Biosci* **14:** 847–871.

Conze D, Krahl T, Kennedy N, Weiss L, Lumsden J, Hess P, Flavell RA, Le Gros G, Davis RJ, Rincon M. 2002. c-Jun NH$_2$-terminal kinase (JNK)1 and JNK2 have distinct roles in CD8$^+$ T cell activation. *J Exp Med* **195:** 811–823.

Copps KD, White MF. 2012. Regulation of insulin sensitivity by serine/threonine phosphorylation of insulin receptor substrate proteins IRS1 and IRS2. *Diabetologia* **55:** 2565–2582.

Cox JS, Walter P. 1996. A novel mechanism for regulating activity of a transcription factor that controls the unfolded protein response. *Cell* **87:** 391–404.

Cox JS, Shamu CE, Walter P. 1993. Transcriptional induction of genes encoding endoplasmic reticulum resident proteins requires a transmembrane protein kinase. *Cell* **73:** 1197–1206.

Coyle SM, Gilbert WV, Doudna JA. 2009. Direct link between RACK1 function and localization at the ribosome in vivo. *Mol Cell Biol* **29:** 1626–1634.

Cretenet G, Le Clech M, Gachon F. 2010. Circadian clock-coordinated 12 hr period rhythmic activation of the IRE1α pathway controls lipid metabolism in mouse liver. *Cell Metab* **11:** 47–57.

Cuadrado A, Nebreda AR. 2010. Mechanisms and functions of p38 MAPK signalling. *Biochem J* **429:** 403–417.

Cuenda A, Rousseau S. 2007. p38 MAP-kinases pathway regulation, function and role in human diseases. *Biochim Biophys Acta* **1773:** 1358–1375.

Cullinan SB, Diehl JA. 2006. Coordination of ER and oxidative stress signaling: The PERK/Nrf2 signaling pathway. *Int J Biochem Cell Biol* **38:** 317–332.

Cullinan SB, Zhang D, Hannink M, Arvisais E, Kaufman RJ, Diehl JA. 2003. Nrf2 is a direct PERK substrate and effector of PERK-dependent cell survival. *Mol Cell Biol* **23:** 7198–7209.

Das M, Sabio G, Jiang F, Rincon M, Flavell RA, Davis RJ. 2009. Induction of hepatitis by JNK-mediated expression of TNF-α. *Cell* **136:** 249–260.

Das M, Garlick DS, Greiner DL, Davis RJ. 2011. The role of JNK in the development of hepatocellular carcinoma. *Genes Dev* **25:** 634–645.

Davis RJ. 2000. Signal transduction by the JNK group of MAP kinases. *Cell* **103:** 239–252.

Delepine M, Nicolino M, Barrett T, Golamaully M, Lathrop GM, Julier C. 2000. EIF2AK3, encoding translation initiation factor 2α kinase 3, is mutated in patients with Wolcott-Rallison syndrome. *Nat Genet* **25:** 406–409.

Deleris P, Trost M, Topisirovic I, Tanguay PL, Borden KL, Thibault P, Meloche S. 2011. Activation loop phosphorylation of ERK3/ERK4 by group I p21-activated kinases (PAKs) defines a novel PAK-ERK3/4-MAPK-activated protein kinase 5 signaling pathway. *J Biol Chem* **286:** 6470–6478.

Deng J, Lu PD, Zhang Y, Scheuner D, Kaufman RJ, Sonenberg N, Harding HP, Ron D. 2004. Translational repression mediates activation of nuclear factor κB by phosphorylated translation initiation factor 2. *Mol Cell Biol* **24:** 10161–10168.

Dong C, Yang DD, Tournier C, Whitmarsh AJ, Xu J, Davis RJ, Flavell RA. 2000. JNK is required for effector T-cell function but not for T-cell activation. *Nature* **405:** 91–94.

Dong C, Davis RJ, Flavell RA. 2002. MAP kinases in the immune response. *Annu Rev Immunol* **20:** 55–72.

Donnelly N, Gorman AM, Gupta S, Samali A. 2013. The eIF2α kinases: Their structures and functions. *Cell Mol Life Sci* **70:** 1252–1256.

Duesbery NS, Webb CP, Leppla SH, Gordon VM, Klimpel KR, Copeland TD, Ahn NG, Oskarsson MK, Fukasawa K, Paull KD, et al. 1998. Proteolytic inactivation of MAP-kinase-kinase by anthrax lethal factor. *Science* **280:** 734–737.

Eguchi K, Manabe I, Oishi-Tanaka Y, Ohsugi M, Kono N, Ogata F, Yagi N, Ohto U, Kimoto M, Miyake K, et al. 2012. Saturated fatty acid and TLR signaling link β cell dysfunction and islet inflammation. *Cell Metab* **15:** 518–533.

Ellgaard L, Helenius A. 2003. Quality control in the endoplasmic reticulum. *Nat Rev Mol Cell Biol* **4:** 181–191.

English A, Zurek N, Voeltz G. 2009. Peripheral ER structure and function. *Curr Opin Cell Biol* **21:** 596–602.

Enslen H, Davis RJ. 2001. Regulation of MAP kinases by docking domains. *Biol Cell* **93:** 5–14.

Fanger GR, Johnson NL, Johnson GL. 1997. MEK kinases are regulated by EGF and selectively interact with Rac/Cdc42. *EMBO J* **16:** 4961–4972.

Fleming Y, Armstrong CG, Morrice N, Paterson A, Goedert M, Cohen P. 2000. Synergistic activation of stress-activated protein kinase 1/c-Jun N-terminal kinase (SAPK1/JNK) isoforms by mitogen-activated protein kinase kinase 4 (MKK4) and MKK7. *Biochem J* **352:** 145–154.

Fonseca SG, Fukuma M, Lipson KL, Nguyen LX, Allen JR, Oka Y, Urano F. 2005. WFS1 is a novel component of the unfolded protein response and maintains homeostasis of the endoplasmic reticulum in pancreatic β-cells. *J Biol Chem* **280:** 39609–39615.

Fonseca SG, Ishigaki S, Oslo CM, Lu S, Lipson KL, Ghosh R, Hayashi E, Ishihara H, Oka Y, Permutt MA, et al. 2010. Wolfram syndrom 1 gene negatively regulate ER stress signaling in rodent and human cells. *J Clin Invest* **120:** 744–755.

Friedlander R, Jarosch E, Urban J, Volkwein C, Sommer T. 2000. A regulatory link between ER-associated protein degradation and the unfolded-protein response. *Nat Cell Biol* **2:** 379–384.

Fu S, Watkins SM, Hotamisligil GS. 2012. The role of endoplasmic reticulum in hepatic lipid homeostasis and stress signaling. *Cell Metab* **15:** 623–634.

Gallo KA, Johnson GL. 2002. Mixed-lineage kinase control of JNK and p38 MAPK pathways. *Nat Rev Mol Cell Biol* **3:** 663–672.

Gardner B, Walter P. 2011. Unfolded proteins are Ire1-activating ligands that directly induce the unfolded protein response. *Science* **333:** 1891–1894.

Garrington TP, Ishizuka T, Papst PJ, Chayama K, Webb S, Yujiri T, Sun W, Sather S, Russell DM, Gibson SB, et al. 2000. MEKK2 gene disruption causes loss of cytokine production in response to IgE and c-Kit ligand stimulation of ES cell-derived mast cells. *EMBO J* **19:** 5387–5395.

Gee H, Noh S, Tang B, Kim K, Lee M. 2011. Rescue of ΔF508-CFTR trafficking via a GRASP-dependent unconventional secretion pathway. *Cell* **146:** 746–760.

Gething MJ, Sambrook J. 1992. Protein folding in the cell. *Nature* **355:** 33–45.

Gonzalez-Teran B, Cortes JR, Manieri E, Matesanz N, Verdugo A, Rodriguez ME, Gonzalez-Rodriguez A, Valverde A, Martin P, Davis RJ, et al. 2013. Eukaryotic elongation factor 2 controls TNF-α translation in LPS-induced hepatitis. *J Clin Invest* **123:** 164–178.

Good M, Tang G, Singleton J, Remenyi A, Lim WA. 2009. The Ste5 scaffold directs mating signaling by catalytically unlocking the Fus3 MAP kinase for activation. *Cell* **136:** 1085–1097.

Good MC, Zalatan JG, Lim WA. 2011. Scaffold proteins: Hubs for controlling the flow of cellular information. *Science* **332:** 680–686.

Gotoh T, Mori M. 2006. Nitric oxide and endoplasmic reticulum stress. *Arterioscler Thromb Vasc Biol* **26:** 1439–1446.

* Green DR, Llambi F. 2014. Cell death signaling. *Cold Spring Harb Perspect Biol* doi: 10.1101/cshperspect.a006080.

Gregor MF, Hotamisligil GS. 2011. Inflammatory mechanisms in obesity. *Annu Rev Immunol* **29:** 415–445.

Gregor MF, Yang L, Fabbrini E, Mohammed BS, Eagon JC, Hotamisligil GS, Klein S. 2009. Endoplasmic reticulum stress is reduced in tissues of obese subjects after weight loss. *Diabetes* **58:** 693–700.

Gupta S, Barrett T, Whitmarsh AJ, Cavanagh J, Sluss HK, Derijard B, Davis RJ. 1996. Selective interaction of JNK protein kinase isoforms with transcription factors. *EMBO J* **15:** 2760–2770.

Han MS, Jung DY, Morel C, Lakhani SA, Kim JK, Flavell RA, Davis RJ. 2013. JNK expression by macrophages promotes obesity-induced insulin resistance and inflammation. *Science* **339:** 218–222.

* Hardie DG. 2012. Organismal carbohydrate and lipid homeostasis. *Cold Spring Harb Perspect Biol* **4:** a006031.

Harding HP, Zhang Y, Ron D. 1999. Protein translation and folding are coupled by an endoplasmic-reticulum-resident kinase. *Nature* **397:** 271–274.

Harding HP, Novoa I, Zhang Y, Zeng H, Wek R, Schapira M, Ron D. 2000. Regulated translation initiation controls stress-induced gene expression in mammalian cells. *Mol Cell* **6:** 1099–1108.

Harding HP, Zeng H, Zhang Y, Jungries R, Chung P, Plesken H, Sabatini DD, Ron D. 2001. Diabetes mellitus and exocrine pancreatic dysfunction in *Perk*[-/-] mice reveals a role for translational control in secretory cell survival. *Mol Cell* **7:** 1153–1163.

Harding HP, Zhang Y, Zeng H, Novoa I, Lu PD, Calfon M, Sadri N, Yun C, Popko B, Paules R, et al. 2003. An integrated stress response regulates amino acid metabolism and resistance to oxidative stress. *Mol Cell* **11:** 619–633.

Hasegawa M, Cuenda A, Spillantini MG, Thomas GM, Buée-Scherrer V, Cohen P, Goedert M. 1999. Stress-activated protein kinase-3 interacts with the PDZ domain of α1-syntrophin. A mechanism for specific substrate recognition. *J Biol Chem* **274:** 12626–12631.

Hatori M, Hirota T, Iitsuka M, Kurabayashi N, Haraguchi S, Kokame K, Sato R, Nakai A, Miyata T, Tsutsui K, et al. 2011. Light-dependent and circadian clock-regulated activation of sterol regulatory element-binding protein, X-box-binding protein 1, and heat shock factor pathways. *Proc Natl Acad Sci* **108:** 4864–4869.

Hayashi T, Rizzuto R, Hajnoczky G, Su TP. 2009. MAM: More than just a housekeeper. *Trends Cell Biol* **19:** 81–88.

Heazlewood CK, Cook MC, Eri R, Price GR, Tauro SB, Taupin D, Thornton DJ, Png CW, Crockford TL, Cornall RJ, et al. 2008. Aberrant mucin assembly in mice causes endoplasmic reticulum stress and spontaneous inflammation resembling ulcerative colitis. *PLoS Med* **5:** e54.

Heo YS, Kim SK, Seo CI, Kim YK, Sung BJ, Lee HS, Lee JI, Park SY, Kim JH, Hwang KY, et al. 2004. Structural basis for the selective inhibition of JNK1 by the scaffolding protein JIP1 and SP600125. *EMBO J* **23:** 2185–2195.

Hinnebusch AG, Lorsch JR. 2012. The mechanism of eukaryotic translation initiation: New insights and challenges. *Cold Spring Harb Perspect Biol* **4:** a011544.

Hirosumi J, Tuncman G, Chang L, Gorgun CZ, Uysal KT, Maeda K, Karin M, Hotamisligil GS. 2002. A central role for JNK in obesity and insulin resistance. *Nature* **420:** 333–336.

Holcik M, Sonenberg N. 2005. Translational control in stress and apoptosis. *Nat Rev Mol Cell Biol* **6:** 318–327.

Hollien J, Weissman JS. 2006. Decay of endoplasmic reticulum-localized mRNAs during the unfolded protein response. *Science* **313:** 104–107.

Holzer RG, Park EJ, Li N, Tran H, Chen M, Choi C, Solinas G, Karin M. 2011. Saturated fatty acids induce c-Src clustering within membrane subdomains, leading to JNK activation. *Cell* **147:** 173–184.

Hotamisligil GS. 2010. Endoplasmic reticulum stress and the inflammatory basis of metabolic disease. *Cell* **140:** 900–917.

Hou SW, Zhi HY, Pohl N, Loesch M, Qi XM, Li RS, Basir Z, Chen G. 2010. PTPH1 dephosphorylates and cooperates with p38γ MAPK to increase ras oncogenesis through PDZ-mediated interaction. *Cancer Res* **70:** 2901–2910.

Hu P, Han Z, Couvillon AD, Kaufman RJ, Exton JH. 2006. Autocrine tumor necrosis factor α links endoplasmic reticulum stress to the membrane death receptor pathway through IRE1α-mediated NF-κB activation and down-regulation of TRAF2 expression. *Mol Cell Biol* **26:** 3071–3084.

Hubner A, Barrett T, Flavell RA, Davis RJ. 2008. Multisite phosphorylation regulates Bim stability and apoptotic activity. *Mol Cell* **30:** 415–425.

Hubner A, Cavanagh-Kyros J, Rincon M, Flavell RA, Davis RJ. 2010. Functional cooperation of the proapoptotic Bcl2 family proteins Bmf and Bim in vivo. *Mol Cell Biol* **30:** 98–105.

Ishihara H, Takeda S, Tamura A, Takahashi R, Yamaguchi S, Takei D, Yamada T, Inoue H, Soga H, Katagiri H, et al. 2004. Disruption of the WFS1 gene in mice causes progressive β-cell loss and impaired stimulus-secretion coupling in insulin secretion. *Hum Mol Genet* **13:** 1159–1170.

Jacobs D, Glossip D, Xing H, Muslin AJ, Kornfeld K. 1999. Multiple docking sites on substrate proteins form a modular system that mediates recognition by ERK MAP kinase. *Genes Dev* **13:** 163–175.

Jaeschke A, Davis RJ. 2007. Metabolic stress signaling mediated by mixed-lineage kinases. *Mol Cell* **27:** 498–508.

Jaeschke A, Czech MP, Davis RJ. 2004. An essential role of the JIP1 scaffold protein for JNK activation in adipose tissue. *Genes Dev* **18:** 1976–1980.

Jiang HY, Wek SA, McGrath BC, Scheuner D, Kaufman RJ, Cavener DR, Wek RC. 2003. Phosphorylation of the α subunit of eukaryotic initiation factor 2 is required for activation of NF-κB in response to diverse cellular stresses. *Mol Cell Biol* **23:** 5651–5663.

Jiang HY, Wek SA, McGrath BC, Lu D, Hai T, Harding HP, Wang X, Ron D, Cavener DR, Wek RC. 2004. Activating transcription factor 3 is integral to the eukaryotic initiation factor 2 kinase stress response. *Mol Cell Biol* **24:** 1365–1377.

Jiao P, Ma J, Feng B, Zhang H, Diehl JA, Chin YE, Yan W, Xu H. 2011. FFA-induced adipocyte inflammation and insulin resistance: Involvement of ER stress and IKKβ pathways. *Obesity* **19:** 483–491.

* Julius D, Nathans J. 2012. Signaling by sensory receptors. *Cold Spring Harb Perspect Biol* **4:** a005991.

Kammoun HL, Chabanon H, Hainault I, Luquet S, Magnan C, Koike T, Ferre P, Foufelle F. 2009. GRP78 expression inhibits insulin and ER stress-induced SREBP-1c activation and reduces hepatic steatosis in mice. *J Clin Invest* **119:** 1201–1215.

Kaneko M, Niinuma Y, Nomura Y. 2003. Activation signal of nuclear factor-κB in response to endoplasmic reticulum stress is transduced via IRE1 and tumor necrosis factor receptor-associated factor 2. *Biol Pharm Bull* **26:** 931–935.

Kant S, Swat W, Zhang S, Zhang ZY, Neel BG, Flavell RA, Davis RJ. 2011. TNF-stimulated MAP kinase activation mediated by a Rho family GTPase signaling pathway. *Genes Dev* **25:** 2069–2078.

Kant S, Barrett T, Vertii A, Noh YH, Jung DY, Kim JK, Davis RJ. 2013. Role of the mixed-lineage protein kinase pathway in the metabolic stress response to obesity. *Cell Rep* **4:** 681–688.

Kars M, Yang L, Gregor MF, Mohammed BS, Pietka TA, Finck BN, Patterson BW, Horton JD, Mittendorfer B, Hotamisligil GS, et al. 2010. Tauroursodeoxycholic Acid may improve liver and muscle but not adipose tissue insulin sensitivity in obese men and women. *Diabetes* **59:** 1899–1905.

Kesavan K, Lobel-Rice K, Sun W, Lapadat R, Webb S, Johnson GL, Garrington TP. 2004. MEKK2 regulates the coordinate activation of ERK5 and JNK in response to FGF-2 in fibroblasts. *J Cell Physiol* **199:** 140–148.

Keyse SM. 2008. Dual-specificity MAP kinase phosphatases (MKPs) and cancer. *Cancer Metastasis Rev* **27:** 253–261.

Kim YK, Kim KS, Lee AS. 1987. Regulation of the glucose-regulated protein genes by β-mercaptoethanol requires de novo protein synthesis and correlates with inhibition of protein glycosylation. *J Cell Physiol* **133:** 553–559.

Klionsky D. 2010. The molecular machinery of autophagy and its role in physiology and disease. *Semin Cell Dev Biol* **21:** 663.

Kokame K, Kato H, Miyata T. 2001. Identification of ERSE-II, a new *cis*-acting element responsible for the ATF6-dependent mammalian unfolded protein response. *J Biol Chem* **276:** 9199–9205.

Kostova Z, Wolf DH. 2003. For whom the bell tolls: Protein quality control of the endoplasmic reticulum and the ubiquitin-proteasome connection. *EMBO J* **22:** 2309–2317.

Kozutsumi Y, Segal M, Normington K, Gething MJ, Sambrook J. 1988. The presence of malfolded proteins in the endoplasmic reticulum signals the induction of glucose-regulated proteins. *Nature* **332:** 462–464.

Kroemer G, Marino G, Levine B. 2010. Autophagy and the integrated stress response. *Mol Cell* **40:** 280–293.

Kuan CY, Whitmarsh AJ, Yang DD, Liao G, Schloemer AJ, Dong C, Bao J, Banasiak KJ, Haddad GG, Flavell RA, et al. 2003. A critical role of neural-specific JNK3 for ischemic apoptosis. *Proc Natl Acad Sci* **100:** 15184–15189.

Ladiges WC, Knoblaugh SE, Morton JF, Korth MJ, Sopher BL, Baskin CR, MacAuley A, Goodman AG, LeBoeuf RC, Katze MG. 2005. Pancreatic β-cell failure and diabetes in mice with a deletion mutation of the endoplasmic reticulum molecular chaperone gene P58^{IPK}. *Diabetes* **54:** 1074–1081.

Lamb JA, Ventura JJ, Hess P, Flavell RA, Davis RJ. 2003. JunD mediates survival signaling by the JNK signal transduction pathway. *Mol Cell* **11:** 1479–1489.

Lamkanfi M, Dixit VM. 2012. Inflammasomes and their roles in health and disease. *Annu Rev Cell Dev Biol* **28:** 137–161.

* Laplante M, Sabatini DM. 2012. mTOR signaling. *Cold Spring Harb Perspect Biol* **4:** a011593.

Lee AH, Iwakoshi NN, Glimcher LH. 2003. XBP-1 regulates a subset of endoplasmic reticulum resident chaperone genes in the unfolded protein response. *Mol Cell Biol* **23:** 7448–7459.

Lee AH, Heidtman K, Hotamisligil GS, Glimcher LH. 2011a. Dual and opposing roles of the unfolded protein response regulated by IRE1α and XBP1 in proinsulin processing and insulin secretion. *Proc Natl Acad Sci* **108:** 8885–8890.

Lee J, Sun C, Zhou Y, Lee J, Gokalp D, Herrema H, Park SW, Davis RJ, Ozcan U. 2011b. p38 MAPK-mediated regulation of Xbp1s is crucial for glucose homeostasis. *Nat Med* **17:** 1251–1260.

Le Guezennec X, Bulavin DV. 2010. WIP1 phosphatase at the crossroads of cancer and aging. *Trends Biochem Sci* **35:** 109–114.

Lei K, Davis RJ. 2003. JNK phosphorylation of Bim-related members of the Bcl2 family induces Bax-dependent apoptosis. *Proc Natl Acad Sci* **100:** 2432–2437.

Lerner AG, Upton JP, Praveen PV, Ghosh R, Nakagawa Y, Igbaria A, Shen S, Nguyen V, Backes BJ, Heiman M, et al. 2012. IRE1α induces thioredoxin-interacting protein to activate the NLRP3 inflammasome and promote programmed cell death under irremediable ER stress. *Cell Metab* **16:** 250–264.

Levine B, Mizushima N, Virgin H. 2011. Autophagy in immunity and inflammation. *Nature* **469:** 323–335.

Li Y, Schwabe RF, DeVries-Seimon T, Yao PM, Gerbod-Giannone MC, Tall AR, Davis RJ, Flavell R, Brenner DA, Tabas I. 2005. Free cholesterol-loaded macrophages are an abundant source of tumor necrosis factor-α and interleukin-6: Model of NF-κB- and map kinase-dependent inflammation in advanced atherosclerosis. *J Biol Chem* **280:** 21763–21772.

Li G, Mongillo M, Chin KT, Harding H, Ron D, Marks AR, Tabas I. 2009. Role of ERO1-α-mediated stimulation of inositol 1,4,5-triphosphate receptor activity in endoplasmic reticulum stress-induced apoptosis. *J Cell Biol* **186:** 783–792.

Li H, Korennykh A, Behrman S, Walter P. 2010. Mammalian endoplasmic reticulum stress sensor IRE1 signals by dynamic clustering. *Proc Natl Acad Sci* **107:** 16113–16118.

★ Lim K-H, Staudt LM. 2013. Toll-like receptor signaling. *Cold Spring Harb Perspect Biol* 5: a011247.

Lu B, Nakamura T, Inouye K, Li J, Tang Y, Lundback P, Valdes-Ferrer SI, Olofsson PS, Kalb T, Roth J, et al. 2012. Novel role of PKR in inflammasome activation and HMGB1 release. *Nature* 488: 670–674.

Lynes E, Simmen T. 2011. Urban planning of the endoplasmic reticulum (ER): How diverse mechanisms segregate the many functions of the ER. *Biochim Biophys Acta* 1813: 1893–1905.

Ma Y, Brewer JW, Diehl JA, Hendershot LM. 2002. Two distinct stress signaling pathways converge upon the CHOP promoter during the mammalian unfolded protein response. *J Mol Biol* 318: 1351–1365.

Marciniak SJ, Ron D. 2006. Endoplasmic reticulum stress signaling in disease. *Physiol Rev* 86: 1133–1149.

Marciniak SJ, Yun CY, Oyadomari S, Novoa I, Zhang Y, Jungreis R, Nagata K, Harding HP, Ron D. 2004. CHOP induces death by promoting protein synthesis and oxidation in the stressed endoplasmic reticulum. *Genes Dev* 18: 3066–3077.

Martinon F, Chen X, Lee A, Glimcher L. 2010. TLR activation of the transcription factor XBP1 regulates innate immune responses in macrophages. *Nat Immunol* 11: 411–418.

Matsukawa J, Matsuzawa A, Takeda K, Ichijo H. 2004. The ASK1-MAP kinase cascades in mammalian stress response. *J Biochem* 136: 261–265.

Matsuzawa A, Nishitoh H, Tobiume K, Takeda K, Ichijo H. 2002. Physiological roles of ASK1-mediated signal transduction in oxidative stress- and endoplasmic reticulum stress-induced apoptosis: Advanced findings from ASK1 knockout mice. *Antioxid Redox Signal* 4: 415–425.

Mazzitelli S, Xu P, Ferrer I, Davis RJ, Tournier C. 2011. The loss of c-Jun N-terminal protein kinase activity prevents the amyloidogenic cleavage of amyloid precursor protein and the formation of amyloid plaques in vivo. *J Neurosci* 31: 16969–16976.

McAlees JW, Sanders VM. 2009. Hematopoietic protein tyrosine phosphatase mediates β2-adrenergic receptor-induced regulation of p38 mitogen-activated protein kinase in B lymphocytes. *Mol Cell Biol* 29: 675–686.

McCullough KD, Martindale JL, Klotz LO, Aw TY, Holbrook NJ. 2001. Gadd153 sensitizes cells to endoplasmic reticulum stress by downregulating Bcl2 and perturbing the cellular redox state. *Mol Cell Biol* 21: 1249–1259.

McGuckin MA, Eri RD, Das I, Lourie R, Florin TH. 2011. Intestinal secretory cell ER stress and inflammation. *Biochem Soc Trans* 39: 1081–1085.

Meusser B, Hirsch C, Jarosch E, Sommer T. 2005. ERAD: The long road to destruction. *Nat Cell Biol* 7: 766–772.

Mighiu PI, Filippi BM, Lam TK. 2012. Linking inflammation to the brain-liver axis. *Diabetes* 61: 1350–1352.

Milanski M, Arruda AP, Coope A, Ignacio-Souza LM, Nunez CE, Roman EA, Romanatto T, Pascoal LB, Caricilli AM, Torsoni MA, et al. 2012. Inhibition of hypothalamic inflammation reverses diet-induced insulin resistance in the liver. *Diabetes* 61: 1455–1462.

Miyake Z, Takekawa M, Ge Q, Saito H. 2007. Activation of MTK1/MEKK4 by GADD45 through induced N-C dissociation and dimerization-mediated trans autophosphorylation of the MTK1 kinase domain. *Mol Cell Biol* 27: 2765–2776.

Morel C, Carlson SM, White FM, Davis RJ. 2009. Mcl-1 integrates the opposing actions of signaling pathways that mediate survival and apoptosis. *Mol Cell Biol* 29: 3845–3852.

Morel C, Standen CL, Jung DY, Gray S, Ong H, Flavell RA, Kim JK, Davis RJ. 2010. Requirement of JIP1-mediated c-Jun N-terminal kinase activation for obesity-induced insulin resistance. *Mol Cell Biol* 30: 4616–4625.

Mori K. 2000. Tripartite management of unfolded proteins in the endoplasmic reticulum. *Cell* 101: 451–454.

Mori K, Ma W, Gething MJ, Sambrook J. 1993. A transmembrane protein with a cdc2$^+$/CDC28-related kinase activity is required for signaling from the ER to the nucleus. *Cell* 74: 743–756.

★ Morrison DK. 2012. MAP kinase pathways. *Cold Spring Harb Perspect Biol* 4: a011254.

Morrison DK, Davis RJ. 2003. Regulation of MAP kinase signaling modules by scaffold proteins in mammals. *Annu Rev Cell Dev Biol* 19: 91–118.

Nakamura K, Johnson GL. 2007. Noncanonical function of MEKK2 and MEK5 PB1 domains for coordinated extracellular signal-regulated kinase 5 and c-Jun N-terminal kinase signaling. *Mol Cell Biol* 27: 4566–4577.

Nakamura T, Furuhashi M, Li P, Cao H, Tuncman G, Sonenberg N, Gorgun CZ, Hotamisligil GS. 2010. Double-stranded RNA-dependent protein kinase links pathogen sensing with stress and metabolic homeostasis. *Cell* 140: 338–348.

Nakatani Y, Kaneto H, Kawamori D, Yoshiuchi K, Hatazaki M, Matsuoka TA, Ozawa K, Ogawa S, Hori M, Yamasaki Y, et al. 2005. Involvement of endoplasmic reticulum stress in insulin resistance and diabetes. *J Biol Chem* 280: 847–851.

★ Newton K, Dixit VM. 2012. Signaling in innate immunity and inflammation. *Cold Spring Harb Perspect Biol* 4: a006049.

Niehrs C, Schäfer A. 2012. Active DNA demethylation by Gadd45 and DNA repair. *Trends Cell Biol* 22: 220–227.

Nikawa J, Yamashita S. 1992. IRE1 encodes a putative protein kinase containing a membrane-spanning domain and is required for inositol phototrophy in *Saccharomyces cerevisiae*. *Mol Microbiol* 6: 1441–1446.

Nikawa J, Akiyoshi M, Hirata S, Fukuda T. 1996. *Saccharomyces cerevisiae* IRE2/HAC1 is involved in IRE1-mediated KAR2 expression. *Nucleic Acids Res* 24: 4222–4226.

Noubade R, Krementsov DN, Del Rio R, Thornton T, Nagaleekar V, Saligrama N, Spitzack A, Spach K, Sabio G, Davis RJ, et al. 2011. Activation of p38 MAPK in CD4 T cells controls IL-17 production and autoimmune encephalomyelitis. *Blood* 118: 3290–3300.

Novoa I, Zeng H, Harding HP, Ron D. 2001 Feedback inhibition of the unfolded protein response by GADD34-mediated dephosphorylation of eIF2α. *J Cell Biol* 153: 1011–1022.

Nozaki J, Kubota H, Yoshida H, Naitoh M, Goji J, Yoshinaga T, Mori K, Koizumi A, Nagata K. 2004. The endoplasmic reticulum stress response is stimulated through the continuous activation of transcription factors ATF6 and XBP1 in *Ins2*$^{+/Akita}$ pancreatic β cells. *Genes Cells* 9: 261–270.

Ogata M, Hino S, Saito A, Morikawa K, Kondo S, Kanemoto S, Murakami T, Taniguchi M, Tanii I, Yoshinaga K, et al. 2006. Autophagy is activated for cell survival after endoplasmic reticulum stress. *Mol Cell Biol* 26: 9220–9231.

Oslowski CM, Hara T, O'Sullivan-Murphy B, Kanekura K, Lu S, Hara M, Ishigaki S, Zhu LJ, Hayashi E, Hui ST, et al. 2012. Thioredoxin-interacting protein mediates ER stress-induced β cell death through initiation of the inflammasome. *Cell Metab* 16: 265–273.

Owens DM, Keyse SM. 2007. Differential regulation of MAP kinase signalling by dual-specificity protein phosphatases. *Oncogene* 26: 3203–3213.

Ozawa K, Miyazaki M, Matsuhisa M, Takano K, Nakatani Y, Hatazaki M, Tamatani T, Yamagata K, Miyagawa J, Kitao Y, et al. 2005. The endoplasmic reticulum chaperone improves insulin resistance in type 2 diabetes. *Diabetes* 54: 657–663.

Ozcan L, Tabas I. 2012. Role of endoplasmic reticulum stress in metabolic disease and other disorders. *Annu Rev Med* 63: 317–328.

Ozcan U, Cao Q, Yilmaz E, Lee AH, Iwakoshi NN, Ozdelen E, Tuncman G, Gorgun C, Glimcher LH, Hotamisligil GS. 2004. Endoplasmic reticulum stress links obesity, insulin action, and type 2 diabetes. *Science* 306: 457–461.

Ozcan U, Yilmaz E, Ozcan L, Furuhashi M, Vaillancourt E, Smith RO, Gorgun CZ, Hotamisligil GS. 2006. Chemical chaperones reduce ER stress and restore glucose homeostasis in a mouse model of type 2 diabetes. *Science* 313: 1137–1140.

Ozcan U, Ozcan L, Yilmaz E, Duvel K, Sahin M, Manning BD, Hotamisligil GS. 2008. Loss of the tuberous sclerosis complex tumor sup-

Cite this chapter as *Cold Spring Harb Perspect Biol* doi: 10.1101/cshperspect.a006072

pressors triggers the unfolded protein response to regulate insulin signaling and apoptosis. *Mol Cell* **29**: 541–551.

Ozcan L, Wong CC, Li G, Xu T, Pajvani U, Park SK, Wronska A, Chen BX, Marks AR, Fukamizu A, et al. 2012. Calcium signaling through CaM-KII regulates hepatic glucose production in fasting and obesity. *Cell Metab* **15**: 739–751.

Palade GE, Porter KR. 1954. Studies on the endoplasmic reticulum: I. Its identification in cells in situ. *J Exp Med* **100**: 641–656.

Patil C, Walter P. 2001. Intracellular signaling from the endoplasmic reticulum to the nucleus: The unfolded protein response in yeast and mammals. *Curr Opin Cell Biol* **13**: 349–355.

Payne DM, Rossomando AJ, Martino P, Erickson AK, Her JH, Shabanowitz J, Hunt DF, Weber MJ, Sturgill TW. 1991. Identification of the regulatory phosphorylation sites in pp42/mitogen-activated protein kinase (MAP kinase). *EMBO J* **10**: 885–892.

Perier C, Bove J, Wu DC, Dehay B, Choi DK, Jackson-Lewis V, Rathke-Hartlieb S, Bouillet P, Strasser A, Schulz JB, et al. 2007. Two molecular pathways initiate mitochondria-dependent dopaminergic neurodegeneration in experimental Parkinson's disease. *Proc Natl Acad Sci* **104**: 8161–8166.

Pfaffenbach KT, Nivala AM, Reese L, Ellis F, Wang D, Wei Y, Pagliassotti MJ. 2010. Rapamycin inhibits postprandial-mediated X-box-binding protein-1 splicing in rat liver. *J Nutr* **140**: 879–884.

Pincus D, Chevalier M, Aragón T, Van Anken E, Vidal S, El-Samad H, Walter P. 2010. BiP binding to the ER-stress sensor Ire1 tunes the homeostatic behavior of the unfolded protein response. *PLoS Biol* **8**: e1000415.

Proud CG. 2009. mTORC1 signalling and mRNA translation. *Biochem Soc Trans* **37**: 227–231.

Purkayastha S, Zhang H, Zhang G, Ahmed Z, Wang Y, Cai D. 2011. Neural dysregulation of peripheral insulin action and blood pressure by brain endoplasmic reticulum stress. *Proc Natl Acad Sci* **108**: 2939–2944.

Puthalakath H, Strasser A. 2002. Keeping killers on a tight leash: Transcriptional and post-translational control of the pro-apoptotic activity of BH3-only proteins. *Cell Death Differ* **9**: 505–512.

Qiu Y, Mao T, Zhang Y, Shao M, You J, Ding Q, Chen Y, Wu D, Xie D, Lin X, et al. 2010. A crucial role for RACK1 in the regulation of glucose-stimulated IRE1α activation in pancreatic β cells. *Sci Signal* **3**: ra7.

Raman M, Earnest S, Zhang K, Zhao Y, Cobb MH. 2007. TAO kinases mediate activation of p38 in response to DNA damage. *EMBO J* **26**: 2005–2014.

Rao RV, Ellerby HM, Bredesen DE. 2004. Coupling endoplasmic reticulum stress to the cell death program. *Cell Death Differ* **11**: 372–380.

Resendez E Jr, Attenello JW, Grafsky A, Chang CS, Lee AS. 1985. Calcium ionophore A23187 induces expression of glucose-regulated genes and their heterologous fusion genes. *Mol Cell Biol* **5**: 1212–1219.

⋆ Rhind N, Russell P. 2012. Signaling pathways that regulate cell division. *Cold Spring Harb Perspect Biol* **4**: a005942.

Robinson MJ, Cobb MH. 1997. Mitogen-activated protein kinase pathways. *Curr Opin Cell Biol* **9**: 180–186.

Rodriguez Limardo RG, Ferreiro DN, Roitberg AE, Marti MA, Turjanski AG. 2011. p38γ activation triggers dynamical changes in allosteric docking sites. *Biochemistry* **50**: 1384–1395.

Ron D, Walter P. 2007. Signal integration in the endoplasmic reticulum unfolded protein response. *Nat Rev Mol Cell Biol* **8**: 519–529.

Rutkowski DT, Kaufman RJ. 2004. A trip to the ER: Coping with stress. *Trends Cell Biol* **14**: 20–28.

Sabio G, Davis RJ. 2010. cJun NH2-terminal kinase 1 (JNK1): Roles in metabolic regulation of insulin resistance. *Trends Biochem Sci* **35**: 490–496.

Salvador JM, Mittelstadt PR, Guszczynski T, Copeland TD, Yamaguchi H, Appella E, Fornace AJ Jr, Ashwell JD. 2005. Alternative p38 activation pathway mediated by T cell receptor-proximal tyrosine kinases. *Nat Immunol* **6**: 390–395.

⋆ Sassone-Corsi P. 2012. The cyclic AMP pathway. *Cold Spring Harb Perspect Biol* **4**: a011148.

Scull CM, Tabas I. 2011. Mechanisms of ER stress-induced apoptosis in atherosclerosis. *Arterioscler Thromb Vasc Biol* **31**: 2792–2797.

Seo HY, Kim YD, Lee KM, Min AK, Kim MK, Kim HS, Won KC, Park JY, Lee KU, Choi HS, et al. 2008. Endoplasmic reticulum stress-induced activation of activating transcription factor 6 decreases insulin gene expression via up-regulation of orphan nuclear receptor small heterodimer partner. *Endocrinology* **149**: 3832–3841.

Sevier C, Kaiser C. 2008. Ero1 and redox homeostasis in the endoplasmic reticulum. *Biochim Biophys Acta* **1783**: 549–556.

Sevier C, Cuozzo J, Vala A, Aslund F, Kaiser C. 2001. A flavoprotein oxidase defines a new endoplasmic reticulum pathway for biosynthetic disulphide bond formation. *Nat Cell Biol* **3**: 874–882.

Shamu CE, Walter P. 1996. Oligomerization and phosphorylation of the Ire1p kinase during intracellular signaling from the endoplasmic reticulum to the nucleus. *EMBO J* **15**: 3028–3039.

Sharma NK, Das SK, Mondal AK, Hackney OG, Chu WS, Kern PA, Rasouli N, Spencer HJ, Yao-Borengasser A, Elbein SC. 2008. Endoplasmic reticulum stress markers are associated with obesity in non-diabetic subjects. *J Clin Endocrinol Metab* **93**: 4532–4541.

Sharma G, Pallesen J, Das S, Grassucci R, Langlois R, Hampton CM, Kelly DF, des Georges A, Frank J. 2013. Affinity grid-based cryo-EM of PKC binding to RACK1 on the ribosome. *J Struct Biol* **181**: 190–194.

Shen X, Ellis RE, Lee K, Liu CY, Yang K, Solomon A, Yoshida H, Morimoto R, Kurnit DM, Mori K, et al. 2001. Complementary signaling pathways regulate the unfolded protein response and are required for *C. elegans* development. *Cell* **107**: 893–903.

Shen J, Chen X, Hendershot L, Prywes R. 2002. ER stress regulation of ATF6 localization by dissociation of BiP/GRP78 binding and unmasking of Golgi localization signals. *Dev Cell* **3**: 99–111.

Shi Y, Vattem KM, Sood R, An J, Liang J, Stramm L, Wek RC. 1998. Identification and characterization of pancreatic eukaryotic initiation factor 2 α-subunit kinase, PEK, involved in translational control. *Mol Cell Biol* **18**: 7499–7509.

Shimazawa M, Ito Y, Inokuchi Y, Hara H. 2007. Involvement of double-stranded RNA-dependent protein kinase in ER stress-induced retinal neuron damage. *Invest Ophthalmol Vis Sci* **48**: 3729–3736.

Shiu RP, Pouyssegur J, Pastan I. 1977. Glucose depletion accounts for the induction of two transformation-sensitive membrane proteins in Rous sarcoma virus-transformed chick embryo fibroblasts. *Proc Natl Acad Sci* **74**: 3840–3844.

Sidrauski C, Walter P. 1997. The transmembrane kinase Ire1p is a site-specific endonuclease that initiates mRNA splicing in the unfolded protein response. *Cell* **90**: 1031–1039.

Somers J, Pöyry T, Willis AE. 2013. A perspective on mammalian upstream open reading frame function. *Int J Biochem Cell Biol* **45**: 1690–1700.

Sumara G, Formentini I, Collins S, Sumara I, Windak R, Bodenmiller B, Ramracheya R, Caille D, Jiang H, Platt KA, et al. 2009. Regulation of PKD by the MAPK p38δ in insulin secretion and glucose homeostasis. *Cell* **136**: 235–248.

Szegezdi E, Logue SE, Gorman AM, Samali A. 2006. Mediators of endoplasmic reticulum stress-induced apoptosis. *EMBO Rep* **7**: 880–885.

Takekawa M, Saito H. 1998. A family of stress-inducible GADD45-like proteins mediate activation of the stress-responsive MTK1/MEKK4 MAPKKK. *Cell* **95**: 521–530.

Talukdar S, Olefsky JM, Osborn O. 2011. Targeting GPR120 and other fatty acid-sensing GPCRs ameliorates insulin resistance and inflammatory diseases. *Trends Pharm Sci* **32**: 543–550.

Tanoue T, Nishida E. 2003. Molecular recognitions in the MAP kinase cascades. *Cell Signal* **15**: 455–462.

Tournier C, Hess P, Yang DD, Xu J, Turner TK, Nimnual A, Bar-Sagi D, Jones SN, Flavell RA, Davis RJ. 2000. Requirement of JNK for stress-induced activation of the cytochrome c-mediated death pathway. *Science* **288**: 870–874.

Tournier C, Dong C, Turner TK, Jones SN, Flavell RA, Davis RJ. 2001. MKK7 is an essential component of the JNK signal transduction pathway activated by proinflammatory cytokines. *Genes Dev* **15:** 1419–1426.

Trusina A, Papa FR, Tang C. 2008. Rationalizing translation attenuation in the network architecture of the unfolded protein response. *Proc Natl Acad Sci* **105:** 20280–20285.

Tu BP, Weissman JS. 2002. The FAD- and O_2-dependent reaction cycle of Ero1-mediated oxidative protein folding in the endoplasmic reticulum. *Mol Cell* **10:** 983–994.

Uehara T, Nakamura T, Yao D, Shi ZQ, Gu Z, Ma Y, Masliah E, Nomura Y, Lipton SA. 2006. S-nitrosylated protein-disulphide isomerase links protein misfolding to neurodegeneration. *Nature* **441:** 513–517.

Urano F, Wang X, Bertolotti A, Zhang Y, Chung P, Harding HP, Ron D. 2000. Coupling of stress in the ER to activation of JNK protein kinases by transmembrane protein kinase IRE1. *Science* **287:** 664–666.

Ventura JJ, Cogswell P, Flavell RA, Baldwin AS Jr, Davis RJ. 2004. JNK potentiates TNF-stimulated necrosis by increasing the production of cytotoxic reactive oxygen species. *Genes Dev* **18:** 2905–2915.

Ventura JJ, Hubner A, Zhang C, Flavell RA, Shokat KM, Davis RJ. 2006. Chemical genetic analysis of the time course of signal transduction by JNK. *Mol Cell* **21:** 701–710.

Walter P, Ron D. 2011. The unfolded protein response: From stress pathway to homeostatic regulation. *Science* **334:** 1081–1086.

Wang XZ, Harding HP, Zhang Y, Jolicoeur EM, Kuroda M, Ron D. 1998. Cloning of mammalian Ire1 reveals diversity in the ER stress responses. *EMBO J* **17:** 5708–5717.

* Ward PS, Thompson CB. 2012. Signaling in control of cell growth and metabolism. *Cold Spring Harb Perspect Biol* **4:** a006783.

Wei Y, Pattingre S, Sinha S, Bassik M, Levine B. 2008. JNK1-mediated phosphorylation of Bcl-2 regulates starvation-induced autophagy. *Mol Cell* **30:** 678–688.

Welch WJ, Garrels JI, Thomas GP, Lin JJ, Feramisco JR. 1983. Biochemical characterization of the mammalian stress proteins and identification of two stress proteins as glucose- and Ca^{2+}-ionophore-regulated proteins. *J Biol Chem* **258:** 7102–7111.

Whinston E, Omerza G, Singh A, Tio CW, Winter E. 2013. Activation of the Smk1 mitogen-activated protein kinase by developmentally regulated autophosphorylation. *Mol Cell Biol* **33:** 688–700.

Whitmarsh AJ, Davis RJ. 2000. Regulation of transcription factor function by phosphorylation. *Cell Mol Life Sci* **57:** 1172–1183.

Whitmarsh AJ, Kuan CY, Kennedy NJ, Kelkar N, Haydar TF, Mordes JP, Appel M, Rossini AA, Jones SN, Flavell RA, et al. 2001. Requirement of the JIP1 scaffold protein for stress-induced JNK activation. *Genes Dev* **15:** 2421–2432.

Wiggins CM, Tsvetkov P, Johnson M, Joyce CL, Lamb CA, Bryant NJ, Komander D, Shaul Y, Cook SJ. 2011. BIM$_{EL}$, an intrinsically disordered protein, is degraded by 20S proteasomes in the absence of polyubiquitylation. *J Cell Sci* **124:** 969–977.

Witzel F, Maddison L, Blüthgen N. 2012. How scaffolds shape MAPK signaling: What we know and opportunities for systems approaches. *Front Physiol* **3:** 475.

Wong HK, Fricker M, Wyttenbach A, Villunger A, Michalak EM, Strasser A, Tolkovsky AM. 2005. Mutually exclusive subsets of BH3-only proteins are activated by the p53 and c-Jun N-terminal kinase/c-Jun

signaling pathways during cortical neuron apoptosis induced by arsenite. *Mol Cell Biol* **25:** 8732–8747.

Wu N, Zheng B, Shaywitz A, Dagon Y, Tower C, Bellinger G, Shen CH, Wen J, Asara J, McGraw TE, et al. 2013 AMPK-dependent degradation of TXNIP upon energy stress leads to enhanced glucose uptake via GLUT1. *Mol Cell* **49:** 1167–1175.

Xia Z, Dickens M, Raingeaud J, Davis RJ, Greenberg ME. 1995. Opposing effects of ERK and JNK-p38 MAP kinases on apoptosis. *Science* **270:** 1326–1331.

Xiao C, Giacca A, Lewis GF. 2011. Sodium phenylbutyrate, a drug with known capacity to reduce endoplasmic reticulum stress, partially alleviates lipid-induced insulin resistance and β-cell dysfunction in humans. *Diabetes* **60:** 918–924.

Yamada T, Ishihara H, Tamura A, Takahashi R, Yamaguchi S, Takei D, Tokita A, Satake C, Tashiro F, Katagiri H, et al. 2006. WFS1-deficiency increases endoplasmic reticulum stress, impairs cell cycle progression and triggers the apoptotic pathway specifically in pancreatic β-cells. *Hum Mol Genet* **15:** 1600–1609.

Yamazaki H, Hiramatsu N, Hayakawa K, Tagawa Y, Okamura M, Ogata R, Huang T, Yao J, Paton A, Paton J, et al. 2009. Activation of the Akt-NF-κB pathway by subtilase cytotoxin through the ATF6 branch of the unfolded protein response. *J Immunol* **183:** 1480–1487.

Yang DD, Kuan CY, Whitmarsh AJ, Rincon M, Zheng TS, Davis RJ, Rakic P, Flavell RA. 1997. Absence of excitotoxicity-induced apoptosis in the hippocampus of mice lacking the Jnk3 gene. *Nature* **389:** 865–870.

Yecies JL, Manning BD. 2011. mTOR links oncogenic signaling to tumor cell metabolism. *J Mol Med* **89:** 221–228.

Yoneda T, Imaizumi K, Oono K, Yui D, Gomi F, Katayama T, Tohyama M. 2001. Activation of caspase-12, an endoplastic reticulum (ER) resident caspase, through tumor necrosis factor receptor-associated factor 2-dependent mechanism in response to the ER stress. *J Biol Chem* **276:** 13935–13940.

Yoshida H, Haze K, Yanagi H, Yura T, Mori K. 1998. Identification of the *cis*-acting endoplasmic reticulum stress response element responsible for transcriptional induction of mammalian glucose-regulated proteins. Involvement of basic leucine zipper transcription factors. *J Biol Chem* **273:** 33741–33749.

Yoshida H, Matsui T, Yamamoto A, Okada T, Mori K. 2001. XBP1 mRNA is induced by ATF6 and spliced by IRE1 in response to ER stress to produce a highly active transcription factor. *Cell* **107:** 881–891.

Yoshida H, Matsui T, Hosokawa N, Kaufman RJ, Nagata K, Mori K. 2003. A time-dependent phase shift in the mammalian unfolded protein response. *Dev Cell* **4:** 265–271.

Zhang P, McGrath B, Li S, Frank A, Zambito F, Reinert J, Gannon M, Ma K, McNaughton K, Cavener DR. 2002. The PERK eukaryotic initiation factor 2 α kinase is required for the development of the skeletal system, postnatal growth, and the function and viability of the pancreas. *Mol Cell Biol* **22:** 3864–3874.

Zhang X, Zhang G, Zhang H, Karin M, Bai H, Cai D. 2008. Hypothalamic IKKβ/NF-κB and ER stress link overnutrition to energy imbalance and obesity. *Cell* **135:** 61–73.

Zhao L, Ackerman SL. 2006. Endoplasmic reticulum stress in health and disease. *Curr Opin Cell Biol* **18:** 444–452.

CHAPTER 19

Cell Death Signaling

Douglas R. Green and Fabien Llambi

Department of Immunology, St. Jude Children's Research Hospital, Memphis, Tennessee 38105

Correspondence: douglas.green@stjude.org

SUMMARY

In multicellular organisms, cell death is a critical and active process that maintains tissue homeostasis and eliminates potentially harmful cells. There are three major types of morphologically distinct cell death: apoptosis (type I cell death), autophagic cell death (type II), and necrosis (type III). All three can be executed through distinct, and sometimes overlapping, signaling pathways that are engaged in response to specific stimuli. Apoptosis is triggered when cell-surface death receptors such as Fas are bound by their ligands (the extrinsic pathway) or when Bcl2-family proapoptotic proteins cause the permeabilization of the mitochondrial outer membrane (the intrinsic pathway). Both pathways converge on the activation of the caspase protease family, which is ultimately responsible for the dismantling of the cell. Autophagy defines a catabolic process in which parts of the cytosol and specific organelles are engulfed by a double-membrane structure, known as the autophagosome, and eventually degraded. Autophagy is mostly a survival mechanism; nevertheless, there are a few examples of autophagic cell death in which components of the autophagic signaling pathway actively promote cell death. Necrotic cell death is characterized by the rapid loss of plasma membrane integrity. This form of cell death can result from active signaling pathways, the best characterized of which is dependent on the activity of the protein kinase RIP3.

Outline

Cite this chapter as *Cold Spring Harb Perspect Biol* doi: 10.1101/cshperspect.a006080

1 INTRODUCTION

Although cell death can happen as a result of overwhelming damage, most cell deaths in animals occur in an active manner, as a consequence of specific signaling events. In general, there are three types of cell death, defined in large part by the appearance of the dying cell: apoptosis (also known as type I cell death), autophagic cell death (type II), and necrosis (type III) (Galluzzi et al. 2007).

Apoptosis is characterized by cell shrinkage, membrane blebbing, and condensation of the chromatin (pyknosis) (Kerr et al. 1972). It can be further defined as cell death accompanied by the activation of caspase proteases (Galluzzi et al. 2012). Two major signaling pathways trigger apoptotic cell death: the mitochondrial (the intrinsic) pathway and the death receptor (the extrinsic) pathway. The latter involves a classical ligand–cell-surface-receptor interaction. For example, cytotoxic lymphocytes can kill infected or transformed cells by expressing ligands for death receptors (DRs), a subset of the tumor necrosis factor (TNF) receptor (TNFR) family. These ligands induce apoptotic cell death of the targeted cells provided they express such DRs. DR-induced cell death in general is critical for immune system function and homeostasis. In contrast, the mitochondrial apoptotic pathway is usually initiated in a cell-autonomous manner. Most cellular stresses, such as DNA damage (induced by genotoxic agents or defects in DNA repair) or endoplasmic reticulum (ER) stress (induced by the accumulation of unfolded proteins), actively engage apoptosis when cells are damaged beyond repair.

Conversely, the lack of a signal, such as those activated by growth factors (e.g., cytokines and neurotrophic factors), can lead to cell death. This mechanism is critical for the development of the nervous system in vertebrates and it is estimated that half of the neurons generated die during this process (Buss et al. 2006). This cell death is due, in part, to the failure of some neuronal precursors to properly migrate or innervate their targets and the consequent lack of neurotrophic factor stimulation. Similarly, during an immune response, cytokine deprivation (together with the DR pathway) is responsible for the acute contraction of the lymphocyte population after clearance of the pathogen. Another example of "loss-of-signal"-induced cell death is the particular form of apoptosis called anoikis, which occurs when epithelial or endothelial cells detach from the extracellular matrix (ECM). In this scenario, unligated ECM receptors of the integrin family cease to induce pro-survival signaling pathways, eventually leading to apoptosis. This mechanism prevents cells shedding from their original location from colonizing elsewhere (a characteristic of metastatic cancer cells). Finally, apoptosis can be induced by oncogenes (e.g., Myc) as a safeguard mecha-

nism against cancer development. This process is controlled in part by a p53-dependent apoptotic pathway, which is activated in response to aberrant mitogenic signals resulting from oncogene overexpression or mutation. As a consequence, evasion of apoptotic cell death is often a requisite to sustain oncogene transformation (see Ch. 21 [Sever and Brugge 2014]).

Autophagic cell death is characterized by the appearance of large intracellular vesicles and engagement of the autophagy machinery. Note that although autophagy (i.e., the membrane engulfment and catabolic degradation of parts of the cytoplasm) is a well-defined process, its function as an active cell death mechanism remains highly controversial. Autophagy is mainly a survival process engaged in response to a metabolic crisis (e.g., low ATP levels and nutrient and amino acid deprivation) or to remove damaged organelles (e.g., mitochondria with low membrane potential) and protein aggregates. As a stress response, autophagy accompanies rather than promotes cell death in most scenarios and merely represents a failed survival attempt (Shen et al. 2012). Nevertheless, there are specific examples in which the autophagy machinery is absolutely required for cell death. During *Drosophila* metamorphosis, obsolete larval tissues such as the midgut and salivary glands regress through massive autophagic cell death, a process triggered by the steroid hormone ecdysone. In this particular case, a deficiency in genes of the autophagic signaling pathway alters the cell death program (Berry and Baehrecke 2007; Denton et al. 2009). Autophagic cell death has also been reported in response to deregulated H-Ras activity and could therefore represent a safeguard mechanism against oncogenic transformation (Elgendy et al. 2011).

Necrosis is characterized by cell swelling and plasma membrane rupture, and a loss of organellar structure without chromatin condensation. Although necrosis can occur as a consequence of irreparable cell damage, at least one pathway of active necrosis exists. This form of cell death, sometimes called necroptosis, is engaged by several signaling pathways that all converge on the activation of receptor-interacting protein kinase 3 (RIP3). RIP3 is activated upon recruitment to macromolecular complexes downstream from various cell-surface receptors: DRs, Toll-like receptors (TLRs), and the T-cell receptor (TCR). Additionally, DNA damage can directly induce the formation of a RIP3-activation platform, independently of cell-surface receptor ligation. Finally, RIP3-dependent necrosis is also triggered by the cytosolic DNA sensor, DNA-dependent activator of interferon (DAI) regulatory factors, following virus infection and the presence, in the cytosol, of double-stranded viral DNA.

Here we are concerned with signaling leading to cell death in vertebrates, and focus on processes that are at least partially understood at the molecular level. We do not discuss the physiology and pathology of cell death in detail or processes involved in the clearance of dying cells. Readers will find a more complete overview of these topics and other aspects of cell death elsewhere (Green 2011b).

2 TYPE I CELL DEATH: APOPTOSIS

Apoptotic cell death is predominantly initiated either by the DR or mitochondrial pathway, although additional pathways exist. DRs—for example, Fas (also known as CD95), Trail receptor (TRAIL-R), or TNFR1—induce apoptosis by directly recruiting a caspase-activation platform upon binding to their respective ligand. The mitochondrial pathway of apoptosis, on the other hand, is triggered upon loss of integrity of the mitochondrial outer membrane, which allows the release of proapoptotic factors (e.g., cytochrome c) from the mitochondria into the cytosol. This process is controlled by the Bcl2 protein family. Once in the cytosol, cytochrome c induces the assembly of a caspase-activation complex: the apoptosome. Both pathways culminate in the activation of caspase proteases and the cleav-

age of intracellular proteins, ultimately leading to the dismantling of the cell. Below, we explore these processes in detail, starting with the end point: caspase activation.

2.1 Caspase Activation, Function, and Regulation

Apoptosis involves the activation of caspases, which orchestrate all of the morphological changes that characterize this form of cell death. Caspases are cysteine proteases with specificity for aspartic acid residues in their substrates. Although at least 17 different caspases exist in mammals, our focus is on only a subset of these for which the activation is at least partially understood and roles in cell death have been established.

The executioner caspases (caspase-3, caspase-6, and caspase-7) effect the destruction and are produced as inactive dimers that lack protein-interaction domains (Fig. 1). Activation is due to proteolytic cleavage between what will be the large and small subunits of the mature enzyme (Salvesen and Riedl 2008). Upon cleavage, the new ends fold into the dimer interface and promote conformational changes to create two active sites in the mature protease.

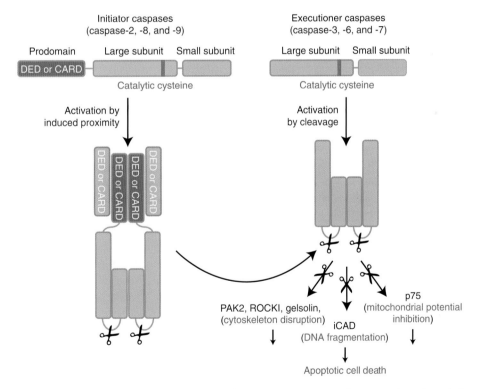

Figure 1. The caspase protein family. Initiator caspases (caspase-2, caspase-8, and caspase-9) are the apical caspases of the apoptotic-signaling cascade. Initiator caspases are produced as inactive zymogens composed of a prodomain (containing a CARD or a death effector domain [DED]) and a large and small subunit. They are recruited through their prodomains into large activation platforms and activated by dimerization. In contrast, executioner caspases (caspase-3, caspase-6, and caspase-7) are activated by cleavage of the zymogen between the large and small subunits and are therefore dependent on initiator caspases for their activation. Catalytically active caspases are composed of a heterotetramer of two small and two large subunits.

Cleavage of caspase-6 is mediated by caspase-3 and caspase-7 (Slee et al. 1999), whereas activation of the latter two caspases is generally the function of "initiator" caspases. It is these initiator caspases, and their activation, that define the apoptotic signaling pathways (see below).

Following activation, the executioner caspases, particularly caspase-3 and caspase-7, can process at least 1000 proteins (Crawford and Wells 2011). The cleavage of these caspase substrates can result in either a gain or a loss of function of these proteins and eventually leads to the cellular changes associated with apoptosis. Notably, caspase proteolysis inactivates components of essential physiological processes. For example, caspase cleavage of the p75 subunit of complex I of the electron transport chain disrupts the mitochondrial transmembrane potential, electron transport, and ATP production during apoptosis (Ricci et al. 2004). Conversely, caspase cleavage can also activate specific pathways. This is the case for caspase-activated nuclease (CAD), a DNAse that cuts chromatin between nucleosomes when its inhibitor, iCAD, is cleaved by caspase-3 (Enari et al. 1998; Sakahira et al. 1998). Additionally, caspases can hijack signaling pathways through constitutive activation of some of their components. For example, the characteristic morphology of apoptotic cells is caused by caspase-mediated activation of several actin cytoskeleton modulators: gelsolin, p21-activated kinase 2 (PAK2), and Rho-associated kinase I (ROCKI). Gelsolin, a calcium-regulated actin-severing protein, becomes constitutively active following caspase processing (Kothakota et al. 1997). PAK2 and ROCK I are serine–threonine kinase effectors of the Rho GTPase family (Rac1, Cdc42, and Rho) that regulate actin polymerization and actin–myosin contractility. Upon caspase cleavage, their kinase activity becomes independent of the GTPases, thus inducing an aberrant reorganization of the actin cytoskeleton and the characteristic membrane blebbing observed in apoptotic cells (Rudel and Bokoch 1997; Coleman et al. 2001; Sebbagh et al. 2001) (for more details on cytoskeleton-modulating proteins, see Ch. 8 [Devreotes and Horwitz 2014]).

Unlike the executioner caspases, initiator caspases (Fig. 1) exist as inactive monomers in cells and are not activated by cleavage. Instead, adaptor molecules that assemble into caspase-activation platforms recruit these initiator caspases, forcing monomers into close proximity and causing conformational changes that result in the formation of active sites. In some (but not all) cases, subsequent autocleavage of the caspase is necessary to stabilize the mature, active enzyme (Pop et al. 2007; Oberst et al. 2010). Note that although cleavage of an executioner caspase is indicative of its activation, that of initiator caspases is not necessarily an indication of activation (McStay et al. 2008).

The interactions between caspase-activation platforms and caspases involve "death fold" domains of the proteins (Kersse et al. 2011). Such death fold elements are present in adaptor proteins and caspases: the caspase-recruitment domain (CARD), and the death effector domain (DED). Other death folds, the death domain (DD) and the pyrin (PYR) domain, are involved in the assembly of some caspase-activation platforms but not present on caspases (Kersse et al. 2011). Death fold domains typically mediate protein–protein interaction through homotypic interaction. They do not share sequence similarity but have similar structures composed of six amphipathic α helices.

2.2 Caspase-9 Activation: The Mitochondrial Pathway of Apoptosis

The mitochondrial pathway of apoptosis, also called the intrinsic pathway, is the most common mechanism of apoptosis in vertebrates. It is activated in response to a variety of cellular stresses, including DNA damage, growth factor deprivation, ER stress, and developmental cues. In this pathway, the executioner caspases are cleaved and activated by caspase-9, which is itself activated by a caspase-activation platform called the apoptosome (Fig. 2) (Bratton et al. 2001; Bratton and Salvesen 2010).

Apoptotic protease-activating factor 1 (APAF1) constitutes the scaffold around which the apoptosome is assembled. During intrinsic apoptosis, cytochrome c is released from mitochondria into the cytosol (see below) and binds to Apaf1 (Zou et al. 1997). This interaction triggers hydrolysis of the Apaf1 cofactor dATP to dADP (Kim et al. 2005). The subsequent exchange of dADP with exogenous dATP allows the oligomerization of seven APAF1–dATP–cytochrome-c units into an active apoptosome. At the center of the apoptosome, exposed CARDs on APAF1 bind to the CARD of caspase-9, thus bringing the inactive caspase-9 monomers into close proximity for activation and autoprocessing (Yu et al. 2005; Yuan et al. 2010). Owing to its higher affinity for the apoptosome, full-length caspase-9 displaces the processed form, creating a continuous cycle of caspase-9 recruitment, activation, processing, and release (Malladi et al. 2009). Because caspase-9 only sustains catalytic activity in this bound state (Rodriguez and Lazebnik 1999; Stennicke et al. 1999; Bratton et al. 2001), the apoptosome functions as a molecular timer in which its lifetime is directly proportional to the amount of unprocessed caspase-9 present (Malladi et al. 2009).

In healthy cells, cytochrome c is found only in the mitochondrial intermembrane space. For it to interact with APAF1, mitochondrial outer membrane permeabilization (MOMP) triggered by apoptotic stimuli must occur (Tait and Green 2010). MOMP induces the release of all soluble

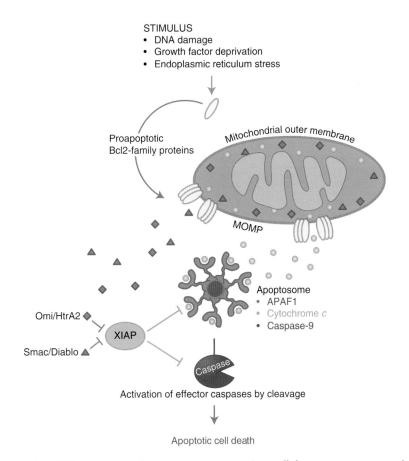

Figure 2. The mitochondrial apoptotic pathway. In response to various cellular stresses, proapoptotic members of the Bcl2 family induce mitochondrial outer membrane permeabilization (MOMP), allowing the release into the cytosol of proapoptotic factors that are normally sequestered in the intermembrane space of the mitochondria (including cytochrome *c*, Smac, and Omi). In the cytosol, cytochrome *c* binds to APAF1 and triggers its oligomerization. Caspase-9 is then recruited and activated by this platform, known as the apoptosome. Catalytically active caspase-9 cleaves and activates the executioner caspases-3 and -7. Upon release in the cytosol, Smac and Omi bind to and inhibit the caspase inhibitor X-linked inhibitor of apoptosis (XIAP). In doing so, they relieve XIAP inhibition of caspase-9, caspase-3, and caspase-7, and potentiate overall caspase activation by the apoptosome.

proteins of the mitochondrial intermembrane space into the cytosol. In addition to cytochrome *c*, two other proapoptotic proteins are released during that process: Smac (also known as Diablo) and Omi (also known as HtrA2). Smac and Omi potentiate the apoptosome activity by antagonizing the caspase inhibitor X-linked inhibitor of apoptosis (XIAP) (Fig. 2). In the absence of Smac and Omi, XIAP binds to, and inhibits the catalytic activity of, the initiator caspase-9 and the executioner caspases-3 and -7. XIAP further dampens intrinsic apoptosis induction through direct ubiquitylation and proteasomal degradation of active caspases (Eckelman et al. 2006).

MOMP is a tightly regulated event controlled by members of the Bcl2 family. These share one or more Bcl2 homology (BH) regions defined by sequence, structure, and function (Fig. 3A). There are three broad classes of Bcl2 proteins: the proapoptotic effector proteins (Bax and Bak),

which are necessary and sufficient for MOMP; the antiapoptotic Bcl2 proteins (e.g., Bcl2, Bcl-xL, and Mcl1), which block MOMP; and the BH3-only proteins (e.g., Bid, Bim, Bad, and Noxa), which activate the proapoptotic effectors and/or neutralize the antiapoptotic Bcl2 proteins. Bcl2 proteins also control several other cellular processes, including mitochondrial fusion, autophagy, and calcium efflux from the ER (Chipuk et al. 2010).

Bax and Bak are directly responsible for the loss of mitochondrial outer membrane integrity (Fig. 3B). Upon activation, they form large oligomers that insert into the mitochondrial outer membrane, disrupting it (Eskes et al. 2000; Korsmeyer et al. 2000; Dewson et al. 2008, 2009). The precise nature of the disruption remains unclear, but it allows the near simultaneous release of all intermembrane space proteins (Goldstein et al. 2000; Munoz-Pinedo et al. 2006).

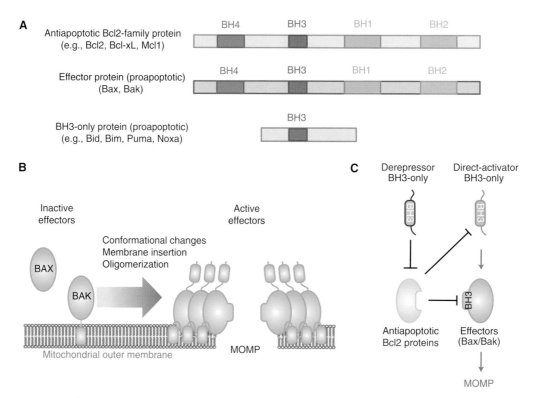

Figure 3. Regulation of mitochondrial outer membrane integrity by the Bcl2 protein family. (*A*) Members of the Bcl2 protein family are characterized by the presence of one or more Bcl2 homology (BH) region. The antiapoptotic Bcl2 proteins (e.g., Bcl2, Bcl-xL, and Mcl1) and the proapoptotic effectors (e.g., Bax and Bak) share four BH regions and a similar globular structure. BH3-only proteins (e.g., Bid, Bim, Bad, and Noxa) are characterized by a single BH region (BH3). (*B*) The proapoptotic effectors reside in cells in inactive forms tethered to the outer mitochondrial membrane (Bak) or soluble in the cytosol (Bax). Upon activation, Bax and Bak oligomerize and further insert into the membrane, causing mitochondrial outer membrane permeabilization (MOMP) and thus apoptosis. (*C*) Bax/Bak activation is triggered following the transient binding of a subset of direct activator BH3-only proteins (e.g., Bid and Bim). Antiapoptotic Bcl2 proteins inhibit MOMP by sequestering the direct activator proteins and/or the effectors. Another group of BH3-only proteins, called sensitizers or derepressors (e.g., Bad and Noxa) promote MOMP by antagonizing antiapoptotic Bcl2 proteins, thereby releasing both direct activator BH3-only proteins and Bax/Bak.

Bax and Bak act redundantly in MOMP and at least one of them is required to permeabilize mitochondria. In living cells these proteins are generally inactive, but become activated in response to upstream events. At least two of the BH3-only proteins (Bim and active Bid) activate Bax and Bak through transient interaction (Fig. 3C), although other conditions (such as heat, changes in pH, and changes in the lipid milieu) may activate the effectors independently of BH3-only proteins (Wei et al. 2000; Kuwana et al. 2002, 2005; Letai et al. 2002).

MOMP is antagonized by the antiapoptotic Bcl2 proteins, which bind to and inhibit both Bax/Bak and the BH3-only proteins by interacting with their BH3 domains (Llambi et al. 2011) (Fig. 3C). Antiapoptotic Bcl2 proteins are also regulated at both the transcriptional and posttranslational levels. In particular, Mcl1 degradation by the ubiquitin-proteasome system participates in apoptosis induction following several cellular stresses. Upon DNA damage, the

BH3-domain-containing protein Mcl1 ubiquitin ligase E3 (MULE) directly binds to Mcl1 and catalyzes its ubiquitylation and degradation (Warr et al. 2005). During antitubulin chemotherapeutic-induced mitotic arrest, Mcl1 is phosphorylated by stress-activated and mitotic kinases such as the MAP kinase (MAPK) Jun amino-terminal kinase (JNK), casein kinase II (CKII), and p38 MAPK. These phosphorylation events unveil a degron on Mcl1 that recruits the SCF-Fbw7 ubiquitin ligase complex, thus targeting Mcl1 for ubiquitylation and degradation (Wertz et al. 2011). Similarly, Mcl1 is degraded following growth factor withdrawal (e.g., IL3) and the subsequent loss of phosphoinositide 3-kinase (PI3K)-Akt signaling. This process relieves glycogen synthase kinase 3 (GSK3) from Akt inhibition. GSK3 then phosphorylates Mcl1, allowing its ubiquitylation by the E3 ligase β-transductin-repeat-containing protein (β-TrCP) and its subsequent degradation (Maurer et al. 2006; Ding et al. 2007). Conversely, cancer cells can increase Mcl1 stability and overall resistance to cellular stress by expressing USP9X, a deubiquitylase that removes polyubiquitin chains from Mcl1 (Schwickart et al. 2010).

Proteins of the Bcl2 family integrate pro- and antiapoptotic signals in healthy and stressed cells and therefore constitute one of the main signaling nodes in the life or death decision. Within this family, BH3-only proteins constitute the main upstream sensors of the mitochondrial apoptotic pathway. A wide variety of signaling pathways converge on the BH3-only family of proteins and regulate their expression level and activity both transcriptionally and posttranscriptionally (see Table 1). For example, Bid is activated upon cleavage by caspase-8 following DR ligation (Li et al. 1998, Luo et al. 1998). In doing so, Bid coordinates the cross-regulation between the extrinsic and intrinsic apoptotic pathways. Additionally, proteolytic activation of Bid can be achieved by granzyme B during the cytotoxic lymphocyte killing process (Heibein et al. 2000; Sutton et al. 2000). The apoptotic response to genotoxic stress is performed, in part, by Puma and Noxa (Jeffers et al. 2003; Villunger et al. 2003). Both are direct transcriptional targets of the tumor suppressor p53 (Oda et al. 2000; Nakano and Vousden 2001; Yu et al. 2003), although other stimuli regulate their expression levels (see Table 1). Similarly, Bim is transcriptionally up-regulated by the forkhead transcription factor FOXO3A upon cytokine deprivation (Dijkers et al. 2000; Gilley et al. 2003; Urbich et al. 2005). Bim is also a major apoptotic factor of the ER stress pathway engaged in response to accumulation of unfolded proteins (Puthalakath et al. 2007). Finally, Bad activity is negatively regulated through phosphorylation by several kinases, such as Akt (Datta et al. 1997; del Peso et al. 1997), which induces its sequestration by 14-3-3 proteins. Upon growth factor deprivation and loss of Akt signaling, Bad is released and antagonizes antiapoptotic Bcl2 proteins (Zha et al. 1996; Datta et al. 1997).

MOMP often condemns a cell to death even if caspase activation is blocked or disrupted (Tait and Green 2010). This is probably the consequence of a catastrophic loss of mitochondrial function that results in bioenergetic failure (Lartigue et al. 2009). Under these conditions, some cells can recover from MOMP, however, and survive (Tait et al. 2010). Therefore, the regulation of caspase activation downstream from MOMP may be important in some settings. For example following cytochrome c binding to APAF1, the failure to replace hydrolyzed dADP by exogenous dATP triggers nonfunctional aggregation and irreversible inactivation of APAF1 (Kim et al. 2005). Nucleotide exchange, and therefore functional apoptosome assembly, is facilitated by a complex composed of heat shock protein 70 (Hsp70), cellular apoptosis susceptibility (CAS) protein, and putative HLA-DR-associated protein I (PHAPI) (Kim et al. 2008). Accordingly, cellular levels of these proteins modulate caspase-activation efficiency during intrinsic apoptosis. tRNA levels also regulate apoptosome activity by binding to cytochrome c, blocking its interaction with APAF1 (Mei et al. 2010). Finally, intrinsic apoptosis can be perturbed downstream from MOMP through phosphorylation and inhibition of caspase-9 (see below) (Allan and Clarke 2009).

2.3 Caspase-8 Activation: The Death Receptor Pathway

Caspase-8 is activated predominantly by the DR pathway of vertebrate apoptosis. DRs are a subset of the TNFR superfamily and include TNFR1, Fas, and TRAIL-R1/2 (Dickens et al. 2012). They contain a DD in their intracellular regions. Through a series of homotypic interactions, these initiate the assembly of large macromolecular complexes that recruit and activate caspase-8 for apoptotic signaling, as well as other signaling molecules that control processes such as inflammation and cell adhesion (Ch. 15 [Newton and Dixit 2012]).

When Fas or a TRAIL-R is bound by its ligand, clustering of the receptors recruits a DD-containing adaptor molecule, FADD, through DD-DD interactions (Fig. 4A). This exposes another death fold domain in FADD, its DED. The complex represents a caspase-activation platform called a death-inducing signaling complex (DISC), as the FADD DEDs bind to DEDs in the prodomain of caspase-8. This brings caspase-8 monomers into close proximity, triggering their protease activity. Caspase-8 then undergoes autocatalytic cleavage both between the large and small subunits and between the large subunit and the prodomain. The first cleavage stabilizes the active dimer (and is required for homodimer activity in this pathway); the second cleavage releases it from the DISC (Dickens et al. 2012).

In some cells (called type I cells), active caspase-8 then promotes apoptosis by cleaving and activating caspase-3 and caspase-7. However, in many cell types (type II cells), the active executioner caspases are inhibited by XIAP, and thus apoptosis is blocked (Jost et al. 2009). In these cases, another caspase-8 substrate comes into play, the BH3-only protein Bid (see above). Caspase-8-mediated cleavage activates Bid, which in turn activates Bax and Bak to promote MOMP. The IAP antagonists Smac and Omi are then released, and these neutralize XIAP to allow apoptosis to proceed.

The signaling pathways engaged downstream from TNFR1 are more complex and potentially lead to three different outcomes: survival, apoptosis or necrosis (Fig. 4B). Upon TNFR1 ligation, a different DD-containing adaptor, TNFR-associated death domain protein (TRADD), is first recruited. TRADD does not directly bind or activate cas-

Table 1. Signaling to BH3-only proteins

BH3-only protein	Stimulus/input	Type of regulation	Signaling pathway	Output	References
Bad	Growth factors and cytokines	Phosphorylation	Akt	Cell survival	Datta et al. 1997, 2002; del Peso et al. 1997
		Phosphorylation	p70 S6K	Cell survival	Harada et al. 2001
		Phosphorylation	PKA	Cell survival	Harada et al. 1999
		Phosphorylation	PIM kinases	Cell survival	Fox et al. 2003; Yan et al. 2003
		Phosphorylation	PAK	Cell survival	Schürmann et al. 2000; Cotteret et al. 2003
		Sequestration	14-3-3	Cell survival	Zha et al. 1996; Datta et al. 2000
	TNF	Phosphorylation	IKK	Cell survival	Yan et al. 2013
	Neuronal activity deprivation	Phosphorylation	Cdc2	Apoptosis	Konishi et al. 2002
	Growth factor or cytokine deprivation	Phosphorylation	JNK	Apoptosis	Donovan et al. 2002
		Dephosphorylation	PP1, PP2A, and PP2C	Apoptosis	Ayllón et al. 2000; Chiang et al. 2001; Klumpp et al. 2003
	Calcium	Dephosphorylation	Calcineurin	Apoptosis	Wang et al. 1999
Bid	Cytotoxic T cell	Cleavage	Granzyme B	Apoptosis	Heibein et al. 2000; Sutton et al. 2000
	Fas/TNF/TRAIL	Cleavage	Caspase-8	Apoptosis	Li et al. 1998; Luo et al. 1998
	Heat shock or ER stress	Cleavage	Caspase-2	Apoptosis	Bonzon et al. 2006; Upton et al. 2008
	Ischemia or cisplatin	Cleavage	Calpain	Apoptosis	Chen et al. 2001; Mandic et al. 2002
	Lysosome permeabilization	Cleavage	Cathepsin	Apoptosis	Stoka et al. 2001; Reiners et al. 2002
Bim	Growth factors or cytokines	Phosphorylation	ERK1/2 and RSK1/2	Cell survival	Ley et al. 2003; Hubner et al. 2008; Dehan et al. 2009
		Ubiquitylation	βTrCP	Cell survival	Akiyama et al. 2003; Ley et al. 2003; Dehan et al. 2009
		mRNA stability	Hsc70	Cell survival	Matsui et al. 2007
	?	mRNA stability	miR-17-92	Cell survival	Xiao et al. 2008
	ER stress	Transcription	CHOP-C/EBPa	Apoptosis	Puthalakath et al. 2007
		Dephosphorylation	PP2A	Apoptosis	Puthalakath et al. 2007
	Glucocorticoid	Transcription	GR	Apoptosis	Erlacher et al. 2005; Ploner et al. 2008
	Growth factor deprivation or DNA damage	Phosphorylation	JNK	Apoptosis	Lei and Davis 2003; Putcha et al. 2003
	Growth factor or cytokine deprivation	Transcription	FOXO3a	Apoptosis	Dijkers et al. 2000; Gilley et al. 2003
Noxa	DNA damage	Transcription	p53	Apoptosis	Oda et al. 2000; Villunger et al. 2003; Naik et al. 2007
	DNA damage	Transcription	p63	Apoptosis	Kerr et al. 2012
	DNA damage	Deubiquitylation	UCH-L1	Apoptosis	Brinkmann et al. 2013
	Glucose deprivation	Transcription	?	Apoptosis	Alves et al. 2006
	Hypoxia	Transcription	HIF1α	Apoptosis	Kim et al. 2004
	Proteasome inhibition	Transcription/protein stabilization	Myc, others	Apoptosis	Fernandez et al. 2005; Qin et al. 2005; Nikiforov et al. 2007
Puma	Cytokine deprivation	Transcription	FOXO3a	Apoptosis	Jeffers et al. 2003; Villunger et al. 2003; You et al. 2006; Ekoff et al. 2007
	DNA damage	Transcription	p53	Apoptosis	Nakano and Vousden 2001; Yu et al. 2001; Jeffers et al. 2003; Villunger et al. 2003
	DNA damage or ER stress	Transcription	p63	Apoptosis	Pyati et al. 2011; Kerr et al. 2012
	Glucocorticoid	Transcription	GR	Apoptosis	Villunger et al. 2003; Erlacher et al. 2005
BIK	DNA damage	Transcription	Smad	Proapoptotic	Spender 2009
	TGFβ	Transcription	E2F1	Proapoptotic	Real 2006
		Ubiquitylation?		Proapoptotic	Nikrad 2005; Zhu 2005
	Proteasome inhibition	Cleavage	RHBDD1	Antiapoptotic	Wang et al. 2008a
		Phosphorylation	Casein kinase	Proapoptotic	Verma 2001; Li 2003
		Transcription	Smad	Proapoptotic	Ramjaun 2007

Continued

Cite this chapter as *Cold Spring Harb Perspect Biol* doi: 10.1101/cshperspect.a006080

Table 1. *Continued*

BH3-only protein	Stimulus/input	Type of regulation	Signaling pathway	Output	References
BMF	TGFβ	Sequestration	DLC2	Antiapoptotic	Puthalakath 2001
		Transcription	HIFα	Proapoptotic	Bruick 2000; Guo 2001
BNIP3	Hypoxia	Transcription	E2F1	Proapoptotic	Yurkova 2008
		Transcription inhibition	NF-κB	Antiapoptotic	Shaw 2006, 2008
		Transcription	Jun	Proapoptotic	Ma 2007
HRK	Potassium deprivation	Transcription	E2F1	Proapoptotic	Hershko 2004
		Transcription inhibition	DREAM	Antiapoptotic	Sanz 2001
	Cytokines	Transcription	HIFα	Proapoptotic	Sowter 2001
NIX	Hypoxia				

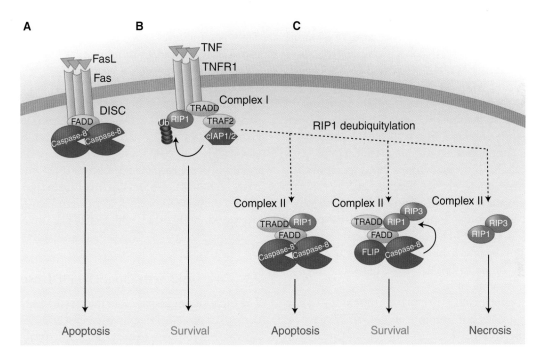

Figure 4. Death receptor signaling pathway. The death receptor signaling pathway is triggered by ligation of a death receptor (TNFR1, Fas, or TRAIL-R1/2) and potentially leads to three different outcomes: survival, apoptosis, or necrosis. (*A*) Upon ligation (by FasL and TRAIL, respectively), Fas/CD95 and TRAIL-Rs assemble a caspase-activation platform called the DISC. This platform recruits and activates caspase-8 via the adaptor protein FADD and thus engages the extrinsic apoptotic pathway. (*B*) Upon ligation to TNF, TNFR1 recruits the adaptor protein TRADD and the kinase RIP1, which in turn mobilizes additional partners, such as the ubiquitin ligases TRAF2 and cIAP1/2. These catalyze the nondegradative ubiquitylation of RIP1 as well as other components of the protein complex, resulting in the stabilization of a prosurvival and proinflammatory signaling platform called complex I. (*C*) The removal of ubiquitin chains of RIP1 by inhibition of cIAP1/2 or through the action of deubiquitylases destabilizes complex I, thus allowing the release of TRADD and RIP1. TRADD and RIP1 then promote the formation of a series of cytosolic complexes (complex II) that can initiate either apoptotic or necrotic cell death. In the proapoptotic configuration of the complex, TRADD and RIP1 recruit FADD and caspase-8, resulting in caspase activation and apoptosis. When FADD or caspase-8 are absent, or when caspase activity is blocked, RIP1 binds to and activates RIP3, thus triggering necrotic cell death. Expression of FLICE-like inhibitory protein (FLIP) inhibits the activity of both complexes. FLIP forms a heterodimer with caspase-8 that cannot induce apoptosis but still displays catalytic activity, and this activity antagonizes the activation of RIP3. Therefore, complexes containing FLIP–caspase-8 heterodimers simultaneously block apoptosis and RIP-dependent necrosis.

pase-8 but instead acts as a membrane-bond scaffold for the recruitment of additional signaling molecules, including a kinase, receptor-interacting protein 1 (RIP1), and the ubiquitin ligases TRAF2 and cIAP1/2. Components of this complex, including RIP1, are modified by a complex series of nondegradative ubiquitylation events performed by both the cIAPs and the linear ubiquitin assembly complex (LUBAC). The assembly and ubiquitylation of this signaling platform (referred to as complex I) culminates in the activation of the NF-κB signaling pathway and an inflammatory response rather than cell death (Ch. 15 [Newton and Dixit 2012]).

TNFR1 ligation leads to cell death when ubiquitylation of RIP1 is compromised either by inhibition of ubiquitin ligases contained in complex I (e.g., cIAP1/2) or through direct de-ubiquitylation of RIP1 by enzymes such as CYLD (and probably other signaling events). RIP1 and TRADD are then released from TNFR1 to form a series of dynamic cytosolic signaling platforms (collectively referred to as complex II) that have the potential to cause both apoptotic and necrotic cell death (Fig. 4C). Cytosolic TRADD recruits FADD (via DD-DD interaction), which in turn can bind to and activate caspase-8 similarly to the DISC (Fig. 4C, left panel) (Christofferson and Yuan 2010; Declercq et al. 2009). In this configuration, complex II formation leads to apoptotic cell death. In many cases, however, apoptosis does not proceed upon TNFR1 ligation, at least in part because the activation of NF-κB induces the expression of the caspase-8-like protein FLICE-like inhibitory protein (FLIP) (Micheau et al. 2001). FLIP has a domain structure similar to that of caspase-8, but it lacks a catalytic cysteine. It preferentially binds to FADD (via DED-DED interactions) and a monomer of caspase-8. The caspase-8–FLIP heterodimer is catalytically active (Micheau et al. 2002; Pop et al. 2011), but does not promote apoptosis. Why this is remains obscure, but may relate to more rapid degradation of the complex (Geserick et al. 2009). Consequently, the assembly of complex II under conditions of active NF-κB signaling, and therefore FLIP expression, leads to a survival outcome (Fig. 4C, middle panel). Finally, complex II formation can engage a necrotic form of cell death (see section below). As we will see below, the activity of the caspase-8–FLIP heterodimer is also important for preventing this necrotic cell death signaling engaged by ligated TNFR1.

2.4 Caspase -1 and Caspase-5/-11 Activation: The Inflammasome Pathway

Caspase-1 and human caspase-5 (caspase-11 in rodents) have large prodomains containing CARDs. These are engaged by caspase-activation platforms called inflammasomes. Generally these platforms form in response to infectious agents (e.g., viruses, bacteria, and fungi) and inert substances that induce inflammation (e.g., uric acid crystals, calcium phosphate crystals, alum, asbestos) (Franchi et al. 2012).

The platforms that engage caspase-1 are established by activated NOD-like receptors (NLRs), which bear either a CARD or, more often, a PYR domain (Fig. 5). In most cases, the NLR engages the adaptor molecule ASC via PYR–PYR interactions. ASC also contains a CARD that binds to and activates caspase-1 (Martinon et al. 2002; Faustin et al. 2007; Franchi et al. 2009).

Another caspase-1 inflammasome involves the sensor AIM2, which binds to cytosolic DNA (e.g., from viruses) and also engages ASC (Fig. 5) (Hornung et al. 2009). Caspase-1 cleaves and promotes the secretion of two inflammatory cytokines, interleukin (IL) 1β and IL18. It can also cleave and activate Bid, as well as caspase-3 and caspase-7 to promote apoptosis. This form of caspase-1-mediated cell death is sometimes called pyroptosis (Brennan and Cookson 2000; Cookson and Brennan 2001; Fink and Cookson 2006; Bergsbaken and Cookson 2007).

Little is known about the activation platform for caspase-5/-11, except that it does not involve ASC or the NLRs that function in caspase-1 activation (Kayagaki et al. 2011). Nevertheless, caspase-11 activation can result in apoptosis (Kayagaki et al. 2011). Caspase-11 can probably also bind to and activate caspase-1 (Green 2011a). In addition, caspase-11, but not caspase-1, is involved in the lethal effects of bacterial lipolysaccharides in mice in vivo (endotoxemia) (Kayagaki et al. 2011), which may also involve one of the executioner caspases, caspase-7 (Lamkanfi et al. 2009).

2.5 Caspase-2 Activation: The PIDDosome Pathway

The functions of caspase-2 in mammalian apoptosis remain somewhat obscure (Krumschnabel et al. 2009). Caspase-2 is activated in response to heat shock, microtubule disruption and DNA damage (Bouchier-Hayes et al. 2009), and it has been implicated in apoptosis in oocytes (Nutt et al. 2005) and degenerating neurons (Troy et al. 1997, 2000, 2001).

The activation platform for caspase-2 includes the adaptor molecule RAIDD (Duan and Dixit 1997; Hofmann et al. 1997; Chou et al. 1998), which bears a CARD that binds to the CARD in the prodomain of the caspase. Like caspase-8, caspase-2 is activated by induced proximity, following which intrachain autocleavage stabilizes the mature enzyme (Baliga et al. 2004).

In addition to a CARD, RAIDD also has a DD. This binds to the DD-containing protein PIDD (Lin et al. 2000; Telliez et al. 2000) to form what has been called the PIDDosome (Tinel and Tschopp 2004). PIDD contains two intein regions that promote its autocleavage. The first cleavage produces PIDD-C, a molecule that functions in

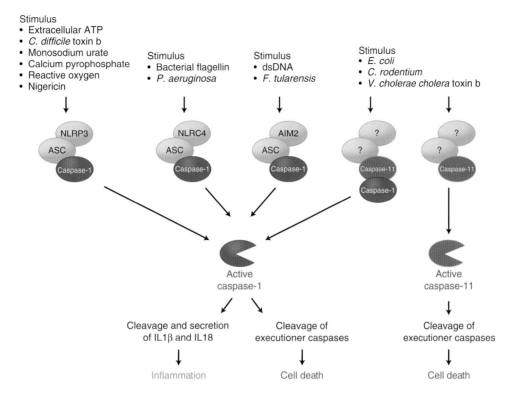

Figure 5. The inflammasomes. The inflammasomes are caspase-activation platforms that assemble in response to infectious agents (e.g., viruses, bacteria, and fungi) and inert substances that induce inflammation (e.g., uric acid crystals, calcium phosphate crystals, alum, asbestos). They recruit and activate the inflammatory caspase-1 and -11. The caspase-1 activation platforms are supported by the NOD-like receptors (NLRs; e.g., NLRP3 and NLRC4) and AIM. Upon activation by inflammatory agents, these proteins recruit the adaptor molecule ASC and caspase-1. The subsequent activation of caspase-1 induces cleavage and secretion of two inflammatory cytokines, interleukin (IL) 1β and IL18.

NF-κB activation (Janssens et al. 2005; Tinel et al. 2007). The second generates PIDD-CC, which binds to RAIDD (Tinel and Tschopp 2004; Tinel et al. 2007). However, caspase-2 activation occurs in the absence of PIDD (Manzl et al. 2009; Ribe et al. 2012), and it is not clear when or if PIDD is required (Manzl et al. 2012). The interaction of caspase-2 with RAIDD is regulated by phosphorylation of the caspase (see below).

2.6 Regulation of Caspase Activation by Kinases

Much of the regulation of apoptosis inevitably occurs upstream of caspase-activation platforms. But other signaling events can modify caspases such that, even when a suitable activation platform forms, their activation is inhibited. XIAP and FLIP are two examples. Below we consider some others.

Regulation of caspase-2 activation is particularly well described. When NADPH is plentiful in the cell (e.g., owing to the activity of the pentose phosphate pathway) calcium/calmodulin-dependent protein kinase II (CaMKII) phosphorylates its CARD (Nutt et al. 2005). This allows the

binding of 14-3-3ζ, which prevents association with the PIDDosome (Nutt et al. 2009). If NADPH levels decrease, protein phosphatase 1 (PP1) dephosphorylates caspase-2, permitting its activation by the PIDDosome (Nutt et al. 2009). SIRT1-mediated acetylation of 14-3-3ζ inhibits its binding to phospho-caspase-2, reinforcing this signal (Andersen et al. 2011). Intriguingly, the *Drosophila* initiator caspase Dronc (see Box 1) is also regulated by NADPH and phosphorylation by CaMKII (Yang et al. 2010; Ch. 6 [Rhind and Russel 2012]).

The cell-cycle regulator cyclin-dependent kinase 1 (CDK1) bound to cyclin B1 also phosphorylates caspase-2 (Andersen et al. 2009). In this case, the phosphorylation is in the region between what will be the large and small subunits of the active caspases, and appears to directly inhibit the generation of the mature, stable enzyme.

Caspase-9 is phosphorylated on a threonine residue in the region between the prodomain and the large subunit by several kinases such as the ERK2 MAPK and CDK1 (Allan and Clarke 2009). Inhibition of caspase-9 activity by ERK2 occurs downstream from Ras in response to prosurvival stimuli such as growth factor stimulation (Allan et al.

BOX 1. CASPASE ACTIVATION IN INVERTEBRATE ORGANISMS

Homologs of mammalian components of the mitochondrial pathway (e.g., caspases, APAF1 and Bcl2 proteins) are found throughout the animal kingdom, including invertebrates. However, in neither of the two invertebrates studied extensively—*Drosophila* and *Caenorhabditis elegans*—has a role for the mitochondrial pathway involving MOMP and cytochrome-*c*-mediated activation of the apoptosome been unambiguously shown. Recent studies indicate that mitochondrial pathway may exist in helminths (Lee et al. 2011; Bender et al. 2012), and cytochrome *c* (Cyt *c*) triggers caspase activation in cytosolic extracts of a helminth and several echinoderms (Bender et al. 2012). Therefore, it is possible that the insect and nematode pathways outlined below are derived from an ancestral mitochondrial pathway resembling that of mammals (see the figure below, part A).

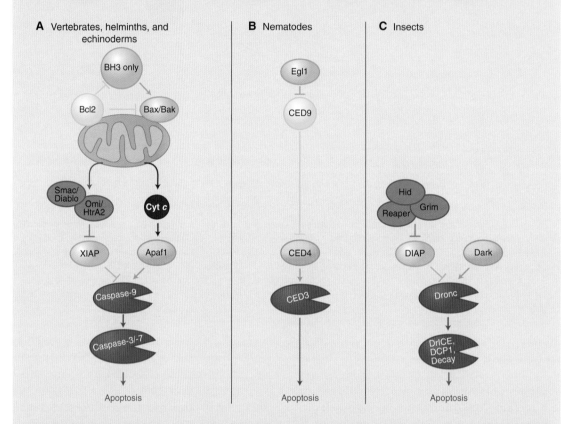

In *C. elegans,* the single caspase involved in apoptosis, CED3, contains a CARD and is activated by a platform composed of the APAF1 ortholog CED4 (Yuan et al. 1993; Chinnaiyan et al. 1997a). Monomeric CED4 is held inactive in living cells by the Bcl2 ortholog CED9 (Chinnaiyan et al. 1997b; Seshagiri and Miller 1997; Spector et al. 1997; Wu et al. 1997a,b). (Recall that this is not the case in mammals, in which Bcl2 proteins do not sequester APAF1, but instead restrict access to cytochrome *c* [Newmeyer et al. 2000].) In cells that are triggered to undergo apoptosis, the BH3-only protein Egl1 binds to CED9, releasing CED4, which then assembles in the caspase-activation platform for CED3 (Conradt and Horvitz 1998; Yan et al. 2004). The control of expression of Egl1 is the major way in which apoptosis in this organism is regulated (see the figure above, part B).

In *Drosophila*, there are several initiator and executioner caspases. The initiator caspase Dronc contains a CARD and is activated on a platform composed of the APAF1 ortholog, Dark (Kanuka et al. 1999; Rodriguez et al. 1999; Zhou et al. 1999). The Dark apoptosome appears to be constitutively active (Yu et al. 2006) but the activity of Dronc is inhibited by an IAP, DIAP1 (Rodriguez et al. 2002). The expression of one of several DIAP1 inhibitors, Reaper, Hid, Grim, or Sickle (Goyal et al. 2000; Lisi et al. 2000; Yoo et al. 2002), leads to displacement of DIAP1, allowing Dronc to cleave and thereby activate the executioner caspases DCP1, drICE, and Decay. Therefore, the expression of the DIAP1 inhibitors appears to directly control apoptosis in flies (see the figure above, part C). The roles of MOMP and cytochrome *c* release in *Drosophila* apoptosis are controversial. Although MOMP occurs, it appears to be predominantly caspase dependent and therefore is probably an effect rather than a cause of caspase activation.

 Cite this chapter as *Cold Spring Harb Perspect Biol* doi: 10.1101/cshperspect.a006080

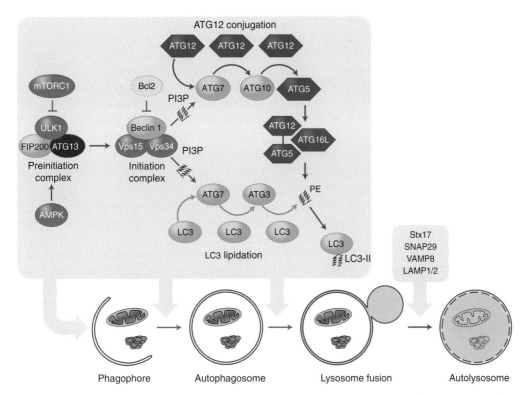

Figure 6. The autophagic signaling pathway. Under metabolic stress, AMPK activation and/or mTORC1 inhibition lead to the activation of the preinitiation complex (ULK1, FIP200, and ATG13). The latter activates the initiation complex (beclin 1, VPS15, and VPS34) that generates PI3P and recruits ATG7 to the phagophore. ATG7 functions similarly to an E1-ubiquitin ligase and initiates two conjugation pathways necessary for membrane elongation and closure of the autophagosome. In the ATG5-ATG12 conjugation pathway, ATG12 is sequentially transferred to ATG7, ATG10, and ATG12. The ATG5-12 conjugate recruits ATG16L and forms a complex necessary to stabilize the phagophore and to complete the second conjugation pathway. In the LC3-PE pathway, LC3 is cleaved by ATG4 and sequentially conjugated to ATG7 and ATG3. The ATG5-ATG12-ATG16L complex carries out the final step by transferring LC3 to PE to form an LC3-PE conjugate (also called LC3-II). LC3-II associates with the autophagosomal membrane and is crucial for the targeting of autophagosomes to lysosomes, as well as for the selective autophagy of organelles and protein aggregates.

2003). CDK1−cyclin-B1 controls caspase-9 phosphorylation and activity during mitosis. This process is thought to dampen the threshold for intrinsic apoptosis signals during the cell cycle, especially upon prolonged mitotic arrest (Allan and Clarke 2007). The mechanism by which this phosphorylation event inhibits caspase-9 activity remains obscure because it does not affect binding to APAF1.

3 TYPE II CELL DEATH AND AUTOPHAGY

3.1 The Autophagy Pathway

Macroautophagy, referred to here simply as autophagy, is a cellular process that is highly conserved in eukaryotes.[1] It is a catabolic process engaged under metabolic stress (such as

nutrient starvation and bioenergetics failure), to ensure availability of critical metabolic intermediates. It is also important for the removal of damaged organelles (including mitochondria), protein aggregates, and infecting organisms (Levine and Kroemer 2008; Kroemer et al. 2010).

Autophagy, in general, is a survival process. Its presence during so-called autophagic (type II) cell death *usually* represents a failed attempt to overcome lethal stress, and disruption of this process promotes rather than inhibits cell death in many cases (but see below). This form of cell death is therefore often referred to as caspase-independent cell death accompanied by autophagy (Kroemer and Levine 2008; Shen et al. 2012).

The autophagy pathway culminates in the formation of a double membrane structure, the autophagosome. This envelops intracellular material and ultimately fuses with lysosomes allowing degradation of the enveloped material (Fig. 6). Autophagosome formation and maturation is reg-

[1]Two other processes, microautophagy and chaperone-mediated autophagy, also exist, but are not further discussed here.

ulated by the sequential function of multiple autophagy-related (ATG) proteins. The process starts with the activation of a preinitiation complex composed of the kinase Unc-51-like kinase 1 (ULK1), FIP200, and ATG13. The activity of this kinase complex is directly regulated by two major metabolic checkpoints: mammalian target of rapamycin complex 1 (mTORC1) and AMP-activated protein kinase (AMPK) (see Ch. 14 [Hardie 2012]; p. 91 [Laplante and Sabatini 2012]). In healthy cells, mTORC1 inhibits the preinitiation complex through phosphorylation of ULK1 and ATG13. Upon metabolic stress (e.g., amino acid deprivation), mTORC1-mediated inhibition of ULK1 and ATG13 is abolished (Ganley et al. 2009; Hosokawa et al. 2009; Jung et al. 2009). In contrast, AMPK activates the preinitiation complex (Egan et al. 2011; Kim et al. 2011). When ATP synthesis is unable to meet the demands of ATP consumption in a cell, AMP and ADP accumulate and activate AMPK. Active AMPK promotes autophagy indirectly through suppression of mTORC1 activity and directly by phosphorylating and activating ULK1. Therefore, conditions under which mTORC1 is inhibited and/or AMPK is activated engage the activity of the preinitiation complex.

The preinitiation complex recruits and activates an initiation complex composed of beclin 1, a class III PI3K (Vps34), and the protein kinase Vps15, which results in the generation of the lipid phosphatidylinositol 3-phosphate (PI3P) (He and Levine 2010). The activity of the initiation complex is negatively regulated by several independent signaling pathways. Akt directly phosphorylates beclin 1, thus allowing it to bind to 14-3-3 (Wang et al. 2012). Beclin-1–14-3-3 complexes are sequestered by vimentin, a component of intermediate filaments. This Akt-mediated cytoskeletal sequestration dampens the lipid kinase activity of the initiation complex. Therefore, growth factors and the PI3K-AKT pathway can inhibit autophagy both directly through beclin 1 and indirectly through the mTOR pathway. Beclin 1 is a scaffold protein that recruits several additional partners (in addition to Vps34 and Vps15) that regulate the lipid kinases activity of the preinitiation complex (for a more comprehensive review of beclin-1-interacting proteins, see Kroemer et al. 2010). For example, beclin-1 binds to autophagy/beclin-1 regulator 1 (AMBRA1), a protein that tethers the beclin-1–Vps34 complex to microtubules through the dynein motor complex (Di Bartolomeo et al. 2010; Fimia et al. 2012). Upon autophagy induction, AMBRA1 is phosphorylated by ULK1 and released from the cytoskeleton to allow autophagosome formation.

Interestingly, members of the Bcl2 family can regulate the initiation complex. Beclin 1 contains a bona fide BH3 domain, which allows its sequestration by antiapoptotic proteins such as Bcl2 and Bcl-xL (Pattingre et al. 2005;

Maiuri et al. 2007b; Oberstein et al. 2007). This dampens the activity of the initiation complex. Under stress conditions (e.g., starvation or hypoxia), the BH3-only proteins Bad and Bnip3 release beclin 1 to permit its participation in the autophagic process (Maiuri et al. 2007a,b; Bellot et al. 2009). Alternatively, dissociation of beclin 1 from Bcl2 and Bcl-xL can be achieved through phosphorylation of the beclin–1 BH3 domain by the death-associated protein kinase (DAPK) (Zalckvar et al. 2009a,b). Finally, kinases of the JNK family disrupt this complex by directly phosphorylating Bcl2, thus decreasing its affinity for beclin 1 (Wei et al. 2008). This process is essential for regulation of exercise-induced autophagy and glucose metabolism in muscles (He et al. 2012).

The elongation and ultimate closure of the autophagosome is regulated by two distinct but complementary ubiquitin-like protein conjugation systems: the ATG5-12 and LC3-PE conjugation pathways (Fig. 6). Both pathways are initiated by a single E1-ligase-like activating enzyme, ATG7, recruited to the phagophore through PI3P generated by the initiation complex. First, a small protein, ATG12, is covalently attached to the active site cysteine residue of ATG7 by a thioester linkage. ATG12 is then transferred via thioester exchange to an E2-like conjugating enzyme, ATG10, which further transfers ATG12 to ATG5. Finally, the ATG5-ATG12 conjugate noncovalently recruits ATG16L to form a large multimeric complex necessary to stabilize the forming phagophore and to complete the second conjugation pathway (Hanada et al. 2007). The LC3-PE conjugation pathway is initiated following cleavage of LC3 by the cysteine protease ATG4. Cleaved LC3 binds to ATG7 and is subsequently transferred to the E2 enzyme ATG3. The ATG5-ATG12-ATG16L complex then acts as an E3 ligase, conjugating LC3 to phosphatidylethanolamine (PE) to form an LC3-PE conjugate (also referred to as LC3-II) (Hanada et al. 2007).

This lipidated, membrane-associated form, LC3-II, is crucial for the targeting of autophagosomes to lysosomes. Once completed, the autophagosome fuses with lysosomes to form an autolysosome, thus degrading the engulfed material. This fusion event is achieved when membranes from both organelles are brought into close proximity through interaction between the autophagosomal protein syntaxin 17 (Stx17), SNAP29, and VAMP8 on the lysosome (Itakura et al. 2012). Additionally, two lysosomal proteins, LAMP1 and LAMP2, are required for the fusion process, and in their absence the autophagosome fails to fuse to lysosomes (Tanaka et al. 2000; Eskelinen et al. 2002, 2004).

During starvation (e.g., amino acid or serum deprivation), large portions of the bulk cytosol are randomly engulfed and degraded by autophagy. Additionally, autophagy can be a selective process and remove specific subcellular structures (e.g., protein aggregates and damaged or excess organelles) in the absence of metabolic stress. During selective autophagy, substrate specificity is conferred by adaptor proteins that tether the targeted structure to the nascent phagophore. Aggregates formed by unfolded proteins are ubiquitylated and linked to the phagophore by proteins such as p62 and neighbor of breast cancer 1 (NBR1) that bind both LC3 and ubiquitin (Shaid et al. 2013). Similarly, autophagic removal of mitochondria (a process called mitophagy) is controlled by mitochondrion-anchored LC3-binding proteins such as NIP-like protein X (NIX, also called Bnip3L) and FUN14-domain-containing 1 (FUNDC1). NIX, a Bcl2-family-related protein, promotes removal of the entire mitochondrial pool during terminal erythrocyte differentiation (Schweers et al. 2007; Sandoval et al. 2008), and FUNDC1 participates in hypoxia-induced mitochondrial clearance (Liu et al. 2012). Additionally, damaged mitochondria that display low membrane potential are also removed by mitophagy following ubiquitylation of mitochondrial proteins by the ubiquitin ligase parkin (Narendra and Youle 2011), although the exact mechanism by which the phagophore is recruited in that scenario is not clear.

It is relatively easy to see how autophagy is engaged under conditions of cellular stress and can ameliorate the stress to protect the cells. Two questions persist: What kills the cells in type II cell death, and *can* autophagy itself ever be a mechanism that promotes cell death? Currently, we do not have a satisfactory answer to the first question but we are closer to answering the second.

3.2 Can Autophagy Kill Cells?

Autophagy often accompanies type II cell death without participating in it. There are cases, however, in which it appears to actively promote cell death. This is best exemplified during *Drosophila* metamorphosis, in which autophagy-dependent cell death drives the involution of the salivary gland (Berry and Baehrecke 2007) and of the midgut (Denton et al. 2009). This physiological form of cell death is triggered by the steroid hormone ecdysone and occurs (in the salivary gland) following growth arrest and inhibition of the PI3K signaling pathway. In both cases, mutation of any of several essential *ATG* genes reduces these developmental cell deaths. Reciprocally, forced expression of the ULK1 ortholog ATG1 is sufficient to drive this form of cell death.

Autophagy-dependent cell death also occurs in mammalian cells transformed with a constitutively active H-Ras (Elgendy et al. 2011). In this setting, Ras-dependent cell death is inhibited if ATG5, ATG7, or beclin 1 are silenced. Intriguingly, this form of autophagic cell death appears to

be regulated by Bcl2 proteins. In particular, displacement of beclin 1 from Mcl1 is thought to be triggered by the proapoptotic BH3-only protein Noxa. Again, how autophagy components promote the death of the cell remains to be elucidated.

Untransformed breast epithelial cells that lose attachment to a substrate can engulf each other, resulting in the death of the engulfed cell by entosis (Overholtzer et al. 2007). The engulfed cell undergoes apoptosis, which can be inhibited by expression of Bcl2, but this does not rescue the cell. However, if the engulfing cell lacks components of the autophagy pathway, the engulfed cell can survive (Krajcovic et al. 2011). Recent studies have shown that components of the autophagy pathway can be recruited to phagosomes that have engulfed organisms or cells, facilitating the degradation of the phagocytosed cargo (Sanjuan et al. 2007; Martinez et al. 2011). The preinitiation complex does not appear to be required. In the case of entosis for example, FIP200 is dispensable (Krajcovic et al. 2011). Thus, in these cases, autophagy may act as a "murder weapon" allowing one cell to efficiently kill another.

4 TYPE III CELL DEATH: NECROSIS

4.1 Types of Necrotic Cell Death

Cellular necrosis or necrotic cell death encompasses a wide variety of cell death processes with one common denominator: the loss of plasma membrane integrity followed by cytoplasmic leakage (Yuan and Kroemer 2010). Necrosis can occur simply as a consequence of such extensive damage that cell integrity is disrupted—for example, at high temperature, following freeze-thaw, or upon mechanical stress. In such cases, cell death is passive and does not require the activation of any particular signaling pathway. Note that a necrotic morphology (i.e., rupture of the plasma membrane) can also be observed at late stages of an apoptotic or autophagic cell death program, when dead cells fail to be cleared from the system by phagocytosis. This process is referred to as secondary necrosis and is independent of any other signaling event than those initially engaged (apoptotic or autophagic). However, necrotic cell death is not always an accidental or passive process and can also be the result of a directed signaling cascade.

The best-characterized form of programmed necrosis is RIP-kinase-dependent necrosis (also referred to as "necroptosis") (Degterev et al. 2005). This requires the kinase activity of RIP3 and can result in a rapid cell death with the features of necrosis (Cho et al. 2009; He et al. 2009; Zhang et al. 2009). Necroptosis can be engaged downstream from TNFR1 (Schulze-Osthoff et al. 1994; Vercammen et al. 1998; Holler et al. 2000; Matsumura et al. 2000). Following

TNFR1 ligation, RIP3 is activated through recruitment to complex II by RIP1 (Fig. 4C). The interaction between RIP1 and RIP3 is mediated by their respective RIP homotypic interaction motifs (RHIMs). In this configuration, complex II induces necrosis rather than apoptosis and is referred to as the RIPoptosome (Feoktistova et al. 2011a; Tenev et al. 2011). The assembly and activation of the necrotic signaling platform are controlled by a series of post-translational modifications of components of this complex. RIPoptosome formation is negatively regulated by ubiquitylation of RIP1. Consequently, treatments with cIAP1/2 inhibitors or expression of the deubiquitylase CYLD promote TNFR1-induced necrosis (Wang et al. 2008b; O'Donnell et al. 2011; Moulin et al. 2012). Additionally, RIP3 activation depends on the kinase activity of RIP1 (Cho et al. 2009; He et al. 2009; Zhang et al. 2009), which can be inhibited by the drug necrostatin (Degterev et al. 2008). Accordingly, RIP-dependent necrosis is blocked by necrostatin (Degterev et al. 2005). In addition, necrosis induction by the RIPoptosome is negatively regulated by the catalytic activity of the caspase-8–FLIP heterodimers (see below).

RIP-dependent necrosis can be triggered by TLR3 and TLR4 (for more details on TLR signaling pathways, see p. 121 [Lim and Staudt 2013]). TLR3 and TLR4 promote the assembly of the RIPoptosome by recruiting the RHIM-containing protein TIR-domain-containing adaptor-inducing interferon-β (TRIF) (Imtiyaz et al. 2006; Feoktistova et al. 2011b). TRIF recruits RIP1 and RIP3 to induce RIP-dependent necrosis (Feoktistova et al. 2011a). TRIF might, however, recruit RIP3 independently of RIP1 through direct RHIM–RHIM interaction. RIP-dependent necrosis can also be engaged during viral infection (Cho et al. 2009; Upton et al. 2010) via the RHIM-containing cytosolic DNA sensor, DNA-dependent activator of interferon regulatory factors (DAI). DAI is activated by double-stranded viral DNA and recruits RIP3 through its RHIM domain (Upton et al. 2012). Finally, DNA damage signaling can also cause RIPoptosome formation independently of any membrane receptor signaling (Tenev et al. 2011).

The mechanism by which RIP3 causes cell death remains obscure. Several mechanisms have been proposed, including the generation of reactive oxygen species (either mitochondrial or via NADPH oxidase) and elevation of metabolic processes to deplete ATP and/or NADH, but the evidence for them is not compelling (Hitomi et al. 2008). Recent studies have implicated the pseudokinase mixed lineage kinase-like (MLKL) (Sun et al. 2012; Zhao et al. 2012) as a critical player in the execution of RIP-dependent necrosis. Phosphorylation by RIP3 triggers MLKL oligomerization and translocation to the plasma membrane (Cai et al. 2014; Chen et al. 2014). The exact

mechanism by which MLKL impairs the plasma membrane integrity is not known but is thought to involve a disruption of ionic homeostasis.

4.2 Control of RIPK-Dependent Necrosis by Caspases

Stimuli that activate RIP1 and the RIPoptosome (e.g., ligation of TNFR1, ligation of TLR3/4, and possibly DNA damage) engage caspase-8 and FLIP. The catalytic activity of caspase-8–FLIP heterodimers antagonizes activation of RIP3 and the ensuing necrosis (Pop et al. 2011). As a result, cells that are treated with TNF or TLR3/4 ligands do not generally undergo cell death. However, if caspase-8–FLIP activity is blocked or disrupted, this can engage necrosis in a manner that is dependent on both RIP1 and RIP3 (Cho et al. 2009; He et al. 2009; Zhang et al. 2009; Oberst et al. 2010). RIP1, RIP3, and CYLD are substrates of caspase-8 (Lin et al. 1999; Feng et al. 2007; Lu et al. 2011; O'Donnell et al. 2011). However, in the absence of FLIP, activation of caspase-8 is not sufficient to prevent RIPK-dependent necrosis, even if apoptosis downstream from caspase-8 activation is blocked (Oberst et al. 2011).

4.3 Other Necrosis Signaling Pathways

Several other forms of necrosis signaling can occur in cells, although, like RIP-dependent necrosis, none of these is fully characterized. Ischemia/reperfusion injury (I/R), for example, causes cell death when a tissue is transiently deprived of blood supply and is then restored. High levels of calcium influx and generation of reactive oxygen species (ROS) have been associated with this form of cell death (Halestrap and Pasdois 2009). These trigger cell death, in part, by causing a mitochondrial permeability transition (mPT), in which the mitochondrial inner membrane becomes permeable to small solutes (Ricchelli et al. 2011). This results in a rapid dissipation of the transmembrane potential, followed by matrix swelling and often rupture of the mitochondrial outer membrane. Cyclophilin D (CypD), a peptidyl-proline trans-isomerase, plays a major role in mPT. Mice lacking CypD display defective mPTs (Baines et al. 2005; Basso et al. 2005; Nakagawa et al. 2005; Schinzel et al. 2005) and display diminished tissue damage under conditions of I/R (Baines et al. 2005; Nakagawa et al. 2005; Schinzel et al. 2005). Importantly, mice lacking CypD have normal apoptosis (Baines et al. 2005; Nakagawa et al. 2005; Schinzel et al. 2005) and RIP-dependent necrosis (Ch'en et al. 2011).

Another form of active necrosis involves NADPH oxidase1 (Nox1) and can be triggered by TNF (Kim et al. 2007). Nox1 generates a ROS burst that may be important.

There may be a relationship between Nox1 and RIP-dependent necrosis (Vanden Berghe et al. 2007), but further evidence supporting this connection is needed.

In neurons, engagement of glutamate receptors can trigger a form of necrotic cell death called excitotoxicity (Wang and Qin 2010). This appears to involve a calcium influx that triggers an ROS burst and also activates calpain (Higuchi et al. 2005; Vosler et al. 2008). Excitotoxicity is associated with I/R and may occur in other pathological conditions (Szydlowska and Tymianski 2010).

5 BEYOND TYPE III—OTHER FORMS OF CELL DEATH

There are additional forms of cell death that appear to require the participation of the cell in its demise. These include mitotic catastrophe (Castedo et al. 2004), ferroptosis (Dixon et al. 2012), and others. Although outside the scope of our discussion, a useful distinction might be made between cellular suicide and cellular sabotage. Suicide mechanisms (including apoptosis, necroptosis, and perhaps autophagic cell death, as outlined above) represent signaling pathways that appear to have evolved to execute the cell. In contrast, sabotage can be thought of as the consequence of disruption of normal cellular processes that have not evolved for the purpose of cell death. Much as a train must be moving if removal of a railroad tie (a *sabot*) is to damage it, these forms of cell death occur when cellular machinery confronts a disruption that results in cellular destruction. For example, if chromosomes are cross-linked, mitosis may sufficiently damage them that the cell dies, even if mechanisms of active cell suicide cannot be engaged (mitotic catastrophe, although it is not clear that mitotic catastrophe proceeds in this way [Castedo et al. 2004]).

6 CONCLUSION

Active or programmed cell death is essential to maintain homeostasis in multicellular organisms as well as for the selective elimination of potentially harmful or infected cells. Accordingly, deregulation of the signaling pathways that trigger cell death can lead to the development of catastrophic diseases such as cancer and autoimmunity (too little cell death) as well as degenerative diseases (too much cell death). Therefore, the existence of tightly controlled and efficient means to induce cell death can be interpreted as the logical consequence of the evolution of multicellular organisms. However, the necessity for so many different death-signaling pathways might appear counterintuitive. Taken together, cell death induction could be viewed as a simple signaling process with multiple inputs/stimuli and

one outcome: the death of the cell. However, the way by which a cell dies has important consequences for neighboring cells and sometimes the entire organism. For example, apoptotic and necrotic cells display divergent inflammatory properties and trigger different immune responses. Additionally, particular death programs include the release of proliferative signals that trigger compensatory proliferation in surrounding tissues. These signals might differ from one type of cell death to the next. Finally, death-signaling pathways are clearly interconnected. For example, autophagic cell death is often potentiated by caspase activation, whereas RIP-dependent necrosis is antagonized by a caspase-dependent activity. Cross talk between these pathways potentially provides numerous backup mechanisms for cell death programs and could explain why inhibition of a single program often has minor consequences for the organism. A better understanding of the impact of each type of cell death on surrounding tissues and of the interplay between these cell death programs might help to answer some of these questions.

REFERENCES

*Reference is in this book.

Akiyama T, Bouillet P, Miyazaki T, Kadono Y, Chikuda H, Chung U-I, Fukuda A, Hikita A, Seto K, Okada T, et al. 2003. Regulation of osteoclast apoptosis by ubiquitylation of proapoptotic BH3-only Bcl-2 family member Bim. *EMBO J* **22:** 6653–6664.

Allan LA, Clarke PR. 2007. Phosphorylation of caspase-9 by CDK1/cyclin B1 protects mitotic cells against apoptosis. *Mol Cell* **26:** 301–310.

Allan LA, Clarke PR. 2009. Apoptosis and autophagy: Regulation of caspase 9 by phosphorylation. *FEBS J* **276:** 6063–6073.

Allan LA, Morrice N, Brady S, Magee G, Pathak S, Clarke PR. 2003. Inhibition of caspase-9 through phosphorylation at Thr 125 by ERK MAPK. *Nat Cell Biol* **5:** 647–654.

Alves NL, Derks IA, Berk E, Spijker R, van Lier RA, Eldering E. 2006. The Noxa/Mcl-1 axis regulates susceptibility to apoptosis under glucose limitation in dividing T cells. *Immunity* **24:** 703–716.

Andersen JL, Johnson CE, Freel CD, Parrish AB, Day JL, Buchakjian MR, Nutt LK, Thompson JW, Moseley MA, Kornbluth S. 2009. Restraint of apoptosis during mitosis through interdomain phosphorylation of caspase 2. *EMBO J* **28:** 3216–3227.

Andersen JL, Thompson JW, Lindblom KR, Johnson ES, Yang CS, Lilley LR, Freel CD, Moseley MA, Kornbluth S. 2011. A biotin switch-based proteomics approach identifies 14–3-3ζ as a target of Sirt1 in the metabolic regulation of caspase 2. *Mol Cell* **43:** 834–842.

Ayllón V, Martínez-A C, García A, Cayla X, Rebollo A. 2000. Protein phosphatase 1α is a Ras-activated Bad phosphatase that regulates interleukin-2 deprivation-induced apoptosis. *EMBO J* **19:** 2237–2246.

Baines CP, Kaiser RA, Purcell NH, Blair NS, Osinska H, Hambleton MA, Brunskill EW, Sayen MR, Gottlieb RA, Dorn GW, et al. 2005. Loss of cyclophilin D reveals a critical role for mitochondrial permeability transition in cell death. *Nature* **434:** 658–662.

Baliga BC, Read SH, Kumar S. 2004. The biochemical mechanism of caspase 2 activation. *Cell Death Differ* **11:** 1234–1241.

Basso E, Fante L, Fowlkes J, Petronilli V, Forte MA, Bernardi P. 2005. Properties of the permeability transition pore in mitochondria devoid of Cyclophilin D. *J Biol Chem* **280:** 18558–18561.

Bellot G, Garcia-Medina R, Gounon P, Chiche J, Roux D, Pouyssegur J, Mazure NM. 2009. Hypoxia-induced autophagy is mediated through hypoxia-inducible factor induction of BNIP3 and BNIP3L via their BH3 domains. *Mol Cell Biol* 29: 2570–2581.

Bender CE, Fitzgerald P, Tait SWG, Llambi F, McStay GP, Tupper DO, Pellettieri J, Sánchez Alvarado A, Salvesen GS, Green DR. 2012. Mitochondrial pathway of apoptosis is ancestral in metazoans. *Proc Natl Acad Sci* 109: 4904–4909.

Bergsbaken T, Cookson BT. 2007. Macrophage activation redirects yersinia-infected host cell death from apoptosis to caspase-1-dependent pyroptosis. *PLoS Pathog* 3: e161.

Berry DL, Baehrecke EH. 2007. Growth arrest and autophagy are required for salivary gland cell degradation in *Drosophila. Cell* 131: 1137–1148.

Bonzon C, Bouchier-Hayes L, Pagliari LJ, Green DR, Newmeyer DD. 2006. Caspase-2-induced apoptosis requires bid cleavage: A physiological role for bid in heat shock-induced death. *Mol Biol Cell* 17: 2150–2157.

Bouchier-Hayes L, Oberst A, McStay GP, Connell S, Tait SWG, Dillon CP, Flanagan JM, Beere HM, Green DR. 2009. Characterization of cytoplasmic caspase-2 activation by induced proximity. *Mol Cell* 35: 830–840.

Bratton SB, Salvesen GS. 2010. Regulation of the Apaf-1-caspase-9 apoptosome. *J Cell Sci* 123: 3209–3214.

Bratton SB, Walker G, Srinivasula SM, Sun XM, Butterworth M, Alnemri ES, Cohen GM. 2001. Recruitment, activation and retention of caspases 9 and 3 by Apaf-1 apoptosome and associated XIAP complexes. *EMBO J* 20: 998–1009.

Brennan MA, Cookson BT. 2000. Salmonella induces macrophage death by caspase-1-dependent necrosis. *Mol Microbiol* 38: 31–40.

Brinkmann K, Zigrino P, Witt A, Schell M, Ackermann L, Broxtermann P, Schull S, Andree M, Coutelle O, Yazdanpanah B, et al. 2013. Ubiquitin C-terminal hydrolase-L1 potentiates cancer chemosensitivity by stabilizing NOXA. *Cell Rep* 3: 881–891.

Bruick RK. 2000. Expression of the gene encoding the proapoptotic Nip3 protein is induced by hypoxia. *Proc Natl Acad Sci* 97: 9082–9087.

Buss RR, Sun W, Oppenheim RW. 2006. Adaptive roles of programmed cell death during nervous system development. *Annu Rev Neurosci* 29: 1–35.

Cai Z, Jitkaew S, Zhao J, Chiang HC, Choksi S, Liu J, Ward Y, Wu LG, Liu ZG. 2014. Plasma membrane translocation of trimerized MLKL protein is required for TNF-induced necroptosis. *Nat Cell Biol* 16: 55–65.

Castedo M, Perfettini JL, Roumier T, Andreau K, Medema R, Kroemer G. 2004. Cell death by mitotic catastrophe: A molecular definition. *Oncogene* 23: 2825–2837.

Chen M, He H, Zhan S, Krajewski S, Reed JC, Gottlieb RA. 2001. Bid is cleaved by calpain to an active fragment in vitro and during myocardial ischemia/reperfusion. *J Biol Chem* 276: 30724–30728.

Ch'en IL, Tsau JS, Molkentin JD, Komatsu M, Hedrick SM. 2011. Mechanisms of necroptosis in T cells. *J Exp Med* 208: 633–641.

Chen X, Li W, Ren J, Huang D, He WT, Song Y, Yang C, Zheng X, Chen P, Han J. 2014. Translocation of mixed lineage kinase domain-like protein to plasma membrane leads to necrotic cell death. *Cell Res* 24: 105–121.

Chiang CW, Harris G, Ellig C, Masters SC, Subramanian R, Shenolikar S, Wadzinski BE, Yang E. 2001. Protein phosphatase 2A activates the proapoptotic function of BAD in interleukin- 3-dependent lymphoid cells by a mechanism requiring 14–3-3 dissociation. *Blood* 97: 1289–1297.

Chinnaiyan AM, Chaudhary D, O'Rourke K, Koonin EV, Dixit VM. 1997a. Role of CED-4 in the activation of CED-3. *Nature* 388: 728–729.

Chinnaiyan AM, O'Rourke K, Lane BR, Dixit VM. 1997b. Interaction of CED-4 with CED-3 and CED-9: A molecular framework for cell death. *Science* 275: 1122–1126.

Chipuk JE, Moldoveanu T, Llambi F, Parsons MJ, Green DR. 2010. The BCL-2 family reunion. *Mol Cell* 37: 299–310.

Cho YS, Challa S, Moquin D, Genga R, Ray TD, Guildford M, Chan FK. 2009. Phosphorylation-driven assembly of the RIP1-RIP3 complex regulates programmed necrosis and virus-induced inflammation. *Cell* 137: 1112–1123.

Chou JJ, Matsuo H, Duan H, Wagner G. 1998. Solution structure of the RAIDD CARD and model for CARD/CARD interaction in caspase-2 and caspase-9 recruitment. *Cell* 94: 171–180.

Christofferson DE, Yuan J. 2010. Necroptosis as an alternative form of programmed cell death. *Curr Opin Cell Biol* 22: 263–268.

Coleman ML, Sahai EA, Yeo M, Bosch M, Dewar A, Olson MF. 2001. Membrane blebbing during apoptosis results from caspase-mediated activation of ROCK I. *Nature Cell Biol* 3: 339–345.

Conradt B, Horvitz HR. 1998. The *C. elegans* protein EGL-1 is required for programmed cell death and interacts with the Bcl-2-like protein CED-9. *Cell* 93: 519–529.

Cookson BT, Brennan MA. 2001. Proinflammatory programmed cell death. *Trends Microbiol* 9: 113–114.

Cotteret S, Jaffer ZM, Beeser A, Chernoff J. 2003. p21-Activated kinase 5 (Pak5) localizes to mitochondria and inhibits apoptosis by phosphorylating BAD. *Mol Cell Biol* 23: 5526–5539.

Crawford ED, Wells JA. 2011. Caspase substrates and cellular remodeling. *Annu Rev Biochem* 80: 1055–1087.

Datta SR, Dudek H, Tao X, Masters S, Fu H, Gotoh Y, Greenberg ME. 1997. Akt phosphorylation of BAD couples survival signals to the cell-intrinsic death machinery. *Cell* 91: 231–241.

Datta SR, Katsov A, Hu L, Petros A, Fesik SW, Yaffe MB, Greenberg ME. 2000. 14-3-3 proteins and survival kinases cooperate to inactivate BAD by BH3 domain phosphorylation. *Mol Cell* 6: 41–51.

Datta SR, Ranger AM, Lin MZ, Sturgill JF, Ma Y-C, Cowan CW, Dikkes P, Korsmeyer SJ, Greenberg ME. 2002. Survival factor-mediated BAD phosphorylation raises the mitochondrial threshold for apoptosis. *Dev Cell* 3: 631–643.

Declercq W, Vanden Berghe T, Vandenabeele P. 2009. RIP kinases at the crossroads of cell death and survival. *Cell* 138: 229–232.

Degterev A, Huang Z, Boyce M, Li Y, Jagtap P, Mizushima N, Cuny GD, Mitchison TJ, Moskowitz MA, Yuan J. 2005. Chemical inhibitor of nonapoptotic cell death with therapeutic potential for ischemic brain injury. *Nat Chem Biol* 1: 112–119.

Degterev A, Hitomi J, Germscheid M, Ch'en IL, Korkina O, Teng X, Abbott D, Cuny GD, Yuan C, Wagner G, et al. 2008. Identification of RIP1 kinase as a specific cellular target of necrostatins. *Nat Chem Biol* 4: 313–321.

Dehan E, Bassermann F, Guardavaccaro D, Vasiliver-Shamis G, Cohen M, Lowes KN, Dustin M, Huang DCS, Taunton J, Pagano M. 2009. βTrCP- and Rsk1/2-mediated degradation of BimEL inhibits apoptosis. *Mol Cell* 33: 109–116.

del Peso L, González-García M, Page C, Herrera R, Nuñez G. 1997. Interleukin-3-induced phosphorylation of BAD through the protein kinase Akt. *Science* 278: 687–689.

Denton D, Shravage B, Simin R, Mills K, Berry DL, Baehrecke EH, Kumar S. 2009. Autophagy, not apoptosis, is essential for midgut cell death in *Drosophila. Curr Biol* 19: 1741–1746.

* Devreotes P, Horwitz AR. 2014. Signaling networks that regulate cell migration. *Cold Spring Harb Perspect Biol* doi: 10.1101/cshperspect.a005959.

Dewson G, Kratina T, Sim HW, Puthalakath H, Adams JM, Colman PM, Kluck RM. 2008. To trigger apoptosis, Bak exposes its BH3 domain and homodimerizes via BH3:Groove interactions. *Mol Cell* 30: 369–380.

Dewson G, Kratina T, Czabotar P, Day CL, Adams JM, Kluck RM. 2009. Bak activation for apoptosis involves oligomerization of dimers via their α6 helices. *Mol Cell* 36: 696–703.

Di Bartolomeo S, Corazzari M, Nazio F, Oliverio S, Lisi G, Antonioli M, Pagliarini V, Matteoni S, Fuoco C, Giunta L, et al. 2010. The dynamic interaction of AMBRA1 with the dynein motor complex regulates mammalian autophagy. *J Cell Biol* 191: 155–168.

Dickens LS, Powley IR, Hughes MA, MacFarlane M. 2012. The "complexities" of life and death: Death receptor signalling platforms. *Exp Cell Res* **318:** 1269–1277.

Dijkers PF, Medema RH, Lammers JW, Koenderman L, Coffer PJ. 2000. Expression of the proapoptotic Bcl-2 family member Bim is regulated by the forkhead transcription factor FKHR-L1. *Curr Biol* **10:** 1201–1204.

Ding Q, He X, Hsu JM, Xia W, Chen CT, Li LY, Lee DF, Liu JC, Zhong Q, Wang X, et al. 2007. Degradation of Mcl-1 by β-TrCP mediates glycogen synthase kinase 3-induced tumor suppression and chemosensitization. *Mol Cell Biol* **27:** 4006–4017.

Dixon SJ, Lemberg KM, Lamprecht MR, Skouta R, Zaitsev EM, Gleason CE, Patel DN, Bauer AJ, Cantley AM, Yang WS, et al. 2012. Ferroptosis: An iron-dependent form of nonapoptotic cell death. *Cell* **149:** 1060–1072.

Donovan N, Becker EBE, Konishi Y, Bonni A. 2002. JNK phosphorylation and activation of BAD couples the stress-activated signaling pathway to the cell death machinery. *J Biol Chem* **277:** 40944–40949.

Duan H, Dixit VM. 1997. RAIDD is a new "death" adaptor molecule. *Nature* **385:** 86–89.

Eckelman BP, Salvesen GS, Scott FL. 2006. Human inhibitor of apoptosis proteins: Why XIAP is the black sheep of the family. *EMBO Rep* **7:** 988–994.

Egan DF, Shackelford DB, Mihaylova MM, Gelino S, Kohnz RA, Mair W, Vasquez DS, Joshi A, Gwinn DM, Taylor R, et al. 2011. Phosphorylation of ULK1 (hATG1) by AMP-activated protein kinase connects energy sensing to mitophagy. *Science* **331:** 456–461.

Ekoff M, Kaufmann T, Engstrom M, Motoyama N, Villunger A, Jonsson JI, Strasser A, Nilsson G. 2007. The BH3-only protein Puma plays an essential role in cytokine deprivation induced apoptosis of mast cells. *Blood* **110:** 3209–3217.

Elgendy M, Sheridan C, Brumatti G, Martin SJ. 2011. Oncogenic Ras-induced expression of Noxa and Beclin-1 promotes autophagic cell death and limits clonogenic survival. *Mol Cell* **42:** 23–35.

Enari M, Sakahira H, Yokoyama H, Okawa K, Iwamatsu A, Nagata S. 1998. A caspase-activated DNase that degrades DNA during apoptosis, and its inhibitor ICAD. *Nature* **391:** 43–50.

Erlacher M, Michalak EM, Kelly PN, Labi V, Niederegger H, Coultas L, Adams JM, Strasser A, Villunger A. 2005. BH3-only proteins Puma and Bim are rate-limiting for γ-radiation- and glucocorticoid-induced apoptosis of lymphoid cells in vivo. *Blood* **106:** 4131–4138.

Eskelinen EL, Illert AL, Tanaka Y, Schwarzmann G, Blanz J, Von Figura K, Saftig P. 2002. Role of LAMP-2 in lysosome biogenesis and autophagy. *Mol Biol Cell* **13:** 3355–3368.

Eskelinen EL, Schmidt CK, Neu S, Willenborg M, Fuertes G, Salvador N, Tanaka Y, Lullmann-Rauch R, Hartmann D, Heeren J, et al. 2004. Disturbed cholesterol traffic but normal proteolytic function in LAMP-1/LAMP-2 double-deficient fibroblasts. *Mol Biol Cell* **15:** 3132–3145.

Eskes R, Desagher S, Antonsson B, Martinou JC. 2000. Bid induces the oligomerization and insertion of Bax into the outer mitochondrial membrane. *Mol Cell Biol* **20:** 929–935.

Faustin B, Lartigue L, Bruey JM, Luciano F, Sergienko E, Bailly-Maitre B, Volkmann N, Hanein D, Rouiller I, Reed JC. 2007. Reconstituted NALP1 inflammasome reveals two-step mechanism of caspase-1 activation. *Mol Cell* **25:** 713–724.

Feng S, Yang Y, Mei Y, Ma L, Zhu DE, Hoti N, Castanares M, Wu M. 2007. Cleavage of RIP3 inactivates its caspase-independent apoptosis pathway by removal of kinase domain. *Cell Signal* **19:** 2056–2067.

Feoktistova M, Geserick P, Kellert B, Dimitrova DP, Langlais C, Hupe M, Cain K, MacFarlane M, Hacker G, Leverkus M. 2011a. cIAPs block Ripoptosome formation, a RIP1/caspase-8 containing intracellular cell death complex differentially regulated by cFLIP isoforms. *Mol Cell* **43:** 449–463.

Feoktistova M, Geserick P, Kellert B, Dimitrova DP, Langlais C, Hupe M, Cain K, MacFarlane M, Häcker G, Leverkus M. 2011b. cIAPs block Ripoptosome formation, a RIP1/caspase-8 containing intracellular cell death complex differentially regulated by cFLIP isoforms. *Mol Cell* **43:** 449–463.

Fernandez Y, Verhaegen M, Miller TP, Rush JL, Steiner P, Opipari AW Jr, Lowe SW, Soengas MS. 2005. Differential regulation of noxa in normal melanocytes and melanoma cells by proteasome inhibition: Therapeutic implications. *Cancer Res* **65:** 6294–6304.

Fimia GM, Corazzari M, Antonioli M, Piacentini M. 2012. Ambra1 at the crossroad between autophagy and cell death. *Oncogene* **32:** 3311–3318.

Fink SL, Cookson BT. 2006. Caspase-1-dependent pore formation during pyroptosis leads to osmotic lysis of infected host macrophages. *Cell Microbiol* **8:** 1812–1825.

Fox CJ, Hammerman PS, Cinalli RM, Master SR, Chodosh LA, Thompson CB. 2003. The serine/threonine kinase Pim-2 is a transcriptionally regulated apoptotic inhibitor. *Genes Dev* **17:** 1841–1854.

Franchi L, Eigenbrod T, Munoz-Planillo R, Nunez G. 2009. The inflammasome: A caspase-1-activation platform that regulates immune responses and disease pathogenesis. *Nat Immunol* **10:** 241–247.

Franchi L, Munoz-Planillo R, Nunez G. 2012. Sensing and reacting to microbes through the inflammasomes. *Nat Immunol* **13:** 325–332.

Galluzzi L, Maiuri MC, Vitale I, Zischka H, Castedo M, Zitvogel L, Kroemer G. 2007. Cell death modalities: Classification and pathophysiological implications. *Cell Death Differ* **14:** 1237–1243.

Galluzzi L, Vitale I, Abrams JM, Alnemri ES, Baehrecke EH, Blagosklonny MV, Dawson TM, Dawson VL, El-Deiry WS, Fulda S, et al. 2012. Molecular definitions of cell death subroutines: Recommendations of the Nomenclature Committee on Cell Death 2012. *Cell Death Differ* **19:** 107–120.

Ganley IG, Lam du H, Wang J, Ding X, Chen S, Jiang X. 2009. ULK1.ATG13.FIP200 complex mediates mTOR signaling and is essential for autophagy. *J Biol Chem* **284:** 12297–12305.

Geserick P, Hupe M, Moulin M, Wong WW, Feoktistova M, Kellert B, Gollnick H, Silke J, Leverkus M. 2009. Cellular IAPs inhibit a cryptic CD95-induced cell death by limiting RIP1 kinase recruitment. *J Cell Biol* **187:** 1037–1054.

Gilley J, Coffer PJ, Ham J. 2003. FOXO transcription factors directly activate *bim* gene expression and promote apoptosis in sympathetic neurons. *J Cell Biol* **162:** 613–622.

Goldstein JC, Waterhouse NJ, Juin P, Evan GI, Green DR. 2000. The coordinate release of cytochrome *c* during apoptosis is rapid, complete and kinetically invariant. *Nat Cell Biol* **2:** 156–162.

Goyal L, McCall K, Agapite J, Hartwieg E, Steller H. 2000. Induction of apoptosis by *Drosophila* reaper, *hid* and *grim* through inhibition of IAP function. *EMBO J* **19:** 589–597.

Green DR. 2011a. Immunology: A heavyweight knocked out. *Nature* **479:** 48–50.

Green DR. 2011b. *Means to an end*. Cold Spring Harbor Laboratory Press, Cold Spring Harbor, NY.

Guo K, Searfoss G, Krolikowski D, Pagnoni M, Franks C, Clark K, Yu KT, Jaye M, Ivashchenko Y. 2001. Hypoxia induces the expression of the pro-apoptotic gene BNIP3. *Cell Death Differ* **8:** 367–376.

Halestrap AP, Pasdois P. 2009. The role of the mitochondrial permeability transition pore in heart disease. *Biochim Biophys Acta* **1787:** 1402–1415.

Hanada T, Noda NN, Satomi Y, Ichimura Y, Fujioka Y, Takao T, Inagaki F, Ohsumi Y. 2007. The Atg12-Atg5 conjugate has a novel E3-like activity for protein lipidation in autophagy. *J Biol Chem* **282:** 37298–37302.

Harada H, Becknell B, Wilm M, Mann M, Huang LJ, Taylor SS, Scott JD, Korsmeyer SJ. 1999. Phosphorylation and inactivation of BAD by mitochondria-anchored protein kinase A. *Mol Cell* **3:** 413–422.

Harada H, Andersen JS, Mann M, Terada N, Korsmeyer SJ. 2001. p70S6 kinase signals cell survival as well as growth, inactivating the proapoptotic molecule BAD. *Proc Natl Acad Sci* **98:** 9666–9670.

* Hardie DG. 2012. Organismal carbohydrate and lipid homeostasis. *Cold Spring Harb Perspect Biol* **4:** a006031.

He C, Levine B. 2010. The Beclin 1 interactome. *Curr Opin Cell Biol* **22:** 140–149.

He S, Wang L, Miao L, Wang T, Du F, Zhao L, Wang X. 2009. Receptor interacting protein kinase-3 determines cellular necrotic response to TNF-α. *Cell* **137:** 1100–1111.

He C, Bassik MC, Moresi V, Sun K, Wei Y, Zou Z, An Z, Loh J, Fisher J, Sun Q, et al. 2012. Exercise-induced BCL2-regulated autophagy is required for muscle glucose homeostasis. *Nature* **481:** 511–515.

Heibein JA, Goping IS, Barry M, Pinkoski MJ, Shore GC, Green DR, Bleackley RC. 2000. Granzyme B-mediated cytochrome *c* release is regulated by the Bcl-2 family members bid and Bax. *J Exp Med* **192:** 1391–1402.

Hershko T, Ginsberg D. 2004. Up-regulation of Bcl-2 homology 3 BH3-only proteins by E2F1 mediates apoptosis. *J Biol Chem* **279:** 8627–8634.

Higuchi M, Tomioka M, Takano J, Shirotani K, Iwata N, Masumoto H, Maki M, Itohara S, Saido TC. 2005. Distinct mechanistic roles of calpain and caspase activation in neurodegeneration as revealed in mice overexpressing their specific inhibitors. *Biol Chem* **280:** 15229–15237.

Hitomi J, Christofferson DE, Ng A, Yao J, Degterev A, Xavier RJ, Yuan J. 2008. Identification of a molecular signaling network that regulates a cellular necrotic cell death pathway. *Cell* **135:** 1311–1323.

Hofmann K, Bucher P, Tschopp J. 1997. The CARD domain: A new apoptotic signalling motif. *Trends Biochem Sci* **22:** 155–156.

Holler N, Zaru R, Micheau O, Thome M, Attinger A, Valitutti S, Bodmer JL, Schneider P, Seed B, Tschopp J. 2000. Fas triggers an alternative, caspase-8-independent cell death pathway using the kinase RIP as effector molecule. *Nat Immunol* **1:** 489–495.

Hornung V, Ablasser A, Charrel-Dennis M, Bauernfeind F, Horvath G, Caffrey DR, Latz E, Fitzgerald KA. 2009. AIM2 recognizes cytosolic dsDNA and forms a caspase-1-activating inflammasome with ASC. *Nature* **458:** 514–518.

Hosokawa N, Hara T, Kaizuka T, Kishi C, Takamura A, Miura Y, Iemura S, Natsume T, Takehana K, Yamada N, et al. 2009. Nutrient-dependent mTORC1 association with the ULK1-Atg13-FIP200 complex required for autophagy. *Mol Biol Cell* **20:** 1981–1991.

Hubner A, Barrett T, Flavell RA, Davis RJ. 2008. Multisite phosphorylation regulates Bim stability and apoptotic activity. *Mol Cell* **30:** 415–425.

Imtiyaz HZ, Rosenberg S, Zhang Y, Rahman ZSM, Hou Y-J, Manser T, Zhang J. 2006. The Fas-associated death domain protein (FADD) is required in apoptosis and TLR-induced proliferative responses in B cells. *J Immunol* **176:** 6852–6861.

Itakura E, Kishi-Itakura C, Mizushima N. 2012. The hairpin-type tail-anchored SNARE syntaxin 17 targets to autophagosomes for fusion with endosomes/lysosomes. *Cell* **151:** 1256–1269.

Janssens S, Tinel A, Lippens S, Tschopp J. 2005. PIDD mediates NF-κB activation in response to DNA damage. *Cell* **123:** 1079–1092.

Jeffers JR, Parganas E, Lee Y, Yang C, Wang J, Brennan J, MacLean KH, Han J, Chittenden T, Ihle JN, et al. 2003. Puma is an essential mediator of p53-dependent and -independent apoptotic pathways. *Cancer Cell* **4:** 321–328.

Jost PJ, Grabow S, Gray D, McKenzie MD, Nachbur U, Huang DC, Bouillet P, Thomas HE, Borner C, Silke J, et al. 2009. XIAP discriminates between type I and type II FAS-induced apoptosis. *Nature* **460:** 1035–1039.

Jung CH, Jun CB, Ro SH, Kim YM, Otto NM, Cao J, Kundu M, Kim DH. 2009. ULK-Atg13-FIP200 complexes mediate mTOR signaling to the autophagy machinery. *Mol Biol Cell* **20:** 1992–2003.

Kanuka H, Sawamoto K, Inohara N, Matsuno K, Okano H, Miura M. 1999. Control of the cell death pathway by Dapaf-1, a *Drosophila* Apaf-1/CED-4-related caspase activator. *Mol Cell* **4:** 757–769.

Kayagaki N, Warming S, Lamkanfi M, Vande Walle L, Louie S, Dong J, Newton K, Qu Y, Liu J, Heldens S, et al. 2011. Non-canonical inflammasome activation targets caspase 11. *Nature* **479:** 117–121.

Kerr JF, Wyllie AH, Currie AR. 1972. Apoptosis: A basic biological phenomenon with wide-ranging implications in tissue kinetics. *Br J Cancer* **26:** 239–257.

Kerr JB, Hutt KJ, Michalak EM, Cook M, Vandenberg CJ, Liew SH, Bouillet P, Mills A, Scott CL, Findlay JK, et al. 2012. DNA damage-induced primordial follicle oocyte apoptosis and loss of fertility require TAp63-mediated induction of *Puma* and *Noxa*. *Mol Cell* **48:** 343–352.

Kersse K, Verspurten J, Vanden Berghe T, Vandenabeele P. 2011. The death-fold superfamily of homotypic interaction motifs. *Trends Biochem Sci* **36:** 541–552.

Kim J-Y, Ahn H-J, Ryu J-H, Suk K, Park J-H. 2004. BH3-only protein Noxa is a mediator of hypoxic cell death induced by hypoxia-inducible factor 1α. *J Exp Med* **199:** 113–124.

Kim HE, Du F, Fang M, Wang X. 2005. Formation of apoptosome is initiated by cytochrome *c*-induced dATP hydrolysis and subsequent nucleotide exchange on Apaf-1. *Proc Natl Acad Sci* **102:** 17545–17550.

Kim YS, Morgan MJ, Choksi S, Liu ZG. 2007. TNF-induced activation of the Nox1 NADPH oxidase and its role in the induction of necrotic cell death. *Mol Cell* **26:** 675–687.

Kim HE, Jiang X, Du F, Wang X. 2008. PHAPI, CAS, and Hsp70 promote apoptosome formation by preventing Apaf-1 aggregation and enhancing nucleotide exchange on Apaf-1. *Mol Cell* **30:** 239–247.

Kim J, Kundu M, Viollet B, Guan KL. 2011. AMPK and mTOR regulate autophagy through direct phosphorylation of Ulk1. *Nat Cell Biol* **13:** 132–141.

Klumpp S, Selke D, Krieglstein J. 2003. Protein phosphatase type 2C dephosphorylates BAD. *Neurochem Int* **42:** 555–560.

Konishi Y, Lehtinen M, Donovan N, Bonni A. 2002. Cdc2 phosphorylation of BAD links the cell cycle to the cell death machinery. *Mol Cell* **9:** 1005–1016.

Korsmeyer SJ, Wei MC, Saito M, Weiler S, Oh KJ, Schlesinger PH. 2000. Proapoptotic cascade activates BID, which oligomerizes BAK or BAX into pores that result in the release of cytochrome *c*. *Cell Death Differ* **7:** 1166–1173.

Kothakota S, Azuma T, Reinhard C, Klippel A, Tang J, Chu K, McGarry TJ, Kirschner MW, Koths K, Kwiatkowski DJ, et al. 1997. Caspase-3-generated fragment of gelsolin: Effector of morphological change in apoptosis. *Science* **278:** 294–298.

Krajcovic M, Johnson NB, Sun Q, Normand G, Hoover N, Yao E, Richardson AL, King RW, Cibas ES, Schnitt SJ, et al. 2011. A non-genetic route to aneuploidy in human cancers. *Nat Cell Biol* **13:** 324–330.

Kroemer G, Levine B. 2008. Autophagic cell death: The story of a misnomer. *Nat Rev Mol Cell Biol* **9:** 1004–1010.

Kroemer G, Marino G, Levine B. 2010. Autophagy and the integrated stress response. *Mol Cell* **40:** 280–293.

Krumschnabel G, Manzl C, Villunger A. 2009. Caspase 2: Killer, savior and safeguard—emerging versatile roles for an ill-defined caspase. *Oncogene* **28:** 3093–3096.

Kuwana T, Mackey MR, Perkins G, Ellisman MH, Latterich M, Schneiter R, Green DR, Newmeyer DD. 2002. Bid, Bax, and lipids cooperate to form supramolecular openings in the outer mitochondrial membrane. *Cell* **111:** 331–342.

Kuwana T, Bouchier-Hayes L, Chipuk JE, Bonzon C, Sullivan BA, Green DR, Newmeyer DD. 2005. BH3 domains of BH3-only proteins differentially regulate Bax-mediated mitochondrial membrane permeabilization both directly and indirectly. *Mol Cell* **17:** 525–535.

Lamkanfi M, Moreira LO, Makena P, Spierings DC, Boyd K, Murray PJ, Green DR, Kanneganti TD. 2009. Caspase-7 deficiency protects from endotoxin-induced lymphocyte apoptosis and improves survival. *Blood* **113:** 2742–2745.

* Laplante M, Sabatini DM. 2012. mTOR signaling. *Cold Spring Harb Perspect Biol* **4:** a011593.

Lartigue L, Kushnareva Y, Seong Y, Lin H, Faustin B, Newmeyer DD. 2009. Caspase-independent mitochondrial cell death results from loss of respiration, not cytotoxic protein release. *Mol Biol Cell* **20:** 4871–4884.

Lee EF, Clarke OB, Evangelista M, Feng Z, Speed TP, Tchoubrieva EB, Strasser A, Kalinna BH, Colman PM, Fairlie WD. 2011. Discovery and molecular characterization of a Bcl-2-regulated cell death pathway in schistosomes. *Proc Natl Acad Sci* **108:** 6999–7003.

Lei K, Davis RJ. 2003. JNK phosphorylation of Bim-related members of the Bcl2 family induces Bax-dependent apoptosis. *Proc Natl Acad Sci* **100:** 2432–2437.

Letai A, Bassik MC, Walensky LD, Sorcinelli MD, Weiler S, Korsmeyer SJ. 2002. Distinct BH3 domains either sensitize or activate mitochondrial apoptosis, serving as prototype cancer therapeutics. *Cancer Cell* **2:** 183–192.

Levine B, Kroemer G. 2008. Autophagy in the pathogenesis of disease. *Cell* **132:** 27–42.

Ley R, Balmanno K, Hadfield K, Weston C, Cook SJ. 2003. Activation of the ERK1/2 signaling pathway promotes phosphorylation and proteasome-dependent degradation of the BH3-only protein, Bim. *J Biol Chem* **278:** 18811–18816.

Li H, Zhu H, Xu CJ, Yuan J. 1998. Cleavage of BID by caspase 8 mediates the mitochondrial damage in the Fas pathway of apoptosis. *Cell* **94:** 491–501.

Li YM, Wen Y, Zhou BP, Kuo H-P, Ding Q, Hung M-C. 2003. Enhancement of Bik antitumor effect by Bik mutants. *Cancer Res* **63:** 7630–7633.

* Lim K-H, Staudt LM. 2013. Toll-like receptor signaling. *Cold Spring Harb Perspect Biol* **5:** a011247.

Lin Y, Devin A, Rodriguez Y, Liu ZG. 1999. Cleavage of the death domain kinase RIP by caspase 8 prompts TNF-induced apoptosis. *Genes Dev* **13:** 2514–2526.

Lin Y, Ma W, Benchimol S. 2000. Pidd, a new death-domain-containing protein, is induced by p53 and promotes apoptosis. *Nat Genet* **26:** 122–127.

Lisi S, Mazzon I, White K. 2000. Diverse domains of THREAD/DIAP1 are required to inhibit apoptosis induced by REAPER and HID in *Drosophila*. *Genetics* **154:** 669–678.

Liu L, Feng D, Chen G, Chen M, Zheng Q, Song P, Ma Q, Zhu C, Wang R, Qi W, et al. 2012. Mitochondrial outer-membrane protein FUNDC1 mediates hypoxia-induced mitophagy in mammalian cells. *Nat Cell Biol* **14:** 177–185.

Llambi F, Moldoveanu T, Tait SWG, Bouchier-Hayes L, Temirov J, McCormick LL, Dillon CP, Green DR. 2011. A unified model of mammalian BCL-2 protein family interactions at the mitochondria. *Mol Cell* **44:** 517–531.

Lu JV, Weist BM, van Raam BJ, Marro BS, Nguyen LV, Srinivas P, Bell BD, Luhrs KA, Lane TE, Salvesen GS, et al. 2011. Complementary roles of Fas-associated death domain (FADD) and receptor interacting protein kinase-3 (RIPK3) in T-cell homeostasis and antiviral immunity. *Proc Natl Acad Sci* **108:** 15312–15317.

Luo X, Budihardjo I, Zou H, Slaughter C, Wang X. 1998. Bid, a Bcl2 interacting protein, mediates cytochrome *c* release from mitochondria in response to activation of cell surface death receptors. *Cell* **94:** 481–490.

Ma C, Ying C, Yuan Z, Song B, Li D, Liu Y, Lai B, Li W, Chen R, Ching Y-P, et al. 2007. dp5/HRK is a c-Jun target gene and required for apoptosis induced by potassium deprivation in cerebellar granule neurons. *J Biol Chem* **282:** 30901–30909.

Maiuri MC, Criollo A, Tasdemir E, Vicencio JM, Tajeddine N, Hickman JA, Geneste O, Kroemer G. 2007a. BH3-only proteins and BH3 mimetics induce autophagy by competitively disrupting the interaction between Beclin 1 and Bcl-2/Bcl-X$_L$. *Autophagy* **3:** 374–376.

Maiuri MC, Le Toumelin G, Criollo A, Rain JC, Gautier F, Juin P, Tasdemir E, Pierron G, Troulinaki K, Tavernarakis N, et al. 2007b. Functional and physical interaction between Bcl-X$_L$ and a BH3-like domain in Beclin-1. *EMBO J* **26:** 2527–2539.

Malladi S, Challa-Malladi M, Fearnhead HO, Bratton SB. 2009. The Apaf-1•procaspase-9 apoptosome complex functions as a proteolytic-based molecular timer. *EMBO J* **28:** 1916–1925.

Mandic A, Viktorsson K, Strandberg L, Heiden T, Hansson J, Linder S, Shoshan MC. 2002. Calpain-mediated Bid cleavage and calpain-independent Bak modulation: Two separate pathways in cisplatin-induced apoptosis. *Mol Cell Biol* **22:** 3003–3013.

Manzl C, Krumschnabel G, Bock F, Sohm B, Labi V, Baumgartner F, Logette E, Tschopp J, Villunger A. 2009. Caspase-2 activation in the absence of PIDDosome formation. *J Cell Biol* **185:** 291–303.

Manzl C, Peintner L, Krumschnabel G, Bock F, Labi V, Drach M, Newbold A, Johnstone R, Villunger A. 2012. PIDDosome-independent tumor suppression by Caspase 2. *Cell Death Differ* **19:** 1722–1732.

Martinez J, Almendinger J, Oberst A, Ness R, Dillon CP, Fitzgerald P, Hengartner MO, Green DR. 2011. Microtubule-associated protein 1 light chain 3α (LC3)-associated phagocytosis is required for the efficient clearance of dead cells. *Proc Natl Acad Sci* **108:** 17396–17401.

Martinon F, Burns K, Tschopp J. 2002. The inflammasome: A molecular platform triggering activation of inflammatory caspases and processing of proIL-β. *Mol Cell* **10:** 417–426.

Matsui H, Asou H, Inaba T. 2007. Cytokines direct the regulation of Bim mRNA stability by heat-shock cognate protein 70. *Mol Cell* **25:** 99–112.

Matsumura H, Shimizu Y, Ohsawa Y, Kawahara A, Uchiyama Y, Nagata S. 2000. Necrotic death pathway in Fas receptor signaling. *J Cell Biol* **151:** 1247–1256.

Maurer U, Charvet C, Wagman AS, Dejardin E, Green DR. 2006. Glycogen synthase kinase-3 regulates mitochondrial outer membrane permeabilization and apoptosis by destabilization of MCL-1. *Mol Cell* **21:** 749–760.

McStay GP, Salvesen GS, Green DR. 2008. Overlapping cleavage motif selectivity of caspases: Implications for analysis of apoptotic pathways. *Cell Death Differ* **15:** 322–331.

Mei Y, Yong J, Liu H, Shi Y, Meinkoth J, Dreyfuss G, Yang X. 2010. tRNA binds to cytochrome *c* and inhibits caspase activation. *Mol Cell* **37:** 668–678.

Micheau O, Lens S, Gaide O, Alevizopoulos K, Tschopp J. 2001. NF-κB signals induce the expression of c-FLIP. *Mol Cell Biol* **21:** 5299–5305.

Micheau O, Thome M, Schneider P, Holler N, Tschopp J, Nicholson DW, Briand C, Grutter MG. 2002. The long form of FLIP is an activator of caspase 8 at the Fas death-inducing signaling complex. *J Biol Chem* **277:** 45162–45171.

Moulin M, Anderton H, Voss AK, Thomas T, Wong WW, Bankovacki A, Feltham R, Chau D, Cook WD, Silke J, et al. 2012. IAPs limit activation of RIP kinases by TNF receptor 1 during development. *EMBO J* **31:** 1679–1691.

Munoz-Pinedo C, Guio-Carrion A, Goldstein JC, Fitzgerald P, Newmeyer DD, Green DR. 2006. Different mitochondrial intermembrane space proteins are released during apoptosis in a manner that is coordinately initiated but can vary in duration. *Proc Natl Acad Sci* **103:** 11573–11578.

Naik E, Michalak EM, Villunger A, Adams JM, Strasser A. 2007. Ultraviolet radiation triggers apoptosis of fibroblasts and skin keratinocytes mainly via the BH3-only protein Noxa. *J Cell Biol* **176:** 415–424.

Nakagawa T, Shimizu S, Watanabe T, Yamaguchi O, Otsu K, Yamagata H, Inohara H, Kubo T, Tsujimoto Y. 2005. Cyclophilin D-dependent mitochondrial permeability transition regulates some necrotic but not apoptotic cell death. *Nature* **434:** 652–658.

Nakano K, Vousden KH. 2001. *PUMA*, a novel proapoptotic gene, is induced by p53. *Mol Cell* **7:** 683–694.

Narendra DP, Youle RJ. 2011. Targeting mitochondrial dysfunction: Role for PINK1 and Parkin in mitochondrial quality control. *Antioxid Redox Signal* **14:** 1929–1938.

Newmeyer DD, Bossy-Wetzel E, Kluck RM, Wolf BB, Beere HM, Green DR. 2000. Bcl-x$_L$ does not inhibit the function of Apaf-1. *Cell Death Differ* **7:** 402–407.

* Newton K, Dixit VM. 2012. Signaling in innate immunity and inflammation. *Cold Spring Harb Perspect Biol* **4:** a006049.

Nikiforov MA, Riblett M, Tang W-H, Gratchouck V, Zhuang D, Fernandez Y, Verhaegen M, Varambally S, Chinnaiyan AM, Jakubowiak AJ, et al. 2007. Tumor cell-selective regulation of NOXA by c-MYC in response to proteasome inhibition. *Proc Natl Acad Sci* **104:** 19488–19493.

Nikrad M, Johnson T, Puthalalath H, Coultas L, Adams J, Kraft AS. 2005. The proteasome inhibitor bortezomib sensitizes cells to killing by death receptor ligand TRAIL via BH3–only proteins Bik and Bim. *Mol Cancer Ther* **4:** 443–449.

Nutt LK, Margolis SS, Jensen M, Herman CE, Dunphy WG, Rathmell JC, Kornbluth S. 2005. Metabolic regulation of oocyte cell death through the CaMKII-mediated phosphorylation of caspase 2. *Cell* **123:** 89–103.

Nutt LK, Buchakjian MR, Gan E, Darbandi R, Yoon SY, Wu JQ, Miyamoto YJ, Gibbons JA, Andersen JL, Freel CD, et al. 2009. Metabolic control of oocyte apoptosis mediated by 14-3-3ζ-regulated dephosphorylation of caspase 2. *Dev Cell* **16:** 856–866.

Oberst A, Pop C, Tremblay AG, Blais V, Denault J-B, Salvesen GS, Green DR. 2010. Inducible dimerization and inducible cleavage reveal a requirement for both processes in caspase-8 activation. *J Biol Chem* **285:** 16632–16642.

Oberst A, Dillon CP, Weinlich R, McCormick LL, Fitzgerald P, Pop C, Hakem R, Salvesen GS, Green DR. 2011. Catalytic activity of the caspase-8-FLIP$_L$ complex inhibits RIPK3-dependent necrosis. *Nature* **471:** 363–367.

Oberstein A, Jeffrey PD, Shi Y. 2007. Crystal structure of the Bcl-X$_L$-Beclin 1 peptide complex: Beclin 1 is a novel BH3-only protein. *J Biol Chem* **282:** 13123–13132.

Oda E, Ohki R, Murasawa H, Nemoto J, Shibue T, Yamashita T, Tokino T, Taniguchi T, Tanaka N. 2000. Noxa, a BH3-only member of the Bcl-2 family and candidate mediator of p53-induced apoptosis. *Science* **288:** 1053–1058.

O'Donnell MA, Perez-Jimenez E, Oberst A, Ng A, Massoumi R, Xavier R, Green DR, Ting AT. 2011. Caspase 8 inhibits programmed necrosis by processing CYLD. *Nat Cell Biol* **13:** 1437–1442.

Overholtzer M, Mailleux AA, Mouneimne G, Normand G, Schnitt SJ, King RW, Cibas ES, Brugge JS. 2007. A nonapoptotic cell death process, entosis, that occurs by cell-in-cell invasion. *Cell* **131:** 966–979.

Pattingre S, Tassa A, Qu X, Garuti R, Liang XH, Mizushima N, Packer M, Schneider MD, Levine B. 2005. Bcl-2 antiapoptotic proteins inhibit Beclin 1-dependent autophagy. *Cell* **122:** 927–939.

Ploner C, Rainer J, Niederegger H, Eduardoff M, Villunger A, Geley S, Kofler R. 2008. The BCL2 rheostat in glucocorticoid-induced apoptosis of acute lymphoblastic leukemia. *Leukemia* **22:** 370–377.

Pop C, Fitzgerald P, Green DR, Salvesen GS. 2007. Role of proteolysis in caspase-8 activation and stabilization. *Biochemistry* **46:** 4398–4407.

Pop C, Oberst A, Drag M, Van Raam BJ, Riedl SJ, Green DR, Salvesen GS. 2011. FLIP$_L$ induces caspase-8 activity in the absence of interdomain caspase-8 cleavage and alters substrate specificity. *Biochem J* **433:** 447–457.

Putcha GV, Le S, Frank S, Besirli CG, Clark K, Chu B, Alix S, Youle RJ, LaMarche A, Maroney AC, et al. 2003. JNK-mediated BIM phosphorylation potentiates BAX-dependent apoptosis. *Neuron* **38:** 899–914.

Puthalakath H, O'Reilly LA, Gunn P, Lee L, Kelly PN, Huntington ND, Hughes PD, Michalak EM, McKimm-Breschkin J, Motoyama N, et al. 2007. ER stress triggers apoptosis by activating BH3-only protein Bim. *Cell* **129:** 1337–1349.

Pyati UJ, Gjini E, Carbonneau S, Lee JS, Guo F, Jette CA, Kelsell DP, Look AT. 2011. p63 mediates an apoptotic response to pharmacological and disease-related ER stress in the developing epidermis. *Dev Cell* **21:** 492–505.

Qin JZ, Ziffra J, Stennett L, Bodner B, Bonish BK, Chaturvedi V, Bennett F, Pollock PM, Trent JM, Hendrix MJ, et al. 2005. Proteasome inhibitors trigger NOXA-mediated apoptosis in melanoma and myeloma cells. *Cancer Res* **65:** 6282–6293.

Ramjaun AR, Tomlinson S, Eddaoudi A, Downward J. 2007. Upregulation of two BH3–only proteins, Bmf and Bim, during TGF β-induced apoptosis. *Oncogene* **26:** 970–981.

Real PJ, Sanz C, Gutierrez O, Pipaon C, Zubiaga AM, Fernandez-Luna JL. 2006. Transcriptional activation of the proapoptotic *bik* gene by E2F proteins in cancer cells. *FEBS Lett* **580:** 5905–5909.

Reiners JJ, Caruso JA, Mathieu P, Chelladurai B, Yin X-M, Kessel D. 2002. Release of cytochrome *c* and activation of procaspase 9 following lysosomal photodamage involves Bid cleavage. *Cell Death Differ* **9:** 934–944.

★ Rhind N, Russel P. 2012. Signaling pathways that regulate cell division. *Cold Spring Harb Perspect Biol* **4:** a005942.

Ribe EM, Jean YY, Goldstein RL, Manzl C, Stefanis L, Villunger A, Troy CM. 2012. Neuronal caspase-2 activity and function requires RAIDD, but not PIDD. *Biochem J* **444:** 591–599.

Ricchelli F, Sileikyte J, Bernardi P. 2011. Shedding light on the mitochondrial permeability transition. *Biochim Biophys Acta* **1807:** 482–490.

Ricci JE, Munoz-Pinedo C, Fitzgerald P, Bailly-Maitre B, Perkins GA, Yadava N, Scheffler IE, Ellisman MH, Green DR. 2004. Disruption of mitochondrial function during apoptosis is mediated by caspase cleavage of the p75 subunit of complex I of the electron transport chain. *Cell* **117:** 773–786.

Rodriguez J, Lazebnik Y. 1999. Caspase 9 and APAF-1 form an active holoenzyme. *Genes Dev* **13:** 3179–3184.

Rodriguez A, Oliver H, Zou H, Chen P, Wang X, Abrams JM. 1999. Dark is a *Drosophila* homologue of Apaf-1/CED-4 and functions in an evolutionarily conserved death pathway. *Nat Cell Biol* **1:** 272–279.

Rodriguez A, Chen P, Oliver H, Abrams JM. 2002. Unrestrained caspase-dependent cell death caused by loss of Diap1 function requires the *Drosophila* Apaf-1 homolog, Dark. *EMBO J* **21:** 2189–2197.

Rudel T, Bokoch GM. 1997. Membrane and morphological changes in apoptotic cells regulated by caspase-mediated activation of PAK2. *Science* **276:** 1571–1574.

Sakahira H, Enari M, Nagata S. 1998. Cleavage of CAD inhibitor in CAD activation and DNA degradation during apoptosis. *Nature* **391:** 96–99.

Salvesen GS, Riedl SJ. 2008. Caspase mechanisms. *Adv Exp Med Biol* **615:** 13–23.

Sandoval H, Thiagarajan P, Dasgupta SK, Schumacher A, Prchal JT, Chen M, Wang J. 2008. Essential role for Nix in autophagic maturation of erythroid cells. *Nature* **454:** 232–235.

Sanjuan MA, Dillon CP, Tait SWG, Moshiach S, Dorsey F, Connell S, Komatsu M, Tanaka K, Cleveland JL, Withoff S, et al. 2007. Toll-like receptor signalling in macrophages links the autophagy pathway to phagocytosis. *Nature* **450:** 1253–1257.

Sanz C, Mellstrom B, Link WA, Naranjo JR, Fernandez-Luna JL. 2001. Interleukin 3-dependent activation of DREAM is involved in transcriptional silencing of the apoptotic Hrk gene in hematopoietic progenitor cells. *EMBO J* **20:** 2286–2292.

Schinzel AC, Takeuchi O, Huang Z, Fisher JK, Zhou Z, Rubens J, Hetz C, Danial NN, Moskowitz MA, Korsmeyer SJ. 2005. Cyclophilin D is a component of mitochondrial permeability transition and mediates neuronal cell death after focal cerebral ischemia. *Proc Natl Acad Sci* **102:** 12005–12010.

Schulze-Osthoff K, Krammer PH, Droge W. 1994. Divergent signalling via APO-1/Fas and the TNF receptor, two homologous molecules involved in physiological cell death. *EMBO J* **13:** 4587–4596.

Schürmann A, Mooney AF, Sanders LC, Sells MA, Wang HG, Reed JC, Bokoch GM. 2000. p21-Activated kinase 1 phosphorylates the death agonist bad and protects cells from apoptosis. *Mol Cell Biol* **20:** 453–461.

Schweers RL, Zhang J, Randall MS, Loyd MR, Li W, Dorsey FC, Kundu M, Opferman JT, Cleveland JL, Miller JL, et al. 2007. NIX is required for programmed mitochondrial clearance during reticulocyte maturation. *Proc Natl Acad Sci* **104:** 19500–19505.

Schwickart M, Huang X, Lill JR, Liu J, Ferrando R, French DM, Maecker H, O'Rourke K, Bazan F, Eastham-Anderson J, et al. 2010. Deubiquitinase USP9X stabilizes MCL1 and promotes tumour cell survival. *Nature* **463:** 103–107.

Sebbagh M, Renvoizé C, Hamelin J, Riché N, Bertoglio J, Bréard J. 2001. Caspase-3-mediated cleavage of ROCK I induces MLC phosphorylation and apoptotic membrane blebbing. *Nature Cell Biol* **3:** 346–352.

Seshagiri S, Miller LK. 1997. *Caenorhabditis elegans* CED-4 stimulates CED-3 processing and CED-3-induced apoptosis. *Curr Biol* **7:** 455–460.

* Sever R, Brugge JS. 2014. Signal transduction in cancer. *Cold Spring Harb Perspect Med* doi: 10.1101/cshperspect.a006098

Shaid S, Brandts CH, Serve H, Dikic I. 2013. Ubiquitination and selective autophagy. *Cell Death Differ* **20:** 21–30.

Shaw J, Zhang T, Rzeszutek M, Yurkova N, Baetz D, Davie JR, Kirshenbaum LA. 2006. Transcriptional silencing of the death gene BNIP3 by cooperative action of NF-κB and histone deacetylase 1 in ventricular myocytes. *Circ Res* **99:** 1347–1354.

Shaw J, Yurkova N, Zhang T, Gang H, Aguilar F, Weidman D, Scramstad C, Weisman H, Kirshenbaum LA. 2008. Antagonism of E2F-1 regulated Bnip3 transcription by NF-κB is essential for basal cell survival. *Proc Natl Acad Sci* **105:** 20734–20739.

Shen S, Kepp O, Kroemer G. 2012. The end of autophagic cell death? *Autophagy* **8:** 1–3.

Slee EA, Harte MT, Kluck RM, Wolf BB, Casiano CA, Newmeyer DD, Wang HG, Reed JC, Nicholson DW, Alnemri ES, et al. 1999. Ordering the cytochrome-*c*-initiated caspase cascade: Hierarchical activation of caspases-2, -3, -6, -7, -8, and -10 in a caspase-9-dependent manner. *J Cell Biol* **144:** 281–292.

Sowter HM, Ratcliffe PJ, Watson P, Greenberg AH, Harris AL. 2001. HIF-1-dependent regulation of hypoxic induction of the cell death factors BNIP3 and NIX in human tumors. *Cancer Res* **61:** 6669–6673.

Spector MS, Desnoyers S, Hoeppner DJ, Hengartner MO. 1997. Interaction between the *C. elegans* cell-death regulators CED-9 and CED-4. *Nature* **385:** 653–656.

Spender LC, O'Brien DI, Simpson D, Dutt D, Gregory CD, Allday MJ, Clark LJ, Inman GJ. 2009. TGF-β induces apoptosis in human B cells by transcriptional regulation of BIK and Bcl-xL. *Cell Death Differ* **16:** 593–602.

Stennicke HR, Deveraux QL, Humke EW, Reed JC, Dixit VM, Salvesen GS. 1999. Caspase 9 can be activated without proteolytic processing. *J Biol Chem* **274:** 8359–8362.

Stoka V, Turk B, Schendel SL, Kim TH, Cirman T, Snipas SJ, Ellerby LM, Bredesen D, Freeze H, Abrahamson M, et al. 2001. Lysosomal protease pathways to apoptosis. Cleavage of bid, not procaspases, is the most likely route. *J Biol Chem* **276:** 3149–3157.

Sun L, Wang H, Wang Z, He S, Chen S, Liao D, Wang L, Yan J, Liu W, Lei X, et al. 2012. Mixed lineage kinase domain-like protein mediates necrosis signaling downstream of RIP3 kinase. *Cell* **148:** 213–227.

Sutton VR, Davis JE, Cancilla M, Johnstone RW, Ruefli AA, Sedelies K, Browne KA, Trapani JA. 2000. Initiation of apoptosis by granzyme B requires direct cleavage of bid, but not direct granzyme B-mediated caspase activation. *J Exp Med* **192:** 1403–1414.

Szydlowska K, Tymianski M. 2010. Calcium, ischemia and excitotoxicity. *Cell Calcium* **47:** 122–129.

Tait SW, Green DR. 2010. Mitochondria and cell death: Outer membrane permeabilization and beyond. *Nat Rev Mol Cell Biol* **11:** 621–632.

Tait SWG, Parsons MJ, Llambi F, Bouchier-Hayes L, Connell S, Muñoz-Pinedo C, Green DR. 2010. Resistance to caspase-independent cell death requires persistence of intact mitochondria. *Dev Cell* **18:** 802–813.

Tanaka Y, Guhde G, Suter A, Eskelinen EL, Hartmann D, Lullmann-Rauch R, Janssen PM, Blanz J, von Figura K, Saftig P. 2000. Accumulation of autophagic vacuoles and cardiomyopathy in LAMP-2-deficient mice. *Nature* **406:** 902–906.

Telliez JB, Bean KM, Lin LL. 2000. LRDD, a novel leucine rich repeat and death domain containing protein. *Biochim Biophys Acta* **1478:** 280–288.

Tenev T, Bianchi K, Darding M, Broemer M, Langlais C, Wallberg F, Zachariou A, Lopez J, MacFarlane M, Cain K, et al. 2011. The Ripoptosome, a signaling platform that assembles in response to genotoxic stress and loss of IAPs. *Mol Cell* **43:** 432–448.

Tinel A, Tschopp J. 2004. The PIDDosome, a protein complex implicated in activation of caspase 2 in response to genotoxic stress. *Science* **304:** 843–846.

Tinel A, Janssens S, Lippens S, Cuenin S, Logette E, Jaccard B, Quadroni M, Tschopp J. 2007. Autoproteolysis of PIDD marks the bifurcation between prodeath caspase 2 and prosurvival NF-κB pathway. *EMBO J* **26:** 197–208.

Troy CM, Stefanis L, Greene LA, Shelanski ML. 1997. Nedd2 is required for apoptosis after trophic factor withdrawal, but not superoxide dismutase (SOD1) downregulation, in sympathetic neurons and PC12 cells. *J Neurosci* **17:** 1911–1918.

Troy CM, Rabacchi SA, Friedman WJ, Frappier TF, Brown K, Shelanski ML. 2000. Caspase-2 mediates neuronal cell death induced by β-amyloid. *J Neurosci* **20:** 1386–1392.

Troy CM, Rabacchi SA, Hohl JB, Angelastro JM, Greene LA, Shelanski ML. 2001. Death in the balance: Alternative participation of the caspase-2 and -9 pathways in neuronal death induced by nerve growth factor deprivation. *J Neurosci* **21:** 5007–5016.

Upton JP, Austgen K, Nishino M, Coakley KM, Hagen A, Han D, Papa FR, Oakes SA. 2008. Caspase-2 cleavage of BID is a critical apoptotic signal downstream of endoplasmic reticulum stress. *Mol Cell Biol* **28:** 3943–3951.

Upton JW, Kaiser WJ, Mocarski ES. 2010. Virus inhibition of RIP3-dependent necrosis. *Cell Host Microbe* **7:** 302–313.

Upton JW, Kaiser WJ, Mocarski ES. 2012. DAI/ZBP1/DLM-1 complexes with RIP3 to mediate virus-induced programmed necrosis that is targeted by murine cytomegalovirus vIRA. *Cell Host Microbe* **11:** 290–297.

Urbich C, Knau A, Fichtlscherer S, Walter DH, Brühl T, Potente M, Hofmann WK, de Vos S, Zeiher AM, Dimmeler S. 2005. FOXO-dependent expression of the proapoptotic protein Bim: Pivotal role for apoptosis signaling in endothelial progenitor cells. *FASEB J* **19:** 974–976.

Vanden Berghe T, Declercq W, Vandenabeele P. 2007. NADPH oxidases: New players in TNF-induced necrotic cell death. *Mol Cell* **26:** 769–771.

Vercammen D, Beyaert R, Denecker G, Goossens V, Van Loo G, Declercq W, Grooten J, Fiers W, Vandenabeele P. 1998. Inhibition of caspases increases the sensitivity of L929 cells to necrosis mediated by tumor necrosis factor. *J Exp Med* **187:** 1477–1485.

Verma S, Zhao LJ, Chinnadurai G. 2001. Phosphorylation of the proapoptotic protein BIK: Mapping of phosphorylation sites and effect on apoptosis. *J Biol Chem* **276:** 4671–4676.

Villunger A, Michalak EM, Coultas L, Mullauer F, Bock G, Ausserlechner MJ, Adams JM, Strasser A. 2003. p53- and drug-induced apoptotic responses mediated by BH3-only proteins puma and noxa. *Science* **302:** 1036–1038.

Vosler PS, Brennan CS, Chen J. 2008. Calpain-mediated signaling mechanisms in neuronal injury and neurodegeneration. *Mol Neurobiol* **38:** 78–100.

Wang Y, Qin ZH. 2010. Molecular and cellular mechanisms of excitotoxic neuronal death. *Apoptosis* **15:** 1382–1402.

Wang HG, Pathan N, Ethell IM, Krajewski S, Yamaguchi Y, Shibasaki F, McKeon F, Bobo T, Franke TF, Reed JC. 1999. Ca^{2+}-induced apoptosis through calcineurin dephosphorylation of BAD. *Science* **284:** 339–343.

Wang Y, Guan X, Fok KL, Li S, Zhang X, Miao S, Zong S, Koide SS, Chan HC, Wang L. 2008a. A novel member of the Rhomboid family, RHBDD1, regulates BIK-mediated apoptosis. *Cell Mol Life Sci* **65:** 3822–3829.

Wang L, Du F, Wang X. 2008b. TNF-α induces two distinct caspase-8 activation pathways. *Cell* **133:** 693–703.

Wang RC, Wei Y, An Z, Zou Z, Xiao G, Bhagat G, White M, Reichelt J, Levine B. 2012. Akt-mediated regulation of autophagy and tumorigenesis through Beclin 1 phosphorylation. *Science* **338:** 956–959.

Warr MR, Acoca S, Liu Z, Germain M, Watson M, Blanchette M, Wing SS, Shore GC. 2005. BH3-ligand regulates access of MCL-1 to its E3 ligase. *FEBS Lett* **579:** 5603–5608.

Wei MC, Lindsten T, Mootha VK, Weiler S, Gross A, Ashiya M, Thompson CB, Korsmeyer SJ. 2000. tBID, a membrane-targeted death ligand, oligomerizes BAK to release cytochrome *c*. *Genes Dev* **14:** 2060–2071.

Wei Y, Pattingre S, Sinha S, Bassik M, Levine B. 2008. JNK1-mediated phosphorylation of Bcl-2 regulates starvation-induced autophagy. *Mol Cell* **30:** 678–688.

Wertz IE, Kusam S, Lam C, Okamoto T, Sandoval W, Anderson DJ, Helgason E, Ernst JA, Eby M, Liu J, et al. 2011. Sensitivity to antitubulin chemotherapeutics is regulated by MCL1 and FBW7. *Nature* **471:** 110–114.

Wu D, Wallen HD, Inohara N, Nunez G. 1997a. Interaction and regulation of the *Caenorhabditis elegans* death protease CED-3 by CED-4 and CED-9. *J Biol Chem* **272:** 21449–21454.

Wu D, Wallen HD, Nunez G. 1997b. Interaction and regulation of subcellular localization of CED-4 by CED-9. *Science* **275:** 1126–1129.

Xiao C, Srinivasan L, Calado DP, Patterson HC, Zhang B, Wang J, Henderson JM, Kutok JL, Rajewsky K. 2008. Lymphoproliferative disease and autoimmunity in mice with increased miR-17-92 expression in lymphocytes. *Nat Immunol* **9:** 405–414.

Yan B, Zemskova M, Holder S, Chin V, Kraft A, Koskinen PJ, Lilly M. 2003. The PIM-2 kinase phosphorylates BAD on serine 112 and reverses BAD-induced cell death. *J Biol Chem* **278:** 45358–45367.

Yan N, Gu L, Kokel D, Chai J, Li W, Han A, Chen L, Xue D, Shi Y. 2004. Structural, biochemical, and functional analyses of CED-9 recognition by the proapoptotic proteins EGL-1 and CED-4. *Mol Cell* **15:** 999–1006.

Yan J, Xiang J, Lin Y, Ma J, Zhang J, Zhang H, Sun J, Danial NN, Liu J, Lin A. 2013. Inactivation of BAD by IKK inhibits TNFα-induced apoptosis independently of NF-κB activation. *Cell* **152:** 304–315.

Yang CS, Thomenius MJ, Gan EC, Tang W, Freel CD, Merritt TJ, Nutt LK, Kornbluth S. 2010. Metabolic regulation of *Drosophila* apoptosis through inhibitory phosphorylation of Dronc. *EMBO J* **29:** 3196–3207.

Yoo SJ, Huh JR, Muro I, Yu H, Wang L, Wang SL, Feldman RM, Clem RJ, Muller HA, Hay BA. 2002. Hid, Rpr and Grim negatively regulate DIAP1 levels through distinct mechanisms. *Nat Cell Biol* **4:** 416–424.

You H, Pellegrini M, Tsuchihara K, Yamamoto K, Hacker G, Erlacher M, Villunger A, Mak TW. 2006. FOXO3a-dependent regulation of Puma in response to cytokine/growth factor withdrawal. *J Exp Med* **203:** 1657–1663.

Yu J, Zhang L, Hwang PM, Kinzler KW, Vogelstein B. 2001. PUMA induces the rapid apoptosis of colorectal cancer cells. *Mol Cell* **7:** 673–682.

Yu J, Wang Z, Kinzler KW, Vogelstein B, Zhang L. 2003. PUMA mediates the apoptotic response to p53 in colorectal cancer cells. *Proc Natl Acad Sci* **100:** 1931–1936.

Yu X, Acehan D, Menetret JF, Booth CR, Ludtke SJ, Riedl SJ, Shi Y, Wang X, Akey CW. 2005. A structure of the human apoptosome at 12.8 A resolution provides insights into this cell death platform. *Structure* **13:** 1725–1735.

Yu X, Wang L, Acehan D, Wang X, Akey CW. 2006. Three-dimensional structure of a double apoptosome formed by the *Drosophila* Apaf-1 related killer. *J Mol Biol* **355:** 577–589.

Yuan J, Kroemer G. 2010. Alternative cell death mechanisms in development and beyond. *Genes Dev* **24:** 2592–2602.

Yuan J, Shaham S, Ledoux S, Ellis HM, Horvitz HR. 1993. The *C. elegans* cell death gene ced-3 encodes a protein similar to mammalian interleukin-1 β-converting enzyme. *Cell* **75:** 641–652.

Yuan S, Yu X, Topf M, Ludtke SJ, Wang X, Akey CW. 2010. Structure of an apoptosome-procaspase-9 CARD complex. *Structure* **18:** 571–583.

Yurkova N, Shaw J, Blackie K, Weidman D, Jayas R, Flynn B, Kirshenbaum LA. 2008. The cell cycle factor E2F-1 activates Bnip3 and the intrinsic death pathway in ventricular myocytes. *Circ Res* **102:** 472–479.

Zalckvar E, Berissi H, Eisenstein M, Kimchi A. 2009a. Phosphorylation of Beclin 1 by DAP-kinase promotes autophagy by weakening its interactions with Bcl-2 and Bcl-X$_L$. *Autophagy* **5:** 720–722.

Zalckvar E, Berissi H, Mizrachy L, Idelchuk Y, Koren I, Eisenstein M, Sabanay H, Pinkas-Kramarski R, Kimchi A. 2009b. DAP-kinase-mediated phosphorylation on the BH3 domain of beclin 1 promotes dissociation of beclin 1 from Bcl-X$_L$ and induction of autophagy. *EMBO Rep* **10:** 285–292.

Zha J, Harada H, Yang E, Jockel J, Korsmeyer SJ. 1996. Serine phosphorylation of death agonist BAD in response to survival factor results in binding to 14-3-3 not BCL-X$_L$. *Cell* **87:** 619–628.

Zhang DW, Shao J, Lin J, Zhang N, Lu BJ, Lin SC, Dong MQ, Han J. 2009. RIP3, an energy metabolism regulator that switches TNF-induced cell death from apoptosis to necrosis. *Science* **325:** 332–336.

Zhao J, Jitkaew S, Cai Z, Choksi S, Li Q, Luo J, Liu Z-G. 2012. Mixed lineage kinase domain-like is a key receptor interacting protein 3 downstream component of TNF-induced necrosis. *Proc Natl Acad Sci* **109:** 5322–5327.

Zhou L, Song Z, Tittel J, Steller H. 1999. HAC-1, a *Drosophila* homolog of APAF-1 and CED-4 functions in developmental and radiation-induced apoptosis. *Mol Cell* **4:** 745–755.

Zhu H, Zhang L, Dong F, Guo W, Wu S, Teraishi F, Davis JJ, Chiao PJ, Fang B. 2005. Bik/NBK accumulation correlates with apoptosis-induction by bortezomib PS-341, Velcade and other proteasome inhibitors. *Oncogene* **24:** 4993–4999.

Zou H, Henzel WJ, Liu X, Lutschg A, Wang X. 1997. Apaf-1, a human protein homologous to *C. elegans* CED-4, participates in cytochrome-*c*-dependent activation of caspase 3. *Cell* **90:** 405–413.

CHAPTER 20

Subversion of Cell Signaling by Pathogens

Neal M. Alto and Kim Orth

UT Southwestern Medical Center, Dallas, Texas 75390

Correspondence: kim.orth@utsouthwestern.edu

SUMMARY

Pathogens exploit several eukaryotic signaling pathways during an infection. They have evolved specific effectors and toxins to hijack host cell machinery for their own benefit. Signaling molecules are preferentially targeted by pathogens because they globally regulate many cellular processes. Both viruses and bacteria manipulate and control pathways that regulate host cell survival and shape, including MAPK signaling, G-protein signaling, signals controlling cytoskeletal dynamics, and innate immune responses.

Outline

1 INTRODUCTION

Viruses and bacteria both produce proteins that hijack cellular signaling machinery to ensure their survival despite the constant negative pressure of their hosts' innate and adaptive immune systems. Studies of these have not only revealed methods by which microbial pathogens cause infectious disease, but also provided critical insights into mechanisms of cellular regulation, particularly in the case of viral oncoproteins (see Box 1). They also yielded valuable tools for dissecting mechanisms involved in regulating signaling.

These proteins help a pathogen achieve its goals of survival, replication, and virulence. For example, they may induce a slow cell death to allow time for replication or cause a rapid cell death so that the pathogen avoids engulfment. Alternatively, they may manipulate the actin cytoskeleton to prevent or accelerate phagocytosis by the host. In the case of viral oncoproteins, they may hijack the cell cycle machinery to promote virus replication.

Here, we do not attempt to be comprehensive, but provide representative examples of how pathogens manipulate host cell signaling, emphasizing the virulence factors produced by bacterial pathogens (Table 1). Targets for these virulence factors include GTPases and their regulators that control the cytoskeleton and vesicular trafficking, kinase cascades involved in intra- and extracellular signaling, and ubiquitin-dependent pathways that regulate signal stability or dictate other outputs.

Bacteria produce different types of virulence factors. Note that these are not always proteins—they may also be peptides or small molecules—but here we confine our discussion to proteins. The first type, called a toxin, is secreted by the pathogen at high concentration and delivered into the host cytoplasm by a variety of mechanisms, including endocytosis or via protein pores formed by the bacterial toxin itself (Fig. 1) (Henkel et al. 2010). Cholera toxin from *Vibrio cholerae*, for example, is an enzyme that enters the cell by endocytosis and ADP-ribosylates the α_s subunit of the heterotrimeric G_s protein. This modification prevents α_s from hydrolyzing GTP, locking it into an active state that constitutively stimulates its downstream effector, adenylyl cyclase (see p. 99 [Sassone-Corsi 2012]). In the intestinal epitheli-

BOX 1. VIRAL ONCOPROTEINS

Studies of viruses have provided critical biological insights into the signal transduction mechanisms of host cells. More than a century ago, Peyton Rous described Rous sarcoma virus (RSV) as a transforming agent that induces solid tumors in chickens. Although this discovery was met with some skepticism at the time, researchers in the 1970s and 1980s isolated and sequenced the viral non-receptor tyrosine kinase-encoding *v-src* gene as the genetic element that causes tumor formation. In fact, *v-src* was the first retroviral oncogene discovered, and this discovery helped define the first proto-oncogene in the vertebrate genome (*c-Src*, now known simply as *Src*). Further investigations into oncoviruses also revealed v-Abl, the retroviral oncoprotein from the Abelson murine leukemia virus (A-MuLV). Like *v-src*, *v-Abl* encodes a nonreceptor tyrosine kinase that is genetically and functionally related to the *ABL* gene in mammals. In humans, the translocation involving chromosomes 22 and 9 (known as the Philadelphia chromosome) causes the first exon in *ABL1* to be replaced by sequences from the *BCR* gene, resulting in expression of the BCR–ABL fusion protein. This genetic translocation results in abnormally high levels of BCR–ABL kinase activity, leading to chronic myeloid leukemia (CML) and a subset of acute lymphocytic leukemia (ALL). Biochemical and structural studies led to one of the successful medicinal treatments for cancer: a specific, small molecule kinase inhibitor referred to as imatinib or GLEEVAC.

Studies of the transforming activity of murine sarcoma viruses were also particularly influential in stimulating work that elucidated small G-protein signaling in human cancer. In the 1960s, the first *Ras* (for Rat sarcoma) genes were identified as transduced oncogenes expressed by the Harvey and Kirsten strains of acutely transforming murine retroviruses. More than a decade later, researchers discovered that these viral Ras proteins interface with guanine nucleotide signaling mechanisms through an intrinsic GTPase activity. Several human genes were subsequently found to display sequence homology to viral *Ras*. The identification of *RAS* as the first human transforming gene set off a global research initiative aimed at discovering the molecular mechanisms of small G proteins.

Interestingly, the major downstream substrate of Ras signal transduction is Raf kinase, encoded by the cellular homolog of the *v-raf* oncogene expressed by the murine retrovirus 3611-MSV. Under normal conditions, binding of extracellular ligands such as growth factors, cytokines, and hormones to cell-surface receptors activates Ras, and this initiates Raf activation. Indeed, both lie downstream from the EGF receptor, which also has a viral oncogene homolog, *v-erbB*. Thus, not only has research on viral oncogenes been an entry point into the molecular genetics of cancer, but it has also revealed critical links between infectious disease mechanisms and eukaryotic signal transduction.

Cite this chapter as *Cold Spring Harb Perspect Biol* doi: 10.1101/cshperspect.a006114

Table 1. Bacterial toxins and effectors and their targets

Pathogen	Toxin	Effector	Target	Activity
Vibrio cholerae	Cholera toxin		Gα_s	ADP ribosylation
Vibrio cholerae	EF edema factor		Calmodulin	Adenylate cyclase
Vibrio cholerae	LF lethal factor		MKK1,2	Metalloprotease
Bordetella pertussis	Pertussus toxin		Gα_i	ADP ribosylation
Clostridium botulinum	C3 botulin toxin		Rho GTPases	ADP ribosylation
Escherichia coli	CNF1		Rho GTPases	Deamination
EPEC/EHEC O157:H7		Tir	Actin	Recruits NCK adaptor
EPEC/EHEC O157:H7		Map	Rho GTPases	GEF
EPEC/EHEC O157:H7		EspFu	N-WASP	Activator of N-WASP
EPEC/EHEC O157:H7		EspG	p21-activated kinase (PAK)	Activator of PAK
EPEC/*Burkholderia* spp.		Cif/CBHP	Ubiquitin, Nedd8	Ubiquitylation inhibitor
Yersinia spp.		YopH	p130Cas	Tyrosine phosphatase
Yersinia spp.		YopE	Rho-like GTPases	GAP
Yersinia spp.		YopT	Rho GTPase	Cysteine protease
Yersinia spp.		YpkA	Gα_q, Rho GTPases	Ser/Thr kinase, GDI
Yersinia spp.		YopJ	MAPKKs, IKK-β	Ser/Thr acetyltransferase
Vibrio parahaemolyticus		VopA/p	MAPKKs	Ser/Thr/Lys acetyltransferase
Vibrio parahaemolyticus		VopS	Rho-GTPases	AMPylation
Vibrio parahaemolyticus		VPA0450	Phosphatidylinositol 4,5-bisphosphate	Lipid phosphatase
Vibrio parahaemolyticus		VopL	Actin	Actin nucleator
Histophilus somni		IbpA	Rho GTPases	Ampylation
Legionella pneumophila		DrrA/SidM	Rab1b	AMPylation, GEF
Legionella pneumophila		SidD	Rab1b	DeAMPylation
Legionella pneumophila		AnkX	Rab1b	Phosphocholination
Shigella spp.		OspF	MAPK	Phosphothreonine lyase
Shigella spp.		IpgD	Phosphatidylinositol 4,5-bisphophate	Lipid phosphatase
Shigella spp.		IpaH9.8	Ste7 MAPK	E3 ubiquitin ligase
Shigella spp.		IpgB2	Rac1 and RhoA	GEF
Salmonella spp.		SopB	Phosphatidylinositol 4,5-bisphophate	Lipid phosphatase
Salmonella spp.		SopE	Cdc42 and Rac GTPases	GEF
Salmonella spp.		SptP	Small GTPase	GAP, tyrosine phosphatase
Pseudomonas aeruginosa		ExoS	Small GTPases	ADP ribosylation, GAP
Listeria monocytogenes		ActA	Arp2/3, actin	Activator of Arp2/3
Viral effector				
Vaccinia virus		A36R	Actin	Adaptor recruits Nck and Grb2
Adenovirus		E1B-55K	p53, Mre11, BLM helicase	Ubiquitin ligase adaptor
Adenovirus		E4orf6	p53, Mre11, BLM helicase	Ubiquitin ligase adaptor
Papillomavirus		E6	E6-AP, p53	E3 ubiquitin ligase
KSHV		RTA	IRF-7	E3 ubiquitin ligase
Gammaherpesviruses		K3 and K5	MHC class 1	E3 ubiquitin ligase
Herpes simplex virus 1		ICP0	PML	E3 ubiquitin ligase

um, the elevated levels of cyclic AMP (cAMP) generated cause an efflux of chloride ions and water, resulting in severe diarrhea and dehydration. Pertussis toxin produced by *Bordetella pertussis*, which causes whooping cough, by contrast, ADP ribosylates the α_i subunit of the heterotrimeric G$_i$ protein so that it remains in the GDP-bound state and cannot be activated by upstream signals. The modified GDP-bound α_i is no longer able to bind to adenylyl cyclase and inhibit production of cAMP. Pertussis toxin causes major trauma in airways during infection because signaling pathways are constitutively activated and results in abnormally

high levels of insulin and histamine sensitivity. These toxins have proven very useful for the elucidation of molecular signaling by G-protein-coupled receptors (GPCRs) and their downstream signaling partners.

Another type of virulence factor, called an effector, is a protein that is directly translocated from bacteria into the host cell by specialized needle-like delivery systems, including the type III and type IV secretion systems (T3SS and T4SS, respectively) (Fig. 1) (Hayes et al. 2010). The effectors are made in the bacterium but appear to be inactive because of association with a chaperone, lack of

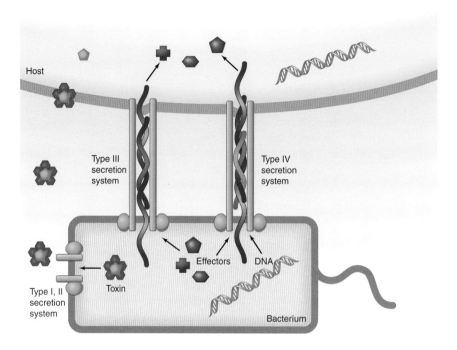

Figure 1. Bacterial secretion systems. Bacteria use several different mechanisms to secrete molecules into the extracellular space and to translocate molecules into a host cell. Toxins, including peptides and proteins, are typically secreted through a type I or II secretion system. Effectors are translocated through a type III or IV secretion system, whereas DNA is only transferred through the type IV secretion system.

appropriate substrate, and/or absence of eukaryotic activators. After delivery into the eukaryotic host, the effectors display very potent activities that manipulate signal transduction pathways. Many of these activities mimic an endogenous activity of the host.

The first such effector analyzed at the molecular level was YopH (*Yersinia* outer protein H) from *Yersinia*, the causal agent of the plague (also known as the Black Death) (Guan and Dixon 1990). This protein contains an unregulated, highly active tyrosine phosphatase domain linked to a leader sequence that both guides its translocation from the bacterium into the host cell and determines its localization after delivery. YopH is translocated into eukaryotic cells through the *Yersinia* T3SS and proceeds to focal adhesions, where it dephosphorylates critical phosphorylated tyrosine residues on protein substrates including p130Cas and Fyb (see Ch. 8 [Devreotes and Horwitz 2014]). The resulting dephosphorylated focal adhesion complex disassembles and, therefore, is unable to promote phagocytosis of the bacterial pathogen. YopH is a typical bacterial effector for the following reasons.

1. Although an extremely active enzyme, YopH has no effect on the pathogen itself, owing to the lack of a substrate.

2. The YopH protein contains information in its amino-terminal domain for both secretion by the T3SS apparatus and localization in the infected host.

3. YopH contains a potent activity that efficiently targets and destroys the Achilles heel of a process, in this case, phagocytosis, by targeting focal adhesions that are regulated by phosphorylated tyrosine residues.

4. YopH plays a critical role in virulence.

Below, we describe several other pathogenic effectors that show these general characteristics, breaking them down by the signaling pathways they target.

2 CORRUPTION OF MAPK SIGNALING

Mitogen-activated protein kinase (MAPK) signaling pathways are cascades of kinases that sequentially activate each other by phosphorylation (p. 81 [Morrison 2012]). A MAPK kinase kinase (MAPKKK) activates a MAPK kinase (MAPKK), which, in turn, activates a MAPK, and there are multiple family members at each stage. Pathogens target these signaling pathways because they regulate many types of cellular behaviors, including cell proliferation, innate immune responses, cell migration, apoptosis, and autophagy.

2.1 *Bacillus anthracis* Lethal Factor Hydrolyzes the MAPKK MKK1/2

Bacillus anthracis, the causal agent of anthrax, releases a multi-subunit complex called anthrax toxin, composed of protective antigen (PA), edema factor (EF), and lethal factor (LF) (Collier 2009). The toxin binds via PA to either anthrax toxin receptor 1 or 2 on the surface of the host cell. After uptake by receptor-mediated endocytosis and acidification of the resulting endosome, EF and LF are released into the cytoplasm. The calcium-binding protein calmodulin binds to cytoplasmic EF, causing a change in conformation that generates an active enzyme that produces cAMP from cellular ATP. The excess cAMP globally disrupts signaling by binding and activating downstream effectors, such as cAMP-dependent kinase (PKA). Cytoplasmic LF is an active metalloprotease that cleaves the amino-terminal extensions from MAPKKs MKK1 and MKK2, producing kinases that can no longer interact with their substrates to activate a proliferative response (Fig. 2). Both of the toxins have an irreversible toxic effect on the infected cell.

2.2 *Yersinia* YopJ Acetylates MAPKK/IKKβ Activation Loop

Yersinia ssp. have a very efficient strategy for disrupting the innate immune response and promoting apoptosis in infected cells, using one molecule, YopJ (also termed YopP). This effector is injected directly into the host's cytoplasm through a T3SS. YopJ blocks all of the MAPK pathways and the NF-κB pathway by preventing the activation of all MAPKKs and IKKβ (but not IKKα) (Fig. 2) (Orth et al. 1999; Hao et al. 2008; p. 81 [Morrison 2012] and p. 121 [Lim and Staudt 2013]). The activity of this 32-kDa effector remained elusive for many years because it contains a catalytic triad similar to that in some cysteine proteases, specifically clan CE proteases, which include adenoviral proteases and ubiquitin-like protein proteases (Orth et al. 2000). However, a classical biochemical approach finally revealed that YopJ does not cleave MAPKKs or any other substrate; instead, it modifies MAPKKs with a small acetyl moiety. This acetyltransferase activity requires an intact catalytic triad and uses acetyl-CoA to modify serine and/or threonine residues in the activation loops of MAPKKs and IKKβ, generating a novel posttranslational modification that competes with phosphorylation (Mittal et al. 2006; Mukherjee et al. 2006).

These findings revealed a new paradigm for signaling, in which serine and threonine residues could be substrates for acetylation (Mukherjee et al. 2007). So how is the catalytic triad of a presumed protease used for acetylation? In fact, YopJ acetyltransferases and clan CE cysteine proteases are both proposed to use the same catalytic "ping-pong" mechanism (Fig. 3). Acetyltransferases containing a catalytic triad react with acetyl-CoA to form a covalent acetyl-enzyme intermediate and release CoA. They then bind their substrate and transfer the acetyl group to the substrate's attacking nucleophile. Cysteine proteases use the

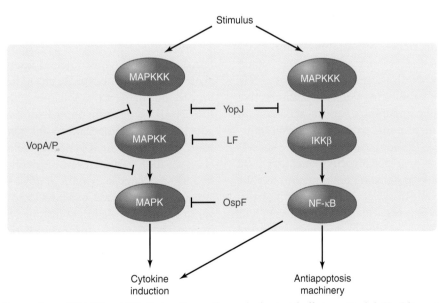

Figure 2. Corruption of MAPK and NFκB signaling pathways by bacterial effectors. *Yersinia* YopJ is an acetyltransferase that acetylates and inhibits MAPKK and IKKβ activation by blocking phosphorylation. Similarly, *Vibrio* VopA/P blocks activation of MAPKK. Lethal factor (LF) from *B. anthracis*, the causal agent of anthrax, is a metalloprotease that inhibits MAPKK by proteolysis. OspF is a phosphothreonine lyase that irreversibly dephosphorylates MAPK by elimination of a phosphate group from activated MAPK.

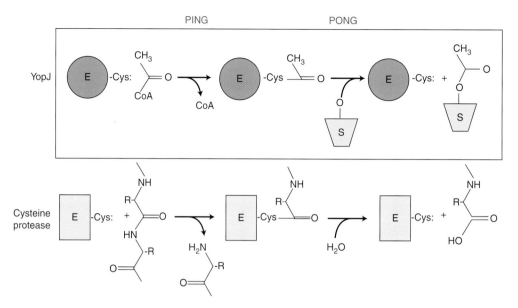

Figure 3. Catalytic triads: same chemistry, different substrates. *Yersinia* YopJ is proposed to use a ping-pong mechanism whereby its catalytic cysteine attacks acetyl-CoA to form a covalent acetyl-enzyme intermediate, followed by a subsequent attack by its second substrate, a hydroxyl on MAPKK, to transfer the acetyl group to MAPKK. A cysteine protease uses the same mechanism with a peptide to form a covalent acetyl-enzyme intermediate. The second substrate in this reaction is water and leads to cleavage of a peptide bond.

same mechanism to form a covalent acyl-enzyme intermediate, release the carboxy-terminal peptide, and then use a second substrate, water, to attack the acyl-enzyme covalent intermediate to complete the hydrolase reaction. Acetyl transferases do not allow water in the catalytic site; otherwise, these enzymes would simply hydrolyze the essential metabolite acetyl-CoA.

During an infection, signals for host survival and apoptosis are induced. In the presence of YopJ, the default pathway will always be death because YopJ blocks the NF-κB survival pathway. By inhibiting signaling pathways that alert the immune system and induce survival signals, YopJ attenuates the immune response to *Yersinia* during infection. In contrast, VopA/P, a YopJ relative from the seafoodborne pathogen *Vibrio parahaemolyticus* that causes food poisoning, inhibits MAPK signaling pathways but not the NF-κB pathway (Fig. 2) (Trosky et al. 2004). Additionally, VopA/P acetylates not only the activation loop of MAPKKs but also a conserved lysine residue in the catalytic loop of MAPKKs that is required for coordination of the γ-phosphate of ATP (Trosky et al. 2007). This inhibits the binding of ATP but not ADP, resulting in an inactive kinase. During infection by *V. parahaemolyticus*, VopA/P efficiently blocks proliferative pathways while allowing activation of survival pathways. Our understanding of this infectious process is in its infancy; thus the activities of the other secreted effectors will need to be uncovered to shed light on the importance

of VopA/P inhibition. Note that this type of serine/threonine acetylation has not yet been observed as an endogenous protein modification in eukaryotes.

2.3 *Shigella* OspF Irreversibly Eliminates a Phosphate

Shigella, the causal agent of bloody dysentery, contains the T3SS effector OspF, which is a phosphothreonine lyase that translocates to the host nucleus upon infection. This enzyme eliminates a phosphate group from a phosphothreonine residue in the activation loop of the MAPKs ERK1 and ERK2 by β-elimination of the hydroxyl moiety, an unprecedented reaction mechanism for removing a phosphate from a protein, which generates dehydroalanine (Fig. 2) (Li et al. 2007). This irreversible dephosphorylation inhibits ERK-mediated activation of downstream mitogen- and stress-activated kinase 1 (MSK1) and MSK2, thereby preventing phosphorylation of histone H3 on S10. This modification is a prerequisite for chromatin reorganization and priming of transcription-factor-binding sites in NF-κB-regulated promoters (Ch. 15 [Newton and Dixit 2012]). Initial studies supported the hypothesis that OspF works to diminish a proinflammatory response during a *Shigella* infection, but other studies implicate an OspF-mediated inhibition of a negative-feedback loop to partially activate immune signaling, which may create an advantageous environment for this intracellular pathogen.

3 MANIPULATION OF G-PROTEIN SIGNALING

Among the Ras superfamily of small G proteins, the Ras, Rho, and Ran subfamilies primarily regulate cell division/differentiation, cytoskeleton remodeling, and nuclear import, respectively (Takai et al. 2001; Ch. 2 [Lee and Yaffe 2014]). Members of the Arf and Rab subfamilies facilitate many aspects of intracellular trafficking (Takai et al. 2001). As in the case of heterotrimeric G proteins, nucleotide binding regulates small G proteins: GTP-bound small G proteins are in an active conformation, and GDP-bound small G proteins are in an inactive conformation. Unsurprisingly, they are important targets of viral and bacterial virulence factors, as are the guanine-nucleotide exchange factors (GEFs) and GTPase-activating proteins (GAPs) that regulate them (Ch. 2 [Lee and Yaffe 2014]). Indeed, the diverse regulatory processes they control make them prime targets, allowing pathogens to subvert host machinery in order to mediate cellular attachment and entry and promote growth, replication, and dissemination of the pathogen within the harsh environment of the host. There is an impressive list of secreted virulence factors that mimic or modify the behavior of host small G proteins (Table 1).

3.1 Bacterial Guanine-Nucleotide Exchange Factor (GEF) Mimics

The pathogenic strategy underlying the subversion of small G proteins during infection depends on the bacterial life cycle. For example, Salmonella, the causal agent of typhoid fever, is intracellular and deploys bacterial GEF proteins such as SopE and SopE2 to activate actin polymerization, which facilitates internalization of the bacterium into host cells. SopE directly activates the Cdc42 and Rac small G proteins to induce membrane ruffling at the site of Salmonella invasion (Fig. 4) (Hardt et al. 1998). SopE2 (Salmonella spp.), BopE (Burkholderia pseudomallei), and CopE (Chromobacterium violaceum) are similar to SopE. A second class of bacterial GEFs shares very low sequence similarity (<15%) with one another, but all contain an invariant WxxxE motif (Alto et al. 2006; Huang et al. 2009). This extends the group of pathogens that directly activate host small G proteins to the facultative intracellular pathogens Shigella spp. and the extracellular attaching/effacing (A/E) Escherichia coli pathogens, including entrohemorrhagic E. coli O157:H7 (EHEC O157:H7) and enteropathogenic E. coli (EPEC).

Bioinformatics analyses reveal that SopE-type and WxxxE-type bacterial GEFs share no overall sequence similarity, yet both classes adopt a conserved V-shape structure and promote the exchange of guanine nucleotides by presenting conserved acidic and amide residues necessary for stabilization of switch I and switch II regions on the target small G proteins (Buchwald et al. 2002; Upadhyay et al. 2008; Huang et al. 2009). However, unlike many forms of pathogenic mimicry, the bacterial GEFs are functional mimics that have structures completely different from those of their eukaryotic counterparts. Indeed, they resemble neither the eukaryotic Dbl-homology (DH) domain nor the dock homology region 2 (DHR2) domain proteins, the two major classes of eukaryotic Rho-family GEFs. Nevertheless, both bacterial and eukaryotic GEFs induce

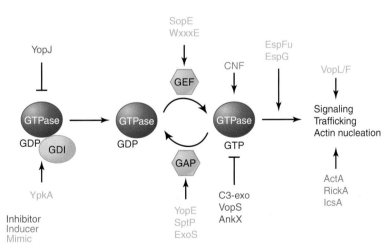

Figure 4. Multilevel regulation of small G proteins by bacterial effectors. Small G proteins cycle on and off membranes, exchange guanine nucleotides (GDP to GTP) for activation, hydrolyze GTP for inactivation, and stimulate downstream signaling pathways. Bacterial pathogens have evolved toxins and effector proteins to usurp nearly every aspect of small G-protein function, using molecular mimicry as well as novel stimulatory and inhibitory mechanisms.

similar nucleotide-free transition states essential for guanine-nucleotide exchange. The catalytic loops of bacterial GEFs are important for making contacts with the switch I and switch II regions of the host small G proteins. Recent studies have shown that these loops are flexible and that proper orientation is important for small G-protein recognition and activation (Klink et al. 2010). Reorientation of the catalytic loop may therefore be a mechanism for bacterial GEFs to control host small G-protein activity in response to bacterial or host stimuli.

3.2 Bacterial GTPase-Activating Protein (GAP) Mimics

Many pathogens have evolved proteins that directly engage small G proteins and stimulate GTP hydrolysis, thus turning the molecular switch off. A key example of this activity is the YopE protein from *Yersinia pseudotuberculosis* (Fig. 4). YopE has in vitro GAP activity toward RhoA, Rac1, and Cdc42 (Black and Bliska 2000; Von Pawel-Rammingen et al. 2000; Andor et al. 2001). It engages the small G proteins directly and presents a critical arginine residue (R144) that stabilizes the β phosphate of the GTP moiety, thus increasing the hydrolytic rate. In fact, mutant bacteria expressing YopE (R144A) that lacks GAP activity are incapable of inducing cytotoxicity. During a *Yersinia* infection, YopE works in concert with YopH to prevent phagocytosis by the host cell. The former effector induces collapse of the actin cytoskeleton and rounding of the host cell, whereas the latter effector disassembles focal adhesions (see above).

Salmonella spp. subvert host small G proteins to gain entry into normally nonphagocytic cells, such as intestinal epithelial cells. It induces its own engulfment by injecting proteins like SopE that reorganize the actin cytoskeleton (discussed above). The type III secreted SptP protein uses a tyrosine phosphatase activity and Rho-GTPase GAP activity in order to return the cytoskeleton back to a preinvasion state (Fu and Galan 1999; Stebbins and Galan 2000). Much like YopE from *Yersinia*, SptP from *Salmonella* has an "arginine finger" that induces efficient GTP hydrolysis by small G proteins. Interestingly, the temporal shift in activity between actin-polymerization-promoting factors such as SopE and actin inhibitors such as SptP is dictated by ubiquitin-dependent degradation (Kubori and Galan 2003). Degradation of SopE allows SptP to rapidly inhibit nonspecific Rho-family GTPase signaling, thus preventing spurious actin assembly so that immune cells no longer detect intracellular *Salmonella*.

Finally, *Pseudomonas*, an opportunistic, extracellular pathogen associated with severe infection in cystic fibrosis patients and burn patients, expresses the effector protein ExoS, which contains two catalytic domains that inhibit small G proteins. The first is an ADP-ribosyltransferase domain that transfers ADP-ribose from NAD^+ to R41 of several Ras-like small G proteins (Coburn et al. 1989; Coburn and Gill 1991). This activity uncouples signaling by preventing GEF-mediated activation of small G proteins. The second is a GAP domain that potently inactivates several host small G proteins (Goehring et al. 1999). Whereas many bacterial effector proteins have evolved to specifically target a particular small G protein, ExoS functions to perturb the activation of multiple GTP-binding proteins and thus the signal transduction pathways they control (Henriksson et al. 2002). Together, the down-regulation of small G-protein signaling by bacterial GAPs dampens actin cytoskeleton dynamics at the bacterium–host interface.

3.3 A *Yersinia* Protease Destroys the Small G-Protein Membrane Anchor

The *Yersinia* effector protein YopT can disrupt the actin cytoskeleton, causing rounding up of host cells, and thereby inhibit phagocytosis (Fig. 4) (Iriarte and Cornelis 1998). YopT is a cysteine protease that cleaves amino-terminally to the prenylated cysteine residue in RhoA, Rac, and Cdc42 (Shao et al. 2002, 2003). This liberates Rho proteins from the membrane and irreversibly inhibits signaling to the actin cytoskeleton. YopT is similar to papain proteases and recognizes basic amino acid residues upstream of the cleavage site as well as both the GDP-bound and GTP-bound forms of RhoA. Thus, cleavage does not depend on the conformational state of the small G protein. When RhoA is no longer tethered to the membrane, actin-mediated cytoskeletal responses during infection are effectively thwarted.

3.4 *Yersinia* YpkA: A Kinase and a GDI

Another *Yersinia* effector protein that regulates host membrane remodeling is YpkA. This three-domain effector contains an amino-terminal kinase domain, central guanine nucleotide dissociation inhibitor (GDI) domain, and a carboxy-terminal actin-binding domain (Juris et al. 2000; Prehna et al. 2006). YpkA preferentially phosphorylates the active GTP-bound form of the α_q heterotrimeric G-protein subunit at position S47 (Navarro et al. 2007). This modification prevents α_q from binding GTP, thereby inhibiting its activity and the subsequent membrane remodeling that would otherwise enhance uptake of bacteria. Interestingly, mice with deficiencies in α_q function have increased bleeding times and defective platelet activation, which is a hallmark of the plague (Offermanns et al. 1997).

The central GDI domain in YpkA from *Yersinia* directly binds Rac1 and mimics the host GDI for Rho small G proteins (Prehna et al. 2006). Consequently, Rac1 cannot undergo GDP for GTP exchange and localization to the membrane, where it would normally function in cell adhesion, migration, and regulation of epithelial cell differentiation.

3.5 Bacterial Effectors Can Modify Small G Proteins

As indicated above, Rho family small G proteins are high-value targets for bacterial effector proteins because they control the eukaryotic cytoskeleton. Therefore, unsurprisingly, some bacterial effector proteins not only mimic the endogenous regulators but can posttranslationally modify eukaryotic small G proteins (Etienne-Manneville and Hall 2002). C3 toxin from *Clostridium botulinum*, more generically known as botulinum toxin, has evolved the ability to ADP-ribosylate and inactivate Rho-family small G proteins, specifically RhoA, RhoB, and RhoC, on an invariant N41 residue (Mohr et al. 1992). Treatment of adherent cells with C3 toxin results in the disassembly of actin microfilaments and rounding up of cells (Chardin et al. 1989). N41 is located at the border of the switch I region of Rho-family proteins, and ADP ribosylation induces close association of the Rho small G protein with its cognate GDI, thereby inhibiting cytosol-to-membrane cycling and preventing activation by GEFs (Sehr et al. 1998). Bacteria with extracellular life cycles, like *E. coli*, must avoid phagocytosis. This can be accomplished through covalent modification of Rho, Rac, and Cdc42. Specifically, cytotoxic necrotizing factor 1 (CNF1) deamidates Q63 of Rho or Q61 of Rac and Cdc42 (Flatau et al. 1997; Schmidt et al. 1997). By removing the functional amine group from the critical catalytic glutamine residue, this renders the small G protein constitutively active. Deamination-induced activation of these small G proteins results in cell ruffling and the formation of stress fibers, focal adhesions, and lamellipodia, as well as extension of filopodia. These effects lead to unregulated membrane protrusions that appear to inhibit normal phagocytic events. An additional interesting feature of CNF1 is its ability to rescue epithelial cells from apoptosis, presumably prolonging the infectivity of the pathogenic bacterium (Miraglia et al. 2007).

Another posttranslational modification used by bacterial effectors is AMPylation, covalent attachment of AMP to a hydroxyl side chain on a protein substrate (Woolery et al. 2010). In the 1960s, Earl Stadtman and colleagues found that *E. coli* glutamine synthetase is regulated by glutamine synthetase adenylyl transferase (GS-ATase), a bifunctional enzyme that catalyzes addition and removal of AMP from glutamine synthetase. This results in inactivation and activation of the enzyme, respectively, which permits cellular regulation of nitrogen metabolism (Brown et al. 1971). VopS, a T3SS effector protein from *V. parahaemolyticus*,

which causes gastroenteritis due to consumption of contaminated raw seafood, has been found to AMPylate the conserved threonine residue in the switch I region of Rho. The AMPylated small G protein can no longer bind to downstream substrates such as PAK and rhotekin, which results in disorganization of the actin cytoskeleton (Yarbrough et al. 2009). This benefits the pathogen because the host cell's actin assembly machinery is compromised, and it can no longer induce phagocytosis.

Bacterial pathogens use AMPylators to disrupt host signaling pathways to promote bacterial survival and replication (for review, see Woolery et al. 2010). IbpA secreted from *Histophilus somni*, like VopS, also modifies RhoA with AMP but on a tyrosine residue instead of a threonine in the switch I region (Worby et al. 2009). Both VopS and IpbA proteins contain a Fic domain (Filamentation induced by cAMP) that mediates this enzymatic activity. The Fic and doc domains share a conserved HPFx[D/E]GN[G/K]R motif, in which the invariant histidine residue is essential for AMPylation activity in Fic proteins (Luong et al. 2010) and cytotoxicity in doc (Garcia-Pino et al. 2008). Indeed, the two domains are grouped in the same family and classified as FIDO domains based on structural similarities (Kinch et al. 2009). Interestingly, Fic domains, found in bacteria, archaea, and metazoans, are thought to function as endogenous signaling elements and, therefore, are regulated so as not to cause harm to the host (Kinch et al. 2009; Engel et al. 2012). Indeed, 90% of Fic domains are regulated by an inhibitory α-helix that prevents constitutive binding of ATP, and binding of a specific substrate is predicted to relieve this inhibition to allow AMPylation (Engel et al. 2012). AMPylators are comparable to kinases in that they both hydrolyze ATP and reversibly transfer a part of this metabolite onto a hydroxyl side chain of the protein substrate.

A more specialized Fic domain can use the substrate CDP-choline, instead of ATP, in a phosphotransfer reaction, producing small G proteins modified by a phosphocholine moiety (Mukherjee et al. 2011). *Legionella*, causal agent of Legionnaires' disease, produces the effector AnkX, which contains an Fic domain that modifies Rab1 (a host small G protein involved in membrane transport) at its switch II region with phosphocholine. The modified Rab1 is no longer recognized by the host GEF, connecdenn, but can bind to and be activated by the *Legionella* GEF DrrA. The inactivation and activation of Rab1 on *Legionella*-containing vacuoles (LCVs) thus become dictated by the pathogen by posttranslational modifications of Rab GTPases.

The adenylyl transferase domain is part of the larger nucleotidyl transferase domain family and, like Fic domains, can catalyze AMPylation. It is characterized by a conserved $Gx_{11}DxD$ motif in which the aspartate residues are essential for the AMPylation activity (Jiang et al. 2007; Muller et al. 2010). This domain has been identified in more than 1400 bacterial proteins among 685 bacterial species, of which the large majority are proteobacteria (Finn et al. 2009).

DrrA (also known as SidM) is a virulence factor secreted from *Legionella*, an intracellular pathogen that survives in LCVs in the cell (Muller et al. 2010). *Legionella* uses a T4SS to secrete proteins from the LCV into the host cell. DrrA is composed of three domains: an amino-terminal adenylyl transferase domain, a GEF domain, and a carboxy-terminal phosphatidylinositol-4-phosphate-binding (P4M) domain (Murata et al. 2006; Brombacher et al. 2009). The first domain of DrrA very closely resembles the carboxy-terminal adenylyl transferase domain of *E. coli* GS-ATase and contains the conserved $Gx_{11}DxD$ motif (Muller et al. 2010). The GEF domain is capable of catalyzing the exchange of GDP for GTP on Rab1, which plays a role in the regulation of vesicular transport from the endoplasmic reticulum (Murata et al. 2006). The P4M domain anchors the effector to the cytoplasmic side of the LCV membrane (Brombacher et al. 2009). DrrA hijacks Rab1 by locking it into its GTP-bound active state, using both the DrrA GEF and adenylyl transferase domains. The GEF domain exchanges GDP for GTP, and the adenylyl transferase domain AMPylates Y77 in the switch II region of Rab1, which blocks its interaction with host GAPs, preventing the hydrolysis of GTP. AMPylated and prenylated GTP-Rab1 localizes with DrrA and targets ER vacuoles to the LCV (Muller et al. 2010). Another virulence factor from *Legionella*, SidD, contains a deAMPylating activity, allowing its recycling (Neunuebel et al. 2011; Tan and Luo 2011). This protein is a phosphodiesterase that, on the basis of structural analysis, resembles a protein phosphatase (Rigden 2011).

3.6 Bacterial Effectors Mimic Host Small G Proteins

Another strategy used by bacterial pathogens is to produce proteins that mimic small G proteins themselves. EHEC O157:H7 secretes the effector protein EspFu (also known as TccP) in order to create actin-based pedestals, allowing attachment of the bacterium to the intestinal epithelium (Campellone et al. 2004; Garmendia et al. 2004). The Wiskott-Aldrich syndrome protein (WASP) family of actin nucleators is normally regulated by Rac and Cdc42, which bind to the Cdc42/Rac1-interaction-binding domain (CRIB) motif within the G-protein-binding domain (GBD) of N-WASP (Ch. 8 [Devreotes and Horwitz 2014]). This releases N-WASP from an autoinhibited conformation that depends on interaction of the GBD and the verprolin-

homology, cofilin-homology, acidic (VCA) domain. Remarkably, EspFu mimics a 17-residue stretch of the VCA motif, thus competing for binding with the GBD, and thereby constitutively activates N-WASP and activates downstream actin polymerization (Cheng et al. 2008; Sallee et al. 2008).

EspFu is not the only bacterial effector to directly activate small G-protein substrates. A second EHEC O157:H7 effector, EspG, is an activator of class I PAK family serine/threonine kinases (Selyunin et al. 2011). Under normal circumstances, PAKs are responsible for regulating cytoskeletal dynamics and are activated by Rac1 or Cdc42 small G proteins in a manner similar to the activation of N-WASP discussed above. The PAKs contain an amino-terminal autoinhibitory domain and a carboxy-terminal kinase domain. Binding of activated Cdc42 or Rac1 to the CRIB domain within the autoinhibitory domain of PAK potently induces kinase activation. In contrast, EHEC EspG binds to and activates PAK but does not recognize the CRIB. Rather, EspG specifically interacts with the $I\alpha3$ helix, which sits upstream of the CRIB domain and serves two primary functions: (1) it occludes the substrate-binding site of the kinase domain, and (2) it positions an inhibitory loop into the kinase catalytic cleft. Binding of EspG to $I\alpha3$ is predicted to unfold the autoinhibitory domain and release the kinase inhibitory loop, leading to PAK activation. Both bacterial virulence factors, EspFu and EspG, thus mimic the basic principles of host small G-protein function to activate their downstream targets by separating the inhibitory domain from the activity-bearing domain, but they use distinct molecular mechanisms to achieve this result.

4 HIGHJACKING LIPID SIGNALING

Phosphoinositides, particularly phosphatidylinositol 4,5-bisphosphate (PIP$_2$), regulate the actin cytoskeleton beneath the plasma membrane, functioning in signaling as well as trafficking by targeting vesicles around the cell. Disruption of phosphoinositide homeostasis at the plasma membrane by bacterial effectors can destabilize actin dynamics and alter the morphology of the membrane. This facilitates the entry of intracellular pathogens or, in the case of extracellular pathogens, can disrupt membrane integrity, which leads to rapid cell lysis in the subsequent stage of infection to facilitate pathogen spreading (Ham et al. 2011).

4.1 The Inositol Polyphosphate 4-Phosphatase *Shigella* IpgD and *Salmonella* SopB Promote Pathogen Entry

IpgD is an effector from the facultative intracellular pathogen *Shigella* that is directly translocated into host cells through a T3SS upon contact with the cell surface (Niebuhr et al. 2000). IpgD is a 4-phosphoinositide phosphatase that hydrolyzes PIP_2 to produce phosphatidylinositol 5-phosphate [PI(5)P] (Niebuhr et al. 2002). Removal of PIP_2 by IpgD decreases the tethering of the plasma membrane to PIP_2-binding cytoskeleton-anchoring proteins, causing extension of membrane filopodia and massive cellular blebbing (observed as bubble-like protrusions) (Charras and Paluch 2008). This reorganization of the actin cytoskeleton at the bacterial entry site promotes the uptake of the pathogen by the host cells.

Like IpgD, the *Salmonella* effector protein SopB hydrolyzes PIP_2 to promote bacterial invasion and establish a niche for its vacuolar life cycle inside the host (Norris et al. 1998; Terebiznik et al. 2002; Hernandez et al. 2004). Thus, modulation of phosphoinositide metabolism appears to be a common strategy for bacterial pathogens to usurp signaling at plasma and vesicular membranes.

4.2 Inositol Polyphosphate 5-Phosphatase *V. parahaemolyticus* VPA0450 Promotes Blebbing

A similar molecular mechanism is used by the T3SS effector VPA0450 from the extracellular pathogen *V. parahaemolyticus* (Broberg et al. 2010). VPA0450 contains catalytic motifs that mimic the activity of the eukaryotic inositol polyphosphate 5-phosphatases (IPP5Cs), which hydrolyze PIP_2 at the membrane surface. In contrast to IpgD, VPA0450 hydrolyzes the D5 phosphate, producing PI(4)P. The removal of PIP_2 disrupts actin dynamics, causing the local detachment of the cortical cytoskeleton from the plasma membrane, which leads to extensive membrane blebbing. Whereas IpgD uses the same molecular mechanism to facilitate internalization of the bacteria, blebbing induced by VPA0450 instead accelerates lysis of the infected host cell (Broberg et al. 2011).

5 OTHER ACTIN REGULATORS TARGETED BY PATHOGENS

The actin cytoskeleton supports focal adhesions and controls cell contraction, cell motility, endocytosis, phagocytosis, and cell division. A characteristic feature of all of these processes is the dynamic transition of cellular actin between its monomeric (G) and polymeric (F) actin states, which is controlled by a myriad of regulatory proteins that act on distinct states of the actin polymer network. For example, the Arp2/3 complex nucleates filaments that grow from the side of existing filaments, creating branched networks, whereas formins and SPIRE nucleate unbranched filaments (Campellone and Welch 2010). A common mechanistic

feature of all three systems is the ability to assemble actin or actin-like proteins into an arrangement that can serve as a template for growth of a new filament. The Arp2/3 complex contains two actin-related subunits, which form a pseudo-actin trimer with an actin monomer provided by activators of the WASP family, such as N-WASP (see above). Formins bind two actin monomers and are thought to position them appropriately for filament growth. SPIRE proteins have multiple repeats of Wiskott-Aldrich homology 2 (WH2) domains that bind to actin and appear to create a three-actin template for filament extension. Bacteria and viruses can hijack these mechanisms to directly regulate actin nucleation, producing the characteristic pathogen motility observed in *Listeria*-, *Shigella*-, *Rickettsia*-, and *Vaccinia*-virus-infected cells. In addition, several bacterial species from the *Vibrio* genus translocate actin-elongation factors into host cells.

5.1 Pathogenic Actin Nucleation Factors

A few bacterial pathogens and viruses use actin polymerization to move around within and between cells. Bacteria from at least three genera (*Listeria*, *Shigella*, and *Rickettsia*) and *Vaccinia* virus all use membrane-anchored proteins to produce actin comet tails that propel the microorganism through the cytoplasm, along the surface of the cell, or through the plasma membrane into a neighboring cell. Although the proteins used by each of these pathogens are unique in structure and function, they all share the common feature of nucleating actin filaments de novo at the membrane surface. For example, ActA directly recruits and activates the Arp2/3 complex at the surface of *Listeria monocytogenes* (Welch et al. 1998). It has, in fact, been an essential tool for studies of various biological processes including cell motility and provided the first physiological evidence for the nucleating activity of the Arp2/3 complex (Welch et al. 1998). Like ActA, *Shigella* VirG/IcsA induces formation of actin comet tails, but this pathogen uses a distinct mechanism. Whereas ActA directly activates Arp2/3, VirG/IcsA on the bacterial surface can recruit N-WASP and induce actin nucleation via Arp2/3 (Egile et al. 1999). Finally, *Vaccinia* virus uses the membrane-anchored protein A36R to facilitate intracellular movement that is strikingly similar. A large domain of A36R on the viral surface is phosphorylated by Src-family tyrosine kinases and then directly interacts with the adaptor protein Nck and subsequently recruits N-WASP. These processes are all essential for the bacteria and viruses to invade systemic tissues of their host organism and therefore represent key virulence factors in a wide range of infectious diseases.

Extracellular pathogens including EPEC also hijack Arp2/3, albeit by a mechanism distinct from that described above. EPEC secretes a cell surface receptor, Tir, which embeds in the host plasma membrane and forms a complex with the bacterial adhesion molecule intimin (Kenny et al. 1997). This causes Tir to cluster at the cell surface, resulting in tyrosine phosphorylation of its cytoplasmic tail and subsequent recruitment of Nck via its SH2 domain (Gruenheid et al. 2001). This recruits N-WASP and the Arp2/3 complex to nucleate branched actin filaments at the EPEC–host interface, resulting in the formation of pedestals. These molecular events are instrumental in the tight attachment of EPEC to the intestinal epithelial cell wall and also induce the characteristic attaching and effacing (A/E) lesion that defines EPEC infections.

5.2 Pathogenic Elongation Factors: *Vibrio* VopL/F

In addition to inducing branched actin networks through activation of Arp2/3, *V. parahaemolyticus* produces VopL, which has three closely spaced WH2 domains that bind actin (Fig. 4) (Liverman et al. 2007). VopL directly induces the nucleation of actin independently of any other eukaryotic factor and is more efficient than its eukaryotic counterparts (Namgoong et al. 2011; Yu et al. 2011). Interspersed with the WH2 domains are three proline-rich motifs (PRMs). PRMs have many potential interacting partners, including WW domains, SH3 domains, and the actin-binding protein profilin. The PRMs in VopL closely resemble those found in the FH1 domains of formins, which are known to bind profilin and profilin–actin complexes (Holt and Koffer 2001). *V. parahaemolyticus* uses VopL to induce unregulated production of stress fibers and thereby disrupts actin homeostasis in the epithelial cells of the gut during infection, resulting in an enterotoxic effect in the intestine. Another T3SS virulence factor from *V. cholera*, VopF, contains a similar WH2/PRM domain architecture and also promotes actin assembly independently of host proteins. VopF induces the formation of small actin protrusions, rather than stress fibers, and may help efficient colonization during infection.

6 TARGETING UBIQUITIN-MEDIATED SIGNAL TRANSDUCTION

Evolutionarily conserved ubiquitylation machinery regulates a diverse set of cellular processes, including development, transcription, replication, cell signaling, and immune function (Pickart 2004; Mukhopadhyay and Riezman 2007; Ch. 2 [Lee and Yaffe 2014]). The versatility of this system to reversibly modify protein function makes it an attractive target for a wide range of pathogens, including viruses and bacteria. These microbes are particularly adept at coopting the ubiquitylation machinery. Indeed, ubiqui-

tin itself is encoded by a large number of viral genomes, and many viruses and bacteria encode the ubiquitin ligases or adaptor proteins required for ubiquitin posttranslational modification (Randow and Lehner 2009; Collins and Brown 2010).

Ubiquitin can be covalently linked to protein substrates as either a single molecule (monoubiquitylation) or a polypeptide chain (polyubiquitylation) (Ch. 2 [Lee and Yaffe 2014]). It is first activated by a ubiquitin-activating enzyme, E1, which involves an ATP-dependent transfer of ubiquitin to the enzyme's catalytic cysteine residue. It is then transferred to the active-site cysteine of an E2 ubiquitin-conjugating enzyme. The ubiquitin residue on the charged E2 enzyme is then targeted to substrates via an E3 ligase. K48-linked chains of ubiquitin mark the substrate for proteasomal degradation. In contrast, monoubiquitylation or K63-linked chains serve as regulatory signals in signal transduction, membrane trafficking, DNA repair, and chromatin remodeling. Below, we highlight a few specific examples of pathogens exploiting the ubiquitin system. These interactions have not only informed us regarding microbial pathogenesis, but also continue to reveal novel mechanisms of ubiquitin regulation in cell signaling.

Many viruses, including baculoviruses, poxviruses, and herpes simplex virus, encode their own ubiquitin molecules but have significantly altered the ubiquitin gene. Human ubiquitin shares only 75% similarity with baculovirus ubiquitin, compared with 96% similarity with yeast ubiquitin (Haas et al. 1996). The viral ubiquitylation machinery may therefore function differently from the host ubiquitylation machinery. Many bacteria also secrete enzymes that modify host ubiquitin or ubiquitin-like molecules (UBLs). Recent studies of EPEC revealed that host ubiquitin is deamidated on Q40 by the bacterial type III effector Cif (Cui et al. 2010). Similarly, the Cif homolog CHBP encoded by *Burkholderia pseudomallei* deamidates Q40 both on ubiquitin and the UBL Nedd8 (Cui et al. 2010; Jubelin et al. 2010; Morikawa et al. 2010). These posttranslational modifications potently inhibit polyubiquitin chain synthesis, resulting in accumulation of host substrates and severe cytopathic effects.

Viruses and bacteria can also encode their own E3 ubiquitin ligases or adaptor proteins that link host E3 enzymes to specific host substrates. Most known eukaryotic E3 ligases belong to one of three types: RING, HECT, and U-box. There are currently no known viral HECT family E3 ubiquitin ligases. Instead, viruses encode RING family or unconventional E3 ligases. Two examples of RING type E3 ligases are the RING-CH family and the Infected Cell Protein 0 (ICP0) family. Initially identified in the murine and human gammaherpes viruses, respectively, these virulence factors down-regulate immune cell surface receptors (Coscoy and Ganem 2000; Ishido et al. 2000; Stevenson et al.

2000; Haque et al. 2001). For example, the K3 and K5 gene products of Kaposi's sarcoma–associated herpesvirus (KSHV) provide immune protection by ubiquitylating MHC class I molecules that present antigen at the cell surface, targeting them for endocytosis and lysosomal degradation. In contrast, ICP0 of herpes simplex virus type I (HSV-1) is required for reactivation of latency and suppression of innate immunity (Everett 2000). The RING domain of ICP0 promotes the accumulation of ubiquitylated proteins and their subsequent proteasomal degradation. In particular, it causes the degradation of RNF8 and RNF168, host cell E3 ligases that are essential for the cellular response to DNA damage. By degrading these proteins, ICP0 blocks the cellular DNA damage response that HSV infection activates, which would otherwise shut off viral transcription.

Another well-characterized example of a virally encoded E3 ligase is adenovirus E4orf6, which, together with adenovirus E1B-55K, substitutes for the substrate recognition subunits of the cullin-EloB-C core complex (Querido et al. 2001a). This host–pathogen complex forms a novel ubiquitin ligase that targets the tumor suppressor p53, Mre11, and the BLM helicase to abrogate the cellular DNA damage response during viral infection (Dobner et al. 1996; Querido et al. 2001b).

The KSHV immediate-early transcription factor RTA shows unconventional E3 ubiquitin ligase activity that targets host immune protein IRF7 for proteasomal degradation (Yu et al. 2005). Similarly, the ubiquitin ligase domain of the IpaH family of bacterial type III effectors is structurally distinct from both the HECT and RING families (Rohde et al. 2007; Singer et al. 2008; Zhu et al. 2008). However, like the HECT-type E3 ligases, IpaH transfers ubiquitin from UbcH5 E2 to substrates by forming a ubiquitin thioester intermediate at a conserved cysteine residue. A series of leucine-rich repeats (LRRs) in IpaH and its family members is responsible for recognizing a diverse array of host substrates and targeting these substrates for ubiquitylation.

Pathogens may also encode adaptor proteins that link E3 ligases to target substrates. In a classic example, the E6 oncoprotein encoded by human papillomavirus (HPC) facilitates ubiquitylation and proteasomal degradation of p53 (Scheffner et al. 1993; Huibregtse et al. 1995). The dimeric E6 forms a complex with human E6-AP, the founding member of the HECT-type E3 ubiquitin ligase family (Huibregtse et al. 1995). The E6–E6-AP complex binds to and targets p53 for ubiquitin-dependent proteolysis, thus interfering with the growth-regulating activities of this tumor suppressor. These discoveries have provided essential insights into cancer caused by high-risk HPV, and have defined an entire class of E3 ubiquitin ligases involved in a myriad of biological processes.

7 CONCLUSION

Virulence factors produced by pathogens have evolved to efficiently manipulate host signaling pathways (Table 1). Mechanisms range from constitutive activation of a pathway, to irreversible inactivation of a critical signaling molecule, to subversion of a whole signaling system to favor the invading pathogen. A major challenge in the future is to determine the enzymatic activities and host substrates for the bacterial and viral virulence factors that show no obvious homology to eukaryotic proteins. Another, even more complex challenge is to understand how these factors work together to orchestrate a successful infection. Temporal and spatial considerations are extremely important for regulating a host cell during infection. Likewise, within the pathogen, determining the regulatory mechanisms that control the activation patterns and spatial dynamics of virulence factors will help reveal how microbial pathogens coopt signal transduction systems during infection. Finally, the use of model organisms to complement studies in mammalian cells will provide valuable insights into the physiological roles of bacterial effector proteins. Such information is essential to gain a system-level view of the infectious disease process and to ultimately design therapeutics that target host–pathogen interactions. Inevitably, by discovering the mechanisms of pathogenic effectors, we have gained a greater understanding into critical steps in eukaryotic signaling. Although a great deal has been learned, given the number and diversity of the yet-to-be-studied bacterial and viral pathogens, much more is left to be discovered.

REFERENCES

*Reference is in this book.

Alto NM, Shao F, Lazar CS, Brost RL, Chua G, Mattoo S, McMahon SA, Ghosh P, Hughes TR, Boone C, et al. 2006. Identification of a bacterial type III effector family with G protein mimicry functions. *Cell* **124:** 133–145.

Andor A, Trulzsch K, Essler M, Roggenkamp A, Wiedemann A, Heesemann J, Aepfelbacher M. 2001. YopE of *Yersinia*, a GAP for Rho GTPases, selectively modulates Rac-dependent actin structures in endothelial cells. *Cell Microbiol* **3:** 301–310.

Black DS, Bliska JB. 2000. The RhoGAP activity of the *Yersinia pseudotuberculosis* cytotoxin YopE is required for antiphagocytic function and virulence. *Mol Microbiol* **37:** 515–527.

Broberg CA, Zhang L, Gonzalez H, Laskowski-Arce MA, Orth K. 2010. A *Vibrio* effector protein is an inositol phosphatase and disrupts host cell membrane integrity. *Science* **329:** 1660–1662.

Broberg CA, Calder TJ, Orth K. 2011. *Vibrio parahaemolyticus* cell biology and pathogenicity determinants. *Microbes Infect* **13:** 992–1001.

Brombacher E, Urwyler S, Ragaz C, Weber SS, Kami K, Overduin M, Hilbi H. 2009. Rab1 guanine nucleotide exchange factor SidM is a major phosphatidylinositol 4-phosphate-binding effector protein of *Legionella pneumophila. J Biol Chem* **284:** 4846–4856.

Brown MS, Segal A, Stadtman ER. 1971. Modulation of glutamine synthetase adenylylation and deadenylylation is mediated by metabolic transformation of the P II -regulatory protein. *Proc Natl Acad Sci* **68:** 2949–2953.

Buchwald G, Friebel A, Galan JE, Hardt WD, Wittinghofer A, Scheffzek K. 2002. Structural basis for the reversible activation of a Rho protein by the bacterial toxin SopE. *EMBO J* **21:** 3286–3295.

Campellone KG, Welch MD. 2010. A nucleator arms race: Cellular control of actin assembly. *Nat Rev* **11:** 237–251.

Campellone KG, Robbins D, Leong JM. 2004. EspFU is a translocated EHEC effector that interacts with Tir and N-WASP and promotes Nck-independent actin assembly. *Dev Cell* **7:** 217–228.

Chardin P, Boquet P, Madaule P, Popoff MR, Rubin EJ, Gill DM. 1989. The mammalian G protein rhoC is ADP-ribosylated by *Clostridium botulinum* exoenzyme C3 and affects actin microfilaments in Vero cells. *EMBO J* **8:** 1087–1092.

Charras G, Paluch E. 2008. Blebs lead the way: How to migrate without lamellipodia. *Nat Rev* **9:** 730–736.

Cheng HC, Skehan BM, Campellone KG, Leong JM, Rosen MK. 2008. Structural mechanism of WASP activation by the enterohaemorrhagic *E. coli* effector EspF(U). *Nature* **454:** 1009–1013.

Coburn J, Gill DM. 1991. ADP-ribosylation of p21^{ras} and related proteins by *Pseudomonas aeruginosa* exoenzyme S. *Infect Immun* **59:** 4259–4262.

Coburn J, Wyatt RT, Iglewski BH, Gill DM. 1989. Several GTP-binding proteins, including p21c–H-ras, are preferred substrates of *Pseudomonas aeruginosa* exoenzyme S. *J Biol Chem* **264:** 9004–9008.

Collier RJ. 2009. Membrane translocation by anthrax toxin. *Mol Aspects Med* **30:** 413–422.

Collins CA, Brown EJ. 2010. Cytosol as battleground: Ubiquitin as a weapon for both host and pathogen. *Trends Cell Biol* **20:** 205–213.

Coscoy L, Ganem D. 2000. Kaposi's sarcoma-associated herpesvirus encodes two proteins that block cell surface display of MHC class I chains by enhancing their endocytosis. *Proc Natl Acad Sci* **97:** 8051–8056.

Cui J, Yao Q, Li S, Ding X, Lu Q, Mao H, Liu L, Zheng N, Chen S, Shao F. 2010. Glutamine deamidation and dysfunction of ubiquitin/NEDD8 induced by a bacterial effector family. *Science* **329:** 1215–1218.

* Devreotes P, Horwitz AR. 2014. Signaling networks that regulate cell migration. *Cold Spring Harb Perspect Biol* doi: 10.1101/cshperspect.a005959.

Dobner T, Horikoshi N, Rubenwolf S, Shenk T. 1996. Blockage by adenovirus E4orf6 of transcriptional activation by the p53 tumor suppressor. *Science* **272:** 1470–1473.

Egile C, Loisel TP, Laurent V, Li R, Pantaloni D, Sansonetti PJ, Carlier MF. 1999. Activation of the CDC42 effector N-WASP by the *Shigella flexneri* IcsA protein promotes actin nucleation by Arp2/3 complex and bacterial actin-based motility. *J Cell Biol* **146:** 1319–1332.

Engel P, Goepfert A, Stanger FV, Harms A, Schmidt A, Schirmer T, Dehio C. 2012. Adenylylation control by intra- or intermolecular active-site obstruction in Fic proteins. *Nature* **482:** 107–110.

Etienne-Manneville S, Hall A. 2002. Rho GTPases in cell biology. *Nature* **420:** 629–635.

Everett RD. 2000. ICP0, a regulator of herpes simplex virus during lytic and latent infection. *Bioessays* **22:** 761–770.

Finn RD, Mistry J, Tate J, Coggill P, Heger A, Pollington JE, Gavin OL, Gunasekaran P, Ceric G, Forslund K, et al. 2009. The Pfam protein families database. *Nucleic Acids Res* **38:** D211–D222.

Flatau G, Lemichez E, Gauthier M, Chardin P, Paris S, Fiorentini C, Boquet P. 1997. Toxin-induced activation of the G protein p21 Rho by deamidation of glutamine. *Nature* **387:** 729–733.

Fu Y, Galan JE. 1999. A *Salmonella* protein antagonizes Rac-1 and Cdc42 to mediate host-cell recovery after bacterial invasion. *Nature* **401:** 293–297.

Garcia-Pino A, Christensen-Dalsgaard M, Wyns L, Yarmolinsky M, Magnuson RD, Gerdes K, Loris R. 2008. Doc of prophage P1 is inhibited by its antitoxin partner Phd through fold complementation. *J Biol Chem* **283:** 30821–30827.

Garmendia J, Phillips AD, Carlier MF, Chong Y, Schuller S, Marches O, Dahan S, Oswald E, Shaw RK, Knutton S, et al. 2004. TccP is an

enterohaemorrhagic *Escherichia coli* O157:H7 type III effector protein that couples Tir to the actin-cytoskeleton. *Cell Microbiol* **6:** 1167–1183.

Goehring UM, Schmidt G, Pederson KJ, Aktories K, Barbieri JT. 1999. The N-terminal domain of *Pseudomonas aeruginosa* exoenzyme S is a GTPase-activating protein for Rho GTPases. *J Biol Chem* **274:** 36369–36372.

Gruenheid S, DeVinney R, Bladt F, Goosney D, Gelkop S, Gish GD, Pawson T, Finlay BB. 2001. Enteropathogenic *E. coli* Tir binds Nck to initiate actin pedestal formation in host cells. *Nat Cell Biol* **3:** 856–859.

Guan KL, Dixon JE. 1990. Protein tyrosine phosphatase activity of an essential virulence determinant in *Yersinia*. *Science* **249:** 553–556.

Haas AL, Katzung DJ, Reback PM, Guarino LA. 1996. Functional characterization of the ubiquitin variant encoded by the baculovirus *Autographa californica*. *Biochemistry* **35:** 5385–5394.

Ham H, Sreelatha A, Orth K. 2011. Manipulation of host membranes by bacterial effectors. *Nat Rev Microbiol* **9:** 635–646.

Hao YH, Wang Y, Burdette D, Mukherjee S, Keitany G, Goldsmith E, Orth K. 2008. Structural requirements for *Yersinia* YopJ inhibition of MAP kinase pathways. *PLoS ONE* **3:** e1375.

Haque M, Ueda K, Nakano K, Hirata Y, Parravicini C, Corbellino M, Yamanishi K. 2001. Major histocompatibility complex class I molecules are down-regulated at the cell surface by the K5 protein encoded by Kaposi's sarcoma-associated herpesvirus/human herpesvirus-8. *J Gen Virol* **82:** 1175–1180.

Hardt WD, Chen LM, Schuebel KE, Bustelo XR, Galan JE. 1998. *S. typhimurium* encodes an activator of Rho GTPases that induces membrane ruffling and nuclear responses in host cells. *Cell* **93:** 815–826.

Hayes CS, Aoki SK, Low DA. 2010. Bacterial contact-dependent delivery systems. *Annu Rev Genet* **44:** 71–90.

Henkel JS, Baldwin MR, Barbieri JT. 2010. Toxins from bacteria. *EXS* **100:** 1–29.

Henriksson ML, Sundin C, Jansson AL, Forsberg A, Palmer RH, Hallberg B. 2002. Exoenzyme S shows selective ADP-ribosylation and GTPase-activating protein (GAP) activities towards small GTPases in vivo. *Biochem J* **367:** 617–628.

Hernandez LD, Hueffer K, Wenk MR, Galan JE. 2004. *Salmonella* modulates vesicular traffic by altering phosphoinositide metabolism. *Science* **304:** 1805–1807.

Holt MR, Koffer A. 2001. Cell motility: Proline-rich proteins promote protrusions. *Trends Cell Biol* **11:** 38–46.

Huang Z, Sutton SE, Wallenfang AJ, Orchard RC, Wu X, Feng Y, Chai J, Alto NM. 2009. Structural insights into host GTPase isoform selection by a family of bacterial GEF mimics. *Nat Struct Mol Biol* **16:** 853–860.

Huibregtse JM, Scheffner M, Beaudenon S, Howley PM. 1995. A family of proteins structurally and functionally related to the E6-AP ubiquitin-protein ligase. *Proc Natl Acad Sci* **92:** 2563–2567.

Iriarte M, Cornelis GR. 1998. YopT, a new *Yersinia* Yop effector protein, affects the cytoskeleton of host cells. *Mol Microbiol* **29:** 915–929.

Ishido S, Wang C, Lee BS, Cohen GB, Jung JU. 2000. Downregulation of major histocompatibility complex class I molecules by Kaposi's sarcoma-associated herpesvirus K3 and K5 proteins. *J Virol* **74:** 5300–5309.

Jiang P, Mayo AE, Ninfa AJ. 2007. *Escherichia coli* glutamine synthetase adenylyltransferase (ATase, EC 2.7.7.49): Kinetic characterization of regulation by PII, PII-UMP, glutamine, and α-ketoglutarate. *Biochemistry* **46:** 4133–4146.

Jubelin G, Taieb F, Duda DM, Hsu Y, Samba-Louaka A, Nobe R, Penary M, Watrin C, Nougayrede JP, Schulman BA, et al. 2010. Pathogenic bacteria target NEDD8-conjugated cullins to hijack host-cell signaling pathways. *PLoS Pathog* **6:** e1001128.

Juris SJ, Rudolph AE, Huddler D, Orth K, Dixon JE. 2000. A distinctive role for the *Yersinia* protein kinase: Actin binding, kinase activation, and cytoskeleton disruption. *Proc Natl Acad Sci* **97:** 9431–9436.

Kenny B, DeVinney R, Stein M, Reinscheid DJ, Frey EA, Finlay BB. 1997. Enteropathogenic *E. coli* (EPEC) transfers its receptor for intimate adherence into mammalian cells. *Cell* **91:** 511–520.

Kinch LN, Yarbrough ML, Orth K, Grishin NV. 2009. Fido, a novel AMPylation domain common to fic, doc, and AvrB. *PLoS ONE* **4:** e5818.

Klink BU, Barden S, Heidler TV, Borchers C, Ladwein M, Stradal TE, Rottner K, Heinz DW. 2010. Structure of *Shigella* IpgB2 in complex with human RhoA: Implications for the mechanism of bacterial guanine nucleotide exchange factor mimicry. *J Biol Chem* **285:** 17197–17208.

Kubori T, Galan JE. 2003. Temporal regulation of *Salmonella* virulence effector function by proteasome-dependent protein degradation. *Cell* **115:** 333–342.

* Lee MJ, Yaffe MB. 2014. Protein regulation in signal transduction. *Cold Spring Harb Perspect Biol* doi: 10.1101/cshperspect.a005918.

Li H, Xu H, Zhou Y, Zhang J, Long C, Li S, Chen S, Zhou JM, Shao F. 2007. The phosphothreonine lyase activity of a bacterial type III effector family. *Science* **315:** 1000–1003.

* Lim K-H, Staudt LM. 2013. Toll-like receptor signaling. *Cold Spring Harb Perspect Biol* **5:** a011247.

Liverman AD, Cheng HC, Trosky JE, Leung DW, Yarbrough ML, Burdette DL, Rosen MK, Orth K. 2007. Arp2/3-independent assembly of actin by *Vibrio* type III effector VopL. *Proc Natl Acad Sci* **104:** 17117–17122.

Luong P, Kinch LN, Brautigam CA, Grishin NV, Tomchick DR, Orth K. 2010. Kinetic and structural insights into the mechanism of AMPylation by VopS Fic domain. *J Biol Chem* **285:** 20155–20163.

Miraglia AG, Travaglione S, Meschini S, Falzano L, Matarrese P, Quaranta MG, Viora M, Fiorentini C, Fabbri A. 2007. Cytotoxic necrotizing factor 1 prevents apoptosis via the Akt/IκB kinase pathway: Role of nuclear factor-κB and Bcl-2. *Mol Biol Cell* **18:** 2735–2744.

Mittal R, Peak-Chew SY, McMahon HT. 2006. Acetylation of MEK2 and IκB kinase (IKK) activation loop residues by YopJ inhibits signaling. *Proc Natl Acad Sci* **103:** 18574–18579.

Mohr C, Koch G, Just I, Aktories K. 1992. ADP-ribosylation by *Clostridium botulinum* C3 exoenzyme increases steady-state GTPase activities of recombinant rhoA and rhoB proteins. *FEBS Lett* **297:** 95–99.

Morikawa H, Kim M, Mimuro H, Punginelli C, Koyama T, Nagai S, Miyawaki A, Iwai K, Sasakawa C. 2010. The bacterial effector Cif interferes with SCF ubiquitin ligase function by inhibiting deneddylation of Cullin1. *Biochem Biophys Res Commun* **401:** 268–274.

* Morrison DK. 2012. MAP kinase pathways. *Cold Spring Harb Perspect Biol* **4:** a011254.

Mukherjee S, Keitany G, Li Y, Wang Y, Ball HL, Goldsmith EJ, Orth K. 2006. *Yersinia* YopJ acetylates and inhibits kinase activation by blocking phosphorylation. *Science* **312:** 1211–1214.

Mukherjee S, Hao YH, Orth K. 2007. A newly discovered post-translational modification—The acetylation of serine and threonine residues. *Trends Biochem Sci* **32:** 210–216.

Mukherjee S, Liu X, Arasaki K, McDonough J, Galan JE, Roy CR. 2011. Modulation of Rab GTPase function by a protein phosphocholine transferase. *Nature* **477:** 103–106.

Mukhopadhyay D, Riezman H. 2007. Proteasome-independent functions of ubiquitin in endocytosis and signaling. *Science* **315:** 201–205.

Muller MP, Peters H, Blumer J, Blankenfeldt W, Goody RS, Itzen A. 2010. The *Legionella* effector protein DrrA AMPylates the membrane traffic regulator Rab1b. *Science* **329:** 946–949.

Murata T, Delprato A, Ingmundson A, Toomre DK, Lambright DG, Roy CR. 2006. The *Legionella pneumophila* effector protein DrrA is a Rab1 guanine nucleotide-exchange factor. *Nat Cell Biol* **8:** 971–977.

Namgoong S, Boczkowska M, Glista MJ, Winkelman JD, Rebowski G, Kovar DR, Dominguez R. 2011. Mechanism of actin filament nucleation by *Vibrio* VopL and implications for tandem W domain nucleation. *Nat Struct Mol Biol* **18:** 1060–1067.

Navarro L, Koller A, Nordfelth R, Wolf-Watz H, Taylor S, Dixon JE. 2007. Identification of a molecular target for the *Yersinia* protein kinase A. *Mol Cell* **26:** 465–477.

Neunuebel MR, Chen Y, Gaspar AH, Backlund PSJr, Yergey A, Machner MP. 2011. De-AMPylation of the small GTPase Rab1 by the pathogen *Legionella pneumophila*. *Science* **333:** 453–456.

* Newton K, Dixit VM. 2012. Signaling in innate immunity and inflammation. *Cold Spring Harb Perspect Biol* **4:** a006049.

Niebuhr K, Jouihri N, Allaoui A, Gounon P, Sansonetti PJ, Parsot C. 2000. IpgD, a protein secreted by the type III secretion machinery of *Shigella flexneri*, is chaperoned by IpgE and implicated in entry focus formation. *Mol Microbiol* **38:** 8–19.

Niebuhr K, Giuriato S, Pedron T, Philpott DJ, Gaits F, Sable J, Sheetz MP, Parsot C, Sansonetti PJ, Payrastre B. 2002. Conversion of PtdIns(4,5)P(2) into PtdIns(5)P by the *S. flexneri* effector IpgD reorganizes host cell morphology. *EMBO J* **21:** 5069–5078.

Norris FA, Wilson MP, Wallis TS, Galyov EE, Majerus PW. 1998. SopB, a protein required for virulence of *Salmonella dublin*, is an inositol phosphate phosphatase. *Proc Natl Acad Sci* **95:** 14057–14059.

Offermanns S, Toombs CF, Hu YH, Simon MI. 1997. Defective platelet activation in Gα(q)-deficient mice. *Nature* **389:** 183–186.

Orth K, Palmer LE, Bao ZQ, Stewart S, Rudolph AE, Bliska JB, Dixon JE. 1999. Inhibition of the mitogen-activated protein kinase kinase superfamily by a *Yersinia* effector. *Science* **285:** 1920–1923.

Orth K, Xu Z, Mudgett MB, Bao ZQ, Palmer LE, Bliska JB, Mangel WF, Staskawicz B, Dixon JE. 2000. Disruption of signaling by *Yersinia* effector YopJ, a ubiquitin-like protein protease. *Science* **290:** 1594–1597.

Pickart CM. 2004. Back to the future with ubiquitin. *Cell* **116:** 181–190.

Prehna G, Ivanov MI, Bliska JB, Stebbins CE. 2006. *Yersinia* virulence depends on mimicry of host Rho-family nucleotide dissociation inhibitors. *Cell* **126:** 869–880.

Querido E, Blanchette P, Yan Q, Kamura T, Morrison M, Boivin D, Kaelin WG, Conaway RC, Conaway JW, Branton PE. 2001a. Degradation of p53 by adenovirus E4orf6 and E1B55K proteins occurs via a novel mechanism involving a Cullin-containing complex. *Genes Dev* **15:** 3104–3117.

Querido E, Morrison MR, Chu-Pham-Dang H, Thirlwell SW, Boivin D, Branton PE. 2001b. Identification of three functions of the adenovirus e4orf6 protein that mediate p53 degradation by the E4orf6–E1B55K complex. *J Virol* **75:** 699–709.

Randow F, Lehner PJ. 2009. Viral avoidance and exploitation of the ubiquitin system. *Nat Cell Biol* **11:** 527–534.

Rigden DJ. 2011. Identification and modelling of a PPM protein phosphatase fold in the *Legionella pneumophila* deAMPylase SidD. *FEBS Lett* **585:** 2749–2754.

Rohde JR, Breitkreutz A, Chenal A, Sansonetti PJ, Parsot C. 2007. Type III secretion effectors of the IpaH family are E3 ubiquitin ligases. *Cell Host Microbe* **1:** 77–83.

Sallee NA, Rivera GM, Dueber JE, Vasilescu D, Mullins RD, Mayer BJ, Lim WA. 2008. The pathogen protein EspF(U) hijacks actin polymerization using mimicry and multivalency. *Nature* **454:** 1005–1008.

* Sassone-Corsi P. 2012. The cyclic AMP pathway. *Cold Spring Harb Perspect Biol* **4:** a011148.

Scheffner M, Huibregtse JM, Vierstra RD, Howley PM. 1993. The HPV-16 E6 and E6-AP complex functions as a ubiquitin-protein ligase in the ubiquitination of p53. *Cell* **75:** 495–505.

Schmidt G, Sehr P, Wilm M, Selzer J, Mann M, Aktories K. 1997. Gln 63 of Rho is deamidated by *Escherichia coli* cytotoxic necrotizing factor-1. *Nature* **387:** 725–729.

Sehr P, Joseph G, Genth H, Just I, Pick E, Aktories K. 1998. Glucosylation and ADP ribosylation of rho proteins: Effects on nucleotide binding, GTPase activity, and effector coupling. *Biochemistry* **37:** 5296–5304.

Selyunin AS, Sutton SE, Weigele BA, Reddick LE, Orchard RC, Bresson SM, Tomchick DR, Alto NM. 2011. The assembly of a GTPase-kinase signalling complex by a bacterial catalytic scaffold. *Nature* **469:** 107–111.

Shao F, Merritt PM, Bao Z, Innes RW, Dixon JE. 2002. A *Yersinia* effector and a *Pseudomonas* avirulence protein define a family of cysteine proteases functioning in bacterial pathogenesis. *Cell* **109:** 575–588.

Shao F, Vacratsis PO, Bao Z, Bowers KE, Fierke CA, Dixon JE. 2003. Biochemical characterization of the Yersinia YopT protease: Cleavage site and recognition elements in Rho GTPases. *Proc Natl Acad Sci* **100:** 904–909.

Singer AU, Rohde JR, Lam R, Skarina T, Kagan O, Dileo R, Chirgadze NY, Cuff ME, Joachimiak A, Tyers M, et al. 2008. Structure of the *Shigella* T3SS effector IpaH defines a new class of E3 ubiquitin ligases. *Nat Struct Mol Biol* **15:** 1293–1301.

Stebbins CE, Galan JE. 2000. Modulation of host signaling by a bacterial mimic: Structure of the *Salmonella* effector SptP bound to Rac1. *Mol Cell* **6:** 1449–1460.

Stevenson PG, Efstathiou S, Doherty PC, Lehner PJ. 2000. Inhibition of MHC class I-restricted antigen presentation by γ2-herpesviruses. *Proc Natl Acad Sci* **97:** 8455–8460.

Takai Y, Sasaki T, Matozaki T. 2001. Small GTP-binding proteins. *Physiol Rev* **81:** 153–208.

Tan Y, Luo ZQ. 2011. *Legionella pneumophila* SidD is a deAMPylase that modifies Rab1. *Nature* **475:** 506–509.

Terebiznik MR, Vieira OV, Marcus SL, Slade A, Yip CM, Trimble WS, Meyer T, Finlay BB, Grinstein S. 2002. Elimination of host cell PtdIns(4,5)P(2) by bacterial SigD promotes membrane fission during invasion by *Salmonella*. *Nat Cell Biol* **4:** 766–773.

Trosky JE, Mukherjee S, Burdette DL, Roberts M, McCarter L, Siegel RM, Orth K. 2004. Inhibition of MAPK signaling pathways by VopA from *Vibrio parahaemolyticus*. *J Biol Chem* **279:** 51953–51957.

Trosky JE, Li Y, Mukherjee S, Keitany G, Ball H, Orth K. 2007. VopA inhibits ATP binding by acetylating the catalytic loop of MAPK kinases. *J Biol Chem* **282:** 34299–34305.

Upadhyay A, Wu HL, Williams C, Field T, Galyov EE, van den Elsen JM, Bagby S. 2008. The guanine-nucleotide-exchange factor BopE from *Burkholderia pseudomallei* adopts a compact version of the *Salmonella* SopE/SopE2 fold and undergoes a closed-to-open conformational change upon interaction with Cdc42. *Biochem J* **411:** 485–493.

Von Pawel-Rammingen U, Telepnev MV, Schmidt G, Aktories K, Wolf-Watz H, Rosqvist R. 2000. GAP activity of the *Yersinia* YopE cytotoxin specifically targets the Rho pathway: A mechanism for disruption of actin microfilament structure. *Mol Microbiol* **36:** 737–748.

Welch MD, Rosenblatt J, Skoble J, Portnoy DA, Mitchison TJ. 1998. Interaction of human Arp2/3 complex and the *Listeria monocytogenes* ActA protein in actin filament nucleation. *Science* **281:** 105–108.

Woolery AR, Luong P, Broberg CA, Orth K. 2010. AMPylation: Something old is new again. *Front Microbiol* **1:** 113.

Worby CA, Mattoo S, Kruger RP, Corbeil LB, Koller A, Mendez JC, Zekarias B, Lazar C, Dixon JE. 2009. The Fic domain: Regulation of cell signaling by adenylylation. *Mol Cell* **34:** 93–103.

Yarbrough ML, Li Y, Kinch LN, Grishin NV, Ball HL, Orth K. 2009. AMPylation of Rho GTPases by *Vibrio* VopS disrupts effector binding and downstream signaling. *Science* **323:** 269–272.

Yu Y, Wang SE, Hayward GS. 2005. The KSHV immediate-early transcription factor RTA encodes ubiquitin E3 ligase activity that targets IRF7 for proteosome-mediated degradation. *Immunity* **22:** 59–70.

Yu B, Cheng HC, Brautigam CA, Tomchick DR, Rosen MK. 2011. Mechanism of actin filament nucleation by the bacterial effector VopL. *Nat Struct Mol Biol* **18:** 1068–1074.

Zhu Y, Li H, Hu L, Wang J, Zhou Y, Pang Z, Liu L, Shao F. 2008. Structure of a *Shigella* effector reveals a new class of ubiquitin ligases. *Nat Struct Mol Biol* **15:** 1302–1308.

CHAPTER 21

Signal Transduction in Cancer

Richard Sever[1] and Joan S. Brugge[2]

[1]Cold Spring Harbor Laboratory, Cold Spring Harbor, New York 11724
[2]Harvard Medical School, Department of Cell Biology, Boston, Massachusetts 02115

Correspondence: joan_brugge@hms.harvard.edu

SUMMARY

Cancer is driven by genetic and epigenetic alterations that allow cells to overproliferate and escape mechanisms that normally control their survival and migration. Many of these alterations map to signaling pathways that control cell growth and division, cell death, cell fate, and cell motility, and can be placed in the context of distortions of wider signaling networks that fuel cancer progression, such as changes in the tumor microenvironment, angiogenesis, and inflammation. Mutations that convert cellular proto-oncogenes to oncogenes can cause hyperactivation of these signaling pathways, whereas inactivation of tumor suppressors eliminates critical negative regulators of signaling. An examination of the PI3K-Akt and Ras-ERK pathways illustrates how such alterations dysregulate signaling in cancer and produce many of the characteristic features of tumor cells.

Outline

1 INTRODUCTION

The development of cancer involves successive genetic and epigenetic alterations that allow cells to escape homeostatic controls that ordinarily suppress inappropriate proliferation and inhibit the survival of aberrantly proliferating cells outside their normal niches. Most cancers arise in epithelial cells, manifesting as carcinomas in organs such as the lung, skin, breast, liver, and pancreas. Sarcomas, in contrast, arise from mesenchymal tissues, occurring in fibroblasts, myocytes, adipocytes, and osteoblasts. Nonepithelial tumors can also develop in cells of the nervous system (e.g., gliomas, neuroblastomas, and medulloblastomas) and hematopoietic tissues (leukemia and lymphoma).

In solid tumors, these alterations typically promote progression from a relatively benign group of proliferating cells (hyperplasias) to a mass of cells with abnormal morphology, cytological appearance, and cellular organization. After a tumor expands, the tumor core loses access to oxygen

and nutrients, often leading to the growth of new blood vessels (angiogenesis), which restores access to nutrients and oxygen. Subsequently, tumor cells can develop the ability to invade the tissue beyond their normal boundaries, enter the circulation, and seed new tumors at other locations (metastasis), the defining feature of malignancy (Fig. 1). This linear sequence of events is clearly an oversimplification of complex cancer-associated events that proceed in distinct ways in individual tumors and between tumor sites; however, it provides a useful framework in which to highlight the critical role of dysregulated signaling in processes associated with the initiation and progression of cancer.

The root cause of cancer is usually genetic or epigenetic alterations in the tumor cells (see below). Progression of the cancer, however, is associated with a complex interplay between the tumor cells and surrounding non-neoplastic cells and the extracellular matrix (ECM). Moreover, the tumor cells develop several well-defined features (Hanahan and Weinberg 2000; Solimini et al. 2007). In addition to

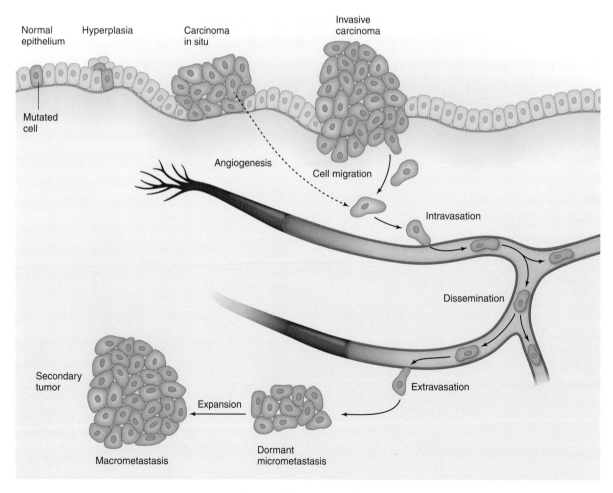

Figure 1. Cancer progression.

increased cell proliferation, these include resistance to apoptosis and other forms of cell death, metabolic changes, genetic instability, induction of angiogenesis, and increased migratory capacity. Dysregulation of cellular signal transduction pathways underlies most of these characteristics.

Here, we describe how tumor cells co-opt signaling pathways to allow them to proliferate, survive, and invade other tissues. To cover all of the signaling molecules involved and their myriad contributions to cancer would require an entire textbook (Weinberg 2013). We therefore focus primarily on two pathways—Ras-ERK (p. 81 [Morrison 2012]) and PI3K-Akt signaling (p. 87 [Hemmings and Restuccia 2012])—that play central roles in multiple processes associated with cancer, while highlighting the involvement of some other key signaling molecules.

2 MUTATIONS AS THE CAUSE OF CANCER

Most tumors arise as a consequence of genetic alterations to cellular genes, which may be inherited or arise spontaneously, for example, as a result of DNA damage induced by environmental carcinogens or mutations arising from replication errors. These alterations confer a selective advantage to the cells, which together with changes in the microenvironment, promote tumor growth and progression. Some are gain-of-function mutations, producing so-called oncogenes that drive tumor formation. Others inactivate tumor suppressor genes that normally ensure that cells do not proliferate inappropriately or survive outside their normal niche.

Tumors can possess tens to hundreds or even thousands of mutations, but many of these are merely so-called "passengers." Typically only two to eight are the "driver mutations" that cause progression of the cancer (Vogelstein et al. 2013). These may be point mutations (such as G12V Ras), deletions (as seen with PTEN), inversions, or amplifications (as seen with Myc). Large-scale rearrangements also occur, for example, the *BCR-ABL* fusions involving chromosomes 9 and 22, which are associated with several leukemias and generate an oncogenic version of the tyrosine kinase Abl. Loss of heterozygosity due to gene conversion or mitotic recombination between normal and mutant parental alleles is another source of genetic alterations that drive cancer. This often affects tumor suppressors such as the retinoblastoma protein (pRB) and p53 (encoded by the *TP53* gene in humans).

Changes in the methylation state of promoters of genes that impact cancer can also play an important role in oncogenesis (Sandoval and Esteller 2012; Suva et al. 2013). Indeed, epigenetic silencing is more common than mutational silencing for some genes, for example, the cyclin-dependent kinase (CDK) inhibitor (CKI) p16 (also known

as CDKN2A or INK4a) and the mismatch repair (MMR) enzyme MLH1. Silencing of MMR enzymes can lead to additional genetic changes because it affects proteins that prevent errors by repairing DNA. Conversely, several mutations associated with cancer affect epigenetic regulators that influence multiple cellular programs, for example, DNMT1 and TET1, which control DNA methylation, and the histone-modifying enzymes EZH2, SETD2, and KDM6A are deleted or mutated in cancer (Delhommeau et al. 2009; Ley et al. 2010; Wu et al. 2012). Interestingly, mutations in the metabolic enzymes isocitrate dehydrogenase (IDH) 1 and IDH2 may promote cancer by generating an "oncometabolite" not present in normal cells that inhibits certain chromatin-modifying enzymes (see below) (Ward and Thompson 2012a).

Finally, in a minority of cancers, infectious agents are the triggers. A few human cancers are triggered by viruses that encode genes that promote tumorigenesis through activation of oncogene pathways or inactivation of tumor suppressors. The human papilloma virus, which is associated with cervical and head and neck cancers, encodes a protein, E6, that promotes degradation of p53, while another viral protein, E7, inactivates pRB and CKIs, among other effects (Munger and Howley 2002). In hepatocellular carcinoma caused by hepatitis B virus, in contrast, it is not clear whether viral proteins themselves are oncogenic, viral integration promotes expression of nearby cellular oncogenes, or cancer is simply a consequence of persistent liver injury and inflammation (Seeger et al. 2013). Epstein−Barr virus (also known as human herpes virus 4) produces a protein called LMP1 that acts as a constitutively active tumor necrosis factor (TNF) receptor, engaging a plethora of signaling pathways including NF-κB, JNK/p38, PI3K, and ERK (Morris et al. 2009). An extreme case of a transmissible cancer is that affecting the Tasmanian devil. All tumors are derived from a founder tumor and are transmitted as allografts from devil to devil during intraspecies facial biting (Murchison et al. 2012; Hamede et al. 2013).

2.1 Cancer-Causing Mutations Affect Signaling Pathways

We can connect the genetic alterations in cancer cells with signaling pathways that control processes associated with tumorigenesis and place these in the context of distortions of wider signaling networks that fuel cancer progression. In each case, the result is dysregulated signaling that is not subject to the normal control mechanisms.

Oncogenic mutations can cause the affected genes to be overexpressed (e.g., gene amplification) or produce mutated proteins whose activity is dysregulated (e.g., point mutations, truncations, and fusions). Examples include proteins involved in signaling pathways that are commonly activated in many physiological responses, such as growth factor receptor tyrosine kinases (RTKs; e.g., the epidermal growth factor receptor, EGFR), small GTPases (e.g., Ras), serine/threonine kinases (e.g., Raf and Akt), cytoplasmic tyrosine kinases (e.g., Src and Abl), lipid kinases (e.g., phosphoinositide 3-kinases, PI3Ks), as well as nuclear receptors (e.g., the estrogen receptor, ER). Components of developmental signaling pathways, such as Wnt, Hedgehog (Hh), Hippo, and Notch can also be affected, as can downstream nuclear targets of signaling pathways, for example, transcription factors (e.g., Myc and NF-κB), chromatin remodelers (e.g., EZH2), and cell cycle effectors (e.g., cyclins).

Alternatively, deletions and other mutations can inactivate negative regulators that normally function as tumor suppressors. Indeed, one of the most commonly mutated genes in cancer is the tumor suppressor p53, the so-called "guardian of the genome." p53 is a critical hub that controls cell proliferation and stress signals such as apoptosis and DNA damage responses (see below). pRB and CKIs such as p16 are other tumor suppressors whose mutation deregulates the cell cycle. Many tumor suppressors function as negative regulators of cytoplasmic signaling, for example, the adenomatous polyposis protein (APC) is a negative regulator of the Wnt pathway, and the lipid phosphatase PTEN is a negative regulator of the PI3K-Akt pathway.

It is worth noting that hyperactivated oncogene pathways can also induce a state of irreversible cell cycle arrest termed senescence (Gorgoulis and Halazonetis 2010; Vargas et al. 2012). This is believed to represent a fail-safe mechanism to inhibit proliferation caused by aberrant activation of oncoproteins in normal cells and is accompanied by changes in cellular structure, chromatin organization, DNA damage, cytokine secretion, and gene expression. Oncogenic transformation requires alterations that abrogate senescence, such as loss of p53 or PTEN.

2.1.1 The PI3K-Akt and Ras-ERK Pathways as Examples of Oncogenic Signaling Pathways

Many of the genes commonly mutated in cancer encode components or targets of the PI3K-Akt and Ras-ERK pathways (Fig. 2). Ordinarily these pathways are transiently activated in response to growth factor or cytokine signaling and ligand occupancy of integrin adhesion receptors, but genetic alterations can lead to constitutive signaling even in the absence of growth factors. The PI3K-Akt pathway can be activated through amplification or activating mutations affecting several PI3K-Akt-pathway proteins—the type I PI3K isoform PIK3CA (p110a), Akt, and the adaptor protein PIK3R1—or through deletion or inactivating mu-

tations in the phosphatases that hydrolyze PI3K products, such as phosphatidylinositol 3,4,5-trisphosphate, the PTEN, and INPP4B tumor suppressors. Further downstream, mutations in the tumor suppressors TSC1 and TSC2 hyperactivate signaling by mTORC1 (p. 91 [Laplante and Sabatini 2012]), an important target of PI3K-Akt signaling. Similarly, the Ras-ERK pathway is activated by mutations in Ras, or its downstream target Raf, that cause constitutive activation of these proteins or by inactivation of GTPase-activating proteins (GAPs), such as NF1 (Cichowski and Jacks 2001), DAB2IP (Min et al. 2010), and RASAL2 (McLaughlin et al. 2013) that stimulate the hydrolysis of GTP bound to Ras that leads to its inactivation. The transcription factor Myc is an important downstream target of Ras-ERK signaling and many other pathways. It is frequently amplified or overexpressed in cancer; interestingly, Myc can not only bind to promoter regions of genes, but also enhance transcriptional elongation of polymerase II, thus extending its effects beyond genes with Myc-binding sites in their promoters. Myc can thus serve as a universal amplifier of expressed genes rather than merely binding to promoters and initiating transcription de novo (Rahl et al. 2010; Lin et al. 2012; Nie et al. 2012).

Oncogenic mutations, amplification, or gene fusions involving upstream tyrosine kinases lead to constitutive signaling through both the Ras-ERK and PI3K-Akt pathways. RTKs including EGFR, ErbB2, fibroblast growth factor receptor (FGFR), and platelet-derived growth factor receptor (PDGFR) are mutated or amplified in a variety of cancers. Similarly, oncogenic mutations in G-protein-coupled receptors (GPCRs) can also activate these pathways.

Finally, it is important to recognize that deregulated synthesis of growth factors themselves plays an important role in many cancers. Inappropriate synthesis of growth factors by cells expressing the appropriate receptor can generate an autocrine loop driving signaling. This can also be achieved through cleavage and release of anchored soluble growth factors by surface ADAM proteases, which are activated downstream from oncogenic signaling pathways (Turner et al. 2009). Alternatively, the growth factor may be synthesized by a neighboring cell (paracrine stimulation). In both cases, signaling via the Ras-ERK and PI3K-Akt pathways may be increased.

3 DYSREGULATION OF CELLULAR PROCESSES BY ONCOGENIC SIGNALING

How, then, does dysregulation of cellular signaling drive cancer progression and produce the characteristic features of tumor cells mentioned above? Below we discuss the role of signal transduction in cancer-associated processes, sur-

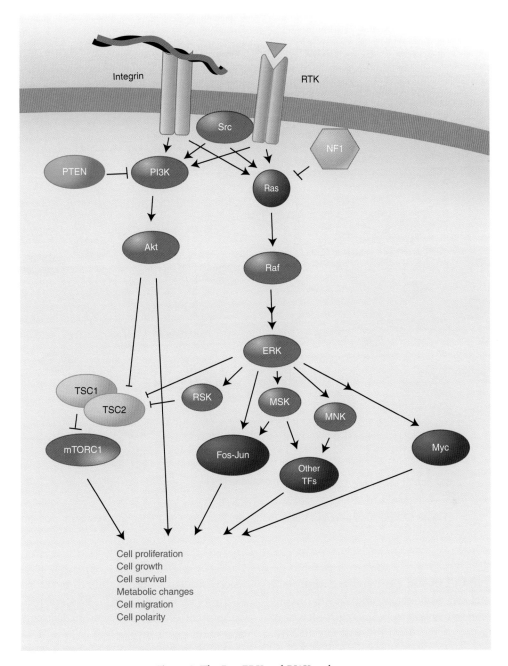

Figure 2. The Ras-ERK and PI3K pathways.

veying the major signals involved and focusing on Ras-ERK and PI3K-Akt signaling to illustrate how their targets influence the behavior of the tumor cells.

4 CELL PROLIFERATION

Excessive cell proliferation is a feature of most cancers. Limited availability of growth factors or nutrients, contact inhibition, and other feedback mechanisms ensure that the pathways that regulate proliferation (see Fig. 3) are normal-

ly tightly controlled. As outlined above, however, mutations in proto-oncogenes and tumor suppressors or inappropriate synthesis of ligands/receptors can hyperactivate these pathways, leading to activation of the cell cycle machinery. Note that signaling targets that represent critical components of cell cycle control mechanism can also undergo genetic alterations in cancer; for example, the genes encoding cyclin D, cyclin E, and CDK4 are amplified in certain cancers and the G1 restriction point inhibitor pRB and p16 can be deleted or mutated as well.

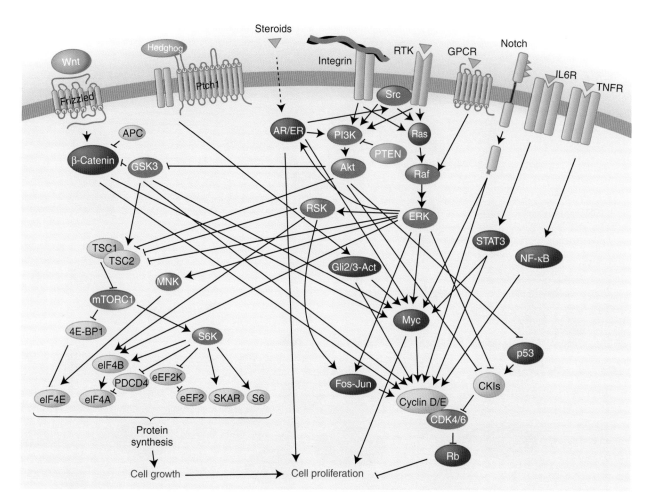

Figure 3. Regulation of cell proliferation by the Ras-ERK and PI3K-Akt pathways.

The Ras-ERK and PI3K-Akt pathways are important regulators of normal cell proliferation and thus their constitutive hyperactivation can lead to excessive proliferation. One important target of the Ras-ERK pathway is Myc, which is phosphorylated by ERK; this leads to its stabilization by suppression of ubiquitylation (Sears et al. 2000). Myc stimulates cell proliferation by inducing numerous genes that promote cell proliferation, including those encoding G_1/S cyclins, CDKs, and the E2F-family transcription factors that drive the cell cycle (Ch. 5 [Duronio and Xiong 2013]). In addition, it represses expression of various cell cycle inhibitors (e.g., CKIs), blocks the activity of transcription factors that promote differentiation (see below), induces genes that enhance translation, and shifts cells to anabolic metabolism. ERK also phosphorylates numerous other transcription factors important for cell proliferation. Elk1, for example, in combination with the SRF transcription factor, induces the immediate early gene FOS,

whose product is also stabilized by ERK phosphorylation (Murphy et al. 2002). FOS, also an oncogene, encodes a component of the transcription factor AP1, which regulates many genes involved in cell proliferation.

Multiple kinases in the ribosomal S6 kinase (RSK), mitogen- and stress-activated kinase (MSK), and mitogen-activated protein kinase (MAPK)-interacting kinase (MNK) families are also phosphorylated by ERK, and these kinases, in turn, phosphorylate transcription factors that regulate cell cycle progression, for example, FOS and CREB (Roux and Blenis 2004). MSKs represent the predominant kinases responsible for the nucleosomal response involving phosphorylation of histone H3 at S10, which is commonly induced by mitotic stimuli (Soloaga et al. 2003). MNKs play an important role regulating translation following mitogenic stimulation by phosphorylating the translation initiation factor eIF4E, and loss of the MNK phosphorylation site completely abrogates its ability to transform cell lines or promote tumors in animal models (Soloaga et al. 2003). Activation of RSK family members by ERK also leads to activation of the mammalian target of the rapamycin (mTOR) pathway through TSC2 phosphorylation and relief of mTOR inhibition. In addition, RSK regulates translation by phosphorylating eIF4B, which increases its interaction with the translation initiation factor eIF3. The promotion of translation by these mechanisms is important for cell growth and, consequently, cell proliferation.

PI3K-Akt signaling controls cell proliferation at various levels. Akt regulates cell growth during cell cycle progression by controlling mTORC1. It inhibits the GAP activity of the TSC1–TSC2 complex toward Rheb, thus allowing GTP-bound Rheb to activate mTORC1. This then phosphorylates eIF4-binding protein, releasing the eIF4E cap-binding factor and allowing it to bind mRNAs, and p70 RSK. This promotes increased protein synthesis, which is critical for enhanced cell growth during cell cycle progression (Richardson et al. 2004). Akt also phosphorylates the kinase GSK3, inhibiting its catalytic activity. Phosphorylation of cyclin D and Myc by GSK3 targets them for degradation; thus, inhibition of this kinase by Akt causes stabilization of these important cell cycle regulators (Diehl et al. 1997; Sears et al. 2000).

In addition, Akt inhibits several cell cycle inhibitors, such as the CKIs p27 (also known as KIP1) and p21 (also known as CIP1); phosphorylation leads to their sequestration in the cytoplasm by 14-3-3 proteins. In the case of p27, phosphorylation also targets it for degradation. Akt-mediated phosphorylation of p21 prevents it from forming a complex with proliferating cell nuclear antigen (PCNA) to inhibit DNA replication, reduces its binding to CDK2/CDK4, and attenuates its inhibitory activity toward CDK2 (Rossig et al. 2001). Furthermore, Akt blocks FoxO-dependent transcription of cell cycle inhibitors such as p27 and RBL2 (retinoblastoma-like protein 2) (Burgering and Medema 2003). It also phosphorylates and activates MDM2 (Ogawara et al. 2002), a ubiquitin ligase that promotes degradation of p53, thereby releasing a key brake on the cell cycle. Later on in the cell cycle, Akt can regulate several enzymes involved in the G_2/M transition (Xu et al. 2012b).

Phosphorylation and consequent inhibition of GSK3 by Akt may, in certain contexts, lead to stabilization and nuclear translocation of the Wnt target β-catenin (Haq et al. 2003; Korkaya et al. 2009; Ma et al. 2013), a transcriptional regulator whose degradation would otherwise be promoted by GSK3 (Polakis 2001; Korkaya et al. 2009). This leads to induction of β-catenin target genes that regulate proliferation, including those encoding Myc and cyclin D. Akt can also phosphorylate β-catenin directly, causing its dissociation from cadherin cell–cell adhesion complexes (see below), thus increasing the pool of β-catenin available and its transcriptional activity (Fang et al. 2007).

Numerous other signaling pathways can, of course, drive cell proliferation in cancer. Cytokine and RTK signaling, for example, activate STAT3, which stimulates synthesis of Myc and cyclin D (p. 117 [Harrison 2012]). Notch, Wnt/β-catenin, and Hedgehog, all of which have been implicated in cancer, also induce Myc and cyclin D (see below). Similarly, the transcription factor NF-κB, which can be activated by TNF and various other signals, also targets cyclin D expression. Cyclin E is induced by several of these signals. Estrogen signaling (see p. 129 [Sever and Glass 2013]) stimulates cell proliferation via activation of the ERα subtype, which induces cyclin D and Myc. Disruption of the balance between ERα and ERβ or mutations in *ERα* that yield truncated proteins or activated proteins can dysregulate this pathway (Thomas and Gustafsson 2011; Li et al. 2013; Robinson et al. 2013; Toy et al. 2013). Note that signaling through ERs and the androgen receptor (AR) is coupled to and enhanced by Ras-ERK and PI3K-Akt signaling (Castoria et al. 2004; Renoir et al. 2013). Growth factor stimulation (e.g., EGF and insulin-like growth factor, IGF) and mutations that activate these pathways increase proliferation of ER/AR-dependent tumors. In addition, these steroid receptors form cytoplasmic complexes with Src and PI3K, which leads to activation of their downstream effectors, and ERK can phosphorylate ERα, which causes its activation in the absence of ligand and stimulation of cell proliferation.

The tumor suppressors that normally hold proliferative signaling in check are obviously also critical. Furthest downstream, pRB normally directly inhibits the transcriptional activity of the E2F proteins until it is deactivated through phosphorylation by CDKs. p53, in contrast, normally blocks cell proliferation in response to stress signals such as DNA damage by inhibiting CDK activity via induction of CKIs. Consequently, mutations in this tumor suppressor deregulate cell proliferation under potentially dangerous, cancer-promoting conditions. The CKIs themselves directly inhibit CDKs and are also inactivated by mutation in many cancers, p16 being the most common example. Further upstream are pathway-specific tumor suppressors, such as the Ras-GAP NF1 and APC, which block Wnt/β-catenin signaling (by promoting GSK3 phosphorylation and, consequently, ubiquitin-dependent destruction of β-catenin). In each case, mutation of the tumor suppressor removes an important brake, allowing cells to proliferate despite signals that would ordinarily restrain them. The Hippo pathway plays a critical role in regulating contact inhibition of proliferation (p. 133

[Harvey and Hariharan 2012]), and disruption of this pathway, which suppresses the transcriptional coactivator YAP, is emerging as a key tumor suppressor pathway in many cancers (Harvey et al. 2013; Lin et al. 2013; Yu and Guan 2013). The Ras-ERK and PI3K-Akt pathways intersect with Hippo pathway components to inactivate its tumor suppressive activity (O'Neill and Kolch 2005; Kim et al. 2010; Collak et al. 2012).

5 CELL SURVIVAL

Cell death functions as a homeostatic mechanism that normally controls cell number. It is also a built-in cancer-protection mechanism that is activated during initial stages of oncogenesis because of stresses associated with unbalanced proliferative signals, excessive cell proliferation, loss of anchorage to natural niches, etc. Mutations that disable cell-death signaling can thus play an important role in cancer. Overexpression of the antiapoptotic protein Bcl2, for example, can occur as a consequence of chromosomal rearrangements in B lymphocytes, and this contributes to follicular lymphoma by preventing cells from undergoing apoptosis. p53 also regulates apoptosis, both by inducing transcription of proapoptotic regulators and binding directly to the proapoptotic protein Bax (Ch. 19 [Green and Llambi 2014]). Loss of this tumor suppressor through mutation can therefore contribute to cancer by reducing cell death, as well as disabling normal cell cycle control. Other cell death regulators that are mutated in cancer include the proapo-ptotic proteins Puma and Bok (which are frequently deleted) and the antiapoptotic proteins Mcl1 and Bcl-xL (whose genes are amplified).

Control of proapoptotic regulators (e.g., Bim and Bad) and antiapoptotic regulators (e.g., Bcl2 and Mcl1) in normal cells ensures that cells undergo apoptosis in the absence of appropriate signals supplied by growth factors or the tissue microenvironment. Hyperactivation of signaling by oncogenic mutations in the Ras-ERK and PI3K-Akt pathways, however, disrupts the balance in favor of antiapoptotic signals, thus contributing to tumor cell survival and abnormal expansion of the cells beyond normal tissue boundaries.

The PI3K-Akt and Ras-ERK pathways regulate cell death in multiple ways (Fig. 4) (review Cagnol and Chambard 2010; Zhang et al. 2011). Akt itself intervenes at several steps in apoptotic signaling from death receptors. It phosphorylates forkhead-family transcription factors such as FoxO3A, which leads to their cytoplasmic sequestration by 14-3-3 proteins, thereby blocking induction of death ligands (e.g., FasL and TRAIL) and the proapoptotic Bcl2-family member Bim. Akt and the ERK-regulated kinase RSK also phosphorylate the proapoptotic Bcl2-family protein Bad,

another target for sequestration by 14-3-3 proteins. In addition, Akt phosphorylates and thereby activates the apoptosis inhibitor XIAP. Akt also activates NF-κB, which regulates multiple survival factors, including antiapoptotic proteins (Bcl2, BCLxl, Mcl1) and the intracellular death receptor inhibitor FLIP (Shen and Tergaonkar 2009). Last, Akt-induced ubiquitylation and degradation of p53 suppresses p53-induced apoptosis (Ogawara et al. 2002).

ERK phosphorylates Bim and the NF-κB inhibitor IκBα (Ghoda et al. 1997), which targets them for degradation. In addition, RSK phosphorylates the caspase-9 scaffolding protein APAF, which impedes the ability of cytochrome c to nucleate apoptosome formation and activate the downstream caspases that drive apoptosis (Kim et al. 2012).

6 CELL METABOLISM

Cell growth needs to be coordinated with metabolic processes involved in the synthesis of macromolecules. Thus, growth factor pathways that regulate both normal and tumor cells impinge on metabolic pathways to program cells to meet the increased need for synthesis of macromolecules to produce new daughter cells (Ch. 7 [Ward and Thompson 2012b]). Activation of oncogenes and loss of tumor suppressors can directly regulate components of metabolic pathways even in the absence of growth factors and, thereby, produce similar metabolic alterations (Fig. 5).

The most common metabolic alteration in cancer cells is increased glucose uptake and glycolysis. At first glance, this might appear a disadvantage because glycolysis generates less ATP than oxidative phosphorylation; however, it allows cells to redirect carbon skeletons from glycolysis to anabolic reactions, such as the pentose phosphate pathway, which leads to nucleotide synthesis and regulates redox homeostasis. These also include the serine/glycine synthesis pathway, which generates several amino acids and charges tetrahydrofolate with a methyl group that is used in pyrimidine synthesis and leads to generation of S-adenosylmethionine, the methyl donor for multiple cellular methyltransferase reactions and methylation of essential molecules such as DNA, RNA, proteins, phospholipids, creatine, and neurotransmitters. Cancer cells show increased glutamine uptake and glutaminolysis to support oxidative phosphorylation and biosynthesis of proteins, lipids, and nucleic acids. They also up-regulate lipid synthesis by redirecting citrate from the Krebs cycle to fatty acid synthesis.

The PI3K-Akt pathway targets numerous substrates to promote these metabolic changes (Plas and Thompson 2005). Regulation of glucose transport and hexokinase by Akt promotes glycolysis, leading to generation of nucleo-

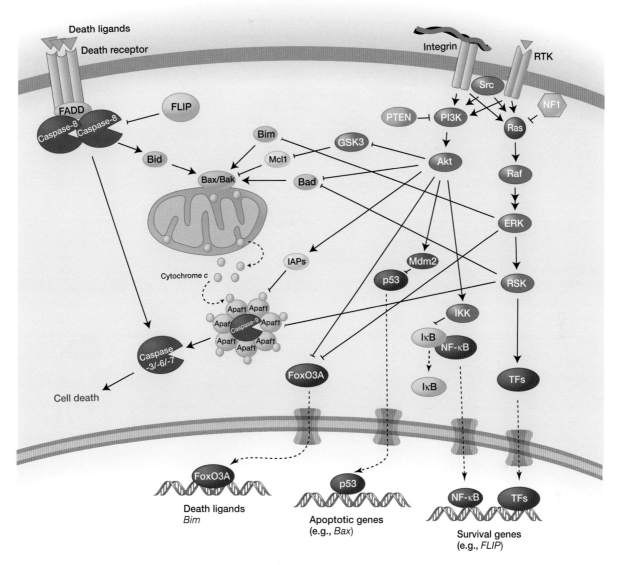

Figure 4. Regulation of cell death by Ras-ERK and PI3K-Akt pathways.

tides and amino acids necessary for cell growth (Engelman et al. 2006). Akt2 regulates glucose transport through multiple mechanisms. Regulation of the glucose transporter GLUT4 by Akt2 is critical for circulating glucose homeostasis. The Akt substrate AS160 plays an undefined role in insulin-stimulated GLUT4 translocation and glucose transport through its Rab-GTPase-activating domain (Miinea et al. 2005), and phosphorylation of the protein synip by Akt2 triggers its dissociation from the trafficking regulator syntaxin 4 and assembly of a protein complex that mediates translocation of GLUT4 vesicles to the plasma membrane (Yamada et al. 2005). Akt2 also regulates transcription, accumulation (Barthel et al. 1999; Jensen et al. 2010), and trafficking of GLUT1, which is the principle

glucose transporter expressed in most cell types (Wieman et al. 2007). Phosphorylation of TSC2 by Akt affects metabolism through mTORC1-mediated regulation of glycolysis; however, the mechanism of regulation is not known. mTORC1 may regulate glycolysis by increasing translation of glycolytic enzymes or their transcriptional regulators, such as Myc (Kim et al. 2004; Sutrias-Grau and Arnosti 2004). Other Akt targets activated by phosphorylation are hexokinase II, whose association with mitochondria is increased (Roberts et al. 2013), and 6-phosphofructo-2-kinase/fructose-2,6-bisphosphatase (Novellasdemunt et al. 2013). Both stimulate glycolysis.

mTORC1 signaling leads to increased synthesis of the transcription factor hypoxia-inducible factor (HIF1). HIF1

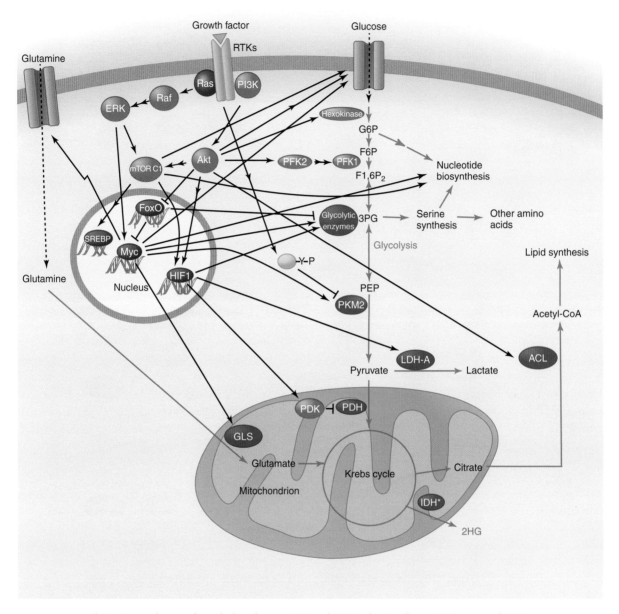

Figure 5. Regulation of metabolism by Ras-ERK and PI3K-Akt signaling. 1DH*, mutated 1DH.

induces glycolytic enzymes and lactate dehydrogenase (LDH-A), providing another means of stimulating glycolysis. In addition, it induces pyruvate dehydrogenase kinase (PDK), which inhibits pyruvate dehydrogenase (PDH) in the mitochondrion and thereby reduces flux from glycolysis into the Krebs cycle. mTORC1 can also stimulate pyrimidine biosynthesis via S6K1 (Ben-Sahra et al. 2013; Nakashima et al. 2013).

Akt/mTORC1 promotes lipid synthesis by activating the transcription factor sterol-response element-binding protein 1 (SREBP), a key regulator of lipid synthesis that is required for tumorigenicity (Bakan and Laplante 2012; Jeon and Osborne 2012; Guo et al. 2013). Loss of SREBP uncouples fatty acid synthase activity from stearoyl-CoA-desaturase-1-mediated desaturation. Another direct target of Akt is ATP-citrate lyase (ACL), an enzyme that converts citric acid to acetyl-CoA, which is required for fatty acid, cholesterol, and isoprenoid synthesis. mTORC1 also regulates amino acid uptake by stimulating translocation of amino acid transporters from intracellular vesicles to the plasma membrane (Berwick et al. 2002; Edinger and Thompson 2002).

Another family of Akt targets that affect cellular and organismal metabolism is FoxO transcription factors.

These are negatively regulated by Akt phosphorylation, which causes their sequestration in the cytoplasm by 14-3-3 proteins. Programs regulated by FoxO transcription factors that increase the cellular capacity for oxidative metabolism are, thus, shut off by active Akt.

Ras-ERK signaling exerts many of its effects on metabolism via Myc. Myc regulates glucose uptake, glycolysis, and the pentose phosphate pathway (Ying et al. 2012) and induces synthesis of glutamine transporters and the enzyme glutaminase (GLS), which converts glutamine into glutamate that can be metabolized in mitochondria (Miller et al. 2012; Dang 2013). It also induces enzymes involved in nucleotide and amino acid synthesis.

The glycolytic enzyme pyruvate kinase is of particular interest in cancer cells. Although glycolysis rates are usually much higher than in noncancer cells, most cancer cells produce an alternative splice form of pyruvate kinase (PKM2) that is less active than the enzyme (PKM1) found in most terminally differentiated cells (Vander Heiden et al. 2009). PKM1 remains active under most physiological conditions, but PKM2 can be turned off by signaling via tyrosine kinases, including the upstream RTKs in the Ras-ERK and PI3K-Akt pathways (Christofk et al. 2008; Hitosugi et al. 2009) and reactive oxygen species (ROS) (Anastasiou et al. 2011). Cancer cells can thus redirect the flux of glycolytic intermediates into anabolic pathways for ribose, serine, and glycine production or production of NADPH and glutathione needed to combat oxidative stress. PKM2 can also enter the nucleus and play a role in gene expression (Luo et al. 2011; Gao et al. 2012). However, deletion of PKM2 accelerates rather than impairs breast tumor formation, which indicates that it is the ability to turn off PKM2 activity that is most critical for tumor growth (Israelsen et al. 2013).

Clearly oncogene activation or loss of tumor suppressors such as PTEN and NF1 can drive these metabolic changes by dysregulating PI3K-Akt and Ras-ERK signaling. Other tumor suppressors also control cell metabolism, however. p53, for example, down-regulates glycolysis by inducing TIGAR, an enzyme that decreases the levels of the glycolytic activator fructose 2,6-bisphosphate. It also stimulates expression of SCO2, which is required for assembly of cytochrome c oxidase and promotes oxidative phosphorylation. Loss of p53 may therefore contribute to the glycolytic phenotype of cancer cells. p53 also regulates glutaminase 2, a metabolic enzyme that controls production of glutamate, which is converted to α-ketoglutarate for mitochondrial respiration and, importantly, glutathione, a critical cellular antioxidant (Hu et al. 2010; Suzuki et al. 2010). Loss of p53 leads to increased levels of ROS and oxidative damage. p53 also regulates the mevalonate pathway that controls cholesterol synthesis and generates inter-

mediates needed for protein geranylgeranylation and farnesylation. Drugs that target this pathway to control cholesterol/ cardiovascular disease have been shown to suppress tumor growth (Shibata et al. 2004; Kubatka et al. 2011).

Similarly, loss of the tumor suppressor LKB1 can lead to metabolic alterations. LKB1 activates AMP-activated protein kinase (AMPK), which acts as a cellular energy regulator and inhibits mTORC1 (Ch. 14 [Hardie 2012]). Loss of LKB1 relieves this inhibition, allowing mTORC1 to promote protein synthesis and lipogenesis. Activators of AMPK, such as metformin, are currently being used in diabetes and cancer therapy.

Finally, mutations associated with cancer can lead to the elevation of metabolites uniquely elevated in cancer cells (Kaelin and McKnight 2013). For example, mutations in IDH1 and IDH2 result in the production of 2-hydroxyglutarate (2HG), a metabolite not present at significant levels in normal cells. 2HG inhibits α-ketoglutarate-dependent enzymes such as the TET family, which regulate DNA methylation, and Jumonji C domain histone demethylases. This leads to epigenetic dysregulation that can drive tumorigenesis. Other oncometabolites may include succinate and fumarate, whose levels can increase because of mutations in succinate dehydrogenase and fumarate hydratase. Both can inhibit the activity of prolyl hydroxylases that control HIF levels, leading to induction of PDK and the other glycolytic enzymes mentioned above.

7 CELL POLARITY AND MIGRATION

As tumors progress toward malignancy, the cancer cells frequently become more migratory and develop the capacity to invade surrounding tissue. This is usually accompanied by changes in adhesion, cell polarity, cytoskeletal dynamics, and morphology. Migration is regulated by growth factors, chemokines, adhesion receptors, and other stimuli (Vicente-Manzanares and Horwitz 2011; Ch. 8 [Devreotes and Horwitz 2014]), many of which are targets for dysregulated signaling in cancer. The PI3K-Akt and Ras-ERK pathways regulate migration and invasion through multiple downstream effectors, including the following (Cain and Ridley 2009):

1. Rho-family GTPases (RhoA, Rac1, Cdc42, ARF6), which control cytoskeletal regulators such as WAVE/WASP-family members, the Arp2/3 complex, formins, the actomyosin contractile machinery, the kinase LIMK, and cofilins (Raftopoulou and Hall 2004);

2. Integrins and associated matrix adhesion proteins (e.g., FAK, paxillin, and calpains) (Ch. 8 [Devreotes and Horwitz 2014]);

3. Extracellular proteases, which degrade ECM proteins, facilitating tumor cell invasion by creating space for cells to move and reducing adhesive contacts that may constrain them, and also release various bioactive molecules anchored in the ECM (see below);

4. Cell–cell adhesion complexes, whose components are regulated through modulation of their stability or protein interactions that affects the strength of adhesion;

5. Transcription factors such as AP1 and Ets2 that regulate expression of many proteins that control migration/ polarity, including matrix metalloproteinases (MMPs), plasminogen activator, cadherins, and actin regulators.

As with other processes regulated by oncoprotein signaling, the outcome of alterations in these pathways is highly context and isoform dependent. For example, Akt1 specifically suppresses migration in many contexts through inhibition of ERK, the transcription factor NFAT, TSC2, or phosphopalladin-induced actin bundling, whereas Akt2 promotes migration through regulation of integrin expression and effects on the epithelial–mesenchymal transition (EMT) (see below; Chin and Toker 2011). Similarly, some isoforms of ERK target RSK to promote cell motility and invasion by altering transcription and integrin activity, whereas others impair cell motility and invasion through effects on the actin cytoskeleton (Sulzmaier and Ramos 2013).

Polarity proteins are critical regulators of tissue architecture. Three protein complexes play central roles in controlling polarity: Scribble, Par, and Crumbs complexes. Through multiple interactions, components of these pathways control signaling pathways that regulate cell polarity and tissue organization. Dysregulation of these pathways is common in tumors and, in some contexts, involves alterations in Ras-ERK and PI3K-Akt signaling. For example, Scribble inhibits ERK activation by functioning as a scaffold to link it with the protein phosphatase PP1γ (Dow et al. 2008; Nagasaka et al. 2013). Loss of Scribble enhances invasion stimulated by H-Ras (Shaikh et al. 1996) and tumor formation promoted by Ras and Myc (Wu et al. 2010). Similarly, loss of the polarity protein Par3 leads to increased invasion in several tumor models (Iden et al. 2012; McCaffrey et al. 2012; Xue et al. 2013) through multiple pathways, including PKC-dependent activation of JAK/STAT3 signaling. This induces expression of a metalloproteinase, MMP9, with subsequent destruction of the ECM and invasion (McCaffrey et al. 2012), and increased Rac activation, leading to decreased cell–cell adhesion (Xue et al. 2013).

Loss of cell polarity is often coupled to cell proliferation because the loss of cell adhesion molecules relieves contact inhibition. One example is the cytoskeletal protein merlin (also known as neurofibromin 2), a tumor suppressor that regulates the Hippo pathway and whose loss is well known to cause increased cell proliferation. Polarity signaling is also coupled to metabolism.

Some subpopulations of epithelial cells in tumors, particularly those at tumor margins, undergo at least a partial EMT. EMTs are associated with various normal physiological processes, for example, wound healing, gastrulation, and branching morphogenesis (Birchmeier and Birchmeier 1995). This developmental process is orchestrated by multiple highly coordinated pathways induced by combinations of different factors including transforming growth factor β (TGFβ), TNF, Wnt, Notch, and some growth factors. EMT is characterized by a loss of apical-basal polarity, down-regulation of E-cadherin cell–cell adhesion molecules, adoption of a more fibroblast-like appearance, and, in some contexts, acquisition of stem- or progenitor-cell phenotypes and anchorage independence, properties that would enhance the cell's ability to invade other tissues and initiate tumors at distant sites.

The Ras-ERK and PI3K-Akt pathways drive the EMT in certain contexts, generally under conditions in which these pathways are hyperactivated together with other pathways implicated in EMT (e.g., TGFβ, Wnt, and Notch signaling) (Larue and Bellacosa 2005). Multiple transcription factors, such as Snail, Slug, Twist, and ZEB, play critical roles driving EMT, and these are regulated by ERK and Akt. For example, Akt can phosphorylate the IκB kinases that regulate NF-κB, a transcription factor that induces Snail. Akt also phosphorylates and inactivates GSK3, which normally promotes ubiquitin-dependent degradation of Snail (Doble and Woodgett 2007); Akt activation will therefore increase Snail stability, further promoting EMT. In addition, Akt2 phosphorylates HNRNP E1, a protein that promotes translational elongation on EMT-promoting transcripts such as those encoding interleukin-like EMT inducer and the adaptor protein DAB2 (Hussey et al. 2011). AP1, which is regulated by the Ras-ERK pathway, can also induce transcription factors that promote EMT as well as other gene expression programs that control phenotypic changes associated with EMT. These include up-regulation of specific integrin heterodimers (e.g., $\alpha_5\beta_1$ and $\alpha_V\beta_6$), vimentin, and fibronectin and down-regulation of cytokeratin, polarity proteins (e.g., Crumbs, PATJ, LGL), and E-cadherin, all of which support cell motility. Interestingly, the polarity protein Scribble maintains cell–cell junctions by suppressing ERK (which stimulates ZEB1) as described above (Elsum et al. 2013; Nagasaka et al. 2013). Dysregulation of both Ras-ERK and PI3K-Akt signaling thus has the potential to play an important role in cancer progression by promoting adoption of an invasive phenotype.

Finally, it is important to note that EMT is not essential for invasion and tumor cell dissemination. Tumor cells can migrate as epithelial sheets within tissues (as occurs during wound healing) or invade by pushing through tissue borders (e.g., basement membrane).

8 CELL FATE AND DIFFERENTIATION

Dysregulation or co-option of developmental signaling pathways is a feature of many cancers. This can disrupt the balance between cell proliferation and differentiation, alter cell fate, and/or inappropriately induce morphogenetic programs such as the EMT (see above) that promote metastasis. Although some oncogenes can directly regulate the developmental state of cells, it is generally believed that cancer progression requires a self-renewing population of "stem-cell-like" cells. These may be induced into a stem-cell-like state by an oncogene(s), or a normal stem/progenitor cell may be the cell-of-origin that sustains the successive mutations that lead to malignancy.

The simplest examples of cancers with dysregulated development are perhaps hematopoietic malignancies in which a differentiation program is stalled before the cells reach their nonproliferative differentiated state. For example, in acute promyelocytic leukemia, a form of acute myelocytic leukemia, myeloblasts fail to differentiate into mature white blood cells because of a translocation that leads to synthesis of a fusion protein combining sequences from a protein called PML and the retinoic acid receptor (RAR). The PML-RAR fusion protein represses RAR-target genes that normally drive differentiation, thereby inactivating the RAR signaling that normally controls this. Subsequently, additional mutations cause overproliferation of the undifferentiated myeloblasts. Inappropriate Wnt signaling has a similar effect in colon cancer. Ordinarily, Wnt signaling via β-catenin (see p. 103 [Nusse 2012]) maintains enterocytes in an undifferentiated state in colon crypts, but is inactivated by APC-induced degradation of β-catenin as cells move up toward the luminal surface of the intestine. Mutation of the APC tumor suppressor in colon cancer, however, means β-catenin is not destroyed and can maintain cells in an undifferentiated state as they move away. Further mutations can then drive neoplasia.

Developmental signals can also drive cancer progression because they stimulate inappropriate cell proliferation (see above). Mutations that activate Notch, for example, contribute to acute lymphocytic leukemia because Notch signaling (p. 109 [Kopan 2012]) can stimulate the cell cycle and also inhibits apoptosis in T cells. Importantly, Notch functions as a tumor suppressor in some other tissues. In others, the concentration of Notch dictates its growth sup-

pressive or stimulatory effects (Mazzone et al. 2010), which illustrates the importance of the signaling context. Activation of the Hedgehog signaling pathway (see p. 107 [Ingham 2012]) by mutations in the Patched receptor occurs in basal cell carcinomas and medulloblastomas and again drives cell proliferation. Hedgehog signaling is also hyperactivated via autocrine loops in many tumors that affect tissues derived from the embryonic gut.

Given that Ras-ERK and PI3K-Akt signaling pathways are activated by growth factors such as EGF, IGF, and fibroblast growth factor (FGF), which play major roles in control of cell fate, they can thus be considered developmental signaling pathways that are hijacked in cancer. Signaling by FGF4/8, for example, activates the Ras-ERK pathway to drive EMT during gastrulation and the Ras-ERK pathway is recapitulated in several cancers (Thiery 2002). The context is important, however; signaling by FGF has the potential to affect cell proliferation, apoptosis, and migration (see above), as well as angiogenesis (see below), but it can also have tumor suppressive effects, maintaining cells in a differentiated, nonproliferative state. For example, whereas FGFR2 is up-regulated in gastric cancers, its expression is reduced in bladder and prostate cancer (Turner and Grose 2010).

9 GENOMIC INSTABILITY

Genomic instability is a common characteristic of cancer cells. Aneuploidy and large-scale DNA rearrangements are frequently observed, and many cancers display elevated mutation rates. Ordinarily, a variety of cellular enzymes repair DNA damage, and checkpoint signaling ensures that DNA replication and cytokinesis are arrested in dividing cells until potentially damaging errors are corrected. Alternatively, checkpoint signaling can induce senescence or apoptosis so that affected cells do not pass on these errors. Whether genomic instability is a cause or a consequence of cancer is still debated, but it clearly reflects a failure of checkpoint signaling and/or DNA repair mechanisms.

DNA damage signals are relayed by the kinases ATM, ATR, Chk1, and Chk2, which stimulate p53, stall the cell cycle, and activate the DNA repair machinery (p. 109 [Kopan 2012]; Ch. 6 [Rhind and Russell 2012]). Downstream of p53, the CKI p21 is induced, and this can halt DNA polymerase if DNA replication has already begun. If the damage cannot be repaired and checkpoint signaling persists, p21 and p53 will induce cells to senesce or undergo apoptosis (see above). Clearly, mutation or epigenetic silencing of these tumor suppressors or upstream kinases can inactivate checkpoint signaling, allowing DNA damage to persist and potentially fuel cancer progression. Indeed, ATM and Chk2 mutations are seen in familial leukemias and colon/breast cancers, respectively, and proteins involved in DNA repair itself are also often mutated, for example, MMR enzymes and BRCA1/2.

The mitotic checkpoint (also known as the spindle assembly checkpoint) ensures that when a cell divides each daughter receives a full complement of chromosomes. A complex containing the proteins Bub1, Bub3, and Mad1-3 monitors attachment of chromosomes to the mitotic spindle, relaying checkpoint signals that block chromosome segregation and subsequent cytokinesis. Once paired, sister chromatids are all attached to microtubules emanating from opposite poles, the signal is switched off, and cells can move from metaphase into anaphase and, ultimately, cytokinesis can proceed (Ch. 6 [Rhind and Russell 2012]). Inactivation of this checkpoint pathway has the potential to lead to aneuploidy, and mutations in Mad1/2 and Bub1 have been observed in cancer (Schvartzman et al. 2010).

Akt has been implicated in multiple aspects of DNA damage responses and genome instability (Xu et al. 2012a). It can inhibit homologous recombinational repair through direct phosphorylation of the checkpoint proteins Chk1 and TopBP1 or indirectly through recruitment of resection factors such as RPA, BRCA1, and Rad51 to sites of double-stranded breaks (DSBs) in DNA. Akt is also activated by DSBs in a DNA-dependent protein-kinase- or ATM/ATR-dependent manner and, in some contexts, can contribute to radioresistance by stimulating DNA repair by nonhomologous end joining. In addition, Akt also inhibits association of BRCA1 with DNA damage foci. As discussed above, dysregulation of the PI3K-Akt pathway suppresses apoptosis through many effectors, thus promoting survival of cells with DNA damage. Because Ras-ERK signaling also inhibits apoptosis, it too could promote survival of damaged cells. Hyperactivation of Ras-ERK signaling has been shown to lead to genomic instability, although the molecular mechanism is unclear (Saavedra et al. 1999). Akt therefore modifies both the response to and repair of genotoxic damage in complex ways that are likely to have important consequences for the therapy of tumors showing deregulation of the PI3K-Akt pathway.

The tumor suppressor PTEN can also regulate chromosome stability, independently of its 3′-phosphatase activity. PTEN regulates the expression of the DNA repair protein RAD51, and loss of PTEN causes extensive centromere breakage and chromosomal translocations (Toda et al. 1993; Liu et al. 2008; p. 109 [Kopan 2012]).

Myc overexpression can induce genomic instability. In mammalian cells and *Drosophila*, overexpression of *Myc* increases the frequency of chromosomal rearrangements (Prochownik and Li 2007; Greer et al. 2013). Multiple mechanisms have been associated with such genomic rearrangements, including ROS-induced DSBs, suppression of

checkpoints that prevent replication of damaged DNA, and telomere clustering.

10 THE TUMOR MICROENVIRONMENT

So far, we have primarily considered how signaling within cancer cells themselves is dysregulated in cancer. However, cancer progression (at least in solid tumors) also depends on the ECM, blood vessels, immune cells, and noncancerous cells such as fibroblasts in the tumor microenvironment, all of which communicate with cancer cells by subverted signaling mechanisms (Fig. 6). Many of the changes in the tumor microenvironment during cancer progression mimic changes that occur during wound healing and/or developmental processes. As tumors evolve, the complexity of their "ecosystem" increases; reciprocal paracrine and juxtacrine interactions between populations of neoplastic cells as well as tumor cells and nonneoplastic cells within the microenvironment control cellular signaling pathways in both positive and negative fashions. Dissecting the roles of individual signaling pathways in these ecosystems is complex because it is difficult to distinguish cell-autonomous and non-cell-autonomous activities.

10.1 The ECM

The ECM is a scaffold that physically supports tissues and provides a substrate for cell adhesion and migration, as well as a source of bioactive molecules. Far from a static structure, it is constantly being remodeled, and its composition plays a critical role in control of cell behavior. Fibronectin, laminin, collagen, and various other ECM components serve as ligands that activate integrin signaling. Integrin signaling leads to activation of canonical pathways such as Ras-ERK, PI3K-Akt, and Src signaling, as well as other proteins, for example, the tyrosine kinase FAK, a scaffold that links integrins with cytoskeletal proteins, adaptors, and enzymes that transduce signals from matrix adhesion complexes. FAK also regulates p53 and members of the miR-200 family of microRNAs, which control apoptosis and epithelial phenotype (Keely 2011).

Heparin sulfate proteoglycans (HSPGs) in the ECM modulate signaling by associating with various ligands and acting as coreceptors (e.g., for FGF and FGFR). In addition, the ECM actively sequesters a variety of growth factors, including TGFβ, vascular endothelial growth factor (VEGF), and platelet-derived growth factor (PDGF), which can be liberated and/or activated by MMPs. Collagens can also be digested and remodeled by proteinases to enhance tumor cell motility.

The ECM changes as cancer progresses (Lu et al. 2011). It stiffens as large quantities of ECM are deposited by can-

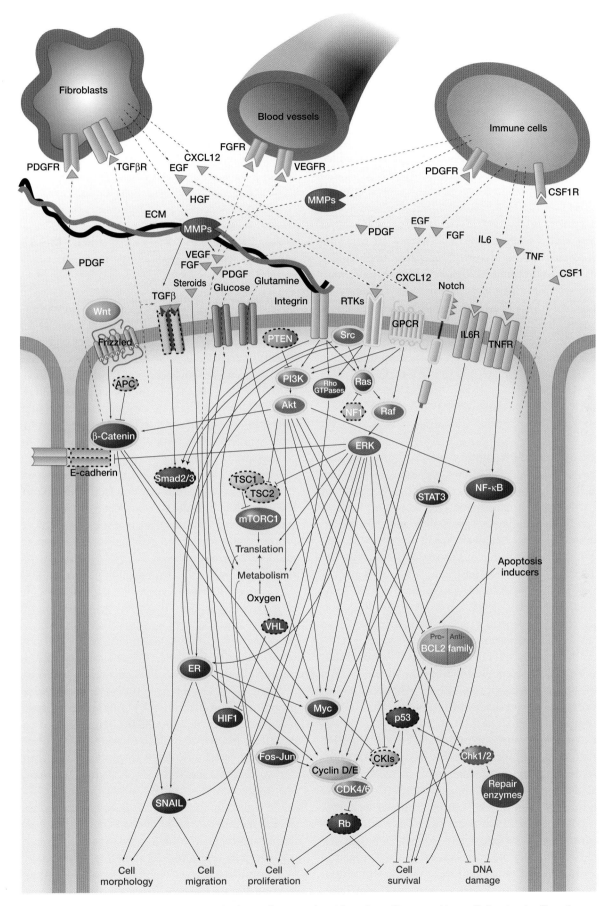

Figure 6. Cancer signaling networks. The figure illustrates the wide variety of intra- and intercellular signals affected in cancer, focusing on Ras-ERK and PI3K-Akt signaling. It is by no means comprehensive; many more pathways are involved and there are other stromal cells involved in paracrine signaling. Oncoproteins are indicated with yellow highlighting; tumor suppressors are indicated with dashed outlines. Arrows do not necessarily indicate direct interactions in this figure.

cer-associated fibroblasts (see below) and collagen fibers become more cross-linked by lysyl oxidases secreted by stromal cells. This increases contractility of the cells, which further fuels stiffening. The changes in stiffness of the ECM promote cell migration and integrin signaling through regulation of Rho family GTPases and other pathways, which synergize with oncogene-activated Ras-ERK and PI3K-Akt pathways to promote invasive growth and cell survival (Keely 2011).

Other changes to the ECM include increased levels of molecules such as tenascin C, a proteoglycan common around developing blood vessels that is induced during inflammation and promotes angiogenesis. MMPs are also up-regulated. These stimulate signaling in various ways and, by degrading the ECM, clear a path for cell migration. Indeed, genes encoding endogenous inhibitors of MMPs such as TIMP3 are known to be targets for hypermethylation in some cancers. HSPGs are also overproduced in cancer and may potentiate oncogenic signaling by FGF, Wnt, and Hedgehog.

10.2 Angiogenesis

Like all tissues, tumors require a blood supply. They acquire this by inducing proliferation and assembly of endothelial cells to form new blood vessels (angiogenesis), co-opting pathways that usually function in wound healing. Central to angiogenesis are signals such as VEGF, PDGF, FGFs, interleukin (IL) 8, and angiopoietin. The PI3K-Akt pathway regulates the induction of angiogenesis as well as vessel integrity (Karar and Maity 2011). Synthesis and secretion of VEGF by cancer cells is induced by HIF1. As mentioned above, HIF1 levels are increased by PI3K-Akt signaling, and hyperactivation of this pathway thus plays an important role in angiogenesis. HIF1 activity can be controlled by the von Hippel–Lindau (VHL) protein, a subunit of an E3 ligase that promotes its ubiquitin-dependent degradation under normoxic conditions when it is proline hydroxylated by proline hydroxylases (see Ch. 7 [Ward and Thompson 2012b]), or through translational control. VHL functions as a tumor suppressor and inactivating VHL mutations occur in a variety of cancers. The PI3K-Akt pathway also modulates the production of other angiogenic factors, such as nitric oxide and angiopoietins. Constitutive endothelial activation of Akt1 has been shown to induce the formation of structurally abnormal blood vessels.

Following its secretion, VEGF is sequestered in the ECM and cannot exert its effects on endothelial cells until it is released by MMPs such as MMP9. These are produced by monocytes and macrophages in the tumor microenvironment (see below), which underscores the importance of immune cells in angiogenesis and existence of the wider signaling network that involves cancer cells, immune cells, and endothelial cells.

Another factor that must be overcome for angiogenesis to occur is inhibitory signals such as thrombospondin 1 (Tsp1). Tsp1 released by various cells normally keeps angiogenesis in check by inducing synthesis of FasL, which causes endothelial cells to undergo apoptosis (see Ch. 19 [Green and Llambi 2014]). Tsp1 is induced by p53, but repressed by Ras, Src, and Myc. It thus represents another control point in angiogenesis that could be activated by dysregulation of the Ras-ERK pathway. Moreover, the gene that encodes Tsp1 is hypermethylated in some cancers.

10.3 Inflammation

Inflammatory cells such as macrophages and neutrophils constitute the first defense against pathogens, but are also involved in tissue remodeling and repair. They are recruited by chemokines secreted by tumor and stromal cells to almost all tumors and secrete various molecules that promote cancer cell proliferation, survival, and migration. In many respects, the contribution of inflammatory cells to tumor progression, like that of angiogenesis, recapitulates their role in wound healing, which also involves these processes.

Signaling via the transcription factor NF-κB (see Ch. 15 [Newton and Dixit 2012]) is important in both cancer cells and tumor-associated inflammatory cells because it can promote cell survival and proliferation and stimulates production of cytokines such as TNF. Oncogenic mutations affecting NF-κB or upstream regulators such as MALT1 and Bcl10 occur in some lymphoid malignancies; however, in most cancers, NF-κB activity is simply increased by cytokine signaling. For example, in colon cancer, TNF produced by macrophages increases NF-κB activity in intestinal epithelial cells, which promotes cell survival; meanwhile, other cytokines such as IL6 and IL11 increase phospho-STAT3 levels (see p. 117 [Harrison 2012]), which promote cell proliferation. A similar phenomenon occurs in hepatocytes in hepatocellular carcinoma, the most common form of liver cancer, and prostate cancer (Karin 2009). NF-κB activation also leads to production of more TNF and synthesis of prostaglandin E2, which further fuels cell proliferation and loss of cell polarity.

In addition to cytokines, inflammatory cells secrete growth factors such as EGF and FGF. These are obviously important regulators of Ras-ERK and PI3K-Akt signaling in cancer cells and will, therefore, dysregulate control of cell proliferation, cell death, metabolism, and cell migration, as discussed above. They also lead to production of colony-stimulating factor 1 (CSF1), a key reciprocal signal that stimulates macrophages, causing them to produce more EGF. Immune cells also produce VEGF and MMPs, which promotes angiogenesis, ECM remodeling, and release of other bioactive molecules (see above). Importantly, all these factors participate in paracrine loops involving immune and cancer cells that sustain chronic inflammation and promote tumor growth and progression.

10.4 Cancer-Associated Fibroblasts

Fibroblasts are present in most tissues, helping to shape organs and control the composition of the ECM. They are activated by tissue injury and, in cancer, produce various factors that stimulate proliferation and migration of cancer cells, along with angiogenesis and inflammation. Release of signals such as TGFβ and PDGF by macrophages and cancer cells activates fibroblasts, which, in turn, can release EGF, HGF, IGF, and chemokines such as CXCR12 (also known as stromal-cell-derived factor 1). These can then further activate Ras-ERK and PI3K-Akt signaling and other pathways in the cancer cells and drive feedback loops that amplify this. Cancer-associated fibroblasts are also responsible for the distinct ECM associated with advanced carcinomas, which also affects signaling within the tumor.

The interplay between cancer cells and these different components of the tumor microenvironment is thus incredibly complex and mimics signaling that tissues display both during development and in normal tissue homeostasis and repair. As cancers metastasize, this extends to other organs. Different cancers are known to seed secondary tumors in particular tissues. Successful colonization depends on the cell surface receptors expressed by the cancer cells and target tissue and the suitability of the microenvironment the latter provides.

11 CONCLUDING REMARKS

Cancer cells show a number of defining characteristics. Underlying these is a dysregulation of cellular signal transduction induced by the genetic and epigenetic changes that drive cancer. This affects not only the cancer cells themselves, but the wider signaling network that encompasses other cells, the ECM, blood vessels, and the immune system. Indeed, metastatic cancer can be considered a systemic disease that affects signaling throughout the affected individual, and systemic effects are ultimately what kill patients in cancer.

Pharmacologic and antibody-based inhibitors that target signaling proteins mutated in tumors or proteins downstream from these have had significant impact as cancer treatments. For example, inhibitors of the nonreceptor tyrosine kinase Abl and RTK ErbB2 dramatically reduce patient mortality in chronic myelogenous leukemia and breast cancer. Other inhibitors, such as those that target B-Raf, EGFR, and the kinase ALK induce remarkable reductions of tumor volume and extend survival in patients with melanoma and nonsmall-cell lung carcinomas; however, the rate of recurrence is high because of the development of drug resistance (Gainor and Shaw 2013; Giroux 2013; Holohan et al. 2013; Lito et al. 2013).

The complexity of the cancer signaling network (see Fig. 6) presents a huge challenge for efforts to develop such anticancer drugs because of the redundancy of pathways that control cell proliferation and survival, cross talk between pathways, and feedback inhibition mechanisms that cause pathway reactivation. The fact that pathways such as Ras-ERK and Akt-PI3K signaling control so many characteristics of cancer cells, and that components of these pathways, or upstream receptors, are so commonly mutated in a variety of cancers gives reason to be optimistic that approaches based on targeting them will be successful. The efficacy of therapies that target these pathways is, however, limited by multiple factors. For example, rewiring of signaling pathways is associated with adaptive responses to inhibition of driver mutations, and this is commonly because of either loss of feedback inhibition or induction of stress pathways (Pratilas and Solit, 2010; Rodrik-Outmezguine et al. 2011; Lito et al. 2013). Moreover, factors from the tumor microenvironment may stimulate alternative pathways that maintain cell viability despite inhibition of the targeted pathways (Castells et al. 2012; Muranen et al. 2012). Alternatively, there can be selection for rare tumor cells that contain drug-resistant variants of the targeted protein or mutations in other pathways that circumvent the dependency on the targeted pathway, and epigenetic or stochastic changes in the state of tumor cells can also activate intrinsic resistance pathways (Holohan et al. 2013; Holzel et al. 2013).

Further complicating matters is the degree of intratumoral genetic heterogeneity. Recent evidence emerging from sequencing of single cells and multiple regions of tumors from individual patients has revealed this is far greater than previously imagined (Navin et al. 2011; Ruiz et al. 2011; Ding et al. 2012; Gerlinger et al. 2012; Xu et al. 2012b; Bashashati et al. 2013) In one study of kidney tumors, only ~45% of mutations were detected in all tumor regions. This heterogeneity also contributes significantly to intratumoral variation in sensitivity to drugs targeting mutated signaling proteins mutated in cancer, and means that single biopsies may not be sufficient to customize patient treatment.

Overcoming these challenges will require a deeper understanding of the nature of resistance mechanisms and how different cellular signaling programs mediate resistant states in heterogeneous populations of tumor cells. Combination therapies that target these should increase the efficacy of targeted therapies. This is a significant challenge, but one that we feel is not insurmountable.

REFERENCES

*Reference is in this book.

Anastasiou D, Poulogiannis G, Asara JM, Boxer MB, Jiang JK, Shen M, Bellinger G, Sasaki AT, Locasale JW, Auld DS, et al. 2011. Inhibition of pyruvate kinase M2 by reactive oxygen species contributes to cellular antioxidant responses. *Science* 334: 1278–1283.

Bakan I, Laplante M. 2012. Connecting mTORC1 signaling to SREBP-1 activation. *Curr Opin Lipidol* 23: 226–234.

Barthel A, Okino ST, Liao J, Nakatani K, Li J, Whitlock JP Jr., Roth RA. 1999. Regulation of GLUT1 gene transcription by the serine/threonine kinase Akt1. *J Biol Chem* 274: 20281–20286.

Bashashati A, Ha G, Tone A, Ding J, Prentice LM, Roth A, Rosner J, Shumansky K, Kalloger S, Senz J, et al. 2013. Distinct evolutionary trajectories of primary high-grade serous ovarian cancers revealed through spatial mutational profiling. *J Pathol* 231: 21–34.

Ben-Sahra I, Howell JJ, Asara JM, Manning BD. 2013. Stimulation of de novo pyrimidine synthesis by growth signaling through mTOR and S6K1. *Science* 339: 1323–1328.

Berwick DC, Hers I, Heesom KJ, Moule SK, Tavare JM. 2002. The identification of ATP-citrate lyase as a protein kinase B (Akt) substrate in primary adipocytes. *J Biol Chem* 277: 33895–33900.

Birchmeier W, Birchmeier C. 1995. Epithelial-mesenchymal transitions in development and tumor progression. *EXS* 74: 1–15.

Burgering BM, Medema RH. 2003. Decisions on life and death: FOXO Forkhead transcription factors are in command when PKB/Akt is off duty. *J Leukoc Biol* 73: 689–701.

Cagnol S, Chambard JC. 2010. ERK and cell death: Mechanisms of ERK-induced cell death—Apoptosis, autophagy and senescence. *FEBS J* 277: 2–21.

Cain RJ, Ridley AJ. 2009. Phosphoinositide 3-kinases in cell migration. *Biol Cell* 101: 13–29.

Castells M, Thibault B, Delord JP, Couderc B. 2012. Implication of tumor microenvironment in chemoresistance: Tumor-associated stromal cells protect tumor cells from cell death. *Int J Mol Sci* 13: 9545–9571.

Castoria G, Lombardi M, Barone MV, Bilancio A, Di Domenico M, De Falco A, Varricchio L, Bottero D, Nanayakkara M, Migliaccio A, et al. 2004. Rapid signalling pathway activation by androgens in epithelial and stromal cells. *Steroids* 69: 517–522.

Chin YR, Toker A. 2011. Akt isoform-specific signaling in breast cancer: Uncovering an anti-migratory role for palladin. *Cell Adh Migr* 5: 211–214.

Christofk HR, Vander Heiden MG, Wu N, Asara JM, Cantley LC. 2008. Pyruvate kinase M2 is a phosphotyrosine-binding protein. *Nature* 452: 181–186.

Cichowski K, Jacks T. 2001. NF1 tumor suppressor gene function: Narrowing the GAP. *Cell* 104: 593–604.

Collak FK, Yagiz K, Luthringer DJ, Erkaya B, Cinar B. 2012. Threonine-120 phosphorylation regulated by phosphoinositide-3-kinase/Akt and mammalian target of rapamycin pathway signaling limits the antitumor activity of mammalian sterile 20-like kinase 1. *J Biol Chem* 287: 23698–23709.

Dang CV. 2013. MYC, metabolism, cell growth, and tumorigenesis. *Cold Spring Harb Perspect Med* 3: a014217.

Delhommeau F, Dupont S, Della Valle V, James C, Trannoy S, Masse A, Kosmider O, Le Couedic JP, Robert F, Alberdi A, et al. 2009. Mutation in TET2 in myeloid cancers. *N Engl J Med* 360: 2289–2301.

*Devreotes P, Horwitz AR. 2014. Signaling networks that regulate cell migration. *Cold Spring Harb Perspect Biol* doi: 10.1101/cshperspect.a005959.

Diehl JA, Zindy F, Sherr CJ. 1997. Inhibition of cyclin D1 phosphorylation on threonine-286 prevents its rapid degradation via the ubiquitin-proteasome pathway. *Genes Dev* 11: 957–972.

Ding L, Ley TJ, Larson DE, Miller CA, Koboldt DC, Welch JS, Ritchey JK, Young MA, Lamprecht T, McLellan MD, et al. 2012. Clonal evolution in relapsed acute myeloid leukaemia revealed by whole-genome sequencing. *Nature* 481: 506–510.

Doble BW, Woodgett JR. 2007. Role of glycogen synthase kinase-3 in cell fate and epithelial-mesenchymal transitions. *Cells Tissues Organs* 185: 73–84.

Dow LE, Elsum IA, King CL, Kinross KM, Richardson HE, Humbert PO. 2008. Loss of human Scribble cooperates with H-Ras to promote cell invasion through deregulation of MAPK signalling. *Oncogene* 27: 5988–6001.

*Duronio RJ, Xiong Y. 2013. Signaling pathways that control cell proliferation. *Cold Spring Harb Perspect Biol* 5: a008904.

Edinger AL, Thompson CB. 2002. Akt maintains cell size and survival by increasing mTOR-dependent nutrient uptake. *Mol Biol Cell* 13: 2276–2288.

Elsum IA, Martin C, Humbert PO. 2013. Scribble regulates an EMT polarity pathway through modulation of MAPK-ERK signaling to mediate junction formation. *J Cell Sci* 126: 3990–3999.

Engelman JA, Luo J, Cantley LC. 2006. The evolution of phosphatidylinositol 3-kinases as regulators of growth and metabolism. *Nat Rev Genet* 7: 606–619.

Fang D, Hawke D, Zheng Y, Xia Y, Meisenhelder J, Nika H, Mills GB, Kobayashi R, Hunter T, Lu Z. 2007. Phosphorylation of β-catenin by AKT promotes β-catenin transcriptional activity. *J Biol Chem* 282: 11221–11229.

Gainor JF, Shaw AT. 2013. Emerging paradigms in the development of resistance to tyrosine kinase inhibitors in lung cancer. *J Clin Oncol* 31: 3987–3996.

Gao X, Wang H, Yang JJ, Liu X, Liu ZR. 2012. Pyruvate kinase M2 regulates gene transcription by acting as a protein kinase. *Mol Cell* 45: 598–609.

Gerlinger M, Rowan AJ, Horswell S, Larkin J, Endesfelder D, Gronroos E, Martinez P, Matthews N, Stewart A, Tarpey P, et al. 2012. Intratumor heterogeneity and branched evolution revealed by multiregion sequencing. *N Engl J Med* 366: 883–892.

Ghoda L, Lin X, Greene WC. 1997. The 90-kDa ribosomal S6 kinase (pp90rsk) phosphorylates the N-terminal regulatory domain of IκBα and stimulates its degradation in vitro. *J Biol Chem* 272: 21281–21288.

Giroux S. 2013. Overcoming acquired resistance to kinase inhibition: The cases of EGFR, ALK and BRAF. *Bioorg Med Chem Lett* 23: 394–401.

Gorgoulis VG, Halazonetis TD. 2010. Oncogene-induced senescence: The bright and dark side of the response. *Curr Opin Cell Biol* 22: 816–827.

*Green DR, Llambi F. 2014. Cell death signaling. *Cold Spring Harb Perspect Biol* doi: 10.1101/cshperspect.a006080.

Greer C, Lee M, Westerhof M, Milholland B, Spokony R, Vijg J, Secombe J. 2013. Myc-dependent genome instability and lifespan in *Drosophila*. *PLoS ONE* 8: e74641.

Guo D, Bell EH, Mischel P, Chakravarti A. 2013. Targeting SREBP-1-driven lipid metabolism to treat cancer. *Curr Pharm Des*. doi: 10.2174/13816128113199990486.

Hamede RK, McCallum H, Jones M. 2013. Biting injuries and transmission of Tasmanian devil facial tumour disease. *J Anim Ecol* 82: 182–190.

Hanahan D, Weinberg RA. 2000. The hallmarks of cancer. *Cell* 100: 57–70.

Haq S, Michael A, Andreucci M, Bhattacharya K, Dotto P, Walters B, Woodgett J, Kilter H, Force T. 2003. Stabilization of β-catenin by a Wnt-independent mechanism regulates cardiomyocyte growth. *Proc Natl Acad Sci* 100: 4610–4615.

* Hardie DG. 2012. Organismal carbohydrate and lipid homeostasis. *Cold Spring Harb Perspect Biol* **4:** a006031.

* Harrison DA. 2012. The Jak/STAT pathway. *Cold Spring Harb Perspect Biol* **4:** a011205.

* Harvey KF, Hariharan IK. 2012. The Hippo pathway. *Cold Spring Harb Perspect Biol* **4:** a011288.

Harvey KF, Zhang X, Thomas DM. 2013. The Hippo pathway and human cancer. *Nat Rev Cancer* **13:** 246–257.

* Hemmings BA, Restuccia DF. 2012. PI3K-PKB/Akt pathway. *Cold Spring Harb Perspect Biol* **4:** a011189.

Hitosugi T, Kang S, Vander Heiden MG, Chung TW, Elf S, Lythgoe K, Dong S, Lonial S, Wang X, Chen GZ, et al. 2009. Tyrosine phosphorylation inhibits PKM2 to promote the Warburg effect and tumor growth. *Sci Signal* **2:** ra73.

Holohan C, Van Schaeybroeck S, Longley DB, Johnston PG. 2013. Cancer drug resistance: An evolving paradigm. *Nat Rev Cancer* **13:** 714–726.

Holzel M, Bovier A, Tuting T. 2013. Plasticity of tumour and immune cells: A source of heterogeneity and a cause for therapy resistance? *Nat Rev Cancer* **13:** 365–376.

Hu W, Zhang C, Wu R, Sun Y, Levine A, Feng Z. 2010. Glutaminase 2, a novel p53 target gene regulating energy metabolism and antioxidant function. *Proc Natl Acad Sci* **107:** 7455–7460.

Hussey GS, Chaudhury A, Dawson AE, Lindner DJ, Knudsen CR, Wilce MC, Merrick WC, Howe PH. 2011. Identification of an mRNP complex regulating tumorigenesis at the translational elongation step. *Mol Cell* **41:** 419–431.

Iden S, van Riel WE, Schafer R, Song JY, Hirose T, Ohno S, Collard JG. 2012. Tumor type-dependent function of the par3 polarity protein in skin tumorigenesis. *Cancer Cell* **22:** 389–403.

* Ingham PW. 2012. Hedgehog signaling. *Cold Spring Harb Perspect Biol* **4:** a011221.

Israelsen WJ, Dayton TL, Davidson SM, Fiske BP, Hosios AM, Bellinger G, Li J, Yu Y, Sasaki M, Horner JW, et al. 2013. PKM2 isoform-specific deletion reveals a differential requirement for pyruvate kinase in tumor cells. *Cell* **155:** 397–409.

Jensen PJ, Gunter LB, Carayannopoulos MO. 2010. Akt2 modulates glucose availability and downstream apoptotic pathways during development. *J Biol Chem* **285:** 17673–17680.

Jeon TI, Osborne TF. 2012. SREBPs: Metabolic integrators in physiology and metabolism. *Trends Endocrinol Metab* **23:** 65–72.

Kaelin WG Jr, McKnight SL. 2013. Influence of metabolism on epigenetics and disease. *Cell* **153:** 56–69.

Karar J, Maity A. 2011. PI3K/AKT/mTOR pathway in angiogenesis. *Front Mol Neurosci* **4:** 51.

Karin M. 2009. NF-κB as a critical link between inflammation and cancer. *Cold Spring Harb Perspect Biol* **1:** a000141.

Keely PJ. 2011. Mechanisms by which the extracellular matrix and integrin signaling act to regulate the switch between tumor suppression and tumor promotion. *J Mammary Gland Biol Neoplasia* **16:** 205–219.

Kim JW, Zeller KI, Wang Y, Jegga AG, Aronow BJ, O'Donnell KA, Dang CV. 2004. Evaluation of myc E-box phylogenetic footprints in glycolytic genes by chromatin immunoprecipitation assays. *Mol Cell Biol* **24:** 5923–5936.

Kim D, Shu S, Coppola MD, Kaneko S, Yuan ZQ, Cheng JQ. 2010. Regulation of proapoptotic mammalian ste20-like kinase MST2 by the IGF1-Akt pathway. *PLoS ONE* **5:** e9616.

Kim J, Parrish AB, Kurokawa M, Matsuura K, Freel CD, Andersen JL, Johnson CE, Kornbluth S. 2012. Rsk-mediated phosphorylation and 14-3-3ε binding of Apaf-1 suppresses cytochrome c-induced apoptosis. *EMBO J* **31:** 1279–1292.

* Kopan R. 2012. Notch signaling. *Cold Spring Harb Perspect Biol* **4:** a011213.

Korkaya H, Paulson A, Charafe-Jauffret E, Ginestier C, Brown M, Dutcher J, Clouthier SG, Wicha MS. 2009. Regulation of mammary stem/progenitor cells by PTEN/Akt/β-catenin signaling. *PLoS Biol* **7:** e1000121.

Kubatka P, Zihlavnikova K, Kajo K, Pec M, Stollarova N, Bojkova B, Kassayova M, Orendas P. 2011. Antineoplastic effects of simvastatin in experimental breast cancer. *Klin Onkol* **24:** 41–45.

* Laplante M, Sabatini DM. 2012. mTOR signaling. *Cold Spring Harb Perspect Biol* **4:** a011593.

Larue L, Bellacosa A. 2005. Epithelial-mesenchymal transition in development and cancer: Role of phosphatidylinositol 3′ kinase/AKT pathways. *Oncogene* **24:** 7443–7454.

Ley TJ, Ding L, Walter MJ, McLellan MD, Lamprecht T, Larson DE, Kandoth C, Payton JE, Baty J, Welch J, et al. 2010. DNMT3A mutations in acute myeloid leukemia. *N Engl J Med* **363:** 2424–2433.

Li S, Shen D, Shao J, Crowder R, Liu W, Prat A, He X, Liu S, Hoog J, Lu C, et al. 2013. Endocrine-therapy-resistant ESR1 variants revealed by genomic characterization of breast-cancer-derived xenografts. *Cell Rep* **4:** 1116–1130.

Lin CY, Loven J, Rahl PB, Paranal RM, Burge CB, Bradner JE, Lee TI, Young RA. 2012. Transcriptional amplification in tumor cells with elevated c-Myc. *Cell* **151:** 56–67.

Lin JI, Poon CL, Harvey KF. 2013. The Hippo size control pathway-ever expanding. *Sci Signal* **6:** pe4.

Lito P, Rosen N, Solit DB. 2013. Tumor adaptation and resistance to RAF inhibitors. *Nat Med* **19:** 1401–1409.

Liu W, Zhou Y, Reske SN, Shen C. 2008. PTEN mutation: Many birds with one stone in tumorigenesis. *Anticancer Res* **28:** 3613–3619.

Lu P, Takai K, Weaver VM, Werb Z. 2011. Extracellular matrix degradation and remodeling in development and disease. *Cold Spring Harb Perspect Biol* **3:** a005058.

Luo W, Hu H, Chang R, Zhong J, Knabel M, O'Meally R, Cole RN, Pandey A, Semenza GL. 2011. Pyruvate kinase M2 is a PHD3-stimulated coactivator for hypoxia-inducible factor 1. *Cell* **145:** 732–744.

Ma L, Zhang G, Miao XB, Deng XB, Wu Y, Liu Y, Jin ZR, Li XQ, Liu QZ, Sun DX, et al. 2013. Cancer stem-like cell properties are regulated by EGFR/AKT/β-catenin signaling and preferentially inhibited by gefitinib in nasopharyngeal carcinoma. *FEBS J* **280:** 2027–2041.

Mazzone M, Selfors LM, Albeck J, Overholtzer M, Sale S, Carroll DL, Pandya D, Lu Y, Mills GB, Aster JC, et al. 2010. Dose-dependent induction of distinct phenotypic responses to Notch pathway activation in mammary epithelial cells. *Proc Natl Acad Sci* **107:** 5012–5017.

McCaffrey LM, Montalbano J, Mihai C, Macara IG. 2012. Loss of the Par3 polarity protein promotes breast tumorigenesis and metastasis. *Cancer Cell* **22:** 601–614.

McLaughlin SK, Olsen SN, Dake B, De Raedt T, Lim E, Bronson RT, Beroukhim R, Polyak K, Brown M, Kuperwasser C, et al. 2013. The RasGAP gene, RASAL2, is a tumor and metastasis suppressor. *Cancer Cell* **24:** 365–378.

Miinea CP, Sano H, Kane S, Sano E, Fukuda M, Peranen J, Lane WS, Lienhard GE. 2005. AS160, the Akt substrate regulating GLUT4 translocation, has a functional Rab GTPase-activating protein domain. *Biochem J* **391:** 87–93.

Miller DM, Thomas SD, Islam A, Muench D, Sedoris K. 2012. c-Myc and cancer metabolism. *Clin Cancer Res* **18:** 5546–5553.

Min J, Zaslavsky A, Fedele G, McLaughlin SK, Reczek EE, De Raedt T, Guney I, Strochlic DE, Macconaill LE, Beroukhim R, et al. 2010. An oncogene-tumor suppressor cascade drives metastatic prostate cancer by coordinately activating Ras and nuclear factor-κB. *Nat Med* **16:** 286–294.

Morris MA, Dawson CW, Young LS. 2009. Role of the Epstein-Barr virus-encoded latent membrane protein-1, LMP1, in the pathogenesis of nasopharyngeal carcinoma. *Future Oncol* **5:** 811–825.

* Morrison DK. 2012. MAP kinase pathways. *Cold Spring Harb Perspect Biol* **4:** a011254.

Munger K, Howley PM. 2002. Human papillomavirus immortalization and transformation functions. *Virus Res* **89:** 213–228.

Muranen T, Selfors LM, Worster DT, Iwanicki MP, Song L, Morales FC, Gao S, Mills GB, Brugge JS. 2012. Inhibition of PI3K/mTOR leads to adaptive resistance in matrix-attached cancer cells. *Cancer Cell* **21:** 227–239.

Murchison EP, Schulz-Trieglaff OB, Ning Z, Alexandrov LB, Bauer MJ, Fu B, Hims M, Ding Z, Ivakhno S, Stewart C, et al. 2012. Genome sequencing and analysis of the Tasmanian devil and its transmissible cancer. *Cell* **148**: 780–791.

Murphy LO, Smith S, Chen RH, Fingar DC, Blenis J. 2002. Molecular interpretation of ERK signal duration by immediate early gene products. *Nat Cell Biol* **4**: 556–564.

Nagasaka K, Seiki T, Yamashita A, Massimi P, Subbaiah VK, Thomas M, Kranjec C, Kawana K, Nakagawa S, Yano T, et al. 2013. A novel interaction between hScrib and PP1γ downregulates ERK signaling and suppresses oncogene-induced cell transformation. *PLoS ONE* **8**: e53752.

Nakashima A, Kawanishi I, Eguchi S, Yu EH, Eguchi S, Oshiro N, Yoshino K, Kikkawa U, Yonezawa K. 2013. Association of CAD, a multifunctional protein involved in pyrimidine synthesis, with mLST8, a component of the mTOR complexes. *J Biomed Sci* **20**: 24.

Navin N, Kendall J, Troge J, Andrews P, Rodgers L, McIndoo J, Cook K, Stepansky A, Levy D, Esposito D, et al. 2011. Tumour evolution inferred by single-cell sequencing. *Nature* **472**: 90–94.

* Newton K, Dixit VM. 2012. Signaling in innate immunity and inflammation. *Cold Spring Harb Perspect Biol* **4**: a006049.

Nie Z, Hu G, Wei G, Cui K, Yamane A, Resch W, Wang R, Green DR, Tessarollo L, Casellas R, et al. 2012. c-Myc is a universal amplifier of expressed genes in lymphocytes and embryonic stem cells. *Cell* **151**: 68–79.

Novellasdemunt L, Tato I, Navarro-Sabate A, Ruiz-Meana M, Mendez-Lucas A, Perales JC, Garcia-Dorado D, Ventura F, Bartrons R, Rosa JL. 2013. Akt-dependent activation of the heart 6-phosphofructo-2-kinase/fructose-2,6-bisphosphatase (PFKFB2) isoenzyme by amino acids. *J Biol Chem* **288**: 10640–10651.

* Nusse R. 2012. Wnt signaling. *Cold Spring Harb Perspect Biol* **4**: a011163.

Ogawara Y, Kishishita S, Obata T, Isazawa Y, Suzuki T, Tanaka K, Masuyama N, Gotoh Y. 2002. Akt enhances Mdm2-mediated ubiquitination and degradation of p53. *J Biol Chem* **277**: 21843–21850.

O'Neill E, Kolch W. 2005. Taming the Hippo: Raf-1 controls apoptosis by suppressing MST2/Hippo. *Cell Cycle* **4**: 365–367.

Plas DR, Thompson CB. 2005. Akt-dependent transformation: There is more to growth than just surviving. *Oncogene* **24**: 7435–7442.

Polakis P. 2001. More than one way to skin a catenin. *Cell* **105**: 563–566.

Pratilas CA, Solit DB. 2010. Targeting the mitogen-activated protein kinase pathway: Physiological feedback and drug response. *Clin Cancer Res* **16**: 3329–3334.

Prochownik EV, Li Y. 2007. The ever expanding role for c-Myc in promoting genomic instability. *Cell Cycle* **6**: 1024–1029.

Raftopoulou M, Hall A. 2004. Cell migration: Rho GTPases lead the way. *Dev Biol* **265**: 23–32.

Rahl PB, Lin CY, Seila AC, Flynn RA, McCuine S, Burge CB, Sharp PA, Young RA. 2010. c-Myc regulates transcriptional pause release. *Cell* **141**: 432–445.

Renoir JM, Marsaud V, Lazennec G. 2013. Estrogen receptor signaling as a target for novel breast cancer therapeutics. *Biochem Pharmacol* **85**: 449–465.

* Rhind N, Russell P. 2012. Signaling pathways that regulate cell division. *Cold Spring Harb Perspect Biol* **4**: a005942.

Richardson CJ, Schalm SS, Blenis J. 2004. PI3-kinase and TOR: PIKTOR-ing cell growth. *Semin Cell Dev Biol* **15**: 147–159.

Roberts DJ, Tan-Sah VP, Smith JM, Miyamoto S. 2013. Akt phosphorylates HK-II at Thr-473 and increases mitochondrial HK-II association to protect cardiomyocytes. *J Biol Chem* **288**: 23798–23806.

Robinson DR, Wu YM, Vats P, Su F, Lonigro RJ, Cao X, Kalyana-Sundaram S, Wang R, Ning Y, Hodges L, et al. 2013. Activating ESR1 mutations in hormone-resistant metastatic breast cancer. *Nat Genet* **45**: 1446–1451.

Rodrik-Outmezguine VS, Chandarlapaty S, Pagano NC, Poulikakos PI, Scaltriti M, Moskatel E, Baselga J, Guichard S, Rosen N. 2011. mTOR kinase inhibition causes feedback-dependent biphasic regulation of AKT signaling. *Cancer Discov* **1**: 248–259.

Rossig L, Jadidi AS, Urbich C, Badorff C, Zeiher AM, Dimmeler S. 2001. Akt-dependent phosphorylation of pp21^{cip1} regulates PCNA binding and proliferation of endothelial cells. *Mol Cell Biol* **21**: 5644–5657.

Roux PP, Blenis J. 2004. ERK and p38 MAPK-activated protein kinases: A family of protein kinases with diverse biological functions. *Microbiol Mol Biol Rev* **68**: 320–344.

Ruiz C, Lenkiewicz E, Evers L, Holley T, Robeson A, Kiefer J, Demeure MJ, Hollingsworth MA, Shen M, Prunkard D, et al. 2011. Advancing a clinically relevant perspective of the clonal nature of cancer. *Proc Natl Acad Sci* **108**: 12054–12059.

Saavedra HI, Fukasawa K, Conn CW, Stambrook PJ. 1999. MAPK mediates RAS-induced chromosome instability. *J Biol Chem* **274**: 38083–38090.

Sandoval J, Esteller M. 2012. Cancer epigenomics: Beyond genomics. *Curr Opin Genet Dev* **22**: 50–55.

Schvartzman JM, Sotillo R, Benezra R. 2010. Mitotic chromosomal instability and cancer: Mouse modelling of the human disease. *Nat Rev Cancer* **10**: 102–115.

Sears R, Nuckolls F, Haura E, Taya Y, Tamai K, Nevins JR. 2000. Multiple Ras-dependent phosphorylation pathways regulate Myc protein stability. *Genes Dev* **14**: 2501–2514.

Seeger C, Zoulim F, Mason W. 2013. Hepsdna viruses. In *Fields virology*, 3rd ed. (ed. Knipe DM, Howley P). Lippincott Williams & Wilkins, Philadelphia.

* Sever R, Glass CK. 2013. Signaling by nuclear receptors. *Cold Spring Harb Perspect Biol* **5**: a016709.

Shaikh S, Collier DA, Sham PC, Ball D, Aitchison K, Vallada H, Smith I, Gill M, Kerwin RW. 1996. Allelic association between a Ser-9-Gly polymorphism in the dopamine D3 receptor gene and schizophrenia. *Hum Genet* **97**: 714–719.

Shen HM, Tergaonkar V. 2009. NFκB signaling in carcinogenesis and as a potential molecular target for cancer therapy. *Apoptosis* **14**: 348–363.

Solimini NL, Luo J, Elledge SJ. 2007. Non-oncogene addiction and the stress phenotype of cancer cells. *Cell* **130**: 986–988.

Soloaga A, Thomson S, Wiggin GR, Rampersaud N, Dyson MH, Hazzalin CA, Mahadevan LC, Arthur JS. 2003. MSK2 and MSK1 mediate the mitogen- and stress-induced phosphorylation of histone H3 and HMG-14. *EMBO J* **22**: 2788–2797.

Sulzmaier FJ, Ramos JW. 2013. RSK isoforms in cancer cell invasion and metastasis. *Cancer Res* **73**: 6099–6105.

Sutrias-Grau M, Arnosti DN. 2004. CtBP contributes quantitatively to Knirps repression activity in an NAD binding-dependent manner. *Mol Cell Biol* **24**: 5953–5966.

Suva ML, Riggi N, Bernstein BE. 2013. Epigenetic reprogramming in cancer. *Science* **339**: 1567–1570.

Suzuki S, Tanaka T, Poyurovsky MV, Nagano H, Mayama T, Ohkubo S, Lokshin M, Hosokawa H, Nakayama T, Suzuki Y, et al. 2010. Phosphate-activated glutaminase (GLS2), a p53-inducible regulator of glutamine metabolism and reactive oxygen species. *Proc Natl Acad Sci* **107**: 7461–7466.

Thiery JP. 2002. Epithelial-mesenchymal transitions in tumour progression. *Nat Rev Cancer* **2**: 442–454.

Thomas C, Gustafsson JA. 2011. The different roles of ER subtypes in cancer biology and therapy. *Nat Rev Cancer* **11**: 597–608.

Toda M, Shirao T, Minoshima S, Shimizu N, Toya S, Uyemura K. 1993. Molecular cloning of cDNA encoding human drebrin E and chromosomal mapping of its gene. *Biochem Biophys Res Commun* **196**: 468–472.

Toy W, Shen Y, Won H, Green B, Sakr RA, Will M, Li Z, Gala K, Fanning S, King TA, et al. 2013. ESR1 ligand-binding domain mutations in hormone-resistant breast cancer. *Nat Genet* **45**: 1439–1445.

Turner N, Grose R. 2010. Fibroblast growth factor signalling: From development to cancer. *Nat Rev Cancer* **10**: 116–129.

Turner SL, Blair-Zajdel ME, Bunning RA. 2009. ADAMs and ADAMTSs in cancer. *Br J Biomed Sci* **66**: 117–128.

Vander Heiden MG, Cantley LC, Thompson CB. 2009. Understanding the Warburg effect: The metabolic requirements of cell proliferation. *Science* **324:** 1029–1033.

Vargas J, Feltes BC, Poloni Jde F, Lenz G, Bonatto D. 2012. Senescence: An endogenous anticancer mechanism. *Front Biosci (Landmark Ed)* **17:** 2616–2643.

Vicente-Manzanares M, Horwitz AR. 2011. Cell migration: An overview. *Methods Mol Biol* **769:** 1–24.

Vogelstein B, Papadopoulos N, Velculescu VE, Zhou S, Diaz LA Jr, Kinzler KW. 2013. Cancer genome landscapes. *Science* **339:** 1546–1558.

Ward PS, Thompson CB. 2012a. Metabolic reprogramming: A cancer hallmark even warburg did not anticipate. *Cancer Cell* **21:** 297–308.

* Ward PS, Thompson CB. 2012b. Signaling in control of cell growth and metabolism. *Cold Spring Harb Perspect Biol* **4:** a006783.

Weinberg R. 2013. *The biology of cancer.* Garland, New York.

Wieman HL, Wofford JA, Rathmell JC. 2007. Cytokine stimulation promotes glucose uptake via phosphatidylinositol-3 kinase/Akt regulation of Glut1 activity and trafficking. *Mol Biol Cell* **18:** 1437–1446.

Wu M, Pastor-Pareja JC, Xu T. 2010. Interaction between Ras^{V12} and scribbled clones induces tumour growth and invasion. *Nature* **463:** 545–548.

Wu G, Broniscer A, McEachron TA, Lu C, Paugh BS, Becksfort J, Qu C, Ding L, Huether R, Parker M, et al. 2012. Somatic histone H3 alterations in pediatric diffuse intrinsic pontine gliomas and non-brainstem glioblastomas. *Nat Genet* **44:** 251–253.

Xu N, Lao Y, Zhang Y, Gillespie DA. 2012a. Akt: A double-edged sword in cell proliferation and genome stability. *J Oncol* **2012:** 951724.

Xu X, Hou Y, Yin X, Bao L, Tang A, Song L, Li F, Tsang S, Wu K, Wu H, et al. 2012b. Single-cell exome sequencing reveals single-nucleotide mutation characteristics of a kidney tumor. *Cell* **148:** 886–895.

Xue B, Krishnamurthy K, Allred DC, Muthuswamy SK. 2013. Loss of Par3 promotes breast cancer metastasis by compromising cell-cell cohesion. *Nat Cell Biol* **15:** 189–200.

Yamada E, Okada S, Saito T, Ohshima K, Sato M, Tsuchiya T, Uehara Y, Shimizu H, Mori M. 2005. Akt2 phosphorylates Synip to regulate docking and fusion of GLUT4-containing vesicles. *J Cell Biol* **168:** 921–928.

Ying H, Kimmelman AC, Lyssiotis CA, Hua S, Chu GC, Fletcher-Sananikone E, Locasale JW, Son J, Zhang H, Coloff JL, et al. 2012. Oncogenic Kras maintains pancreatic tumors through regulation of anabolic glucose metabolism. *Cell* **149:** 656–670.

Yu FX, Guan KL. 2013. The Hippo pathway: Regulators and regulations. *Genes Dev* **27:** 355–371.

Zhang X, Tang N, Hadden TJ, Rishi AK. 2011. Akt, FoxO and regulation of apoptosis. *Biochim Biophys Acta* **1813:** 1978–1986.

CHAPTER 22

Outlook

Jeremy Thorner[1], Tony Hunter[2], Lewis C. Cantley[3], and Richard Sever[4]

[1]Department of Molecular and Cell Biology, University of California, Berkeley, California 94720-3202

[2]Molecular and Cell Biology Laboratory, Salk Institute for Biological Studies, La Jolla, California 92037

[3]Cancer Center, Weill Cornell Medical College, New York, New York 10065

[4]Cold Spring Harbor Laboratory, Cold Spring Harbor, New York 11724

Correspondence: jthorner@berkeley.edu

SUMMARY

We have come a long way in the 55 years since Edmond Fischer and the late Edwin Krebs discovered that the activity of glycogen phosphorylase is regulated by reversible protein phosphorylation. As the contents of this book attest, many of the other fundamental molecular mechanisms that operate in biological signaling have since been characterized and the vast web of interconnected pathways that make up the cellular signaling network has been mapped in considerable detail. Nonetheless, it is important to consider how fast this field is still moving and the issues at the current boundaries of our understanding. One must also appreciate what experimental strategies have allowed us to attain our present level of knowledge. Therefore, we summarize here some key issues (both conceptual and methodological), raise unresolved questions, discuss potential pitfalls, and highlight areas in which our understanding is still rudimentary. We hope these wide-ranging ruminations will be useful to investigators who carry studies of signal transduction forward during the rest of the 21st century.

Outline

1 SIGNALING IN THE ATOMIC AGE

Our perceptions about signal transduction processes and the functions of the molecules involved have blossomed as a consequence of technological advances in genomics (Rogne and Taskén 2013), mass spectrometry (Bensimon et al. 2012), structure determination (Chiu et al. 2006), and computational power (Chakraborty and Das 2010). The increase in structural biology, in particular, has been extremely important. It is often helpful, especially for newcomers, to illustrate information flow, protein–protein interactions, and other basic concepts by using schematic diagrams. However, the acquisition of structural information about signaling molecules at atomic resolution has given us unprecedented insights into the way signaling proteins operate as nano-machines to control cellular processes. This level of detail is vital to convey to the advanced student, the practitioner at the bench, and those attempting to develop effective therapeutics. Starting more than 20 years ago with acquisition of the first three-dimensional picture of a protein kinase, cAMP-dependent protein kinase (also known as protein kinase A, PKA) (Knighton et al. 1991) and elucidation of crystal structures of the first noncatalytic folds found in many signaling proteins—SH2 domains (which recognize phosphorylated tyrosine residues in various sequence contexts) (Waksman et al. 1992), and SH3 domains (which recognize PxxP motifs and variants thereof) (Musacchio et al. 1992)—the pace of advance has been truly remarkable.

Structural information at atomic resolution has, for example, turned our rather naïve initial views about how ligand-induced dimerization of receptor-tyrosine kinases leads to activation into a highly nuanced view of the articulation and dynamics of these molecules (Jura et al. 2011; Bessman and Lemmon 2012; Endres et al. 2013). The same is true for the cytosolic protein-tyrosine kinases (Bradshaw 2010; Kiu and Nicholson 2012). Similarly, deconvolution of the structures of polytopic integral membrane proteins, in particular G-protein-coupled receptors (GPCRs), which seemed an insurmountable technological challenge less than 10 years ago, has yielded to new methodology and produced a flood of high-resolution structures for these molecules (Stenkamp et al. 2005; Granier and Kobilka 2012; Katritch et al. 2013), and even a four-protein complex of one GPCR with its cognate heterotrimeric G protein (Rasmussen et al. 2011). However, some molecules critical for signaling remain recalcitrant to this approach. For example, although structural information at atomic resolution has had an enormous impact on our level of understanding of the function and regulation of certain classes of ion channels (MacKinnon 2003), many that have critical roles in vision, taste, mechanical sensation (hearing and touch), pain perception, and other aspects of neurotransmission and neurosensation have not yet been visualized in their native full-length form (Li et al. 2011; Lau et al. 2012; Kumar and Mayer 2013). Nonetheless, arguably, the depth of our understanding of detailed molecular mechanisms at the atomic level has come farther and faster for proteins involved in signal transduction than for proteins in many other areas of biology. That pace needs to be continued.

2 SEEING IS BELIEVING

Technological advances in imaging also deserve special mention because of the critically important information that direct visualization provides about the localization and movement of signaling molecules in individual cells. The demonstration that a green fluorescent protein (GFP) from a jellyfish could be fused to a protein of interest and the resulting chimera used to observe the concentration and subcellular distribution of signaling molecules in live cells in real time via fluorescence microscopy was a seminal breakthrough (Chalfie et al. 1994). The discovery of GFP provoked a hunt for additional proteins capable of generating a fluorophore internally, such as DsRed (Miyawaki 2002), or by coupling to endogenous chromophores such as biliverdin (Shu et al. 2009; Piatkevich et al. 2013). Successful mutational strategies allowed modification of the spectral properties of GFP itself (Zacharias and Tsien 2006; Tsien 2009) and its relatives (Shaner et al. 2008; Piatkevich and Verkhusha 2010). Similarly, mutagenesis of small single-chain enzymes has yielded variants that self-label covalently with a fluorophore substituent in the course of one catalytic turnover when provided with an appropriate cell-permeable, dye-derivatized substrate. This approach produced the so-called SNAP (Corrêa et al. 2013), CLIP (Gautier et al. 2008), and Halo (Encell et al. 2012) tags. It is important to emphasize, however, that studies tracking hybrid proteins must be confirmed by independent approaches (e.g., genetic complementation) that show they retain biological function; sadly, such constructs are often used and conclusions drawn without this necessary validation.

There are additional tactics for bioorthogonal labeling of proteins (Debets et al. 2013; Herner et al. 2013). For example, unnatural amino acid (UAA) technology (Liu and Schultz 2010; Chin 2014) offers another route for genetically encoding a protein with a site-specific fluorescent tag. By extending the genetic code and making the necessary changes to the translation machinery, one can incorporate at any specified position in a protein a 21st amino acid (either a UAA with a fluorescent side chain [Kang et al. 2013; Niu and Guo 2013] or a UAA whose side chain can be readily coupled to a cell-permeable fluorophore [Borr-

Cite this chapter as *Cold Spring Harb Perspect Biol* doi: 10.1101/cshperspect.a022913

mann et al. 2012], or even two different UAAs at distinct sites in the same protein [Xiao et al. 2013a]). Together, all of these methods provide a researcher with a broad palette for fluorescent labeling, permitting simultaneous interrogation of the locations of and/or conformational changes in multiple signaling proteins in the same cell.

Such labels have become even more useful because of concomitant advances in fluorescence imaging that have taken us well beyond conventional epifluorescence microscopy. These include dramatic improvements to the quality of the microscopes themselves and, most notably, the advent of various optical and computational methods to determine the centroid of a point source of emitted fluorescence, such as photoactivated localization microscopy (PALM), stochastic optical reconstruction microscopy (STORM), and stimulated emission-depletion microscopy (STED) (Sengupta et al. 2012; Coelho et al. 2013). These techniques (collectively, referred to as super-resolution fluorescence microscopy) achieve a spatial resolution better than the diffraction limit of the illuminating light ($\sim$200 nm), providing the capability to discriminate between molecules separated by only $\sim$30 nm (Huang et al. 2009). Spinning-disk confocal microscopy, deconvolution microscopy, and two-photon microscopy allow image reconstruction in three dimensions, providing spatial information. Depending on the nature of the molecule, the process, and the cellular region being examined, techniques such as total internal reflection fluorescence microscopy or selective plane illumination microscopy can be used to follow the movement of single molecules by illuminating only a very thin region of the cell.

Fluorescence speckle microscopy and photoactivation microscopy use photoconvertible forms of fluorescent protein tags to permit monitoring of the motion of individual molecules by limiting the number of detectable molecules in the field of view. Fluorescence recovery after photobleaching and fluorescence lifetime imaging, meanwhile, permit measurement of the dynamics of the population of fluorescently labeled molecules in the cell. Förster resonance energy transfer (FRET) between two different proteins tagged with distinct fluorescent labels with appropriate spectral overlap can be used to determine if they ever approach each other more closely than 10 nm (Schmid and Birbach 2007; Miyawaki 2011; Zhou et al. 2012). A related method is bimolecular fluorescence complementation, in which the amino-terminal half of a fluorescent protein is tethered to one protein, whereas the carboxy-terminal half is tethered to another; if the two proteins interact, they stabilize association of the two halves of the fluorescent reporter, allowing acquisition of its fluorescent state (Kerppola 2009; Filonov and Verkhusha 2013). Collectively, application of such methods can provide crucial information about the spatial and temporal behavior of molecules responsible for signaling. Microfluidic devices in which individual cells are restrained provide a convenient means to carefully manipulate their conditions and monitor their resulting responses by such techniques (Cheong et al. 2009).

The ability to tag a protein that undergoes a marked conformational change on ligand binding with a fluorescent protein that undergoes a concomitant spectral change, or tag a protein at each end with two compatible fluorophores that undergo a FRET change in response to ligand binding or a posttranslational modification, has permitted construction of in vivo biosensors that report responses in individual cells, such as an increase in intracellular calcium ion concentration (Akerboom et al. 2012; Tong et al. 2013) or activation of a particular protein kinase (Kunkel et al. 2007; Harvey et al. 2008; González-Vera 2012). Similarly, fusion of fluorescent labels to ion channels (Guerrero and Isacoff 2001; Mutoh et al. 2011) and photoinduced electron transfer to membrane-seeking chemical probes (Miller et al. 2012) provide sensitive readouts for voltage changes in individual neurons.

Conversely, fluorescently tagged ion channels whose functional properties can be switched reversibly between an active and inactive state in response to laser illumination allow nondestructive manipulation of cell behavior by light of a particular wavelength (Gorostiza and Isacoff 2008; Kramer et al. 2013). Using genetically encoded light-controlled proteins to monitor and manipulate the behavior of live cells in real time has been termed optogenetics (Lima and Miesenböck 2005; Miller 2006; Fenno et al. 2011). Other molecules that undergo dramatic conformational changes on light absorption have also been exploited, such as flavin adenine dinucleotide-binding light-, oxygen-, or voltage-sensing (LOV) domains (Christie et al. 2012; Renicke et al. 2013). Likewise, following pioneering work of Peter Quail (Leivar and Quail 2011) and Clark Lagarias (Rockwell et al. 2006) on how plant and cyanobacterial phytochromes recruit protein cofactors (PIFs) in a light-dependent manner to regulate transcription, the light-gated dimerization of a protein fused to an approximately 900-residue amino-terminal chromophore-containing fragment of a phytochrome and another protein fused to an approximately 100-residue fragment of a PIF has been exploited to control the initiation of signaling in animal cells (Levskaya et al. 2009; Toettcher et al. 2013). Note, however, that the necessary segment of the phytochrome can be difficult to work with because of its large size (nearly 100,000 kDa) and the requirement to supply exogenously its cognate linear tetrapyrrole chromophore (either phytochromobilin or phycocyanobilin), which does not readily enter all cell types. These approaches owe much to methods for small-molecule-driven protein–protein association

(Crabtree and Schreiber 1996; Farrar et al. 1996), for example, those using the immunosuppressant (and mTORC1 inhibitor) rapamycin. In the presence of rapamycin, a protein fused to the rapamycin-binding protein FKBP12 will interact with a protein fused to the FKBP12–rapamycin-complex-binding domain of mTOR, and both fusion proteins can also be marked with fluorescent tags.

Advances in instrumentation have also made possible the imaging of bioluminescent reporters, such as luciferase, in intact tissues and whole animals (Sadikot and Blackwell 2008; Close et al. 2011). Microscopes equipped with ultra-sensitive charge-coupled device cameras optimized for the near infrared range and upright parcentered and parfocal optical configurations that provide a very long working distance and high numerical aperture make the capture of this light possible. This form of optical imaging is low cost and noninvasive and facilitates real-time in situ visualization of signaling. For example, luciferase, either intact or split, when fused to a protein(s) of interest can be used for analyzing where a protein is located or in what cells a protein–protein interaction occurs, respectively (Stynen et al. 2012); and a luciferase gene placed downstream of another gene's promoter can serve as a transcriptional reporter to identify the cells that respond to a signal (Ravnskjaer et al. 2013).

Improvements in imaging have not been limited to light microscopy. Advances in electron microscopy (EM), including electron diffraction, cryo-EM, electron tomography, and single-particle reconstruction, allow visualization of the arrangement of the constituent polypeptides in large multiprotein complexes (Baumeister and Steven 2000; Frank 2009; Lander et al. 2012). Development of enzymic probes compatible with EM fixation and sectioning methods, such as miniSOG (Shu et al. 2011) and APEX (Martell et al. 2012), provide means to determine where a given signaling protein is localized at the ultrastructural level. An additional advantage of miniSOG is that this flavoprotein can also be visualized by light microscopy in living cells, which can then be processed at any given time during a signaling process for subsequent analysis at the EM level (Butko et al. 2012).

Our ability to obtain a snapshot of what is happening to individual proteins in single cells in response to an external cue is not confined to microscopy. Recently developed mass spectrometry methods allow protein constituents in a single cell to be quantified, using heavy-atom-labeled antibodies against those proteins and/or specific phospho-sites on those proteins. This method has been dubbed "mass cytometry" because fluorophore-tagged antibody reporters are replaced with isotopically tagged ones, and depends on the use of an inductively coupled-plasma time-of-flight mass spectrometer, which can resolve up to 100 different rare-earth-metal-labeled antibody probes with single-cell

sensitivity (Bandura et al. 2009; Zivanovic et al. 2013). This technology has been used successfully to define the state of 31 different proteins in particular cell types of the hematopoietic lineage in bone marrow (Bendall et al. 2011), examine the effect of 27 small-molecule protein kinase inhibitors on 14 different phosphorylation sites in human peripheral blood mononuclear cells from eight donors (Bodenmiller et al. 2012), and identify T cells specific for particular epitopes among 77 candidate rotaviral antigens (Newell et al. 2013). However, the sophisticated instrumentation necessary is expensive and not widely available. Moreover, the cells being analyzed are fixed, "stained" with the antibodies, and then destroyed in the process. Thus, cells assigned to a given signaling state by their mass cytometry signature cannot be recovered and followed subsequently (unlike in conventional cytometry, in which the cells can remain viable and be separated on the basis of their fluorescence signature by a fluorescence-activated cell sorter).

Additional technological leaps make feasible observation of single signaling molecules in action in vitro. Standard biochemical measurements yield an ensemble average and, even for a purified preparation, it is frequently difficult to discern what fraction of the molecules present are properly folded and functional. In contrast, the use of optical traps ("molecular tweezers") allows measurement of the biophysical properties of an individual protein in operation (Greenleaf et al. 2007; Moffitt et al. 2008). Similarly, in silico molecular dynamic simulations based on crystal structure coordinates and/or nuclear magnetic resonance constraints now allow us to predict the accessible conformational states of a signaling protein (Baker 2006; Friedland and Kortemme 2010), or how it might dock with another protein or small molecule (Kolb et al. 2009). This approach has been made accessible by construction of computers with the necessary speed and capacity (Dror et al. 2012) (and/or linking many computers together via the Internet) (Beauchamp et al. 2012) and corresponding improvements in algorithms for calculating the necessary energetics (Brooks et al. 2009; Lane et al. 2013).

3 SIGNALING IN THE POSTGENOMIC ERA

Genetic methods have been just as important for the progress made in deciphering signal transduction pathways. These biological regulatory mechanisms have roots deep in evolutionary time. Thus, genetic studies in single-celled eukaryotic microbes (e.g., budding [Thorner 2006] and fission [Otsubo and Yamamato 2008] yeast) and experimentally tractable invertebrates (e.g., fruit flies [Thompson 2010] and nematodes [Bargmann and Kaplan 1998]), as well as in model vertebrates (including zebrafish [Moro et al. 2013] and the mouse [Nardella et al. 2010; Guerra and Barbacid 2013]), have been invaluable. Genetic analysis of these organisms has supplied critical information about gene products that represent, generate, and/or propagate signals that control cell metabolism, growth, size, division, differentiation, shape, and motility, and that, when defective, cause disease. The development of instrumentation for facile DNA sequence determination (Shendure et al. 2004; Ståhl and Lundeberg 2012) and effective algorithms for comparative genomics (Venter et al. 2003; Dewey and Pachter 2006; Jiang et al. 2009; Washietl et al. 2012) have permitted rapid decoding of entire genomes. With this comprehensive information, the genetic findings made in model organisms can illuminate related processes in all other living things, including humans, because of the high degree of evolutionary conservation in major signaling pathways. Moreover, interrogating the entire biosphere will continue to be as important for understanding human biology and disease as studies of human beings themselves. This viewpoint is readily defensible especially when one considers that many clinically useful drugs that modulate signaling are natural products. For example, rapamycin was discovered only because someone chose to isolate a new aerobic Gram-positive filamentous bacterium (*Streptomyces hygroscopicus*) from the soil of Easter Island (Rapa Nui in Polynesian) (Vézina et al. 1975).

With the rise in genetics and genomics, sophisticated new tools for genome engineering are allowing analysis of the function of genes in human cells, including those involved in signaling, by many of the paradigmatic methods initially developed for use in studies of model organisms. These methods exploit nucleases and DNA- and RNA-binding proteins derived from yet other microorganisms (Urnov et al. 2010; Carroll 2011; Gaj et al. 2013; Wei et al. 2013). At the same time, computational methods for identifying conserved structural domains and other sequence motifs in proteins have also advanced greatly (Doolittle 1995; Copley et al. 2002; Galperin and Koonin 2012; Huang et al. 2013).

4 MODULARITY IN SIGNALING

One revelation derived from the explosion of sequence and structural information is the extent to which signaling proteins appear to have arisen during evolution by the shuffling and assembly of readily identifiable modules. These stably folded domains, joined by flexible linkers, frequently serve as recognition elements that mediate specific protein–protein interactions that link them together in specific complexes. Understanding the dynamics of the assembly of such complexes, how they direct signal propagation, enhance signaling efficiency, and insulate pathways

against inadvertent stimulation continues to be an area of ongoing research.

Another important consequence of the modular nature of the proteins involved in cellular regulation is that such an architecture allows the constituent domains to evolve discrete and separable functions, which we are just beginning to uncover and appreciate. For example, the p110β isoform of the catalytic subunit of Class 1A phosphoinositide 3-kinase (PI3K) can generate phosphatidylinositol 3,4,5-trisphosphate (PIP$_3$) at the plasma membrane via its carboxy-terminal kinase domain in response to growth-factor-initiated signaling by receptor-tyrosine kinases or GPCR activation (Vadas et al. 2011; Dbouk et al. 2012; see also p. 87 [Hemmings and Restuccia 2012]). However, on growth factor withdrawal, p110β dissociates from receptors, interacts via its so-called helical domain with the small GTPase Rab5 and helps stabilize the GTP bound state of Rab5 (Dou et al. 2013). This interaction stimulates autophagy because increasing the amount of Rab5-GTP that decorates internal membranes results in recruitment of the class III PI3K hVps34, which generates the phosphatidylinositol 3-phosphate necessary for assembly of preautophagosomes (Parzych and Klionsky 2013). The mitogen-activated protein kinase (MAPK) kinase MEK1 provides another example. It was thought to be simply a dedicated component of the Ras-Raf-MEK-ERK signaling cascade (English and Cobb 2002; Roskoski 2012; see also p. 81 [Morrison 2012]); however, we now know that once feedback phosphorylated on its amino-terminal extension by the kinase ERK, MEK1 forms a ternary complex with a multidomain adaptor protein called MAGI-1, which is necessary for membrane recruitment of the PIP$_3$-specific phosphatase PTEN; MEK1 thereby promotes down-regulation of the PIP$_3$-dependent protein kinase Akt (Zmajkovicova et al. 2013). A related issue is that inherently disordered regions in some proteins adopt alternative structures when associated with different interaction partners, leading to different outcomes. The p53 transcription factor provides a particularly dramatic example of this (Dunker et al. 2008; Joerger and Fersht 2008; Freed-Pastor and Prives 2012).

The phenomenon whereby a protein has multiple distinct functions has been dubbed "moonlighting" (Jeffery 2009). The potential for this evolving is greatest in multidomain proteins, but not restricted to them. For example, one of the splice variants of the muscle form of the glycolytic enzyme pyruvate kinase, PKM2 (Hitosugi et al. 2009), appears to have other roles. On the one hand, PKM2, when proline-hydroxylated by prolyl hydroxylase 3, seems to associate with the transcription factor HIF1α and act as a coactivator that promotes expression of HIF1α-dependent genes (Luo et al. 2011). On the other hand, once tyrosine phosphorylated in response to growth factors, PKM2 may undergo a switch in both oligomerization state (from a tetramer to a dimer) and catalytic function (from its glycolytic role to a protein kinase), and affect transcription by phosphorylating both histones (e.g., T11 on histone H3, which promotes acetylation at K9, a modification that stimulates transcription) (Yang et al. 2012) and transcription factors (e.g., Y705 in STAT3, which promotes its dimerization and transactivator function) (Gao et al. 2012).

Such instances of moonlighting in signaling proteins may help explain how the intricacies of human biology are achieved with only 21,000 or so protein-coding genes (just four times as many as a yeast cell). Thus, investigators need to be alert to the possibility that moonlighting could contribute in unanticipated ways to the biological complexity observed in a signaling process, above and beyond alternative pre-mRNA splicing, differential protein processing, and other mechanisms for generating protein diversity that we already understand.

5 POSTTRANSLATIONAL MODIFICATIONS AND SIGNALING

For many of the stably folded domains that mediate specific protein–protein interactions in signal transduction (see Ch. 2 [Lee and Yaffe 2014]), target recognition depends on a posttranslational modification of an amino acid side chain, from phosphorylation to methylation, acetylation, and ubiquitylation (Pawson and Nash 2003; Bhattacharyya et al. 2006; Nash 2012; Sadowski and Taylor 2013). Indeed, proteins can be covalently modified to form all sorts of other adducts and more than 300 types are known (Krishna and Wold 1993; Sims and Reinberg 2008; Farley and Link 2009; Hart et al. 2011; Scarpa et al. 2013). The protein–protein interactions dependent on such modifications are generally dynamic because these groups are installed and removed by enzymes whose own activity often depends on signaling—in particular, protein phosphorylation (Hunter 2012; Jin and Pawson 2012)—which makes the study of protein kinases and phosphoprotein phosphatases central to our understanding of signal transduction (Cohen 2002; Fischer 2013). This area of research has driven development of new tools for globally interrogating both the kinome (Knight et al. 2013) and the phosphoproteome (Leitner et al. 2011), including position-oriented, combinatorial, synthetic peptide libraries (Turk et al. 2006; Arsenault et al. 2011) and immobilized whole proteome arrays ("protein chips") (Ptacek and Snyder 2006) for delineating kinase-substrate specificity, ever more specific small-molecule kinase inhibitors (Cohen and Alessi 2013), genetic approaches to uniquely sensitize a given kinase to inhibition (Elphick et al. 2007; Knight and Shokat 2007; Feldman and Shokat 2010; Kliegman et al. 2013), selective chemical tags to covalently label particular kinases and phosphatases or their substrates (Allen et al. 2007; Patricelli et al. 2007; Hertz et al. 2010; Sadowsky et al. 2011; Miller et al. 2013), yeast two-hybrid screens (Cook et al. 1996; Fukada and Noda 2007; Sopko and Andrews 2008), and other strategies to trap particular kinases and phosphatases in complexes with their targets (Blanchetot et al. 2005; Boubekeur et al. 2011), as well as sophisticated mass spectrometry instrumentation and corresponding methods for detecting and cataloging phosphoproteins (Cohen and Knebel 2006; Chi et al. 2007; Gevaert and Vandekerckhove 2009; Palumbo et al. 2011; Engholm-Keller and Larsen 2013; Loroch et al. 2013; Roux and Thibault 2013).

In signaling that regulates cell growth and division (Ch. 6 [Rhind and Russell 2012] and Ch. 5 [Duronio and Xiong 2013], respectively), the actions of protein kinases are pivotal (Morgan 2007). The chemical constraints on how these enzymes operate have underappreciated implications for pathway logic and dynamics. Every catalytic turnover of a protein kinase requires that the phosphoacceptor sequence and ADP dissociate from the jaws of the active site to permit entry of a fresh molecule of ATP into the back of the catalytic pocket for the next phosphotransfer event (Tarrant and Cole 2009; Lassila et al. 2011). How then can the multifarious protein kinases present in a cell at any given time avoid adventitious modification of inappropriate targets? Likewise, how can a protein kinase phosphorylate a given substrate processively at several sites, as is often the case?

Part of the answer may lie in condition-, developmental-stage-, and tissue-specific expression of the genes encoding these protein kinases and their corresponding substrates, which could ensure that they are available only in the right cells under the right circumstances. RNA-seq experiments (for this method, see McGettigan 2013) and studies using beads containing a mixture of protein-kinase-binding inhibitors followed by identification of the bound enzymes by mass spectrometry show, however, that at least certain immortalized cell lines and some breast tumor tissue express the majority (60%–75%) of the kinome repertoire (Duncan et al. 2012). But, kinase targets may be more cell-type specific. For example, arachidonic acid, the precursor for eisocanoids, is released from membrane phospholipids in tissues such as spleen, gut, white fat, and macrophages by the action of the known kinase target cytosolic phospholipase A2 (cPLA2). In humans, this enzyme exists as six isoforms (cPLA2α, cPLA2β, cPLA2γ, cPLA2δ, cPLA2ε, and cPLA2ζ) encoded by distinct genes. The most ubiquitously expressed and well-studied isoform, cPLA2α, is phosphorylated at multiple sites by various different protein kinases. Depending on the cell type and agonist examined, phosphorylation seems to regulate both membrane binding and catalytic activity of cPLA2α (Ghosh et al. 2006;

Dennis et al. 2011). The biochemical properties and tissue distributions of the other five isoforms suggest that regulation of their phosphorylation is distinct from that of cPLA2α and may also involve other mechanisms. Moreover, although cPLA2 is responsible for arachidonate generation in many tissues, this fatty acid is supplied in brain, liver, and lung via the action of a separate class of serine acylhydrolase, MAGL, which further extends the regulatory possibilities (Savinainen et al. 2012; Mulvihill and Nomura 2013).

There are many other mechanisms that can dictate how a protein kinase achieves its target specificity (see Ch. 2 [Lee and Yaffe 2014]). On the one hand, various molecular matchmaker strategies (see below) have evolved to ensure that a protein kinase encounters its proper substrates with high probability and efficiency (Endicott et al. 2012). On the other hand, for subsequent events to unfold, the rate of phosphorylation must exceed the rate of dephosphorylation by ever-present and more promiscuous phosphoprotein phosphatases (Cohen 1992). For example, because of just this sort of antagonism, during the response to stimulation by epidermal growth factor (EGF), the phosphorylated tyrosine sites on the cytosolic domain of the EGF receptor turn over 100–1000 times before maximal receptor phosphorylation is achieved (Kleiman et al. 2011). Thus, any elevation in phosphorylation detected in response to a stimulus is not simply caused by modifications that are installed and remain until the signal is terminated. Hence, the hydrolytic activity of phosphatases ensures that inefficient or inadvertent phosphorylation events have no physiological consequence.

Protein kinases that act on targets at a cellular membrane are either integral membrane proteins (e.g., receptor tyrosine kinases) or possess domains that permit association with receptors or receptor-associated proteins (e.g., the JAK family of protein tyrosine kinases bound to the cytosolic segments of cytokine receptors) and/or are posttranslationally modified with substituents (e.g., *N*-myristoyl and *S*-palmitoyl groups) that strongly promote partition into the membrane (e.g., the Src family of protein tyrosine kinases) (Groves and Kuriyan 2010). In other instances, association of the kinase and its substrate with a third partner (a scaffold, linker, or anchor protein) brings about the necessary propinquity and, in addition, enhances reaction rate by achieving a high local concentration of the reactants (Ferrell 2000; Kuriyan and Eisenberg 2007). Thus, if biochemical analysis of the proteins associated with a given protein kinase, or genetic analysis of a process, reveals another gene in which mutation yields a phenotype similar to that resulting from loss of the kinase, and the gene product in question has no obvious catalytic function itself, then it may serve such a scaffolding role. Moreover, there

is emerging evidence that at least one mammalian protein involved in the innate immune response to RNA virus infection serves as just such a signal-activating platform when, in prion-like fashion, it is converted into an amyloid-like fibril (Hou et al. 2011; Wu 2013).

In some cases, phosphorylation of a substrate by one protein kinase converts a sequence motif into a high-affinity binding site for, and permits its subsequent phosphorylation by, another protein kinase. The first of such two-step modifications is termed "priming." For example, the prior phosphorylation of substrates by the cell cycle kinase cyclin-dependent kinase (CDK) 1 can generate a phospho-epitope recognized by the Polo box domains of the Polo family of protein kinases, which execute later cell cycle events (Lowery et al. 2005; Strebhardt 2010). This feature can be exploited in other ways. For example, a peptide library approach (Turk and Cantley 2003) indicates GSK3β is a protein kinase that strongly prefers to phosphorylate a primed substrate, one that possesses a serine or threonine residue in a sequence with a phosphorylated serine residue at position +4. When S9 in GSK3β is phosphorylated by Akt, however, this modification creates a substrate-like sequence that moors its own amino terminus in its active site, thereby blocking GSK3β action on other primed substrates (Frame and Cohen 2001; Weston and Davis 2001). For both kinetic and thermodynamic reasons, intramolecular interactions are favored over intermolecular associations. Hence, investigators should always be aware that phosphorylation sites present in a kinase may serve such roles.

In many cases, a protein kinase recognizes and interacts with its substrate via at least one other association distinct and physically distant from the active-site–phosphoacceptor-sequence interaction. Such secondary points of enzyme-substrate binding are termed docking sites (Bardwell and Thorner 1996; Reményi et al. 2006; Goldsmith et al. 2007). The more docking interactions between a kinase and its target, the higher the probability that the substrate will not dissociate from the enzyme in the time it takes for the next catalytic turnover event. Thus, multiple docking interactions ensure that the enzyme will stay bound to the same substrate molecule (and phosphorylate another phosphoacceptor site, if there is one) rather than jump to a new substrate. Moreover, such docking interactions supply additional binding energy that makes it possible for substrate phosphorylation to occur at noncanonical (suboptimal, lower affinity) phosphorylation sites. Such considerations help explain why an investigator may find, perhaps perplexingly, nonconsensus sites in mass spectrometry data, even when they are using a purified substrate and protein kinase.

A particularly illustrative example of the vital role of such docking interactions in ensuring processive multisite phosphorylation that is now well understood at the mechanistic level is CDK1-dependent phosphorylation of the yeast CDK inhibitor Sic1 (Nash et al. 2001). In the case of Sic1, the phosphomodifications are a prelude to its timely ubiquitylation by an SCF-type ubiquitin ligase (Silverman et al. 2012) whose substrate recognition subunit (Cdc4) interacts with phosphoepitopes (Tang et al. 2012). In Sic1, there is a sequence motif (VLLPP) that binds with high affinity to a site in the G_1 cyclin (Cln2) in Cln2-CDK1-Cks1 complexes; similarly, there are four interspersed RxL motifs that bind to a site in the S phase cyclin (Clb5) in Clb5–CDK1-Cks1 complexes. The obligatory Cks1 subunit in both CDK complexes contains a pocket that preferentially binds phospho-TP sites (CDKs are generally highly selective for -SP- and -TP- sites); hence, once phosphorylated at any such site, the phosphoepitope–Cks1 interaction provides yet another docking interaction. After a sufficient amount of the Cln2-bound CDK1-Cks1 complex builds up during G_1 phase, it holds onto Sic1 via two contacts (active-site–phosphoacceptor-site binding and VLLPP-motif–Cln2-docking-pocket association); following the first phosphorylation, there are now three such interactions possible (active-site–phosphoacceptor-site binding, VLLPP-motif–Cln2-docking-pocket association, and phosphoepitope–Cks1 interaction), which explains initiation and establishment of processive phosphorylation (Kõivomägi et al. 2011). Similarly, as the amount of Clb5-bound CDK1-Cks1 builds up in late-G_1/early-S phase, maintenance and completion of processive multisite phosphorylation by the Clb5-CDK1-Cks1 enzyme is very efficient, given that it has four times the probability of engaging an RxL motif than the Cln2-CDK1-Cks1 complex does a single VLLPP motif (Venta et al. 2012).

6 INTEGRATION OF CELL METABOLISM AND SIGNALING

Intermediates in metabolism may have been the first intracellular (and intercellular) signaling molecules, acting as feedback or feed-forward regulators (either allosteric effectors or covalent modifiers) of enzymes in metabolic pathways and transcription factors that controlled expression of those enzymes. Our focus on other levels of regulation has perhaps diverted us from exploring these more ancestral control mechanisms because, in metazoans, GPCR and protein kinase signaling in response to hormones and growth factors override the selfish metabolic needs of any given cell, in favor of the needs of the organism (see Ch. 14 [Hardie 2012]).

So, just as proteins can have moonlighting functions, we need to better understand the manifold functions in signaling of what were formerly considered merely metabolic intermediates. For example, until the stimulatory function of histone lysine ε-N-acetylation in chromatin remodeling and gene expression was uncovered (Racey and Byvoet 1971; Turner 1991), acetyl-CoA was presumed to have only the less glamorous role of conveying carbon from glycolytically generated pyruvate, or the breakdown of fatty acids, to the tricarboxylic acid (TCA) cycle for energy generation in the mitochondrion. We now know that acetyl-CoA has other roles that impinge critically on the capacity of a cell to signal (Lin et al. 2013). For example, in neurons and some other cells, it is needed for synthesis of the central neurotransmitter acetylcholine. Moreover, in all cells, three molecules of acetyl-CoA can condense to form HMG-CoA for the mevalonate pathway to make both isoprenoid compounds (including the farnesyl and geranylgeranyl moieties attached to the carboxy-terminal CAAX motifs in many small GTPases involved in signaling, such as H-Ras and K-Ras) (Berndt et al. 2011; Resh 2012) and cholesterol (which is not only important for the architecture and fluidity of the membrane in which receptors reside, but is also attached to the carboxyl terminus of an important developmental signaling protein, Hedgehog; Creanga et al. 2012; p. 107 [Ingham 2012]). Yet another role for acetyl-CoA that affects signaling has recently turned up. The α-N-acetylation of Met1 in the Rub1/Nedd8-specific E2 enzyme Ubc12 (although an irreversible modification, once installed) increases the avidity of its binding to a hydrophobic pocket in its cognate E3 ligase Dcn1 and thereby stimulates Nedd8-ylation of the Cul1 scaffold protein (Monda et al. 2013), which is necessary, in turn, for assembly and activity of the SCF-type E3 ligases that control the ubiquitin-dependent degradation of numerous signaling proteins (Deshaies et al. 2010). It is worth noting that, simply through substrate availability (and dependent on the relative K_m values of the various ε-N-lysine acetyltransferases), the level of acetyl-CoA will dictate the rate of acetylation of histones and other proteins with concomitant consequences for gene expression and other cellular functions (Starai et al. 2004; Londoño Gentile et al. 2013; Zhang et al. 2013).

Acetyl-CoA is just one prominent nexus linking cell metabolism and signaling. The levels of other key metabolites have multiple effects that influence both metabolism and cell signaling in ways that we are just beginning to understand. For example, 2-oxoglutarate (2OG, also known as α-ketoglutarate) is a central intermediate in the TCA cycle, but also needed for amino acid interconversion and breakdown by transamination, which may affect the activity level of a major amino acid sensor and growth regulator, mTORC1 (Bar-Peled et al. 2012). In addition, 2OG is a critical cofactor for the Jumonji class of histone methyl-N-lysine demethylases (Hou and Yu 2010), TET family 5-methylcytosine hydroxylases (Wu and Zhang 2011), and EglN-type prolyl-4-hydroxylases (Freeman et al. 2003), all of which likely affect the level of expression of genes encoding signaling proteins. Moreover, neomorphic mutations in isocitrate dehydrogenase (both IDH1 and IDH2) that prevent oxidative decarboxylation of isocitrate to 2OG, but catalyze instead NADPH-dependent reduction of 2OG to (R)-2-hydroxyglutarate (Dang et al. 2009), an antagonist of 2OG-dependent protein and DNA demethylation, are strongly associated with certain cancers, clearly implicating altered metabolism in tumorigenesis (Losman and Kaelin 2013).

Unsurprisingly, the major cellular methyl donor, S-adenosylmethionine (AdoMet), is important not only for methylation of DNA bases and epigenetic silencing (Guibert and Weber 2013), but also for protein methylation that influences signaling. For example, AdoMet-dependent N-methylation of an arginine residue in the inhibitory factor Smad6 dissociates it from activated bone morphogenetic protein (BMP) receptors, alleviating inhibition and enabling the type I subunit of the BMP receptor complex to phosphorylate the transcription factors Smad1 and Smad5, which permits their nuclear import and activates their gene regulatory function (Xu et al. 2013). Similarly, succinyl-CoA is another central intermediate in the TCA cycle, and lysine succinylation and cysteine succination have both been recently identified as posttranslational modifications of proteins that influence the properties of the enzymes to which they are attached (Zhang et al. 2011; Lin et al. 2012). An open-ended question for future study is how many other cellular metabolites lie at the crossroads between central metabolism and cell signaling or are involved in the fine-tuning of biological processes that impinge on signaling.

7 SIGNAL DIVERSITY

During evolution, mechanisms have arisen that allow diverse cell types to sense and respond to various stimuli. In vision, light photons are absorbed by the rhodopsins (chromophore-containing GPCRs) and converted into intracellular chemical and then electrochemical changes in the rod and cone cells in the retinas of our eyes. Hearing depends on conversion of sound waves into intracellular signals via the opening and closing of stretch-activated ion channels that are mechanically coupled to ciliary bundles in specialized hair cells in our ears. In a similar way, touch depends on conversion of mechanical pressure or thermal differences into conformational changes in ion channels that open to elevate the level of intracellular calcium ions in specialized nerve fibers in our skin. Of course,

other stimuli to which our senses respond are chemical in nature. Taste depends on conversion of the binding of various soluble compounds into intracellular changes in the gustatory-receptor-containing cells in the papillae on our tongues. Smell depends on conversion of the binding of various volatile compounds into intracellular signals in the olfactory-receptor-containing cells on the roof of our nasal cavity. Skin irritation caused by the common nettle in our gardens has a similar source; its trichomes inject chemicals normally made in animal cells, such as the neurotransmitter serotonin and the immune mediator histamine, thus aggravating our nerves and provoking inflammation.

Indeed, the variety of chemical signals generated by cells and to which they are able to respond is rather staggering. Many of the classical endocrine and pituitary hormones were discovered and chemically identified in the 1800s and early 1900s. For example, Frederick Banting announced the isolation of insulin in 1921 and Fred Sanger determined its structure in 1953 (Nobel Prizes being awarded for both accomplishments). However, new types of hormones that impact cell and organismal physiology and development are still being discovered at a surprising rate, which requires that their cognate receptors and downstream mediators be identified. For example, adipose-derived leptin and its receptor (Isse et al. 1995; Tartaglia et al. 1995), neuropeptides orexin-A/hypocretin-1 and andorexin-B/hypocretin-2 and their receptors (de Lecea et al. 1998; Sakurai et al. 1998), and stomach-derived ghrelin and its receptor (Howard et al. 1996; Kojima et al. 1999), which have such critical roles in controlling the interrelated processes of energy metabolism and obesity, wakefulness and appetite, and hunger and growth, were not characterized until the late 1990s. Indeed, there are still many GPCRs and nuclear receptors that are "orphans," in the sense that their physiological ligands have not yet been determined (Civelli 2012; Pearen and Muscat 2012).

The constellation of known signaling molecules continues to expand in new and unanticipated directions. For example, it has recently been appreciated that different bacterial species, some of which are intracellular pathogens, synthesize unusual cyclic dinucleotides that control their own transcription, including cyclic-di-AMP, cyclic-di-GMP, and the mixed cyclic-GMP-AMP, in all cases linked $3'$-$5'$ (Kalia et al. 2013). The presence of such compounds in mammalian cells is sensed by their binding to an endoplasmic reticulum (ER)-localized protein called STING (Woodward et al. 2010). Once activated, STING stimulates a protein kinase (TANK-binding kinase 1), which, in turn, phosphorylates and activates a transcription factor, IRF3, that regulates interferon production (Chin et al. 2013). However, STING is activated much more potently and elicits a more efficacious interferon response when foreign DNA enters cells. The difference is due to the fact that cytosolic DNA binds to the regulatory domain of a mammalian enzyme, cGAMP synthase (cGAS), thereby stimulating production of cyclic-GMP-AMP in which the linkage is cyclic-[G(2′-5′)pA(3′-5′)p] (Kranzusch et al. 2013; Shaw and Liu 2014). This endogenously generated signal potently activates diverse hSTING variants, whereas not all of them respond well to bacterial cyclic-[G(3′-5′)pA(3′-5′)p] (Diner et al. 2013). Thus, mammalian cells have evolved a mechanism to generate a novel signal that permits the innate immune system to distinguish infiltration by a naked DNA from entry of a bacterial invader. Such interkingdom signaling, mediating interplay between viruses, bacteria, and mammalian cells (Lim et al. 2009; Marks et al. 2013; Pluznick et al. 2013), is clearly prevalent, has important consequences for human health, and warrants continued exploration (see also Ch. 20 [Alto and Orth 2012]).

How a signal is deployed or displayed can also have information content that needs to be considered. Extracellular signaling ligands not only are released from cells via the classical secretory pathway in an autocrine, juxtacrine, or endocrine manner, but can also pass "through" cells via the process of paracytophagy and the generation of so-called argosomes (Greco et al. 2001). Such mechanisms are also used for the entry and cell-to-cell passage of many prokaryotic intracellular pathogens (Portnoy 2012). Some mammalian cell types can even engulf an entire other cell by a macroendocytic process dubbed entosis (Overholtzer et al. 2007; Florey and Overholtzer 2012). We now appreciate that cells can also generate "exosomes" (small ∼100-nm-diameter vesicles) as a means for quantal export of ligands and other classes of informational molecules, including miRNAs, because, once released into extracellular fluids, they can be taken up by other cells (Bang and Thum 2012; Briscoe and Thérond 2013; Choi et al. 2013).

Cells know their place, at least in part, by making contacts with adjacent cells and components of the extracellular matrix. However, many cells types erect a single specialized projection, the primary cilium (Garcia-Gonzalo and Reiter 2012; Nozawa et al. 2013), which constrains to this one location certain classes of signaling receptors (such as the receptors for the Hedgehog family of ligands) (Wong and Reiter 2008; Ch. 10 [Perrimon et al. 2012]), presumably to confer the capacity to respond only to a highly polarized or localized signal source. In contrast, other cells extend ultrafine processes and specialized filopodial extensions (also referred to as cytonemes) that can mediate cell-to-cell contacts and long-range transport of signal molecules over a distance of many cell lengths (Roy et al. 2011; Sanders et al.

2013). Similarly, juxtaposed cells in many epithelial layers are connected by gap junctions that act as portholes through which certain intracellular signals, such as an increase in cytosolic calcium ion concentration, can be spread from cell to cell (Goodenough and Paul 2009). Clearly, we still have much to learn about the interplay between signaling molecules and all levels of cellular organization.

8 PROSPECTUS

For the foreseeable future, signal transduction research remains a field confronted with a still vast frontier. As the contents of this book and issues raised here make clear, unraveling signal transduction processes is a multidisciplinary enterprise. Ultimately, understanding signaling will require an appreciation for, understanding of, and the means to usefully grasp the seamless interconnectedness among all the bits of information gleaned by what were formerly considered disparate branches of the biological, chemical, and physical sciences. Fortunately, as we highlight in this book, there are recurrent themes, general mechanisms, common strategies, and ubiquitous reactions in cell signaling that allow the complexity to be parsed out productively. Indeed, we can already discern general design principles that are allowing us to reengineer cellular responses to stimuli different from those evolved in nature (Pryciak 2009; Burrill et al. 2011; Blount et al. 2012; Lim et al. 2013).

Nonetheless, many unexpected discoveries that provide novel insights and fresh paradigms will continue to be made, and these will open up new avenues for understanding cell signaling. Some recent examples highlight this point. Remarkably, proteins of the previously uncharacterized Fam20C family turn out to be secreted atypical protein kinases that phosphorylate casein and other extracellular substrates that have important physiological roles in bone mineralization (Tagliabracci et al. 2013; Xiao et al. 2013b). This discovery raises interesting questions about how these enzymes acquire ATP in the lumen of the Golgi and whether such enzymes are made and have important roles in the nervous system, in which ATP is stored in synaptic vesicles (Zimmermann 2008) and released extracellularly to stimulate purinergic receptors (Khakh and North 2012). Likewise, it has been appreciated for decades that blood platelets also store ATP and other compounds in storage vesicles that are released extracellularly when platelets are activated at a site of injury (Da Prada et al. 1971; Higashii et al. 1985). Does this released ATP also act, in part, through Fam20C-like extracellular protein kinases? Another example is that very high serum levels of high-molecular-weight hyaluronan dramatically protect the naked mole rat against cancer (Tian et al. 2013). Hyaluronan is an extracellular proteoglycan and can bind to various cell surface receptors. How does it suppress malignant growth and/or enhance immune surveillance of precancerous tissue? And, do the same mechanisms operate in humans?

It is also clear that we have just began to scratch the surface of how microRNAs (Martinez and Gregory 2010; Mendell and Olson 2012) and other RNAs encoded in our genomes (Hancks and Kazazian 2012; Batista and Chang 2013) coordinately influence the levels of gene products involved in signaling and are themselves controlled by signaling processes. Likewise, an area of traditional biochemistry that clearly intersects with signaling in many ways is the function of various classes of proteases; yet, we understand the roles of many of these enzymes only very superficially and know even less about their physiologically relevant enzyme–substrate relationships. For example, in 2012, a new, circulating, exercise-induced regulatory hormone, dubbed irisin, was described that converts white fat into more thermogenic beige fat (Boström et al. 2012). However, irisin is identical to the ectodomain of a small (212-residue) cell surface protein, fibronectin type III domain-containing protein 5 (FNDC5), which is anchored in the plasma membrane by a single carboxy-terminal hydrophobic transmembrane segment. In fact, protease-mediated shedding of the extracellular domains of transmembrane signaling proteins as separate entities with distinct functions is a common phenomenon (Horiuchi 2013). Exercise stimulates FNDC5 expression in skeletal muscle and the cleavage and release of its uniquely structured amino-terminal fibronectin-III-like ectodomain as irisin (Erickson 2013; Schumacher et al. 2013); however, the protease responsible for this shedding, and whether it too is under any sort of regulation, is unknown.

Thus, if we take the long view, studies of signal transduction are still in exponential phase with many important discoveries to come. Moreover, we anticipate continued development of evermore sophisticated experimental tools—from improvements in automated deep sequencing to characterize the global transcriptome (Malone and Oliver 2011), to new mass spectrometry instrumentation to catalog the cellular metabolome (Rubakhin et al. 2013), to further refinement of mathematical, statistical, and computational theories and methods to assist with display, interpretation, and modeling of the complex networks of relationships involved in intra- and intercellular signaling (Janes and Lauffenburger 2013; see also Ch. 4 [Azeloglu and Iyengar 2014]). Continued advances of this sort will allow us to address questions at an ever greater level of detail and resolution, providing answers at the molecular level to long-standing mechanistic questions about the myriad processes that comprise cell signaling.

REFERENCES

*Reference is in this book.

Akerboom J, Chen TW, Wardill TJ, Tian L, Marvin JS, Mutlu S, Calderón NC, Esposti F, Borghuis BG, Sun XR, et al. 2012. Optimization of a GCaMP calcium indicator for neural activity imaging. *J Neurosci* **32:** 13819–13840.

Allen JJ, Li M, Brinkworth CS, Paulson JL, Wang D, Hübner A, Chou WH, Davis RJ, Burlingame AL, Messing RO, et al. 2007. A semi-synthetic epitope for kinase substrates. *Nat Methods* **4:** 511–516.

* Alto NM, Orth K. 2012. Subversion of cell signaling by pathogens. *Cold Spring Harb Perspect Biol* **4:** a006114.

Arsenault R, Griebel P, Napper S. 2011. Peptide arrays for kinome analysis: New opportunities and remaining challenges. *Proteomics* **11:** 4595–4609.

* Azeloglu EU, Iyengar R. 2014. Signaling networks: Information flow, computation, and decision making. *Cold Spring Harb Perspect Biol* doi: 10.1101/cshperspect.a005934.

Baker D. 2006. Prediction and design of macromolecular structures and interactions. *Philos Trans R Soc Lond B Biol Sci* **361:** 459–463.

Bandura DR, Baranov VI, Ornatsky OI, Antonov A, Kinach R, Lou X, Pavlov S, Vorobiev S, Dick JE, Tanner SD. 2009. Mass cytometry: Technique for real time single cell multitarget immunoassay based on inductively coupled plasma time-of-flight mass spectrometry. *Anal Chem* **81:** 6813–6822.

Bang C, Thum T. 2012. Exosomes: New players in cell-cell communication. *Int J Biochem Cell Biol* **44:** 2060–2064.

Bardwell L, Thorner J. 1996. A conserved motif at the amino termini of MEKs might mediate high-affinity interaction with the cognate MAPKs. *Trends Biochem Sci* **21:** 373–374.

Bargmann CI, Kaplan JM. 1998. Signal transduction in the *Caenorhabditis elegans* nervous system. *Annu Rev Neurosci* **21:** 279–308.

Bar-Peled L, Schweitzer LD, Zoncu R, Sabatini DM. 2012. Ragulator is a GEF for the rag GTPases that signal amino acid levels to mTORC1. *Cell* **150:** 1196–1208.

Batista PJ, Chang HY. 2013. Long noncoding RNAs: Cellular address codes in development and disease. *Cell* **152:** 1298–1307.

Baumeister W, Steven AC. 2000. Macromolecular electron microscopy in the era of structural genomics. *Trends Biochem Sci* **25:** 624–631.

Beauchamp KA, McGibbon R, Lin YS, Pande VS. 2012. Simple few-state models reveal hidden complexity in protein folding. *Proc Natl Acad Sci* **109:** 17807–17813.

Bendall SC, Simonds EF, Qiu P, Amir el-AD, Krutzik PO, Finck R, Bruggner RV, Melamed R, Trejo A, Ornatsky OI, et al. 2011. Single-cell mass cytometry of differential immune and drug responses across a human hematopoietic continuum. *Science* **332:** 687–696.

Bensimon A, Heck AJ, Aebersold R. 2012. Mass spectrometry-based proteomics and network biology. *Annu Rev Biochem* **81:** 379–405.

Berndt N, Hamilton AD, Sebti SM. 2011. Targeting protein prenylation for cancer therapy. *Nat Rev Cancer* **11:** 775–791.

Bessman NJ, Lemmon MA. 2012. Finding the missing links in EGFR. *Nat Struct Mol Biol* **19:** 1–3.

Bhattacharyya RP, Reményi A, Yeh BJ, Lim WA. 2006. Domains, motifs, and scaffolds: The role of modular interactions in the evolution and wiring of cell signaling circuits. *Annu Rev Biochem* **75:** 655–680.

Blanchetot C, Chagnon M, Dubé N, Hallé M, Tremblay ML. 2005. Substrate-trapping techniques in the identification of cellular PTP targets. *Methods* **35:** 44–53.

Blount BA, Weenink T, Ellis T. 2012. Construction of synthetic regulatory networks in yeast. *FEBS Lett* **586:** 2112–2121.

Bodenmiller B, Zunder ER, Finck R, Chen TJ, Savig ES, Bruggner RV, Simonds EF, Bendall SC, Sachs K, Krutzik PO, et al. 2012. Multiplexed mass cytometry profiling of cellular states perturbed by small-molecule regulators. *Nat Biotechnol* **30:** 858–867.

Borrmann A, Milles S, Plass T, Dommerholt J, Verkade JM, Wiessler M, Schultz C, van Hest JC, van Delft FL, Lemke EA. 2012. Genetic encoding of a bicyclo[6.1.0]nonyne-charged amino acid enables fast cellular protein imaging by metal-free ligation. *Chembiochem* **13**: 2094–2099.

Boström P, Wu J, Jedrychowski MP, Korde A, Ye L, Lo JC, Rasbach KA, Boström EA, Choi JH, Long JZ, et al. 2012. A PGC1-α-dependent myokine that drives brown-fat-like development of white fat and thermogenesis. *Nature* **481**: 463–468.

Boubekeur S, Boute N, Pagesy P, Zilberfarb V, Christeff M, Issad T. 2011. A new highly efficient substrate-trapping mutant of protein tyrosine phosphatase 1B (PTP1B) reveals full autoactivation of the insulin receptor precursor. *J Biol Chem* **286**: 19373–19380.

Bradshaw JM. 2010. The Src, Syk, and Tec family kinases: Distinct types of molecular switches. *Cell Signal* **22**: 1175–1184.

Briscoe J, Thérond PP. 2013. The mechanisms of Hedgehog signalling and its roles in development and disease. *Nat Rev Mol Cell Biol* **14**: 416–429.

Brooks BR, Brooks CL III, Mackerell ADJ, Nilsson L, Petrella RJ, Roux B, Won Y, Archontis G, Bartels C, Boresch S, et al. 2009. CHARMM: The biomolecular simulation program. *J Comput Chem* **30**: 1545–1614.

Burrill DR, Boyle PM, Silver PA. 2011. A new approach to an old problem: Synthetic biology tools for human disease and metabolism. *Cold Spring Harb Symp Quant Biol* **76**: 145–154.

Butko MT, Yang J, Geng Y, Kim HJ, Jeon NL, Shu X, Mackey MR, Ellisman MH, Tsien RY, Lin MZ. 2012. Fluorescent and photo-oxidizing Time-STAMP tags track protein fates in light and electron microscopy. *Nat Neurosci* **15**: 1742–1751.

Carroll D. 2011. Genome engineering with zinc-finger nucleases. *Genetics* **188**: 773–782.

Chakraborty AK, Das J. 2010. Pairing computation with experimentation: A powerful coupling for understanding T cell signalling. *Nat Rev Immunol* **19**: 59–71.

Chalfie M, Tu Y, Euskirchen G, Ward WW, Prasher DC. 1994. Green fluorescent protein as a marker for gene expression. *Science* **263**: 802–805.

Cheong R, Wang CJ, Levchenko A. 2009. Using a microfluidic device for high-content analysis of cell signaling. *Sci Signal* **2**: pl2.1–pl2.18.

Chi A, Huttenhower C, Geer LY, Coon JJ, Syka JE, Bai DL, Shabanowitz J, Burke DJ, Troyanskaya OG, Hunt DF. 2007. Analysis of phosphorylation sites on proteins from *Saccharomyces cerevisiae* by electron transfer dissociation (ETD) mass spectrometry. *Proc Natl Acad Sci* **104**: 2193–2198.

Chin JW. 2014. Expanding and reprogramming the genetic code of cells and animals. *Annu Rev Biochem* doi: 10.1146/annurev-biochem-060713-035737.

Chin KH, Tu ZL, Su YC, Yu YJ, Chen HC, Lo YC, Chen CP, Barber GN, Chuah ML, Liang ZX, et al. 2013. Novel c-di-GMP recognition modes of the mouse innate immune adaptor protein STING. *Acta Crystallogr D Biol Crystallogr* **69**: 352–366.

Chiu W, Baker ML, Almo SC. 2006. Structural biology of cellular machines. *Trends Cell Biol* **16**: 144–150.

Choi DS, Kim DK, Kim YK, Gho YS. 2013. Proteomics, transcriptomics and lipidomics of exosomes and ectosomes. *Proteomics* **13**: 1554–1571.

Christie JM, Gawthorne J, Young G, Fraser NJ, Roe AJ. 2012. LOV to BLUF: Flavoprotein contributions to the optogenetic toolkit. *Mol Plant* **5**: 533–544.

Civelli O. 2012. Orphan GPCRs and neuromodulation. *Neuron* **76**: 12–21.

Close DM, Xu T, Sayler GS, Ripp S. 2011. In vivo bioluminescent imaging (BLI): Noninvasive visualization and interrogation of biological processes in living animals. *Sensors* **11**: 180–206.

Coelho M, Maghelli N, Tolic-Nørrelykke IM. 2013. Single-molecule imaging in vivo: The dancing building blocks of the cell. *Integr Biol* **5**: 748–758.

Cohen P. 1992. Signal integration at the level of protein kinases, protein phosphatases and their substrates. *Trends Biochem Sci* **17**: 408–413.

Cohen P. 2002. The origins of protein phosphorylation. *Nat Cell Biol* **4**: E127–E130.

Cohen P, Alessi DR. 2013. Kinase drug discovery—What's next in the field? *ACS Chem Biol* **8**: 96–104.

Cohen P, Knebel A. 2006. KESTREL: A powerful method for identifying the physiological substrates of protein kinases. *Biochem J* **393**: 1–6.

Cook JG, Bardwell L, Kron SJ, Thorner J. 1996. Two novel targets of the MAP kinase Kss1 are negative regulators of invasive growth in the yeast *Saccharomyces cerevisiae*. *Genes Dev* **10**: 2831–2848.

Copley RR, Doerks T, Letunic I, Bork P. 2002. Protein domain analysis in the era of complete genomes. *FEBS Lett* **513**: 129–134.

Corrêa IR Jr, Baker B, Zhang A, Sun L, Provost CR, Lukinavicius G, Reymond L, Johnsson K, Xu MQ. 2013. Substrates for improved live-cell fluorescence labeling of SNAP-tag. *Curr Pharm Des* **19**: 5414–5420.

Crabtree GR, Schreiber SL. 1996. Three-part inventions: Intracellular signaling and induced proximity. *Trends Biochem Sci* **21**: 418–422.

Creanga A, Glenn TD, Mann RK, Saunders AM, Talbot WS, Beachy PA. 2012. Scube/You activity mediates release of dually lipid-modified Hedgehog signal in soluble form. *Genes Dev* **26**: 1312–1325.

Dang L, White DW, Gross S, Bennett BD, Bittinger MA, Driggers EM, Fantin VR, Jang HG, Jin S, Keenan MC, et al. 2009. Cancer-associated IDH1 mutations produce 2-hydroxyglutarate. *Nature* **462**: 739–744.

Da Prada M, Pletscher A, Tranzer JP. 1971. Storage of ATP and 5-hydroxytryptamine in blood platelets of guinea-pigs. *J Physiol* **217**: 679–688.

Dbouk HA, Vadas O, Williams RL, Backer JM. 2012. PI3Kβ downstream of GPCRs—Crucial partners in oncogenesis. *Oncotarget* **3**: 1485–1486.

Debets MF, van Hest JC, Rutjes FP. 2013. Bioorthogonal labelling of biomolecules: New functional handles and ligation methods. *Org Biomol Chem* **11**: 6439–6455.

de Lecea L, Kilduff TS, Peyron C, Gao X, Foye PE, Danielson PE, Fukuhara C, Battenberg EL, Gautvik VT, Bartlett FS, et al. 1998. The hypocretins: Hypothalamus-specific peptides with neuroexcitatory activity. *Proc Natl Acad Sci* **95**: 322–327.

Dennis EA, Cao J, Hsu YH, Magrioti V, Kokotos G. 2011. Phospholipase A₂ enzymes: Physical structure, biological function, disease implication, chemical inhibition, and therapeutic intervention. *Chem Rev* **111**: 6130–6185.

Deshaies RJ, Emberley ED, Saha A. 2010. Control of cullin-ring ubiquitin ligase activity by nedd8. *Subcell Biochem* **54**: 41–56.

Dewey CN, Pachter L. 2006. Evolution at the nucleotide level: The problem of multiple whole-genome alignment. *Hum Mol Genet* **15**: R51–R56.

Diner EJ, Burdette DL, Wilson SC, Monroe KM, Kellenberger CA, Hyodo M, Hayakawa Y, Hammond MC, Vance RE. 2013. The innate immune DNA sensor cGAS produces a noncanonical cyclic dinucleotide that activates human STING. *Cell Rep* **3**: 1355–1361.

Doolittle RF. 1995. The multiplicity of domains in proteins. *Annu Rev Biochem* **64**: 287–314.

Dou Z, Pan JA, Dbouk HA, Ballou LM, DeLeon JL, Fan Y, Chen JS, Liang Z, Li G, Backer JM, et al. 2013. Class IA PI3K p110β subunit promotes autophagy through Rab5 small GTPase in response to growth factor limitation. *Mol Cell* **50**: 29–42.

Dror RO, Dirks RM, Grossman JP, Xu H, Shaw DE. 2012. Biomolecular simulation: A computational microscope for molecular biology. *Annu Rev Biophys* **41**: 429–452.

Duncan JS, Whittle MC, Nakamura K, Abell AN, Midland AA, Zawistowski JS, Johnson NL, Granger DA, Jordan NV, Darr DB, et al. 2012. Dynamic reprogramming of the kinome in response to targeted MEK inhibition in triple-negative breast cancer. *Cell* **149**: 307–321.

Dunker AK, Silman I, Uversky VN, Sussman JL. 2008. Function and structure of inherently disordered proteins. *Curr Opin Struct Biol* **18**: 756–764.

* Duronio RJ, Xiong Y. 2013. Signaling pathways that control cell proliferation. *Cold Spring Harb Perspect Biol* **5:** a008904.

Elphick LM, Lee SE, Gouverneur V, Mann DJ. 2007. Using chemical genetics and ATP analogues to dissect protein kinase function. *ACS Chem Biol* **2:** 299–314.

Encell LP, Friedman Ohana R, Zimmerman K, Otto P, Vidugiris G, Wood MG, Los GV, McDougall MG, Zimprich C, Karassina N, et al. 2012. Development of a dehalogenase-based protein fusion tag capable of rapid, selective and covalent attachment to customizable ligands. *Curr Chem Genomics* **6:** 55–71.

Endicott JA, Noble MA, Johnson LN. 2012. The structural basis for control of eukaryotic protein kinases. *Annu Rev Biochem* **81:** 587–613.

Endres NF, Das R, Smith AW, Arkhipov A, Kovacs E, Huang Y, Pelton JG, Shan Y, Shaw DE, Wemmer DE, et al. 2013. Conformational coupling across the plasma membrane in activation of the EGF receptor. *Cell* **152:** 543–556.

Engholm-Keller K, Larsen MR. 2013. Technologies and challenges in large-scale phosphoproteomics. *Proteomics* **13:** 910–931.

English JM, Cobb MH. 2002. Pharmacological inhibitors of MAPK pathways. *Trends Pharmacol Sci* **23:** 40–45.

Erickson HP. 2013. Irisin and FNDC5 in retrospect: An exercise hormone or a transmembrane receptor? *Adipocyte* **2:** 289–293.

Farley AR, Link AJ. 2009. Identification and quantification of protein posttranslational modifications. *Methods Enzymol* **463:** 725–763.

Farrar MA, Alberol-Ila J, Perlmutter RM. 1996. Activation of the Raf-1 kinase cascade by coumermycin-induced dimerization. *Nature* **383:** 178–181.

Feldman ME, Shokat KM. 2010. New inhibitors of the PI3K-Akt-mTOR pathway: Insights into mTOR signaling from a new generation of Tor kinase domain inhibitors (TORKinibs). *Curr Top Microbiol Immunol* **347:** 241–262.

Fenno L, Yizhar O, Deisseroth K. 2011. The development and application of optogenetics. *Annu Rev Neurosci* **34:** 389–412.

Ferrell JE Jr. 2000. What do scaffold proteins really do? *Sci STKE* **2000:** e1.

Filonov GS, Verkhusha VV. 2013. A near-infrared BiFC reporter for *in vivo* imaging of protein-protein interactions. *Chem Biol* **20:** 1078–1086.

Fischer EH. 2013. Cellular regulation by protein phosphorylation. *Biochem Biophys Res Commun* **430:** 865–867.

Florey O, Overholtzer M. 2012. Autophagy proteins in macroendocytic engulfment. *Trends Cell Biol* **22:** 374–380.

Frame S, Cohen P. 2001. GSK3 takes centre stage more than 20 years after its discovery. *Biochem J* **359:** 1–16.

Frank J. 2009. Single-particle reconstruction of biological macromolecules in electron microscopy—30 years. *Q Rev Biophys* **42:** 139–158.

Freed-Pastor WA, Prives C. 2012. Mutant p53: One name, many proteins. *Genes Dev* **26:** 1268–1286.

Freeman RS, Hasbani DM, Lipscomb EA, Straub JA, Xie L. 2003. SM-20, EGL-9, and the EGLN family of hypoxia-inducible factor prolyl hydroxylases. *Mol Cells* **16:** 1–12.

Friedland GD, Kortemme T. 2010. Designing ensembles in conformational and sequence space to characterize and engineer proteins. *Curr Opin Struct Biol* **20:** 377–384.

Fukada M, Noda M. 2007. Yeast substrate-trapping system for isolating substrates of protein tyrosine phosphatases. *Methods Mol Biol* **365:** 371–382.

Gaj T, Gersbach CA, Barbas CF III. 2013. ZFN, TALEN, and CRISPR/Cas-based methods for genome engineering. *Trends Biotechnol* **31:** 397–405.

Galperin MY, Koonin EV. 2012. Divergence and convergence in enzyme evolution. *J Biol Chem* **287:** 21–28.

Gao X, Wang H, Yang JJ, Liu X, Liu ZR. 2012. Pyruvate kinase M2 regulates gene transcription by acting as a protein kinase. *Mol Cell* **45:** 598–609.

Garcia-Gonzalo FR, Reiter JF. 2012. Scoring a backstage pass: Mechanisms of ciliogenesis and ciliary access. *J Cell Biol* **197:** 697–709.

Gautier A, Juillerat A, Heinis C, Corrêa IRJ, Kindermann M, Beaufils F, Johnsson K. 2008. An engineered protein tag for multiprotein labeling in living cells. *Chem Biol* **15:** 128–136.

Gevaert K, Vandekerckhove J. 2009. Reverse-phase diagonal chromatography for phosphoproteome research. *Methods Mol Biol* **527:** 219–227.

Ghosh M, Tucker DE, Burchett SA, Leslie CC. 2006. Properties of the Group IV phospholipase A_2 family. *Prog Lipid Res* **45:** 487–510.

Goldsmith EJ, Akella R, Min X, Zhou T, Humphreys JM. 2007. Substrate and docking interactions in serine/threonine protein kinases. *Chem Rev* **107:** 5065–5081.

González-Vera JA. 2012. Probing the kinome in real time with fluorescent peptides. *Chem Soc Rev* **41:** 1652–1664.

Goodenough DA, Paul DL. 2009. Gap junctions. *Cold Spring Harb Perspect Biol* **1:** a002576.

Gorostiza P, Isacoff EY. 2008. Optical switches for remote and noninvasive control of cell signaling. *Science* **322:** 395–399.

Granier S, Kobilka B. 2012. A new era of GPCR structural and chemical biology. *Nat Chem Biol* **8:** 670–673.

Greco V, Hannus M, Eaton S. 2001. Argosomes: A potential vehicle for the spread of morphogens through epithelia. *Cell* **106:** 633–645.

Greenleaf WJ, Woodside MT, Block SM. 2007. High-resolution, single-molecule measurements of biomolecular motion. *Annu Rev Biophys Biomol Struct* **36:** 171–190.

Groves JT, Kuriyan J. 2010. Molecular mechanisms in signal transduction at the membrane. *Nat Struct Mol Biol* **17:** 659–665.

Guerra C, Barbacid M. 2013. Genetically engineered mouse models of pancreatic adenocarcinoma. *Mol Oncol* **7:** 232–247.

Guerrero G, Isacoff EY. 2001. Genetically encoded optical sensors of neuronal activity and cellular function. *Curr Opin Neurobiol* **11:** 601–607.

Guibert S, Weber M. 2013. Functions of DNA methylation and hydroxymethylation in mammalian development. *Curr Top Dev Biol* **104:** 47–83.

Hancks DC, Kazazian HHJ. 2012. Active human retrotransposons: Variation and disease. *Curr Opin Genet Dev* **22:** 191–203.

* Hardie DG. 2012. Organismal carbohydrate and lipid homeostasis. *Cold Spring Harb Perspect Biol* **4:** a006031.

Hart GW, Slawson C, Ramirez-Correa G, Lagerlof O. 2011. Cross talk between O-GlcNAcylation and phosphorylation: Roles in signaling, transcription, and chronic disease. *Annu Rev Biochem* **80:** 825–858.

Harvey CD, Ehrhardt AG, Cellurale C, Zhong H, Yasuda R, Davis RJ, Svoboda K. 2008. A genetically encoded fluorescent sensor of ERK activity. *Proc Natl Acad Sci* **105:** 19264–19269.

* Hemmings BA, Restuccia DF. 2012. The PI3K-PKB/Akt pathway. *Cold Spring Harb Perspect Biol* **4:** a011189.

Herner A, Nikic I, Kállay M, Lemke EA, Kele P. 2013. A new family of bioorthogonally applicable fluorogenic labels. *Org Biomol Chem* **11:** 3297–3306.

Hertz NT, Wang BT, Allen JJ, Zhang C, Dar AC, Burlingame AL, Shokat KM. 2010. Chemical genetic approach for kinase-substrate mapping by covalent capture of thiophosphopeptides and analysis by mass spectrometry. *Curr Protoc Chem Biol* **2:** 15–36.

Higashii T, Isomoto A, Tyuma I, Kakishita E, Uomoto M, Nagai K. 1985. Quantitative and continuous analysis of ATP release from blood platelets with firefly luciferase luminescence. *Thromb Haemost* **53:** 65–69.

Hitosugi T, Kang S, Vander Heiden MG, Chung TW, Elf S, Lythgoe K, Dong S, Lonial S, Wang Z, Chen GZ, et al. 2009. Tyrosine phosphorylation inhibits PKM2 to promote the Warburg effect and tumor growth. *Sci Signal* **2:** ra73.71–ra73.78.

Horiuchi K. 2013. A brief history of tumor necrosis factor α-converting enzyme: An overview of ectodomain shedding. *Keio J Med* **62:** 29–36.

Hou H, Yu H. 2010. Structural insights into histone lysine demethylation. *Curr Opin Struct Biol* **20:** 739–748.

Hou F, Sun L, Zheng H, Skaug B, Jiang QX, Chen ZJ. 2011. MAVS forms functional prion-like aggregates to activate and propagate antiviral innate immune response. *Cell* **146:** 448–461.

Howard AD, Feighner SD, Cully DF, Arena JP, Liberator PA, Rosenblum CI, Hamelin M, Hreniuk DL, Palyha OC, Anderson J, et al. 1996. A receptor in pituitary and hypothalamus that functions in growth hormone release. *Science* **273:** 974–977.

Huang B, Bates M, Zhuang X. 2009. Super-resolution fluorescence microscopy. *Annu Rev Biochem* **78:** 993–1016.

Huang IK, Pei J, Grishin NV. 2013. Defining and predicting structurally conserved regions in protein superfamilies. *Bioinformatics* **29:** 175–181.

Hunter T. 2012. Why nature chose phosphate to modify proteins. *Philos Trans R Soc Lond B Biol Sci* **367:** 2513–2516.

* Ingham PW. 2012. Hedgehog signaling. *Cold Spring Harb Perspect Biol* **4:** a011221.

Isse N, Ogawa Y, Tamura N, Masuzaki H, Mori K, Okazaki T, Satoh N, Shigemoto M, Yoshimasa Y, Nishi S, et al. 1995. Structural organization and chromosomal assignment of the human obese gene. *J Biol Chem* **270:** 27728–27733.

Janes KA, Lauffenburger DA. 2013. Models of signalling networks—What cell biologists can gain from them and give to them. *J Cell Sci* **126:** 1913–1921.

Jeffery CJ. 2009. Moonlighting proteins—An update. *Mol Biosyst* **5:** 345–350.

Jiang Z, Rokhsar DS, Harland RM. 2009. Old can be new again: HAPPY whole genome sequencing, mapping and assembly. *Int J Biol Sci* **5:** 298–303.

Jin J, Pawson T. 2012. Modular evolution of phosphorylation-based signalling systems. *Philos Trans R Soc Lond B Biol Sci* **367:** 2540–2555.

Joerger AC, Fersht AR. 2008. Structural biology of the tumor suppressor p53. *Annu Rev Biochem* **77:** 557–582.

Jura N, Zhang X, Endres NF, Seeliger MA, Schindler T, Kuriyan J. 2011. Catalytic control in the EGF receptor and its connection to general kinase regulatory mechanisms. *Mol Cell* **42:** 9–22.

Kalia D, Merey G, Nakayama S, Zheng Y, Zhou J, Luo Y, Guo M, Roembke BT, Sintim HO. 2013. Nucleotide, c-di-GMP, c-di-AMP, cGMP, cAMP, (p)ppGpp signaling in bacteria and implications in pathogenesis. *Chem Soc Rev* **42:** 305–341.

Kang JY, Kawaguchi D, Coin I, Xiang Z, O'Leary DD, Slesinger PA, Wang L. 2013. In vivo expression of a light-activatable potassium channel using unnatural amino acids. *Neuron* **80:** 358–370.

Katritch V, Cherezov V, Stevens RC. 2013. Structure-function of the G protein-coupled receptor superfamily. *Annu Rev Pharmacol Toxicol* **53:** 531–556.

Kerppola TK. 2009. Visualization of molecular interactions using bimolecular fluorescence complementation analysis: Characteristics of protein fragment complementation. *Chem Soc Rev* **38:** 2876–2886.

Khakh BS, North RA. 2012. Neuromodulation by extracellular ATP and P2X receptors in the CNS. *Neuron* **76:** 51–69.

Kiu H, Nicholson SE. 2012. Biology and significance of the JAK/STAT signalling pathways. *Growth Factors* **30:** 88–106.

Kleiman LB, Maiwald T, Conzelmann H, Lauffenburger DA, Sorger PK. 2011. Rapid phospho-turnover by receptor tyrosine kinases impacts downstream signaling and drug binding. *Mol Cell* **43:** 723–737.

Kliegman JI, Fiedler D, Ryan CJ, Xu YF, Su XY, Thomas D, Caccese MC, Cheng A, Shales M, Rabinowitz JD, et al. 2013. Chemical genetics of rapamycin-insensitive TORC2 in *S. cerevisiae*. *Cell Rep* **5:** 1725–1736.

Knight ZA, Shokat KM. 2007. Chemical genetics: Where genetics and pharmacology meet. *Cell* **128:** 425–430.

Knight JD, Pawson T, Gingras AC. 2013. Profiling the kinome: Current capabilities and future challenges. *J Proteomics* **81:** 43–55.

Knighton DR, Zheng JH, Ten Eyck LF, Ashford VA, Xuong NH, Taylor SS, Sowadski JM. 1991. Crystal structure of the catalytic subunit of cyclic adenosine monophosphate-dependent protein kinase. *Science* **253:** 407–414.

Kõivomägi M, Valk E, Venta R, Iofik A, Lepiku M, Balog ER, Rubin SM, Morgan DO, Loog M. 2011. Cascades of multisite phosphorylation control Sic1 destruction at the onset of S phase. *Nature* **480:** 128–131.

Kojima M, Hosoda H, Date Y, Nakazato M, Matsuo H, Kangawa K. 1999. Ghrelin is a growth-hormone-releasing acylated peptide from stomach. *Nature* **402:** 656–660.

Kolb P, Ferreira RS, Irwin JJ, Shoichet BK. 2009. Docking and chemoinformatic screens for new ligands and targets. *Curr Opin Biotechnol* **20:** 429–436.

Kramer RH, Mourot A, Adesnik H. 2013. Optogenetic pharmacology for control of native neuronal signaling proteins. *Nat Neurosci* **16:** 816–823.

Kranzusch PJ, Lee AS, Berger JM, Doudna JA. 2013. Structure of human cGAS reveals a conserved family of second-messenger enzymes in innate immunity. *Cell Rep* **3:** 1362–1368.

Krishna RG, Wold F. 1993. Post-translational modification of proteins. *Adv Enzymol Relat Areas Mol Biol* **67:** 265–298.

Kumar J, Mayer ML. 2013. Functional insights from glutamate receptor ion channel structures. *Annu Rev Physiol* **75:** 313–337.

Kunkel MT, Toker A, Tsien RY, Newton AC. 2007. Calcium-dependent regulation of protein kinase D revealed by a genetically encoded kinase activity reporter. *J Biol Chem* **282:** 6733–6742.

Kuriyan J, Eisenberg D. 2007. The origin of protein interactions and allostery in colocalization. *Nature* **450:** 983–990.

Lander GC, Saibil HR, Nogales E. 2012. Go hybrid: EM, crystallography, and beyond. *Curr Opin Struct Biol* **22:** 627–635.

Lane TJ, Shukla D, Beauchamp KA, Pande VS. 2013. To milliseconds and beyond: Challenges in the simulation of protein folding. *Curr Opin Struct Biol* **23:** 58–65.

Lassila JK, Zalatan JG, Herschlag D. 2011. Biological phosphoryl-transfer reactions: Understanding mechanism and catalysis. *Annu Rev Biochem* **80:** 669–702.

Lau SY, Procko E, Gaudet R. 2012. Distinct properties of Ca^{2+}-calmodulin binding to N- and C-terminal regulatory regions of the TRPV1 channel. *J Gen Physiol* **140:** 541–555.

* Lee ML, Yaffe MB. 2014. Protein regulation in signal transduction. *Cold Spring Harb Perspect Biol* doi: 10.1101/cshperspect.a005918.

Leitner A, Sturm M, Lindner W. 2011. Tools for analyzing the phosphoproteome and other phosphorylated biomolecules. *Anal Chim Acta* **703:** 19–30.

Leivar P, Quail PH. 2011. PIFs: Pivotal components in a cellular signaling hub. *Trends Plant Sci* **16:** 19–28.

Levskaya A, Weiner OD, Lim WA, Voigt CA. 2009. Spatiotemporal control of cell signalling using a light-switchable protein interaction. *Nature* **461:** 997–1001.

Li M, Yu Y, Yang J. 2011. Structural biology of TRP channels. *Adv Exp Med Biol* **704:** 1–23.

Lim JH, Kim HJ, Komatsu K, Ha U, Huang Y, Jono H, Kweon SM, Lee J, Xu X, Zhang GS, et al. 2009. Differential regulation of *Streptococcus pneumoniae*-induced human MUC5AC mucin expression through distinct MAPK pathways. *Am J Transl Res* **1:** 300–311.

Lim WA, Lee CM, Tang C. 2013. Design principles of regulatory networks: Searching for the molecular algorithms of the cell. *Mol Cell* **49:** 202–212.

Lima SQ, Miesenböck G. 2005. Remote control of behavior through genetically targeted photostimulation of neurons. *Cell* **121:** 141–152.

Lin H, Su X, He B. 2012. Protein lysine acylation and cysteine succination by intermediates of energy metabolism. *ACS Chem Biol* **7:** 947–960.

Lin R, Tao R, Gao X, Li T, Zhou X, Guan KL, Xiong Y, Lei QY. 2013. Acetylation stabilizes ATP-citrate lyase to promote lipid biosynthesis and tumor growth. *Mol Cell* **51:** 506–518.

Liu CC, Schultz PG. 2010. Adding new chemistries to the genetic code. *Annu Rev Biochem* **79:** 413–444.

Londoño Gentile T, Lu C, Lodato PM, Tse S, Olejniczak SH, Witze ES, Thompson CB, Wellen KE. 2013. DNMT1 is regulated by ATP-citrate lyase and maintains methylation patterns during adipocyte differentiation. *Mol Cell Biol* **33:** 3864–3878.

Loroch S, Dickhut C, Zahedi RP, Sickmann A. 2013. Phosphoproteomics—More than meets the eye. *Electrophoresis* **34:** 1483–1492.

Losman JA, Kaelin WG Jr. 2013. What a difference a hydroxyl makes: Mutant IDH, (*R*)-2-hydroxyglutarate, and cancer. *Genes Dev* **27:** 836–852.

Lowery DM, Lim D, Yaffe MB. 2005. Structure and function of Polo-like kinases. *Oncogene* **24:** 248–259.

Luo W, Hu H, Chang R, Zhong J, Knabel M, O'Meally R, Cole RN, Pandey A, Semenza GL. 2011. Pyruvate kinase M2 is a PHD3-stimulated coactivator for hypoxia-inducible factor 1. *Cell* **145:** 732–744.

MacKinnon R. 2003. Potassium channels. *FEBS Lett* **555:** 62–65.

Malone JH, Oliver B. 2011. Microarrays, deep sequencing and the true measure of the transcriptome. *BMC Biol* **9:** 34.31–34.39.

Marks LR, Davidson BA, Knight PR, Hakansson AP. 2013. Interkingdom signaling induces *Streptococcus pneumoniae* biofilm dispersion and transition from asymptomatic colonization to disease. *mBio* **4:** e00438-13.

Martell JD, Deerinck TJ, Sancak Y, Poulos TL, Mootha VK, Sosinsky GE, Ellisman MH, Ting AY. 2012. Engineered ascorbate peroxidase as a genetically encoded reporter for electron microscopy. *Nat Biotechnol* **30:** 1143–1148.

Martinez NJ, Gregory RI. 2010. MicroRNA gene regulatory pathways in the establishment and maintenance of ESC identity. *Cell Stem Cell* **7:** 31–35.

McGettigan PA. 2013. Transcriptomics in the RNA-seq era. *Curr Opin Chem Biol* **17:** 4–11.

Mendell JT, Olson EN. 2012. MicroRNAs in stress signaling and human disease. *Cell* **148:** 1172–1187.

Miller G. 2006. Optogenetics. Shining new light on neural circuits. *Science* **314:** 1674–1676.

Miller EW, Lin JY, Frady EP, Steinbach PA, Kristan WB Jr, Tsien RY. 2012. Optically monitoring voltage in neurons by photo-induced electron transfer through molecular wires. *Proc Natl Acad Sci* **109:** 2114–2119.

Miller RM, Paavilainen VO, Krishnan S, Serafimova IM, Taunton J. 2013. Electrophilic fragment-based design of reversible covalent kinase inhibitors. *J Am Chem Soc* **135:** 5298–5301.

Miyawaki A. 2002. Green fluorescent protein-like proteins in reef Anthozoa animals. *Cell Struct Funct* **27:** 343–347.

Miyawaki A. 2011. Development of probes for cellular functions using fluorescent proteins and fluorescence resonance energy transfer. *Annu Rev Biochem* **80:** 357–373.

Moffitt JR, Chemla YR, Smith SB, Bustamante C. 2008. Recent advances in optical tweezers. *Annu Rev Biochem* **77:** 205–228.

Monda JK, Scott DC, Miller DJ, Lydeard J, King D, Harper JW, Bennett EJ, Schulman BA. 2013. Structural conservation of distinctive N-terminal acetylation-dependent interactions across a family of mammalian NEDD8 ligation enzymes. *Structure* **21:** 42–53.

Morgan DO. 2007. *The cell cycle: Principles of control.* New Science, London.

Moro E, Vettori A, Porazzi P, Schiavone M, Rampazzo E, Casari A, Ek O, Facchinello N, Astone M, Zancan I, et al. 2013. Generation and application of signaling pathway reporter lines in zebrafish. *Mol Genet Genomics* **288:** 231–242.

* Morrison DK. 2012. MAP kinase pathways. *Cold Spring Harb Perspect Biol* **4:** a011254.

Mulvihill MM, Nomura DK. 2013. Therapeutic potential of monoacylglycerol lipase inhibitors. *Life Sci* **92:** 492–497.

Musacchio A, Noble M, Pauptit R, Wierenga R, Saraste M. 1992. Crystal structure of a Src-homology 3 (SH3) domain. *Nature* **359:** 851–855.

Mutoh H, Perron A, Akemann W, Iwamoto Y, Knöpfel T. 2011. Optogenetic monitoring of membrane potentials. *Exp Physiol* **96:** 13–18.

Nardella C, Carracedo A, Salmena L, Pandolfi PP. 2010. Faithful modeling of PTEN loss-driven diseases in the mouse. *Curr Top Microbiol Immunol* **347:** 135–168.

Nash PD. 2012. Why modules matter. *FEBS Lett* **586:** 2572–2574.

Nash P, Tang X, Orlicky S, Chen Q, Gertler FB, Mendenhall MD, Sicheri F, Pawson T, Tyers M. 2001. Multisite phosphorylation of a CDK inhibitor sets a threshold for the onset of DNA replication. *Nature* **414:** 514–521.

Newell EW, Sigal N, Nair N, Kidd BA, Greenberg HB, Davis MM. 2013. Combinatorial tetramer staining and mass cytometry analysis facilitate T-cell epitope mapping and characterization. *Nat Biotechnol* **31:** 623–629.

Niu W, Guo J. 2013. Expanding the chemistry of fluorescent protein biosensors through genetic incorporation of unnatural amino acids. *Mol Biosyst* **9:** 2961–2970.

Nozawa YI, Lin C, Chuang PT. 2013. Hedgehog signaling from the primary cilium to the nucleus: An emerging picture of ciliary localization, trafficking and transduction. *Curr Opin Genet Dev* **23:** 429–437.

Otsubo Y, Yamamato M. 2008. TOR signaling in fission yeast. *Crit Rev Biochem Mol Biol* **43:** 277–283.

Overholtzer M, Mailleux AA, Mouneimne G, Normand G, Schnitt SJ, King RW, Cibas ES, Brugge JS. 2007. A nonapoptotic cell death process, entosis, that occurs by cell-in-cell invasion. *Cell* **131:** 966–979.

Palumbo AM, Smith SA, Kalcic CL, Dantus M, Stemmer PM, Reid GE. 2011. Tandem mass spectrometry strategies for phosphoproteome analysis. *Mass Spectrom Rev* **30:** 600–625.

Parzych KR, Klionsky D. 2013. An overview of autophagy: Morphology, mechanism and regulation. *Antioxid Redox Signal* **20:** 460–473.

Patricelli MP, Szardenings AK, Liyanage M, Nomanbhoy TK, Wu M, Weissig H, Aban A, Chun D, Tanner S, Kozarich JW. 2007. Functional interrogation of the kinome using nucleotide acyl phosphates. *Biochemistry* **46:** 350–358.

Pawson T, Nash P. 2003. Assembly of cell regulatory systems through protein interaction domains. *Science* **300:** 445–452.

Pearen MA, Muscat GE. 2012. Orphan nuclear receptors and the regulation of nutrient metabolism: Understanding obesity. *Physiology* **27:** 156–166.

* Perrimon N, Pitsouli C, Shilo B-Z. 2012. Signaling mechanisms controlling cell fate and embryonic patterning. *Cold Spring Harb Perspect Biol* **4:** a005975.

Piatkevich KD, Verkhusha VV. 2010. Advances in engineering of fluorescent proteins and photoactivatable proteins with red emission. *Curr Opin Chem Biol* **14:** 23–29.

Piatkevich KD, Subach FV, Verkhusha VV. 2013. Engineering of bacterial phytochromes for near-infrared imaging, sensing, and light-control in mammals. *Chem Soc Rev* **42:** 3441–3452.

Pluznick JL, Protzko RJ, Gevorgyan H, Peterlin Z, Sipos A, Han J, Brunet I, Wan LX, Rey F, Wang T, et al. 2013. Olfactory receptor responding to gut microbiota-derived signals plays a role in renin secretion and blood pressure regulation. *Proc Natl Acad Sci* **110:** 4410–4415.

Portnoy DA. 2012. Yogi Berra, Forrest Gump, and the discovery of Listeria actin comet tails. *Mol Biol Cell* **23:** 1141–1145.

Pryciak PM. 2009. Designing new cellular signaling pathways. *Chem Biol* **16:** 249–254.

Ptacek J, Snyder M. 2006. Charging it up: Global analysis of protein phosphorylation. *Trends Genet* **22:** 545–554.

Racey LA, Byvoet P. 1971. Histone acetyltransferase in chromatin. Evidence for in vitro enzymatic transfer of acetate from acetyl-coenzyme A to histones. *Exp Cell Res* **64:** 366–370.

Rasmussen SG, DeVree BT, Zou Y, Kruse AC, Chung KY, Kobilka TS, Thian FS, Chae PS, Pardon E, Calinski D, et al. 2011. Crystal structure of the β_2 adrenergic receptor-Gs protein complex. *Nature* **477:** 549–555.

Ravnskjaer K, Hogan MF, Lackey D, Tora L, Dent SY, Olefsky J, Montminy M. 2013. Glucagon regulates gluconeogenesis through KAT2B- and WDR5-mediated epigenetic effects. *J Clin Invest* **123:** 4318–4328.

Reményi A, Good MC, Lim WA. 2006. Docking interactions in protein kinase and phosphatase networks. *Curr Opin Struct Biol* **16:** 676–685.

Renicke C, Schuster D, Usherenko S, Essen LO, Taxis C. 2013. A LOV2 domain-based optogenetic tool to control protein degradation and cellular function. *Chem Biol* **20:** 619–626.

Resh MD. 2012. Targeting protein lipidation in disease. *Trends Mol Med* **18:** 206–214.

★ Rhind N, Russell P. 2012. Signaling pathways that regulate cell division. *Cold Spring Harb Perspect Biol* **4:** a005942.

Rockwell NC, Su YS, Lagarias JC. 2006. Phytochrome structure and signaling mechanisms. *Annu Rev Plant Biol* **57:** 837–858.

Rogne M, Taskén K. 2013. Cell signalling analyses in the functional genomics era. *New Biotechnol* **30:** 333–338.

Roskoski RJ. 2012. MEK1/2 dual-specificity protein kinases: Structure and regulation. *Biochem Biophys Res Commun* **417:** 5–10.

Roux PP, Thibault P. 2013. The coming of age of phosphoproteomics—From large data sets to inference of protein functions. *Mol Cell Proteomics* **12:** 3453–3464.

Roy S, Hsiung F, Kornberg TB. 2011. Specificity of *Drosophila* cytonemes for distinct signaling pathways. *Science* **332:** 354–358.

Rubakhin SS, Lanni EJ, Sweedler JV. 2013. Progress toward single cell metabolomics. *Curr Opin Biotechnol* **24:** 95–104.

Sadikot RT, Blackwell TS. 2008. Bioluminescence: Imaging modality for *in vitro* and *in vivo* gene expression. *Methods Mol Biol* **477:** 383–394.

Sadowski MI, Taylor WR. 2013. Prediction of protein contacts from correlated sequence substitutions. *Sci Prog* **96:** 33–42.

Sadowsky JD, Burlingame MA, Wolan DW, McClendon CL, Jacobson MP, Wells JA. 2011. Turning a protein kinase on or off from a single allosteric site via disulfide trapping. *Proc Nat Acad Sci* **108:** 6056–6061.

Sakurai T, Amemiya A, Ishii M, Matsuzaki I, Chemelli RM, Tanaka H, Williams SC, Richardson JA, Kozlowski GP, Wilson S, et al. 1998. Orexins and orexin receptors: A family of hypothalamic neuropeptides and G protein-coupled receptors that regulate feeding behavior. *Cell* **92:** 573–585.

Sanders TA, Llagostera E, Barna M. 2013. Specialized filopodia direct long-range transport of SHH during vertebrate tissue patterning. *Nature* **497:** 628–632.

Savinainen JR, Saario SM, Laitinen JT. 2012. The serine hydrolases MAGL, ABHD6 and ABHD12 as guardians of 2-arachidonoylglycerol signalling through cannabinoid receptors. *Acta Physiol* **204:** 267–276.

Scarpa ES, Fabrizio G, Di Girolamo M. 2013. A role of intracellular mono-ADP-ribosylation in cancer biology. *FEBS J* **280:** 3551–3562.

Schmid JA, Birbach A. 2007. Fluorescent proteins and fluorescence resonance energy transfer (FRET) as tools in signaling research. *Thromb Haemost* **97:** 378–384.

Schumacher MA, Chinnam N, Ohashi T, Shah RS, Erickson HP. 2013. The structure of irisin reveals a novel intersubunit β-sheet fibronectin type III (FNIII) dimer: Implications for receptor activation. *J Biol Chem* **288:** 33738–33744.

Sengupta P, Van Engelenburg S, Lippincott-Schwartz J. 2012. Visualizing cell structure and function with point-localization superresolution imaging. *Dev Cell* **23:** 1092–1102.

Shaner NC, Lin MZ, McKeown MR, Steinbach PA, Hazelwood KL, Davidson MW, Tsien RY. 2008. Improving the photostability of bright monomeric orange and red fluorescent proteins. *Nat Methods* **5:** 545–551.

Shaw N, Liu ZJ. 2014. Role of the HIN domain in regulation of innate immune responses. *Mol Cell Biol* **34:** 2–15.

Shendure J, Mitra RD, Varma C, Church GM. 2004. Advanced sequencing technologies: Methods and goals. *Nat Rev Genet* **5:** 335–344.

Shu X, Royant A, Lin MZ, Aguilera TA, Lev-Ram V, Steinbach PA, Tsien RY. 2009. Mammalian expression of infrared fluorescent proteins engineered from a bacterial phytochrome. *Science* **324:** 804–807.

Shu X, Lev-Ram V, Deerinck TJ, Qi Y, Ramko EB, Davidson MW, Jin Y, Ellisman MH, Tsien RY. 2011. A genetically encoded tag for correlated light and electron microscopy of intact cells, tissues, and organisms. *PLoS Biol* **9:** e1001041.

Silverman JS, Skaar JR, Pagano M. 2012. SCF ubiquitin ligases in the maintenance of genome stability. *Trends Biochem Sci* **37:** 66–73.

Sims RJ III, Reinberg D. 2008. Is there a code embedded in proteins that is based on post-translational modifications? *Nat Rev Mol Cell Biol* **9:** 815–820.

Sopko R, Andrews BJ. 2008. Linking the kinome and phosphorylome—A comprehensive review of approaches to find kinase targets. *Mol Biosyst* **4:** 920–933.

Ståhl PL, Lundeberg J. 2012. Toward the single-hour high-quality genome. *Annu Rev Biochem* **81:** 359–378.

Starai VJ, Takahashi H, Boeke JD, Escalante-Semerena JC. 2004. A link between transcription and intermediary metabolism: A role for Sir2 in the control of acetyl-coenzyme A synthetase. *Curr Opin Microbiol* **7:** 115–119.

Stenkamp RE, Teller DC, Palczewski K. 2005. Rhodopsin: A structural primer for G-protein coupled receptors. *Arch Pharm* **338:** 209–216.

Strebhardt K. 2010. Multifaceted polo-like kinases: Drug targets and antitargets for cancer therapy. *Nat Rev Drug Discov* **9:** 643–660.

Stynen B, Tournu H, Tavernier J, Van Dijck P. 2012. Diversity in genetic *in vivo* methods for protein-protein interaction studies: From the yeast two-hybrid system to the mammalian split-luciferase system. *Microbiol Mol Biol Rev* **76:** 331–382.

Tagliabracci VS, Xiao J, Dixon JE. 2013. Phosphorylation of substrates destined for secretion by the Fam20 kinases. *Biochem Soc Trans* **41:** 1061–1065.

Tang X, Orlicky S, Mittag T, Csizmok V, Pawson T, Forman-Kay JD, Sicheri F, Tyers M. 2012. Composite low affinity interactions dictate recognition of the cyclin-dependent kinase inhibitor Sic1 by the SCF^{Cdc4} ubiquitin ligase. *Proc Natl Acad Sci* **109:** 3287–3292.

Tarrant MK, Cole PA. 2009. The chemical biology of protein phosphorylation. *Annu Rev Biochem* **78:** 797–825.

Tartaglia LA, Dembski M, Weng A, Deng N, Culpepper J, Devos R, Richards GJ, Campfield LA, Clark FT, Deeds J, et al. 1995. Identification and expression cloning of a leptin receptor, OB-R. *Cell* **83:** 1263–1271.

Thompson BJ. 2010. Developmental control of cell growth and division in *Drosophila*. *Curr Opin Cell Biol* **22:** 788–794.

Thorner J. 2006. Signal transduction. In *Landmark papers in yeast biology* (ed. Linder P, et al.), pp. 193–210. Cold Spring Harbor Laboratory Press, Cold Spring Harbor, NY.

Tian X, Azpurua J, Hine C, Vaidya A, Myakishev-Rempel M, Ablaeva J, Mao Z, Nevo E, Gorbunova V, Seluanov A. 2013. High-molecular-mass hyaluronan mediates the cancer resistance of the naked mole rat. *Nature* **499:** 346–349.

Toettcher JE, Weiner OD, Lim WA. 2013. Using optogenetics to interrogate the dynamic control of signal transmission by the ras/erk module. *Cell* **155:** 1422–1434.

Tong X, Shigetomi E, Looger LL, Khakh BS. 2013. Genetically encoded calcium indicators and astrocyte calcium microdomains. *Neuroscientist* **19:** 274–291.

Tsien RY. 2009. Constructing and exploiting the fluorescent protein paintbox. *Angew Chem Int Ed Engl* **48:** 5612–5626.

Turk BE, Cantley LC. 2003. Peptide libraries: At the crossroads of proteomics and bioinformatics. *Curr Opin Chem Biol* **7:** 84–90.

Turk BE, Hutti JE, Cantley LC. 2006. Determining protein kinase substrate specificity by parallel solution-phase assay of large numbers of peptide substrates. *Nat Protoc* **1:** 375–379.

Turner BM. 1991. Histone acetylation and control of gene expression. *J Cell Sci* **99:** 13–20.

Urnov FD, Rebar EJ, Holmes MC, Zhang HS, Gregory PD. 2010. Genome editing with engineered zinc finger nucleases. *Nat Rev Genet* **11:** 636–646.

Vadas O, Burke JE, Zhang X, Berndt A, Williams RL. 2011. Structural basis for activation and inhibition of class I phosphoinositide 3-kinases. *Sci Signal* **4:** re.2.1–re.2.12.

Venta R, Valk E, Kõivomägi M, Loog M. 2012. Double-negative feedback between S-phase cyclin-CDK and CKI generates abruptness in the G1/S switch. *Front Physiol* **3**: 459.

Venter JC, Levy S, Stockwell T, Remington K, Halpern A. 2003. Massive parallelism, randomness and genomic advances. *Nat Genet* **33**: 219–227.

Vézina C, Kudelski A, Sehgal SN. 1975. Rapamycin (AY-22,989), a new antifungal antibiotic. I. Taxonomy of the producing streptomycete and isolation of the active principle. *J Antibiot (Tokyo)* **28**: 721–726.

Waksman G, Kominos D, Robertson SC, Pant N, Baltimore D, Birge RB, Cowburn D, Hanafusa H, Mayer BJ, Overduin M, et al. 1992. Crystal structure of the phosphotyrosine recognition domain SH2 of v-src complexed with tyrosine-phosphorylated peptides. *Nature* **358**: 646–653.

Washietl S, Will S, Hendrix DA, Goff LA, Rinn JL, Berger B, Kellis M. 2012. Computational analysis of noncoding RNAs. *Wiley Interdiscip Rev RNA* **3**: 759–778.

Wei C, Liu J, Yu Z, Zhang B, Gao G, Jiao R. 2013. TALEN or Cas9—Rapid, efficient and specific choices for genome modifications. *J Genet Genomics* **40**: 281–289.

Weston CR, Davis RJ. 2001. Signaling specificity—A complex affair. *Science* **292**: 2439–2440.

Wong SY, Reiter JF. 2008. The primary cilium at the crossroads of mammalian hedgehog signaling. *Curr Top Dev Biol* **85**: 225–260.

Woodward JJ, Iavarone AT, Portnoy DA. 2010. c-di-AMP secreted by intracellular Listeria monocytogenes activates a host type I interferon response. *Science* **328**: 1703–1705.

Wu H. 2013. Higher-order assemblies in a new paradigm of signal transduction. *Cell* **153**: 287–292.

Wu H, Zhang Y. 2011. Mechanisms and functions of Tet protein-mediated 5-methylcytosine oxidation. *Genes Dev* **25**: 2436–2452.

Xiao H, Chatterjee A, Choi SH, Bajjuri KM, Sinha SC, Schultz PG. 2013a. Genetic incorporation of multiple unnatural amino acids into proteins in mammalian cells. *Angew Chem Int Ed Engl* **52**: 14080–14083.

Xiao J, Tagliabracci VS, Wen J, Kim SA, Dixon JE. 2013b. Crystal structure of the Golgi casein kinase. *Proc Natl Acad Sci* **110**: 10574–10579.

Xu J, Wang AH, Oses-Prieto J, Makhijani K, Katsuno Y, Pei M, Yan L, Zheng YG, Burlingame A, Brückner K, et al. 2013. Arginine methylation initiates BMP-induced Smad signaling. *Mol Cell* **51**: 5–19.

Yang W, Xia Y, Hawke D, Li X, Liang J, Xing D, Aldape K, Hunter T, Alfred Yung WK, Lu Z. 2012. PKM2 phosphorylates histone H3 and promotes gene transcription and tumorigenesis. *Cell* **150**: 685–696.

Zacharias DA, Tsien RY. 2006. Molecular biology and mutation of green fluorescent protein. *Methods Biochem Anal* **47**: 83–120.

Zhang Z, Tan M, Xie Z, Dai L, Chen Y, Zhao T. 2011. Identification of lysine succinylation as a new post-translational modification. *Nat Chem Biol* **7**: 58–63.

Zhang M, Galdieri L, Vancura A. 2013. The yeast AMPK homolog SNF1 regulates acetyl coenzyme A homeostasis and histone acetylation. *Mol Cell Biol* **33**: 4701–4717.

Zhou X, Herbst-Robinson KJ, Zhang J. 2012. Visualizing dynamic activities of signaling enzymes using genetically encodable FRET-based biosensors from designs to applications. *Methods Enzymol* **504**: 317–340.

Zimmermann H. 2008. ATP and acetylcholine, equal brethren. *Neurochem Int* **52**: 634–648.

Zivanovic N, Jacobs A, Bodenmiller B. 2013. A practical guide to multiplexed mass cytometry. *Curr Top Microbiol Immunol* doi: 10.1007/82_2013_335.

Zmajkovicova K, Jesenberger V, Catalanotti F, Baumgartner C, Reyes G, Baccarini M. 2013. MEK1 is required for PTEN membrane recruitment, AKT regulation, and the maintenance of peripheral tolerance. *Mol Cell* **50**: 43–55.

Index